MATHEMATICS, GAME THEORY AND ALGEBRA COMPENDIUM. VOLUME 1

MATHEMATICS, GAME THEORY AND ALGEBRA COMPENDIUM. VOLUME 1

JACOB H. MATHIAS
EDITOR

Nova Science Publishers, Inc.
New York

For permission to use material from this book please contact us:
Telephone 631-231-7269; Fax 631-231-8175
Web Site: http://www.novapublishers.com

LIBRARY OF CONGRESS CATALOGING-IN-PUBLICATION DATA

Elias, Bartholomew.

Federal Aviation Administration reauthorization / Bart Elias, John W. Fischer, Robert S. Kirk et al., authors.

p. cm.

ISBN 978-1-60456-659-8 (softcover)

1. Aeronautics--Technological innovations--Government policy--United States. 2. United States. Federal Aviation Administration--Appropriations and expenditures. 3. Aeronautics--Technological innovations--Finance--Government policy--United States. 4. Aeronautics--Cost control--Research--United States. I. Fischer, John W. II. Kirk, Robert S. III. Title.

TL521.E43 2008

354.790973--dc22

2008014424

Published by Nova Science Publishers, Inc. ✦ New York

Contents

PREFACE

This book is devoted to new advances in all branches of mathematics, game theory and applications, and pure and applied algebra and geometry including Field Theory and Polynomials; Commutative Rings and Algebras; Algebraic Geometry; Linear and Multilinear Algebras; Matrix Theory; Associative Rings and Algebras; Nonassociative Rings and Algebras; K-Theory; Group Theory and Generalizations; Topological Groups, Lie Groups; Geometry; Differential Geometry.

In: Mathematics, Game Theory and Algebra Compendium ISBN 978-1-60456-659-8
Editor: Jacob H. Mathias, pp. 1-5

Chapter 1

ON PROPERTY $(E.A)$ IN GAUGE SPACES

Adnan Alhomaidan
Department of Mathematics, King Abdulaziz University,
P.O. Box 80203, Jeddah 21589, Saudi Arabia

Abstract

We study mappings satisfying the property $(E.A)$ in gauge spaces. Our results include recent results of Aamri and El Moutawakil [1].

1. Introduction and Preliminaries

Recently, Aamri and El Moutawakil [1] defined the property $(E.A)$ in metric spaces, which generalizes the concept of noncompatible mappings, and gave some common fixed point theorems under strict contractive conditions for mappings satisfying the property $(E.A)$. This concept was extended to multivalued mappings independently by Kamran [5] and Singh and Hashim [6]. Frigon [3] obtained the Banach contraction principle in complete guage spaces. The aim of this paper to extend the property $(E.A)$ to gauge spaces and obtain common fixed points of mappings satisfying this property. Our results include recent results of Aamri and El Moutawakil [1].

Let $E = (E, \{d_\alpha\}_{\alpha\in\Lambda})$ be a gauge space endowed with a gauge structure $\{d_\alpha : \alpha \in \Lambda\}$; here Λ is a directed set (see [2, pp. 198, 308]).

Definition 1 *Let F and G be two self-mappings on a gauge space E. We say that F and G satisfy the property $(E.A)$ if there exists a sequence $\{x_n\}$ such that*

$$\lim_{n\to\infty} d_\alpha(Fx_n, t) = 0 = \lim_{n\to\infty} d_\alpha(Gx_n, t)$$

for all $\alpha \in \Lambda$ and some $t \in E$.

Definition 2 *Two selfmappings F and G are said to be weakly compatible [4] if $Fa = Ga$ for some $a \in E$, then $FGa = GFa$.*

2. Main Results

Theorem 3 *Let E be a gauge space and F and G be weakly compatible self-mappings on E such that*
(i) F and G satisfy the property $(E.A)$.
(ii) for each $\alpha \in \Lambda$ and $x, y \in E$ such that $x \neq y$,

$$\begin{aligned} d_\alpha(Gx, Gy) &< \max\{d_\alpha(Fx, Fy), \lambda[d_\alpha(Gx, Fx) + d_\alpha(Gy, Fy)], \\ &\quad \lambda[d_\alpha(Gy, Fx) + d_\alpha(Gx, Fy)]\}, \end{aligned}$$

where $0 \leq \lambda \leq \frac{1}{2}$ (iii) $F(E)$ or $G(E)$ is a closed subspace of E such that $G(E) \subset F(E)$. Then F and G have a unique common fixed point.

Proof Since F and G satisfy the property (EA), then there exists a sequence $\{x_n\}$ in E satisfying

$$\lim_{n\to\infty} d_\alpha(Fx_n, t) = 0 = \lim_{n\to\infty} d_\alpha(Gx_n, t)$$

for all $\alpha \in \Lambda$ and some $t \in E$.
Suppose that $F(E)$ is closed. Then $t = \lim_{n\to\infty} Fx_n \in F(E)$ and so there exists $a \in E$ such that $t = Fa$. As a result we have

$$\lim_{n\to\infty} d_\alpha(Fx_n, Fa) = 0 = \lim_{n\to\infty} d_\alpha(Gx_n, Fa)$$

We claim that $Fa = Ga$. For this we prove that $d_\alpha(Fa, Ga) = 0$ for all $\alpha \in \Lambda$. Suppose that there exists some $\alpha_0 \in \Lambda$ such that $d_{\alpha_0}(Fa, Ga) \neq 0$.

Now, condition (ii) implies that

$$\begin{aligned} d_{\alpha_0}(Gx_n, Ga) &< \max\{d_{\alpha_0}(Fx_n, Fa), \lambda[d_{\alpha_0}(Gx_n, Fx_n) + d_{\alpha_0}(Ga, Fa)], \\ &\quad \lambda[d_{\alpha_0}(Ga, Fx_n) + d_{\alpha_0}(Gx_n, Fa)]\}. \end{aligned}$$

On taking the limit as $n \to \infty$,

$$\begin{aligned} d_{\alpha_0}(Fa, Ga) &\leq \max\{d_{\alpha_0}(Fa, Fa), \lambda[d_\alpha(Fa, Fa) + d_\alpha(Ga, Fa)], \\ &\quad \lambda[d_{\alpha_0}(Ga, Fa) + d_{\alpha_0}(Fa, Fa)]\} \\ &= \lambda d_{\alpha_0}(Ga, Fa) \leq d_{\alpha_0}(Ga, Fa)/2, \end{aligned}$$

which is a contradiction. Hence $Fa = Ga$. Since F and G are weakly compatible, $F(Ga) = G(Fa)$, and so $G(Ga) = G(Fa) = F(Ga) = F(Fa)$.

Finally, we show that $b := Ga$ is a common fixed point of G and F. Suppose that there exists some α_0 such that $d_{\alpha_0}(b, Gb) \neq 0$. Then

$$\begin{aligned} d_{\alpha_0}(b, Gb) &= d_{\alpha_0}(Ga, GGa) \\ &< \max\{d_{\alpha_0}(Fa, FGa), \lambda[d_{\alpha_0}(Ga, Fa) + d_{\alpha_0}(GGa, FGa)], \\ &\quad \lambda[d_{\alpha_0}(GGa, Fa) + d_{\alpha_0}(Ga, FGa]\} \\ &\leq \max\{d_{\alpha_0}(Ga, GGa), 2\lambda d_{\alpha_0}(GGa, Ga)\} \\ &= d_{\alpha_0}(b, Gb), \end{aligned}$$

which is a contradiction. Thus $d_\alpha(b, Gb) = 0$ for all $\alpha \in \Lambda$. Hence $b = Gb$ and $Fb = F(Ga) = G(Ga) = Gb = b$.

The proof is similar when $G(E)$ is a closed subspace of E since $G(E) \subset F(E)$.

To prove the uniqueness of the common fixed point, we suppose that $c \in E$ is another common fixed point of F and G, i.e., $c = Fc = Gc$ and suppose that, for some $\alpha_0 \in \Lambda$, $d_{\alpha_0}(b, c) \neq 0$. Then

$$\begin{aligned} d_{\alpha_0}(b,c) &= d_{\alpha_0}(Fb, Gc) \\ &< \max\{d_{\alpha_0}(Fb, Fc), \lambda[d_{\alpha_0}(Gb, Fb) + d_{\alpha_0}(Gc, Fc)], \\ &\quad \lambda[d_{\alpha_0}(Gc, Fb) + d_{\alpha_0}(Gb, Fc)]\} \\ &= \max\{d_{\alpha_0}(b, c), \lambda[d_{\alpha_0}(b, b) + d_{\alpha_0}(c, c)], \\ &\quad \lambda[d_{\alpha_0}(c, b) + d_{\alpha_0}(b, c)]\} \\ &= \max\{d_{\alpha_0}(b, c), 2\lambda d_{\alpha_0}(b, c) \\ &= d_{\alpha_0}(b, c), \end{aligned}$$

which is a contradiction. Hence $d_{\alpha_0}(b, c) = 0$ for all $\alpha \in \Lambda$. Hence, the common fixed point is unique.

Corollary 4 *(Aamri and El Moutawakil [1, Theorem 1]) Let $E = (E, d)$ be a metric space and F and G be weakly compatible self-mappings of E such that*
(i) F and G satisfy the property $(E.A)$.
(ii) for each $x, y \in E$ such that $x \neq y$,

$$\begin{aligned} d(Gx, Gy) &< \max\{d(Fx, Fy), [d(Gx, Fx) + d(Gy, Fy)]/2, \\ &\quad [d(Gy, Fx) + d(Gx, Fy)]/2\}, \end{aligned}$$

(iii) $G(E) \subset F(E)$.
If $F(E)$ or $G(E)$ is a complete subspace of E, then F and G have a unique common fixed point.

Theorem 5 *Let A, B, F and G be self-mappings of a gauge space E. Suppose that for each $\alpha \in \Lambda$, there exists a nondecreasing function $\phi_\alpha : [0, \infty) \to [0, \infty)$ satisfying $\phi_\alpha(t) < t$ for $t > 0$ such that for $x, y \in E$ we have such that*

$$d_\alpha(Ax, By) \leq \phi_\alpha(\max\{d_\alpha(Fx, Gy), d_\alpha(Fx, By), d_\alpha(Gy, By)\}), \tag{1}$$

(4) (A, F) and (B, G) are weakly compatible,
(5) (A, F) or (B, G) satisfies the property $(E.A)$,
(6) $A(E) \subseteq G(E)$ and $E(E) \subseteq F(E)$.
If the range of the one of the mappings A, B, G or F is a closed subspace of E, then A, B, F and G have a unique common fixed point.

Proof Suppose that (B, G) satisfies the property $(E.A)$. Then there exists a sequence $\{x_n\}$ in E such that

$$\lim_{n\to\infty} d_\alpha(Bx_n, t) = 0 = \lim_{n\to\infty} d_\alpha(Gx_n, t)$$

for all $\alpha \in \Lambda$ and for $t \in E$. Since $B(E) \subseteq F(E)$, there exists a sequence $\{y_n\} \subset E$ such that$Bx_n = Fy_n$, for $n \in \mathbf{N}$. So (2.), we have $\lim_{n\to\infty} d_\alpha(Fy_n, t) = 0$, for all $\alpha \in \Lambda$. We now show that $\lim_{n\to\infty} d_\alpha(Ay_n, t) = 0$, for all $\alpha \in \Lambda$. Fix $\alpha \in \Lambda$. Notice, from (1), we have that

$$\begin{aligned} d_\alpha(Ay_n, Bx_n) &\le \phi_\alpha(\max\{d_\alpha(Fy_n, Gx_n), d_\alpha(Fy_n, Bx_n), d_\alpha(Gx_n, Bx_n)\}) \\ &\le \phi_\alpha(d_\alpha(Gx_n, Bx_n)) \\ &\le d_\alpha(Gx_n, Bx_n) \end{aligned}$$

Letting $n \to \infty$ and using (2.), we get $\lim_{n\to\infty} d_\alpha(Ay_n, Bx_n) = 0$. Using the triangle inequality, we have

$$d_\alpha(Ay_n, t) \le d_\alpha(Ay_n, Bx_n) + d_\alpha(Bx_n, t),$$

which, on taking the limit as $n \to \infty$, gives $\lim_{n\to\infty} d_\alpha(Ay_n, t) = 0$. Suppose that $F(E)$ is closed and $t \in F(E)$.Then there exists some $u \in E$ such that $t = Fu$. As a result we have,

$$\lim_{n\to\infty} d_\alpha(Bx_n, Fu) = \lim_{n\to\infty} d_\alpha(Gx_n, Fu) = \lim_{n\to\infty} d_\alpha(Fy_n, Fu).$$

for all $\alpha \in \Lambda$ Fix $\alpha \in \Lambda$. We show that $Au = Fu$. Notice that

$$d_\alpha(Au, Bx_n) \le \phi_\alpha(\max\{d_\alpha(Fu, Gx_n), d_\alpha(Fu, Bx_n), d_\alpha(Gx_n, Bx_n)\}).$$

Letting $n \to \infty$, we get $d_\alpha(Au, Fu) = 0$ for all $\alpha \in \Lambda$, Hence $Au = Fu$. By the weak compatibility of A and F, we have $AFu = FAu$ and so $AAu = AFu = FAu = FFu$. Now, since $A(E) \subset G(E)$, there exists $v \in E$ such that $Au = Gv$. We show that $Gv = Bv$. Using (1), we have

$$\begin{aligned} d_\alpha(Au, Bv) &\le \phi_\alpha(\max\{d_\alpha(Fu, Gv), d_\alpha(Fu, Bv), d_\alpha(Gv, Bv\}) \\ &\le \phi_\alpha(d_\alpha(Au, Bv)) \end{aligned}$$

for all $\alpha \in \Lambda$, which implies that $Au = Fu = Gv = Bv$. Since B and G are weakly compatible, we have $BGv = GBv$, and so $GGv = GBv = BGv = BBv$.

Next we show that Au is a common fixed point of A, B, F and G. Using (1), we have

$$\begin{aligned} d_\alpha(Au, AAu) &= d_\alpha(AAu, Bv) \\ &\le \phi_\alpha(\max\{d_\alpha(FAu, Gu), d_\alpha(FAu, Bv), d_\alpha(Gv, Bv)\}) \\ &\le \phi_\alpha(d_\alpha(AAu, Au)) \end{aligned}$$

for all $\alpha \in \Lambda$. Therefore, $Au = AAu = FAu$ and Au is a common fixed point of A and F. Similarly we prove that Bv is a common fixed point of B and G, and since $Au = Bv$, then Au is a common fixed point of A, B, F and G.

The proof is similar when $G(E)$ is a closed subspace of E. The cases in which $A(E)$ or $B(E)$ is closed are similar to the cases in which $G(E)$ or $F(E)$ is closed, since $A(E) \subset G(E)$ and $B(E) \subset F(E)$.

If $Au = Bu = Fu = Gu = u$ and $Av = Bv = Fv = Gv = v$, then from (1), we have

$$\begin{aligned} d_\alpha(u,v) &= d_\alpha(Au,Bv) \\ &\le \phi_\alpha(\max\{d_\alpha(Fu,Gv), d_\alpha(Fu,Bv), d_\alpha(Gv,Bv\}) \\ &\le \phi_\alpha(d_\alpha(Au,Bv)) \end{aligned}$$

for all $\alpha \in \Lambda$. This implies that $u = v$ and so the common fixed point is unique.

Corollary 6 *(Aamri and El Moutawakil [1, Theorem 2]) Let A, B, F and G be self-mappings of a metric space (E, d). Suppose that there exists a nondecreasing function $\phi : [0,\infty) \to [0,\infty)$ satisfying $\phi(t) < t$ for $t > 0$ such that for $x, y \in E$ we have such that*

$$d(Ax, By) \le \phi(\max\{d(Fx, Gy), d(Fx, By), d(Gy, By)\}), \tag{2}$$

(4) (A, F) and (B, G) are weakly compatible,
(5) (A, F) or (B, G) satisfies the property $(E.A)$,
(6) $A(E) \subseteq G(E)$ and $E(E) \subseteq F(E)$.
If the range of the one of the mappings A, B, G or F is a complete subspace of E, then A, B, F and G have a unique common fixed point.

References

[1] M. Aamri and D. El-Moutawakil, Some new common fixed point theorems under strict contractive conditions, *J. Math. Anal. Appl.*, **270** (2002), 181-188.

[2] J. Dugundji, Topology, *Ally and Bacon*, Boston, 1966.

[3] M. Frigon, Fixed point results for generalized contraction in gauge spaces and applications, *Proc. Amer. Math. Soc.*, **128**, (2000), 2957–2965.

[4] G. Jungck and B. E. Rhoades, Fixed points for set valued functions without continuity, *Indian J. Pure App. Math.*, **29** (1998), 227-238.

[5] T. Kamran, Coincidence and fixed points on hybrid strict contractions, *J. Math. Anal. Appl.*, **299** (2004), 235-241.

[6] S. L. Singh and A. M. Hashim, New coincidence and fixed point theorems for strictly contractive hybrid maps, *Austral. J. Math. Anal. Appl.*, **2** (2005), 1-7.

In: Mathematics, Game Theory and Algebra Compendium
Editor: Jacob H. Mathias, pp. 7-13
ISBN 978-1-60456-659-8

Chapter 2

STRUCTURE OF CERTAIN PERIODIC NEAR RINGS

Asma Ali[*]
Department of Mathematics
Aligarh Muslim University
Aligarh-202002 (India)

Abstract

Let R be a left near ring and let $m = m(x, y) \geq 1$, $n = n(x, y) \geq 1$ be positive integers depending on x and y in R. Further, let $p(xy)$ be an element of R which is finite sum of powers $(xy)^k$; $k \geq 2$ and additive inverses of such powers. In the present paper we establish some decomposition theorems for a near ring R satisfying any one of the conditions, (i) $xy = (xy)^n p(xy)$, (ii) $xy = (yx)^n p(xy)$, (iii) $xy = (yx)^n p(yx)$, (iv) $xy = x^m y^n p(xy)$, (v) $xy = y^m x^n p(xy)$, (vi) $xy = x^m p(xy) x^n$ and (vii) $xy = y^m p(xy) y^n$.

AMS Subject Classifications (1991): 16W25, 16U80.

Key Words and Phrases: Distributively generated (d-g)-near ring, D-near ring, zero-symmetric near ring and zero-commutative near ring.

1. Preliminaries

Throughout the paper R will denote a left near ring with multiplicative center $Z(R)$. We shall denote by N the set of nilpotent elements and by P the set of potent elements that is, the set $\{x \in R \mid x^n = x,$ for some positive integer $n(x) > 1\}$. If $P = R$, we shall call R a J-ring. By the well known $x = x^{n(x)}$ theorem of Jacobson [6], J-rings are necessarily commutative. The set of commutators is denoted by C. An element x of R is said to be distributive if $(y + z)x = yx + zx$, for all $y, z \in R$. If every element of R is distributive, then R is said to be a distributive near ring. A near ring is said to be distributively generated $(d-g)$ if it contains a multiplicative subsemigroup of distributive elements which generates

*E-mail address: asma786@postmark.net

the additive group $(R,+)$. A near ring R is called zero-commutative if for all $x, y \in R$, $xy = 0$, implies that $yx = 0$, and if for all $x \in R$, $0x = 0$, R is called zero-symmetric (we may recall that the left distributivity of R yields $x0 = 0$). A near ring R is called periodic if for every $x \in R$ there exist distinct positive integers $m = m(x)$, $n = n(x)$ such that $x^m = x^n$. A sufficient condition for a ring R to be periodic is Chacron's criterion : For each $x \in R$ there exists an integer $m = m(x) \geq 1$ and a polynomial $f(X) \in Z[X]$, the ring of polynomials in X such that $x^m = x^{m+1}f(x)$ ([4]).

For each element x of a near ring R, the subnear ring generated by x will be denoted by $< x >$. Note that if x is a distributive element, then the element $t(x) \in < x >$ (or of $< 1, x >$ if R has unity 1) may be assumed to be a finite sum of powers of x.

An ideal of a near ring R is a normal subgroup I of R^+ such that (i) $RI \subseteq I$ and (ii) $(x+i)y - xy \in I$ for all $x, y \in R$ and $i \in I$.

A near ring R is called a D-near ring if every nonzero homomorphic image T of R satisfies the following conditions:

(i) T has a nonzero right distributive element.

(ii) $(T,+)$ is abelian implies that $(T,+,.)$ is a ring.

It is evident by definition that all distributive near rings are examples of D-near rings. However, the example 2.5 # 6 of [5] illustrates that the class of D- near rings is larger than the class of $d - g$ near rings.

2. Introduction

The property $x^n = x$ has been among the favourites of many ring theorists over the last few decades since Jacobson[6] first studied the commutativity of rings satisfying this condition in order to generalize the classical Wedderburn theorem [10]. Later, this property has been weakened by Searcoid and MacHale [9] who proved that a ring satisfying the condition $(xy)^{n(x,y)} = xy$ must be commutative. Ligh and Luh [7] established that, a ring R satisfying the above condition is a direct sum of a J-ring and a zero ring. Later, Bell and Ligh [3] obtained structure of rings satisfying the related properties like $xy = (xy)^2p(xy)$ and $xy = (yx)^2p(xy)$, for $p(XY) \in \mathbb{Z}[X,Y]$, the ring of polynomials in two noncommuting indeterminates X and Y.

It is natural to question whether the analogous hypotheses yield the direct sum decomposition in the case of near rings. An example due to Clay (cf. [5, Example 2.5 # 29]) shows that it is not possible to obtain such a decomposition.

Example 2.1. Consider the non-abelian additive group $(R,+)$, isomorphic to the symmetric group S_3 and define addition and multiplication in R as follows :

+	0	a	b	c	u	v
0	0	a	b	c	u	v
a	a	o	v	u	c	b
b	b	u	0	v	a	c
c	c	v	u	0	b	a
u	u	b	c	a	v	0
v	v	c	a	b	0	u

.	0	a	b	c	u	v
0	0	0	0	0	0	0
a	0	a	a	a	0	0
b	0	a	a	a	0	0
c	0	a	a	a	0	0
u	0	0	0	0	0	0
v	0	0	0	0	0	0

It is easy to see that $(R, +, \ \cdot)$ is a commutative (distributive) near ring satisfying $xy = (xy)^2 = xy^2x$, for all $x, y \in R$. However $P = \{0, a\}$ is not an ideal of R.

Hence, following [3], we define a weaker notion of orthogonal sum. A near ring R is an orthogonal sum of subnear rings A and B denoted by $R = A \uplus B$ if $AB = BA = \{0\}$ and each element of R has a unique representation in the form $a + b$ with $a \in A$ and $b \in B$.

In this paper we shall obtain an orthogonal sum decomposition for a near ring R satisfying either of the following conditions, where $p(xy)$ denotes an element of R which is finite sum of powers $(xy)^k$; $k \geq 2$ and additive inverses of such powers :

(P_1) For every pair of elements $x, y \in R$, there exists a positive integer $n = n(x, y) \geq 1$ such that $xy = (yx)^n p(yx)$.

(P_2) For every pair of elements $x, y \in R$, there exists a positive integer $n = n(x, y) \geq 1$ such that $xy = (yx)^n p(xy)$.

(P_3) For every pair of elements $x, y \in R$, there exist positive integers $m = m(x, y) \geq 1$ and $n = n(x, y) \geq 1$ such that $xy = y^m x^n p(xy)$.

(P_4) For every pair of elements $x, y \in R$, there exist positive integers $m = m(x, y) \geq 1$ and $n = n(x, y) \geq 1$ such that $xy = y^m p(xy) y^n$.

(P_5) For every pair of elements $x, y \in R$, there exists a positive integer $n = n(x, y) \geq 1$ such that $xy = (xy)^n p(xy)$.

(P_6) For every pair of elements $x, y \in R$, there exist positive integers $m = m(x, y) \geq 1$ and $n = n(x, y) \geq 1$ such that $xy = x^m y^n p(xy)$.

(P_7) For every pair of elements $x, y \in R$, there exist positive integers $m = m(x, y) \geq 1$ and $n = n(x, y) \geq 1$ such that $xy = x^m p(xy) x^n$.

3. Main Results

Theorem 3.1. Let R be a D-near ring satisfying condition (P_1). Then R is periodic and $R = P \uplus N$, where P is a subring, N is a subnear ring with trivial multiplication.

Theorem 3.2. Let R be a zero-symmetric D-near ring satisfying either of the conditions $(P_2) - (P_4)$. Then R is periodic and $R = P \uplus N$, where P is a subring, N is a subnear ring with trivial multiplication.

Theorem 3.3. Let R be a D-near ring satisfying condition (P_7). If idempotent elements of R are multiplicatively central, then R is periodic. Moreover, $R = P \uplus N$, where P is a subring and N is a subnear ring with trivial multiplication.

Theorem 3.4. Let R be a zero-symmetric D-near ring satisfying either of the conditions (P_5) and (P_6). If idempotent elements of R are central, then R is periodic. Moreover, $R = P \uplus N$, where P is a subring and N is a subnear ring with trivial multiplication.

Before starting the proof of the above theorems we state the following lemmas which are essentially proved in [1], [2], [3] and [8] respectively.

Lemma 3.1. Let R be a zero-symmetric near ring satisfying the following properties :

(i) For each x in R, there exists a positive integer $n(x) > 1$ such that $x^{n(x)} = x$.

(ii) Every non trivial homomorphic image of R contains a nonzero central idempotent.

Then $(R, +)$ is commutative.

Lemma 3.2. Let R be a near ring with unity 1. Then for every $x \in R$, $< x >= x < 1, x >$.

Lemma 3.3. If R is a zero-commutative periodic near ring , then $R = P + N$.

Lemma 3.4. Let R be a near ring in which idempotents are multiplicatively central. If e and f are any idempotents, there exists an idempotent g such that $ge = e$ and $gf = f$.

Lemma 3.5. If R is a zero-commutative near ring, then N is an ideal of R.

The following Lemma was obtained for $d.g$ near ring by Bell [2, Theorem 12]. Now we shall extend the result for D-near ring.

Lemma 3.6. Let R be a D-near ring satisfying $x^n = x^n p(x)$, where $n = n(x)$ a positive integer and $p(x) \in < x >$. If $N \subseteq Z(R)$, then R/N is a periodic and commutative ring.

Proof. Since $N \subseteq Z(R)$, N is an ideal by Lemma 3.5. Consider the near ring R/N. Since R/N can be written as a subdirect product of near rings without zero divisors, we may assume R/N has no nonzero divisors of zero. Let d be a nonzero distributive element of R/N. Note that in view of the property that $x^n = x^n p(x)$, $d = dt(d)$ for some $t(d) \in < d >$. It follows that $e = t(d)$ is a nonzero idempotent in R/N. Since $e(er - r) = 0$, for every $r \in R/N$, e is the left identity in R/N. Considering arbitrary $x, y \in R/N$ and using the fact that e commutes with d and $d = de$, we have, $0 = (x + y)de - (xde - yde) = (x + y)e - (xe - ye)d$. Again since R/N has no nonzero divisors of zero, e is a distributive element of R/N. Hence e is the multiplicative identity in R/N. If x is an arbitrary nonzero element of R/N, then $x = t(x)$, where $t(x) \in < x >$. Using Lemma 3.2, we can write $t(x) = xt'(x)$ for some $t'(x) \in < 1, x >$. Now $x = x^2 t'(x)$ yields that $xt'(x) = 1$ i.e., R/N is a division near ring. Thus R/N is additively commutative and is a ring. HenceR/N is periodic by Chacron's criterion [4] and a commutative ring by [2, Theorem 2].

The proof of the above lemma runs on the same lines if we replace the property $x^n = x^n p(x)$ by $x^n = x^n p(x) x^n$.

Lemma 3.7. Let R be a near ring satisfying either of the conditions (P_1) and (P_7). Then $RN = NR = \{0\}$.

Proof. Clearly R satisfying (P_1) is zero-commutative. Hence N is an ideal by Lemma 3.5. Replacing y by x in (P_1), we find that

$$x^2 = x^{2m} p(x^2) \text{ ; for } m = m(x) \geq 1. \tag{3.1}$$

If $u \in N$, then making repeated use of (3.1), we have $u^2 = 0$. Now for any $u \in N$ by condition (P_1), we have $ux = (xu)^l p(xu) = 0$, for $l = l(u, x) \geq 1$. Zero-commutativity yields that $xu = 0$, for $u \in N$ and $x \in R$. Hence $RN = NR = \{0\}$.

The proof runs on the parallel lines for condition (P_7).

Proceeding on the same lines we can prove the following lemma :

Lemma 3.8. Let R be a zero-symmetric near ring satisfying either of the conditions $(P_2) - (P_6)$. Then $RN = NR = \{0\}$.

Lemma 3.9. Let R be a D-near ring satisfying condition (P_1). Then idempotent elements of R are central.

Proof. Let e be an idempotent and $x \in R$. Then by condition (P_1) there exists an integer $t = t(x, e) \geq 1$ such that $xe = (ex)^t h(ex)$. Multiplying by e on the left we get $exe = xe$. Application of Lemma 3.6 and Lemma 3.7 yields that $C \subseteq N \subseteq Z(R)$ and we have $e(xe - ex) = 0$, for all $x \in R$. Hence $ex = xe$, for all $x \in R$.

In the similar manner we can prove the following :

Lemma 3.10. Let R be a zero-symmetric D-near ring satisfying either of the conditions $(P_2) - (P_4)$. Then idempotent elements of R are central.

Proof of Theorem 3.1. In view of Lemma 3.6 we see that for each $x \in R$, there exist distinct positive integers $m = m(x), n = n(x)$ such that $x^m - x^n \in N$. Hence using Lemma 3.7, we have $x^{m+1} = x^{n+1}$, for each $x \in R$ and R is periodic.

Next we show that P is a subring. Let $a, b \in P$ and choose integers $p = p(a) > 1$ and $q = q(b) > 1$ such that $a^p = a$ and $b^q = b$. Let $r = (p-1)q-(p-2) = (q-1)p-(q-2)$. Then it is clear that $a^r = a$ and $b^r = b$. Note that $e = a^{r-1}$, $f = b^{r-1}$ are idempotents with $ea = a$ and $fb = b$. Since $C \subseteq N \subseteq Z(R)$, we can write $a^2b = aba$ for all a, b in P using Lemma 3.7. Obviously $ab = a^r b^r = (ab)^r$, hence $ab \in P$ for all a, b in P. Moreover since R/N has the $x^l = x$ property we have an integer $j > 1$ such that

$$(a-b)^j = a - b + u \; ; \; u \in N. \tag{3.2}$$

Using Lemma 3.4 we can choose an idempotent g for which $ge = e$ and $gf = f$. Therefore $ga = a$ and $gb = b$. Now multiplying (3.2) by g we have $(a-b)^j = a - b$ i.e. $a - b \in P$. Also by Lemma 3.1 $(P, +)$ is abelian. Hence P is a subring.

Trivially $P \cap N = \{0\}$. Let $a + u = b + v$ where $a, b \in P$ and $u, v \in N$. Then $a - b = v - u \in P \cap N = \{0\}$, which yields $a = b$ and $v = u$. Hence $R = P \uplus N$.

The proof of Theorem 3.2 - 3.4 can be obtained in the similar manner.

References

[1] H. E. Bell, Near rings in which each element is a power of itself, *Bull. Austral. Math. Soc.* **2** (1970), 363-368.

[2] H. E. Bell, On commutativity of periodic rings and near rings, *Acta Math. Acad. Sci. Hungar.* **36** (1980), 293-302.

[3] H. E. Bell and S. Ligh , Some decomposition theorems for periodic rings and near rings, *Math. J. Okayama Univ.* **31** (1989), 93-99.

[4] M. Chacron, On a theorem of Herstein, *Canad. J. Math.* **21** (1969), 1348-1353.

[5] J. R. Clay , The near rings on groups of low order, *Math Z.* **104** (1968), 364-371.

[6] N. Jacobson, Structure theory of algebraic algebras of bounded degree, *Ann. of Math.* **46** (1945),695-707.

[7] S. Ligh and J. Luh, Direct sum of J-rings and zero rings, *Amer. Math. Monthly* **96** (1989), 40-41.

[8] Y. F. Reddy and C. V. L. N. Murty, Semi-symmetric ideals in near rings, *Indian J. Pure Appl. Math.* **16** (1985), 17-21.

[9] M. O. Searcoid and D. MacHale, Two elementary generalizations of Boolean rings, *Amer. Math. Monthly* **93** (1986), 121-122.

[10] J.H.M. Wedderburn, A theorem on finite algebras, *Trans. Math. Soc.* **6** (1905), 349-352.

In: Mathematics, Game Theory and Algebra Compendium ISBN 978-1-60456-659-8
Editor: Jacob H. Mathias, pp. 15-21

Chapter 3

EXTENDED METHOD OF QUASILINEARIZATION FOR A THREE-POINT BOUNDARY VALUE PROBLEM WITH GENERAL NONLINEAR NONCONVEX BOUNDARY CONDITIONS

Ahmed Alsaedi * ***and Mohammed Alhuthali*** †
Department of Mathematics,
Faculty of science, King Abdul Aziz University,
P.O.Box. 80257, Jeddah 21589, Saudi Arabia

Abstract

In this paper, We apply the generalized quasilinearization technique to obtain a monotone sequence of approximate solutions converging monotonically and quadratically to the unique solution of a second order three-point boundary value problem with general nonlinear nonconvex boundary conditions.

Key Words and Phrases: Quasilinearization, three-point boundary value problem, quadratic convergence.

AMS Subject Classifications (2000): 34B10, 34B15.

1. Introduction

The method of quasilinearization initiated by Bellman and Kalaba [1], and generalized by Lakshmikantham [2-3] has been studied and extended in several diverse disciplines. In fact, it is generating a rich history and a comprehensive description of this method can be found in [4-10].

Multi-point nonlinear boundary value problems, which refer to a different family of

*E-mail address: aalsaedi@hotmail.com
†E-mail address: moadth@yahoo.com

boundary conditions in the study of disconjugacy theory [11], have been addressed by many authors, for example, [12-14]. In particular, Eloe and Gao [15] discussed the quasilinearization method for a three-point boundary value problem.
The aim of this paper is to relax the assumption of convexity/concavity on the nonlinear general boundary conditions involved in the second order three-point boundary value problem and discuss the extended method of quasilinearization for this problem. In fact, we develop a sequence of approximate solutions converging monotonically and quadratically to a solution of the problem.

2. Preliminary Results

We consider the three-point boundary value problem with general nonlinear boundary conditions

$$x''(t) = f(t, x(t)), \tag{1}$$

$$px(0) - qx'(0) = a, \qquad px(1) + qx'(1) = g(x(\frac{1}{2})), \tag{2}$$

where $f \in C[[0,1] \times R, R]$, $p, q > 0$ with $p > 1$ and $g : R \longrightarrow R$ is continuous. By Green's function method, the solution, $x(t)$ of (1)-(2) can be written as

$$x(t) = a(\frac{-t}{p+2q} + \frac{p+q}{p^2+2pq}) + g(x(\frac{1}{2}))[\frac{t}{p+2q} + \frac{q}{p^2+2pq}] + \int_0^1 G(t,s) f(s, x(s)) ds,$$

where the Green's function $G(t, s)$ for the general three-point boundary value problem is given by

$$G(t,s) = \frac{1}{(p^2+2pq)} \begin{cases} (pt+q)(p(s-1)-q), & \text{if } 0 \le t \le s \le 1 \\ (p(t-1)-q)(ps+q), & \text{if } 0 \le s \le t \le 1 \end{cases}$$

Notice that $G(t, s) < 0$ on $(0,1) \times (0,1)$.

We say that $\alpha \in C^2[0,1]$ is a lower solution of the boundary value problem (1)-(2) if

$$\alpha''(t) \ge f(t, \alpha),\ t \in [0,1],$$

$$p\alpha(0) - q\alpha'(0) \le a, \qquad p\alpha(1) + q\alpha'(1) \le g(\alpha(\frac{1}{2})),$$

and $\beta \in C^2[0,1]$ be an upper solution of the boundary value problem (1)-(2) if

$$\beta''(t) \le f(t, \beta),\ t \in [0,1],$$

$$p\beta(0) - q\beta'(0) \ge a, \qquad p\beta(1) + q\beta'(1) \ge g(\beta(\frac{1}{2})).$$

Now, we state the following theorems which play a pivotal role in the proof of the main result (for the proof of these theorems, see [16]).

Theorem 1. Assume that f is continuous with $f_x > 0$ on $[0,1] \times R$ and g is continuous with $0 \leq g' < 1$ on R. Let β and α be the upper and lower solutions of (1)-(2) respectively. Then $\alpha(t) \leq \beta(t)$, $t \in [0,1]$.

Theorem 2. Assume that f is continuous on $[0,1] \times R$ and g is continuous on R satisfying $0 \leq g' < 1$. Further, we assume that there exist an upper solution β and a lower solution α of (1)-(2) such that $\alpha(t) \leq \beta(t),\ t \in [0,1]$. Then there exists a solution $x(t)$ of (1)-(2) satisfying $\alpha(t) \leq x(t) \leq \beta(t),\ t \in [0,1]$.

3. Main Result

Theorem 3. Assume that

($\mathbf{A_1}$) $f(t,x) \in C([0,1] \times R)$ such that $\frac{\partial f}{\partial x}(t,x) > 0,\ \frac{\partial^2}{\partial x^2}(f(t,x) + \phi(t,x)) \leq 0$, where $\frac{\partial^2}{\partial x^2}\phi(t,x) \leq 0$ for some continuous function $\phi(t,x)$.

($\mathbf{A_2}$) $\alpha, \beta \in C^2[0,1]$ are lower and upper solutions of (1.1)-(1.2) respectively.

($\mathbf{A_3}$) $g(x), g'(x)$ are continuous on R with $0 \leq g' \leq 1$ and $g''(x) + \psi''(x) \geq 0$ for some continuous function $\psi(x)$ satisfying $\psi''(x) \geq 0$.

Then there exists a monotone sequence of solutions $\{w_n\}$ converging quadratically to the unique solution, x of (1)-(2).

Proof. Define $F : [0,1] \times R \longrightarrow R$ by

$$F(t,x) = f(t,x) + \phi(t,x).$$

Using (A_1) and (A_3), we obtain

$$f(t,x) \leq F(t,y) + F_x(t,y)(x-y) - \phi(t,x), \tag{3}$$

and

$$g(x) \geq \chi(y) + \chi'(y)(x-y) - \psi(x), \tag{4}$$

where $\chi(x) = g(x) + \psi(x)$ and $\alpha \leq y \leq x \leq \beta$.
Define

$$F^*(t,x,y) = F(t,y) + F_x(t,y)(x-y) - \phi(t,x),$$

and

$$h(x,y) = \chi(y) + \chi'(y)(x-y) - \psi(x).$$

We observe that

$$f(t,x) = \min_y F^*(t,x,y), \qquad f(t,x) = F^*(t,x,x) \tag{5}$$

$$g(x) = \max_y h(x,y), \qquad g(x) = h(x,x). \tag{6}$$

In view of the fact that $f_x(t,x) > 0$, we find that $F_x^*(t,x,y) > 0$ which implies that $F_x^*(t,x,y)$ is increasing in x for each $(t,y) \in [0,1] \times R$. Similarly, $g'(x) = h_x(x,y)$ and by (A_3), we have $0 \le h_x(x,y) \le 1$. Select $\alpha = w_0$ and consider the following BVP

$$x''(t) = F^*(t, x(t), w_0(t)),\ t \in [0,1], \tag{7}$$

$$px(0) - qx'(0) = a, \qquad px(1) + qx'(1) = h(x(\frac{1}{2}), w_0(\frac{1}{2})). \tag{8}$$

Using (A_3), (5) and (6), we obtain

$$w_0'' \ge f(t, w_0) = F^*(t, w_0, w_0),\ t \in [0,1],$$

$$pw_0(0) - qw_0'(0) \le a, \qquad pw_0(1) + qw_0'(1) \le g(w_0(\frac{1}{2})) = h(w_0(\frac{1}{2}), w_0(\frac{1}{2})),$$

and

$$\beta'' \le f(t,\beta) \le F^*(t, \beta, w_0),\ t \in [0,1],$$

$$p\beta(0) - q\beta'(0) \ge a, \qquad p\beta(1) + q\beta'(1) \ge g(\beta(\frac{1}{2})) \ge h(\beta(\frac{1}{2}), w_0(\frac{1}{2})),$$

which imply that w_0 and β are lower and upper solutions of (7)-(8) respectively. It follows by Theorems 1 and 2 that there exists a unique solution, w_1 of (7)-(8) such that

$$w_0(t) \le w_1(t) \le \beta(t),\ t \in [0,1].$$

Now, we consider the BVP

$$x'' = F^*(t, x(t), w_1(t)),\ t \in [0,1], \tag{9}$$

$$px(0) - qx'(0) = a, \qquad px(1) + qx'(1) = h(x(\frac{1}{2}), w_1(\frac{1}{2})). \tag{10}$$

Again, using (A_3), (5) and (6), we get

$$w_1'' = F^*(t, w_1, w_0) \ge F^*(t, w_1, w_1),\ t \in [0,1],$$

$$pw_1(0) - qw_1'(0) \le a, \qquad pw_1(1) + qw_1'(1) = h(w_1(\frac{1}{2}), w_0(\frac{1}{2})) \le h(w_1(\frac{1}{2}), w_1(\frac{1}{2})),$$

and

$$\beta'' \le f(t,\beta) \le F^*(t, \beta, w_1),\ t \in [0,1],$$

$$p\beta(0) - q\beta'(0) \ge a, \qquad p\beta(1) + q\beta'(1) \ge g(\beta(\frac{1}{2})) \ge h(\beta(\frac{1}{2}), w_1(\frac{1}{2})),$$

implying that w_1 and β are lower and upper solutions of $(9) - (10)$ respectively. By the earlier arguments, there exists a solution, w_2 of $(9) - (10)$ such that

$$w_1(t) \le w_2(t) \le \beta(t),\ t \in [0,1].$$

Continuing this process successively, we obtain a monotone sequence $\{w_n\}$ of solutions satisfying

$$w_0(t) \le w_1(t) \le w_2(t) \le ... \le w_n(t) \le \beta(t),\ t \in [0,1],$$

where each element w_n of the sequence is a solution of the BVP

$$x'' = F^*(t, x(t), w_{n-1}(t)),\ t \in [0,1],$$

$$px(0) - qx'(0) = a, \qquad px(1) + qx'(1) = h(x(\frac{1}{2}), w_{n-1}(\frac{1}{2})),$$

and is given by

$$\begin{aligned} w_n(t) &= a(\frac{-t}{p+2q} + \frac{p+q}{p^2+2pq}) + h(w_n(\frac{1}{2}), w_{n-1}(\frac{1}{2}))[\frac{t}{p+2q} + \frac{q}{p^2+2pq}] \\ &+ \int_0^1 G(t,s)F^*(s, w_n(s), w_{n-1}(s))ds. \end{aligned} \tag{11}$$

Employing the fact that $[0,1]$ is compact and the monotone convergence is pointwise, it follows that the convergence of the sequence is uniform. If $x(t)$ is the limit point of the sequence, then passing onto the limit $n \to \infty$, (11) gives

$$\begin{aligned} x(t) &= a(\frac{-t}{p+2q} + \frac{p+q}{p^2+2pq}) + h(x(\frac{1}{2}), x(\frac{1}{2}))[\frac{t}{p+2q} + \frac{q}{p^2+2pq}] \\ &+ \int_0^1 G(t,s)F^*(s, x(s), x(s))ds \\ &= a(\frac{-t}{p+2q} + \frac{p+q}{p^2+2pq}) + g(x(\frac{1}{2}))[\frac{t}{p+2q} + \frac{q}{p^2+2pq}] \\ &+ \int_0^1 G(t,s)f(s, x(s))ds. \end{aligned}$$

Hence, $x(t)$ is the solution of (1)-(2).
Now, we show that the convergence of the sequence of iterates is of order $k (k \geq 2)$. For that, we define $e_n(t) = x(t) - w_n(t),\ t \in [0,1]$ and note that $e_n(t) \geq 0$. Further

$$pe_n(0) - qe_n'(0) = 0,\ pe_n(1) + qe_n'(1) = g(x(\frac{1}{2})) - h(w_n(\frac{1}{2}), w_{n-1}(\frac{1}{2})).$$

Using the generalized mean value theorem, we have

$$\begin{aligned} e''_{n+1}(t) &= x'' - w''_{n+1} \\ &= F(t,x) - \phi(t,x) - [F(t,w_n) + F_x(t,w_n)(w_{n+1} - w_n) - \phi(t, w_{n+1})] \\ &= F_x(t,c_1)(x - w_n) - F_x(t,w_n)(x - w_n) \\ &+ [F_x(t,w_n) - \phi_x(t,c_2)](x - w_{n+1}) \\ &= F_{xx}(t,c_3)(c_1 - w_n)(x - w_n) + [F_x(t,w_n) - \phi_x(t,c_2)](x - w_{n+1}) \\ &\geq F_{xx}(t,c_3)e_n^2 + f_x(t,c_2)e_{n+1} \\ &\geq -M\|e_n\|^2, \end{aligned}$$

where M is a bound on $F_{xx}(t,x)$ for $t \in [0,1]$, $w_n < c_3 < c_1 < x$, $w_{n+1} < c_2 < x$ and $\|e_n\| = \max\{|e_n(t)| : t \in [0,1]\}$. Thus, we have

$$\begin{aligned}
e_{n+1}(t) &= [g(x(\frac{1}{2})) - h(w_{n+1}(\frac{1}{2}), w_n(\frac{1}{2})))](\frac{t}{p+2q} + \frac{q}{p^2+2pq}) + \int_0^1 G(t,s)e''_{n+1}(s) \\
&\leq [\chi(x(\frac{1}{2})) - \chi(w_n(\frac{1}{2})) - (\psi(x(\frac{1}{2})) - \psi(w_{n+1}(\frac{1}{2})))\\
&- \chi'(w_n(\frac{1}{2}))(w_{n+1}(\frac{1}{2}) - w_n(\frac{1}{2}))][\frac{t}{p+2q} + \frac{q}{p^2+2pq}] + M\|e_n\|^2 \int_0^1 |G(t,s)| \\
&= [\chi'(c_4)(x(\frac{1}{2}) - w_n(\frac{1}{2})) - \psi'(c_5)(x(\frac{1}{2}) - w_{n+1}(\frac{1}{2}))\\
&- \chi'(w_n(\frac{1}{2}))(w_{n+1}(\frac{1}{2}) - w_n(\frac{1}{2}))][\frac{t}{p+2q} + \frac{q}{p^2+2pq}] + M\|e_n\|^2 \int_0^1 |G(t,s)| \\
&\leq [\chi''(c_6)e_n^2(\frac{1}{2}) + g'(w_n(\frac{1}{2}))e_{n+1}(\frac{1}{2})][\frac{t}{p+2q} + \frac{q}{p^2+2pq}] + M_1\|e_n\|^2,
\end{aligned}$$

where M_1 provides a bound for $M\int_0^1 |G(t,s)|ds$, $w_n(\frac{1}{2}) < c_6 < c_4 < x(\frac{1}{2})$, $w_n(\frac{1}{2}) < c_5 < x(\frac{1}{2})$. Taking maximum over the interval $[0,1]$, we get

$$\|e_{n+1}\| < M_5\|e_n\|^2 + \lambda_1\|e_{n+1}\|, \tag{12}$$

where M_2 provides a bound for $|\chi''|$ on $[w_n(\frac{1}{2}), x(\frac{1}{2})]$, $|g'| \leq \lambda < 1$, $M_5 = M_4 + M_1$, $M_4 = M_2M_3$, $\lambda_1 = \lambda M_3$ and $M_3 = \frac{1}{p+2q} + \frac{q}{p^2+2pq}$. Solving (12) algebraically, we obtain

$$\|e_{n+1}\| \leq \frac{M_5}{1-\lambda_1}\|e_n\|^2.$$

This completes the proof.

References

[1] R. Bellman and R. Kalaba, Quasilinearization and Nonlinear Boundary Value Problems, Amer. Elsevier, New York, 1965.

[2] V. Lakshmikantham, An extension of the method of quasilinearization, *J. Optim. Theory Appl.*, **82**(1994), 315-321.

[3] V. Lakshmikantham, Further improvement of Generalized quasilinearization, *Nonlinear Anal.*, **27**(1996), 223-227.

[4] V. Lakshmikantham and A.S. Vatsala, Generalized Quasilinearization for Nonlinear Problems, Kluwer Academic Publishes, Dordrecht, 1998.

[5] J.J. Nieto, Generalized quasilinearization method for a second order differential equation with Dirichlet boundary conditions, *Proc. Amer. Math. Soc.*, **125**(1997), 2599-2604.

[6] A. Cabada, J.J. Nieto and Rafael Pita-da-veige, A note on rapid convergence of approximate solutions for an ordinary Dirichlet problem, *Dynam. Contin. Discrete Impuls. systems*, **4**(1998), 23-30.

[7] A. Cabada and J.J. Nieto, Quasilinearization and rate of convergence for higher order nonlinear periodic boundary value problems, *J. Optim. Theory Appl.*, **108**(2001), 97-107.

[8] Bashir Ahmad, J.J. Nieto and N. Shahzad, The Bellman-Kalaba-Lakshmikantham quasilinearization method for Neumann problems, *J. Math. Anal. Appl.*, **257**(2001), 356-363.

[9] Bashir Ahmad, J.J. Nieto and N. Shahzad, Generalized quasilinearization method for mixed boundary value problems, *App. Math. Comput.*, **133**(2002), 423-429.

[10] Bashir Ahmad, Ahmed Alsaedi and S. Sivasundaram, Approximation of the solution of nonlinear second order integro-differential equations, *Dynamic Systems Appl.* (To appear).

[11] W. Coppel, Disconjugacy, Lecture Notes in Mathematics, Vol.220, Springer-Verlag, NewYork/Berlin (1971).

[12] I.T. Kiguradze and A.G. Lomtatidze, In certain boundary value problems for second order linear ordinary differential equations with singularities, *J. Math. Anal. Appl.*, **101**(1984), 325-347 .

[13] C.P. Gupta, A second order m-point boundary value problem at resonance, *Nonlinear Anal.*, **24**(1995), 1483-1489.

[14] C.P. Gupta and S. Trofimchuck, A priori estimates for the existence of a solution for a multi-point boundary value problem, *J. Inequal. Appl.*, **5**(2000), 351-365 .

[15] P. Eloe and Y. Gao, The method of quasilinearization and a three-point boundary value problem, *J. Korean Math. Soc.*, **39**(2002), 319-330 .

[16] Bashir Ahmad and Tagreed G. Sogati, A second order three-point boundary value problem with mixed nonlinear boundary conditions, *Methods Appl. Anal.* **11**(2004), 295-302.

In: Mathematics, Game Theory and Algebra Compendium ISBN 978-1-60456-659-8
Editor: Jacob H. Mathias, pp. 23-29

Chapter 4

THE JONES POLYNOMIAL OF TWIST KNOTS

Yılmaz Altun*
Yuzuncu Yil University, Faculty of Art and Science,
Department of Mathematics, 65080 Van, Turkey

Abstract

In this paper, some interesting invariants of twist knots are calculated. Especially, the Jones polynomial and the Conway polynomial of them are given. In addition, the Arf invariant of twist knots is calculated.

Key Words: Twist knot, Jones polynomial, Conway polynomial.

2000 Subject Classification: 57M25

1. Introduction

Twist knots [4] are in certain sense generalizations of the trefoil knot, the figure-eight knot and stevedore's knot. A twist knot with n-half-twists is denoted by T_n. Knots T_n are alternating knots and the number of half twists determining them. All twist knots are invertible. All of them T_n expect T_0 and T_2 are not amphicherial [3](See,[13, p.300 Corollary3]).

It is well-known from the relevant literature that, first in 1985, Jones [8] discovered a new polynomial invariant of oriented knots and links in S^3. The invariant known as Jones polynomial is denoted $V_K(t)$ for an oriented knot K. Later, Kauffman [10] constructed a state model for the Jones polynomial. In this modal, bracket polynomial of the diagram of the knot and the sum of the sings of the crosing of this diagram are considered. (See, [10, p.399,400]). Thus the Jones polynomial of a knot can be easily determined by using the bracket polynomial, for example, it can be easily seen that $V_{T_1}(t) = -t^{-4} + t^{-3} + t^{-1}$.

A formula related with the Jones polynomial of the $(n, m)-$ torus knot (with $(n, m) = 1$) has been given in [9]. Besides, Landvoy [11] constructed a formula for the Jones polynomial of pretzel knots and links. In this note we constructed a formula for both $V_{T_{2n}}$ and $V_{T_{2n+1}}$ (See, [12,p.30, Exercise 2]).

*E-mail address: artvin2003@yahoo.com

Definition 1.1. (See, [8]). If K is a knot, define $W_K(t)$ to be the Laurent polynomial

$$W_K(t) = (1 - V_K(t)) / \left(1 - t^3\right)(1 - t) \tag{1.1}$$

It should be noted that $1 - V_K(t)$ is divisible by $\left(1 - t^3\right)(1 - t)$, (See, [9]).

Definition 1.2. (See, [8]). Arf invariant of K is defined by

$$ArfK = W_K(i). \tag{1.2}$$

It can be shown that the Alexander polynomial [1] $\Delta(t)$ of a knot may be written as a polynomial with integral coefficients in $u = t + t^{-1} - 2$, $\Delta(t) = f(u)$. Hence $\Delta(t^2)$ is a polynomial in $z = \left(t - t^{-1}\right)$. (It is even a polynomial in z^2, see, for instance [5]).

In [6], Conway defined a polynomial $\nabla(z)$ with integral coefficients for oriented links. Let $\nabla_K(t)$ be the Conway polynomial of knot K. Then it can be written

$$\Delta_K(t^2) = \nabla_K\left(t - t^{-1}\right) \tag{1.3}$$

(See, [14]). Therefore if $\Delta_K(t)$ is known, $\nabla_K(t)$ can be easily calculated.

2. The Jones Polynomial of T_n

Theorem 2.1. *For* $n = 1, 2, ...$

$$V_{T_{2n}}(t) = t^{-2n} - t^{-2n+1} + t^{-2n+2} - 2t^{-2n+3} + 2t^{-2n+4}$$

$$-2t^{-2n+5} + ... + 2t^{-4} - 2t^{-3} + 2t^{-2} - 2t^{-1} + 2 - t + t^2 \tag{2.1}$$

(See, Table1).

Proof. The proof of theorem will be performed by using induction method. Now, we prove that for every positive integer n,

$$V_{T_{2n}}(t) = t^{-2n} - t^{-2n+1} + t^{-2n+2} - 2t^{-2n+3} + 2t^{-2n+4}$$

$$-2t^{-2n+5} + ... + 2t^{-4} - 2t^{-3} + 2t^{-2} - 2t^{-1} + 2 - t + t^2$$

Let

$$S = \left\{n \in N : V_{T_{2n}}(t) = t^{-2n} - t^{-2n+1} + ... + 2 - t + t^2\right\}.$$

$1 \in S$, since

$$V_{T_2}(t) = t^{-2} - t^{-1} + 1 - t + t^2.$$

If $k \in S$, then

$$V_{T_{2k}}(t) = t^{-2k} - t^{-2k+1} + t^{-2k+2} - 2t^{-2k+3} + 2t^{-2k+4}$$

$$-2t^{-2k+5} + ... + 2t^{-4} - 2t^{-3} + 2t^{-2} - 2t^{-1} + 2 - t + t^2. \tag{2.2}$$

To show that $k+1 \in S$, it suffices to prove that

$$V_{T_{2k+2}}(t) = t^{-2k-2} - t^{-2k-1} + t^{-2k} - 2t^{-2k+1}$$
$$+2t^{-2k} + ... + 2t^{-4} - 2t^{-3} + 2t^{-2} - 2t^{-1} + 2 - t + t^2. \quad (2.3)$$

To drive (6) from (5), add

$$t^{-2k-2} - t^{-2k-1} - t^{-2k+1} + t^{-2k+2}$$

to each side of (5) to obtain

$$V_{T_{2k}}(t) + t^{-2k-2} - t^{-2k-1} - t^{-2k+1} + t^{-2k+2}$$
$$= t^{-2k} - t^{-2k+1} + t^{-2k+2} - 2t^{-2k+3} + 2t^{-2k+4} - 2t^{-2k+5} + ... + 2t^{-4}$$
$$-2t^{-3} + 2t^{-2} - 2t^{-1} + 2 - t + t^2 + t^{-2k-2} - t^{-2k-1} - t^{-2k+1} + t^{-2k+2}$$
$$= t^{-2k-2} - t^{-2k-1} + t^{-2k} - 2t^{-2k+1} + 2t^{-2k+2} + ... + 2t^{-4}$$
$$-2t^{-3} + 2t^{-2} - 2t^{-1} + 2 - t + t^2$$
$$= V_{T_{2k+2}}(t).$$

Hence, $k+1 \in S$. Then, $S = N$, and this completes the proof. □

Corollary 2.2.

$$V_{T_{2n}}(t) = \begin{cases} 1 & , \quad n = 0 \\ V_{T_{2n-2}}(t) + t^{-2n} - t^{-2n+1} - t^{-2n+3} + t^{-2n+4} & , \quad n = 1, 2, 3, ... \end{cases}$$

Theorem 2.3. *For* $n = 1, 2, ...$

$$V_{T_{2n+1}}(t) = -t^{-2n-4} + t^{-2n-3} - t^{-2n-2} + 2t^{-2n-1} - 2t^{-2n}$$
$$+2t^{-2n+1} - 2t^{-2n+2} + ... + 2t^{-5} - 2t^{-4} + 2t^{-3} - t^{-2} + t^{-1}. \quad (2.4)$$

(See, Table2).

Proof. The proof of Theorem 2.3 is similar that of Theorem 2.1, and hence it is omitted. □

Corollary 2.4.

$$V_{T_{2n+1}}(t) = \begin{cases} -t^{-4} + t^{-3} + t^{-1} & , \quad n = 0 \\ V_{T_{2n-1}}(t) - t^{-2n-4} + t^{-2n-3} + t^{-2n-1} - t^{-2n} & , \quad n = 1, 2, 3, ... \end{cases}$$

Corollary 2.5. *The Laurent polynomial of* T_{2n} *can be calculated by*

$$W_{T_{2n}}(t) = -\left(t^{-2n} + t^{-2n+2} + t^{-2n+4} + ... + t^{-6} + t^{-4} + t^{-2}\right), (n = 1, 2, 3, ...). \quad (2.5)$$

(See, Table3).

Table 1. The Jones polynomial table for T_{2n}

T_0														**1**		
T_2												1	-1	**1**	-1	1
T_4										1	-1	1	-2	**2**	-1	1
T_6								1	-1	1	-2	2	-2	**2**	-1	1
T_8						1	-1	1	-2	2	-2	2	-2	**2**	-1	1
...	.	.	.	.	.	.	.	.	.	.	.	.	.	**.**	.	.
T_{2n}	1	-1	1	-2	2	...	...	-2	2	-2	2	-2	2	−2	**2**	-1	1
				$n-1$		...		4		3		2		1			

Table 2. The Jones polynomial table for T_{2n+1}

T_1													-1	1	0	1
T_3											-1	1	-1	2	-1	1
T_5									-1	1	-1	2	-2	2	-1	1
T_7							-1	1	-1	2	-2	2	-2	2	-1	1
T_9					-1	1	-1	2	-2	2	-2	2	-2	2	-1	1
...	.	.	.	.	.	.	.	.	.	.	.	.	.	.	.	.	
T_{2n+1}	-1	1	-1	2	-2	...	2	-2	2	-2	2	-2	2	-2	2	-1	1
				$n-1$		...	4		3		2		1				

Proof. Clearly, the relation in (2.5) can be obtained by using (1.1) and (2.1). □

Remark. We note that since $V_{T_0} = 1$, it follows easily that $W_{T_0} = 0$.

Corollary 2.6. *The Laurent polynomial of T_{2n+1} can be calculated by*

$$W_{T_{2n+1}}(t) = t^{-2n-4} + t^{-2n-2} + t^{-2n} + ... + t^{-8} + t^{-6} + t^{-4}, (n = 0, 1, 2, ...) . \quad (2.6)$$

(See, Table4).

Proof. The proof of this corollary can be completed taking into consideration (1.1) and Corollary 2.4. □

Corollary 2.7. *The Arf invariant of T_{2n} can be written as*

$$ArfT_{2n} = \begin{cases} 0 & , \quad n \ even \\ 1 & , \quad n \ odd \end{cases}$$

Proof. Since there exist $n-$terms in (2.5), then we have from (1.2) and (2.5) that

$$ArfT_{2n} = \begin{cases} 0 & , \quad n \ even \\ 1 & , \quad n \ odd \end{cases}$$

□

Table 3. The Laurent polynomial table for T_{2n}

T_0													0
T_2													-1
T_4											-1	0	-1
T_6									-1	0	-1	0	-1
T_8							-1	0	-1	0	-1	0	-1
...	.	.	.	.	.	.	.	.	.	.	.	.	.
T_{2n}	-1	0	...	...	-1	0	-1	0	-1	0	-1	0	-1
	$n-1$		...	...	4		3		2		1		

Table 4. The Laurent polynomial table for T_{2n+1}

T_1													1
T_3											1	0	1
T_5									1	0	1	0	1
T_7									1	0	1	0	1
T_9							1	0	1	0	1	0	1
...	.	.	.	.	.	.	.	.	.	.	.	.	.
T_{2n+1}	1	0	...	...	1	0	1	0	1	0	1	0	1
	n		...	...	4		3		2		1		

Corollary 2.8. *The Arf invariant of T_{2n+1} can be written as*

$$ArfT_{2n+1} = \begin{cases} 0 & , \quad n \ odd \\ 1 & , \quad n \ even. \end{cases}$$

Proof. Since there exist $n+1-$terms in (2.6), then we can conclude the result from (1.2) and (2.6). □

3. The Conway Polynomial of T_n

Theorem 3.1. *The Alexander polynomials of T_{2n} and T_{2n+1} are different and are as follows*

$$\Delta_{2n}(t) = nt^2 - (2n+1)t + n$$

and

$$\Delta_{2n+1}(t) = (n+1)^2 t^2 - (2n+1)t + (n+1),$$

respectively.

Proof. See [2].

Recall that

$$\Delta_{2n}(t) = 1 - n\left(t + t^{-1} - 2\right),$$

$$\Delta_{2n+1}(t) = 1 + (n+1)\left(t + t^{-1} - 2\right).$$

Those are normalized Alexander polynomials of T_{2n} and T_{2n+1} respectively. □

Theorem 3.2. *The Conway polynomials of* T_{2n} *and* T_{2n+1} *are different and are as follows, respectively,*

$$\nabla_{2n}(t) = 1 - nt^2,$$
$$\nabla_{2n+1}(t) = 1 + (n+1)t^2.$$

Proof. Since

$$\Delta_{2n}(t) = 1 - n\left(t + t^{-1} - 2\right),$$

then it is clear that

$$\Delta_{2n}\left(t^2\right) = 1 - n\left(t - t^{-1}\right)^2. \tag{3.1}$$

But it is known that by (1.3),

$$\Delta_{T_{2n}}\left(t^2\right) = \nabla_{T_{2n}}\left(t^2\right)\left(t - t^{-1}\right). \tag{3.2}$$

Now, it follows from (3.1) and(3.2) that

$$\nabla_{T_{2n}}\left(t - t^{-1}\right) = 1 - n\left(t - t^{-1}\right)^2.$$

Thus,

$$\nabla_{2n}(t) = 1 - nt^2.$$

Similarly, for $\nabla_{2n+1}(t)$, we can easily find

$$\nabla_{T_{2n+1}}(t) = 1 + (n+1)t^2.$$

This completes the proof of the theorem. □

Corollary 3.3.

$$ArfT_{2n} = \begin{cases} 0 & , \quad \Delta_{T_{2n}}(-1) \equiv 1(mod8) \\ 1 & , \quad \Delta_{T_{2n}}(-1) \equiv 5(mod8). \end{cases}$$

Proof. The proof this corollary can be seen from the fact $\Delta_{T_{2n}}(-1) = 4n + 1$. (Compare, [8, Corollary 20]). □

Corollary 3.4.

$$ArfT_{2n+1} = \begin{cases} 0 & , \quad \Delta_{T_{2n+1}}(-1) \equiv 3(mod8) \\ 1 & , \quad \Delta_{T_{2n+1}}(-1) \equiv 5(mod8) \end{cases}$$

Proof. Similarly, by considering the fact $\Delta_{T_{2n+1}}(-1) = 4n + 3$, ones can complete the proof. (Compare, [12,Theorem 10.7]). □

References

[1] Alexander, J. W., Topological invarints of knots and links. *Trans. Amer.Math. Soc.* **30** (1928), 275-306.

[2] Altun, Y., Bozhüyük, M. E., Some properties of shepherd's knots.(in Turkish), *Doga Seri* A_1, **8** (1984), 1-5.

[3] Altun, Y., Bozhüyük, M. E., The group of twist knots, *Yuzuncu Yil University, Jour. of Fac. of Education*, Vol 1, no. 2 (1996), 89-102.

[4] Bing, R. A., Martin, J. M., Cubes with knoted holes, *Trans. of the Amer. Math. Soc.* **155** (1971), 217-231.

[5] Burde, G., Zieschang, H., Knots, Berlin-Newyork de Gruyter (1985).

[6] Conway, J. H., An enumeration of knots and links, *Comp. problems in abstract algebra,* Pergamon, New york (1970) 329-358.

[7] Crowel, R. H., Fox, R. H., Introduction to knot theory, *Blaisdell-Ginn*, New york (1963).

[8] Jones, V., A polynomial invariant for knots via von Neumann algebras, *Bull. AMS* **12** (1985) 103-111.

[9] Jones, V., Hecke algebra representations of braid groups and link polynomials, *Annals of Math.,* **126** (1987) 335-388.

[10] Kauffman, L. H., State models and the Jones polynomial, *Topology* **26**(1987) 395-407.

[11] Landvoy, R. A., The Jones polynomial of pretzel knots and links, *Topology and its Appl.* **83** (1998) 135-147.

[12] Lickorish, W. B. R., An introduction to knot theory, *Springer-Verlag New york Inc.,* (1997).

[13] Thistlethwaite,M. B., A spanning tree expansion of the Jones polynomial, *Topology,* Vol.26, No.3, (1987) 297-309.

[14] Zieschang, H., On the Alexander and Jones polynomial, M. E. Bozhüyük (ed.), Topics in knot theory, *Kluwer-Acad. Publishers,* Printed in Netherlands, (1993) 229-257.

In: Mathematics, Game Theory and Algebra Compendium ISBN 978-1-60456-659-8
Editor: Jacob H. Mathias, pp. 31-41

Chapter 5

Abstract Ideal Theory: Multiplicative Lattices and x-Systems*

D. D. Anderson†
Department of Mathematics, The University of Iowa ,
Iowa City, IA 52242
Communicated by Ayman Badawi

Abstract

We survey some aspects of abstract ideal theory including multiplicative lattices and especially the work of K. E. Aubert concerning x-systems.

2000 AMS Mathematics Subject Classification: 13-02, 13A15.

This paper is a slight elaboration of a talk given at the International Symposium on Commutative Rings and Monoids dedicated to the 60th birthday of Prof. Franz Halter-Koch held September 29-October 1, 2004 at the Institute of Mathematics, The University of Graz, Graz, Austria. The author would like to take this opportunity to thank the organizers of this conference, especially Alfred Geroldinger.

There are at least three different approaches to abstract ideal theory: (1) multiplicative lattices, (2) x-systems and r-systems, and (3) partially ordered abelian groups. In this talk we briefly address (1) - a fuller account is given in [5] and cover (2), especially K. E. Aubert's work, in more detail. Later work on (2) may be found in F. Halter-Koch's text [21]. We do not discuss (3) which is fully elaborated by P. Jafford [22] and J. Močkoř [26].

By the late 1800's Dedekind (1831–1916) had introduced the notion of an ideal. By the early 1900's ideal theory, at least for number rings and polynomial rings, was well established. For example, in 1905 E. Lasker (1868–1941), a world chess champion, gave primary decomposition for $k[X_1, \cdots, X_n]$, k a field. But ideal theory in general (Noetherian) rings really got its start in 1921 with E. Noether's (1882–1935) classic paper [27] and

*To Franz Halter-Koch on his sixtieth birthday
†E-mail address: dan-anderson@uiowa.edu

later in the work of W. Krull (1899–1971). In fact, by 1935 we have Krull's *Idealtheorie* [24]. However, already by 1924 abstract ideal theory had started with Krull's paper [23].

The setting for Krull's work was a lattice with a multiplication, that is, a multiplicative lattice. What is (or should) a multiplicative lattice $\mathcal{L}$ be? $\mathcal{L}$ is a lattice with join $\vee$ and meet $\wedge$, often (usually) is complete with least element O and greatest element I, and usually is modular ($B \leq A \Rightarrow A \wedge (B \vee C) = B \vee (A \wedge C)$). $\mathcal{L}$ has a commutative, associative multiplication with $A\,(B \vee C) = AB \vee AC$ ($A\,(\bigvee B_\alpha) = \bigvee AB_\alpha$ when $\mathcal{L}$ is complete), and $IA = A$ when I exists. It is often assumed that $\mathcal{L}$ is compactly generated and that I is compact. (Recall that A is compact if $A \leq \bigvee A_\alpha \Rightarrow A \leq A_{\alpha_1} \vee \cdots \vee A_{\alpha_n}$ for some finite subset.)

For example, for a commutative ring R, the lattice $\mathcal{L}\,(R)$ of ideals of R is a multiplicative lattice. Here an ideal is compact as an element of $\mathcal{L}\,(R)$ if and only if it finitely generated.

Actually, Krull's work had little influence. The real start was papers by M. Ward (1901–1963), R. P. Dilworth (1914–1993), and Ward and Dilworth, especially the 1939 Ward/Dilworth paper [29]. At this stage they already saw the connection between the ideal theory of commutative rings and semigroups [30]. (Ward was a student of E. T. Bell and Dilworth was a student of Ward. Dilworth's thesis involved multiplicative lattices. At one time, Bell, Ward, Dilworth, and Dilworth's student R. S. Pierce were all on the faculty of Cal Tech.)

Interestingly, in the same year 1939 P. Lorenzen (1915–1994) introduced another setting for abstract ideal theory, the r-system. The r-system and its generalization the x-system of K. E. Aubert will be our main focus. But first back to multiplicative lattices.

To get deeper results in a multiplicative lattice we need the analog of a principal ideal. Ward/Dilworth [29] called an element A "principal" if for $B \leq A$, $B = CA$ for some C. So a principal ideal (a) of a commutative ring R is a principal element of $\mathcal{L}\,(R)$. They proved the following theorem.

Theorem 1. *Let $\mathcal{L}$ be a modular multiplicative lattice with ACC in which every element is a (finite) join of "principal" elements. Then each element ($\neq I$) of $\mathcal{L}$ is a finite meet of primary elements.*

However, this form of "principal element" was too weak to obtain deeper results. Twenty-three years later in 1972 Dilworth remedied this in his fundamental paper [18].

He defined an element a of a multiplicative lattice $\mathcal{L}$ to be *principal* if (1) $b \wedge ca = ((b{:}a) \wedge c)\,a$ and (2) $b \vee (c{:}a) = (ba \vee c){:}a$ for all $b, c \in \mathcal{L}$. Here $b{:}a = \bigvee \{c \mid ca \leq b\}$. An element satisfying (1) (resp., (2)) is now called *meet principal* (resp., *join principal*). Observe that (1) and (2) are dual where we interchange $\bigvee$ and $\bigwedge$ and $\cdot$ and :. For results on principal elements the reader is referred to [4].

An ideal A of a commutative ring R is a principal element of $\mathcal{L}\,(R)$ if and only if A is finitely generated and locally principal [4].

Dilworth defined a *Noether lattice* to be a complete modular multiplicative lattice that satisfies ACC in which every element is a (finite) join of principal element. He showed that a Noether lattice satisfies (1) the normal decomposition theory (of primary elements), (2) the Krull Intersection Theorem, and (3) the Generalized Principal Ideal Theorem.

There is an extensive literature on Noether lattices and "modules" over them. There is a highly developed theory of completion, localization, dimension theory and Hilbert functions, and distributive Noether lattices have been classified.

For the non-Noetherian case we have the notion of an r-lattice introduced in [2]. An *r-lattice* is a complete modular multiplicative lattice that is compactly generated, principally generated, and has greatest element I compact.

Examples of Multiplicative Lattices

(1) R a commutative ring with 1. $\mathcal{L}(R)$ is an r-lattice and $\mathcal{L}(R)$ is a Noether lattice $\Leftrightarrow R$ is Noetherian.

(2) R a commutative ring with 1. Let
$\mathcal{L}_f(R) = \{\text{faithful ideals of } R\} \cup \{0\}$
$\mathcal{L}_r(R) = \{\text{regular ideals of } R\} \cup \{0\}$
$\mathcal{L}_{rg}(R) = \{\text{regularly generated ideals of } R\} \cup \{0\}$.
These are all multiplicative lattices, but need not be r-lattices. $\mathcal{L}_r(R)$ is an r-lattice if R is Marot or Prüfer.

(3) R a graded ring. Then $\mathcal{L}_h(R)$, the lattice of homogeneous ideals of R, is an r-lattice.

(4) R a commutative semiring with 1 (same axioms as a commutative ring except $(R, +)$ is only a monoid). Here $\mathcal{L}(R)$ is a compactly generated, principally generated multiplicative lattice that need not be modular. (For example, $\mathcal{L}(R[X])$ is modular $\Leftrightarrow R$ is a ring.) See [1] for results on the lattice of ideals of a semiring.

(5) S a commutative semigroup; here we might as well assume S has a 0 and 1. Let $\mathcal{L}(S)$ be the lattice of ideals of S. Here a semigroup ideal is a nonempty subset I of S with the property that $s \in S$ and $i \in I \Rightarrow si \in I$. So $I = \bigcup\{Si \mid i \in I\}$. In $\mathcal{L}(S)$, $\bigvee I_\alpha = \bigcup I_\alpha$ and $\bigwedge I_\alpha = \bigcap I_\alpha$. So $\mathcal{L}(S)$ is distributive. Also, S (or $\mathcal{L}(S)$) is quasilocal, and the principal ideal Sa while compact need not be a principal element (it is meet principal). $\mathcal{L}(S)$ is an r-lattice $\Leftrightarrow ab = ac \neq 0 \Rightarrow b = \lambda c$, λ a unit. Call such an S an *r-semigroup*. See [3] for the ideal theory of commutative semigroups and its relation to abstract ideal theory.

$\mathcal{L}(S)$, S a semigroup, or more generally any distributive multiplicative lattice, has the odd property that for $\{P_a\}$ a set of prime elements, $\bigvee P_a$ is either prime or I, the greatest element. Thus we have no analog of the important ring property that if an ideal $J \subseteq P_1 \cup \cdots \cup P_n$, a finite union of prime ideals, then $J \subseteq$ some P_i. This is due to "the lack of an addition" and is also a problem for x-systems. So sometimes we assume that a multiplicative lattice $\mathcal{L}$ satisfies (1) the *weak union condition*: given $A, B, C \in \mathcal{L}$ with $A \nleq B$ and $A \nleq C$, there exists a principal element $E \leq A$ with $E \nleq B$ and $E \nleq C$, or (2) the *union condition on primes*: given $A, P_1, \cdots, P_n \in \mathcal{L}$ with $P_1, \cdots, P_n$ prime and each $A \nleq P_i$, there exists a principal element $E \leq A$ with $E \nleq P_i$ for each i.

As is to be expected, more is known about multiplicative lattices in the distributive case. The following is a sample.

Theorem 2.

(1) *Let $\mathcal{L}$ be a quasilocal distributive r-lattice. Then $\mathcal{L} \approx \mathcal{L}(S)$ for some r-semigroup S.*
(2) *A distributive r-lattice is embeddable into the lattice of ideals of a commutative ring.*
(3) *Let $\mathcal{L}$ be an r-lattice domain. Then the following are equivalent: (a) $\mathcal{L}$ is distributive and satisfies the weak union condition, (b) for each maximal element M of $\mathcal{L}$, $\mathcal{L}_M$ is totally ordered, (c) $A(B \wedge C) = AB \wedge AC$ for all $A, B, C \in \mathcal{L}$, (d) every compact element of $\mathcal{L}$ is principal, and (e) $\mathcal{L} \approx \mathcal{L}(R)$ for R a Prüfer domain.*

Some additional references for multiplicative lattices are [4],[5],[19], and [21]. We have kept the discussion of multiplicative lattices brief since they are already surveyed in [5]. Workers in multiplicative lattices include D. D. Anderson and his students, Ward, Dilworth and his student K. P. Bogart, several students of L. J. Ratliff, especially the Johnson brothers E. W. Johnson and J. A. Johnson and their students, K. E. Aubert, I. Fleischer, P. J. McCarthy, H. M. Nakkar, and C. Jayaram.

Let me end our discussion of multiplicative lattices with two open questions.

(1) Let $\mathcal{L}$ be a Noether lattice or an r-lattice. When is $\mathcal{L}$ embeddable into the lattice of ideals of a commutative ring? While false in general, this is true in the distributive case.
(2) Let $\mathcal{L}$ be a regular local Noether lattice. Does $\mathcal{L}$ have unique factorization? This is the case for $\dim \mathcal{L} \leq 2$ or if $\mathcal{L}$ is a *small* regular local Noether lattice, that is, $\mathcal{L}$ has only finitely many height-one primes. Note that $\mathcal{L}$ is small regular local of dimension $n \Leftrightarrow \mathcal{L}$ is distributive regular local of dimension $n \Leftrightarrow \mathcal{L} \approx \mathcal{L}(S)$, S a free commutative monoid on n generators, $\approx$ the lattice of monomial ideals in an R-sequence of length n.

What is an ideal? Now in a multiplicative lattice $\mathcal{L}$, an "ideal" is just an element of $\mathcal{L}$. Now Dedekind introduced ideals to restore unique factorization. Later the central role of ideals became apparent when it was realized that ideals are kernels of ring homomorphisms.

Already in the late 1920's we have the notion of a v-ideal. In 1929 J. Arnold wrote on ideal theory in commutative semigroups.

In 1932 we have the important paper [28] by H. Prüfer (1896-1934). Prüfer looks at the general notion of a (fractional) "ideal" generated by a set. The abstraction of an ideal generated by a set will lead to a second abstract ideal theory. Let D be an integral domain with quotient field K. For $X \subseteq K$, we want the "$\star$-ideal" $X^\star$ generated by X (with some conditions on X). What properties should $X \longrightarrow X^\star$ have? Prüfer postulated the following four conditions.

(1) $X \subseteq X^\star$,
(2) $X \subseteq Y^\star \Rightarrow X^\star \subseteq Y^\star$
$\}$ $\star$ is a closure operation
(3) $X^\star$ should be a fractional ideal (so this puts the condition on X that there exists a nonzero $r \in D$ with $rX \subseteq D$: So $\star : B(K) \longrightarrow B(K)$ where $B(K) =$ bounded sets of K.)
(4) $\{a\}^\star = aD$ for $0 \neq a \in K$, so $(a)^\star = (a)$, and
(5) $(aX)^\star = aX^\star$.

Note that (1) and (3) give that $(X) \subseteq X^\star$ and $X^\star = (X)^\star$, so we could just as well define $\star$ on $F(D)$, the set of nonzero fractional ideals of D. This is what is usually done today for rings and we get the notion of a star-operation. A *star-operation* $\star$ on $F(D)$ is a function $\star : F(D) \longrightarrow F(D)$ satisfying (1) $A \subseteq A^\star$, $(A^\star)^\star = A^\star$ (so $\star$ is a closure operation), (2) $(aA)^\star = aA^\star$ for $a \in K^* = K - \{0\}$, and (3) $(a)^\star = (a)$ (or equivalently $D^\star = D$ by (2)). See [20] for an introduction to star-operations.

Examples 3. *Let D be an integral domain with quotient field K and let $X \in B(K)$.*

(1) $X_d = (X) = \{r_1x_1 + \cdots + r_nx_n \mid r_i \in D, x_i \in X\}$, the (usual) ideal generated by X. (Here d stands for Dedekind.)

(2) $X_v = (X)_v = ((X)^{-1})^{-1} = \bigcap\{(z) \mid X \subseteq (z), z \in K^*\}$.

(3) $X_t = (X)_t = \bigcup\{Y_v \mid \phi \neq Y \subseteq X, |Y| < \infty\}$.

Prüfer's work was put in an abstract setting by P. Lorenzen [25] via the notion of an r-system.

Let H be a cancellative commutative monoid with quotient group G. Let $B(H) = \{\phi \neq X \subseteq G \mid \text{there exists } h \in H \text{ with } hX \subseteq H\}$. An *$r$-system* on H is an operation $r{:}B(H) \longrightarrow B(H)$ that satisfies

(1) $X \subseteq X_r$,
(2) $X \subseteq Y_r \Rightarrow X_r \subseteq Y_r$, } r is a closure operation
(3) $\{a\}_r = aH$, and
(4) $aX_r = (aX)_r$.

Examples 4.

(1) Let D be an integral domain with quotient field K. Take $H = D^* = D - \{0\}$, $G = K^* = K - \{0\}$. If $\star$ is a star-operation on $F(D)$, then $\star$ gives an r-system on G by defining $X_r = (X)^\star$. In particular, we have the d-system, the v-system, and the t-system.

(2) H a commutative cancellative monoid and $X_s = \bigcup\{Hx \mid x \in H\}$. So X_s is the semigroup ideal generated by X.

We will give more examples later. Note the Lorenzen's framework is rather restrictive in the H must be a cancellation semigroup with identity. So the theory does not apply to rings with zero divisors or without identity. Also, the condition $\{a\}_r = Ha$ is somewhat restrictive. Of course, the first objection is easy to overcome. Take H to be a commutative semigroup and let $G = \{h/k \mid h, k \in H, ka = kb \Rightarrow a = b, a, b \in H\}$ - the total quotient semigroup of H.

To overcome these restrictions, especially the condition that $\{a\}_r = Ha$, K. E. Aubert (1924–1991) introduced the notion of an x-system. This work which was part of his 1957 dissertation [7] was first announced in [6] and finally published as the massive work [8]. (K. E. Aubert once told the author that it had been a mistake to publish the long paper [8]; the work would have been more widely read if it has been presented in a series of shorter papers.)

Definition of an (Integral) x-System

Let S be a commutative semigroup. There is defined an *x-system on S* if for every $\phi \neq A \subseteq S$, there exists $A_x \subseteq S$ satisfying

(1) $A \subseteq A_x$ (so $S_x = S$),
(2) $A \subseteq B_x \Rightarrow A_x \subseteq B_x$, and $\Big\}$ x is a closure operation
(3) $AB_x \subseteq B_x \cap (AB)_x$ (or equivalently, $aB_x \subseteq B_x \cap (aB)_x$ for all $a \in S$)
Here (3) is equivalent to
(3.1) $AB_x \subseteq B_x$ or $SB_x \subseteq B_x$, that is, B_x is a semigroup ideal, and
(3.2) $AB_x \subseteq (AB)_x$ (or equivalently, $aB_x \subseteq (aB)_x$) (The *Continuity Axiom*)

We could assume that $0, 1 \in S$ and then replace (1) by $A \cup \{0\} \subseteq A_x$.

Generalizing Lorenzen, an (*integral*) r-*system* on a semigroup S (usually with $0, 1$) is a mapping that sends each nonempty subset $A \subseteq S$ to $A_r \subseteq S$ satisfying

(1) $A \subseteq A_r$ (or $A \cup \{0\} \subseteq A_r$),
(2) $A \subseteq B_r \Rightarrow A_r \subseteq B_r$,
(3) $\{a\}_r = aS^1$, $S^1 = S \cup \{1\}$ if needed, and
(4) $aA_r = (aA)_r$.

Note that $b \in B \Rightarrow bS^1 = \{b\}_r \subseteq B_r$ so (3) gives that B_r is a semigroup ideal of S. So (3.1) holds and of course (4)$\Rightarrow$(3.2). If S (or $S - \{0\}$) is a group, then $aB_x \subseteq (aB)_x$ for all $a, B \Leftrightarrow (aB)_x \subseteq aB_x$ for all $a, B \Leftrightarrow aB_x = (aB)_x$ for all a, B. So in this case the notions of x-system and r-system coincide.

Let x be an x-system on a semigroup S. Define for subsets of S the x-*product* $A \circ B = (AB)_x$ and the x-*sum* $A \cup_x B = (A \cup B)_x$ or more generally, $\bigcup_x A_i = (\bigcup A_i)_x$. Note that we always have $\bigcap_i (A_i)_x = (\bigcap_i (A_i)_x)_x$. The x-ideals form a complete lattice with respect to $\subseteq$.

Aubert was very proud of the Continuity Axiom. He believed it provided exactly the right level of generality to do abstract ideal theory. This is partially justified by his following theorem and also as we will see later by his partial converse of Krull's Theorem.

Theorem 5. *For a semigroup S and closure operation $A \longrightarrow A_x$, the following conditions are equivalent.*

(1) The Continuity Axiom: $AB_x \subseteq (AB)_x$ for all $A, B \subseteq S$.
(2) $A \circ B = A \circ B_x$ (or $A \circ B = A_x \circ B_x$) for all $A, B \subseteq S$.
(3) $A(B \cup_x C) \subseteq AB \cup_x AC$ for all $A, B, C \subseteq S$.
(4) $A \circ (B \cup_x C) = A \circ B \cup_x A \circ C$ for all $A, B, C \subseteq S$.
(5) $(A_x{:}B)_x = A_x{:}B$ for all $A, B \subseteq S$.
(6) $A_x{:}\cup_x B_i = \bigcap (A_x{:}B_i)$ for all $A, B_i \subseteq S$.
If S has an identity, we can further add
(7) $(A \circ B) \circ C = A \circ (B \circ C)$ for all $A, B, C \subseteq S$, and
(8) $(A_x{:}B){:}C = A_x{:}(B \circ C)$ for all $A, B, C \subseteq S$.

So the set of x-ideals of an x-system (S, x) (where S has a 0) forms a multiplicative lattice with $A_x \vee B_x = A_x \cup_x B_x$ and $A_x \wedge B_x = A_x \cap B_x$ which is complete ($\bigvee A_x^{(i)} = \bigcup_x A_x^{(i)}$ and $\bigwedge A_x^{(i)} = \bigcap A_x^{(i)}$ and has greatest element S and least element 0). Note that the Continuity Axiom gives that $A(\bigvee_\alpha B_\alpha) = \bigvee_\alpha AB_\alpha$ and $A \circ B = (A_x B_x)_x$ is associative. However, the lattice need not be modular as seen by the lattice of ideals of a semiring.

Aubert remarks that the Continuity Axiom can not be formulated in a multiplicative lattice.

Example 6. *(An x-system NOT) A commutative ring R is a* differential ring *if there is a map $\delta: R \longrightarrow R$ with $\delta(a+b) = \delta(a) + \delta(b)$ and $\delta(ab) = a\delta(b) + \delta(a)b$ for all $a, b \in R$. A* δ^*-ideal *is an ideal A of R that satisfies $a \in A \Rightarrow \delta(a) \in A$. Then $X \longrightarrow X_{\delta^*} = \delta^*$-ideal generated by X is* not *an x-system. For take $R = Z[x]$ with δ the usual derivative. Then the Continuity Axiom fails. Here $x\{x\}_{\delta^*} \subsetneq \{x^2\}_{\delta^*}$. Now $1 = \delta(x) \in \{x\}_{\delta^*} \Rightarrow \{x\}_{\delta^*} = \mathbb{Z}[x] \Rightarrow x\{x\}_{\delta^*} = x\mathbb{Z}[x]$. However, $x^2, 2 \in \{x^2\}_{\delta^*}$ and $(x^2, 2)$ is a δ^*-ideal; so $\{x^2\}_{\delta^*} = (x^2, 2)$. Hence $x\mathbb{Z}[x] \nsubseteq (x^2, 2)$.*

However, we do get an x-system for the differential ring R and $X \longrightarrow X_\delta$ =the perfect δ^*-ideal generated by X =smallest radical δ^*-ideal $\supseteq X$. This x-system is called the δ-system.

Call a closure operation $A \longrightarrow A_{x^*}$ on a semigroup S an *x^*-system* if (3.1) $aB_{x^*} \subseteq B_{x^*}$ holds.

Let (S, x) be an x-system. Then x has *finite character* if $A_x = \bigcup\{B_x \mid B \subseteq A, |B| < \infty\}$. As for rings, an x-ideal $P_x \neq S$ is *prime* if $ab \in P_x \Rightarrow a \in P_x$ or $b \in P_x$, or equivalently, $A_x \circ B_x \subseteq P_x \Rightarrow A_x \subseteq P_x$ or $B_x \subseteq P_x$ and $\operatorname{rad} A_x = \{s \in S \mid s^n \in A_x \text{ for some } n \geq 1\}$. Aubert generalized Krull's theorem that the nilradical of an ideal (in a commutative ring) is the intersection of the minimal primes containing it.

Theorem 7 (Krull-Stone Theorem). *Let (S, x) be a finite character x-system. Then* $\operatorname{rad}(A_x) = \bigcap\{P_x \mid P_x \text{ is a minimal prime} \supseteq A_x\}$.

As a corollary we have the following.

Corollary 8 (Ritt-Raudenbush Theorem (1934))**.** *Let R be a differential ring. Then each perfect (=radical) differential ideal of R is an intersection of prime differential ideals.*

Aubert also gave a partial converse to the Krull-Stone Theorem. Call an x-system a *radical x-system* if each A_x is a radical ideal. The family of radical x-ideals of an x-system is also an x-system. The next result gives further evidence of the central role of the Continuity Axiom.

Theorem 9. *Let x^* be a finite character x^*-system on a semigroup S. Then x^* satisfies the Continuity Axiom (that is, x^* is an x-system) if and only if x^* satisfies the Krull-Stone Theorem.*

We next look at congruences on an x-system. Let (S, x) be an x-system and $A_x \subsetneq S$ an x-ideal. Define $b \equiv c\,(A_x)$ if $(A_x, b)_x = (A_x, c)_x$. While natural, this congruence does *not* reduce to the usual one in ring theory $b \equiv c \mod A_x \Leftrightarrow b - c \in A_x$. For example, in $\mathbb{Z}$ with $A_x = 3\mathbb{Z}$, $b \equiv c\,(3\mathbb{Z}) \Leftrightarrow (3, b) = (3, c)$. So $1 \equiv 2\cdot(3\mathbb{Z})$. Here there are two classes $\{n \mid n \equiv 0 \bmod 3\}$ and $\{n \mid n \equiv 1, 2 \bmod 3\}$.

To remedy this Aubert introduced the "additivity axiom" in [9].

Now the congruence relation $\equiv (A_x)$ leads to the residue class monoid S/A_x and a canonical multiplicative homomorphism $\phi{:}S \longrightarrow S/A_x$ given by $\phi(x) = \{y \in S \mid x \equiv y\,(A_x)\}$. (Note: The Continuity Axiom gives that $\equiv$ is actually a congruence. Assume $b \equiv c\,(A_x)$, so $(A_x, b)_x = (A_x, c)_x$. Then $db \in d\,(A_x, b)_x = d\,(A_x, c)_x \subseteq (dA_x, dc)_x \subseteq (A_x, dc)_x$. Likewise $dc \in (A_x, bc)_x$, so $(A_x, bd)_x = (A_x, cd)_x$. Hence $bd \equiv cd\,(A_x)$.)

Note that S/A_x is a monoid with 0. The 0 of S/A_x is the class A_x. We have an x-system on S/A_x, the x_ϕ-system, whose x_ϕ ideals are the subsets B of S/A_x with $\phi^{-1}(B)$ a x-ideal. So there is a one-to-one correspondence between the x-ideals of S containing A_x and the x_ϕ-ideals of S/A_x. For $\equiv (3\mathbb{Z})$ on $\mathbb{Z}$, the d-ideals containing $3\mathbb{Z}$ are $3\mathbb{Z}$ and $1\mathbb{Z} = \mathbb{Z}$. Here $\mathbb{Z}/(3\mathbb{Z})$ has two classes $\bar{0} = \phi((0)) = 3\mathbb{Z}$ and $\bar{1} = \phi((1)) = \{n \mid n \equiv 1, 2 \bmod 3\}$ and two d_ϕ-ideals $(\bar{0})$ and $(\bar{1})$. So $\mathbb{Z}/((3))$ is just the two element monoid $\{\bar{0}, \bar{1}\}$ with zero $\bar{0}$ and identity $\bar{1}$.

In what follows S is a commutative monoid with 0 and x is a finite character x-system on S. The x-system is *additive* if for $c \in A_x \cup_x B_x$, there exists $b \in B_x$ with $c \equiv b\,(A_x)$. Observe that the d-system on rings is additive: $c \in A_d + B_d \Rightarrow c = a + b$, $a \in A_d$, $b \in B_d$. So $(A_d, c) = (A_d, b)$, that is $c \equiv b\,(A_d)$. Aubert proved the following theorem [9].

Theorem 10. *Let x be a finite character x-system on a commutative monoid S with 0. Then the following conditions are equivalent.*

(1) x is an additive ideal system.
(2) For each x-ideal A_x, $\phi{:}S \longrightarrow S/A_x$ is the canonical map, then $\phi(A_x \cup_x B_x) = \phi(B_x)$ for all x-ideals B_x.
(3) $\phi^{-1}(\phi(B_x)) = A_x \cup_x B_x$ with notation as in (2).
(4) $\phi(B_x) = (\phi(B))_{x_\phi}$ with notation as in (2).
(5) $\phi(B_x \cup_x C_x) = \phi(B_x) \cup_{x_\phi} \phi(C_x)$ as in (2).
(6) $\phi(B \cup C)_{x_\phi} = \phi(B) \cup_{x_\phi} \phi(C)$ as in (2).
(7) $\phi(B_x \circ C_x) = \phi(B_x) \circ_{x_\phi} \phi(C_x)$ as in (2).
(8) $\phi(B \circ C) = \phi(B) \circ_{x_\phi} \phi(C)$ as in (2).
(9) The canonical map $B_x/A_x \cap B_x \longrightarrow (A_x \cup_x B_x)/A_x$ is surjective (or equivalently, bijective, or an isomorphism).

The proof of the equivalence of (1) and (9) uses the modular law. In fact, we have the following result, again from [9].

Theorem 11. *The lattice of x-ideals of an additive ideal system is modular.*

We have not yet discussed primary decomposition in Noetherian x-systems. As in the multiplicative lattice case, an irreducible x-ideal in a Noetherian semigroup need not be primary. But if the lattice of x-ideals is modular, we do have primary decomposition. Hence an additive Noetherian x-system does have primary decomposition.

Aubert has several other papers on x-systems involving localization of x-systems [10], modules over x-systems [17], tensor products of ideal systems [16], and ideal systems and lattice theory [11]-[13]. Finally, the reader may consult Aubert's collected works [15].

Let me end with the following quotes.

H. Weyl (1940) [31] "Therefore, when one widens the realm of elements to that of ideals in a given ring, one sometimes gains and sometimes loses. One gets the impression that, generally speaking, the truth lies halfway: if the domain of integers in many cases is too narrow, the domain of ideals is in most cases too wide."

The next quote by Aubert (1983) [14] reflects I believe Franz Halter-Koch's philosophy.

"...the notion of divisibility appears as a strictly multiplicative theory. ...a somewhat...irrelevant additive ingredient was...brought in...from the start.

The true multiplicative liberation came from Lorenzen's thesis in 1939. The purpose of this paper is to continue and revive some of the work of Lorenzen... Only Jaffard's monograph (our reference [23]) seems to us to do full justice to Lorenzen's ideas. This is a very fine book, but it is written in a style and uses terminology which may have prevented many from reading it... The present paper revolve(s) around... t-ideals. In fact, ... t-ideals represent "the truth that lies halfway" that is alluded to by ...Herman Weyl."

Let me end by saying that Aubert's statement that only Jaffard's monograph does justice to Lorenzen's theory is *no* longer true. Indeed, we now have [21] F. Halter-Koch, *Ideal Systems: An Introduction to Multiplicative Ideal Theory*, Dekker, New York, 1998.

Happy 60th Birthday Franz!

References

[1] F. E. Alarcón and D. D. Anderson, Commutative semirings and their lattices of ideals, *Houston J. Math.* **20** (1994), 571–590.

[2] D. D. Anderson, Abstract commutative ideal theory without chain condition, *Algebra Universalis* **6** (1976), 131–145.

[3] D. D. Anderson and E. W. Johnson, Ideal theory in commutative semigroups, *Semigroup Forum* **30** (1984), 127–158.

[4] D. D. Anderson and E. W. Johnson, Dilworth's principal elements, *Algebra Universalis* **36** (1996), 392–404.

[5] D. D. Anderson and E. W. Johnson, Abstract ideal theory from Krull to the present, *Ideal Theoretic Methods in Commutative Algebra* (eds. D. D. Anderson and I. Papick), Marcel Dekker, New York, 2001, 27–47.

[6] K. E. Aubert, Généralisation de la théorie des r-idéaux de Prüfer-Lorenzen, *C. R. Acad. Sci. Paris* **238** (1954), 2214–2216.

[7] K. E. Aubert, Contribution à la théorie des idéaux et à la théorie des valuations, Thesis, DSC, Paris, 1957.

[8] K. E. Aubert, The theory of x-ideals, *Acta Math.* **107** (1962), 1–52.

[9] K. E. Aubert, Additive ideal systems, *J. Algebra* **18** (1971), 511–528.

[10] K. E. Aubert, Localisation dans les systèmes d'idéaux, *R. R. Acad. Sci. Paris* **272** (1971), 465–468.

[11] K. E. Aubert, Ideal systems and lattice theory, I., *Algebra Universalis* **1** (1971), 204–213.

[12] K. E. Aubert, Ideal systems and lattice theory, II., *J. Reine Angew. Math.* **298** (1978), 32–42.

[13] K. E. Aubert, Ideal systems and lattice theory, III., *Math. Japon.* **24** (1979), 247–252.

[14] K. E. Aubert, Divisors of finite character, *Ann. Math. pure et app.* **33** (1983), 327–361.

[15] K. E. Aubert, *Collected Papers of Karl Egil Aubert* (edited by P. Ribenboim), *Queen's Papers in Pure and Applied Mathematics*, vol. 89, Queen's University, Kingston, ON, 1992.

[16] K. E. Aubert and I. Fleischer, Tensor products of ideal systems and their modules, *Pacific J. Math.* **74** (1978), 285–296.

[17] K. E. Aubert and E. R. Hansén, Systèmes de modules, *C. R. Acad. Sci. Paris* **272** (1971), 525–528.

[18] R. P. Dilworth, Abstract commutative ideal theory, *Pacific J. Math.* **12** (1962), 481–498.

[19] R. P. Dilworth, *The Dilworth Theorems* (edited by K. Bogart, R. Freese, and J. Kung), Birkhäuser, Boston, 1990.

[20] R. Gilmer, *Multiplicative Ideal Theory*, Queen's Papers in Pure and Applied Mathematics, vol. 90, Queen's University, Kingston, ON, 1992.

[21] F. Halter-Koch, *Ideal Systems: An Introduction to Multiplicative Ideal Theory*, Marcel Dekker, New York, 1998.

[22] P. Jaffard, *Les Systèmes d'Ideaux*, Dunod, Paris, 1960.

[23] W. Krull, Axiomatische Begründung der allgemeinen Idealtheorie, *Sitzungsbericht der physikalisch-medicischen Societat zu Erlangen* **56** (1924), 47–63.

[24] W. Krull, *Idealtheorie*, Springer-Verlag, Berlin, 1935.

[25] P. Lorenzen, Abstrakte Begründung der multiplikativen Idealtheorie, *Math. Z.* **45** (1939), 533–553.

[26] J. Močkoř, *Groups of Divisibility*, D. Reidel Publishing Company, Dordrecht/Boston/Lancaster, 1983.

[27] E. Noether, Idealtheorie in Ringbereichen, *Math. Ann.* **83** (1921), 24–66.

[28] H. Prüfer, Untersuchungen über Teilbarkeitseigenschaften in Korpen, *J. Reine Angew. Math* **168** (1932), 1–36.

[29] M. Ward and R. P. Dilworth, Residuated lattices, *Trans. Amer. Math. Soc.* **45** (1939), 335–354.

[30] M. Ward and R. P. Dilworth, The lattice theory of ova, *Ann. of Math.* **40** (1939), 600–608.

[31] H. Weyl, *The Algebraic Theory of Numbers*, Princeton University Press, Princeton, 1940.

In: Mathematics, Game Theory and Algebra Compendium
Editor: Jacob H. Mathias, pp. 43-51
ISBN 978-1-60456-659-8

Chapter 6

COMMUTATIVITY OF RINGS WITH CONSTRAINTS ON NILPOTENTS AND THE JACOBSON RADICAL

Mohammad Ashraf*
Department of Mathematics
Faculty of Science, King Abdulaziz University
P.O. Box. 80203, Jeddah 21589
Saudi-Arabia

Abstract

Let R be an associative ring with identity 1, J the Jacobson radical of R, and N the set of nilpotent elements of R. Suppose that $n \geq 1$ is a fixed positive integer. In the present paper, it is shown that R is commutative if and only if it satisfies the property $(xy)^n = y^n x^n$ for all $x, y \in R \backslash (N \cup J)$ and commutators in R are $n(n+1)$-torsion free. Further, similar result has been obtained for ring R satisfying the property $(xy)^n = (yx)^n$, for all $x, y \in R \setminus (N \cup J)$.

AMS Subject Classifications (1991): 16U80.

Key Words and Phrases: Jacobson radical, nilpotent elements, torsion free rings, centre, commutativity, polynomial identity.

1. Introduction

Throughout the paper R will denote an associative ring with identity 1, Z the centre of R, J the Jacobson radical of R, N the set of nilpotent elements of R, and C the commutator ideal of R. As usual, for any $x, y \in R$ we write $[x, y] = xy - yx$. For a fixed positive integer $n > 1$, and a non-empty subset A of R, consider the following ring properties:

$P_1(n, A)$ $\quad (xy)^n = x^n y^n$, for all $x, y \in A$.

*E-mail address: mashraf80@hotmail.com

$P_2(n, A)$	$(xy)^n = y^n x^n$, for all $x, y \in A$.
$P_3(n, A)$	$(xy)^n = (yx)^n$, for all $x, y \in A$.
$Q(n)$	For all $x, y \in R, n[x, y] = 0$ implies that $[x, y] = 0$.

The property $P_1(n, R)$ has been among the favourites of many ring theorists over the last few decades since Herstein [12], first studied this property and proved that a ring (may be without identity 1) satisfying $P_1(n, R)$ must have nil commutator ideal (for a fairly complete bibliography see Bell [10]). One of the interesting generalization of this result is due to Abu-Khuzam [1], which states that an $n(n-1)$-torsion free ring R satisfying $P_1(n, R)$ is commutative. Further, Bell and Yaqub [11,Theorem 2] established commutativity of R satisfying $P_1(n, R \setminus N)$ (resp.$P_3(n, R \setminus N)$) together with the property $Q(n(n-1))$ (resp. $Q(n)$). They also proved that the conclusion of their result remains true if the property $P_1(n, R \setminus N)$ (resp. $P_3(n, R \setminus N)$ is replaced by $P_1(n, R \setminus J)$ (resp.$P_3(n, R \setminus J)$. Similar kind of results are also available in [6] for rings satisfying either of the property $P_2(n, R \setminus N)$ or $P_2(n, R \setminus J)$. In a very recent paper Abu-Khuzam [3] improved the result of Bell and Yaqub and established that a ring satisfying $P_1(n, R \setminus (N \cup J))$ and $Q(n(n-1))$ must be commutative. Now, it is natural to question that whether the above result remains true if the property $P_1(n, R \setminus (N \cup J))$ is replaced by $P_2(n, R \setminus (N \cup J))$ or $P_3(n, R \setminus (N \cup J))$. In the present paper, we have investigated this problem and proved that a ring satisfying $P_2(n, R \setminus (N \cup J))$ (resp. $P_3(n, R \setminus (N \cup J))$) together with the property $Q(n(n+1))$ (resp. $Q(n)$) is commutative. Thus we generalize many well-known commutativity theorems.

2. Main Results

The main results of the present paper are as follows:

Theorem 2.1. A ring R is commutative if and only if R satisfies the properties $P_2(n, R \setminus (N \cup J))$ and $Q(n(n+1))$.

Theorem 2.2. A ring R is commutative if and only if R satisfies the properties $P_3(n, R \setminus (N \cup J))$ and $Q(n)$.

The following lemmas are essential in order to prove the above theorems:

Lemma 2.1([14, page 221]). If x, y are elements of a ring R (may be without identity 1) such that $[x, [x, y]] = 0$, then $[x^m, y] = mx^{m-1}[x, y]$, for all positive integers $m \geq 1$.

Lemma 2.2([10, Lemma]). Suppose that f is a polynomial function of two variables on R with the property that $f(x + 1, y) = f(x, y)$ for all $x, y \in R$. If there exists a positive integer m such that $x^m f(x, y) = 0$, for all $x, y \in R$, then $f(x, y) = 0$, for all $x, y \in R$.

A similar result holds if we assume $f(x, y)x^m = 0$ instead.

Lemma 2.3([15, Theorem]). Let f be a polynomial in n non-commuting indeterminates $x_1, x_2, \cdots, x_n$ with relatively prime integral coefficients. Then the following are equivalent:

(i) For any ring satisfying the polynomial identity $f = 0$, C is a nil ideal.

(ii) For every prime p, $G(F(p))_2$ fails to satisfy $f = 0$.

Proof of Theorem 2.1. Let u, v be units in R. By our hypothesis, we have

$$(uvu^{-1})^n = u^{-n}(uv)^n = u^{-n}v^n u^n.$$

On the other hand $(uvu^{-1})^n = uv^n u^{-1}$. Thus $uv^n u^{-1} = u^{-n}v^n u^n$ and hence $[u^{n+1}, v^n] = 0$, for all $u, v \in U$. This yields that $[u^{n(n+1)}, v^n] = 0$ and $[u^{n(n+1)}, v^{n+1}] = 0$, for all $u, v \in U$. Combining of these two relations gives that

$$[u^{n(n+1)}, v] = 0, \text{ for all } u, v \in U. \tag{2.1}$$

Assume that $x \in N \cup J$. Then $1 - x \in U$ and application of (2.1) yields that

$$[u^{n(n+1)}, x] = 0, \text{ for all } u \in U \text{and } x \in N \cup J. \tag{2.2}$$

Further, if $ux \in N \cup J$, then (2.2) implies that $[u^{n(n+1)}, ux] = 0$, and hence

$$[u^{n(n+1)}, x] = 0, \text{ for all } u \in U \text{ and } ux \in N \cup J. \tag{2.3}$$

Now suppose that $x \notin N \cup J$ and $ux \notin N \cup J$. Then by our hypothesis $(uxu^{-1})^n = u^{-n}(ux)^n = u^{-n}x^n u^n$. Since, $(uxu^{-1})^n = ux^n u^{-1}$, we find that $ux^n u^{-1} = u^{-n}x^n u^n$, and hence

$$[u^{n(n+1)}, x^n] = 0, \text{ for all } x \notin N \cup J \text{ and } ux \notin N \cup J. \tag{2.4}$$

Combining of (2.2),(2.3) and (2.4) yields that

$$[u^{n(n+1)}, x^n] = 0, \text{ for all } x \in R, \text{ and } u \in U. \tag{2.5}$$

In view of (2.5), we find that

$$[u^k, x^k] = 0, (k = n(n+1)) \text{ for all } x \in R, \text{ and } x \in U. \tag{2.6}$$

Now our objective is to prove that

$$(x^{k-1}u^k x^{-1})^{nk} = x^{nk^2}u^{nk^2}, \text{ for all} x \in R, \text{ and} u \in U. \tag{2.7}$$

Let $x^{k-1} \in N \cup J$ or $u^k x \in N \cup J$. Then $1 - x^{k-1} \in U$ or $1 - u^k x \in U$. Hence by (2.1), $[u^k, 1 - x^{k-1}] = 0$ or $[u^k, 1 - u^k x] = 0$. Thus in either case $[u^k, x^{k-1}] = 0$. Now combine this fact together with (2.6), to get

$$
\begin{aligned}
(x^{k-1}u^kx)^{nk} &= (u^kx^k)^{nk} \\
&= \underbrace{u^kx^k.u^kx^k\cdots\cdots u^kx^k}_{nk-times} \\
&= \underbrace{x^k.x^k\cdots x^k}_{nk-times}\underbrace{u^k.u^k\cdots u^k}_{nk-times} \\
&= x^{nk^2}u^{nk^2}
\end{aligned}
$$

-i.e. $(x^{k-1}u^kx)^{nk} = x^{nk^2}u^{nk^2}$, for all $x^{k-1} \in N \cup J$ or $u^kx \in N \cup J$.

On the other hand, if $x^{k-1} \notin N \cup J$ and $u^kx \notin N \cup J$, then we distinguish two cases.

Case(i). Let $x \in N \cup J$. Then by (2.2), $[u^k, x] = 0$. This implies that $[u^k, x^k] = 0, (k = n(n+1))$, and hence repeated application of these two relations yields that

$$
\begin{aligned}
(x^{k-1}u^kx)^{nk} &= ((x^ku^k)^n)^k \\
&= (\underbrace{x^ku^k.x^ku^k\cdots\cdots x^ku^k}_{n-times})^k \\
&= (x^{nk}u^{nk})^k \\
&= \underbrace{x^{nk}u^{nk}.x^{nk}u^{nk}\cdots\cdots x^{nk}u^{nk}}_{k-times} \\
&= x^{nk^2}u^{nk^2}
\end{aligned}
$$

This proves (2.7) in the present case.

Case(ii).Let $x \notin N \cup J$. Then applications of $P_2(n, R \setminus N \cup J)$, (2.4) and (2.6) give that

$$
\begin{aligned}
(x^{k-1}u^kx)^{nk} &= ((x^{k-1}u^kx)^n)^k \\
&= ((u^kx)^nx^{n(k-1)})^k \\
&= (x^nu^{nk}x^{n(k-1)})^k \\
&= (x^{nk}u^{nk})^k \\
&= \underbrace{x^{nk}u^{nk}.x^{nk}u^{nk}\cdots\cdots x^{nk}u^{nk}}_{k-times} \\
&= x^{nk^2}u^{nk^2}
\end{aligned}
$$

This establishes(2.7) in all cases.

For any $x \in R$ and $u \in U$, application of (2.6) yields that

$$
\begin{aligned}
(x^{k-1}u^kx)^{nk} &= \underbrace{x^{k-1}u^kx.x^{k-1}u^kx\cdots\cdots x^{k-1}u^kx}_{nk-times} \\
&= x^{k-1}(u^kx^k)^{nk-1}u^kx \\
&= x^{k-1}(x^ku^k)^{nk-1}u^kx \\
&= x^{k-1}x^{nk^2-k}u^{nk^2-k}u^kx \\
&= x^{nk^2-1}u^{nk^2-1}x
\end{aligned}
$$

Combine the above relation with (2.7), to get $x^{nk^2-1}[u^{nk^2}, x] = 0$, for all $x \in R$, $u \in U$. Now in view of Lemma 2.2, we have

$$[u^{nk^2}, x] = 0, \text{ for all } x \in R, u \in U. \tag{2.8}$$

Now our objective is to show that nilpotents in R are central. Let $a \in N$. Then there exists a minimal positive integer t such that

$$[a^m, x] = 0, \text{ for all } m \geq t. \tag{2.9}$$

If $t = 1$, then $a \in Z$. Therefore, assume that $t > 1$. Now replace u by $1 + a^{t-1}$ in (2.8), to get $[(1 + a^{t-1})^{nk^2}, x] = 0$, for all $x \in R$. Hence application of (2.9) gives that $nk^2[a^{t-1}, x] = 0$, for all $x \in R$ and $a \in N$. Since $k = n(n+1)$ and R has the property $Q(k)$, it has also the the property $Q(nk^2)$. Thus $[a^{t-1}, x] = 0$, for all $x \in R$. This contradicts the minimality of t in (2.9). Hence, $t = 1$ and in view of (2.9), we get

$$N \subseteq Z \tag{2.10}$$

Our next objective is to show that the commutator ideal of R is nil.

If $x \notin N \cup J$ and $y \notin N \cup J$, then by our hypothesis $(xy)^n = y^n x^n$. Since $(xy)^n x = x(yx)^n$, we find that $y^n x^{n+1} = x^{n+1} y^n$ -i.e.

$$[x^{n+1}, y^n] = 0, \text{ for all } x \notin N \cup J \text{and } y \notin N \cup J. \tag{2.11}$$

Now if $x \in N \cup J$, then $1 - x \in U$, and in view of (2.8),we get

$$[(1-x)^{nk^2}, y] = 0, \text{ for all } y \in R \text{and } x \in N \cup J. \tag{2.12}$$

Similarly, if $y \in N \cup J$, then $1 - y \in U$ and hence

$$[x, (1-y)^{nk^2}] = 0, \text{ for all } x \in R \text{and } y \in N \cup J. \tag{2.13}$$

Combine (2.11), (2.12) and (2.13), to get

$$[[[x^{n+1}, y^n], (1-x)^{nk^2}], (1-y)^{nk^2}] = 0, \text{ for all } x, y \in R. \tag{2.14}$$

We see that (2.14) is a polynomial identity satisfied by all elements of R. But if we consider $x = e_{11}, y = e_{11} + e_{12}$, we find that no ring of 2×2 matrices over $GF(p), p$ a prime satisfies this equality. Hence, it follows by Lemma 2.3 that the commutator ideal of R is nil. Combining this fact together with (2.10), we have

$$C \subseteq N \subseteq Z. \tag{2.15}$$

Thus Lemma 2.1 is applicable in the present situation and in view of (2.8), we get $nk^2 u^{nk^2-1}[u, x] = 0$, for all $x \in R$ and $u \in U$. This implies that $nk^2[u, x] = 0$. But R satisfied the property $Q(nk^2)$ and hence $[u, x] = 0$, for all $x \in R$ -i.e.

$$U \subseteq Z. \tag{2.16}$$

If $x \in N \cup J$, then in view of (2.16), $1 - x \in U \subseteq Z$ and hence

$$x \in Z \text{ and } (xy)^n = y^n x^n, \text{ for all } y \in R, \ x \in N \cup J. \tag{2.17}$$

Similarly, if $y \in N \cup J$, then $1 - y \in U \subseteq Z$ and we get

$$y \in Z \text{ and } (xy)^n = y^n x^n, \text{ for all } x \in R, \ y \in N \cup J. \tag{2.18}$$

Combining of (2.17) and (2.18) together with the property $P_2(n, R \setminus (N \cup J))$ gives that

$$(xy)^n = y^n x^n \text{ for all } x, y \in R. \tag{2.19}$$

Since $(xy)^n x = x(yx)^n$ for all $x, y \in R$, application of (2.19) gives that $[x^{n+1}, y^n] = 0$ for all $x, y \in R$. Now apply Lemma 2.1, to get $(n+1)x^n[x, y^n] = 0$. This in view of Lemma 2.2 gives that $(n+1)[x, y^n] = 0$. Since R has the property $Q(n+1)$ also,we find that $[x, y^n] = 0$, for all $x, y \in R$. Further application of Lemma 2.1 gives that $ny^{n-1}[x, y] = 0$. Now similar arguments as above,leads to $[x, y] = 0$, for all $x, y \in R$. This completes the proof of our theorem.

Corollary 2.1([6, Theorem 1]). A ring R is commutative if and only if R satisfies the properties $P_2(n, R \setminus N)$ and $Q(n(n+1))$.

Corollary 2.2([6, Theorem 2]). A ring R is commutative if and only if R satisfies the properties $P_2(n, R \setminus J)$ and $Q(n(n+1))$.

Proof of Theorem 2.2. Let $u, v \in U$.By our hypothesis $P_3(n, R \setminus N \cup J)$, $(uvu^{-1})^n = (u^{-1}uv)^n = v^n$. Also, $(uvu^{-1})^n = uv^n u^{-1}$. Thus $uv^n u^{-1} = v^n$ and hence

$$[u, v^n] = 0, \text{ for all } x, y \in U. \tag{2.20}$$

Using similar techniques with necessary variations as used to get (2.2)—(2.5) from (2.1), we have the following.

$$[u^n, x] = 0, \text{ for all } u \in U \text{and } x \in N \cup J. \tag{2.21}$$

$$[u^n, x] = 0, \text{ for all } u \in U \text{and } ux \in N \cup J. \tag{2.22}$$

$$[u, x^n] = 0, \text{ for all } x \notin N \cup J \text{and } ux \notin N \cup J. \tag{2.23}$$

$$[u^n, x^n] = 0, \text{ for all } x \in R \text{ and } u \in U. \tag{2.24}$$

Now our aim is to show that

$$(x^{n-1}u^n x)^n = x^{n^2} u^{n^2}, \text{ for all } x \in R, u \in U. \tag{2.25}$$

Suppose that $x^{n-1} \in N \cup J$ or $u^n x \in N \cup J$. Then $1 - x^{n-1} \in U$, or $1 - u^n x \in U$ and hence by (2.20), we find that $[1 - x^{n-1}, u^n] = 0$ or $[1 - u^n x, u^n] = 0$. Thus in both the cases $[x^{n-1}, u^n] = 0$. Now application of (2.24) gives that

$$(x^{n-1}u^n x)^n = (u^n x^n)^n = \underbrace{u^n x^n . u^n x^n \cdots\cdots u^n x^n}_{n-times} = x^{n^2} u^{n^2}.$$

On the other hand if $x^{n-1} \notin N \cup J$, then we distinguish two cases.

Case(i). Let $x \in N \cup J$. Then by (2.21), $[u^n, x] = 0$, and hence

$$(x^{n-1}u^n x)^n = (x^n u^n)^n = \underbrace{x^n u^n . x^n u^n \cdots\cdots x^n u^n}_{n-times} = x^{n^2} u^{n^2}.$$

Case(ii). Let $x \notin N \cup J$. Then by our hypothesis $P_3(n, R \setminus (N \cup J))$ and (2.24), we get

$$(x^{n-1}(u^n x))^n = (u^n x^n)^n = x^{n^2} u^{n^2}$$

Thus (2.5) is proved in all cases.
For all $x \in R$ and $u \in U$, application of (2.24)yields that

$$\begin{aligned}(x^{n-1}u^n x)^n &= \underbrace{x^{n-1}u^n x . x^{n-1}u^n x \cdots\cdots x^{n-1}u^n x}_{n-times} \\ &= x^{n-1}(u^n x^n)^{n-1} u^n x \\ &= x^{n^2-1} u^{n^2} x\end{aligned}$$

Combining this fact together with (2.25) we find that $x^{n^2-1}[x, u^{n^2}] = 0$ and hence by Lemma 2.2, we have

$$[u^{n^2}, x] = 0, \text{ for all } x \in R, u \in U. \tag{2.26}$$

Now using similar arguments as used to get (2.10) from (2.8), we find that

$$N \subseteq Z. \tag{2.27}$$

Further, it is easy to see that$R$ satisfies the property $[(xy)^n, x] = 0$, for all $x \notin N \cup J$ *and* $y \notin N \cup J$,. Now repetition of arguments which led to (2.14) from (2.11) yield that R satisfies the polynomial identity $[[[(xy)^n, x], (1-x)^{n^2}], (1-y)^{n^2}] = 0$, for all $x, y \in R$. Hence the similar arguments as used in the proof of previous theorem gives that

$$C \subseteq N \subseteq Z. \tag{2.28}$$

Thus Lemma 2.1 is applicable in the present situation and hence in view of (2.26) we find that $n^2 u^{n^2-1}[u, x] = 0$. This implies that $n^2[u, x] = 0$. Since R has the property $Q(n)$, it has also the property $Q(n^2)$, and hence $[u, x] = 0$ for all $x \in R$. This yields that

$$U \subseteq Z. \tag{2.29}$$

Now repetition of arguments which led to (2.19) from (2.16) yield that R satisfies the property $P_3(n, R)$. Hence our theorem follows from the proof of [9, Theorem 5].

Corollary 2.3. ([11, Theorem 1]). A ring R is commutative if and only if R satisfies the properties $P_3(n, R \setminus N)$ and $Q(n)$.

Corollary 2.4. ([11, Theorem 1]). A ring R is commutative if and only if R satisfies the properties $P_3(n, R \setminus J)$ and $Q(n)$.

Remark 2.1. The above theorems are true for the case $n = 1$ also. To see this, observe that any ring R with identity 1 for which $N \cup J$ and $R \setminus (N \cup J)$ are both commutative sets is necessarily commutative. In fact, if $n = 1$, then in both the cases R satisfies the property $[x, y] = 0$, for all $x, y \in R \setminus (N \cup J)$. Now let $a, b \in N \cup J$. Then $1 + a, 1 + b$ are both units and hence $0 = [1 + a, 1 + b] = [a, b]$. Thus $R \setminus (N \cup J)$ and $N \cup J$ are both commutative sets. If $a \in R \setminus (N \cup J)$ and $b \in N \cup J$, then $1 + b$ is a unit and $0 = [a, 1 + b] = [a, b]$.

References

[1] H. Abu-khuzam, A commutativity theorem for rings, *Math. Japonica* **25** (1980), 593-595.

[2] H. Abu-khuzam, Commutativity results for rings, *Bull. Austral. Math. Soc.* **38** (1988), 191-195.

[3] H. Abu-khuzam, *A commutativity theorem for rings with constraints involving nilpotents and the Jacobson radical*, Algebra and its applicatios, Narosa Publishing House, New Delhi(2000), 1-7.

[4] H. Abu-khuzam and A. Yaqub, Commutativity of rings satisfying some polynomial constraints, *Acta Math. Hungar.* **67** (1995), 207-217.

[5] H. Abu-khuzam, H.E. Bell and A. Yaqub, Commutativity of rings satisfying certain polynomial identities, *Bull. Austral. Math. Soc.* **44** (1991), 63-69.

[6] M. Ashraf and M.A. Quadri, Commutativity of rings with some polynomial constraints, *Bull. Austral. Math. Soc.* **41** (1990), 201-206.

[7] M. Ashraf, On commutativity of rings with constraints on a subset, *Internat. J. Math., Game Theory and Algebra*, Vol 11, No. 6(2001), 1-13 .

[8] H.E. Bell, On some commutativity theorems of Herstein, *Arch. Math.* **24** (1973), 34-38.

[9] H.E. Bell, On rings with commuting powers, *Math. Japonica* **24** (1979), 473-478.

[10] H.E. Bell, The identity $(XY)^n = X^n Y^n$: does it buy commutativity ? *Math. Mag.* **55** (1982), 165-170.

[11] H.E. Bell and A. Yaqub, Commutativity of rings with certain constraints, *Math. Japonica* **4** (1987), 511-519.

[12] I.N. Herstein, Power maps in rings, *Michigan Math. J.* **8** (1961), 29-32.

[13] M. Hongan and H. Tominaga, Some commutativity theorems for semiprime rings, *Hokkaido Math. J.* **10** (1981), 271-277.

[14] N. Jacobson, Structure of rings, (*Amer. Math. Soc. Colloq. Publ. Providence* 1956).

[15] T.P. Kezlan A note on commutativity of semiprime PI-rings, *Math. Japonica* **27** (1982), 297-268.

In: Mathematics, Game Theory and Algebra Compendium ISBN 978-1-60456-659-8
Editor: Jacob H. Mathias, pp. 53-57

Chapter 7

DECOMPOSABILITY OF ITERATED EXTENSIONS*

***V.K. Bhat*†**
80/345, Patel Marg, Mansarover,
Jaipur, India-304022

Abstract

It is known that R[x,s], $R[x, x^{-1}, s]$ and R[x,d] (where s is an automorphism of a right Noetherian ring and d is a derivation of a right Noetherian Q-algebra) are decomposable. In this paper we prove that iterated extensions of these rings are decomposable, and hence they have primary decomposition such that each primary component has an artinian quotient ring.

AMS Subject Classification: Primary 16-XX, Secondary 16P40, 16P50, 16U20

Keywords: Automorphism, derivation. Ore-extensions, primary decomposition, regular elements, quotient ring

Introduction

In [1] it has been proved that R[x,s], $R[x, x^{-1}, s]$ and R[x,d] (where s is an automorphism of a right Noetherian ring and d is a derivation of a right Noetherian Q-algebra) are decomposable.In this paper we introduce decomposable rings.We say a Noetherian ring R is decomposable if in R there exist ideals Ij, j=1,2,,n such that $0 = \cap Ij$ and each ring R/Ij has a right artinian quotient ring.We prove that all the iterated Ore-extensions of a commutative Noetherian ring R, described below, are decomposable. One notable feature of this paper is that we show that the above mentioned decomposability is actually extended from the primary decomposition of the coefficient rings. The rings that we deal with are $S_t(R) = R[x_1, s_1][x_2, s_2][x_t, s_t]$, the iterated skew-polynomial ring where each s_i is an automorphism of $S_{i-1}(R)$; $L_t(R) = R[x_1, x_1^{-1}, s_1][x_2, x_2^{-1}, s_2][x_t, x_t^{-1}, s_t]$, the

*Communicated by Ayman Badawi
†E-mail address: vijaykumarbhat2000@yahoo.com

iterated skew-Laurant polynomial ring, where each s_i is an automorphism of $L_{i-1}(R)$ and the iterated differential operator ring $D_t(R) = R[x_1, d_1][x_2, d_2][x_t, d_t]$, where each d_i is a derivation of $D_{i-1}(R)$. We denote by S(R), L(R) and D(R) respectively the rings R[x,s], $R[x, x^{-1}, s]$, and R[x,d], where s is an automorphism of R and d is a derivation of R. $A_t(R)$ denotes any of $S_t(R)$ or $L_t(R)$ where $s_i \mid R$ is an automorphism of R. Throughout Q denotes the field of rational numbers and Z denotes the ring of integers unless otherwise stated. Let R be a ring. Then the set of associated prime ideals of R is denoted by Ass(R). C(I) denotes the set of elements regular modulo I, where I is an ideal of R.

1. Invariance of Symbolic Powers of Certain Ideals

Definition 1.1. Let R be a commutative Noetherian ring and P a semiprime ideal of R. Let $k \geq 1$ be an integer then the symbolic power of P is denoted by $P^{(k)}$ and is defined as $P^{(k)} = \{a \in R$ such that there exists $d \in C(P)$ such that $ad \in P^k\}$

Proposition 1.2. *Let R be a commutative Noetherian ring and P be a semiprime ideal of R then $P^{(k)}$ is an ideal of R.*

Proof. let a, $b \in P(k)$ then there exist $d_1, d_2 \in C(P)$ such that $ad_1 \in P^k$ and $bd_2 \in P^k$. So $ad_1d_2 \in P^k$ and $bd_1d_2 \in P^k$; i.e. $(a - b)d_1d_2 \in P^k$ and since $d_1d_2 \in C(P)$ so $(a - b) \in P^{(k)}$. Now let $a \in P^{(k)}$ and $r \in R$. Then there exists $d \in C(P)$ such that $ad \in P^k$. So $ard \in P^k$. Since $d \in C(P)$, $ar \in P^{(k)}$. Hence $P^{(k)}$ is an ideal of R. □

Proposition 1.3. *Let P be a semiprime ideal of a commutative Noetherian ring R and s be an automorphism of R with s(P)=P,then $s(P^{(K)} = P^{(k)}$.*

Proof. See Proposition (3) of [1] □

Proposition 1.4. *Let R be commutative Noetherian ring and let d be a derivation of R. If P be a semiprime ideal of R such that $d(P) \subseteq P$, then $d(P^{(k)}) \subseteq P^{(k)}$.*

Proof. See Theorem (7) of [1] □

Proposition 1.5. *Let R be a commutative Noetherian ring and $A_t(R)$ be the usual skew polynomial ring. Let $Ass(R) = \{P_j\ , 1 \leq j \leq n$. Let $s_i^{m_i}(P_j) = P_j$, $m_i \geq 1$ for all j, $1 \leq j \leq n$. Let $P_{1j} = \cap s_1^k(P_j)$, $1 \leq k \leq m_1$ and $P_{tj} == \cap s_t^u(P_{t-1j})$, $1 \leq u \leq m_t$, $t \geq 2$ for all j, $1 \leq j \leq n$, then:*

(1) $s_i(P_{tj}^{(k)}) = P_{tj}^{(k)}$ for all i,j; $1 \leq j \leq n$, $1 \leq i \leq t$, $k \geq 1$.

(2) $R/(P_{tj}^{(k)})$ has an artinian quotient ring for all j, $1 \leq j \leq n$.

Proof. **(1)** we will use induction on t. For t = 1, clearly $s_1(P_{1j}) = P_{1j}$ Suppose $s_i(P_{t-1j}) = P_{t-1j}$, $1 \leq i \leq t - 1$. We now show that $s_i(P_{tj}) = P_{tj}$, $1 \leq i \leq t$. Note that we have an automorphism At of R such that $s_t \mid R$ is same as At and At can be extended to $S_1(R)$, $S_2(R)$,...,$S_{t-1}(R)$ such that $At(x_i) = x_i$, $1 \leq i \leq t - 1$. Now $s_i(P_{tj}) = s_i(\cap s_t^u(P_{t-1j})) = s_i(\cap At^u(P_{t-1j})) = s_i(\cap At^u(L_{i-1}(P_{t-1j}) \cap R))$, $1 \leq u \leq m_t$. So

$s_i(P_{tj}) = s_i(\cap At^u(L_{i-1}(P_{t-1j}) \cap R)$ by (10.6.3) of [5]. So $s_i(P_{tj}) = (\cap At^u(P_{t-1j})) = \cap s_t^u(P_{t-1j})) = P_{tj}$.

(2) $N(R/P_{tj}^{(k)}) = (\cap s_j^i(P_{tj}))/P_{tj}^{(k)}$, $1 \leq i \leq m_j$, where j=t+1 and $s_j^{m_j}(P_{tj}) = P_{tj}$. Let $N_1 = (R/P_{tj}^{(k)})$ and let $a + P_{tj}^{(k)} \in C(N_1)$ then $a \in \cap s_j^i(P_{tj})$, $1 \leq i \leq m_j$. So $a \in C(s_j^{m_j}(P_{tj})) = C(P_{tj})$. Now let $b \in R$ such that $ab \in P_{tj}^{(k)}$. Then there exists $d \in C(P_{tj})$ such that $abd \in P_{tj}^k$ i.e. $bad \in P_{tj}^k$. Since $ad \in C(P_{tj})$, $b \in P_{tj}^{(k)}$. So by Small's Theorem, namely Theorem (10.9) of [4], $R/P_{tj}^{(k)}$ has an artinian quotient ring. □

Proposition 1.6. *Let R be a Noetherain Q-algebra and $D_t(R)$ as usual the t-th different operator ring then:*

(1) $d_i(P_j) \subseteq P_j$ implies $d_i(P_j^{(k)}) \subseteq P_j^{(k)}$ for all i, j, $1 \leq i \leq t$, $1 \leq j \leq n$, $k \geq 1$.

(2) $R/P_j^{(k)}$ has an artinian quotient ring.

Proof. **(1)** The proof is obvious using 1.4

(2) $N(R/P_j^{(k)}) = \cap s_1^i(P_j))/P_j(k)$, $1 \leq i \leq m_t$, where $s_1^{m_1}(P_j) = P_j$. Let $N_1 = N(R/P_j^{(k)})$ and let $a + P_j^{(k)} \in C(N_1)$. Then $a \in C(\cap s_1^i(P_j))$, so $a \in s_1^i(P_j)$ for all i, $1 \leq i \leq m_t$. So $a \in C(s_1^{m_1}(P_j)) = C(P_j)$. Now let $ab \in P_j^{(k)}$, $b \in R$. Then there exists $d \in C(P_j)$ such that $abd \in P_j^k$;i.e; $bad \in P_j^k$. Since $ad \in C(P_j)$, $b \in P_j^{(k)}$. So by Small's Theorem ,namely Theorem (10.9) of [4], $R/P_j^{(k)}$ has an artinian quotient ring. □

2. Decomposability of the Iterated Extension Rings

The followings two Propositions are well known in commutative ring theory. One can see [6].

Proposition 2.1. *Let R be a commutative Noetherian ring, let $0 = \cap Ij$, $1 \leq j \leq n$ be a reduced primary decomposition of R, and let $\sqrt{Ij} = P_j$. Then there exists an integer $k \geq 1$ such that $P_j^{(k)} \subseteq Ij$.*

Proposition 2.2. *Let R be a commutative Noetherian ring. Then R is a decomposable ring.*

Proposition 2.3. *Let R be a ring and S(R) be as usual and $S_2(R) = S(R)[x_2, s_2]$, let each $s_i \mid R$ be an automorphism of R for i=1,2, and let I be an ideal of R such that $s_1(I) = I$ and $s_2(I) = I$. Then $s_2(S(I)) = S(I)$.*

Proof. Let $f = x^m a_m + ... + a_0 \in S(I)$, $a_i \in I$. Then $s_2(f) = s_2(x^m a_m + ... + a_0) = s_2(x^m a_m) + ... + s_2(a_0) = s_2(x^m)s_2(a_m) + ... + s_2(a_0) = g_m b_m + ... + b_0$ where $s_2(x^i) = g_i \in S_1(R)$ and $s_2(a_i) = b_i \in I$. So $s_2(f) \in S_1(I)$. Hence $s_2(S(I)) \subseteq S(I)$. Replacing s_2 by s_2^{-1} ,we get that $s_2(S(I)) = S(I)$. □

Corollary 2.4. *Proposition 2.3 holds if S(R) is replaced by L(R).*

Proposition 2.5. *Let R be a Noetherian Q-algebra and $Ass(R) = \{P_j, 1 \leq j \leq n\}$. Consider $D_t(R)$, then $d_i(D_{i-1}(P_j^{(k)})) \subseteq D_{i-1}(P_j^{(k)})$ for all i, j, $1 \leq i \leq n$, $1 \leq i \leq t$.*

Proof. $Ass(R) = \{P_j, 1 \leq j \leq n\}$. Now $d_1(P_j) \subseteq P_j$ by Theorem (2.5) of [2]and so $d_1(P_j^{(k)}) \subseteq P_j^{(k)}$ by Proposition 1.4. Now by Theorem (2.6) of [2], $D(P_j) = P_j[x_1, d_1] \in AssD(R)$. So $d_2(D(P_j)) \subseteq D(P_j)$ by Theorem (2.6) of [2]. We will show that $d_2(D(P_j^{(k)})) \subseteq D(P_j^{(k)})$. Let $f_1 = \sum x_1^i a_i \in D(P_j^{(k)})$, $a_i \in P_j^{(k)}$, $0 \leq i \leq s$. Then there exists $u_i \in C(P_j)$ such that $a_i u_i \in (P_j^k)$. Let $u = u_0.u_1....u_s$. Then $a_i u \in P_j^k$. So $g_1 = \sum x_1^i a_i u \in D(P_j^k)$ and $d_2(g_1) \in D(P_j^k)$; i.e; $d_2(f_1 u) \in D(P_j^k)$ or $d_2(f_1)u + f_1 d_2(u) \in D(P_j^k) \subseteq D(P_j^{(k)})$. So $d_2(f_1)u \in D(P_j^{(k)})$ as $f_1 \in D(P_j^{(k)})$. Let $d_2(f_1) = \sum x^i b_i$, $0 \leq i \leq m$. Then $b_i u \in P_j^{(k)}$. So there exists $v_i \in C(Pi)$ such that $b_i u v_i \in P_j^k$. Let $v = v_0 v_1 ... v_m$. Now $b_i uv \in P_j^k$. Since $uv \in C(P_j)$, so $b_i \in P_j^{(k)}$, $d(D(P_j^{(k)})) \subseteq D(P_j^{(k)})$. With the same process in a finite number of steps by easy induction it can be seen that $d(D_{i-1}(P_j^{(k)})) \subseteq D_{i-1}(P_j^{(k)})$ for all $i \geq 3$. □

Theorem 2.6. *Let R be a commutative Noetherian ring and let $A_t(R)$ and $D_t(R)$ be as usual (in case of $D_t(R)$, R is moreover a Q-algebra). Then $A_t(R)$ and $D_t(R)$ are decomposable rings.*

Proof. Let $0 = \cap Ij$, $1 \leq j \leq n$ be a reduced primary decomposition of R and let $\sqrt{Ij} = P_j$. Then $Ass(R) = \{P_j, 1 \leq j \leq n\}$ and there exists an integer $k \geq 1$ such that $P_j^{(k)} \subseteq Ij$. So $\cap P_j^{(k)} = 0, 1 \leq j \leq n$.

$A_t(R)$ case : Let P_{tj} be as in Proposition 1.5 then $R/P_{tj}^{(k)}$ has an artinian quotient ring by Proposition 1.5 and $s_i(P_{tj}^{(k)}) = P_{tj}^{(k)}$ by 1.5 for all i, $1 \leq i \leq t$. So by [3] and Proposition 2.3 $A_t(R)/A_t(P_{tj}^{(k)})$ has an artinian quotient ring. Now $\cap A_t(P_{tj}^{(k)}) = 0$, $1 \leq j \leq n$. So $A_t(R)$ is a decomposable ring.

$D_t(R)$ case : $R/P_j^{(k)}$ has an artinian quotient ring by Proposition 1.6. Now by Theorem 2.5 $d_i(D_{i-1}(P_{tj}^{(k)})) \subseteq D_{i-1}(P_{tj}^{(k)})$ for all i, $1 \leq i \leq t$ and for all j, $1 \leq j \leq n$. So by [3] $D_t(R)/D_t(P_j(k))$ has an artinian quotient ring. Now $\cap D_t(P_j^{(k)}) = 0$, $1 \leq j \leq n$. Thus $D_t(R)$ is a decomposable ring. □

References

[1] V.K.Bhat, Decomposability of skew-polynomial rings, To appear, *Int. Jl. Of Comm. Rings.*

[2] V.K.Bhat, Structure of associated prime ideals of certain Ore-extensions. Communicated.

[3] W. D. Blair and L. W. Small, Embedding differential and skew-polynomial rings into artinain rings, *Proc. Amer, Math. Soc.* **109**(4) 1990, 881-886.

[4] K.R. Goodearl and R.B. Warfield Jr. , An introduction to non-commutative Noetherian rings, *Cambridge Uni. Press* 1989.

[5] J.C.McConnell and J.C.Robson, Noncommutative Noetherian Rings,Wiley(1987); revised edition:*American Math.* Society (2001).

[6] O. Zariski and P. Samuel Commutative Algebra Vol 1, *D Van Nostrand Company Inc.* 1967.

Received: August 2004
Revised: December 2004

In: Mathematics, Game Theory and Algebra Compendium
Editor: Jacob H. Mathias, pp. 59-78
ISBN: 978-1-60456-659-8

Chapter 8

SOME ASPECTS OF A PURE THEORY OF BARGAINING: PLAYERS, INFORMATION, EQUILIBRIUM AND VOTING POWER

Volker Bieta, Siegfried Gelbhaar
Trier University, Germany
Martin Straub
Bielefeld University, Germany

Abstract

Players A,C and possibly B as an intermediary are playing a bargaining game on a position. The players are characterized by their types respectively their quadratic loss functions differing only by their bliss points. A's type is chosen randomly, in the case of incomplete information ex post revealed to A only. A has a limited possibility of signalling his type. We prove equivalent conditions regarding B's type under which the game with B leads to a separating equilibrium (with signalling) superior for both, A and C, to the pooling equilibrium (without signalling) which results in the game without B. This is a positive argument for an intermediary from a strategic perspective.

Keywords: AMS subject classification 91-xx.:game theory, economics, social and behavioral sciences

1. Introduction

We analyze the problem of intermediation in a stylized game theoretic framework with signalling. Two or three players are playing a one step bargaining game *(Güth 1996/Rubinstein 1982/1985)* with complete or incomplete (asymmetric) information on a position (outcome, payoff). The object of the paper is a strategic point of view on intermediation. The middleman or intermediary B is regarded as a third player, who changes the original two person game A, C into a three person game A,B,C. Drawing on the Bayesian equilibrium concept *(Harsanyi 1967/Kreps 1990/ Myerson 1991)* we prove in the case of asymmetric information that by introducing an intermediary the equilibrium can be Pareto improved for A and C *(Rubinstein /Wolinski 1987)*. The possibility of Pareto improvement arises from the indirect information transfer from A to C, which becomes possible by the mere existence of a third player. By B signalling from A to C may become useful, and a separating equilibrium may arised.

Our point of view is rather different from most of the existing literature on middlemen, which concentrates mainly on financial intermediation *(Arshadi/Karels 1997)*.[1] Even though an asymmetric information setting is also considered in *(Leland/Pyle 1977)* and an information transfer takes also place by delegated monitoring in *(Diamond 1984)*, these models are quite different from our's.

The intermediary B in our setting has not the task of monitoring or of another direct information transfer but acts as an independent third player.[2] In the two person asymmetric information game without B the loss functions of A and C differ only in the bliss points characterizing their types. A's type is chosen randomly and ex post known to A only. C proposes an outcome or position x and A accepts or rejects according to his type. In the case of rejection the outcome $x=0$ results. Even though by institution A has the possibility of signalling information on his type to C before C's proposal x, this possibility cannot be used strategically (section 4.1, result 2). The intuitive reason for this is the limitation of information transfer from A to C to two intervals of types (section 2, the model (3)).

Allowing for binding pre-game-commitments the equilibrium will be expected in general to be Pareto improved (for A and C). But such free pre-game-commitments are not given here by institution.[3] The only allowed commitment consists in accepting unanimously or rejecting a third player B – equal to A by his veto power with respect to x and to C by his information and different from A by his given and known type. The pre-game-commitments on B are not explicitly included in the games analyzed, but are the background for analyzing separately the games with or without B. [4]

For the game with B (with veto power) we characterize a separating equilibrium as sequential equilibrium (section 4.2, result 3) and prove this to be Pareto superior for A and C to the pooling equilibrium (section 4.2, result 4). Finally we characterize the set of types B, which are Pareto optimal – i. e. such that A and C cannot be both made better of by another type of B (section 4.2, result 5).

The paper is organized as follows: After the introduction (1) we describe in section (2) the game theoretic model of the bargaining process. The games with or without intermediary are analyzed in section (3) for the case of complete information. Section (4) constitutes the central body of the paper. Here we derive in the case of incomplete (asymmetric) information the main results as indicated above. Section (5) closes with a brief conclusion.

---

[1] Therefore we keep the citation of this literature rather small. For an overview see also *(Spulber 1996)*.

[2] As a "real" example for our model we may see the 1994/1996 conflict between Chechnia and Russia as players A and C and General Lebed as independent third player respectively.

[3] We do not allow for any precommitment of A such as to reject a „too bad" offer ($x < 2x_A$). We also do not consider the possibility of an „outside option" of A, which might also give A the possibility of rejecting „too bad" offers ($x < 2x_A$). We do not discuss here the connection of these concepts to our model, which may be the issue of further research.

[4] One may think of B also as a controller of, or mediator between A and C introduced by the government and endowed with veto power.

2. The Model

We begin by describing the game with three players A, B, C – i. e. including a third player B with veto power as intermediary or middleman in an asymmetric (incomplete) information setting. By small variations of this game model at the end of the section, we get the case of complete information and the case without B.

The game with three players A, B, C and asymmetric information is based on the following assumptions:

(1) $x \in X = [0, x_*] \cup [x^*, 1] \subset [0,1]$ denotes a position or outcome or payoff.[5]

(2) A, B, C have quadratic loss functions denoted by u_A, u_B, u_C:

$$u_A = -(x - x_A)^2 \,;\ u_B = -(x - x_B)^2 \,;\ u_C = -(x - x_C)^2$$

with $0 < x_A < 1,\ 0 < x_B < 1,\ x_C = 1$. We call $t_A = 2x_A,\ t_B = 2x_B,\ t_C = 2x_C := 2$ the types of A, B, C respectively.[6]

(3) The game proceeds as follows:

–Nature draws a type $t_A \in\,]0, \sigma_2] \cap T,\ T = 2 \cdot X$, with a given $0 < \sigma_2 < 2$ at random according to a uniform distribution on $]0, \sigma_2] : t_A \sim E\,]0, \sigma_2]$. The situation, before t_A is drawn, (σ_2, t_B, t_C), is common knowledge to all players, but the type t_A after it is drawn, is known only to A.[7]

–A sends a signal $s = s(t_A) \in X = [0, x_*] \cup [x^*, 1] \subset [0,1]$ (regarding his type t_A), which is observed by C.

–Having observed A's signal, C responds by a proposal $x \in X$ observed by A and B.

–If both, A and B, accept *x*, each player gets respectively the payoff $u_A(x), u_B(x), u_C(x)$. If A or B rejects *x*, then the outcome $x = 0$ is realized, and each player gets respectively $u_A(0)$, $u_B(0)$, $u_C(0)$.[8]

---

[5] The intervals $[0, x_*]$ resp. $[x^*, 1]$ with $0 < x_* < x^* < 1$ will be defined later in section 4.1. The definition of x_*, x^* is crucial for result 2 in section 4.1. Indeed it turns out $x^* = 3^{-1}\left[2 + \sigma_2 - 3^{-1}\left(\sigma_2^2 - 2\sigma_2 + 4\right)\right]^{2^{-1}}$, $x_* = 1 - \left[2^{-1} + 2^{-1}\left(1 - x^*\right)^{2^{-1}}\right]$ but because this does not tell much at this stage, the reader may think of arbitrary $0 < x_* < x^* < 1$ meanwhile.

[6] The quadratic loss functions are used e. g. in *(Persson/Tabellini 1990)*, but are not in general common in bargaining literature. The non-monotonicity of the loss functions is an essential feature in our model, but the quadratic form does not seem to be essential for our results.

[7] It does not matter, if we allow that the signal is observed by B too.

[8] We call the game a bargaining game with asymmetric information on the position.

We describe some variations:

–In the game without B the above description applies omitting B. In the last step of (3) A alone suffices to accept x.

–In the game with B without veto power, B is not omitted. But in the last step of (3) A is again alone sufficient to accept x.[9]

–In the game with perfect information (with or without B) in (3), step 1, the type t_A of A is ex post (after having been drawn by nature) common knowledge for all players. The signalling procedure in (3), step 2, can be maintained but has no more any meaning in this case and could be omitted as well.

Our object is the game theoretic analysis of the described bargaining model.[10] The focus is to compare the games with or without intermediary. Besides the framework of incomplete information we analyze first the simpler case of complete information *(Cothren/Loewenstein ND)* in the following section.

3. The Complete Information Setting

In this section we consider the case of full information – i. e. after A's type is chosen randomly his type is known to all players. Besides the existence of a sequential equilibrium in both games – with or without veto player B – we are also interested whether A or C are better off with or without B. The answer is given in

Result 1

In the complete information setting in both games – with or without B – a unique sequential equilibrium exists. Ex ante A resp. C is better off resp. worse off in the game with B compared with the game without B. Ex post A resp. C is better off resp. worse off or indifferent in the game with B compared with the game without B.[11]

Proof

In the game without B, C is ex post fully informed about A's type t and his proposal $x \in X$ will be as high as possible, such that A does accept. Hence C will propose:

[9] It is obvious that this game is equivalent to the game without B.

[10] Here we use the notion of sequential equilibrium as defined in Mas-Colell et. al. in a mathematical strong form (definition 9. C. 4). We give a description of the used equilibrium concepts which is more intuitively. We denote an equilibrium as separating if several types of a player choose different actions, thus revealing their types. We denote an equilibrium as pooling, if each type of a player chooses the same action, thus providing no information about his type.

[11] Ex ante resp. ex post means before resp. after the random choice of the type t of A.

$$\dot{x} = \begin{cases} t & t \le x_* \\ x_* & x_* < t < x^* \\ t & x^* \le t \end{cases}$$

and A will accept the offer.

In the game with B is given veto power, C's proposal $x \in X$ will be as high as possible, such that A and B do accept, i. e. $x \le t \wedge 2x_B$:

$$\dot{x} = \begin{cases} t \wedge 2x_B & t \wedge 2x_B \le x_* \\ x_* & x_* < t < x^* \\ t \wedge 2x_B & x^* \le t \wedge 2x_B \end{cases}$$

and A and B will accept the offer.

To answer the question whether or not the game without B or with B is given veto power is ex post (given *t*) better, resp. worse, for A resp. C, we remark:

If $2x_B \ge t$ the solution of the game is identical with the solution of the game without B. Hence A and C are ex post indifferent in the game with or without B.

If $2x_B < t$ the solution is:

$$\dot{x} = \begin{cases} 2x_B & 2x_B \le x_* \\ x_* & x_* < 2x_B < x^* \\ 2x_B & x^* \le 2x_B \end{cases}$$

offered by C and A and B accepting. Then A resp. C is ex post better resp. worse off in the game with B is given veto power compared with the game without B.

The statement ex ante follows immediately from the results ex post by taking expectations. ♦

Denoting I_1 resp. I_2 the (ex ante) utility of A in the game without resp. with B is given veto power, we know $I_2 \ge I_1 \forall x_B$. Moreover we answer the question, which state of B is best for A. A's utility is:

– without B:

$$I_1 = -\sigma_2^{-1}\left\{ \int_0^{\sigma_2 \wedge x_*} \left(t - t2^{-1}\right)^2 dt + \int_{x_*}^{\sigma_2 \wedge x^*} \left(x_* - t2^{-1}\right)^2 dt + \int_{x^*}^{\sigma_2 \vee x^*} \left(t - t2^{-1}\right)^2 dt \right\}$$

$$\text{With B}: I_2 = -\sigma_2^{-1}\left\{\int_0^{2x_B \wedge x_*} \left(t - t2^{-1}\right)^2 dt + \int_{x_*}^{\sigma_2 \wedge x^*} \left(x_* - t2^{-1}\right)^2 dt + \int_{2x_B \vee x^*}^{\sigma_2} \left(2x_B - t2^{-1}\right)^2 dt\right\}$$

We find from $\frac{\partial I_2}{\partial x_B}=0$ and $\frac{\partial^2 I_2}{\partial x_B^2}<0$ the maximizer $x_B = \sigma_2 6^{-1}$. I_2, at which $x_B = 0$ and $x_B = \sigma_2 2^{-1}$ imply $I_1 = I_2$.

However, we find the complete information assumption not very realistic. In the following section we address the question under which conditions the intermediary B should be given the right to intervene when asymmetric information prevails.

4. The Incomplete Information Setting

Let C now be uninformed about A's type $t \in T$. Then C's analysis of signals A sends regarding his type gains strategic importance. C has to account for A not telling the truth about his true type. A reports truthfully whenever the message he sends coincides with his true type, otherwise he lies (see for example *Ben-Porath/Dekel* 1992; *Osborne/Rubinstein* 1990). The decoding of A's signals is of central importance because C's beliefs regarding A's type have to account for the risk that A strategically sends wrong messages (see for example *Matthews* 1989; *Lax/Sebenius* 1991).

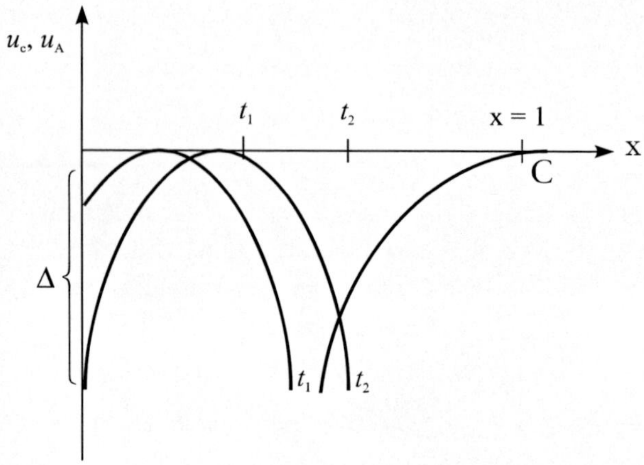

Figure 1. Illustrates C's decision problem when receiving different type signals t_1 and t_2, $t_2 > t_1$. A type t_2 player will prefer an offer $x = t_1$ to $x = t_2$. Since $u_c(t_2) > u_c(t_1)$, C will respond to a signal t_1 with $x = t_1$ and to a signal t_2 with $x = t_2$, given that A reports truthfully.

However, signalling truthfully his type t as a strategy for A and proposing $x=t$ as a strategy for C is no equilibrium. As figure 1 shows a type t_2 will always prefer $x = t_1$ to $x = t_2$ since that yields a surplus in payoff of Δ. A type t_2 player will therefore pretend to be a type t_1 player. That means that in a sequential equilibrium A is not signalling truthfully. Indeed as shown in 4.1 below the possibility of limited signalling two intervals of types is not used in the resulting pooling equilibrium. As shown in 4.2 a third player may be able to improve upon the outcome (*Rubinstein/Wolinsky* 1987) by the use of signalling in the separating equilibrium which results in this case.

4.1. The Incomplete Information Setting without Intermediary

In this section we consider the game with incomplete information without veto player B – i. e. after A's type is chosen randomly this type is known to A only. As a by-product of the result 2 below we get the values of x^*, x_* :

$$x^* = 3^{-1}\left[2+\sigma_2-3^{-1}\left(\sigma_2^2-2\sigma_2+4\right)\right]^{2^{-1}}, \; x_* = 1-\left[2^{-1}+2^{-1}\left(1-x^*\right)^{2^{-1}}\right]$$

at which $0<x_*<x^*<1$, $T=\left[0,x_*\right]\cup\left[x^*,1\right]$ holds.

Result 2

In the incomplete information game without B there exists a unique sequential equilibrium. The equilibrium strategies of C resp. A consist in offering resp. accepting x^* (pooling equilibrium).[12]

Proof

We make the preliminary assumption that A has no ability of signalling. Therefore C makes an offer based on the known distribution of the types of A, at which holds:

–for offers $x>t$, a *t*-type of A will always reject C's offer (with probability 1);
–for offers $0\le x\le t$, a *t*-type of A will always accept C's offer (with probability 1).

Then C's expected utility from an offer *x* is given by:

$$Eu_C(x) = -\left(\sigma_2 - x\right)\sigma_2^{-1}\left(x-1\right)^2 - \left[1-\left(\sigma_2 - x\right)\sigma_2^{-1}\right].$$

[12] We remark that by footnote 6 "*X*" is disconnected, which is important for our result and the cause of the definition of "*X*" in (1), section (2).

From $\dfrac{\partial Eu_C}{\partial x} = \left(3x^2 - 4x - 2\sigma_2 x + 2\sigma_2\right)\sigma_2^{-1} = 0$ and $\dfrac{\partial^2 Eu_C\left(x^*\right)}{\partial x^2} < 0$ we get C's expected utility maximizing offer:

$$x^* = 3^{-1}\left[2 + \sigma_2 - 3^{-1}\left(\sigma_2^2 - 2\sigma_2 + 4\right)\right]^{2^{-1}}.$$

Any type of A on the interval $\left[x^*, \sigma_2\right]$ will accept x^*, while any other type on $\left[0, x^*\right]$ will reject. The probability of A accepting x^* is given by $\left(\sigma_2 - x^*\right)\sigma_2^{-1}$, and the probability of rejecting by $x^*\sigma_2^{-1}$, with expected utility for A of:

$$I_3 = \sigma_2^{-1}\left\{-\int_0^{x^*}\left(0 - t2^{-1}\right)^2 dt - \int_{x^*}^{\sigma_2}\left(x^* - t2^{-1}\right)^2 dt\right\}$$

at which $I_1 \le I_3$ holds obviously.

Then, under the preliminary assumption that A cannot send any signals to C, we have found a pooling equilibrium in a setting of incomplete information and when B cannot intervene decisions (*Cho/Kreps* 1987).

Now we show that even if A can send signals, this pooling equilibrium to be the unique sequential equilibrium. For that it suffices to show that any strategy deviating from the pooling equilibrium C might apply in response to A's signal, cannot be a best response to A's strategy according to the pooling equilibrium.

Suppose that, after observing signal s, C reacts with $x(s) < x^*$. Applying that strategy and given $t \in \left[x(s), x^*\right[$, C can generate gains which in expectation gives:

$$\Delta u_C = \left[1 - \left(1 - x(s)\right)^2\right]\left[x^* - x(s)\right]\sigma_2^{-1}; (> 0).$$

But for $t \in \left[2x(s), x(s) + x^*\right[$ A will gain by sending s, making C worse off. In expectation C will gain:

$$\overline{\Delta}u_C = \left[\left(1 - x^*\right)^2 - \left(1 - x(s)\right)^*\right]\left[x^* - x(s)\right]\sigma_2^{-1}; (< 0).$$

Together, C will gain – or loose –:

$$\Delta u_C + \overline{\Delta}u_C = \left[1 - \left(1 - x(s)\right)^2 + \left(1 - x^*\right)^2 - \left(1 - x(s)\right)^2\right]\left[x^* - x(s)\right]\sigma_2^{-1}.$$

The first bracket is monotonically increasing in $x(s)$. We define x_* by $x_* := x(s)$ such as to make this bracket zero (0). Therefore $x(s) \in [0, x_*]$ and therefore also $x(s) \in T = [0, x_*] \cup [x^*, 1]$ with $x(s) < x^*$ implies $\Delta u_C + \overline{\Delta} u_C < 0$. Consequently the offer $x(s) < x^*$ cannot be an equilibrium.

The reaction $x(s) > x^*$ is excluded too. An equilibrium strategy of A requires $t > x(s) + x^*$, which leads to the contradiction $x(s) > x(s) + x^*$ and, hence $x(s) > x^*$ can also not be an equilibrium.♦

4.2. Incomplete Information Setting and Intermediary Given Veto Power

In the game with B as veto player, C has to make offer $x \in [0, 2x_B]$ since independent of his beliefs about A's type $t \in T$ B will reject any offer $x \notin [0, 2x_B]$. If C believes that $t \in [2x_B, \sigma_2]$ he will offer $x = 2x_B$, which is his best response and which will guarantee that any player A and B will accept. For C's ability to make a type inference from A's type report it then becomes important that A's and B's objectives are compatible.

A truthful type revelation by A is driven by how attractive an offer $x \in [0, 2x_B]$ is to A. If an offer $x \in [2x_B, t]$ will sufficiently offset a false revelation for any A, then A might as well report truthfully to C and accept an offer $x = 2x_B$.

However, there is a risk involved for A that C might not believe his signal and propose the pooling offer $x \in [0, 2x_B[$. Any such offer will be accepted by the intermediary but will make A worse off among all other possible offers. Then, in addition to the difference in bargaining objectives A and B might have, an important question in three person bargaining games becomes if A can use the intermediary as a bargaining cushion so as to trigger the pooling offer or whether B as an independent player can arrive at a solution that makes A better off.

In this three person bargaining signalling game we determine the sequential equilibrium (*Banks/ Sobel* 1987; *Wilson* 1982). Thereby $\hat{t} \in T$ denotes a critical type, which will allow C to propose type separating offers. We then obtain a separating equilibrium with a bargaining solution in which A truthfully reports some information about his type and which dominates any pooling outcome. That gives an answer to the question, under which conditions it is optimal for A to allow some intermediary to intervene in the bargaining.

Before stating result 3 we give some notations:

$$x^0 \equiv (6x_B)^{-1}\left\{4x_B^2 + 2x_B + \sigma_2 - \left[\left(4x_B^2 - 4x_B + \sigma_2\right)^2 + 12x_B^2\right]^{2^{-1}}\right\}$$

$$x_0 \equiv 1 - \left\{2^{-1} + 2^{-1}\left(1 - x_0\right)^{2^{-1}}\right\} < x^0$$

$$\hat{x} \equiv 3^{-1}\left\{2+\hat{t}-3^{-1}\left(\hat{t}^2-2\hat{t}+4\right)^{2^{-1}}\right\} \text{ and } \hat{t} \equiv 2x_B + \hat{x}.$$

Furthermore we assume that $s_1 \leq \hat{t}$, $s_2 > \hat{t}$ are arbitrarily chosen constants and in step 3 below we use two constants $\underline{x}_B$, $\overline{x}_B$ at which $0 < \underline{x}_B < \overline{x}_B < 1$ holds.

Result 3

In the game with incomplete information and intermediary with veto power the sequential equilibrium is uniquely determined.

For $x_B < \overline{x}_B$, $x^* < x^0$, $x^* \leq \hat{x} < x^0$, the sequential equilibrium is given by:

$$\alpha(s) = \begin{cases} s_1 \leq \hat{t} & \text{if } s = s_1 \leq \hat{t} \\ s_2 > \hat{t} & \text{if } s = s_s > \hat{t} \end{cases},$$

$$\mu(s) = \begin{cases} E[0,t] & \text{if } s = s_1 \leq \hat{t} \\ E[\hat{t},\sigma_2] & \text{if } s = s_s > \hat{t} \end{cases},$$

$$x = x(s) = \begin{cases} \hat{x} & \text{if } s = s_1 \leq \hat{t} \\ 2x_B & \text{if } s = s_s > \hat{t} \end{cases}.$$

For $x_B < \overline{x}_B$, $x^* \leq x^0$, $\hat{x} < x^*$ the sequential equilibrium is given by:

$$\alpha(s) = \begin{cases} s_1 \leq \hat{t} & \text{if } s = s_1 \leq \hat{t} \\ s_2 > \hat{t} & \text{if } s = s_s > \hat{t} \end{cases},$$

$$\mu(s) = \begin{cases} E[0,t] & \text{if } s = s_1 \leq \hat{t} \\ E[\hat{t},\sigma_2] & \text{if } s = s_s > \hat{t} \end{cases},$$

$$x = x(s) = \begin{cases} x^* & \text{if } s = s_1 \leq \hat{t} \\ 2x_B & \text{if } s = s_s > \hat{t} \end{cases}.$$

For $x_B < \overline{x}_B$, $x^0 < x^*$, $\overline{x}_B < x_B < \sigma_2 2^{-1}$ the sequential equilibrium is given by the pooling equilibrium $x = x(s) = x^*$.

Proof

First we remark that the game with B coincides with the game without B in the case $\sigma_2 \le 2x_B$. In the following, we consider the case $2x_B < \sigma_2$. The game with B is then equivalent to the game without B but no more with equally distributed types of A but with a distribution $F(t \le x) = 2xx_B\sigma_2^{-1} + 1_{\{x \le 2x_B\}}(x)(\sigma_2 - 2x_B)\sigma_2^{-1}$.

In the following we exclude signalling and determine a pooling equilibrium.

C's expected utility from an offer $x \le 2x_B$ is given by:

$$E\,u_C(x) = -(x-1)^2 F(x) - [1 - F(x)] = -(x-1)^2\{(2x_B - x)2x_B\sigma_2^{-1} + (\sigma_2 - 2x_B)\sigma_2^{-1}\} - 2xx_B\sigma_2^{-1}$$

From

$$\frac{\partial E}{\partial x}u_C(x) = 6(x-1)^2 x_B\sigma_2 - 2(x-1)\{4x_B^2 - 4x_B + \sigma_2\}\sigma_2^{-1} - 2x_B\sigma_2^{-1} = 0$$

and

$$\frac{\partial^2 E}{\partial x^2}u_C(x) < 0$$

we get:

$$x = x^0 = (6x_B)^{-1}\left\{4x_B + 2x_B + \sigma_2 - \left[(4x_B^2 - 4x_B + \sigma_2)^2 + 12x_B\right]^{2^{-1}}\right\}.$$

Because T is restricted by $T = [0, x_*] \cup [x^*, 1] \ne X = [0,1]$ we consider the set of propositions x where signalling may be worthwhile.

Let C react with $x(s) < x^0$ after observing signal s. In analogy to result 2 – with x^0 instead of x^* – we can derive an interval $[x_0, x^0]$ with $x_0 = 1 - \left[2^{-1} + 2^{-1}(1 - x^0)^{2^{-1}}\right] < x^0$ such that signalling does not occur outside of $[x_0, x^0]$.

We remark

– x_0 resp. x_* is monotonically increasing with x^0 resp. x^*.

We have to consider the cases:

$$x^0 < x^* \Rightarrow x_0 < x_* \Rightarrow \left[x_0, x^0\right] \cap T = \left[x_0, x_*\right],$$

$$x^0 > x^* \Rightarrow x_0 > x_* \Rightarrow \left[x_0, x^0\right] \cap T = \left[x^*, x^0\right],$$

$$x^0 = x^* \Rightarrow x_0 = x_* \Rightarrow \left[x_0, x^0\right] \cap T = \left[x_*, x^*\right]$$

–all these cases may occur as the following consideration shows:

$$x^0 \begin{pmatrix} \leq \\ > \end{pmatrix} \Leftrightarrow 2x_B x^0 \begin{pmatrix} \leq \\ > \end{pmatrix} \Leftrightarrow (2x_B - \sigma_2)(2x_B - 1) \begin{pmatrix} \leq \\ > \end{pmatrix} \left[\left(4x_B^2 - 4x_B + \sigma_2\right)^2 + 12x_B^2\right]^{2^{-1}} - 3^{-1}\left[\sigma_2^2 - 2\sigma_2 + 4\right]^{2^{-1}}$$

at which holds:

$$\sigma_2 = 2x_B \to 0 \Rightarrow 0 \begin{pmatrix} \leq \\ > \end{pmatrix} - 23^{-1} \Rightarrow x^0 > x^*,$$

$$\sigma_2 = 2x_B \to 1 \Rightarrow 0 \begin{pmatrix} \leq \\ > \end{pmatrix} 13^{2^{-1}} - 3^{-1}3^{2^{-1}} > 0 \Rightarrow x^0 \leq x^*.$$

By derivation x^0 is the pooling equilibrium in the unrestricted game with $T = [0,1]$. In the restricted game with $T = [0, x_*] \cup [x^*, 1]$ and in the case $x^0 < x^*$ instead of x^0 the pooling equilibrium becomes x^* and we have no signalling. In the case $x^0 > x^*$ the pooling equilibrium x^0 remains in the game with B. In this case signalling may be useful for getting offers from the interval $\left[x^*, x^0\right]$.

In the following reasonings of the proof the restrictions on T are neglected – i.e. $T = X = [0,1]$ is assumed. After having determined sequential equilibrium – pooling or separating – we modify it in order to get a sequential equilibrium of the restricted game with $T = [0, x_*] \cup [x^*, 1]$.

The sequential equilibrium of the unrestricted game is determined in the following five steps:

Step 1

C observes signals. Any signal $s_1 \leq \hat{t}$ uniquely identifies a player A on $t \in [0, \hat{t}]$; any signal $s_2 > \hat{t}$ identifies a type $t \in]\hat{t}, \sigma_2]$. Given his observation C holds beliefs about what A could have send a particular signal according to $t \sim \mu(s) = \begin{cases} E[0,\hat{t}] & if\ s = s_1 \leq \hat{t} \\ E[\hat{t},\sigma_2] & if\ s = s_2 > \hat{t} \end{cases}$. The choice of s_1, s_2, given $s_1 \leq \hat{t}$ and $s_2 > \hat{t}$ as the only relevant information.

Step 2

Given his expectation about possible types C determines his bargaining strategy. If he observes $s_2 > \hat{t}$, he must believe that A is from the interval $]\hat{t}, \sigma_2]$ which corresponds to $[2x_B, \sigma_2]$. Given his strategy he will make an offer $x = 2x_B$ for any $t \in]\hat{t}, \sigma_2]$. That is his best response given his beliefs, since for $x = 2x_B$ B will always accept the offer. If C believes that A belongs to $[0, 2x_B]$ he can guarantee that B and any A will accept his offer only if he proposes $x \leq 2x_B$. Then C must have observed a signal $s_1 \leq \hat{t}$ corresponding to types on the interval $[0, \hat{t}]$.

For any offer $x \leq 2x_B$ C's expected utility given the signals and the responses is given by $Eu_C(x) = -(\hat{t} - x)\hat{t}^{-1}(x-1)^2 + (1-(\hat{t}-x)\hat{t}^{-1})(0-1)^2$ at which $(\hat{t} - x)\hat{t}^{-1}$ denotes the probability that his offer will be accepted and $1-(\hat{t}-x)\hat{t}^{-1}$ denotes the probability that his offer will be rejected. $\frac{\partial Eu_C(x)}{\partial x} = 0$ gives with $\hat{x} = \left[2 + \hat{t} - \left(\hat{t}^2 - 2\hat{t} + 4\right)^{2^{-1}}\right]3^{-1}$ C's utility maximizing offer, with $\frac{\partial^2 Eu_C(\hat{x})}{\partial x^2} < 0$ and $\hat{x} < 2x_B$, which implies $\hat{t} \geq 2x_B$. Then any player A of type $\hat{t}$ is indifferent between offers $\hat{x}$ and $2x_B$, which implies $\hat{t} = 2x_B + \hat{x}$.

Step 3

Step 2 separates types t and hence C's offers s.t.:

- $t \in [0, \hat{x}]$: t-types of A will reject the offer $\hat{x}$ by C. Although B accepts the offer and the solution is equal to $x = q = 0$
- $t \in [\hat{x}, \hat{t}]$: t-types of A will accept the offer $\hat{x}$ by C and so will B
- $t \in]\hat{t}, \sigma_2]$: t-types of A in $x \in [0, 2x_B]$ accept the offer $2x_B$, which is barely being accepted by B

Therefore

$$I_4 = \sigma_2^{-1}\left\{-\int_0^{\hat{x}}\left(0-t2^{-1}\right)^2 dt - \int_{\hat{x}}^{\hat{t}}\left(\hat{x}-t2^{-1}\right)^2 dt - \int_{\hat{t}}^{\sigma_2}\left(2x_B - t2^{-1}\right)^2 dt\right\} = f\left(x_B\right)$$

is A's expected utility. The maximal value of I_4 in $\left[\underline{x}_B, \overline{x}_B\right]$ is given by $\frac{\partial I_4}{\partial x_B} = 0$ and $\frac{\partial^2 I_4}{\partial x_B^2} < 0$ and as results we get: $I_4 < I_3$ for $x_B < \underline{x}_B$, $I_4 = I_3$ for $x_B = \underline{x}_B$, $I_4 > I_3$ for $\underline{x}_B < x_B < \overline{x}_B$ and $I_4 = I_3$ for $x_B \geq \overline{x}_B$.

Figure 2 illustrates the interval Δ for which a player A will prefer the intervening intermediary where the magnitude of Δ will depend on the intermediary's bargaining objective.

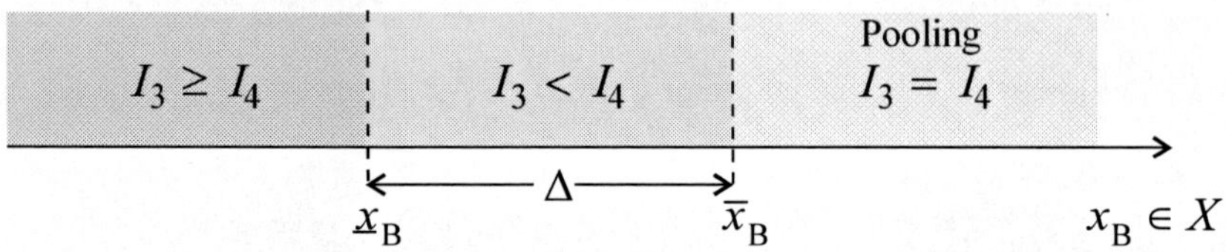

Figure 2.

We see that on the interval $x_B \geq \overline{x}_B$ and $I_3 = I_4$ any x_B^* can result in a pooling equilibrium. With the assumption of uniformly distributed types C will always offer $x = x^*$. Then B ceases to improve final allocations, he no longer has the role as an information transmitter to A.

Step 4

By definition of a sequential equilibrium for $x_B < \overline{x}_B$ we need strategies for A s.t.

$$\alpha(t) = \begin{cases} s_1 \ \ if\ t \in \left[0, \hat{t}\right] \\ s_2 \ \ if\ t \in \left]\hat{t}, \sigma_2\right] \end{cases}$$

with arbitrary chosen $s_1 \leq \hat{t}$, $s_2 > \hat{t}$ and A has no incentive to deviate. $\alpha(t)$ will satisfy $x_B < \overline{x}_B$ because only the information $t \leq \hat{t}$ or $t > \hat{t}$ is relevant for C. If a type $t \in \left[0, \hat{t}\right[$ will send $s_2 > \hat{t}$ rather than $s_1 \leq \hat{t}$ he will be offered $x = 2x_B$. But any type $t \in \left[0, \hat{x}\right[$ will reject $x = 2x_B$ and therefore $x = \hat{x}$ will be rejected by C. For $t \in \left[\hat{x}, \hat{t}\right[$, C will prefer $x = \hat{x}$ to $x = 2x_B$.

Hence, no type $t \in [0,\hat{t}[$ gains from misreporting his type and signalling s_2. Likewise, no type $t \in]\hat{t},\sigma_2]$ has any incentive to signal s_1 when his true type is s_2 since $u_{A,t}(2x_B) > u_{A,t}(\hat{x})$. Then $\alpha(t)$ is an equilibrium strategy for A and beliefs of C are rational.

In analogy to result 2 we conclude that there does not exist any type separating strategy for sequential equilibria with more than the two signals $s_1 \neq s_2$. Therefore from step 2 above we obtain a unique sequential equilibrium since for a given $\hat{t}$, $\hat{x}$ can be uniquely determined by $\hat{x} = \left[2 + \hat{t} - \left(\hat{t}^2 - 2\hat{t} + 4\right)^{2^{-1}}\right]3^{-1}$. From $\hat{t} = 2x_B + \hat{x}$, $\hat{t}$ is the unique positive solution to a quadratic equation.

Step 5

From the equilibrium strategies of types of A and the equilibrium beliefs of C we obtain the equilibrium offer by C: $x = x(s) = \begin{cases} \hat{x} & if\ s = s_1 \leq \hat{t} \\ 2x_B & if\ s = s_2 > \hat{t} \end{cases}$, given that $x_B \leq \bar{x}_B$.

We modify the solution of the unrestricted game to get a solution of the restricted game:

–In the case $x^0 \leq x^*$ the game reduces to the game without B with x^* as unique pooling equilibrium without signalling.

–In the case $x^0 > x^*$, $\hat{x} \in T$ and hence $x^* \leq \hat{x} < x^0$ we maintain the equilibrium offer of C from step 5:

$$x = x(s) = \begin{cases} \hat{x} & if\ s = s_1 \leq \hat{t} \\ 2x_B & if\ s = s_2 > \hat{t} \end{cases}.$$

–In the case $x^0 > x^*$, $\hat{x} \notin T$ and hence $\hat{x} < x^*$ we get the equilibrium:

$$x = x(s) = \begin{cases} x^* & if\ s = s_1 \leq \hat{t} \\ 2x_B & if\ s = s_2 > \hat{t} \end{cases}. \blacklozenge$$

Result 3 shows that in an incomplete information setting an intermediary can change the bargaining solution. B's strategy ultimately affects the offers, $x = \hat{x}$, $x = 2x_B$ and $x = x^*$, respectively, C is going to propose. It is obvious that B's the intermediary's role is that of an information transmitter. Any type A will find it optimal to reveal information on his type truthfully and consequently C submits a type separating offer. If A does not expect to gain from B's involvement he will prefer the game without B.

The institutional or technical instrument of signalling types is necessary for the described scenario of separating types. If this instrument is not given, then there scenarios only the pooling scenario, in which C forms his beliefs about A's type according to the known uniform distribution of types t of A. In the following we deal with the question which of the two scenarios is better for A or C.

Result 4

For $x_B \in \left]\underline{x}_B, \overline{x}_B\right[$ with $x^* \leq \hat{x} < x^0$ any players A and C prefer ex ante the game with B with the type separating equilibrium to the game without B with the pooling equilibrium and also the pooling equilibrium arising in the game without signalling but with B.

Proof

With B, C will make offers according to $\tilde{x} = \min\left(x^*, 2x_B\right)$, at which holds:

–for any $x^* > 2x_B$ C expects that B will reject x^*, and hence offers $\tilde{x} = 2x_B$. Any A in $t \in \left[0, 2x_B\right]$ will reject the offer while any A in $t \notin \left[0, 2x_B\right]$ will accept. A's expected utility is I_2.

–for any $x^* \leq 2x_B$ B accepts any offer $\tilde{x} = x^*$. A's expected utility is I_3.

For $I_2 < I_3 < I_4$, $x \in \left]\underline{x}_B, \overline{x}_B\right[$ any A will prefer the equilibrium which is type separating to a pooling solution. $I_5 = \sigma_2^{-1}\left\{-\int_0^{\tilde{x}}(0-1)^2\,dt - \int_{\tilde{x}}^{\sigma_2}(\tilde{x}-1)^2\,dt\right\}$ will be C's expected utility from a pooling solution, whereas C's expected utility form the separating equilibrium will be $I_6 = \sigma_2^{-1}\left\{-\int_0^{\hat{x}}(0-1)^2\,dt - \int_{\hat{x}}^{\hat{t}}(\hat{x}-1)^2\,dt - \int_{\hat{t}}^{\sigma_2}(2x_B-1)^2\,dt\right\}$ at which $I_6 \geq I_5$ holds obviously. Then also C prefers the type separating scenario ex ante. ♦

In the following we deal the question how in the type separating scenario the ex ante utility of A or C depends on the type $t_B = 2x_B$ of B. We call B' dominated by B if as well A and C are ex ante better off with B than without B'.

The answer is given by:

Result 5

The set $\left[\tilde{x}_B, \bar{x}_B\right] \cap \left\{x^* \leq \hat{x} < x^0\right\} \subset \left[\underline{x}, \bar{x}\right] \cap \left\{x^* \leq \hat{x} < x^0\right\}$ denotes the set of Pareto-optimal x_B with respect to A and C. Within that interval both players have different objectives regarding the choice of x_B while outside the interval – i.e. $\left[\underline{x}_B, \tilde{x}_B\right[$ – the utility of both players increases with higher x_B.

Proof

Since types $t \in \left[0, \hat{x}\right]$ of A reject offer $\hat{x}$ in a separating scenario, the probability of A accepting is given by $\varphi_1 = \left(\sigma_2 - \hat{x}\right)\sigma_2^{-1}$ and the probability of A rejecting by $\varphi_2 = \hat{x}\sigma_2^{-1}$. From $\hat{x}$ it follows that φ_1 increases if x_B decreases which implies that B's and C's bargaining objectives – i.e x_B resp. $x_C = 1$ – will have to widely separate in order for successful solutions to arise. Then the question becomes if decreasing $x_B \in \left[\underline{x}_B, \bar{x}_B\right]$ is always in the interest of player A. I.e. we have to show $\frac{\partial I_4}{\partial x_B} < 0$.

At first we notice that A's utility I_4 attains a maximum at $\tilde{x}_B$ which is an interior point on the interval $\left[\underline{x}_B, \bar{x}_B\right]$. Then it is true that decreasing x_B will decrease A's utility for offers $x_B \in \left[\underline{x}_B, \tilde{x}_B\right[$ and increase for offers $x_B \in \left]\tilde{x}_B, \bar{x}_B\right[$. From $\frac{\partial I_4}{\partial x_B} = 0$ we obtain an optimum at: $\tilde{x}_B = \left[2 + 2\sigma_2 - \left(\sigma_2^2 + 2\sigma_2 + 4\right)^{2^{-1}}\right]6^{-1}$. Furthermore, $\left[\tilde{x}_B, \bar{x}_B\right] \subset \left[\underline{x}_B, \bar{x}_B\right]$ denotes the Pareto-optimal set which is the set of x_B chosen by A and C for which no other x_B can make one player A or C better off without leaving some other player worse off.

For the proof it suffices to show that $\frac{\partial I_6}{\partial x_B} > 0$ for $x_B \in \left]\tilde{x}_B, \bar{x}_B\right[$, because of $\frac{\partial I_4}{\partial x_B} < 0$ for $x_B \in \left]\tilde{x}_B, \bar{x}_B\right[$.

From $\hat{t} = 2x_B + \hat{x}$ it follows that $\hat{t}$ is the unique solution of the quadratic equation $\hat{t} - 2x_B = \left[2 + \hat{t} - \left(\hat{t}^2 - 2\hat{t} + 4\right)^{2^{-1}}\right]3^{-1}$ which is given by $\hat{t} = 4x_B + 1 - \left(4x_B^2 + 1\right)^{2^{-1}}$.

From this argumentation we get $\frac{\partial \hat{t}}{\partial x_B} = 4 - 4\left(4x_B^2 + 1\right)^{2^{-1}} > 0$ and because $2x_B \leq \sigma_2$ it follows

$$\frac{\partial \hat{x}}{\partial x_B} = \frac{\partial \hat{t}}{\partial x_B} - 2 = 2 - 4\left(4x_B^2 + 1\right)^{2^{-1}} < 0 .$$

From our earlier equation for I_6 we obtain

$$\frac{\partial I_6}{\partial x_B} = -\sigma_2^{-1}\left\{4x_B(1-x_B)\frac{\partial \hat{x}}{\partial x_B} + 2(\hat{x}-1)^2 - 2(2x_B-1)^2 - 4(\sigma_2-\hat{t})(1-2x_B)\right\} > 0. \;\blacklozenge$$

Result 5 then gives an answer to the question we posed at the beginning: whether an intermediary should be involved in the bargaining given his power to intervene decisions and hence improve solutions for players A and C.

5. Conclusion

In this paper we discuss a one step bargaining game on a position. Two or three players (A, C and possibly B as intermediary) interact in the context of full or asymmetric information. The player's quadratic loss functions differ only in their bliss points characterizing their types. C's and possibly B's type are given and common knowledge. A's type is chosen randomly, ex post known to A only in the context of asymmetric information. A has a limited possibility of signalling his type to C. The information transfer is limited to two intervals of types. C proposes a position x and A and possibly B have a veto power against x. In case of rejection the position $x=0$ results. We compare the games with or without intermediary in the case of full or asymmetric information.

After a brief discussion of the setting of complete information we concentrate on the case of asymmetric information. The game without intermediary always results in a pooling equilibrium (section 4.1, result 2). In the game with B a separating equilibrium results for an interval of types of B (section 4.2, result 3). For types of B in a bit smaller interval this separating equilibrium is proven to be Pareto superior (for A and C) to the pooling equilibrium arising in the game without intermediary (section 4.2, result 4). Finally we characterize an interval yet a bit smaller as the set of types of B, where A and C cannot be both made better of by another type of B (section 4.2, result 5). This interval constitutes the set of Pareto optimal types of intermediaries.

The message of our paper is to analyze in the framework of a one step bargaining game on a position with asymmetric information and the possibility of a limited information transfer from A to C by signalling two intervals of types, how an additional third veto player may improve the equilibrium for both other players. This is achieved by signalling from A to C made possible by the third player B acting in this way indirectly as an information transmitter or intermediary or middleman. Furthermore our non-cooperative game model sets the stage for the more advanced analysis of Pareto improvements in the context of bargaining solutions in cooperative games.[13]

[13] We remark that in our Chechnia example the third player „General Lebed" is only Pareto improving the situation, if he is endowed with veto power.

Summary

In a two/three player bargaining game with asymmetric information and signalling we give equivalent conditions that the existence of an intermediary can Pareto improve the bargaining solution. With intermediary actively involved we get a Pareto superior separating equilibrium (with signalling) in contrast to a Pareto inferior pooling equilibrium (without signalling) when no intermediary is involved.

References

Arshadi, N.; Karels, G. V. (1997): Modern Financial Intermediaries and Markets, NY

Banks J.; Sobel,S. (1987): Equilibrium selection in signalling games, *Econometrica*, Vol. 55, p. 647-661

Ben-Porath, E.; Dekel, E. (1992): Signaling Future Actions and the Potential for Sacrifics, *Journal of Economic Theory*, Vol. 57, p. 36-51

Cho, I. K.; Kreps, D. M. (1987): Signalling Games and Stable Equilibria, *Quarterly Journal of Economics*, Vol. 102, p. 179-221

Cothren, R.; Loewenstein, *M. A.* (N D): Quality Signals and Asymmetric Information in a Sequential Bargaining Game, unpublished paper, Virginia Polytechnic Institute/State University

Cremer, J. (1989): Commitment Through Third Parties, *Quarterly Journal of Economics*, Vol. 104, p. 74 - 89

Diamond, D. (1984): Financial Intermediation and Delegated Monitoring, *Review of Economic Studies,* Vol. 51, p. 393 - 414

Güth, W.; Steffen, H.; Ockenfels, P. (1996): Two-Level Ultimatum Bargaining with Incomplete Information-An Experimental Study, *The Economic Journal*, Vol. **106**, p. 593-604

Harsanyi J. C. (1967): Games.with incomplete information played by Bayesian players: Part I, II, III, *Management Science*, p. 159-182, p. 320-334, p. 486-502

Kreps, D. M. (1990): *Game Theory and Economic Modelling*, Oxford

Kreps, D. M.; Wilson, R. (1982): *Sequential Equilibria, Econometrica*, Vol. 50, p. 863-894

Lax, D.; Sebenius, J. (1991): Negotiating through an agent, *Journal of Conflict Resolution*, Vol. 37, p. 474-493

Leland, H. E.; Pyle, D. H. (1977): Informational Asymmetries, Financial Structure, and Financial Intermediation, *Journal of Finance*, Vol. 32, p. 371 – 387

Mas-Colell, A.; Whinston, M.D.; Green, J.A. (1995): Microeconomic Theory, Oxford-NY

Matthews, S. (1989): Veto Threats: Rhetoric in an Bargaining Game, *Quarterly Journal of Economics*, Vol. 104, p. 347-369

Myerson, R. B. (1991): Game Theory-Analysis of Conflict, Cambridge

Osborne, M. J.; Rubinstein, A. (1990): Bargaining and Markets, San Diego

Persson, T.; Tabellini G. (1990): Macroeconomic Policy, Credibility and Politics, Paris

Rubinstein, A. (1982): Perfect equilibrium in a bargaining model, *Econometrica*, Vol. 50, p. 97-109

Rubinstein, A. (1985): A bargaining model with incomplete information about time preferences, *Econometrica*, Vol. 53, p. 1151-1172

Rubinstein, A; Wolinky, A. (1987): Middlemen, *Quarterly Journal of Economics*, Vol. 102, p. 581-593

Spulber, D. (1996): Market Microstructure and Intermediation, *Journal of Economic Perspectives*, Vol 10, p 135-152

Van Damme, E. (1991): *Stability and Perfection of Nash Equilibria*, Heidelberg -New York

In: Mathematics, Game Theory and Algebra Compendium ISBN 978-1-60456-659-8
Editor: Jacob H. Mathias, pp. 79-83

Chapter 9

LIMITS AT INFINITY OF GENERALIZED RIESZ POTENTIALS

Inan Cinar*
100. Yil University, Faculty of Science and Art,
Department of Mathematics,
65080 Van, Turkey

Abstract

In this paper we study the existence of limits of generalized Riesz potentials depending on λ-distance a long the sets parallel to some coordinate axis. Our main results is the generalization of Mizuta theorem for classical Riesz potentials.

AMS Subject Classification: 31B10, 42B20, 47B06

Keywords: Riesz potentials, non-isotropic Riesz Potentials

Mizuta [1] studied the limits of classical Riesz potentials on the sets parallel to the x_1-axis,...,x_l-axis and showed that if $l < \alpha p < n$ and $f \in L_p(\mathbb{R}^n)$ is the nonnegative function then there exist a set $E \subset \mathbb{R}^{n-l}$ with $C_{\alpha,p}(\{0\} \times E) = 0$ such that for all $x' \in \mathbb{R}^{n-l}$

$$\lim_{|x'| \longrightarrow \infty} \int |x - y|^{\alpha - n} f(y) dy = 0$$

where $x = (x', x'')$ and $C_{\alpha,p}(B)$ is a Bessel capacity of the set B ([2][3]).

In this work we study the limits at infinity of generalized Riesz potentials depending on λ-distance and for this potentials established the analogous of Mizuta theorem.

We first defined the λ-distance. Let $\lambda = (\lambda_1, ..., \lambda_n)$ where λ_i are positive real numbers and $|\lambda| = \lambda_1 + ... + \lambda_n$. For any $x \in \mathbb{R}^n$ its λ-distance from origin is defined by the following way [4].

$$\|x\|_\lambda = (|x_1|^{1/\lambda_1} + ... + |x_n|^{1/\lambda_n})^{\frac{|\lambda|}{n}}$$

*E-mail address: inancinar@mynet.com

In ([4], [5]) singular integrals depending on λ-distance investigated. Riesz potentials depending on this λ-distance has the form

$$R^{\lambda}_{\alpha,l} f(x) = \int_{\mathbb{R}^n} \frac{f(y)}{\|x-y\|_{\lambda}^{n-\alpha}} dy \tag{9.1}$$

Some properties of this potentials studied in [4]. It is known form the definition of Bessel capacity that this capacity closely connected with Riesz potentials [2]. By analogues we are defined there the capacities connected with the potentials (9.1). Since all distance defined in $\mathbb{R}^n$ are equivalents this capacities also equivalents to Bessel capacity.

Let $E \subset \mathbb{R}^n$ We are defined the following functional set

$$E_{\lambda,g} = \{g \geq 0,\ g \in L_p(\mathbb{R}^n) : \int_{\mathbb{R}^n} \|x-y\|_{\lambda}^{\alpha-n} g(y)dy \geq 1,\ x \in E\}$$

Bessel capacity of the set E connected with the λ-distance we denote by $C^{\lambda}_{\alpha p}(E)$ and defined it as

$$C^{\lambda}_{\alpha p}(E) = \inf_{f \in E_{\lambda,g}} \|f\|_p^p .$$

It easy seen that if $C_{\alpha p}(E) = \|f^*\|_p^p$ then there exist a constant C_1 such that $C_1 f^* \in E_{\lambda,g}$ and therefore

$$C^{\lambda}_{\alpha p}(E) \leq C_1 C_{\alpha p}(E).$$

It maybe shown that the inverse inequality also holds. That is there exist a constant C_2 for which $C_{\alpha p}(E) \leq C_2 C^{\lambda}_{\alpha p}(E)$. This means that two capacities are equivalent and the both maybe used in investigation of generalized Riesz potentials. From this equivalentness and the Mizuta theorems [6] it follows that if E_1 is the projection of set E to the set $\{0\} \times \mathbb{R}^{n-l}$ then

$$C^{\lambda}_{\alpha p}(E_1) \leq C^{\lambda}_{\alpha p}(E).$$

Our main result is following.

Theorem 9.1. *Let* $|\lambda|_{k,q} = \lambda_k + ... + \lambda_q$, $1 \leq k < q \leq n$, $1 \leq p < \infty$, $\frac{1}{p} + \frac{1}{p'} = 1$, $l - \frac{2|\lambda|_{1,l}}{p'} < \alpha < n - \frac{2|\lambda|}{p'}$ *and* $f \in L_p(\mathbb{R}^n)$ *is nonnegative function. Then there exist a set* $E \subset \mathbb{R}^{n-l}$ *which* $C^{\lambda}_{\alpha p}(\{0\} \times E) = 0$ *such that for all* $x' \in \mathbb{R}^{n-l} - E$

$$\lim_{\|x'\|_{\lambda} \longrightarrow \infty} \int_{\mathbb{R}^n} \|x-y\|_{\lambda}^{\alpha-n} f(y)dy = 0$$

where $x = (x', x'')$.

Proof. Let $B_{\lambda}(x_0, r)$ is the λ-ball with radius r, centered in x_0. That is

$$x \in B_{\lambda}(x_0, r) \Rightarrow \|x - x_0\|_{\lambda} \leq r.$$

Then for $\|x\|_{\lambda} \geq 2r$ the inequality

$$\int_{B_{\lambda}(0,r)} \|x-y\|_{\lambda}^{\alpha-n} f(y)dy \tag{9.2}$$

$$\leq \|f\|_p \left(\int_{\mathbb{R}^n - B_\lambda(x,r)} \|x-y\|_\lambda^{p'(\alpha-n)} \, dy \right)^{\frac{1}{p'}}$$

holds. Passing to spherical coordinates by formulas $y_k = (x_k + \rho U_k)^{2\lambda_k}$ $k = 1, 2, ..., n$ where U_k are the ordinary angles coordinate we can see that this transform has the Jacobian

$$I_\lambda = \rho^{2\|\lambda\|-1} \omega_k(\theta_1, ..., \theta_{n-1}),$$

where ω_k depend only on angles coordinates [5]. By this reason

$$\left(\int_{\mathbb{R}^n - B_\lambda(x,r)} \|x-y\|_\lambda^{p'(\alpha-n)} \, dy \right)^{\frac{1}{p'}} \leq M_1 r^{\alpha - (n - \frac{2\|\lambda\|}{p'})}$$

where M_1 is a constant depending on λ and p.

Using this in (9.2) we have

$$\int_{B_\lambda(0,r)} \|x-y\|_\lambda^{\alpha-n} f(y) dy \leq M_1 \|f\|_p \, r^{\alpha - (n - \frac{2\|\lambda\|}{p'})} \tag{9.3}$$

By the definition of $C_{\alpha p}^\lambda$ capacity

$$C_{\alpha p}^\lambda \left(\left\{ x : \int_{\mathbb{R}^n - B_\lambda(0,r)} \|x-y\|_\lambda^{\alpha-n} f(y) dy \geq 1 \right\} \right) \leq \int_{\mathbb{R}^n - B_\lambda(0,r)} f^p(y) dy$$

From this we obtain

$$C_{\alpha p}^\lambda \left(\left\{ x : \int_{\mathbb{R}^n - B_\lambda(0,r)} \|x-y\|_\lambda^{\alpha-n} f(y) dy \geq \frac{1}{m} \right\} \right) \tag{9.4}$$

$$\leq m^p \int_{\mathbb{R}^n - B_\lambda(0,r)} f^p(y) dy$$

By (9.3) for sufficiently large $r > r(m)$ and $x \in \mathbb{R}^n - B_\lambda(0, 2r)$ the inequality

$$\int_{B_\lambda(0,r)} \|x-y\|_\lambda^{\alpha-n} f(y) dy \leq \frac{1}{m}$$

holds.

Therefore we have from (9.4)

$$C_{\alpha p}^\lambda \left(\left\{ x : \int_{\mathbb{R}^n - B_\lambda(0,r)} \|x-y\|_\lambda^{\alpha-n} f(y) dy \geq \frac{1}{m} \right\} \right) \leq 2^{-m} \tag{9.5}$$

From (9.5) follows the inequality

$$C_{\alpha p}^\lambda \left(\left\{ x : \int_{\mathbb{R}^n - B_\lambda(0,r)} \|x-y\|_\lambda^{\alpha-n} f(y) dy \geq \frac{2}{m} \right\} \right) \leq 2^{-m}$$

Denote

$$\Omega_{m,r}=\left\{x\in\mathbb{R}^n-B_\lambda(0,2r):\int_{\mathbb{R}^n}\|x-y\|_\lambda^{\alpha-n}f(y)dy\geq\frac{2}{m}\right\}$$

Then form obvious embedding

$$\left\{x:\int_{\mathbb{R}^n-B_\lambda(0,2r)}\|x-y\|_\lambda^{\alpha-n}f(y)dy\geq\frac{2}{m}\right\}$$

$$\subset\left\{x:\int_{\mathbb{R}^n-B_\lambda(0,r)}\|x-y\|_\lambda^{\alpha-n}f(y)dy\geq\frac{2}{m}\right\}.$$

We see that

$$C_{\alpha p}^\lambda(\Omega_{m,r})\leq 2^{-m}.$$

Now if F'' is the projection of set F to the set $\{0\}\times\mathbb{R}^{n-l}$ we set $E_j=\bigcup_{m=j}^\infty\Omega_{m,r(m)}$ and $E=\bigcap_{j=1}^\infty E_j''$ then as mentioned above

$$C_{\alpha p}^\lambda(E_j'')\leq C_{\alpha p}^\lambda(E_j).$$

By the properties of capacities [3]

$$C_{\alpha p}^\lambda(E_j)=C_{\alpha p}^\lambda\left(\bigcup_{m=j}^\infty\Omega_{m,r(m)}\right)$$

$$\leq\sum_{m=j}^\infty C_{\alpha p}^\lambda(\Omega_{m,r(m)})\leq\sum_{m=j}^\infty\frac{1}{2^m}=\frac{1}{2^{j-1}}$$

and consequently

$$C_{\alpha p}^\lambda(E_j'')\leq\frac{1}{2^{j-1}}.$$

From this we obtain $C_{\alpha p}^\lambda(E)=0$. In outside of the set E the integral

$$\int_{\mathbb{R}^n-B_\lambda(0,r)}\|x-y\|_\lambda^{\alpha-n}f(y)dy$$

is bounded and by (9.3) the generalized Riesz potential is bounded too. Therefore $\|x-y\|_\lambda^{\alpha-n}f(y)\in L_1(\mathbb{R}^n)$. Now for $x=(x',x'')$, $x'\in\mathbb{R}^l$, we have

$$\|x-y\|_\lambda^{\alpha-n}=\left[\|x'-y'\|_\lambda^{\frac{n-l}{|\lambda|_l l}}+\|x''-y''\|_\lambda^{\frac{n-l}{|\lambda|_{l_n}}}\right]^{(\alpha-n)\frac{|\lambda|}{n}}.$$

From this $\|x-y\|_\lambda^{\alpha-n}f(y)\longrightarrow 0$ as $\|x'\|_\lambda\longrightarrow\infty$.
Applying Lebesque dominated convergence theorem we see that for $x''\in\mathbb{R}^{n-l}-E$

$$\lim_{\|x'\|_\lambda\longrightarrow\infty}\int_{\mathbb{R}^n}\|x-y\|_\lambda^{\alpha-n}f(y)dy=0.$$

The proof is completed. □

References

[1] Y. Mizuta. On the existence of spherical limits of Beppo Levi functions. *J. London Math. Soc.*, (2) **41**, 89-99, (1990).

[2] N. G. Meyers. A theory of capacities for potentials in Lebesgue classes. *Math. Scand.*, **8** , 255 - 292, (1970).

[3] N. G. Meyers. Continuity properties of potentials. *Duke Math. J.*, **42**, 157 -166, (1975).

[4] I. Cinar, H. Duru, On Continuity Properties of Potentials Depending on λ - Distance. *J. of Appl. Math. and Comput.*, **139**: 531-534, (2003).

[5] I. Cinar, On nonizotropic Riesz Potentials. *Indian J. Pure Appl. Math.*, **28**(9): 1197-1205, (1997).

[6] Y. Mizuta. On the differentiability of Riesz potentials of functions. *Hiroshima Math. J.*, **17**, 355 - 359, (1987).

In: Mathematics, Game Theory and Algebra Compendium ISBN 978-1-60456-659-8
Editor: Jacob H. Mathias, pp. 85-102

Chapter 10

PHASE OPERATOR ON A DEFORMED HILBERT SPACE

P.K. Das*
Physics and Applied Mathematics Unit
Indian Statistical Institute
203,B.T.Road,Kolkata-700108, India

Abstract

In this paper we propose a phase operator and study its distribution on a deformed Hilbert space. Phase measurement statistics through quantum estimation theory is also explored.

1. Introduction

To deal with the fluctuating fields we introduce a distribution for the complex field amplitude in classical coherence theory. By integrating over the strength of the field we then obtain the phase distribution. The description of the phase in quantum mechanical terms has been influenced by the difficulty of ascribing an operator to it in the quantum sense. To define a Hermitian phase operator in the quantum mechanical description of phase goes back to the work of Dirac [1], who attempted a defination of a phase operator with the help of polar decomposition of the annihilation operator in radiation field. But a polar decomposition of the one-mode field complex amplitude operator does not give a unitary operator exponential of the phase. Thereafter, Susskind and Glogower[2], Carruthers and Nieto[3], Pegg and Barnett[4], Shapiro and co-workers[5] have studied further in this topic. Susskind and Glogower modified Dirac's phase operator though it is one-sided unitary operator. Nevertheless, their phase operator has been extensively used in quantum optics. Shapiro and co-workers introduced phase measurement statistics through quantum estimation theory[6]. Pegg and Barnett carried out a polar decomposition of the annihilation operator in a truncated Hilbert space of dimension $s + 1$, and defined a Hermitian phase operator in this finite-dimensional space. Now, given a state in the finite-dimensional Hilbert space one

*E-mail address: daspk@isical.ac.in

first computes the expectation value with the restricted state to the $s+1$-dimensional space. It is natural now to take the limit s to infinity and recover an Hermitian phase operator on the full Hilbert space. However, in this limit the PB phase operator does not converge to an Hermitian phase operator, but the distribution does converge to the SG phase distribution. Thus it appears to be computationally advantageous to describe the quantum-mechanical phase via a phase distribution rather than through a phase operator. This view was manifested in the work of Shapiro and co-workers. Agarwal and co-workers[7] adopted this point of view in investigating the quantum-mechanical phase properties of the nonlinear optical phenomena.

Taking the ideas of Susskind and Glogower in mind I [9, 11, 12, 13] recently, described a phase operator in a deformed Hilbert space and studied phase distribution of Kerr vectors and phase measurement problems. But it appears that the phase operator is to be modified to an appropriate one [15]. In this paper we propose the modified phase operator on the deformed Hilbert space and study its phase distribution.

Another interesting problem worth studying is the phase measurement statistics through quantum estimation theory by the Susskind-Glogower type operator. In this direction we have the following points to note:

Refractive index variations due to various environmental effects causes phase shifts to optical beams which we can observe interferometrically. In a coherent state (shot-noise limited) phase sensing interferometer, the root-mean-square phase error is proportional to $1/\sqrt{N}$, where N is the average number of detected photons. If we use squeezed states in a single frequency Michelson interferometer gravity-wave detector then root- mean-square phase error is proportional to $1/N$. This means that squeezed state interferometer requires fewer photons than does a conventional coherent-state interferometer to reach the standard quantum limit on position measurement accuracy. But the squeezed state interferometry does not represent the ultimate quantum limit on the measurement of optical phase. Better performance can be achieved in phase measurement problem by using quantum estimation theory. As the statistics of a quantum measurement depend on both the measurement operator and the input state, one studies the phase measurement problem on a single mode radiation field optimizing both the quantum state and the quantum measurement and arrives a local accuracy $\delta\phi \sim 1/N^2$ which is the reciprocal peak likelihood error and not the root-mean-square error.

In this direction Shapiro and others studied phase measurements on a single- mode radiation field of annihilation operator a on Boson-Fock space. They further studied probability operator measures(POM) and had shown that POM generated by the Susskind-Glogower operator yields the maximum-likelihood phase estimation for an arbitrary input state.

Taking the view point of Shapiro and others we study here the same problem in the deformed Hilbert space setting and observe that their result can be improved further in the new setting.

The work is organized as follows. In section 1, we give a brief introduction to the subject we wish to study. In section 2, we give preliminaries and notations that will be used throughout the paper. In section 3, we propose the phase operator in the deformed Hilbert space. In section 4, we discuss first phase vectors and its completeness. Then discuss phase distribution in the deformed Hilbert space and give a few illustrative examples. In fact, we elaborate how phase distribution will look like when we take incoherent vector, coherent vector, coherent phase vector in the deformed space. In section 5, we consider phase measurement problem after a brief discussion of probability operator measure in the deformed space and finally in section 6, we give a conclusion.

2. Preliminaries and Notations

We consider the set

$$H_q = \{f : f(z) = \sum a_n z^n \text{where} \sum [n]! \, |a_n|^2 < \infty\},$$

where $[n] = \dfrac{1-q^n}{1-q}$, $0 < q < 1$.

For $f, g \in H_q$, $f(z) = \sum_{n=0}^{\infty} a_n z^n$, $g(z) = \sum_{n=0}^{\infty} b_n z^n$ we define addition and scalar multiplication as follows:

$$f(z) + g(z) = \sum_{n=0}^{\infty} (a_n + b_n) z^n \tag{1}$$

and

$$\lambda \, o \, f(z) = \sum_{n=0}^{\infty} \lambda \, a_n z^n. \tag{2}$$

It is easily seen that H_q forms a vector space with respect to usual pointwise scalar multiplication and pointwise addition by (1) and (2). We observe that $e_q(z) = \sum_{n=0}^{\infty} \dfrac{z^n}{[n]!}$ belongs to H_q.

Now we define the inner product of two functions $f(z) = \sum a_n z^n$ and $g(z) = \sum b_n z^n$ belonging to H_q as

$$(f \, , \, g) = \sum [n]! \, \bar{a_n} \, b_n. \tag{3}$$

Corresponding norm is given by

$$\|f\|^2 = (f \, , \, f) = \sum [n]! \, |a_n|^2 < \infty \, .$$

With this norm derived from the inner product it can be shown that H_q is a complete normed space.Hence H_q forms a Hilbert space.

In a recent paper [14, 16] we have proved that the set $\{\dfrac{z^n}{\sqrt{[n]!}}, n = 0, 1, 2, 3, \ldots\}$ forms a complete orthonormal set. If we consider the following actions on H_q:

$$\begin{aligned} T f_n &= \sqrt{[n]} f_{n-1} \\ T^* f_n &= \sqrt{[n+1]} f_{n+1} \end{aligned} \tag{4}$$

where T is the backward shift and its adjoint T^* is the forward shift operator on $H_q$and $f_n(z) = \dfrac{z^n}{\sqrt{[n]!}}$ then we have shown [14, 16] that the solution of the following eigenvalue equation

$$T f_\alpha = \alpha f_\alpha \tag{5}$$

is given by

$$f_\alpha = e_q(|\alpha|^2)^{-\frac{1}{2}} \sum_{n=0}^{\infty} \frac{\alpha^n}{\sqrt{[n]!}} f_n. \tag{6}$$

We call f_α a **coherent vector** in H_q.

3. Phase Operator

Before going to define the phase operator we observe that

$$TT^* = [N+1], T^*T = [N] \tag{7}$$

where the operator N is such that

$$Nf_n = nf_n. \tag{8}$$

Also we can verify that

$$NT - TN = -T, NT^* - T^*N = T^* \tag{9}$$

and

$$TT^* - T^*T = q^N. \tag{10}$$

We can also show that q^N commutes with both T^*T and TT^*.

Now, analogous to the idea of Carruthers and Nieto [3], we initially proposed the phase operator to be

$$P = (q^N + T^*T)^{-1/2}T \tag{11}$$

where N is given by (8).

Now because of the relation

$$q^n + [n] = [n+1] \tag{12}$$

our phase operator(11) does not produce anything new but the phase distribution produced by Susskind-Glogower phase operator.

To circumvent this situation we propose our phase operator to be

$$P = (q^{N+1} + T^*T)^{-1/2}T \tag{13}$$

where N is given by (8).

The operator P (13) is not unitary but is one-sided unitary as we can easily verify

$$PP^* = I, P^*P \neq I. \tag{14}$$

4. Phase Distribution

In this section we describe the phase distribution in the deformed Hilbert space. To do this we introduce the phase vector and obtain its distributions in details.

4.1. Phase Vector

To obtain the phase vector we consider first the Susskind-Glogower type **phase operator** $P = (q^{N+1} + T^*T)^{-1/2}T$ as discussed above(13).

Now the phase vector is obtained by solving the eigenvalue equation

$$Pf_\beta = \beta f_\beta \tag{15}$$

where $f_\beta(z) = \sum_{n=0}^{\infty} a_n z^n = \sum_{n=0}^{\infty} a_n \sqrt{[n]!}\, f_n(z)$.That is,

$$f_\beta = \sum_{n=0}^{\infty} a_n \sqrt{[n]!}\, f_n. \tag{16}$$

Then

$$\begin{aligned} Pf_\beta &= \sum_{n=0}^{\infty} a_n \sqrt{[n]!}\,(q^{N+1} + T^*T)^{-1/2} T f_n \\ &= \sum_{n=1}^{\infty} a_n \sqrt{[n]!}\,(q^{N+1} + T^*T)^{-1/2} \sqrt{[n]} f_{n-1} \\ &= \sum_{n=1}^{\infty} a_n \sqrt{[n]!}\sqrt{[n]}(q^n + [n-1])^{-1/2} f_{n-1} \\ &= \sum_{n=0}^{\infty} a_{n+1} \sqrt{[n+1]!}\sqrt{[n+1]}(q^{n+1} + [n])^{-1/2} f_n \end{aligned} \tag{17}$$

and

$$\beta f_\beta = \beta \sum_{n=0}^{\infty} a_n \sqrt{[n]!} f_n. \tag{18}$$

From (15), (16), (17) and (18) we observe that a_n satisfies the following difference equation:

$$a_{n+1}\sqrt{[n+1]!}\sqrt{[n+1]}(q^{n+1} + [n])^{-1/2} = \beta a_n \sqrt{[n]!}. \tag{19}$$

That is,

$$a_{n+1} = \frac{\beta a_n (q^{n+1} + [n])^{1/2}}{[n+1]}. \tag{20}$$

Hence,

$$\begin{aligned} a_1 &= \frac{\beta (q + [0])^{1/2} a_0}{[1]}. \\ a_2 &= \frac{\beta a_1 (q^2 + [1])^{1/2}}{[2]} = \frac{\beta^2 a_0 \sqrt{(q + [0])(q^2 + [1])}}{[2]!}. \\ a_3 &= \frac{\beta a_2 (q^3 + [2])^{1/2}}{[3]} = \frac{\beta^3 a_0 \sqrt{(q + [0])(q^2 + [1])(q^3 + [2])}}{[3]!}. \end{aligned}$$

and so on. Thus,

$$a_n = \frac{\beta^n a_0 \sqrt{(q + [0])(q^2 + [1])(q^3 + [2]) \ldots (q^n + [n-1])}}{[n]!}.$$

Hence,

$$\begin{aligned} f_\beta &= \sum_{n=0}^{\infty} a_n \sqrt{[n]!} f_n \\ &= a_0 \sum_{n=0}^{\infty} \beta^n \sqrt{\frac{(q+[0])(q^2+[1])(q^3+[2])\dots(q^n+[n-1])}{[n]!}} f_n. \end{aligned}$$

where $\beta = |\beta| e^{i\theta}$ is a complex number.These vectors are normalizable in a strict sense only for $|\beta| < 1$.

Now, if we take $a_0 = 1$ and $|\beta| = 1$ we have

$$f_\beta = \sum_{n=0}^{\infty} e^{in\theta} \sqrt{\frac{(q+[0])(q^2+[1])(q^3+[2])\dots(q^n+[n-1])}{[n]!}} f_n. \tag{21}$$

Henceforth,we shall denote this vector as

$$f_\theta = \sum_{n=0}^{\infty} e^{in\theta} \sqrt{\frac{(q+[0])(q^2+[1])(q^3+[2])\dots(q^n+[n-1])}{[n]!}} f_n, \tag{22}$$

$0 \leq \theta \leq 2\pi$ and call f_θ a **phase vector** in H_q.

4.2. Completeness of Phase Vectors

The phase vectors f_θ are neither normalizable nor orthogonal.The completeness relation

$$I = \frac{1}{2\pi} \int_X \int_0^{2\pi} d\nu(x,\theta) |f_\theta >< f_\theta| \tag{23}$$

where

$$d\nu(x,\theta) = d\mu(x) d\theta \tag{24}$$

may be proved as follows:

Here we consider the set X consisting of the points $x = 0, 1, 2, \dots$ and $\mu(x)$ is the measure on X which equals

$$\mu_n \equiv \frac{[n]!}{(q+[0])(q^2+[1])\dots(q^n+[n-1])}$$

at the point $x = n$ and θ is the Lebesgue measure on the circle.

Define the operator

$$|f_\theta >< f_\theta| : H_q \to H_q \tag{25}$$

by

$$|f_\theta >< f_\theta| f = (f_\theta, f) f_\theta \tag{26}$$

with $f(z) = \sum_{n=0}^{\infty} a_n z^n$

Now,

$$\begin{aligned}
&(f_\theta, f)\\
&= \sum_{n=0}^{\infty}[n]!\frac{e^{-in\theta}}{\sqrt{[n]!}}\sqrt{\frac{(q+[0])(q^2+[1])(q^3+[2])\dots(q^n+[n-1])}{[n]!}}a_n \qquad (27)\\
&= \sum_{n=0}^{\infty} e^{-in\theta}\sqrt{(q+[0])(q^2+[1])(q^3+[2])\dots(q^n+[n-1])}a_n.
\end{aligned}$$

Then,

$$\begin{aligned}
&(f_\theta, f)f_\theta\\
&= \sum_{n=0}^{\infty}\sum_{m=0}^{\infty} a_n\, e^{i(m-n)\theta}\sqrt{\frac{(q+[0])(q^2+[1])\dots(q^m+[m-1])}{[m]!}} \qquad (28)\\
&\times \sqrt{(q+[0])(q^2+[1])\dots(q^n+[n-1])}f_m.
\end{aligned}$$

Using

$$\int_0^{2\pi} d\theta e^{i(m-n)\theta} = 2\pi\delta_{mn} \qquad (29)$$

we have

$$\begin{aligned}
&\frac{1}{2\pi}\int_X\int_0^{2\pi} d\nu(x,\theta)|f_\theta >< f_\theta|f\\
&= \int_X d\mu(x)\sum_{n=0}^{\infty}\sum_{m=0}^{\infty} a_n\, f_m\sqrt{\frac{(q+[0])(q^2+[1])\dots(q^m+[m-1])}{[m]!}}\\
&\times \sqrt{(q+[0])(q^2+[1])\dots(q^n+[n-1])}\frac{1}{2\pi}\int_0^{2\pi} e^{i(m-n)\theta}d\theta\\
&= \sum_{n=0}^{\infty} a_n f_n\int_X\frac{(q+[0])(q^2+[1])\dots(q^n+[n-1])}{\sqrt{[n]!}}d\mu(x) \qquad (30)\\
&= \sum_{n=0}^{\infty} a_n f_n\frac{(q+[0])(q^2+[1])\dots(q^n+[n-1])}{\sqrt{[n]!}}\\
&\times \frac{[n]!}{(q+[0])(q^2+[1])\dots(q^n+[n-1])}\\
&= \sum_{n=0}^{\infty}\sqrt{[n]!}a_n f_n\\
&= f.
\end{aligned}$$

Thus,(23) follows.

4.3. Distribution

We use the vectors f_θ to associate, to a given density operator ρ, a phase distribution as follows:

$$\begin{aligned}
P(\theta) &= \tfrac{1}{2\pi}(f_\theta, \rho f_\theta)\\
&= \tfrac{1}{2\pi}\textstyle\sum_{m,n=0}^{\infty}\sqrt{\tfrac{(q+[0])\dots(q^m+[m-1])}{[m]!}}.\sqrt{\tfrac{(q+[0])\dots(q^n+[n-1])}{[n]!}}.e^{i(n-m)}.(f_m, \rho f_n)
\end{aligned} \qquad (31)$$

The $P(\theta)$ as defined in (31) is positive, owing to the positivity of ρ, and is normalized

$$\int_X \int_0^{2\pi} P(\theta) d\nu(x,\theta) = 1 \tag{32}$$

where

$$d\nu(x,\theta) = d\mu(x) d\theta \tag{33}$$

for,

$$\begin{aligned}
\int_X \int_0^{2\pi} P(\theta) d\nu(x,\theta) &= \int_X d\mu(x) \sum_{m,n=0}^{\infty} \sqrt{\frac{(q+[0])...(q^m+[m-1])}{[m]!}} . \sqrt{\frac{(q+[0])...(q^n+[n-1])}{[n]!}} . \\
&\quad \frac{1}{2\pi} \int_0^{2\pi} e^{i(m-n)\theta} d\theta . (f_m, \rho f_n) \\
&= \int_X d\mu(x) \sum_{n=0}^{\infty} \frac{(q+[0])...(q^n+[n-1])}{[n]!} . (f_n, \rho f_n) \\
&= \sum_{n=0}^{\infty} (f_n, \rho f_n) \\
&= 1.
\end{aligned} \tag{34}$$

In particular, the **phase distribution** over the window $0 \leq \theta \leq 2\pi$ for any vector f is then defined by

$$\begin{aligned}
P(\theta) &= \frac{1}{2\pi} (f_\theta, |f><f| f_\theta) \\
&= \frac{1}{2\pi} |(f_\theta, f)|^2 .
\end{aligned} \tag{35}$$

4.4. Examples

We now consider some specific vectors in the Hilbert space H_q and compute their corresponding phase distributions.

4.4.1. Incoherent Vectors

For the incoherent vectors we take the density operator to be

$$\rho = \sum_{n=0}^{\infty} p_n |f_n >< f_n|, \tag{36}$$

with

$$p_n \geq 0 \text{ and } \sum_{n=0}^{\infty} p_n = 1.$$

Now we calculate the phase distribution $P(\theta)$ as

$$\begin{aligned}
P(\theta) &= \frac{1}{2\pi} (f_\theta, \rho f_\theta) \\
&= \frac{1}{2\pi} \sum_{n=0}^{\infty} p_n (f_\theta, |f_n >< f_n| f_\theta) \\
&= \frac{1}{2\pi} \sum_{n=0}^{\infty} p_n |(f_\theta, f_n)|^2 \\
&= \frac{1}{2\pi} \sum_{n=0}^{\infty} p_n . \frac{(q+[0])...(q^n+[n-1])}{[n]!}
\end{aligned} \tag{37}$$

4.4.2. Coherent Vectors

For the coherent vectors $f_\alpha(6)$,

$$f_\alpha = e_q(|\alpha|^2)^{-\frac{1}{2}} \sum_{n=0}^{\infty} \frac{\alpha^n}{\sqrt{[n]!}} f_n. \tag{38}$$

we take the density operator to be

$$\rho = |f_\alpha >< f_\alpha|, \ \alpha = |\alpha| e^{i\theta_0} \tag{39}$$

and calculate the phase distribution $P(\theta)$ as

$$\begin{aligned} P(\theta) &= \tfrac{1}{2\pi}(f_\theta, \rho f_\theta) \\ &= \tfrac{1}{2\pi}(f_\theta, |f_\alpha >< f_\alpha| f_\theta) \\ &= \tfrac{1}{2\pi}|(f_\theta, f_\alpha)|^2 \\ &= \tfrac{1}{2\pi}|\textstyle\sum_{n=0}^{\infty} e^{in(\theta_0-\theta)}. \frac{|\alpha|^n}{\sqrt{[n]!}}.e_q(|\alpha|^2)^{-\frac{1}{2}}.\sqrt{\frac{(q+[0])\ldots(q^n+[n-1])}{[n]!}}|^2 \end{aligned} \tag{40}$$

4.4.3. Coherent Phase Vectors

The coherent phase vectors $\{f_\beta\}$ are the unit length eigenvectors of the operator P whose eigenvalues β are complex numbers,namely,

$$Pf_\beta = \beta f_\beta.$$

From the previous section, 4.1, we see that

$$\begin{aligned} & f_\beta \\ = & \sum_{n=0}^{\infty} a_n \sqrt{[n]!} f_n \\ = & a_0 \sum_{n=0}^{\infty} \beta^n \sqrt{\frac{(q+[0])(q^2+[1])(q^3+[2])\ldots(q^n+[n-1])}{[n]!}} f_n. \end{aligned} \tag{41}$$

Here, $|\beta| < 1$ is necessary in order to ensure normalization.For,

$$\|f_\beta\|^2 = |a_0|^2 \sum_{n=0}^{\infty} |\beta|^{2n} \frac{(q+[0])(q^2+[1])(q^3+[2])\ldots(q^n+[n-1])}{[n]!}. \tag{42}$$

The above series (42) is convergent for $|\beta| < 1$.Thus,when f_β is normalized we have $|a_0| = \Phi(|\beta|^2)^{-1/2}$ where

$$\Phi(|\beta|^2) = \sum_{n=0}^{\infty} |\beta|^{2n} \frac{(q+[0])(q^2+[1])(q^3+[2])\ldots(q^n+[n-1])}{[n]!}. \tag{43}$$

Thus, f_β (41) takes the form

$$f_\beta = \Phi(|\beta|^2)^{-1/2} \sum_{n=0}^{\infty} \beta^n \sqrt{\frac{(q+[0])(q^2+[1])(q^3+[2])\ldots(q^n+[n-1])}{[n]!}} f_n. \tag{44}$$

where $|\beta| < 1$.

It is proper to call these vectors as **coherent phase vectors**,because they are eigenvectors of the **phase operator** P and they are **coherent** with respect to P.

We have that the coherent phase vectors are normalized,and nonorthogonal,

$$(f_\alpha, f_\beta) = \frac{\Phi(\bar{\alpha}\beta)}{\sqrt{\Phi(|\alpha|^2)\Phi(|\beta|^2)}}. \tag{45}$$

If we ignore the normalization factor $\Phi(|\beta|^2)^{-1/2}$ then we get

$$f_\beta \to \sum_{n=0}^{\infty} e^{in\theta} \sqrt{\frac{(q+[0])(q^2+[1])(q^3+[2])\ldots(q^n+[n-1])}{[n]!}} f_n = f_\theta. \tag{46}$$

for $\beta = |\beta| e^{i\theta}$ with $|\beta| \to 1$.

Now we take the density operator to be

$$\rho = |f_\beta >< f_\beta| \tag{47}$$

and calculate the phase distribution $P(\theta)$ as

$$\begin{aligned} P(\theta) &= \tfrac{1}{2\pi}(f_\theta, \rho f_\theta) \\ &= \tfrac{1}{2\pi}(f_\theta, |f_\beta >< f_\beta| f_\theta) \\ &= \tfrac{1}{2\pi}|(f_\theta, f_\beta)|^2 \\ &= \tfrac{1}{2\pi}|\Phi(|\beta|^2)^{-1/2} \sum_{n=0}^{\infty} \beta^n e^{-in\theta} \frac{(q+[0])\ldots(q^n+[n-1])}{[n]!}|^2 \\ &= \tfrac{1}{2\pi} \frac{|\Phi(e^{-i\theta}\beta)|^2}{\Phi(|\beta|^2)}. \end{aligned} \tag{48}$$

5. Phase Measurement Statistics

In this section we discuss phase measurement statistics in the deformed Hilbert space setting. We firstly discuss probability operator measure in this space.

5.1. Probability Operator Measure

A discrete spectrum **probability operator measure(POM)** consists of a set of Hermitian,positive semidefinite operators, $\{\Pi_n : n \in I\!N\}$, which resolves the identity

$$I = \sum_{n \in I\!N} \Pi_n. \tag{49}$$

Measurement of this POM , by definition , gives a discrete , classical random variable with probability distribution

$$P(n, g) = (g, \Pi_n g) \quad \text{for } n \in I\!N \tag{50}$$

where g is any vector of unit norm in H_q.

In order that the laws of classical probability be satisfied , it is necessary and sufficient that

$$0 \leq P(n,g) \leq 1\,,\ \sum_{n=0}^{\infty} P(n,g) = 1 \tag{51}$$

are satisfied for arbitrary g of unit norm in H_q.

We know that the sequence $f_n = \frac{z^n}{\sqrt{[n]!}}$ forms a complete orthonormal sequence in H_q and are eigen vectors of the operator $N = T^*T$ such that

$$Nf_n = [n]f_n. \tag{52}$$

Measurement of N for any arbitrary vector $g \in H_q$ of unit norm yields a discrete -valued , classical random variable with probability distribution

$$P(f_n, g) = |(f_n, g)|^2 \quad \text{for } n = 0, 1, 2, \ldots \tag{53}$$

In order that the law of classical probability be satisfied it is necessary and sufficient that

$$0 \leq P(f_n,g) \leq 1\,,\ \sum_{n=0}^{\infty} P(f_n,g) = 1 \tag{54}$$

for arbitrary $g \in H_q$ of unit norm.

The completeness of $\{f_n\}$ guarantees that the prescription in equation (53)obeys equation (54).For,if we expand the arbitrary vector g of unit norm in terms of f_n we have

$$\begin{aligned} g &= \sum_{n=0}^{\infty}(f_n, g)f_n \\ &= \sum_{n=0}^{\infty}|f_n >< f_n|g \end{aligned} \tag{55}$$

Where we define the operator

$$|f_n >< f_n| : H_q \rightarrow H_q$$

by

$$|f_n >< f_n| = (f_n, g)f_n.$$

Equation (54) is now easily verified from equation (53) and equation (55).

Thus , N operator measurement is equivalent to the POM

$$\{\Pi_n = |f_n >< f_n| : n = 0, 1, 2, \ldots\}. \tag{56}$$

Similarly,a continuous spectrum POM consists of a set of Hermitian, positive semidefinite differential operators,$\{d\Pi(\beta) : \beta \in \mathbb{C}\}$,which resolve the identity,

$$I = \int_{\beta \in \mathbb{C}} d\Pi(\beta). \tag{57}$$

The result of measuring this POM is,by definition,a continuous,classical random variable whose probability density function is given by

$$p(\beta, g) = \frac{(g, d\Pi(\beta)g)}{d\beta} \quad \text{for } \beta \in \mathbb{C} \tag{58}$$

where g is any vector of unit norm in H_q.

We know that the backwardshift T has eigenvectors -the coherent vectors f_α (6).These vectors are not orthogonal but they form a resolution of the identity

$$I = \frac{1}{2\pi} \int_{\alpha \in \mathbb{C}} d\mu(\alpha) \, |f_\alpha >< f_\alpha| \tag{59}$$

where

$$d\mu(\alpha) = e_q(|\alpha|^2) e_q(-|\alpha|^2) d_q |\alpha|^2 d\theta$$

with $\alpha = re^{i\theta}$, which defines a T-POM

$$d\Pi(\alpha) \equiv |f_\alpha >< f_\alpha| \, \frac{d\mu(\alpha)}{2\pi} \quad \text{for } \alpha \in \mathbb{C}. \tag{60}$$

The outcome of the T-POM is a complex-valued,continuous classical random variable with probability density function

$$p(\alpha, g) = \frac{(g, d\Pi(\alpha)g)}{d\mu(\alpha)} = \frac{1}{2\pi} |(f_\alpha, g)|^2 \quad \text{for } \alpha \in \mathbb{C} \tag{61}$$

where g is any vector of unit norm in H_q.

Because of (59),it follows that

$$p(\alpha, g) \geq 0 \ , \quad \int_{\alpha \in \mathbb{C}} d\mu(\alpha) \, p(\alpha, g) = 1 \tag{62}$$

hold for any vector g of unit norm in H_q.

5.2. Phase Estimation

Once we have the POM information we are ready to discuss the phase estimation problem.Without loss of generality,we assume that $0 \leq \theta \leq 2\pi$.The class of POM's we must optimize over in order to find the best phase estimate is taken to be

$$\{d\hat{\Pi}(\theta) : 0 \leq \theta \leq 2\pi\},$$

where

$$d\hat{\Pi}(\theta) = d[\hat{\Pi}(\theta)]^\dagger \text{ and } I = \int_0^{2\pi} d\hat{\Pi}(\theta). \tag{63}$$

The conditional probability density,given the phase operator

$$P = (q^n + T^*T)^{-1/2} T,$$

for obtaining a phase value θ from this POM is

$$p(\theta, P) = \frac{(g, d\hat{\Pi}(\theta)g)}{d\nu(x,\theta)}, \text{ for } 0 \leq \theta \leq 2\pi \text{ , } x \text{ an integer} \tag{64}$$

where g is a vector of unit norm in H_q.

We choose the POM,$d\hat{\Pi}(\theta)$,and the input vector g to optimize our estimate of the phase shift P.For a given POM and the input vector, equation (64) supplies the PDF needed to perform classical maximal likelihood estimation.The observed phase value θ is our estimate of P.In order for this estimate to be one of maximum likelihood,we restrict our attention to the POM's satisfying

$$P_{ML}(\theta) = arg\max_{\theta} p(\theta, P), \text{ for } 0 \leq \theta \leq 2\pi \tag{65}$$

and optimize our estimate over $d\hat{\Pi}$ and g by maximizing the peak likelihood - minimizing $\delta\theta \equiv \frac{1}{p(\theta, P)}$.

For the input vector

$$g = \sum_{n=0}^{\infty}(f_n, g)f_n$$

where

$$(f_n, g) = |(f_n, g)|e^{ik_n} \text{ , } n = 0, 1, 2, \ldots, \tag{66}$$

$\delta\theta$ is minimized by the following POM

$$d\hat{\Pi}(\theta) = |f_\theta^g >< f_\theta^g|\frac{d\nu(x,\theta)}{2\pi} \tag{67}$$

where

$$d\nu(x,\theta) = d\mu(x)d\theta, 0 \leq \theta \leq 2\pi$$

as in (24) and

$$f_\theta^g \equiv \sum_{n=0}^{\infty} e^{in\theta + ik_n}\sqrt{\frac{(q+[0])(q^2+[1])(q^3+[2])\ldots(q^n+[n-1])}{[n]!}}f_n. \tag{68}$$

To calculate reciprocal peak likelihood $\delta\theta$ with this optimum POM to estimate P we observe first

$$\begin{aligned} p(\theta, P) &= \frac{(g, d\hat{\Pi}(\theta)g)}{d\nu(x,\theta)} \\ &= \frac{|(f_\theta^g, g)|^2}{2\pi} \\ &= \frac{1}{2\pi}|\sum_{n=0}^{\infty} e^{-in\theta}\sqrt{\frac{(q+[0])(q^2+[1])\ldots(q^n+[n-1])}{[n]!}}|(f_n, g)||^2 \end{aligned} \tag{69}$$

Hence a suitable peak likelihood $\delta\theta$ for maximum $p(\theta, P)$ can be [6],

$$\begin{aligned} \delta\theta &= 2\pi|(f_\theta^g, g)|^{-2} \\ &= 2\pi|\sum_{n=0}^{\infty}\sqrt{\frac{(q+[0])(q^2+[1])\ldots(q^n+[n-1])}{[n]!}}|(f_n, g)||^{-2} \end{aligned} \tag{70}$$

which is independent of the phases $\{k_n\}$. In fact , $p(\theta, P)$ is independent of the phases $\{k_n\}$.

As peak likelihood $\delta\theta$ is independent of $\{k_n\}$ we can assume , without loss of generality , that the input vector $g = \sum_{n=0}^{\infty}(f_n, g)f_n$ has positive real coefficient (f_n, g).Equation (68) then reduces to

$$f_\theta^g = f_\theta = \sum_{n=0}^{\infty} e^{in\theta}.\sqrt{\frac{(q+[0])\dots(q^n+[n-1])}{[n]!}}f_n \tag{71}$$

for $0 \leq \theta \leq 2\pi$ which is the solution of the eigenvalue equation (15)

$$Pf_\theta = e^{i\theta}f_\theta.$$

Now we consider the operator

$$U = \sum_{n=0}^{\infty} e^{-ik_n}|f_n >< f_n|. \tag{72}$$

Observe that

$$UU^* = U^*U = I.$$

Thus,U is a unitary transformation.

Now,for an arbitrary input vector,g,the optimum POM from equation (67) is equivalent to performing the unitary transformation U followed by the POM

$$d\Pi(\theta) = |f_\theta >< f_\theta|\frac{d\nu(x,\theta)}{2\pi} \tag{73}$$

where

$$d\nu(x,\theta) = d\mu(x)d\theta, 0 \leq \theta \leq 2\pi$$

as in (23) and (24) for,

$$\begin{aligned} Uf_\theta^g &= \sum_{n=0}^{\infty} e^{in\theta+ik_n}.e^{-ik_n}\sqrt{\frac{(q+[0])\dots(q^n+[n-1])}{[n]!}}f_n \\ &= \sum_{n=0}^{\infty} e^{in\theta}.\sqrt{\frac{(q+[0])\dots(q^n+[n-1])}{[n]!}}f_n \\ &= f_\theta \end{aligned} \tag{74}$$

where f_θ^g is given by (68).

Shifting the input vector's phase by the phase operator P amounts to

$$(f_n, g) \to e^{in\theta_0}(f_n, g) \text{ for n=0,1,2, \dots} \tag{75}$$

By rotating out the input phases k_n with the U transformation we get the transformed input as

$$e^{in\theta_0}(f_n, g) \xrightarrow{U} e^{in\theta_0}|(f_n, g)| \tag{76}$$

The effect of POM in equation (73) to this transformed vector

$$g' = \sum_{n=0}^{\infty} e^{in\theta_0}|(f_n, g)|f_n \tag{77}$$

gives the classical phase with PDF

$$\begin{aligned} p(\theta, P) &= \frac{(g', d\Pi(\theta)g')}{d\nu(x,\theta)} \\ &= \frac{1}{2\pi}(g', |f_\theta >< f_\theta|g') \\ &= \frac{|(f_\theta, g'|^2}{2\pi} \\ &= \frac{1}{2\pi}|\sum_{n=0}^{\infty} e^{in(\theta_0-\theta)}\sqrt{\frac{(q+[0])\ldots(q^n+[n-1])}{[n]!}}|(f_n, g)||^2 \end{aligned} \tag{78}$$

From the above equation it is clear that ML estimate obeys $P_{ML}(\theta) = \theta$.

Thus,the POM in equation (23) leads to the ML phase estimate for all vectors in H_q. Thus, to achieve our goal of jointly optimizing phase estimate performance over both the measurement and the input vector,it remains for us to minimize $\delta\theta$ from equation (70),by appropriate choice of input vector. Specifically,the coefficients $\{(f_n, g)\}$ for the input vector must minimize the right side of the equation (70) subject to the normalization constraint

$$\sum_{n=0}^{\infty} |(f_n, g)|^2 = 1 \tag{79}$$

and the average number constraint

$$\sum_{n=0}^{\infty} [n]|(f_n, g)|^2 = N_0 \tag{80}$$

where $N_0 = (g, T^*Tg)$.

Without loss of generality,we shall assume that (f_n, g) are positive real.

Now , Maximize

$$\begin{aligned} L(g, \lambda_1, \lambda_2) &\equiv \frac{1}{2\pi}[\sum_{n=0}^{\infty}(f_n, g)]^2 + \lambda_1[\sum_{n=0}^{\infty}(f_n, g)^2 - 1] \\ &+ \lambda_2[\sum_{n=0}^{\infty}[n](f_n, g)^2 - N_0], \end{aligned} \tag{81}$$

where λ_1 and λ_2 are Lagrange's multipliers.

It is straight forward to show that

$$(f_n, g) = \frac{c}{k+[n]} \text{ for } n = 0, 1, 2, \ldots \tag{82}$$

achieves the required stationary point for L,where c and k are positive constants depending on the Lagrange's multipliers.For brevity we shall chose $k = 1$.

As we know that $[n] \geq n$ for $q > 0$ we have

$$\frac{c/(1+[n])}{1/n} \leq \frac{c}{1/n+1}$$

Hence we see that

$$\lim_{n\to\infty} \frac{c/(1+[n])}{1/n} \leq c.$$

Thus,the series $\sum_{n=0}^{\infty} \frac{c}{1+[n]}$ and $\sum_{n=0}^{\infty} \frac{1}{n}$ converge or diverge together. But $\sum_{n=0}^{\infty} \frac{1}{n}$ diverges. Hence,we must introduce a truncation parameter in equation (82).That is, we have

$$\begin{aligned} (f_n, g) &= \frac{c}{1+[n]} \text{ for, } n = 0, 1, 2, \ldots s \\ &= 0 \text{ for, } n > s \end{aligned} \tag{83}$$

Now,we have

$$\begin{aligned} N_0 &= \sum_{n=0}^{s} [n].|(f_n, g(\alpha))|^2 \\ &= \sum_{n=0}^{s} [n].\frac{c^2}{(1+[n])^2} \\ &= \sum_{n=0}^{s} \frac{c^2}{1+[n]} - 1 \end{aligned} \tag{84}$$

where we have used equations (80) , (81) and (83) with the truncation point s.

Then,

$$\begin{aligned} \delta\theta &= 2\pi[\sum_{n=0}^{s} \sqrt{\frac{(q+[0])(q^2+[1])\ldots(q^n+[n-1])}{[n]!}}(f_n, g)]^{-2} \\ &= 2\pi c^2.[\sum_{n=0}^{s} \sqrt{\frac{(q+[0])(q^2+[1])\ldots(q^n+[n-1])}{[n]!}} \frac{c^2}{1+[n]}]^{-2} \\ &< 2\pi c^2 A.[\sum_{n=0}^{s} \frac{c^2}{1+[n]}]^{-2} \\ &= \frac{2\pi c^2 A}{(N_0+1)^2} \\ &\approx \frac{2\pi c^2 A}{N_0^2} \end{aligned} \tag{85}$$

for $N_0 \gg 1$.Here A is a constant.

6. Conclusion

Thus, with the proposed phase operator we describe a phase distribution to calculate phase distribution for specific vectors in the deformed space. Also we know [5] that ML phase estimation with optimized state leads to $\delta\theta \sim \frac{1}{N_0^2}$ reciprocal peak likelihood performance where we are interested in the behaviour at high average photon number,namely $N_0 \gg 1$.In this paper we show that in the deformed case $\delta\theta$ can even less than $\frac{1}{N_0^2}$.

References

[1] P.A.M.Dirac, Proc.R.Soc.London, *Ser.A*, vol.114, 243 (1927).

[2] L.Susskind and J.Glogower, *Physics* **1**, 49(1964).

[3] P.Carruthers and M.M.Nieto, Phase and angle variables in Quantum Mechanics, *Rev.Mod.Phys.* **40**, 411(1968).

[4] D.T.Pegg and S.M.Barnett,Unitary phase operator in Quantum Mechanics, *Europhys.Lett.* **6**, 483(1989).

[5] J.H.Shapiro and S.R.Shepard, Quantum phase measurement:A system-theory perspective, *Phys.Rev.A* **43**, 3795(1991).

[6] C.W.Helstrom, Quantum detection and estimation theory, *Academic Press*, New York, 1976.

[7] G.S.Agarwal, S.Chaturvedi, K.Tara and V.Srinivasan, Classical phase changes in nonlinear processes and their quantum counterparts, *Phys.Rev.A* **45**, 4904(1992).

[8] G.S.Agarwal, Negative binomial states of the field-operator representation and production by state reduction in optical processes, *Phys.Rev.A*, **45**, 1787 (1992).

[9] P.K.Das, Phase distribution of Kerr vectors in a deformed Hilbert space, *International Journal of Theoretical Physics*, vol.38, no.6, 1807, (1999).

[10] P.K.Das, Erratum: Phase distribution of Kerr vectors in a deformed Hilbert space, *International Journal of Theoretical Physics*, vol.39, no.4, (2000).

[11] P.K.Das, Probability Operator Measure and Phase Measurementin a Deformed Hilbert Space, *International Journal of Theoretical Physics*, vol.39, no.4, (2000).

[12] P.K.Das, Squeezed Vector and Its Phase Distribution in a Deformed Hilbert Space, *International Journal of Theoretical Physics*, vol.40, no.4, 205, (2001).

[13] P.K.Das, *Nonlinear Phase Changes in a Deformed Hilbert Space, International Journal of Theoretical Physics*, vol.40, no.4, 217, (2001).

[14] P.K.Das, Eigenvectors of Backwardshift on a Deformed Hilbert Space, *International Journal of Theoretical Physics*, vol.37, no.9, 2363, (1998).

[15] P.K.Das, Erratum: Correction to phase operator on a Deformed Hilbert Space, *International Journal of Theoretical Physics*, vol.41, no.02, 371, (2002).

[16] P. K. Das, Coherent vectors as eigenvectors of a backwardshift on a deformed Hilbert space., *International Journal of Mathematics*, Game Theory and Algebra, vol. 11, no.02, 81, (2001).

In: Mathematics, Game Theory and Algebra Compendium ISBN 978-1-60456-659-8
Editor: Jacob H. Mathias, pp. 103-109

Chapter 11

GALOIS THEORY OF GRADED FIELDS*

L. El Fadil
Department of Mathematics, Faculty of Sciences
Dhar-Mahraz, P.O. Box 1796, Fes-Atlas, Fes-Morocco

Abstract

In this paper we study galois theory of graded field extensions. In particular, we take a closer look at the galois theory studied by Hwang, Wadsworth, and Boulagouaz. On the other hand, the galois theory notions studied by Boulagouaz do not cover many situations, for example group algebras. We show that these notions are simple cases of galois extensions of commutative rings.

Introduction

Let $(\Gamma, +)$ be an abelian group and R a unit commutative graded ring with respect to Γ, i.e., $R = \oplus_{\sigma\in\Gamma} R_\sigma$ such that R_σ is a R_0-submodule of R and $R_\sigma R_\tau \subset R_{\sigma+\tau}$ for every $(\sigma,\tau) \in \Gamma^2$. Set $\Gamma_R = \{\sigma \in \Gamma \mid R_\sigma \neq 0\}$ and $R^h = \cup_{\sigma\in\Gamma} R_\sigma$ the set of homogeneous elements of R. For every nonzero homogeneous element $x \in R_\sigma$, we write $deg(x) = \sigma$ and we call it the degree of x. The graded ring R is said to be a graded field if $1 \neq 0$ and every nonzero homogeneous element in R is invertible. When this occurs, Γ_R is a subgroup of Γ, called the grade group of R, R_0 is a field and for every $\sigma \in \Gamma$, R_σ is a R_0-vector space of dimension 1.

Throughout this paper, S/R is a graded field extension with grade groups $\Gamma_R = \Gamma$ and $\Gamma_S = \Delta$, i.e., $\Gamma \subset \Delta$ and $R_\sigma \subset S_\sigma$ for every $\sigma \in \Gamma$. In section 1, we show every homogeneous element of S which is integral over R has its minimal polynomial over R. we characterize the separability of a such element via its minimal polynomial. In the second section, we investigate galois theory of graded field extensions. In particular, we state galois theory results given in [1]. Moreover the galois notions defined in [2] do not cover many situations, for example group algebras. We show that these notions are simple cases of galois extensions of commutative rings.

*Communicated by Ayman Badawi

1. Preliminaries

Let S be a free R-algebra of finite rank. Then every $x \in S$ induces a R-homomorphism l_x of S defined by $l_x(s) = xs$. Define $T_{S/R}(x) = tr(l_x)$ the trace of x and $N_{S/R}(x) = det(l_x)$ the norm of x, where $tr(l_x)$ is the trace of l_x and $det(l_x)$ is the determinant of l_x. $T_{S/R}$ is a linear form of S which induces a bilinear form of S defined by $T_{S/R}(x, y) = T_{S/R}(xy)$ for every $(x, y) \in S^2$. The determinant of the bilinear form $T_{S/R}$ with respect to a R-basis $(e_1, ..., e_n)$ of S is denoted by $D(e_1, ..., e_n)$, and called the discriminant of $(e_1, ..., e_n)$. The discriminant ideal of the R-algebra S is the principal ideal generated by $D(e_1, ..., e_n)$, where $(e_1, ..., e_n)$ is a R-basis of S. The discriminant plays a key role to characterize separability of commutative algebras. For more details see [4].

Let $P = \sum_{i=0}^{n} a_i X^i$ be a polynomial of $R[X]$ of degree n and $\sigma \in \Delta$. P is said to be a σ-homogeneous polynomial if every $a_i \neq 0$, a_i is an homogeneous element of R and for every (i, j) such that $a_i \neq 0$ and $a_j \neq 0$, $deg(a_i) + i\sigma = deg(a_j) + j\sigma$. Let $\lambda = deg(a_n) + n\sigma$, called the grade of the polynomial P. For every $\lambda \in \Gamma < \sigma >$, let $R[X]_\lambda$ be the set of σ-homogeneous polynomials of grade λ. Then $R[X] = \sum_{\lambda \in \Gamma<\sigma>} R[X]_\lambda$ is a graded ring with respect to $\Gamma < \sigma >$, which will be denoted $R[X]^{(\sigma)}$. In that way $P(X)$ is an homogeneous element of degree λ in the graded ring $R[X]^{(\sigma)}$. In particular, every polynomial of $R[X]$ splits as a sum of σ-homogeneous polynomials of $R[X]^{(\sigma)}$.

Exemple 1.1 *Let $s \in R$ be an homogeneous element of degree σ and $n \in \mathbb{N}^*$. Denote $t\mathbb{Z}$ the free abelian group with basis t and $G = \Gamma < t >= \Gamma \oplus t\mathbb{Z}$. Put $\Delta = \Gamma < t > / < nt - \sigma >$, where $< nt - \sigma >$ is the subgroup of G generated by $nt - \sigma$. Then Δ is an abelian group which contains Γ.*
Indeed, let $g \in \Gamma$ such that $\bar{g} = 0$. Then there exists $k \in \mathbb{Z}$ such that $g = k(nt - \sigma)$. So, $k = 0$.
Let $\tau = \bar{t}$ be in Δ. Then $n\tau = \sigma$ and $P(X) = X^n - s$ is a τ-homogeneous polynomial of grade σ in $R[X]^{(\tau)}$.

Lemma 1.2 *Let S/R be a graded field extension and $s \in S_\sigma$ a nonzero homogeneous element which is integral over R. Then $R[s]$ is a graded field with grade group $\Gamma < \sigma >$.*

Proof. Since s is integral over R, $R[s]$ is a finitely generated R-module. Let $P(X) = X^n + ... + a_0$ be a monic polynomial of minimal degree of $R[X]$, which annihilate s and $Q(X)$ the $n\sigma$-homogeneous component of $P(X)$ in the graded ring $R[X]^{(\sigma)}$. Then there exist $q_0, ..., q_{n-1}$ homogeneous elements of R such that $Q(s) = s^n + q_{n-1}s^{n-1} + ... + q_0 = 0$. Since s is invertible and $Q(X)$ is a polynomial of minimal degree of $R[X]$ which annihilate s, q_0 is a nonzero homogeneous element of R. So, q_0 is invertible in R. Hence $s^{-1} = -q_0^{-1}(s^{n-1} + q_{n-1}s^{n-2} + ... + q_1) \in R[s]$. Consequently, $R[s] = \sum_{g \in \Gamma<\sigma>} R[s]_g$, where $R[s]_g = \sum_{\tau + n\sigma = g} R_\tau s^n$. Therefore every homogeneous component of $R[s]$ contains an invertible element. Hence it suffices to show that $R[s]_0$ is a field. First, $R[s]_0$ is a subring of the domain S_0. Moreover $R[s]_0$ is a R_0-vector space of finite dimension. Then $R[s]_0$ is a field. Finally, $R[s]$ is a graded field with grade group $\Gamma < \sigma >$. ∎

Theorem 1.3 *Let S/R be a graded field extension and $s \in S_\sigma$ a nonzero homogeneous element which is integral over R. Let d be the cardinal order of $\Gamma < \sigma > /\Gamma$, a a nonzero homogeneous element of R of degree $d\sigma$ and $[R[s] : R] = n$. Then $\mu_s(X) = a^{\frac{n}{d}} H(a^{-1}X^d)$ is the minimal polynomial of s over R, where $H(X) \in R_0[X]$ is the minimal polynomial of $a^{-1}s^d$ over R_0. In particular, $R[s]_0 = R_0[a^{-1}s^d]$ and if $s^d \in R$, then $\mu_s(X) = X^d - s^d$.*

Proof. Since s is a nonzero homogeneous element which is integral over R, $R[s]/R$ is a graded field extension. So, $R[s]$ is a free R-module such that $[R[s] : R] = [R[s]_0 : R_0][\Gamma < \sigma >: \Gamma]$. Let $\mu_s(X)$ be the characteristic polynomial of the endomorphism $l_s : R[s] \longrightarrow R[s]$ defined by $l_s(x) = sx$. Since $\mu_s(X)$ is a monic polynomial lies in $R[X]$ of degree $[R[s] : R]$, $\mu_s(X)$ is the minimal polynomial of s over R, and then s has its minimal polynomial over R. Consider $Q(X)$ the $n\sigma$-homogeneous component of $\mu_s(X)$ in the graded ring $R[X]^{(\sigma)}$. Then $Q(s) = 0$. Since $Q(X)$ and $\mu_s(X)$ are monic polynomials of $R[X]$ with the same degree, the uniqueness of the minimal polynomial implies that $Q(X) = \mu_s(X)$. So, $\mu_s(X)$ is an homogeneous polynomial of $R[X]^{(\sigma)}$. Let (i, j) such that $i \neq j$, $q_i \neq 0$, and $q_j \neq 0$. Since $(i-j)\sigma = deg(q_j) - deg(q_i) \in \Gamma$, d divides $i-j$. The fact that s is invertible implies that $q_0 \neq 0$. So, d divides every i, and then $\mu_s(X) = \sum_{i=0}^{m} q_{di} X^{di}$, where $m = \frac{n}{d}$. Therefore, $\mu_s(X) = b \sum_{i=0}^{m} b^{-1} a^i q_{di} (a^{-1}X^d)^i = bP((a^{-1}X^d))$, where $b = a^{\frac{n}{d}}$. In the other hand, since $deg(b^{-1}a^i q_{di}) = (i.d - n)\sigma + deg(q_{di}) = 0$, $P(X) \in R_0[X]$. Moreover the fact that $deg(a^{-1}s^d) = -d\sigma + d\sigma = 0$ implies that $a^{-1}s^d \in R[s]_0$. Hence $a^{-1}s^d$ is algebraic over R_0. Let $H(X)$ be the minimal polynomial of $a^{-1}s^d$ over R_0. Then $H(a^{-1}X^d)$ annihilates s, and hence $\mu_s(X)$ divides $bH((a^{-1}X^d))$. By an account of degree we have $P(X) = H(X)$. ■

Corollary 1.4 *Every homogeneous element of S, which is integral over R has its minimal polynomial over R.*

Example. Let K be a field, $L = K(X)$, $R = L[T, T^{-1}]$ and $S = R[s]$, where s is a root of $Y^4 - XT^2$. Since $Y^4 - XT^2$ is a $\frac{1}{2}$-homogeneous polynomial of grade 2, s is a homogeneous element of degree $\frac{1}{2}$. On the other hand, since $Y^4 - XT^2 = T((T^{\frac{-1}{2}}Y^2)^2 - TX)$, $R[s]_0 = L[T^{\frac{-1}{2}}s^2]$.

Proposition 1.5 *Let S/R be a graded field extension and $s \in S_\sigma$ a nonzero homogeneous element which is integral over R. Let d be the cardinal order of $\Gamma < \sigma > /\Gamma$. Then $R[s]/R$ is separable if and only if d is invertible in R_0 and $R[s]_0/R_0$ is a separable field extension.*

Proof. Let $F = R[s]$ and $\mu_s(X) = bH((a^{-1}X^d))$ be the minimal polynomial of s over R. Then $F \simeq R[X]/(\mu_s(X))$, and then $D_R(F) = N_{F/R}(\mu_s'(s))R$. So, F/R is separable if and only if $a^{-n}d^n(N_{F/R}(s))^{d-1}N_{F/R}(H'((a^{-1}s^d)))$ is invertible in R. Since $a^{-n}N_{F/R}(s)^{d-1}$ is invertible in R, F/R is separable if and only if d is invertible in R_0 and $N_{F/R}(H'((a^{-1}s^d)))$ is invertible in R. The fact that $a^{-1}s^d \in F_0$ and F is a free F_0-algebra implies that $N_{F/R}(H'((a^{-1}s^d))) = (N_{F_0/R}(H'((a^{-1}s^d))))^d$. Therefore, $N_{F/R}(H'((a^{-1}s^d)))$ is invertible in R if and only if $N_{F_0/R}(H'((a^{-1}s^d)))$, i.e., F_0/R_0 is a separable extension of fields. ■

Corollary 1.6 *Let S/R be a graded field extension and $s \in S_\sigma$ a nonzero homogeneous element which is integral over R. Then $R[s]/R$ is separable if and only if s is a simple root of its minimal polynomial over R.*

Indeed, let $\mu_s(X) = X^n + a_{n-1}X^{n-1} + ... + a_0$ be the minimal polynomial of s over R. Since $\mu_s(X)$ is a σ-homogeneous polynomial, for every i such that $a_i \neq 0$, $deg(a_i) + i\sigma = n\sigma$. So, for every i such that $a_i \neq 0$, $deg(a_i) + (i-1)\sigma = (n-1)\sigma$. Hence $\mu'_s(X)$ is a σ-homogeneous polynomial. Consequently, $\mu'_s(s)$ is a homogeneous element of S, and then $N_{R[s]/R}(\mu'_s(s))$ is a homogeneous element of R. Therefore, $R[s]/R$ is separable if and only if $\mu'_s(s) \neq 0$.

2. Galois Extensions

In this section, we investigate galois theory of graded field extensions. We characterize simple galois extension of graded fields by its ramification index and its homogeneous components of degree 0. In particular, we state galois theory results given in [1].

Let A be commutative free R-algebra and G a finite subgroup of $Aut_R(A)$ such that $A^G = R$. Denote $A^{(G)}$ the A-algebra with basis $\{e_\sigma | \sigma \in G\}$ made up by orthogonal idempotents. The extension A/R is said to be a G-galois extension if the homomorphism h is an isomorphism of A-algebras

$$\begin{array}{llll} h: & A \otimes A & \longrightarrow & A^{(G)} \\ & x \otimes y & \longmapsto & \sum_{\sigma \in G} x\sigma(y)e_\sigma \end{array}$$

The following Propositions give us a characterization of simple galois extensions of graded fields.

Proposition 2.1 *Let T/R be a totally ramified graded field extension with respect to torsion free abelian groups Γ and Δ, respectively and let $s \in T$ be a nonzero homogeneous element of degree σ, which is integral over R. Set n the cardinal order of the group $\Gamma < \sigma > /\Gamma$. Then $R[s]/R$ is a galois extension if and only if n is invetible in R_0 and R_0 contains ζ_n, a nth primitive root of $X^n - 1$.*

Proof. Set $a = s^n$. Then $deg(a) = nx \in \Gamma$, and then there exists $v \in R[s]_0$ such that $a = vu_{nx}$. Since $S_0 = R_0$, $R[s]_0 = R_0$, and then $v \in R$. Consequently, $a \in R$, and then $\mu_s(X) = X^n - a$ is the minimal polynomial of s over R. Since $R[s]/R$ is a galois extension, it is separable. So, n is invertible in R_0. Since the cardinal order of G is equal to n, $\mu_s(X)$ splits in $R[s]$ with simple roots $s_1, ..., s_n$. Hence $\frac{s_1}{s}, ..., \frac{s_n}{s}$ are the roots of $X^n - 1$. Therefore, $\zeta_n \in R[s]_0 = R_0$.
Conversely, assume that n is invetible in R_0 and R_0 contains ζ_n. Hence $s, \zeta_n s, ..., \zeta_n^{n-1}s$ are the distinct roots of $\mu_s(X)$. Set $G =< \sigma >$, where $\sigma(s) = \zeta_n s$. Let $y = \sum_{i=0}^{n-1} r_i s^i \in R[s]$. For $0 \leq k \leq n-1$, $\sigma^k(y) = r_0 + \sum_{i=1}^{n-1} r_i \zeta_n^{ik} s^i$. Hence if $y \in R[s]^G$, then $y = r_0 \in R$. Now we compute $det(h)$, the determinant of the homomorphism h of $R[s] \otimes R[s]$ into $R[s]^{(G)}$ with respect to $R[s]$−bases $\mathcal{B} = (1, s, ..., s^{n-1})$ and G respectively. We obtain

$$det(h) = \begin{vmatrix} 1 & s & \cdots & s^{n-1} \\ 1 & \zeta_n s & \cdots & (\zeta_n s)^{n-1} \\ \vdots & \vdots & & \vdots \\ 1 & \zeta_n^{n-1} s & \cdots & \zeta_n^{(n-1)^2} s^{n-1} \end{vmatrix} = s^{n(n+1)/2} \prod_{1 \leq i < j \leq n} (\zeta_n^i - \zeta_n^j).$$ Since s is an invertible element and $\prod_{1 \leq i < j \leq n} (\zeta_n^i - \zeta_n^j) \in R$ is a nonzero homogeneous element of degree 0, $det(h)$ is an invertible element of R. So, $R[s]/R$ is a $G-$galois extension.

Corollary 2.2 *Let T/R be a totally ramified graded field extension with respect to torsion free abelian groups Γ and Δ, respectively and let $s \in T_\sigma$ be a nonzero homogeneous element, which is integral over R. Then $R[s]/R$ is a galois extension if and only if the minimal polynomial of s over R splits square free in $R[s]$.*

Proof. If $R[s]/R$ is a galois extension with galois group G, then $\mu_s(X) = X^n - s^n = \prod_{i=0}^{n-1}(X - \zeta_n^i s)$ splits square free in $R[s]$, where $n = [R[s] : R]$.
Conversely, $\mu_s(X) = X^n - s^n$ splits square free in $R[s]$, then n is invetible in R_0 and R_0 contains ζ_n.

For two commutative free R-algebras A and B of finite ranks, let G_1 (resp. G_2) be a finite subgroup of $\mathcal{A}ut_R(A)$ (resp. of $\mathcal{A}ut_R(B)$) such that $A^{G_1} = R$ and $B^{G_2} = R$. Then $G_1 \times G_2$ is a finite subgroup of $\mathcal{A}ut_R(A \otimes_R B)$ such that $\mathcal{A}ut_R(A \otimes_R B)^{G_1 \times G_2} = R$.

Lemma 2.3 *Under the above hypotheses, $A \otimes_R B$ is a $G_1 \times G_2$-galois extension over R if and only if A/R is a G_1-galois extension and B/R is a G_2-galois extension.*

Proof. Consider the following homomorphisms,

$$\begin{array}{rccc} h_A: & A \otimes_R A & \longrightarrow & A^{(G_1)} \\ & x \otimes y & \longmapsto & \sum_{\sigma \in G_1} x\sigma(y)e_\sigma. \end{array}$$

$$\begin{array}{rccc} h_B: & B \otimes_R B & \longrightarrow & B^{(G_2)} \\ & x \otimes y & \longmapsto & \sum_{\sigma \in G_2} x\sigma(y)e_\sigma. \end{array}$$

$$\begin{array}{rccc} h: & A \otimes_R B \otimes_R A \otimes_R B & \longrightarrow & A \otimes_R B^{(G_1 \times G_2)} \\ & (a \otimes b) \otimes (a' \otimes b') & \longmapsto & \sum_{(\sigma,\tau) \in G_1 \times G_2} a\sigma(a') \otimes b\tau(b').e_{\sigma,\tau}. \end{array}$$

Then h is an isomorphism if and only if h_A and h_B are isomorphisms. Indeed, set $\mathcal{A} = (a_1, ..., a_n)$ and $\mathcal{B} = (b_1, ..., b_m)$ two R-bases of A and B respectively. Then $\mathcal{F} = (a_i \otimes b_j)_{i,j}$ is a R-basis of $A \otimes B$. Let $M = (\alpha_{ij})_{1 \leq i,j \leq n}$ (resp. $N = (\beta_{ij})_{1 \leq i,j \leq m}$) be the matrix of h_A with respect to bases $\mathcal{A}$ and G_1 (resp. the matrix of h_B with respect to the bases $\mathcal{B}$ and G_2). Then $H = ((\alpha_{ik}\beta_{jl})_{(i,j)(k,l)})$ is the matrix of h with respect to bases $\mathcal{F}$ and G. Hence $H = M \otimes N$ and $det(M \otimes N) = (det(M))^r(det(N))^n$. ■

Now, we give a new proof of [1, Proposition 3.3].

Corollary 2.4 *Let T/R be a totally ramified graded field extension with respect to torsion free abelian groups Γ and Δ, respectively and let l be the exponent of Δ/Γ. Then T/R is a galois extension if and only if l is invertible in R_0, which contains ζ_l a primitive n^{th}-root of 1.*

Proof. Set $\triangle/\Gamma =< \bar{\sigma_1} > ... < \bar{\sigma_r} >$ and let $s_1, ..., s_r$ be homogeneous elements of T such that $deg(s_i) = \sigma_i$ for each i. Then the multiplication of T induces an isomorphism of R-algebras $\mu : R[s_1] \otimes ... \otimes R[s_r] \longrightarrow T$. Consequently, T/R is a galois extension if and only if $R[s_i]/R$ is a galois extension for every $1 \leq i \leq r$. That is equivalently to l is invertible in R_0, which contains ζ_l. ■

3. Concluding Remarks

Let S/R be a graded field extension such that R is a domain with quotient field K.
1) Since $D_R(S)$ is generated by an homogeneous element of R, of degree 0, S/R is separable if and only if $D_R(S) \neq 0$, i.e., $D_K(KS) \neq 0$. Hence S/R is separable if and only if KS/K is separable too.
2) If $\triangle$ is a torsion free abelian group, then R and S are domains. Since S is a connected ring, S/R is a galois extension if and only if S/R is separable and $S^G = R$, where $G = Aut_R(S)$ (see [7, Theorem 2.1, page 7]). From [1, corollary 1.3, page 824] , S is an integrally closed domain. Then $S^G = R$ if and only if $(KS)^G = K$. Consequently, S/R is a galois extension if and only if KS/K is a galois extension. In particular, we find Theorem 3.11, given in [1].

Corollary 3.1 *Let S/R be a graded field extension with torsion free abelian groups and $s \in S$ a homogeneous element which is integral over R. Then $R[s]/R$ is a galois extension if and only if $\mu_s(X)$ splits square free in $R[s]$.*

Indeed, since $R[s]/R$ is a graded field extension with torsion free abelian groups, $R[s]/R$ is a galois extension if and only if $K[s]/K$ is a galois extension, where K is the quotient field of R, i.e., if $\mu_s(X)$ splits square free in $R[s]$ (because $R[s]$ is an integrally closed domain). ■

The following Theorem allows us to show that the notions of galois extension, defined in [2], are simple cases of galois extensions of commutative rings.

Theorem 3.2 *Let S/R be a graded field extension such that $\triangle$ is a torsion free abelian group and S is finitely generated as R-module. Then the following statements are equivalent :*
1) S/R is a galois extension.
2) Every homogeneous element $s \in S$ is a simple root of its minimal polynomial which splits in S.

Proof. First, since $\triangle$ is a torsion free abelian group, R and S are domains with quotient fields K and $L = KS$ respectively. Let $s \in S_\sigma$ be an homogeneous element. Then $R[s]/R$ is a graded field extension. Since S/R is a galois extension, L/K is a galois extension. So, $\mu_s(X)$, the minimal polynomial of s splits square free in L. Since S is an integrally closed domain, $\mu_s(X)$, splits square free in S.
Conversely, to show that S/R is a galois extension it suffices to show that L/K is a galois extension. Since L is the K-algebra generated by S and S is generated by its homogeneous elements, it suffices to show that for every homogeneous element $s \in S$, $R[s]/R$ is a galois extension. ■

References

[1] Y. S. Hwang and A. R. Wadsworth, Algebraic extensins of graded and valued fields, *Comm. in Algebra*, **27**(2)(1999), 821-840.

[2] M. Boulagouaz, An introduction to the galois theory for graded fields, Lect. Notes in *Pure and Appl. math.*, Marcel Dekker, v : (208)(1999), 21-31.

[3] M. A. Knus and M. Ojanguren, Théorie de la descente et algèbres d'Azummaya, *Lect. Notes in Math.*, v. 389, Springer-Verlag Berlin, Heidelberg, New York, 1974.

[4] L. El Fadil and M. Charkani. Generalization of the discriminant and applications. (to appear in the *Arabian Journal for Science and Engineering*, **(1A)**29, 2004.

[5] N. Bourbaki, *Eléments de Mathématique Ch. 1 à 3*, Hermman, 1970.

[6] L. El fadil, Separability of graded fields. (Submitted)

[7] C. Greither, Cyclic Galois extensions of commutative rings, *Lect. Notes in Math.*, v. 1534, Springer-Verlag Berlin Heidelberg, New York, 1992.

Received: September 2004
Revised: December 2004

In: Mathematics, Game Theory and Algebra Compendium ISBN: 978-1-60456-659-8
Editor: Jacob H. Mathias, pp. 111-122

Chapter 12

SELF SIMILAR ISOTHERMAL EXPANSION OF MAGNETO-HYDRODYNAMIC GAS BEHIND A SPHERICAL SHOCK WAVE WITH SELF GRAVITATION

Ashok Ganguly
Department of Applied Mathematics and Computational Science
Shri G.S. Institute of Technology and Science 23, Park Road ,
Indore (M.P.) – 452003, India
Pankaj Sharma
Department of Applied Mathematics Jawaharlal
Institute of Technology Borawan
(Khargone)(M.P.) – 451228, India

Abstract

Self –similar power driven isothermal flow behind a spherical Shock has been discussed in a heterogeneous ,magnetic ,self – gravitating medium. The Shock radius, the density distribution and the internal heat generation are assumed to vary with some power of time .The variation of flow variables behind the shock with respect to distance have been displayed graphically and comparison has been made for the cases of presence and absence of magnetic field.

Introduction

Self –similar solutions to problems in gas dynamics are of considerable interest in many fields. Spherically symmetric self –similar flows of a polytropic gas in the absence of gravity have been studied by Sedov[1946],Courant and Friedreich [1948], Taylor[1950] and Rogers [1957]. They have profound influence in the study of flare generated blast waves ,e.g. Parker [1963] ,Hundhausen [1972] ,Dryer [1975] and Holzer [1979].

Stanyukovich [1960] formulated some self –similar non-adiabatic gas expansion problems with very specific source terms including the propulsion of a piston by the non-instantaneous release of energy in the gas behind it but did not obtain any general solution. In a series of papers ,Nemchinov [1961][1964] considered the expansion of a gas driven by internal heat generation of a very general form and obtained an analytic solution for a special case .However, his source term was too general to admit a closed form solution. During the period of novae outburst , instantaneous release of large amount energy and propagation of Shock waves from the centre layer to periphery in a non- homogeneous medium of density have been observed. The solution for the isothermal expansion of a gas into vaccum for temperature depending on the time obeying some power law was obtained by Farnsworth [1980] while Caporaso [1982] has studied self –similar solutions for the non-adiabatic expansion of a gas in plane , cylindrical and spherical symmetries owing to a general source function.

In this paper we have investigated a self-similar solution for power driven isothermal expansion headed by a spherical Shock surface in a heterogeneous medium with the effect of magnetic field and self gravitation .To cover wide range of physical situations we take the source term Q which has dimensions of energy per unit mass per unit time , in the form (Caporaso[1982])

$$Q = k\, r^{\nu} \rho^{\mu} p^{\sigma} t^{\delta} h^{\eta} \tag{1}$$

For isothermal expansion of the gas ,the internal heat generation per unit mass is identical for all elements of the fluid behind surface of discontinuity and is only a function of time , so

$$\eta = \nu = \mu = \sigma = 0 \ \text{ and } \ Q = k\, t^{\delta} \tag{2}$$

The medium in which expansion takes place is heterogeneous and the Shock radius is a function of time. Thus , the density and the Shock radius are assumed to obey the following power law

$$\rho_0 = A\, t^{\alpha} \tag{3}$$

$$R = B\, t^{\beta} \tag{4}$$

Where ρ_0 is the density of the ambient medium and R is the Shock radius. And A ,B, α and β are constants. Also the magnetic field in the undisturbed medium is taken as

$$h_0 = c\, R^{\lambda} \tag{5}$$

where h_0 is assumed to be directed tangentially to the advancing Shock front and c is a constant.

Basic Equations and Boundary Conditions

The fundamental equations for the MHD gas in a spherically symmetric self-gravitating radial flow are ,

$$\frac{\partial \rho}{\partial t}+u\frac{\partial \rho}{\partial r}+\rho\left[\frac{\partial u}{\partial r}+\frac{2u}{r}\right]=0 \tag{6}$$

$$\frac{\partial u}{\partial t}+u\frac{\partial u}{\partial r}+\frac{1}{\rho}\frac{\partial p}{\partial r}+\frac{Gm}{r^2}+\frac{h}{\rho r}\frac{\partial}{\partial r}(hr)=0 \tag{7}$$

$$\frac{\partial h}{\partial t}+\frac{\partial}{\partial r}(uh)+\frac{hu}{r}=0 \tag{8}$$

$$\frac{\partial e}{\partial t}+u\frac{\partial e}{\partial r}-\frac{p}{\rho^2}\left[\frac{\partial \rho}{\partial t}+u\frac{\partial \rho}{\partial r}\right]=Q \tag{9}$$

$$\frac{\partial m}{\partial r}=4\pi\rho r^2 \tag{10}$$

where

$$e=\frac{p}{\rho(\gamma-1)} \tag{11}$$

and ρ , u , p , G , m ,e , h and γ represent density, flow velocity, pressure, gravitational constant , mass ,energy per unit mass of the fluid behind the spherical shock , transverse magnetic field and the ratio of specific heats of gas respectively.

The shock conditions across the shock wave heading the expanding gas are

$$u_1=\frac{2D}{(\gamma+1)}\left(1-\frac{1}{M^2}\right) \tag{12}$$

$$p_1=\frac{p_0}{(\gamma+1)}\left[2\gamma M^2-(\gamma-1)\right] \tag{13}$$

$$\rho_1=\frac{\rho_0(\gamma+1)M^2}{(\gamma-1)M^2+2} \tag{14}$$

$$h_1 = \frac{h_0 (\gamma + 1) M^2}{(\gamma - 1) M^2 + 2} \tag{15}$$

$$m_1 = m_0 \tag{16}$$

where suffices 0 and 1 denote state of flow just ahead and behind the shock surface and D = dR/dt.

At equilibrium state, the pressure and mass are given by

$$p_0 = -\frac{2\pi\beta^2 G R^2 \rho_0^2}{(\alpha + 3\beta)(\alpha + \beta)} - \frac{(\lambda + 1) h_0^2}{2\lambda} \tag{17}$$

$$m_0 = \frac{4\pi\beta R^3 \rho_0}{(\alpha + 3\beta)} \tag{18}$$

For positive p_0 and m_0 we have $(\alpha+3\beta) > 0$ and $(\alpha+\beta) < 0$ (19)

and

$$-1 < \lambda < 0 \tag{20}$$

Similarity Solutions

Let us consider solution of the equations in the form

$$u = D\, U(\xi) \tag{21}$$

$$\rho = \rho_0 V(\xi) \tag{22}$$

$$p = \rho_0 D^2 W(\xi) \tag{23}$$

$$h = \sqrt{\rho_0} D Z(\xi) \tag{24}$$

$$m = D^2 R\, Y(\xi) \tag{25}$$

where ξ is a non- dimensional parameter defined by

$$\xi = \frac{r}{R(t)} \tag{26}$$

Transformation of the shock conditions (12) – (16) in non-dimensional form We get ,

$$U(1) = \frac{2}{(\gamma+1)}\left(1 - \frac{1}{M^2}\right) \tag{27}$$

$$W(1) = \frac{1}{(\gamma+1)}\left(2 - \frac{(\gamma-1)}{\gamma M^2}\right) \tag{28}$$

$$V(1) = \frac{(\gamma+1)\ M^2}{(\gamma-1)\ M^2 + 2} \tag{29}$$

$$Y(1) = 1 \tag{30}$$

$$Z(1) = \frac{1}{M_A}\left(\frac{(\gamma+1)M^2}{(\gamma-1)M^2 + 2}\right) \tag{31}$$

where Alf 'ven Mach number M_A and the usual Mach number M are defined as

$$M_A^{\ 2} = \frac{\rho_0 D^2}{h_0^{\ 2}}$$

$$M^2 = \frac{\rho_0 D^2}{\gamma\, p_0}$$

respectively.

The equations of motion (6) – (10) may be reduced to the following ordinary differential equations with the help of non dimensional variables (21) - (25).

$$(U-\xi)V'\xi + V\left[\xi\left(\frac{\alpha}{\beta} + U'\right) + 2U\right] = 0 \tag{32}$$

$$(U-\xi)U' + \left(1 - \frac{1}{\beta}\right)U + \frac{GY}{\xi^2} + \frac{1}{V}\left[W' + \frac{Z^2}{\xi} + ZZ'\right] = 0 \tag{33}$$

$$(U-\xi)Z'+\left[\frac{\left(\frac{\alpha}{2}+\beta-1\right)}{\beta}+U'+\frac{U}{\xi}\right]Z=0 \tag{34}$$

$$\frac{W}{(\gamma-1)}\left[1+\frac{2(\beta-1)}{\alpha}+\frac{\beta W'}{\alpha W}(U-\xi)\right]-\frac{W\gamma}{(\gamma-1)}\left[1+\frac{\beta V'}{\alpha V}(U-\xi)\right]$$

$$=\frac{K}{\alpha}\beta^{2\sigma+\eta-2}\xi^{\nu}B^{\nu+2\sigma+\eta-2}A^{\mu+\sigma+\frac{\eta}{2}}V^{\mu+1}W^{\sigma}Z^{\eta}t^{\alpha(\mu+\sigma)-2\sigma+\delta+\frac{\alpha\eta}{2}-\eta+3+\beta(\nu+2\sigma+\eta-2)} \tag{35}$$

$$Y'=\frac{4\pi AV\xi^{2}}{\beta^{2}}t^{(\alpha+2)} \tag{36}$$

Noting the fact that motion is self-similar, we have from (35) and (36)

$$\alpha(\mu+\sigma)-2\sigma+\delta+\frac{\alpha\eta}{2}-\eta+3+\beta(\nu+2\sigma+\eta-2)=0 \tag{37}$$

and

$$\alpha=-2$$

and so,

$$\beta=\frac{\left[-\alpha(\mu+\sigma)+2\sigma-\delta-\frac{\alpha\eta}{2}+\eta-3\right]}{(\nu+2\sigma+\eta-2)} \tag{39}$$

For an isothermal flow

$$\beta=\frac{\delta+3}{2} \tag{40}$$

$$[\because \nu=\mu=\sigma=\eta=0]$$

With the help of (19) , (20) , (38) & (40) we have

$$\frac{-5}{3}<\delta<1 \quad , \quad -1<\lambda<0 \quad , \quad \frac{2}{3}<\beta<2 \tag{41}$$

To analyze the variation of flow variables behind the shock surface the equations (32)-(36) may be manipulated in the forms

$$U' = \frac{\left[\frac{V(\gamma-1)}{\beta^3} - W\left\{\frac{(\alpha+2\beta-2)}{\beta} + \frac{2\gamma U}{\xi}\right\} + (U-\xi)V\left\{U\left(1-\frac{1}{\beta}\right) + \frac{GY}{\xi^2}\right\} - Z^2\left\{\frac{\alpha+4\beta-2}{2\beta}\right\}\right]}{\left[Z^2 - V(U-\xi)^2 + W\gamma\right]} \tag{42}$$

$$Z' = -\frac{\left[\frac{(\alpha+2\beta-2)}{2\beta} + U' + \frac{U}{\xi}\right]Z}{(U-\xi)} \tag{43}$$

$$V' = -V\frac{\left[\xi\left(\frac{\alpha}{\beta} + U'\right) + 2U\right]}{(U-\xi)\xi} \tag{44}$$

$$W' = \frac{Z^2\left[\frac{(\alpha+2\beta-2)}{2\beta} + \frac{U}{\xi}\right] + U'\left[Z^2 - V(U-\xi)^2\right]}{(U-\xi)} - \frac{(\beta-1)}{\beta}UV - \frac{(GYV + Z^2\xi)}{\xi^2} \tag{45}$$

$$Y' = \frac{LV\xi^2}{\beta^2} \tag{46}$$

Where L = 4πA.
Under the Equilibrium conditions , we have

$$G = -\frac{2(\alpha+3\beta)\,(\alpha+\beta)}{L}\left[\frac{1}{\gamma M^2} + \frac{(\lambda+1)}{2\lambda M_A^2}\right] \tag{47}$$

The boundary conditions are given by

$$Z(1) = 1\ ,\ V(1) = 1\ ,\ W(1) = 1\ ,Y(1) = 1 \tag{48}$$

$$U(1) = \frac{2}{(\gamma+1)}\left[1 - \frac{1}{M^2}\right]$$

for ξ =1.

Results and Discussions

The set of differential equations (42)- (46) have been integrated numerically with the boundary conditions (48) by the well known Runga-Kutta method for β =0.7 and 1. 5 , γ =1. 33 and 1.4 , for M^2 = 5 and $M_A{}^2$ =10 , λ = - 0. 5 . We have plotted the graphs showing variations of various flow parameters with distance in the presence and absence of magnetic field. Two cases have been considered for the values of β =0.7 and 1. 5.

Case I

For β = 0.7 , the line patterns used in the figures (1A) - (4A) stand for the following respective cases.

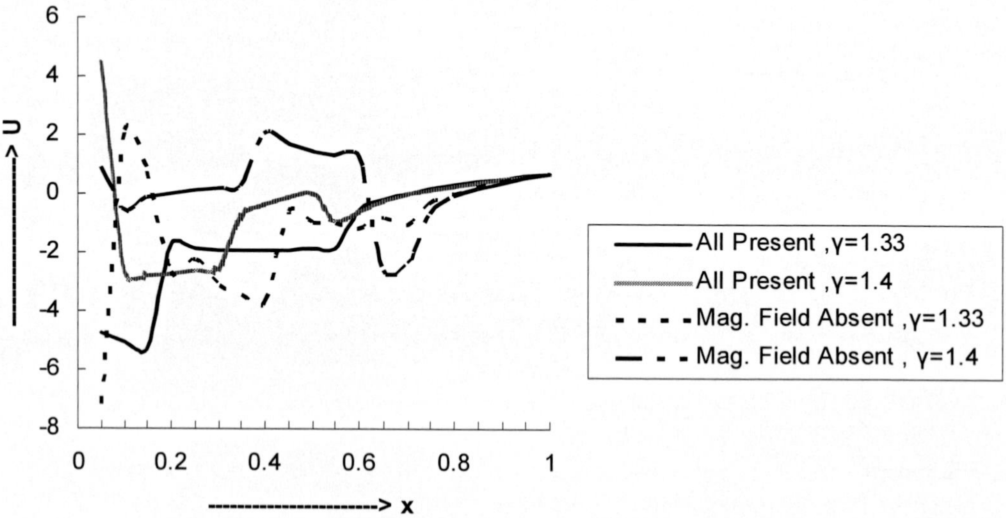

Figure 1A. Variation of Velocity with Distance (For ß=0.7).

In Case of Fig. I(A) , velocity starts from a lower value and tries to increase and then drops and finally increases and tends to attain a uniform pattern as we move away from the shock –centre. The variation for the case γ = 1.4 is noticeable .Here the velocity tries to re-increase as we approach to the centre of the shock , more so when magnetic field is present.

In figures 2A and 3A , the variations of density and pressure , respectively , are similar. In general , from a dormant, low value at the origin of shock , it tries to build up and attain a uniform value near zero as one goes away from the centre of shock. However ,for γ = 1. 33 and magnetic field present , pressure and density drop considerably from the centre of shock and then attain a uniform value . In fact , for this case a significant decrease in the variation from the centre of the shock is observed.

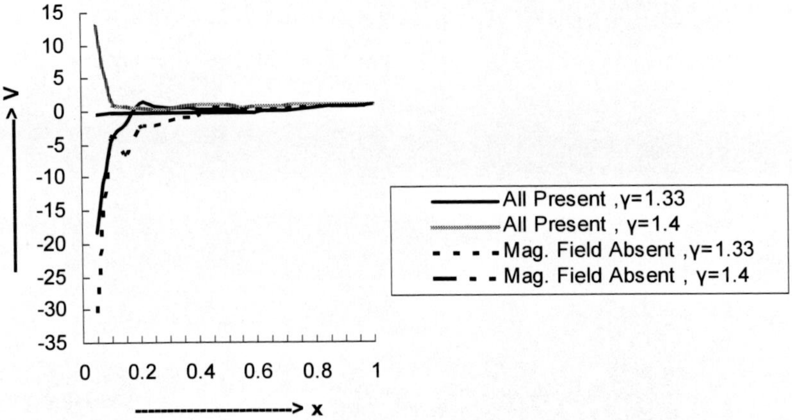

Figure 2A. Variaton of Velocity with Distance (for β=0.7).

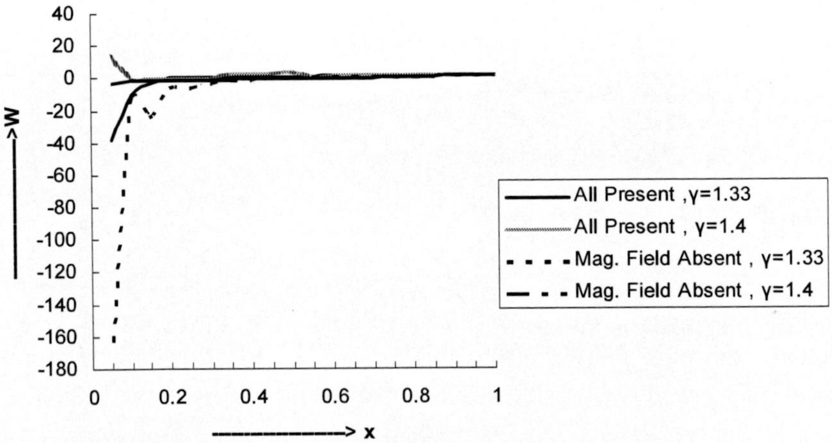

Figure 3A.Variation of Velocity with Distance (For ß=0.7).

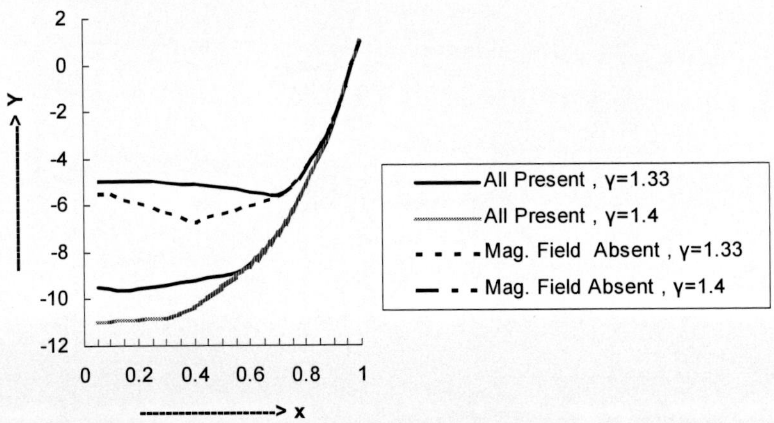

Figure 4A.Variation of Magnetic Field with Distance (For ß=0.7)

In fig. 4A , mass decreases , as we move towards the centre of the shock. However ,the behavior for two cases of γ is very significant , especially around the beginning of the shock .

Case II

For β = 1.5 (hence δ =0) ,the line patterns used in figures (1B) – (4B) stand for the respective cases as given below.

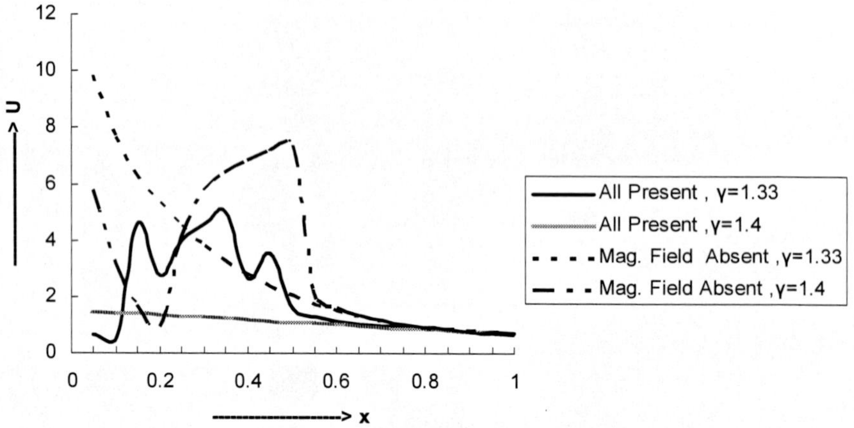

Figure 1B. Variaton of Velocity with Distance (for β=1.5).

In fig. I (B) , velocity decreases as we move away from the centre of the shock. This is the distinction from case I. This is obvious as here δ =0 , therefore $Q = k\ t^{\delta}$ becomes uniformly constant , as in (2). Again there is a marked difference in the behavior for γ =1. 33 and γ =1.4 . In figures 2B and 3B , while density and pressure drop very fast as we move away from the centre of the shock , they gradually attain a uniform value near x = 1.

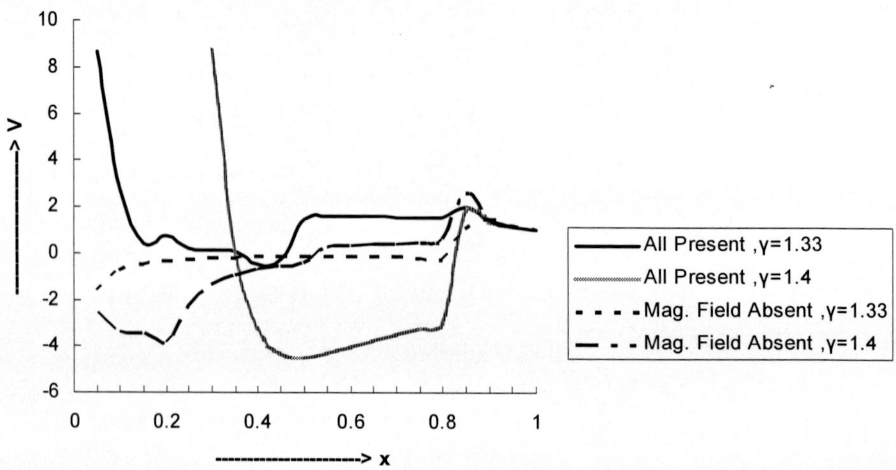

Figure 2B. Variation of Velocity with Distance (For ß=1.5).

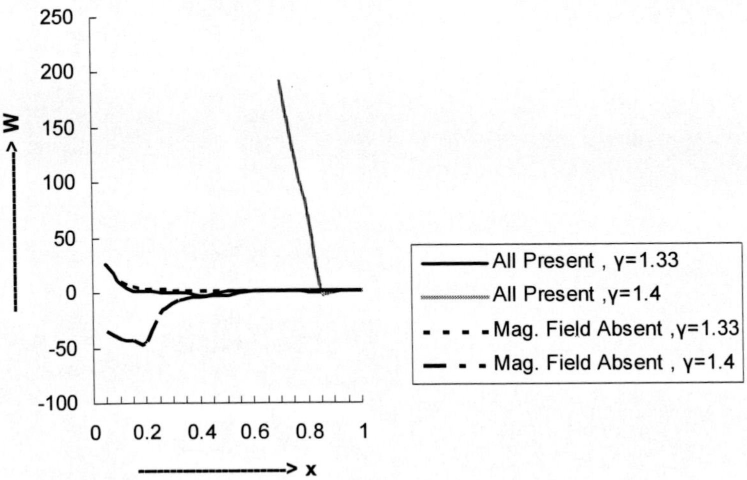

Figure 3B. Variation of Velocity with Distance (For ß=1.5).

However , a transfer phenomenon is noticeable in both the cases. The variations for γ =1.33 are highly significant in comparison with γ =1.4.

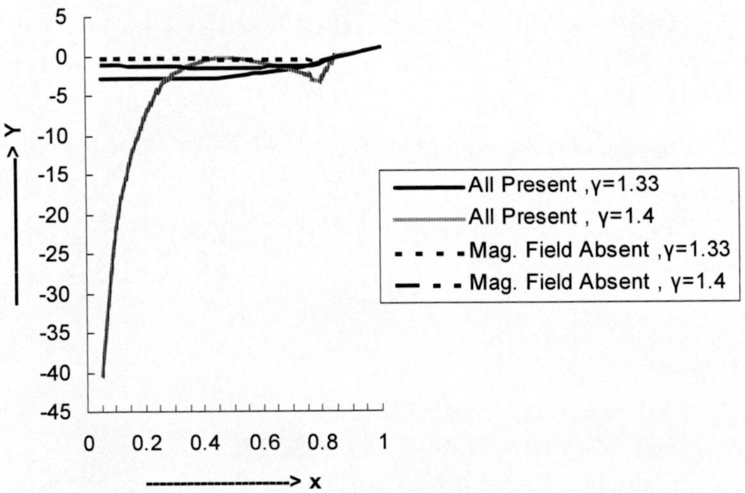

Figure 4B. Variation of Mass with Distance (For ß=1.5).

In fig. 4B , there is a uniform transfer of mass throughout the shock process. However , abnormality is noticeable for γ =1.4 and magnetic field present. It results in a blast and transfers large amount of mass particles on account of shock at β = 1. 5. This is also distinctly different from case I for β = 0.7.

Finally , we observe the variations in magnetic field in fig. 5. The line patterns used in fig. 5 stand for the respective cases as given below.

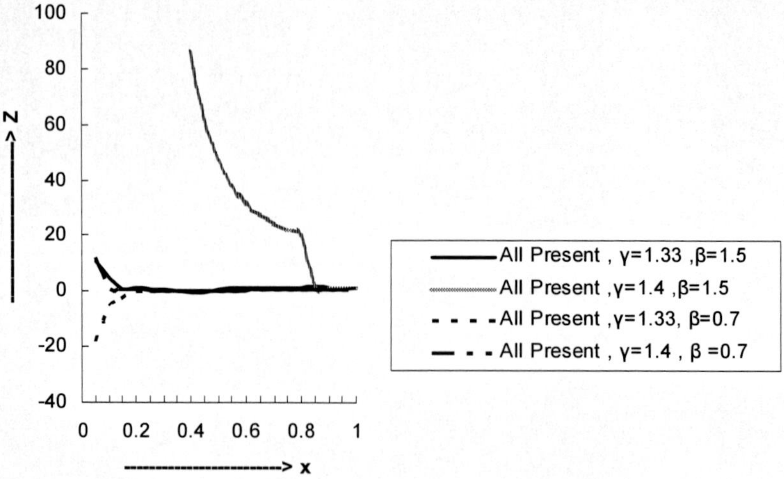

Figure 5. Variation of Magnetic Field with Distance.

Magnetic field variation seems to have , in the main , no effect in the self- similar isothermal expansion of a gas. However , abnormal behavior is noticed for the cases $\beta = 0.7$ and $\gamma = 1.33$ and $\beta = 1.5$, $\gamma = 1.4$. This may be due to extreme values of β as it lies between 2/3 and 2 , see[cf. (41)]. As our values of β are close to extreme values , may be due to this abnormality is seen for $\gamma = 1.33$ (for lower values of β) and $\gamma = 1.4$ (for higher extreme values of β)

References

Rogers , M.H. : 1957 , *Ap. J.* **125** , 478.
Dryer , M. : 1975 , *space science Review* , **17**, 277.
Farnsworth , A.V. : 1980 , *Phys . Fluids* , **23** ,1496.
Caporaso , G .I .: 1982 , *Phys . Fluids* , **25** ,436.
Nemchinov, I.V. : 1961 , *Zh. Prikl . Mekh.Tekh.Fiz.* **1** , 17.
Nemchinov, I.V. : 1964 , *Zh. Prikl . Mekh.Tekh.Fiz.* **5** , 18.
Courant , R. and Friedrichs , K.O. : 1948 ,Supersonic flow and shock waves , New York, *Inter science.*
Parker , E . N. : 1963 , *Inter Planetary Dynamical Process* ,New York , Ins. Pu.
Hundhausen , A.J. : 1972 , Coronal expansion and Solar wind , *Springer Verlag* ,New York.
Holzer ,T.E. : 1979 , *The Solar wind and related Astrophysical phenomena* Vol. 1, Amsterdam North , Holland .
Stanyukovich , K . P. : 1960 , Unsteady motion of continuous media ,Pergamon Press.
Taylor , G. I. : 1950 , *Proc. Roy. Soc. London* , A ; **201** , 1975.
Sedov , L. I. : 1946 , *Prikl Mat. Mekh.* , **10** , 241.

In: Mathematics, Game Theory and Algebra Compendium ISBN 978-1-60456-659-8
Editor: Jacob H. Mathias, pp. 123-136

Chapter 13

CLASSIFICATION OF THE 5-DIMENSIONAL POWER-ASSOCIATIVE 2nd-ORDER BERNSTEIN ALGEBRAS

***P. Vicente**[*] **and H. Guzzo Jr.**[†]*
Universidad de León Universidade de São Paulo
Departamento de Matemáticas Instituto de Matemática e Estatística
Campus de Vegazana Caixa Postal 66281
24071, León - Spain 05315-970, São Paulo - Brazil

Abstract

In this paper we classify the 5-dimensional Power-Associative 2nd-order Bernstein algebras

1. Preliminary Results

In this paper all algebras considered will be commutative, non necessarily associative and finite dimensional over an infinite field K of char$(K) \neq 2$.

As usual in non-associative algebras we will consider right (left) principal powers and plenary powers of an element $x \in A$, defined by $x^1 = x$ and $x^{n+1} = x^n x$ (${}^1x = x$ and ${}^{n+1}x = x({}^n x)$) for $n \geq 1$ and $x^{[1]} = x,\ x^{[n+1]} = x^{[n]}x^{[n]}$, $n \geq 1$ respectively.

If ω: $A \longrightarrow K$ is a nonzero homomorphism, then the ordered pair (A, ω) will be called a baric algebra over K and ω its weight function. For $x \in A$, $\omega(x)$ is called the weight of x. The set $N = \{x \in A \mid w(x) = 0\}$ is a two-sided ideal of A of codimension 1. If $e \in A$ has weight 1 and $e^2 = e$, then e will be called idempotent element of A. If $X \subseteq A$, then $K(X)$ is the vector subspace of A generated by X.

A nth*-order Bersntein algebra* is a commutative baric algebra (A, ω) satisfying:

$$x^{[n+2]} = (\omega(x))^{2^n} x^{[n+1]}, \text{ for all } x \in A$$

[*]E-mail address: demmvm@unileon.es
[†]E-mail address: guzzo@ime.usp.br

and n is the smallest one with such property. If $n = 2$, that is, $x^{[4]} = \omega(x)^4 x^{[3]}$, then A is a 2nd*-order Bernstein algebra.*

Some general results about n^{th}-order Bernstein algebras are the following (see [5], [7]).

Proposition 1.1 *Let* (A, ω) *be a* n^{th}*-order Bernstein algebra, then:*

1. *The weight homomorphism is unique.*

2. *The set of idempotent elements of A is* $\Im(A) = \{x^{[n+1]} \mid x \in A \text{ and } \omega(x) = 1\}$.

3. *If* $\mathrm{e} \in A$ *is an idempotent element, then A has a Peirce decomposition,* $A = K\mathrm{e} \bigoplus N$, *where* $N = U_{\mathrm{e}} \bigoplus V_{\mathrm{e}}$, $U_{\mathrm{e}} = \{x \in A \mid \mathrm{e}x = \frac{1}{2}x\}$, $V_{\mathrm{e}} = \{x \in A \mid L_{\mathrm{e}}^n(x) = 0\}$ *and* $L_{\mathrm{e}}(x) = \mathrm{e}x$.

4. $U_{\mathrm{e}}^2 \subseteq V_{\mathrm{e}}$, $\dim U_{\mathrm{e}}$ *(and so* $\dim V_{\mathrm{e}}$*) does not depend on the particular idempotent element, and we can define* $\mathrm{type}\,(A) = (1 + r, s)$*, where* $r = \dim U$ *and* $s = \dim V$.

2. Jordan and Power-Associative 2nd-order Bernstein Algebras

An algebra A is power-associative if $x^i x^j = x^{i+j}$ for every $i, j \in N$ and $x \in A$. A commutative algebra A is a Jordan algebra if $x(x^2 y) = x^2(xy)$ for all $x, y \in A$. Some general results about Jordan or power-associative and 2nd-order Bernstein are the following (see [4], [6], [8], [9]):

If (A, ω) is a baric algebra, then (A, ω) is a power-associative 2nd-order Bernstein algebra if and only if there exists an idempotent element e of A such that, $A = K\mathrm{e} \bigoplus U_{\mathrm{e}} \bigoplus V_{\mathrm{e}}$, where $U_{\mathrm{e}} = \{x \in A \mid \mathrm{e}x = \frac{1}{2}x\}$, $V_{\mathrm{e}} = \{x \in A \mid \mathrm{e}x = 0\}$, $U_{\mathrm{e}}^2 \subseteq V_{\mathrm{e}}$, $U_{\mathrm{e}}V_{\mathrm{e}} \subseteq U_{\mathrm{e}}$, $0 \neq V_{\mathrm{e}}^2 \subseteq V_{\mathrm{e}}$ and $u^3 = 0$, $u^2 v = 2u(uv)$, $uv^2 = 2v(uv)$, $v^{[3]} = v^4 = 0$, for all $u \in U_{\mathrm{e}}$, $v \in V_{\mathrm{e}}$. Then, the same decomposition exists for an arbitrary element and satisfies the same properties.

In the case of power-associative 2nd-order Bernstein algebras the set of idempotents and the relations between decompositions associated to two idempotent elements are known and are given by the following:

Proposition 2.1 *Let* (A, ω) *be a power-associative* 2nd*-order Bernstein algebra and* e *an idempotent element of A, then:*

1. *The set of idempotent elements of A is given by* $\Im(A) = \{\mathrm{e} + u + u^2 \mid u \in U_{\mathrm{e}}\}$.

2. *If* $\mathrm{e}' = \mathrm{e} + u + u^2$ *is other idempotent element of A, then* $U_{\mathrm{e}'} = \{u' + 2uu' \mid u' \in U_{\mathrm{e}}\}$ *and* $V_{\mathrm{e}'} = \{v - 2uv \mid v \in V_{\mathrm{e}}\}$.

3. $\mathrm{Ann}_N N$, $\mathrm{Ann}_{U_{\mathrm{e}}} U_{\mathrm{e}}$, $U_{\mathrm{e}} + U_{\mathrm{e}}^2$, $\dim(U_{\mathrm{e}}^2 + V_{\mathrm{e}}^2)$, $\dim U_{\mathrm{e}} V_{\mathrm{e}}$, $\dim U_{\mathrm{e}}^2$, $\dim U_{\mathrm{e}}^3$ *are invariants of A, that is, they do not depend on the choice of the idempotent element.*

Remark 1 *If* (A, ω) *is a power-associative* 2nd*-order Bernstein algebra, then* $\dim V_{\mathrm{e}}^2 \geq 1$.

Similarly, if (A, ω) is a Jordan 2nd-order Bernstein algebra, then there exists an idempotent element of A such that, $A = Ke \bigoplus U_e \bigoplus V_e$, where $U_e = \{x \in A \mid ex = \frac{1}{2}x\}$, $V_e = \{x \in A \mid ex = 0\}$, $U_e^2 \subseteq V_e$, $U_e V_e \subseteq U_e$, $0 \neq V_e^2 \subseteq V_e$, and $\forall u \in U_e$, $\forall v, w \in V_e$, $u^3 = 0$, $u^2 v = 2u(uv)$, $uv^2 = 2v(uv)$, $v^{[3]} = v(v^2 w) = 0$.

Again the same decomposition with the same properties is true for every idempotent element of A.

Every Jordan algebra is power-associative. The converse, in a context of 2nd-order Bernstein algebras is true if some additional condition is satisfied:

Corollary 2.1 *Let (A, ω) be a 2nd-order Bernstein algebra. The following conditions are equivalent:*

1. *(A, ω) is a power-associative algebra and $v(v^2 w) = 0$, for every $v, w \in V$.*

2. *(A, ω) is a Jordan algebra.*

Some relations satisfied by products of elements in the Peirce components of a Jordan 2nd-order Bernstein algebra (respectively power-associative 2nd-order Bernstein algebras in which $v^3 = 0$ for all $v \in V_e$, where e is an idempotent element of A) where found in [4]:

Lemma 2.1 *If (A, ω) is a Jordan 2nd-order Bernstein algebra, then we have the following relations for every $u \in U$ and $v, w \in V$,*

$$v(v^2 w) = 0 \tag{1}$$

$$v^2(vw) = 0 \tag{2}$$

$$v^2(uv) = v(uv^2) = 2v(v(uv)) \tag{3}$$

$$uv^3 = 2v^2(uv) \tag{4}$$

$$v^3 w + 2v(v(vw)) = 0 \tag{5}$$

$$v^2 w^2 + 2(vw)^2 = 0 \tag{6}$$

$$v^2(u^2 v) = v^2(u(uv)) = 0 \tag{7}$$

$$v(uv)^2 + u(v^2(uv)) = 0 \tag{8}$$

Lemma 2.2 *If (A, ω) is a power-associative 2nd-order Bernstein algebra with an idempotent element e such that $v^3 = 0$ for all $v \in V_e$, then we have the following relations for every $u \in U_e$ and $v, w \in V$:*

$$v^2(uv) = v(uv^2) = v(v(uv)) = 0 \tag{9}$$

$$v(uv)^2 = 0 \tag{10}$$

Corollary 2.2 *If (A, ω) is a power-associative 2nd-order Bernstein algebra, then the following conditions are equivalent:*

1. *There exists an idempotent element e such that $v^3 = 0$, $\forall v \in V_e$.*

2. *For every idempotent element* e', $v'^3 = 0$, $\forall v' \in V_{e'}$.

Proof: (1) $\Rightarrow$ (2) Let e, e' be idempotent elements of A. We know that $e' = e + u + u^2$, with $u \in U_e$. For every $v' \in V_{e'}$, $v' = v - 2uv$, with $v \in V_e$. Applying $v^3 = u^3 = 0$, (9) and (10), we get $v'^3 = 0$.

(2) $\Rightarrow$ (1) Trivial. ■

We have proved that in our context of power-associative 2nd-order Bernstein algebras, the condition $v^3 = 0$, $\forall v \in V_e$ does not depend on the particular idempotent element, that is, it is an invariant. We will find other invariants.

Let E be a vector subspace of A, we define N_E as the linear span of elements of square zero, that is, $N_E = K(x \in E \mid x^2 = 0)$,

Proposition 2.2 *Let* (A, ω) *be a power-associative* 2nd*-order Bernstein algebra. If* $e, e' \in \Im(A)$, *then* $\dim N_{U_e} = \dim N_{U_{e'}}$.

Proof: If $e, e' \in \Im(A)$ then $U_{e'} = \{u' + 2uu' \mid u' \in U_e\}$ where $e' = e + u + u^2$ with $u \in U_e$. The function $f: U_e \longrightarrow U_{e'}$ defined by $f(u') = u' + 2uu'$ is a linear bijection. Applying $u^2v = 2u(uv)$, and $u^3 = 0$, we have $4(uu')^2 = 4(uu')(uu') = 4u(u'(uu')) + 4u'(u(uu')) = -2u(uu'^2) - 2u'(u'u^2) = -2u(uu'^2) - u'^2u^2 = -4u(uu'^2)$. So $(u' + 2uu')^2 = u'^2 + 4u'(uu') + 4(uu')^2 = u'^2 - 2u'^2u - 4u(uu'^2)$. If $u'^2 = 0$, then $(u' + 2uu')^2 = 0$ and if $(u' + 2uu')^2 = 0$, then $uu'^2 = 0$ and $u'^2 - 4u(uu'^2) = 0$. So we have $u'^2 = 0$. Therefore $u'^2 = 0$ if and only if $(f(u'))^2 = 0$ and $\dim N_{U_e} = \dim N_{U_{e'}}$. ■

Proposition 2.3 *Let* (A, ω) *be a power-associative* 2nd*-order Bernstein algebra such that* $v^3 = 0$, $\forall v \in V$. *If* $e, e' \in \Im(A)$, *then* $\dim N_{V_e} = \dim N_{V_{e'}}$.

Proof: If $e, e' \in \Im(A), e' = e + u + u^2$ with $u \in U_e$, then $V_{e'} = \{v - 2uv \mid v \in V_e\}$. The function $g: V_e \longrightarrow V_{e'}$ defined by $g(v) = v - 2uv$ is a linear bijection. Applying $u^2v = 2u(uv), uv^2 = 2v(uv)$, and $v^3 = 0$, we have $4(uv)^2 = 4(uv)(uv) = 4(u(uv))v - 4u(v(uv)) = 2(u^2v)v - 2u(uv^2) = -4u(uv^2)$. So, $(v - 2uv)^2 = v^2 - 4v(uv) + 4(uv)^2 = v^2 - 2uv^2 - 4u(uv^2)$. If $v^2 = 0$, then $(v - 2uv)^2 = 0$ and if $(v - 2uv)^2 = 0$, then $uv^2 = 0$, $v^2 - 4u(uv^2) = 0$ and so we have $v^2 = 0$. Therefore $v^2 = 0$ if and only if $(g(v))^2 = 0$ and $\dim N_{V_e} = \dim N_{V_{e'}}$. ■

The following two results have been proved in [4].

Proposition 2.4 *Let* (A, ω) *be a* 2nd*-order Bernstein algebra with an idempotent element* e *such that* $v^3 = 0$, *for all* $v \in V_e$. *Then* (A, ω) *is a Jordan algebra if and only if* (A, ω) *is a power-associative algebra.*

Proposition 2.5 *Let* (A, ω) *be a* 2nd*-order Bernstein algebra with* $\dim V_e < 5$. *Then* (A, ω) *is a Jordan algebra if and only if* (A, ω) *is a power-associative algebra.*

Clearly, if (A, ω) is a Jordan and 2nd-order Bernstein algebra, $\dim V_e = s$, then $V_e^{s+1} = 0$. It follows immediatly from the fact that V_e is nilpotent (see [1])

Lemma 2.3 *Let* (A, ω) *be a Jordan* 2nd*-order Bernstein. If* $\dim V_e = s$*, then* $V_e^{s+1} = 0$.

Proof: By [1], V_e is nilpotent. So $\dim V_e^2 \leq s - 1$, $\dim V_e^3 \leq s - 2$ and so on, hence $\dim V_e^{s+1} = 0$. ■

Lemma 2.4 *Let* (A, ω) *be a power-associative* 2nd*-order Bernstein algebra. If* $\dim V_e = 4$*, then we obtain the following relations for every* $v, w \in V_e$*:*

$$v^2(v^2 w) = 0 \tag{11}$$

$$v^3(vw) = 0; \tag{12}$$

$$v^3 w^2 = 0. \tag{13}$$

Proof: Linearizing $v(v^2 w) = 0$ we have $v'(v^2 w) + 2v((vv')w) = 0$. Doing $v' = v^2$ and applying (5), we get $v^2(v^2 w) = -2v(v^3 w) = 4v(v(v(vw))) = 0$ since $V_e^5 = 0$.

By (5) $v^3(vw) = -2v(v(v(vw))) = 0$; $v^3 w^2 = -2v(v(vw^2)) = 0$. ■

Remark 2 *The relation (11) also fields if* (A, ω) *is a power-associative* 2nd*-order Bernstein algebra with* $v^3 = 0$, $\forall v \in V_e$.

Proposition 2.6 *Let* (A, ω) *be a power-associative* 2nd*-order Bernstein algebra and* $v^3 = 0$ *for all* $v \in V_e$*. If* v^2, w^2 *are linearly independent, then* v, v^2, w, w^2 *are linearly independent too.*

Proof: If $v = \alpha v^2 + \beta w^2$, then $v^2 = 2\alpha\beta v^2 w^2 = 4\alpha^2\beta^2 w^2(v^2 w^2) = 0$ by Remark 2, but this is a contradiction, therefore v, v^2, w^2 are linearly independent.

If $w = \alpha v + \beta v^2 + \gamma w^2$, then $w^2 = \alpha^2 v^2 + 2\alpha\gamma v w^2 + 2\beta\gamma v^2 w^2$. Multiplying by v^2 and using (2) and (11) we have $v^2 w^2 = 0$. Multiplying now by v and by (5) we have $vw^2 = 2\alpha\gamma v(vw^2) = 4\alpha^2\gamma^2 v(v(vw^2)) = 0$. We get $w^2 = \alpha^2 v^2$ and this is a contradiction, therefore v, v^2, w, w^2 are linearly independent. ■

Proposition 2.7 *Let* (A, ω) *be a power-associative* 2nd*-order Bernstein algebra. If A has type* $(1 + r, 3)$ *and* $v^3 = 0$*, for all* $v \in V_e$*, then* $\dim V_e^2 = 1$ *and* $V_e^3 = 0$.

Proof: We know that $\dim V_e^2 \leq 2$ because (A, ω) is a Jordan 2nd-order Bernstein algebra and V_e is nilpotent by Proposition 2.4 and Lemma 2.3. If $\dim V_e^2 = 2$, then there exist $v, w \in V_e$ such that v^2, w^2 are linearly independent and by Proposition 2.6 v, v^2, w, w^2 are linearly independent too, but it is impossible since $\dim V_e = 3$. ■

Proposition 2.8 *Let* (A, ω) *be a power-associative* 2nd*-order Bernstein algebra of type* $(1 + r, 4)$*:*

1. *If* $v^3 = 0$*, for all* $v \in V_e$*, then* $\dim V_e^2 \leq 2$ *and* $V_e^3 = 0$.
2. *If there exists* $v \in V_e$ *such that* $v^3 \neq 0$*, then* $V_e^2 = K(v^2, v^3)$ *and* $V_e^4 = 0$.

Proof: By [2] we have $V_e^3 = 0$, so $\dim V_e^2 \leq 3$. If $\dim V_e^2 = 3$, then exist $v, w \in V_e$ such that v, v^2, w^2 are linearly independent. By Proposition 2.6$\{v, v^2, w, w^2\}$ is a basis of V_e. So $vw = \alpha v + \beta v^2 + \gamma w + \delta w^2$ and multiplying by v we have $0 = \alpha v^2 + \gamma v w \alpha v^2 + \gamma(\alpha v + \beta v^2 + \gamma w + \delta w^2)$, so $\alpha = \gamma = 0$ and therefore $V_e^2 \subseteq K(v^2, v^3)$ but that is imposible since we have supposed $\dim V_e^2 = 3$.

Assertion 2 follows from [2], since V_e is a commutative power-associative nilagebra of nilindex 4 and dimension 4. ■

Proposition 2.9 *Let* (A, ω) *be a power-associative* 2nd*-order Bernstein algebra. If there is an idempotent element* e *such that* $U_eV_e = 0$*, then* $U_{e'}V_{e'} = 0$, $V_e = V_{e'}$, $U_e^2 = U_{e'}^2$ *for every* $e' \in \Im(A)$.

Proof: Let e' be other idempotent element, then $e' = e + u + u^2$ with $u \in U_e$, $U_{e'} = \{u' + 2uu' \mid u' \in U_e\}$ and $V_{e'} = \{v - 2uv \mid v \in V_e\}$. Since $\dim U_eV_e$ is an invariant of A we have $U_{e'}V_{e'} = 0$. So $V_e = V_{e'}$. If $u' + 2uu', u'' + 2uu'' \in U_{e'}$ with $u', u'' \in U_e$, then $(u' + 2uu')(u'' + 2uu'') = u'u''$ since $u'(uu'') \in U_eV_e$ and $(uu')(uu'') \in U_e^2 V_e \subseteq U_e(U_eV_e) = 0$. Therefore $U_e^2 = U_{e'}^2$. ■

Proposition 2.10 *Let* (A, ω) *be a power-associative* 2nd*-order Bernstein algebra. If* type $(A) = (2, s)$, *then* $U_eV_e = 0$, $V_e = V_{e'}$, *and* $U_e^2 = U_{e'}^2$ *for every* $e, e' \in \Im(A)$.

Proof: Let $U_e = K(u)$. For every $v \in V_e$ we will have $uv = \lambda u$, because $U_eV_e \subseteq U_e$. So $uv^2 = 2v(uv) = 2v(\lambda u) = 2\lambda^2 u$ and $0 = (v^2)^2 u = 2v^2(v^2 u) = 4\lambda^2 v^2 u = 8\lambda^4 u$. Therefore $\lambda = 0$ and $uv = 0$. ■

Proposition 2.11 *Let* (A, ω) *be a power-associative* 2nd*-order Bernstein algebra. If* $\dim V_e = 3$ *and* $U_eV_e = 0$ *and there exists* $v \in V_e$ *such that* $v^3 \neq 0$*, then* $U_e^2 \subseteq V_e^3$.

Proof: By $u^2 v = 2u(uv)$ we have $U_e^2 V_e = 0$. By $v^3 \neq 0$, $V_e = K(v, v^2, v^3)$ and $uu' = \alpha v + \beta v^2 + \gamma v^3$ for all $u, u' \in U_e$. Multiplying twice by v we have $0 = v(v(uu')) = \alpha v^3$, then $\alpha = 0$; $0 = v(uu') = \beta v^3$, then $\beta = 0$. So $uu' = \gamma v^3 \in V_e^3$. ■

Proposition 2.12 *If* (A, ω) *is a power-associative* 2nd*-order Bernstein algebra such that* $\dim A \leq 5$, *then* $\dim N_{V_e}$ *is an invariant.*

Proof: If $\dim A \leq 5$, then $v^3 = 0$, for all $v \in V_e$ or $U_eV_e = 0$ since if there exists $v \in V_e$ such that $v^3 \neq 0$, then $\dim V_e \geq 3$ and if $U_eV_e \neq 0$, then $\dim U_e \geq 2$. Applying Proposition 2.3 and Proposition 2.10 we have $\dim N_{V_e}$ is an invariant. ■

3. Classification in Dimension 5

Let $A = Ke \oplus U_e \oplus V_e$ be a power-associative 2nd-order Bernstein algebra with $\dim A = 5$ and type (A) $= (1 + r, s)$, where $r + s = 4$. Let $\{e, u_1, \ldots u_r, v_1, \ldots, v_s\}$ be a basis of A with $u_i \in U_e, v_j \in V_e$, $i = 1, 2, ..., r$, $j = 1, 2, ..., s$. Since $\dim V_e \leq 4$, by Proposition 2.5, (A, ω) will be a Jordan 2nd-order Bernstein algebra. Since $\dim V_e \geq 2$, then type (A) $=$ $(1, 4)$ or $(2, 3)$ or $(3, 2)$.

3.1. Power-associative 2nd-order Bernstein Algebras of Type (1,4)

Theorem 3.1 *Let $A = Ke + K(v_1, v_2, v_3, v_4)$ be a power-associative 2nd-order Bernstein algebra with* type (A) $= (1,4)$. *Then A is isomorphic to an algebra whose nonzero products are given by one of the following types:*

1. $A(1)(\alpha,\beta)$: $e^2 = e \quad v_2^2 = v_1 \quad v_3^2 = \alpha v_1 \quad v_4^2 = \beta v_1$.
   In this case, $v^3 = 0,\ \forall v \in V_e,\ \dim V_e^2 = 1$ *and* $V_e^3 = 0$.

2. $A(2)(\alpha)$: $e^2 = e \quad v_2^2 = v_1 \quad v_4^2 = v_3 \quad v_2v_4 = v_1 + \alpha v_3, \quad \alpha \in K^* = K \setminus \{0\}$.
   Now, $v^3 = 0,\ \forall v \in V_e,\ \dim V_e^2 = 2$ *and* $V_e^3 = 0$.

3. $A(3)(\alpha,\beta)$: $e^2 = e \quad v_2v_3 = v_1 \quad v_3^2 = v_2 \quad v_4^2 = \alpha v_1 + \beta v_2$.
   Let's notice that, $\exists v \in V_e,\ v^3 \neq 0,\ \dim V_e^2 = 2$ *and* $\dim V_e^3 = 1$.

4. $A(4)(\alpha,\beta)$: $e^2 = e \quad v_2v_3 = v_2v_4 = v_1 \quad v_3^2 = v_2 \quad v_4^2 = \alpha v_1 + \beta v_2$.
   Here again, $\exists v \in V_e,\ v^3 \neq 0,\ \dim V_e^2 = 2$ *and* $\dim V_e^3 = 1$.

Proof: We will consider three cases:

1. $v^3 = 0\ \forall v \in V_e,\ \dim V_e^2 = 1$ and $V_e^3 = 0$.
   In this case $V_e^2 = K(z_1)$, $V_e = K(z_1, z_2, z_3, z_4)$. So $z_1^2 = z_1z_2 = z_1z_3 = z_1z_4 = 0$. Let f be the bilinear symmetric form over $V' = K(z_2, z_3, z_4)$ given by $xy = f(x,y)z_1$. Since this bilinear form is diagonalizable, we can find a basis $\{w_2, w_3, w_4\}$ of V' with $w_2w_3 = w_2w_4 = w_3w_4 = 0$ $w_2^2 = az_1$ $w_3^2 = bz_1$ $w_4^2 = cz_1$ and $a \in K^*$. So our algebra A is isomorphic to $A(1)(a^{-1}b, a^{-1}c)$. Conversely every algebra $A(1)(\alpha,\beta)$ is a power-associative 2nd-order Bernstein algebra with type (A) $= (1,4)$, $v^3 = 0\ \forall v \in V_e,\ \dim V_e^2 = 1$ and $V_e^3 = 0$.

2. $v^3 = 0\ \forall v \in V_e,\ \dim V_e^2 = 2$ and $V_e^3 = 0$.
   Now $V_e^2 = K(v_2^2, v_4^2)$, $V_e = K(v_2^2, v_2, v_4^2, v_4)$. If $v_1 = v_2^2$ and $v_3 = v_4^2$ we have v_1v_j, $v_3v_j \in V_e^3 = 0$ and $v_2v_4 \in V_e^2$, so $v_2v_4 = \lambda v_1 + \mu v_3$. If $\lambda \neq 0$ and $\mu \neq 0$ doing $v_1' = \lambda^2 v_1$, $v_2' = \lambda v_2$, $v_3' = v_3$ and $v_4' = v_4$ we have $A(2)(\lambda\mu)$. In the other cases we obtain an algebra $A(2)(\frac{1-\gamma^2}{4})$ with $\gamma \in K \setminus \{0,-1,1\}$. Conversely, every algebra $A(2)(\alpha)$ is a power-associative 2nd-order Bernstein algebra with type (A) $= (1,4)$, $v^3 = 0\ \forall v \in V_e,\ \dim V_e^2 = 2$ and $V_e^3 = 0$.

3. $\exists v \in V_e$ such that $v^3 \neq 0,\ \dim V_e^2 = 2$ and $\dim V_e^3 = 1$.
   In this case $V_e^2 = K(v^2, v^3)$, $V_e^3 = K(v^3)$, $V_e = K(v, v^2, v^3, w)$, $vw = av^2 + bv^3$. If $w' = w - av - bv^2$ then $vw' = 0$, $w'^2 = \alpha v^2 + \beta v^3$ and $v^2w' = \gamma v^3$. If $\gamma = 0$ and $v_1 = v^3$, $v_2 = v^2$, $v_3 = v$, $v_4 = w'$ we have $A(3)(\alpha,\beta)$. If $\gamma \neq 0$ and $v_1 = v^3$, $v_2 = v^2$, $v_3 = v$, $v_4 = \gamma^{-1}w'$ we have $A(4)(\alpha,\beta)$. Conversely algebras $A(3)(\alpha,\beta)$ and $A(4)(\alpha,\beta)$ are power-associative 2nd-order Bernstein algebras with type (A) $= (1,4)$, $\exists v \in V_e$ such that $v^3 \neq 0,\ \dim V_e^2$ 2 and $\dim V_e^3 = 1$. ■

Proposition 3.1 *With the notation of Theorem 3.1:*

1. *If* $\alpha \neq 0$*, then* $A(1)(0,0) \not\simeq A(1)(\alpha,\beta)$.
2. $A(1)(\alpha,0) \simeq A(1)(0,\alpha)$.
3. $A(1)(\alpha,0) \simeq A(1)(\alpha',0) \Leftrightarrow \alpha' = \gamma^2\alpha$ *with* $\gamma \in K^*$.
4. *If* $\alpha, \alpha', \beta \in K^*$*, then* $A(1)(\alpha,0) \not\simeq A(1)(\alpha',\beta')$.
5. *If* $\alpha' = \gamma^2\alpha$ *and* $\beta' = \delta^2\beta$ *with* $\gamma, \delta \in K^*$*, then* $A(1)(\alpha,\beta) \simeq A(1)(\alpha',\beta')$.
6. *If* $\alpha \neq 0$ *or* $\beta \neq 0$*, then* $A(3)(0,0) \not\simeq A(3)(\alpha,\beta)$.
7. *If* $\alpha \neq 0$*, then* $A(3)(\alpha,0) \simeq A(3)(1,0)$.
8. *If* $\beta' \neq 0$*, then* $A(3)(\alpha,0) \not\simeq A(3)(\alpha',\beta')$.
9. *If* $\beta, \beta' \in K^*$*, then* $A(3)(\alpha,\beta) \simeq A(3)(\alpha',\beta') \Leftrightarrow \beta' = \gamma^2\beta$ *with* $\gamma \in K^*$.
10. $A(4)(\alpha,\beta) \not\simeq A(3)(0,0)$.
11. $A(4)(\alpha,0) \simeq A(4)(0,0)$.
12. *If* $\beta \neq 0$*, then* $A(4)(\alpha,\beta) \not\simeq A(4)(0,0)$.
13. *If* $\beta, \beta' \in K^*$*, then* $A(4)(\alpha',\beta') \simeq A(3)(0,\beta) \Leftrightarrow \beta' \neq -1$ *and* $\beta' = \gamma^2\beta$ *with* $\gamma \in K^*$.

Proof: We use the invariant $\mathrm{Ann}_N N$. If $\alpha = \beta = 0$, then $\mathrm{Ann}_N N = K(v_1, v_3, v_4)$. If $\alpha \neq 0$ and $\beta = 0$, then $\mathrm{Ann}_N N = K(v_1, v_4)$ and if $\alpha \neq 0$ and $\beta \neq 0$, then $\mathrm{Ann}_N N = K(v_1)$.

If $\{e, v_1, v_2, v_3, v_4\}$ is a basis of $A(1)(\alpha,0)$, then $\{e, v_1, v_2, v_4, v_3\}$ is a basis of $A(1)(0,\alpha)$.

Let us suppose $A(1)(\alpha,0) \simeq A(1)(\alpha',0)$. Since $U_e = 0$ then there is only one idempotent element e. Let $\{e, v_1, v_2, v_3, v_4\}$ be a basis of $A(1)(\alpha,0)$ then they must exist $w_i = \sum_{j=1}^{4} a_{ij}v_j$, $i = 1,2,3,4$ such that $\{e, w_1, w_2, w_3, w_4\}$ is a basis of $A(1)(\alpha',0)$. The relation $w_2^2 = w_1$ implies $a_{22}^2 + \alpha a_{23}^2 = a_{11} \neq 0$ and $a_{12} = a_{13} = a_{14} = 0$. The relation $w_3^2 = \alpha' w_1$ implies $a_{32}^2 + \alpha a_{33}^2 = \alpha' a_{11}$ and finally the relation $w_2 w_3 = 0$ implies $a_{22}a_{32} + \alpha a_{23}a_{33} = 0$. If $a_{22} = 0$ then $\alpha a_{23}^2 = a_{11} \neq 0$, $a_{33} = 0$, $a_{32}^2 = \alpha' a_{11} \neq 0$ and so $\alpha' = \frac{a_{32}^2}{a_{11}} = \frac{a_{32}^2}{\alpha a_{23}^2} = \left(\frac{a_{32}}{\alpha a_{23}}\right)^2 \alpha = \gamma^2\alpha$ with $\gamma \in K^*$. If $a_{22} \neq 0$, $(a_{22}a_{32})^2 = (\alpha a_{23}a_{33})^2$; $(a_{11} - \alpha a_{23}^2)(\alpha' a_{11} - \alpha a_{33}^2) = \alpha^2 a_{23}^2 a_{33}^2$. So $\alpha' a_{11} - \alpha a_{33}^2 - \alpha\alpha' a_{23}^2 = 0$, and replacing a_{11} we have $\alpha' a_{22}^2 - \alpha a_{33}^2 = 0$. So again $\alpha' = \gamma^2\alpha$ with $\gamma \in K^*$.

Conversely, if $\alpha' = \gamma^2\alpha$ with $\gamma \in K^*$ and $\{e, v_1, v_2, v_3, v_4\}$ is a basis of $A(1)(\alpha,0)$, then $\{e, v_1, v_2, \gamma v_3, v_4\}$ is a basis of $A(1)(\alpha',0)$.

It's sufficient to see $\mathrm{Ann}_N N$ to prove that $A(1)(\alpha,0) \not\simeq A(1)(\alpha',\beta')$.

If $\alpha' = \gamma^2\alpha$ and $\beta' = \delta^2\beta$ with $\gamma, \delta \in K^*$ and $\{e, v_1, v_2, v_3, v_4\}$ is a basis of $A(1)(\alpha,\beta)$, then $\{e, v_1, v_2, \gamma v_3, \delta v_4\}$ is a basis of $A(1)(\alpha',\beta')$.

In $A(3)(0,0)$, $\mathrm{Ann}_N N = K(v_1, v_4)$ and if $\alpha \neq 0$ or $\beta \neq 0$ in $A(3)(\alpha, \beta)$, $\mathrm{Ann}_N N = K(v_1)$ so if $\alpha \neq 0$ or $\beta \neq 0$, then $A(3)(0,0) \not\simeq A(3)(\alpha, \beta)$.

If $\alpha \neq 0$ and $\{e, v_1, v_2, v_3, v_4\}$ is a basis of $A(3)(\alpha, 0)$, then $\{e, \frac{1}{\alpha^3}v_1, \frac{1}{\alpha^2}v_2, \frac{1}{\alpha}v_3, \frac{1}{\alpha^2}v_4\}$ is a basis of $A(3)(1,0)$.

Let us suppose that $A(3)(\alpha, \beta) \simeq A(3)(\alpha', \beta')$ with $\beta' \neq 0$. If $\{e, v_1, v_2, v_3, v_4\}$ be a basis of $A(3)(\alpha, \beta)$, then they will exist $w_i = \sum_{j=1}^{4} a_{ij} v_j$, $i = 1, 2, 3, 4$ such that $\{e, w_1, w_2, w_3, w_4\}$ is a basis of $A(3)(\alpha', \beta')$. So $w_3^2 = w_2$; $w_2 w_3 = w_1$; $w_2 w_4 = 0$ and $w_4^2 = \alpha' w_1 + \beta' w_2$. They imply $2a_{32}a_{33} + a_{34}^2\alpha = a_{21}$, $a_{33}^2 + a_{34}^2\beta = a_{22}$, $a_{23} = a_{24} = 0$; $a_{22}a_{33} = a_{11} \neq 0$, $a_{12} = a_{13} = a_{14} = 0$; $a_{43} = 0$; $a_{44}^2\alpha = \alpha' a_{11} + \beta' a_{21}$, $a_{44}^2\beta = \beta' a_{22}$. If $\beta = 0$ then $\beta' = 0$. So if $\beta' \neq 0$, then $A(3)(\alpha, 0) \not\simeq A(3)(\alpha', \beta')$. If $\beta' \neq 0$ then $a_{44}^2\beta \neq 0$ and so we have $\beta \neq 0$ and $a_{44} \neq 0$. Moreover $w_3 w_4 = 0$ implies $a_{34} = 0$. So $a_{22} = a_{33}^2$ and finally $\beta' = \left(\frac{a_{44}}{a_{33}}\right)^2 \beta = \gamma^2\beta$ with $\gamma \in K^*$.

Conversely if $\beta, \beta' \neq 0$ and $\beta' = \gamma^2\beta$ with $\gamma \neq 0$ and $\{e, v_1, v_2, v_3, v_4\}$ is a basis of $A(3)(\alpha, \beta)$, then $\{e, w_1, w_2, w_3, w_4\}$ be a basis of $A(3)(\alpha', \beta')$, where $w_1 = v_1$, $w_2 = \frac{1}{\beta'}(\gamma^2\alpha - \alpha')v_1 + v_2$, $w_3 = \frac{1}{2\beta'}(\gamma^2\alpha - \alpha')v_2 + v_3$ and $w_4 = \gamma v_4$.

In $A(4)(\alpha, \beta)$, $\mathrm{Ann}_N N = K(v_1)$ and $\mathrm{Ann}_N N = K(v_1, v_4)$ in $A(3)(0,0)$, then we have $A(4)(\alpha, \beta) \not\simeq A(3)(0,0)$.

If $\{e, v_1, v_2, v_3, v_4\}$ be a basis of $A(4)(\alpha, 0)$, then $\{e, v_1, v_2 + \alpha v_1, v_3 + \frac{\alpha}{2}v_2, v_4 - \frac{\alpha}{2}v_2\}$ be a basis of $A(4)(0,0)$.

If $A(4)(\alpha, -1) \simeq A(4)(0,0)$ they will exist $\{e, v_1, v_2, v_3, v_4\}$, and $\{e, w_1, w_2, w_3, w_4\}$ basis of $A(4)(\alpha, -1)$ and $A(4)(0,0)$ respectively, with $w_i = \sum_{j=1}^{4} a_{ij} v_j$, $i = 1, 2, 3, 4$ verifying $w_3^2 = w_2$, $w_2 w_3 = w_1$, $w_2 w_4 = w_1$, $w_3 w_4 = 0$ and $w_4^2 = 0$. This conditions imply $2a_{32}(a_{33} + a_{34}) + a_{34}^2\alpha = a_{21}$, $a_{33}^2 - a_{34}^2 = a_{22}$, $a_{23} = a_{24} = 0$; $a_{22}(a_{33} + a_{34}) = a_{11} \neq 0$, $a_{12} = a_{13} = a_{14} = 0$; $a_{22}(a_{43} + a_{44}) = a_{11} \neq 0$, $a_{33}a_{43} - a_{34}a_{44} = 0$; $a_{43}^2 - a_{44}^2 = 0$. Of this relation as $a_{11} \neq 0$ we obtain $a_{43} = a_{44}$ and so $a_{11} = 2a_{22}a_{43} \neq 0$, $a_{33} = a_{34}$ and $a_{22} = 0$, but this is a contradiction, so $A(4)(\alpha, -1) \not\simeq A(4)(0,0)$.

Since in $A(4)(0,0)$, $N_{V_e} = K(v_1, v_2, v_4)$ and if $\beta = -\gamma^2$with $\gamma \in K - \{0, 1, -1\}$, then $N_{V_e} = K(v_1, v_2, v_3, v_4)$ and if $\beta \notin -K^2$, then $N_{V_e} = K(v_1, v_2)$, we have $A(4)(\alpha, \beta) \not\simeq A(4)(0,0)$, in the others cases.

Can be proved similarly assertion 13.

The following table shows the results obtained in Theorem 3.1 and Proposition 3:1

	d_1	d_2	$\mathrm{Ann}_N N$	N_{V_e}	v_2^2	v_3^2	v_4^2	v_2v_3	v_2v_4
$A(1)(0,0)$			$K(v_1,v_3,v_4)$	$K(v_1,v_3,v_4)$	v_1	0	0	0	0
$A(1)(-1,0)$	1	0	$K(v_1,v_4)$	$K(v_1,v_2,v_3,v_4)$	0	0	0	v_1	0
$A(1)(\alpha,0)\ -\alpha\notin K^2$			$K(v_1,v_4)$	$K(v_1,v_4)$	v_1	αv_1	0	0	0
$A(1)(\alpha,\beta)\ \alpha,\beta\neq 0$			$K(v_1)$		v_1	αv_1	βv_1	0	0
$A(2)(\alpha)\ 1-4\alpha\notin K^2$	2	0	$K(v_1,v_3)$	$K(v_1,v_3)$	v_1	0	v_3	0	$v_1+\alpha v_3$
$A(2)\left(\frac{1}{4}\right)$			$K(v_1,v_3)$	$K(v_1,v_3,v_4-2v_2)$	v_1	0	v_3	0	$v_1+\frac{1}{4}v_3$
$A(3)(0,0)$			$K(v_1,v_4)$	$K(v_1,v_2,v_4)$	0	v_2	0	v_1	0
$A(3)(1,0)$			$K(v_1)$	$K(v_1,v_2)$	0	v_2	v_1	v_1	0
A(3)(0,-1)			$K(v_1)$	$K(v_1,v_2,v_3,v_4)$	0	v_2	$-v_2$	v_1	0
$A(3)(0,\beta)\ -\beta\notin K^2$	2	1	$K(v_1)$	$K(v_1,v_2)$	0	v_2	βv_2	v_1	0
$A(4)(0,0)$			$K(v_1)$	$K(v_1,v_2,v_4)$	0	v_2	0	v_1	v_1
$A(4)(0,-1)$			$K(v_1)$	$K(v_1,v_2,v_3,v_4)$	0	v_2	$-v_2$	v_1	v_1
$A(4)(\alpha,-1)\ \alpha\neq 0$			$K(v_1)$	$K(v_1,v_2,v_3+v_4)$	0	v_2	αv_1-v_2	v_1	v_1

where $d_1 = \dim V^2$ and $d_2 = \dim V^3$.

These algebras are nonisomorphic and $A(1)(\alpha,0)$ and $A(1)(\alpha',0)$ are isomorphic if and only if $\alpha' = \gamma^2\alpha$ with $\gamma \in K^*$. Similarly $A(3)(0,\beta)$ and $A(3)(0,\beta')$ are isomorphic if and only if $\beta' = \gamma^2\beta$ with $\gamma \in K^*$. ■

3.2. Power-associative 2nd-order Bernstein Algebras of Type (2,3)

Theorem 3.2 *Let $A = Ke + K(u) + K(v_1,\ v_2,\ v_3)$ be a power-associative 2nd-order Bernstein algebra with* type (A) $= (2,3)$. *Then A is isomorphic to an algebra whose nonzero products are given by one of the following types:*

1. *$A(5)(\alpha)$: $e^2 = e,\quad eu = \frac{1}{2}u,\quad v_2^2 = v_1,\quad v_3^2 = \alpha v_1$.*
   Here $v^3 = 0,\ \forall v \in V_e,\ \dim V_e^2 = 1,\ V_e^3 = 0$ and $U_e^2 = 0$.

2. *$A(6)$: $e^2 = e,\quad eu = \frac{1}{2}u,\quad u^2 = v_3,\quad v_2^2 = v_1$.*

3. *$A(7)(\alpha,\beta)$: $e^2 = e,\quad eu = \frac{1}{2}u,\quad u^2 = v_1,\quad v_2^2 = \alpha v_1,\quad v_3^2 = \beta v_1,\quad (\alpha \neq 0)$.*
   In these two cases $v^3 = 0,\ \forall v \in V_e,\ \dim V_e^2 = 1,\ V_e^3 = 0$ and $U_e^2 \neq 0$.

4. *$A(8)(\alpha)$: $e^2 = e,\quad eu = \frac{1}{2}u,\quad u^2 = \alpha v_1,\quad v_2v_3 = v_1,\quad v_3^2 = v_2$.*
   Let's notice that $\exists v \in V_e,\ v^3 \neq 0,\ \dim V_e^2 = 2$ and $\dim V_e^3 = 1$.

Proof: By Proposition 2.10 we have $U_eV_e = 0$, so $U_eV_e^2 = U_e^2V_e = 0$ too. We consider two cases: 1) $v^3 = 0\ \forall v \in V_e$, and 2) $\exists v \in V_e$ such that $v^3 \neq 0$.

1. In the first case by Proposition 2.7 $\dim V_e^2 = 1$ and $V_e^3 = 0$. So $V_e^2 = K(z_1)$, $V_e = K(z_1,z_2,z_3)$ with $z_1^2 = z_1z_2 = z_1z_3 = 0$. Let f be the bilinear symmetric form over $V' = K(z_2,z_3)$ given by $xy = f(x,y)z_1$. Since this bilinear form is diagonalizable, we can find a basis $\{w_2,w_3\}$ of V' with $w_2w_3 = 0$, $w_2^2 = \alpha z_1$, $w_3^2 = \beta z_1$ and $\alpha \in K^*$. If $w_1 = \alpha z_1$ then $V_e = K(w_1,w_2,w_3)$, $U_e = K(u)$,

$0 = uw_i = w_1w_i = w_2w_3$, $i = 1,2,3$, $w_2^2 = w_1$, $w_3^2 = \alpha^{-1}\beta w_1$. If $U_e^2 = 0$ and $v_i = w_i$, A is isomorphic to $A(5)(\alpha^{-1}\beta)$.

If $U_e^2 \neq 0$, $u^2 = aw_1 + bw_2 + cw_3$. Multiplying respectively by w_2 and w_3 and using $U_e^2V_e = 0$ we get $b = 0$ and $c\beta = 0$. If $c \neq 0$, $v_1 = w_1$, $v_2 = w_2$, $v_3 = aw_1 + cw_3$, we have $A(6)$.

If $c = 0$ then $a \neq 0$. Taking $v_1 = aw_1, v_2 = aw_2$ and $v_3 = w_3$, we get $A(7)(a,(a\alpha)^{-1}\beta)$.

2. In the second case $V = K(v, v^2, v^3)$, $V_e^2 = K(v^2, v^3)$ and $V_e^3 = K(v^3)$. By Proposition 2.11 $U_e^2 \subseteq V_e^3$, therefore $u^2 = \alpha v^3$. Taking $v_1 = v^3$, $v_2 = v^2$, and $v_3 = v$, we have $A(8)(\alpha)$. ■

Proposition 3.2 *With the notation of Theorem 3.2 we have:*

1. $A(5)(\alpha) \simeq A(5)(\alpha') \Leftrightarrow \alpha' = \gamma^2\alpha$ with $\gamma \in K^*$.
2. $A(7)(\alpha,\beta) \not\simeq A(6)$.
3. If $\beta \neq 0$, then $A(7)(\alpha,\beta) \not\simeq A(7)(\alpha,0)$.
4. $A(7)(\alpha,0) \simeq A(7)(\alpha',0) \Leftrightarrow \alpha' = \gamma^2\alpha$ with $\gamma \in K^*$.
5. $A(7)(\alpha,\beta) \simeq A(7)(\alpha',\beta') \Rightarrow \alpha'\beta' = \gamma^2\alpha\beta$ with $\gamma \in K^*$.
6. If $\alpha' = \gamma^2\alpha$ and $\beta' = \delta^2\beta$ with $\gamma, \delta \in K^*$ then $A(7)(\alpha,\beta) \simeq A(7)(\alpha',\beta')$.
7. If $\beta' = \gamma^2\alpha$ and $\alpha' = \delta^2\beta$ with $\gamma, \delta \in K^*$ then $A(7)(\alpha,\beta) \simeq A(7)(\alpha',\beta')$.
8. If $\beta \neq 0$, then $A(7)(\alpha,\beta) \simeq A(7)(1,-1) \Leftrightarrow \beta = -\gamma^2\alpha$, with $\gamma \in K^*$.
9. If $\alpha \neq 0$, then $A(8)(\alpha) \not\simeq A(8)(0)$.
10. If $\alpha' = \gamma^2\alpha$ with $\gamma \in K^* \Rightarrow A(8)(\alpha) \simeq A(8)(\alpha')$.

Proof: The proof of 1, 4 and 5 is similar to the assertion 3 of Proposition 3.1

If $\beta \neq 0$, in $A(7)(\alpha,\beta)$ we have $\operatorname{Ann}N = K(v_1)$ and in $A(6)$ $\operatorname{Ann}N = K(v_1, v_3)$. So if $\beta \neq 0$, $A(7)(\alpha,\beta) \not\simeq A(6)$.

Let us see now that $A(7)(\alpha,0) \not\simeq A(6)$. Let us suppose that $A(7)(\alpha,0) \simeq A(6)$, then it will exist one basis $\{e,u,v_1,v_2,v_3\}$ of $A(7)(\alpha,0)$ and other basis $\{e',u',w_1,w_2,w_3\}$ of $A(6)$, where $e' = e + \delta u + \delta^2 v_1$, $u' = \lambda u + 2\lambda\delta v_1$, $w_i = \sum_{j=1}^{3} a_{ij}v_j$, $i = 1,2,3$. The conditions $u'^2 = w_3$ and $w_2^2 = w_1$ imply $0 = a_{32} = a_{33} = a_{12} = a_{13}$, but that is a contradiction because w_1 and w_3 are linearly independent.

Since in $A(7)(\alpha,0)$, $\operatorname{Ann}N = K(v_1,v_3)$ we have if $\beta \neq 0$, $A(7)(\alpha,\beta) \not\simeq A(7)(\alpha,0)$

If $\alpha' = \gamma^2\alpha$ and $\beta' = \delta^2\beta$ with $\gamma,\delta \in K^*$ and $\{e,u,v_1,v_2,v_3\}$ is a basis of $A(7)(\alpha,\beta)$, then $\{e,u,v_1,\gamma v_2,\delta v_3\}$ is a basis of $A(7)(\alpha',\beta')$. Therefore $A(7)(\alpha,\beta) \simeq A(7)(\alpha',\beta')$.

The proof is similar if $\beta' = \gamma^2\alpha$ and $\alpha' = \delta^2\beta$ with $\gamma,\delta \in K^*$.

If $\beta \neq 0$, and $A(7)(\alpha,\beta) \simeq A(7)(1,-1)$, by assertion 5, $-1 = \delta^2\alpha\beta$ with $\delta \in K^*$ and so $\beta = -((\delta\alpha)^{-1})^2\alpha$. Conversely if $\beta = -\gamma^2\alpha$, with $\gamma \in K^*$, and $\{e,u,v_1,v_2,v_3\}$

is a basis of $A(7)(\alpha,\beta)$, then $\{e,u,v_1,v_2',v_3'\}$ with $v_2' = \frac{1}{2\beta}[\gamma(\beta-1)v_2+(\beta+1)v_3]$ and $v_3' = \frac{1}{2\beta}[\gamma(\beta+1)v_2+(\beta-1)v_3]$ is a basis of $A(7)(1,-1)$.

$A(8)(\alpha) \not\simeq A(8)(0)$ with $\alpha \neq 0$ because $\dim U^2$ is different.

If $\alpha' = \gamma^2\alpha$ with $\gamma \in K^*$ and $\{e,u,v_1,v_2,v_3\}$ is a basis of $A(8)(\alpha)$ then $\{e,\gamma u,v_1,v_2,v_3\}$ is a basis of $A(8)(\alpha')$. ■

The following table shows the results obtained in Theorem 3.2 and Proposition 3.2

	d_1	d_2	d_3	$\mathrm{Ann}_N N$	N_{V_e}	u^2	v_2^2	v_3^2	v_2v_3
$A(5)(0)$				$K(u,v_1,v_3)$	$K(v_1,v_3)$	0	v_1	0	0
$A(5)(-1)$	0	1	0	$K(u,v_1)$	$K(v_1,v_2,v_3)$	0	0	0	v_1
$A(5)(\alpha)$ $-\alpha \notin K^2$				$K(u,v_1)$	$K(v_1)$	0	v_1	0	αv_1
$A(6)$				$K(v_1,v_3)$	$K(v_1,v_3)$	v_3	v_1	0	0
$A(7)(\alpha,0)$ $\alpha \neq 0$	1	1	0	$K(v_1,v_3)$	$K(v_1,v_3)$	v_1	αv_1	0	0
$A(7)(1,-1)$				$K(v_1)$	$K(v_1,v_2,v_3)$	v_1	0	0	v_1
$A(7)(\alpha,\beta)$ $-\frac{\beta}{\alpha} \notin K^2, \alpha,\beta \neq 0$				$K(v_1)$	$K(v_1)$	v_1	αv_1	βv_1	0
$A(8)(0)$	0	2	1	$K(v_1)$	$K(v_1,v_2)$	0	0	v_2	v_1
$A(8)(\alpha)$ $\alpha \neq 0$	1	2	1	$K(v_1)$	$K(v_1,v_2)$	αv_1	0	v_2	v_1

where $d_1 = \dim U^2, d_2 = \dim V^2$ and $d_3 = \dim V^3$

3.3. Power-associative 2nd-order Bernstein Algebras of Type (3,2)

Theorem 3.3 *Let* $A = Ke + K(u_1,\ u_2) + K(v_1,\ v_2)$ *be a power-associative* 2nd*-order Bernstein algebra with* type (A) $= (3,2)$. *Then* A *is isomorphic to one of the followings:*

1. $A(9)(\alpha,\beta)$: $U_eV_e = 0$

   $e^2 = e, \quad eu_i = \frac{1}{2}u_i, \quad u_1^2 = \alpha v_2, \quad u_2^2 = \beta v_2, \quad v_1^2 = v_2.$

2. $A(10)(\alpha)$: $U_eV_e \neq 0$

   $e^2 = e, \quad eu_i = \frac{1}{2}u_i, \quad u_1^2 = \alpha v_2, \quad u_1v_1 = u_2, \quad v_1^2 = v_2.$

Proof: El caso $U_eV_e \neq 0$ is known by [6]. So let's assume that $U_eV_e = 0$. By [6] $U_e^2 \subseteq V_e^2$. Since $\exists v \in V_e$ with $v^2 \neq 0$, $V_e = K(v,v^2)$ and $V_e^2 = K(v^2)$. Let f be the bilinear symmetric form over $U_e = K(\widetilde{u_1},\widetilde{u_2})$ given by $uu' = f(u,u')v^2$. Since this bilinear form is diagonalizable, we can find a basis $\{u_1,u_2\}$ of U_e with $u_1u_2 = 0$, $u_1^2 = \alpha v^2$, $u_2^2 = \beta v^2$. Taking $v_1 = v$ and $v_2 = v^2$, we get $A(9)(\alpha,\beta)$. ■

Proposition 3.3 *With the notation of Theorem 3.3 we have:*

1. *If* $\alpha \neq 0$ *or* $\beta \neq 0$, *then* $A(9)(0,0) \not\simeq A(9)(\alpha,\beta)$.
2. $A(9)(\alpha,\beta) \simeq A(9)(\beta,\alpha)$.
3. $A(9)(\alpha,0) \simeq A(9)(\alpha',0)$ *if and only if* $\alpha' = \gamma^2\alpha$ *with* $\gamma \in K^*$.
4. $A(9)(\alpha,\beta) \simeq A(9)(\alpha',\beta') \Rightarrow \alpha'\beta' = \gamma^2\alpha\beta$ *with* $\gamma \in K^*$.

5. *If* $\alpha' = \gamma^2\alpha$ *and* $\beta' = \delta^2\beta$ *with* $\gamma, \delta \in K^*$ *then* $A(9)(\alpha, \beta) \simeq A(9)(\alpha', \beta')$.

6. *If* $\beta \neq 0$, *then* $A(9)(\alpha, \beta) \simeq A(9)(1, -1) \Leftrightarrow \beta = -\gamma^2\alpha$, *with* $\gamma \in K^*$.

7. *If* $\alpha \neq 0$, *then* $A(10)(0) \not\simeq A(10)(\alpha)$.

8. $A(10)(\alpha) \simeq A(10)(\alpha') \Leftrightarrow \alpha' = \gamma^2\alpha$ *with* $\gamma \in K^*$.

Proof: It sufficient to see the $\mathrm{Ann}_N N$. In $A(9)(\alpha, \beta)$ we have: if $\alpha = \beta = 0$, then $\mathrm{Ann}N = K(u_1, u_2, v_2)$, if $\alpha = 0$ and $\beta \neq 0$, then $\mathrm{Ann}N = K(u_1, v_2)$ and if $\alpha \neq 0$ and $\beta = 0$, then $\mathrm{Ann}_N N = K(u_2, v_2)$.

If $\{e, u_1, u_2, v_1, v_2\}$ is a basis of $A(9)(\alpha, \beta)$, then $\{e, u_2, u_1, v_1, v_2\}$ is a basis of $A(9)(\beta, \alpha)$.

In $A(10)(0)$, $N_{U_e} = K(u_1, u_2)$ and if $\alpha \neq 0$ in $A(10)(\alpha)$, $N_{U_e} = K(u_2)$, so $A(10)(0) \not\simeq A(10)(\alpha)$.

The proof of the other assertions follows the same lines of Propositions 3.1, and 3.2. ■

We can recopilate the previous results in the following table:

	d_1	d_2	$\mathrm{Ann}_N N$	N_{U_e}	N_{V_e}	u_1^2	u_2^2	u_1u_2	u_1v_1	v_1^2
$A(9)(0,0)$		0	$K(u_1, u_2, v_2)$	$K(u_1, u_2)$	$K(v_2)$	0	0	0	0	v_2
$A(9)(\alpha, 0)\ \alpha \neq 0$	0		$K(u_2, v_2)$	$K(u_2)$	$K(v_2)$	αv_2	0	0	0	v_2
$A(9)(1, -1)$		1	$K(v_2)$	$K(u_1, u_2)$	$K(v_2)$	0	0	v_2	0	v_2
$A(9)(\alpha, \beta)\ -\frac{\beta}{\alpha} \notin K^2, \alpha, \beta \neq 0$			$K(v_2)$	0	$K(v_2)$	αv_2	βv_2	0	0	v_2
$A(10)(0)$	1	0	$K(u_2, v_2)$	$K(u_1, u_2)$	$K(v_2)$	0	0	0	u_2	v_2
$A(10)(\alpha)\ \alpha \neq 0$	1	1	$K(u_2, v_2)$	$K(u_2)$	$K(v_2)$	αv_2	0	0	u_2	v_2

where $d_1 = \dim UV$ and $d_2 = \dim U^2$

References

[1] M.A Garcia Muñiz, Derivaciones en álgebras de Bernstein de orden 2, Ph.D. thesis, *University of Oviedo*, Spain (1998).

[2] M. Gerstenhaber, H.C. Myung, On commutative power-associative nilalgebras of low dimension, *Proc.Amer.Math. Soc.* **48** (1975) 412-417

[3] S. González, J.C. Gutiérrez and C. Martínez, On Bernstein algebras of nth-order, presented at *Third International Conference on Non-associative Algebra and Its Applications*, Oviedo, Spain (1993).

[4] S. González, C. Martínez and P. Vicente, Power-Associative and Jordan 2nd-order Bernstein algebras, *Nova Journal of Algebra and Geometry*, Vol. 2, No. 4, (1993) 367-381.

[5] J.C. Gutiérrez, Clasificación de Algebras de Bernstein en función del tipo y la dimensión, Ph.D. thesis, *University of Oviedo*, Spain (1995).

[6] A. Labra, C. Mallol and A. Suazo, A characterization of Power-Associative Bernstein algebras of order 2, *Nova Journal of Algebra and Geometry*, Vol. 3, No. 1, (1994) 83-96.

[7] C. Mallol, A propos del algèbres de Bernstein, Ph.D. Thesis, *Université des Sciencies et Techniques du Languedoc, Montpellier* II (1989).

[8] M. Ouattara, Sur les algèbres de Bernstein d'ordre 2, *Linear Algebra and its Applications*, (1991) 144:29-38.

[9] S. Walcher, Bernstein's algebras which are Jordan algebras, *Arch. Math.*, **50**, (1988) 218-222.

In: Mathematics, Game Theory and Algebra Compendium ISBN 978-1-60456-659-8
Editor: Jacob H. Mathias, pp. 137-143

Chapter 14

A Dual Approach to Alpha-Regularity

R.J. Higgs*
School of Mathematical Sciences,
University College Dublin,
Belfield, Dublin 4, Ireland
J.F. Humphreys†
Department of Mathematical Sciences,
University of Liverpool, Peach Street,
Liverpool L69 7ZL, United Kingdom

Abstract

Let G be a group and let $M(G)$ be its Schur multiplier. The normal problem associated with the subject of α-regularity is to determine the subset of G that consists of the α-regular elements for a given $[\alpha] \in M(G)$. We present a new idea here: Let A be any subset of G. Then we consider the subset $M_A(G)$ of $M(G)$ that consists of those $[\alpha] \in M(G)$ for which every element of A is α-regular. The advantage of this dual approach, which we exploit in a number of applications, is that $M_A(G)$ is always a subgroup of $M(G)$.

1. Introduction

Fuller details of the concepts mentioned in this paper are given in Karpilovsky [6], also to avoid repetition G will always denote a finite group throughout this paper.

Definition. A 2-*cocycle* α of G over the field of complex numbers $\mathbb{C}$ is a function $\alpha : G \times G \to \mathbb{C}^{\times}$ such that

$$\alpha(x, yz)\alpha(y, z) = \alpha(x, y)\alpha(xy, z)$$

for all $x, y, z \in G$.

*E-mail address: russell.higgs@ucd.ie
†E-mail address: su01@liverpool.ac.uk

The set of all such 2-cocycles of G forms an infinite abelian group $Z^2(G, \mathbb{C}^\times)$ under the operation of multiplication.

Definition. Let $\alpha, \beta \in Z^2(G, \mathbb{C}^\times)$. Then α and β are *cohomologous* if there exists a function $\delta : G \to \mathbb{C}^\times$ such that

$$\beta(x,y) = \frac{\delta(x)\delta(y)}{\delta(xy)}\alpha(x,y)$$

for all $x, y \in G$.

Cohomology defines an equivalence relation on $Z^2(G, \mathbb{C}^\times)$ and the set of all such cohomology classes $[\alpha]$ forms the finite abelian group $H^2(G, \mathbb{C}^\times)$, which in this context is more normally denoted by $M(G)$, the so called *Schur multiplier* of G.

Definition. Let $\alpha \in Z^2(G, \mathbb{C}^\times)$. Then an element $x \in G$ is *α-regular* if $\alpha(x,y) = \alpha(y,x)$ for all $y \in C_G(x)$.

Let Proj(G, α) denote the set of all projective characters of irreducible projective representations of G with 2-cocycle α. Then we state as facts the following:
(1) $x \in G$, is α-regular if and only if x is β-regular for any $\beta \in [\alpha]$;
(2) the set, S_α, of α-regular elements of G is not in general a subgroup of G;
(3) any conjugate of an α-regular element of G is itself α-regular, so that we may refer to the α-regular conjugacy classes of G;
(4) $|\text{Proj}(G, \alpha)|$ equals the number of α-regular conjugacy classes of G.

In this paper we will adopt a different approach to dealing with the concept of α-regularity. In section 2 we outline this new approach by considering any subset A of G and then defining the subgroup $M_A(G)$ of $M(G)$ to be $\{[\alpha] \in M(G) :$ every element of A is α-regular$\}$. The normal problem in dealing with α-regularity is to find S_α given α. From our new viewpoint we are given a subset A of G and then are faced with the equally difficult task of determining $M_A(G)$, however, one advantage of our dual approach is that we only have to find a set of generators for $M_A(G)$. The easiest case to deal with using the traditional approach is when G is abelian, for in this case $S_\alpha \leq G$. Not surprisingly this is also the easiest case for our dual approach, and in the main result in section 3 we determine $M_A(G)$, when G is abelian and $A \leq G$. In section 3 we also illustrate the potential usefulness of our dual approach by finding (as an example) a lower bound for the order of the Schur multiplier of a certain group of order p^9, which has the property that it has a non-trivial cohomology class $[\alpha]$ for which every element of the group is α-regular.

2. Subgroups of the Schur Multiplier

We begin this section by defining new subgroups of the Schur multiplier and then investigating their properties in a series of elementary lemmas.

Definition. Let $A \subseteq G$, and let $M \leq M(G)$. Then we define

$$M_A = \{[\alpha] \in M : \text{ every element of } A \text{ is } \alpha\text{-regular}\}.$$

The normal case in which the above definition will be applied is when $M = M(G)$, and in this case we will accordingly write $M_A(G)$ for M_A.

Lemma 2.1. *With notation as above, $M_A \leq M$.*

Proof. Let $[\alpha_1]$ denote the cohomology class of the trivial 2-cocycle α_1. Then $[\alpha_1] \in M_A$ for any A, since every element of G is α_1-regular. We also observe that $M_\emptyset = M$.

Now suppose that $A \neq \emptyset$. Let $[\alpha], [\beta] \in M_A$. Then $\alpha(x, y) = \alpha(y, x)$ and $\beta(x, y) = \beta(y, x)$ for all $y \in C_G(x)$ and all $x \in A$. Thus

$$\alpha\beta^{-1}(x, y) = \alpha\beta^{-1}(y, x)$$

for all $y \in C_G(x)$ and all $x \in A$. Thus every element of A is $\alpha\beta^{-1}$-regular.

We note in passing that $M_{\{1\}} = M$. Of course the formula in the proof of lemma 2.1 indicates that an element of G could be $\alpha\beta^{-1}$-regular without being either α or β-regular, this is essentially the basis of the next result.

Proposition 2.2. *Let $A, B \subseteq G$, and let $M \leq M(G)$. Then*

$$M_A M_B \leq M_{A \cap B};$$

and

$$M_A \cap M_B = M_{A \cup B}.$$

Proof. Both results hold with equality if either A or $B = \emptyset$, so we assume that this is not the case.

Let $[\alpha] \in M_A$, and let $[\beta] \in M_B$. If possible let $x \in A \cap B$, then x is $\alpha\beta$-regular as in the proof of lemma 2.1.

Let $[\alpha] \in M_A \cap M_B$. Then every element of A is α-regular and every element of B is α-regular. Thus every element of A or B is α-regular, that is $[\alpha] \in M_{A \cup B}$. The reverse inclusion is equally simple.

We thus deduce from the Factor of a Factor theorem:

Corollary 2.3. *With notation as above*

$$M_A / M_{A \cup B} \cong M_A M_B / M_B,$$

and in particular

$$|M_A M_B| = \frac{|M_A||M_B|}{|M_{A \cup B}|},$$

so that $|M_{A \cap B}|$ is divisible by this number.

This very simple result will be the tool that we will use repeatedly in the first part of section 3.

Now suppose A is a non-empty subset of G such that $\langle A \rangle \subseteq S_\alpha$ for all $[\alpha] \in M_A$. Then $M_A = M_{\langle A \rangle}$. This condition does occur in certain circumstances, for example Read [7, Lemma 3.3] proved that

$$U_\alpha(G) := S_\alpha \cap Z(G)$$

is a subgroup of G for any $\alpha \in Z^2(G, \mathbb{C}^\times)$. Let $\alpha, \beta \in Z^2(G, \mathbb{C}^\times)$, then

$$U_\alpha(G) \cap U_\beta(G) \leq U_{\alpha\beta}(G),$$

and if A and B are non-empty subsets of $Z(G)$,

$$M_{\langle A \rangle} \cap M_{\langle B \rangle} = M_{\langle A, B \rangle}.$$

3. Applications

In Higgs [3], an example of a group G of order p^9 for which $M_G(G)$ is non-trivial was constructed, as summarized below. (The reader should note that the notation $M_E(G)$ was used by Higgs in this earlier paper.) Before we can investigate the Schur multiplier of this group we need two preliminary results, the first of which is inherently contained in Higgs [3, Proposition 1.5 and Corollary 1.10].

Proposition 3.1. *Let N be a normal abelian subgroup of G such that $G/C_G(N)$ is cyclic. Then the inflation homomorphism from $M(G/N)$ into $M_N(G)$ is surjective.*

Let $K \leq Z(G)$. Then the image of the transgression homomorphism from the dual group of K into $M(G/K)$ is isomorphic to $G' \cap K$. Hence interpreting Read [7, Theorem 5.1] using this isomorphism we obtain:

Corollary 3.2. $M_K(G) \cong M(G/K)/(G' \cap K)$.

It is of interest to apply this result to a representation group of G.

Proposition 3.3. *Let H be a representation group of G, so that H has a subgroup A with $A \leq Z(H) \cap H'$, $H/A \cong G$ and $A \cong M(G)$. Now let B be a subgroup of H such that $A \leq B \leq Z(H)$. Then $M_B(H)$ is trivial and $M(H/B) \cong H' \cap B$. In particular if $Z(H) \cap H' \leq B \leq Z(H)$, then $M(H/B) \cong Z(H) \cap H'$.*

Proof. By definition $A \cong M(H/A)$ and $A \leq Z(H) \cap H'$, so that $M_A(H)$ is trivial from corollary 3.2. Now $M_B(H)$ is trivial, since $M_B(H) \leq M_A(H)$. Thus from corollary 3.2, $M(H/B) \cong H' \cap B$.

Example. It is well-known that the Schur multiplier of C_p^4, an elementary abelian p-group of order p^4, is C_p^6. One presentation for a representation group H of C_p^4 is

$$H = \langle X_1, X_2, X_3, X_4 : X_i^p = 1 = [X_i, X_j, X_k] \text{ for } 1 \leq i, j, k \leq 4 \rangle,$$

in which the centre of H is generated by the commutators $[X_i, X_j]$ for $1 \leq i < j \leq 4$.

We now define G to be $H/\langle [X_1, X_2][X_3, X_4] \rangle$. Thus G is a group of order p^9 generated by x_1, x_2, x_3, x_4, where $x_i = X_i \langle [X_1, X_2][X_3, X_4] \rangle$. The centre of G has order p^5 generated by the elements $z_{ij} = [x_i, x_j]$ for $1 \leq i < j \leq 4$ but we note that $z_{34} = z_{12}$.

Now $G' = Z(G)$ and $G/G' \cong C_p^4$. Thus by corollary 3.2,

$$M_{Z(G)}(G) \cong M(G/G')/G' \cong C_p.$$

However, by construction $M_G(G)$ has order divisible by p and hence we conclude that

$$M_G(G) = M_{Z(G)}(G).$$

We next show for illustrative purposes how to use the results in section 2 to find a lower bound for the order of $M(G)$.

Let A_i for $i = 1, \ldots, 4$ be the central subgroups of G of order p^4 generated by z_{12} and

$$(1)\ z_{13}, z_{14}, z_{23},\ (2)\ z_{13}, z_{14}, z_{24},\ (3)\ z_{13}, z_{23}, z_{24},\ (4)\ z_{14}, z_{23}, z_{24},$$

respectively. Then $G/A_i \cong C_p^2 \times P$ for $i = 1, \ldots, 4$, where P is a non-abelian group of order p^3 generated by two elements of order p. Thus P has exponent p for p odd, and is the dihedral group of order 8 for $p = 2$. Thus for $i = 1, \ldots, 4$

$$M(G/A_i) \cong M(C_p^2) \times M(P) \times (C_p^2 \otimes P) \cong \begin{cases} C_p^7, & \text{if } p \text{ is odd;} \\ C_p^6, & \text{if } p = 2. \end{cases}$$

We assume henceforward that p is odd, so by corollary 3.2,

$$M_{A_i}(G) \cong M(G/A_i)/A_i \cong C_p^3.$$

Thus

$$|M_{A_i}(G)M_{A_j}(G)| = p^5,$$

for $1 \leq i < j \leq 4$, since $A_iA_j = Z(G)$ for $1 \leq i < j \leq 4$.

We now define $B_{ij} = A_i \cap A_j$ for $1 \leq i < j \leq 4$, so that p^5 divides $|M_{B_{ij}}(G)|$.

Now for $((i,j),(s,t)) = ((1,2),(3,4)), ((1,3),(2,4)), ((1,4),(2,3))$, we have that $|M_{B_{ij}}(G)M_{B_{st}}(G)|$ is divisible by p^9, since $B_{ij}B_{st} = Z(G)$ in these three cases.

Finally $B_{ij} \cap B_{st} = \langle z_{12}\rangle$ in any of the three cases above, and we have shown that p^9 divides $|M_{\langle z_{12}\rangle}(G)|$.

In the case $p = 2$ we may just adjust the powers of p that occur in the above argument to give that 2^5 divides $|M_{\langle z_{12}\rangle}(G)|$.

To complete the example we briefly consider the central subgroup A of G generated by z_{13}, z_{14}, z_{23} and z_{24}. Now $G/A \cong Q$, where Q is an extra-special p-group of order p^5, so that $M(Q) \cong C_p^5$. Thus $M_A(G) \cong C_p$ from corollary 3.2 and we conclude that $M_A(G) = M_{Z(G)}(G)$. Unfortunately, further consideration of A does not lead to an improvement in the lower bounds obtained above.

We next turn our attention to abelian groups, but first we have a more general result. The greatest common divisor of the degrees of the elements of $\mathrm{Proj}(G,\alpha)$ is described in Higgs [2].

Proposition 3.4. *Let $[\alpha] \in M(G)$, and let c denote the greatest common divisor of the degrees of the elements of* $\mathrm{Proj}(G,\alpha)$. *Let k denote the number of conjugacy classes of G. Then*

(1) *$o([\alpha])$ in $M(G)$ divides c;*

(2) $k \geq |G/G'| + (|G'| - 1)(|Z(G)|/|G'|)$*;*

(3) *if $[\alpha] \in M_G(G)$, then $c^2k \leq |G|$ and $c^2 \leq |G : Z(G)|$.*

Proof. (1) is well-known. Now let $[\alpha] \in M_G(G)$, and let $\xi \in \mathrm{Proj}(G,\alpha)$. Then, from Clifford's theorem for projective characters, $\xi_{Z(G)} = \xi(1)\lambda$ for some $\lambda \in \mathrm{Proj}(Z(G), \alpha_{Z(G)})$, consequently $\xi(1)$ is the degree of some irreducible projective representation of $G/Z(G)$. Thus $\xi(1)^2 \leq |G : Z(G)|$. This in particular applies when $[\alpha]$ is trivial, so that G has at least $(|G| - |G/G'|)/|G : Z(G)|$ non-linear irreducible characters.

We note that (3) confirms a result of Higgs [4, Corollary 1.4] that $M_G(G)$ is trivial when G is abelian. We may also rearrange the inequalities in (2) and (3) to give that

$$c^2 \leq \frac{|G : Z(G)||G'|}{|G : Z(G)| + (|G'| - 1)}$$

for $[\alpha] \in M_G(G)$. Thus $c^2 < \min\{|G'|, |G : Z(G)|\}$ if G is non-abelian and $[\alpha] \in M_G(G)$. Notice that any p-group with $|G'|$ or $|G : Z(G)|$ dividing p^2 has $M_G(G)$ trivial, and in particular extra-special p-groups possess this property as was noted in Higgs [4, Proposition 2.4].

Let $R(m,n)$ denote the group under matrix addition of all $n \times n$ alternating matrices over $\mathbb{Z}_m$ for $m > 1$, so that $|R(m,n)| = m^{n(n-1)/2}$.

Corollary 3.5. *Let $G \cong C_{s_1} \times \cdots \times C_{s_n}$ for $n \geq 1$, where $s_i = p^{a_i}$ for p a prime number and $a_i \geq a_{i+1} > 0$. Now let $A \leq G$, so that $A \cong C_{k_1} \times \cdots \times C_{k_n}$, where $k_i = p^{b_i}$, $a_i \geq b_i \geq 0$ and $b_i \geq b_{i+1}$. Then, setting $m = s_1$,*

$$M_A(G) \cong \{C \in R(m,n) : c_{ij} \in p^{a_1 - r_{ij}}\mathbb{Z}_m \text{ for } i \neq j\},$$

where c_{ij} denotes the (i,j)th entry of C and $r_{ij} = \min\{a_i - b_i, a_j - b_j\}$. In particular if π is a permutation of $\{1,\ldots,n\}$ such that $a_{\pi(i)} - b_{\pi(i)} \geq a_{\pi(i+1)} - b_{\pi(i+1)}$ for all i, then

$$|M_A(G)| = \prod_{i=1}^{n} (s_{\pi(i)}/k_{\pi(i)})^{n-\pi(i)}.$$

Proof. Suppose $k_r \neq 0$, but $k_{r+1} = 0$. Then we let $y_1,\ldots,y_r$ be generators of A such that $\langle y_i \rangle \cong C_{k_i}$ (and set $y_{r+1},\ldots,y_n$ equal to the identity element). Now let $x_1,\ldots,x_n$ be generators of G such that $x_i^{t_i} = y_i$, where $t_i = p^{a_i - b_i}$.

Let ω be a primitive mth root of unity in $\mathbb{C}$, and let $\alpha \in Z^2(G,\mathbb{C}^\times)$. Then $\alpha(x_i,x_j)/\alpha(x_j,x_i) = \omega^{c_{ij}}$ for all $i,j \in \{1,\ldots,n\}$. We define the matrix $C(\alpha)$ that represents $[\alpha]$ to be the $n \times n$ alternating matrix with entries from $\mathbb{Z}_m$ whose entries are c_{ij}. It is explained in Frucht [1] that $[\alpha] \mapsto C(\alpha)$ defines a monomorphism from $M(G)$ into $R(m,n)$. The image of this map is the set of matrices $C(\alpha)$ in $R(m,n)$ with $c_{ij} \in p^{a_1 - a_j}\mathbb{Z}_m$ for $i \neq j$. Moreover an element $x = x_1^{d_1} \cdots x_n^{d_n}$ is α-regular if and only if $C(\alpha)\vec{d} = \vec{0}$, where the transpose of $\vec{d}$ equals $[d_1,\ldots,d_n]$. Applying this result here, we have that y_i is α-regular if and only if $t_i C(\alpha)\vec{e}_i = \vec{0}$, that is if and only if $t_i C_i = \vec{0}$, where C_i denotes the ith column of $C(\alpha)$. Thus y_i is α-regular if and only if all the entries in the ith column (and row) of $C(\alpha)$ are in $(m/t_i)\mathbb{Z}_m$. We conclude that

$$M_{\langle y_i \rangle}(G) = \{C \in R(m,n) : c_{ij} \in p^{a_1 - s_{ij}}\mathbb{Z}_m \text{ for } i \neq j\},$$

where $s_{ij} = \min\{a_j, a_i - b_i\}$. Finally from the remarks at the end of section 2 and proposition 3.4,

$$M_A(G) = M_{\langle y_1,\ldots,y_r \rangle}(G) = \bigcap_{i=1}^{r} M_{\langle y_i \rangle}(G).$$

In particular since $c_{ji} = -c_{ij}$ this yields that if y_i and y_j are both α-regular for $i \neq j$, then $c_{ij} \in p^{a_1 - r_{ij}}\mathbb{Z}_m$, where $r_{ij} = \min\{a_i, a_j, a_i - b_i, a_j - b_j\}$.

We notice that setting $b_1 = a_1 - a_2, b_2 = b_3 = \cdots = b_n = 0$ in corollary 3.5 (for $n \geq 2$) gives that $M_{\langle y_1 \rangle}(G) = M(G)$. In fact in this setting Higgs and Humphreys [5, Corollary 3.8] proved that $\langle y_1 \rangle$ is precisely the subgroup of G that consists of those elements that are α-regular for all $[\alpha] \in M(G)$.

The most important case of corollary 3.5 occurs when G is homocyclic.

Corollary 3.6. *Let G be the direct product of $n \geq 1$ copies of C_m, where $m = p^a$ for p a prime number and $a \geq 1$. Let A be a non-trivial subgroup of G, so that $A \cong C_{k_1} \times \cdots \times C_{k_r}$ for some $r \leq n$, and $k_i = p^{a_i}$ for $a \geq a_1 \geq a_2 \geq \ldots \geq a_r > 0$. Then*

(1) *$M_A(G)$ is isomorphic to the subgroup of $R(m,n)$, which has its ith column (and row) in $k_i\mathbb{Z}_m$ for i taken in the order $1,\ldots,r$. In particular*

$$|M_A(G)| = |R(m,n-r)| \prod_{i=1}^{r} (m/k_i)^{n-i}.$$

(2) $M_A(G) \cong R(m,n-r)$ *if* $a=1$.

Acknowledgments

The authors would like to thank the referee for his role in improving the presentation of this paper.

References

[1] R. Frucht, Über die Darstellung endlicher Abelscher Gruppen durch Kollineationen, *J. Reine angew. Math.* **166** (1931), 16–29.

[2] R.J. Higgs, On the degrees of projective representations, *Glasgow Math. J.* **30** (1988), 133–135.

[3] ______, Projective characters of degree one and the inflation-restriction sequence, *J. Austral. Math. Soc. Ser. A* **46** (1989), 272–280.

[4] ______, Subgroups of the Schur multiplier, *J. Austral. Math. Soc. Ser. A* **48** (1990), 497–505.

[5] R.J. Higgs and J.F. Humphreys, Projective characters of direct products of certain p-groups, *Comm. Algebra* **27** (1999), 4493–4514.

[6] G. Karpilovsky, Projective representations of finite groups, Monographs and Textbooks in Pure and Applied Mathematics, vol. 94, *Marcel Dekker Inc.*, New York, 1985.

[7] E.W. Read, On the centre of a representation group, *J. London Math. Soc.* (2) **16** (1977), 43–50.

In: Mathematics, Game Theory and Algebra Compendium ISBN 978-1-60456-659-8
Editor: Jacob H. Mathias, pp. 145-155

Chapter 15

LIMIT CYCLE IN FRACTIONAL DIFFERENTIAL SYSTEMS

Rabha W. Ibrahim*
Department of Mathematics,
University of Pune,
Ganeshkhind, Pune−411007, India

Abstract

In this paper we mention the link between Melnikove integral function and Riemann-Liouville integral function for two-dimensional ordinary differential systems of degree $n \geq 1$, and show that the solutions of ordinary differential systems reduced to the solutions of fractional differential systems.From this solution, we show that the limit cycle for the planner system is disappear after reduced to fractional differential system.

Key Words: Limit cycle,Melnikove integral function,Riemann-Liouville integral function,Abelian integral function,fractional differential system .

1. Introduction

Melnikove integrals function or Poincare- Melnikove integrals function; [3], or Poincare-Pontryagin integrals function; [6], is an analytic tool to determined the solution of dynamic systems. Especially determining the number and nature of critical points, the separatrix structure and the number and location of closed trajectories (limit cycles). The question is still the main open problem in the qualitative theory of real differential systems.In the study of two-dimensional systems,

$$x^{\cdot} = P(x, y),\ y^{\cdot} = Q(x, y), \tag{1}$$

where P and Q are real polynomials in the variables x and y of degree $n \geq 1$. The determination of the number and position of limit cycles is known as the second part of the Hilbert

*E-mail address: rabhaibrahim@yahoo.com

$16-th$ problem (see [12]). The importance of the Melnikove integrals function comes from the importance of this problem (one of the two methods for solving this problem) ;[2] ,[9]. These integrals introduced by Melnikove in the early $1960^{,}s$; [7], [9]. Some recent results of Mardesic ;[8], for polynomially perturbed systems

$$x^{\cdot} = P(x, y, \mu), \; y^{\cdot} = Q(x, y, \mu) \tag{2}$$

where μ is a parameter in R, show that there is a direct correspondence between the maximum number of limit cycles of such a system and the maximum number of zeros of a certain Melnikove integral functions as well as Abelian integral function, later Blows and Perko in [2], show that Abelian integral function and Melnikove integral functions for such a system are equivalence by using divergence-type integrals (divergence theorem).

System (2) can reduce to the perturbed Hamiltonian system(see [3]) of the form

$$x^{\cdot} = H_y + p(x, y, \mu), \; y^{\cdot} = -H_x + q(x, y, \mu) \tag{3}$$

where $H(x, y) = \frac{x^2+y^2}{2}$, Hamiltonian polynomial then we have the system

$$x^{\cdot} = y + p(x, y, \mu), \; y^{\cdot} = -x + q(x, y, \mu) \tag{4}$$

where p and q of degree $n > 1$. Note that system (4) has many applications;[9]. In this paper we show that Melnikove integral function is equivalent to another kind of integral function call it Riemann-Liouville integral function, where Abelian integral function is a special case of the Riemann-Liouville integral function.

2. Preliminaries

Definition 2.1[2]. The Melnikove function for a system of the form

$$x^{\cdot} = f(x) + g(x, \mu),$$

with $x \in R^2$ and $\mu \in R^n$ along the close orbit (cycle) of the unperturbed system

$$x^{\cdot} = f(x)$$

$$L_\alpha : x = \ell_\alpha(t), \; 0 \leq t \leq T_\alpha$$

is define by the integral function :

$$M(\alpha, \mu) = \int_0^{T_\alpha} \exp(-\int_0^t \nabla.f(\ell_\alpha(s))ds).f \wedge g(\ell_\alpha(t), 0, \mu)dt. \tag{5}$$

Where the wedge product of two vectors $x, y \in R^2$ is defined by $x \wedge y = x_1y_2 - y_1x_2$.

Also another definition for Melnikov function for system (3) in sense of [6],

$$M(h) = \int_{H=h} q(x, y)dx - p(x, y)dy. \tag{6}$$

Where h is a Hamiltonian value in a level set $H = h$. The feature of equation (6) is called Abelian integral function. Note that the equations (5) and (6) are equivalent (see[2]).

Definition 2.2. The Abelian integral function is given in many senses for system (3), first in the sense of [2],

$$I(h,\mu) = -\int_{L(h)} F(x,y,\mu) \wedge dX,$$

where $h \in J$, an open set in R, $L(h)$ is a cycle (close orbit) of the unperturbed Hamiltonian system $(H_y, -H_x)^T$ and $F(x,y,\mu) = (p(x,y,\mu), q(x,y,\mu))^T$ and $dX = (dx, dy)^T$. Second in the sense of [3], for system of the form

$$x^{\cdot} = V(x,y)H_y + p(x,y,\mu), y^{\cdot} = -V(x,y)H_x + q(x,y,\mu),$$

$$I(h,\mu) = -\int_{H=h} \frac{F(x,y,\mu) \wedge dX}{V(x,y)} \tag{7}$$

define on a region

$$D_{[h_o,h_1]} := H(x,y) = h : h_0 \leq h \leq h_1 \subset R^2$$

where $V(x,y)$ does not vanish and the Hamiltonian system $(H_y, -H_x)$ has all their periodic solutions.

In the sense of [10], Abelion integral function is defined for one variable (t) :

$$I(t) = \frac{1}{\Gamma(\alpha)} \int_0^t \frac{f(\tau)d\tau}{(t-\tau)^{1-\alpha}} \tag{8}$$

where $t > 0$ and $0 < \alpha < 1$. Comparing equation (8) with (7), we find that $V(t) = \frac{1}{\Gamma(\alpha)(t-\tau)^{1-\alpha}}$. We can make a generalization for equation (8) for two variables:

$$\Pi_{i=1}^2 I_i(t_1,t_2) = \frac{1}{\Gamma(\alpha)} \Pi_{i=1}^2 \int_0^{t_2} \int_0^{t_1} \frac{f_i(\tau_1,\tau_2)}{(t_i-\tau_i)^{1-\alpha}} d\tau_1 d\tau_2.$$

Abelian integral function helps us to find a solution of many applied problems lead to integral equations. However, their transformation to the form of Abelian integral function may often be convenient for obtaining the solutions for system of differential equations (see [10],[11]).

From equation (8) we obtain two- dimensional system of solution for systems of the form (1,2), containing Abelian integral functions I_1 and I_2. In the next section ,we prove that Abelian integral function and Melnikove integral function are equivalent for all $V(X) = V(x,y)$ with application systems reduced their solutions to Abelian integral function. Note that the proof of equivalently in [2], is hold when $V(X) \equiv 1$.The next definition is called the Riemann-Liouville integral function such that Abelian integral function is a special case of it.

Definition 2.3. The Riemann-Liouville integral function of order $\alpha > 0$ is defined for one variable (t) as in [1], [4] and [5] :

$$\ell^{\alpha} f(t) = \frac{1}{\Gamma(\alpha)} \int_0^t (t-\tau)^{\alpha-1} f(\tau) d\tau. \tag{9}$$

Comparing equation (9) with (7), we find $V(t) = \frac{(t-\tau)^{\alpha-1}}{\Gamma(\alpha)}$ where $\alpha > 0$.As in definition(2.2) , equation (9) has a generalization for two variables t_1, t_2 :

$$\Pi_{i=1}^{2} \ell_i^{\alpha} = L^{\alpha} = \frac{1}{\Gamma(\alpha)} \Pi_{i=1}^{2} \int_0^{t_2} \int_0^{t_1} (t_i - \tau_i)^{\alpha-1} f_i(\tau_1, \tau_2) d\tau_1 d\tau_2$$

where $\alpha > 0$.

3. Melnikove Function Reduce to Abelian Function or Riemann-Liouville Function

In this section we prove equivalently between Melnikove function ; equations (5,6) and Abelian function ; equation (7) for all $V(x, y)$. For $h \in J$ an open set in R, and $L(h)$ is a cycle of the unperturbed Hamiltonian system

$$x^{\cdot} = G(x, y) \tag{10}$$

where $G(x, y) = (H_y, -H_x)^T$ and $dX = (dx, dy)^T$. L_{α} is a periodic orbit of the system (10), where $\alpha(h)$ is a monotone function, then we have from equation (7)

$$I(h, \mu) = -\int_{H=h} \frac{F(X, \mu) \wedge dX}{V(x, y)}$$

$$= -\int_{H=h} \frac{p(X, \mu)dy - q(X, \mu)dx}{V(X)}$$

$$= -\int_{H=h} \frac{p(X, \mu)dy}{V(X)} - \frac{q(X, \mu)dx}{V(X)}$$

$$= \int_{H=h} Q(x, y, \mu)dx - P(x, y, \mu)dy$$

where $P(X, \mu) = \frac{p(X,\mu)}{V(X)}$ and $Q(X, \mu) = \frac{q(X,\mu)}{V(X)}$. Now let $\overline{F}(X, \mu) = (P, Q)^T$ then we obtain

$$I(h, \mu) = -\int_0^{T_{\alpha}} \overline{F}(\ell_{\alpha}(t), \mu) \wedge \ell^{\cdot}_{\alpha}(t) dt$$

$$= \int_0^{T_{\alpha}} G(\ell_{\alpha}(t)) \wedge \overline{F}(\ell_{\alpha}(t), \mu) dt$$

$$= M(\alpha(h), \mu).$$

From above we obtain that the Melnikove integral function is equal to Abelian integral function for all $V(x, y)$ in (7) and that led consequently to Riemann-Liouville integral function for all $\alpha > 0$ in (9) and this recently led to some important new results on $Hilbert's16-thproblem$ for perturbed systems of the form (2) . Later we illustrate number of examples , show that the Melnikov integral function reduced to Abelian integral function or Riemann-Liouville integral function, that is the solutions of two dimensional planner systems reduce to the solutions of fractional differential systems by illustrating number of substitutions.

Note that all examples are for two-dimensional systems (perturbed or un-perturbed).

Example 3.1. Consider the un-perturbed system

$$x^{\cdot} = H_y + xf(x^2 + y^2)$$

$$y^{\cdot} = -H_x + yf(x^2 + y^2)$$

The Melnikov integral function for such a system is

$$M(h) = \int_{H=h} yf(x^2 + y^2)dx - xf(x^2 + y^2)dy$$

then we have the integral equation :

$$\int_{-\infty}^{\infty} yf(x^2 + y^2)dx = g(y)$$

or

$$\int_{0}^{\infty} f(x^2 + y^2)dx = \frac{g(y)}{2y}$$

substitution of variables $\chi = x^2, \eta = y^2$ gives

$$\int_{0}^{\infty} f(\chi, \eta)\frac{d\chi}{\sqrt{\chi}} = \frac{g(\sqrt{\eta})}{\sqrt{\eta}}$$

then the further substitution $\tau = \frac{1}{\chi+\eta}$ leads to the integral equation

$$\int_{0}^{\frac{1}{\eta}} (\frac{1}{\eta} - \tau)^{-1/2}\tau^{-3/2}f(\tau)d\tau = g(\sqrt{\eta})$$

and denoting by $t_1 = \frac{1}{\eta}$, and $F(\tau) = \tau^{-3/2}f(\tau)$ then we have the integral equation :

$$\int_0^{t_1} (t_1 - \tau)^{-1/2} F(\tau) d\tau = g(\frac{1}{\sqrt{t_1}}). \tag{11}$$

Comparing with (8) we have $\alpha = \frac{1}{2}$ and with the help of equation 10.4 in [10], (the solution of Abelian integral function); the solution of (11) becomes:

$$F(\tau) = \frac{1}{\Gamma(1/2)} D_{t_1}^{1/2} g(\frac{1}{\sqrt{t_1}})$$

where $D^{1/2}$ is an operator of differential fractional(see [10]). Similarly, from the integral of second term of the Melnikov integral function , we have the solution

$$F(\tau) = \frac{1}{\Gamma(1/2)} D_{t_2}^{1/2} g(\frac{1}{\sqrt{t_2}})$$

where $t_2 = \frac{1}{\chi}$. From above we find that the Melnikov function of unperturbed two dimensional system reduced to Abelian function.

Example 3.2 Consider the system

$$x^{\cdot} = H_y + f(x^n + y^n)$$

$$y^{\cdot} = -H_x + f(x^n + y^n)$$

The Melnikov integral function for such a system is

$$M(h) = \int_{H=h} f(x^n + y^n) dx - f(x^n + y^n) dy$$

to reduce this integral function to system of Abelian integral function, we need some substitutions as in the previous example. For the first integral

$$\int_0^{\infty} f(x^n + y^n) dx = g(y)/2$$

change the variables $\chi = x^n, \eta = y^n$ leads to the function

$$\int_0^{\infty} f(\chi + \eta) \chi^{\frac{1}{n}-1} d\chi / n = \frac{g(\eta^{1/n}}{2})$$

then the substitution $\tau = \frac{1}{\chi+\eta}$ gives the integral function :

$$\int_0^{\frac{1}{\eta}} -f(\frac{1}{\tau})(\frac{1}{\eta} - \tau)^{\frac{1}{n}-1} . (\frac{1}{\tau})^{\frac{1}{n}-1} . (\frac{d\tau}{\tau^2}) = \frac{g(\eta^{1/n})}{2\eta^{\frac{1}{n}-1}}$$

denoting $F(\tau) = -f(\frac{1}{\tau})((\frac{1}{\tau})^{\frac{1}{n}-1} . (\frac{1}{\tau^2})$ and $G(\eta) = \frac{g(\eta^{1/n})}{2\eta^{\frac{1}{n}-1}}$ with $t_1 = \frac{1}{\eta}$ then we obtain the Abelian integral function of degree $\alpha = \frac{1}{n}$:

$$\int_0^{t_1}(t-\tau)^{\frac{1}{n}-1}F(\tau)d\tau = G(\frac{1}{t_1^{1/n}})$$

with the solution :

$$F(\tau) = \frac{1}{\Gamma(1/n)}.D_{t_1}^{1/n}G(\frac{1}{t_1^{1/n}})$$

where $n \geq 1$. Similarly for the integral equation

$$\int_{H=h} f(x^n + y^n)dy = g(x)$$

the solution is

$$F(\tau) = \frac{1}{\Gamma(1/n)}D_{t_2}^{1/n}G(\frac{1}{t_2^{1/n}})$$

where $t_2 = 1/\chi$.

Example 3.3 suppose the system of the form :

$$x^{\cdot} = H_y + xf((x^2 + y^2)^k)$$

$$y^{\cdot} = -H_x + yf((x^2 + y^2)^k)$$

the Melnikov integral function for this system is :

$$M(h) = \int_{H=h} yf((x^2 + y^2)^k)dx - xf((x^2 + y^2)^k)dy$$

As in previous examples, we illustrate the above substitutions : $\chi = x^2, \eta = y^2$ with $\tau = \frac{1}{\chi+\eta}$ and at last $t_1 = \frac{1}{\eta}$, we have the integral function

$$\int_0^{t_1}(t_1 - \tau^{1/k})^{-1/2}F(\tau)d\tau = G(\frac{1}{\sqrt{t_1}})$$

denoting $\overline{\tau} = \tau^{1/k}$, lead to Abelian integral function :

$$\int_0^{t_1}(t_1 - \overline{\tau})^{-1/2}F(\overline{\tau})d\overline{\tau} = G(\frac{1}{\sqrt{t_1}})$$

with degree $\alpha = 1/2$, then the solution for this system is:

$$F(\overline{\tau}) = \frac{1}{\Gamma(1/2)}.D_{t_1}^{1/2}G(\frac{1}{\sqrt{t_1}})$$

and

$$F(\overline{\tau}) = \frac{1}{\Gamma(1/2)}.D_{t_2}^{1/2} G(\frac{1}{\sqrt{t_2}})$$

where $t_1 = \frac{1}{\eta}$ and $t_2 = \frac{1}{\chi}$.

The next example is a perturbed system where $\mu \in R^n$, where $n \geq 1$. All these examples have many applications in the qualitative theory of differential equations especially in the theory of center problem (find all sufficient and necessary conditions to make the system has a center). Also one of the most important application of these systems in the theory of limit cycles(find the numbers, positions and their multiplicities) which corresponding to the zeros of Melnikove integral function, Riemann-Liouville integral function, and Abelian integral function (see [2] and [9]).

Example 3.4. Consider the perturbed system

$$x^{\cdot} = H_y + xf((x^2 + y^2 \pm \mu)^k)$$

$$y^{\cdot} = -H_x + yf((x^2 + y^2 \pm \mu)^k)$$

where μ is a parameter in R^n. The Melnikov integral function is reduced to Abelian integral function of degree $\alpha = 1/2$.

$$M(h) = \int_{H=h} yf((x^2 + y^2 \pm \mu)^k)dx - xf((x^2 + y^2 \pm \mu)^k)dy$$

the same substitution in above examples holds then we have for the integral function

$$\int_0^{\infty} f((x^2 + y^2 \pm \mu)^k)dx = \frac{g(y)}{2y}$$

the Abelian integral function

$$\int_0^{t_1} (t_1 - \tau)^{-1/2} F(\tau, \mu) d\tau = g(\frac{1}{\sqrt{t_1}})$$

where $F(\tau) = -f((\frac{1}{\tau} \pm \mu)^k).\tau^{-3/2}$ and $t_1 = \frac{1}{\eta}$ with $\alpha = 1/2$. Then the solution of this integral equation is:

$$F(\tau, \mu) = \frac{1}{\Gamma(1/2)}.D_{t_1}^{1/2} g(\frac{1}{\sqrt{t_1}})$$

Similarly for the integral

$$\int_{H=h} xf((x^2 + y^2 \pm \mu)^k)dy$$

the solution is

$$F(\tau,\mu) = \frac{1}{\Gamma(1/2)}.D_{t_2}^{1/2} g(\frac{1}{\sqrt{t_2}})$$

with $t_2 = \frac{1}{\chi}$.

4. Limit Cycles in Fractional Differential Systems

In the qualitative theory of differential equations, research on limit cycles(isolated close orbit) is an interesting and difficult part. Limit cycles appear in the plane autonomous differential systems :(1),(2) and (3). It is worthwhile to mention that, owing to research developments from other fields such biology and chemistry, research on limit cycles, especially in the theory of polynomial systems (of integer order $n \geq 1$,) has become more and more important. Now, the question is : what happen to the limit cycle in fractional differential system?or if systems (1),(2)and (3) have limit cycles, are these limit cycles still remain when these systems reduce to fractional differential systems? In this section, we try to answer these questions by illustrate an example show that the limit cycles of systems (1),(2)and (3) disappear when these systems reduce to fractional differential systems.

Example 4.1. Consider the planner analytic system:

$$x^{\cdot} = y + x(x^2 + y^2 - \mu) \tag{12}$$

$$y^{\cdot} = -x + y(x^2 + y^2 - \mu)$$

where μ is a parameter in R. There are two methods for determining the existence and the number of limit cycles.The first one is called Hopf bifurcation [9], the other one is called Melnikove integral function [2].Frome the first method, the system (12) has unique limit cycle when $\mu > 0$ and this limit cycle bifurcate from a center (type of the critical point at the origin for the linear part of system (12)). Also by using Melnikove integral function for the system (12) :

$$M(\alpha,\mu) = 2\pi\alpha^2(\alpha^2 - \mu),$$

the system(12) has unique limit cycle of multiplicity one at $\mu > 0$ [2]. The question is: what happen to this limit cycle when system(12) reduce to system of fractional order α_i where $0 < \alpha_i \leq 1$?

Suppose system (12) reduce to fractional differential system of the form :

$$D^{(\alpha_i)}x - \lambda x = F(x) \tag{13}$$

where $0 < \alpha_i \leq 1$,and $x \in R^2, \lambda = (\lambda_1, \lambda_2)^T, andF(x) = (f_1, f_2)^T$. Then system (12) is a special case of system in example (3.4) where

$$f((x^2 + y^2 \pm \mu)^k) = x^2 + y^2 - \mu$$

then the system (12), can reduce to system of fractional order $\alpha_i = 1/2$, where $i = 1, 2$, so we obtain the system :

$$D_{t_1}^{1/2} g(t_1) = F(\tau, \mu)\Gamma(1/2)$$

$$D_{t_2}^{1/2} g(t_2) = F(\tau, \mu)\Gamma(1/2)$$

this system is equivalent to the system :

$$D_{t_1}^{1/2} t_1 = F_1(t_1, t_2, \mu) \tag{14}$$

$$D_{t_2}^{1/2} t_2 = F_2(t_1, t_2, \mu)$$

where F_1, F_2 are not linear function in t_1, t_2 (see example 3.4).Compering system (14) with system (13), we find that the linear part was disappear that is $\lambda \equiv 0$ this lead us to reality that the limit cycle is disappear. From the above example we have this result : The existence of limit cycle in planner analytic system of the form (1),(2) and (3), need not to be exists when these systems reduced to systems of fractional order.

References

[1] I.Ali, V.Kiryakova, and S.Kalla, Solutions of fractional multi-order integral and differential equations using a Poisson-type transform, *J.M.A.A.* **269**, 172-199 (2002).

[2] T.R.Blows and L.M.Perko, Bifurcation of limit cycles from centers and separatrix cycles of planner analytic systems, *S.I.A.M.* **36**, 341-376 (1994).

[3] B.Coll, A.Gasull, R.Prohens, Bifurcation of limit cycles from two families of centers. *C.R.M.preprint* **491** (2002).

[4] V.Daftardar - Gejji and A.Babakhani, *Analysis of system of fractional differential equations*. to appear.

[5] K.Diethelm and N.Ford, Analysis of fractional differential equations, *J.M.A.A.* **265**, 229-248 (2002).

[6] L.Gavrilov and I.Iliev, Second-order analysis in polynomially perturbed reversible quadratic Hamiltonian systems, *Ergod.th. and dynam.syc.* **20**, 1671-1686 (2000).

[7] J.Guckenheimer and P.Holmes, Nonlinear oscillations, dynamical systems, and bifurcation of vector fields, *Appl.Math.Sci.* **42**, Springer - Verlag, Berlin and New York, 1983.

[8] P.Mardesic, An explicit bound for the multiplicity of zeros of generic Abelian integrals, *Nonlinearity* , **4**, 845-852(1991).

[9] L.M.Perko, Differential equations and dynamical systems, *Text in Appl.Math.7 Springer - Verlag*, Berlin and New York, 1991.

[10] I.Podlubny, Fractional Differential Equations, *Acad.Press*, London, 1999.

[11] H.M.Srivastava and R.C.Buschman, Theory and Applications of convolution Integral Equations, *Kluwer Acad.Publishers*, Dordrecht - Bosten - London, 1992.

[12] H.Zoladek, The XVI-th Hilbert problem about limit cycles, *Panoramas of Math.* **34**, 167-174 (1995).

In: Mathematics, Game Theory and Algebra Compendium ISBN 978-1-60456-659-8
Editor: Jacob H. Mathias, pp. 157-163

Chapter 16

THE HALF-FACTORIAL PROPERTY IN THE RING A+XI[X] WHERE A IS A UFR

Murat Alan [*] ***and A. Göksel Ağargün*** [†]
Yildiz Technical University
Faculty of Arts and Sciences, Mathematics Department
Davutpasa Campus, 34210, Esenler, Istanbul, Turkey

Abstract

In this paper we give necessary and sufficient conditions for A+XI[X] to be a U-HFR where A is a UFR .

1. Introduction

The purpose of this paper is to study the half-factorial property in the context of commutative rings with zero divisors by using U-decompositions and to give a generalization of the result due to Gonzalez, Pellerin and Robert in [9] which states the integral domain $A+XI[X]$ where A is a UFD and I is a nonzero proper ideal of A is an HFD if and only if I is a prime ideal of A.

Let R be a commutative ring with identity and group of units $U(R)$. Two elements a and b are *associate*, denoted $a \sim b$, if $a \mid b$ and $b \mid a$, that is, $(a) = (b)$. A nonunit $a \in R$ is an *irreducible* (or an *atom*) if $a = bc \Rightarrow a \sim b$ or $a \sim c$. Here we allow a to be 0 and 0 is irreducible if and only if R is an integral domain. R is *atomic* if each nonzero, nonunit element of R is a finite product of irreducible elements. R is said to be *présimplifiable* if $x = xy \Rightarrow x = 0$ or $y \in U(R)$. An integral domain or a quasilocal ring is a présimplifiable ring. A principal ideal ring (PIR) is called a *special principal ideal ring* ($SPIR$) if it has only one prime ideal $P \neq R$ and P is nilpotent, that is, $P^n = (0)$ for some integer $n > 0$. Fletcher [6] introduced a new type of irreducible factorization in a commutative ring, called a U-decomposition, which eliminates the bad behavior of a factorization because of non trivial idempotents. A *U-decomposition* of $r \in R$ is a

* E-mail address: alan@yildiz.edu.tr
† E-mail address: agar@yildiz.edu.tr

factorization $r = (p_1' \dots p_{k'}')(p_1 \dots p_k)$ such that (i) the p_i''s and p_i's are irreducible (ii) $p_i'(p_1 \dots p_k) = (p_1 \dots p_k)$ for $i = 1, \dots, k'$ and (iii) $p_i(p_1 \dots \widehat{p_i} \dots p_k) \neq (p_1 \dots \widehat{p_i} \dots p_k)$ for $i = 1, \dots, k$, where $\widehat{p_i}$ means the omission of this element from the product. In a U-decomposition $r = (p_1' \dots p_{k'}')(p_1 \dots p_k)$, the product $p_1 \dots p_k$ is called the *relevant part* and the other is the *irrelevant part*. If a nonunit $r \in R$ has a U-decomposition then clearly r has an irreducible factorization and the converse is also true [6, Proposition 2]. Two U-decompositions of $r \in R$, $r = (p_1' \dots p_{k'}')(p_1 \dots p_k) = (q_1' \dots q_{k'}')(q_1 \dots q_l)$ are *associates* if (i) $k = l$ and (ii) after a suitable change in the order of the factors in the relevant parts, we have p_i and q_i are associates for $i = 1, \dots, k$. A ring R is called a *unique factorization ring* (UFR) if each nonunit has a U-decomposition and any two U-decompositions of a nonunit element of R are associates. Fletcher shows that R is a UFR if and only if R is a finite direct product of UFD's and $SPIR$'s.

Let D be an integral domain. Although the classical situation of factorization in D is unique factorization, there are some other factorization properties which are weaker than unique factorization. For example D is a *half-factorial domain* (HFD) if D is atomic and for each nonzero, nonunit $x \in D$, if $x = x_1 \dots x_n = y_1 \dots y_m$ with each x_i, y_j irreducible in D, then $n = m$. This concept was introduced by Zaks in [13]. For a detailed study of these factorization properties see [2, 3]. In [7], Axtell and et al. study *U-half-factorial rings* (*U-HFR*) without the assumption of atomicity. In Section 2, we briefly review this factorization property.

In [11] the authors investigated the domains of the form $A+XI[X]$ where A is a UFD and I is a nonzero proper ideal of A. One of the most interesting results in their paper is the result that $A + XI[X]$ is an HFD if and only if I is a prime ideal of A. This result raises the question of whether this is still true for commutative rings. In Section 3, we investigate this question and give some positive results.

Throughout this paper, all rings will be a commutative ring with identity and for any undefined terminology or notation, see [10].

2. Half-Factorial Property

In this section we briefly review the half-factorial property in commutative rings which we use them in Section 3.

Let R be a commutative ring with identity and $a \in R$ be a nonunit. By a factorization of a we mean $a = a_1 a_2 \dots a_n$ where each a_i is a nonunit. Another type of factorization is a U-factorization which was introduced in [1] and extensively examined in [6,7,12].

Definition 1 *Let $r \in R$ a nonunit. If $r = a_1 a_2 \dots a_n b_1 b_2 \dots b_m$ is a factorization, then $r = a_1 a_2 \dots a_n [b_1 b_2 \dots b_m]$ is a **U-factorization** of r if*

(1) $a_i(b_1 b_2 \dots b_m) = (b_1 b_2 \dots b_m)$ for $1 \leq i \leq n$, and

(2) $b_j(b_1 \dots \hat{b}_j \dots b_m) \neq (b_1 b_2 \dots \hat{b}_j \dots b_m)$ *for* $1 \leq j \leq m$, *where* $\hat{b}_m$ *means the omission of this element from the product.*

The b_j's are called the essential divisors of this particular U-factorization of r, and the a_i's are the inessential divisors of this particular U-factorization of r.

Hence every U-decomposition of a nonunit $x \in R$ is a U-factorization of x. A ring R is called *U-atomic* if every nonzero, nonunit has a U-factorization in which all the essential divisors are irreducible [6]. Even though it is not required in the definition of U-atomic, if R is U-atomic then we can always find a U-factorization of 0 in which all the essential divisors are irreducible. If R is présimplifiable then there are no inessential divisors in any U-factorization of a nonzero, nonunit $r \in R$. A ring R is a *U-half-factorial ring* (U-HFR) if R is U-atomic and if $r = a_1a_2 \dots a_n[b_1b_2 \dots b_m] = c_1c_2 \dots c_l[d_1d_2 \dots d_s]$ are two U-factorizations with b_i, d_j irreducible, then $m = s$. We use the following theorem later.

Theorem 2 [*7, Theorem 5.3*] $R = R_1 \times \dots \times R_n$ *is a U-HFR if and only if* R_i *is a U-HFR for* $1 \leq i \leq n$.

R is a $half\text{-}factorial\ ring$ (HFR) if R is atomic and for each nonzero, nonunit $x \in R$, if $x = x_1 \dots x_n = y_1 \dots y_m$ are two irreducible factorizations of x, then $n = m$. Although HFR is also a generalization of HFD to commutative rings, HFR and $U\text{-}HFR$ need not implies each other. Z_6 is an example of an atomic $U\text{-}HFR$ that is not an HFR. Example 9 shows that R an HFR need not imply R is a $U\text{-}HFR$ even though R is présimplifiable. Note that R a UFR need not imply R is an HFR (e.g. consider Z_6).

Theorem 3 *Let R be a commutative ring with identity. Then the following conditions are equivalent.*

(1) R is an atomic $U\text{-}HFR$ *and présimplifiable.*

(2) R is an HFR *and* 0 *has the U-half-factorial property.*

Proof. (1) $\Rightarrow$ (2) : This follows immediately from the fact that in a présimplifiable ring any U-factorization of a nonzero nonunit in which all the essential divisors are irreducible is the same as its irreducible factorization. (2) $\Rightarrow$ (1) : It is clear. ■

3. The Rings A+XI[X]

Suppose A is an $SPIR$ and $P = (p)$ is the unique prime ideal of A, where $P^n = (0)$ and n is the smallest integer which satisfies $P^n = (0)$. If $n = 1$ then A is a field which is not an interesting case. So from now on, we assume that $n > 1$ and A, P and n will be as above. Note that for any ideal I of A, $A + XI[X]$ is always Noetherian and hence always atomic and for any proper ideal I of A, $U(A + XI[X]) = U(A) + XI[X]$.

Proposition 4 *If A is an* $SPIR$ *then* $A[X]$ *is présimplifiable.*

Proof. Suppose $f(X) \neq 0$. So $f(X)(1 - g(X)) = 0$ gives that all the coefficients of $1 - g(X)$ are zero divisors. Hence $1 - g(X) \in P[X]$ and thus is nilpotent. So $g(X) \in U(A[X])$. ■

Lemma 5 *Let* $A \subset B$ *be an extension of commutative rings and* $U(B) \cap A = U(A)$. *If B is présimplifiable then A is also.*

Proof. Let $a, b \in A$ and $a = ab$. Suppose $a \neq 0$. Then $b \in U(B) \Rightarrow b \in U(A)$. ■

Corollary 6 *If A is an $SPIR$ then for any ideal I of A, $R = A+XI[X]$ is présimplifiable.*

Proof. $R \subset A[X]$ and $U(A[X]) \cap R = U(R)$. ■

Now for any proper ideal I of A we determine all the irreducible elements of $R = A + XI[X]$. Let A be an $SPIR$. If I is any proper ideal of A then I is of the form (p^k) where $1 \leq k \leq n$. So we may take a general element of $R = A + XI[X]$ as $f(X) = p^{\alpha_0}a_0 + p^{k+\alpha_1}a_1X + ... + p^{k+\alpha_s}a_sX^s$, $\alpha_i \geq 0, a_i \in U(A)$. Here $f(X) \in U(R) \Leftrightarrow \alpha_0 = 0$.

Proposition 7 *Let A be an $SPIR$ and $I = (p^k)$, $1 \leq k \leq n$, be a proper ideal of A. Suppose $f(X) = p^{1+\alpha_0}a_0 + p^{k+\alpha_1}a_1X + ... + p^{k+\alpha_s}a_sX^s$ ($\alpha_i \geq 0$, $a_i \in U(A)$) is a nonunit element of $R = A + XI[X]$. Then $f(X)$ is irreducible in R if and only if α_i is 0 for some $i \in \{0, 1, ..., s\}$.*

Proof. Suppose $\alpha_i \geq 1$ for all i. Then $f(X) = p(p^{\alpha_0}a_0 + p^{k+\alpha_1-1}a_1X + ... + p^{k+\alpha_s-1}a_sX^s)$ is not irreducible in R since R is présimplifiable. Now suppose that for some $t \in \{0, 1, ..., s\}$, $\alpha_t = 0$ and $f(X) = g(X)h(X)$ where $g(X) = p^{\beta_0}b_0 + p^{k+\beta_1}b_1X + ... + p^{k+\beta_r}b_rX^r$, $h(X) = p^{\gamma_0}c_0 + p^{k+\gamma_1}c_1X + ... + p^{k+\gamma_l}c_lX^l \in R$. If $\alpha_0 = 0$ then clearly $f(X)$ is irreducible. Suppose $t \geq 1$. Then

$$p^{k+\alpha_t}a_t = p^{\beta_0+\gamma_t+k}b_0c_t + p^{\beta_t+k+\gamma_0}b_tc_0 + \sum p^{\beta_j+\gamma_{t-j}+2k}b_jc_{t-j}\,,$$
$$\beta_0, \gamma_0 \geq 0,\ \beta_j, \gamma_{t-j} \geq 0.$$

If $\beta_0, \gamma_0 \geq 1$ then for some $r \in A$, $p^ka_t = p^{k+1}r \Rightarrow p^k = p^{k+1}ra_t^{-1} \Rightarrow p^{n-1} = 0$ which is a contradiction. Thus we have either $\beta_0 = 0$ or $\gamma_0 = 0$, that is, either $g(X) \in U(R)$ or $h(X) \in U(R)$. Hence $f(X)$ is irreducible. ■

Corollary 8 *Let I be a nonzero proper ideal of A. If I is not the prime ideal of A then $R = A + XI[X]$ can not be a U-HFR.*

Proof. Since $(0) \neq I$ is not a prime ideal, I is of the form $I = (p^k)$ where $2 \leq k < n$. Since $k \geq 2$, $p^{2k}X^2$ has two U-decompositions $(\ )(p^kX.p^kX) = (\ )(p^k.p^kX^2)$ one of them has length 2 and the other has length $k + 1$. ■

Even though $A + XI[X]$ is présimplifiable, in this corollary we can not replace U-HFR with HFR. The next example shows that $A + XI[X]$ can be an HFR for a non prime ideal I. Example 9 also shows that in a commutative ring every nonzero, nonunit having the half-factorial property need not imply that 0 does.

Example 9 *Let $R = Z_8 + X(4)[X]$. R is an HFR but not a U-HFR. 0 has two nonassociate U-decomposition as $0 = (\ )(2.2.2) = (\ )(4X.4X)$. Hence R is not a U-HFR. But R is an HFR. To see this, let $f \in R$ be a nonzero, nonunit element of R. Suppose f is irreducible in R and $f = gh$. Then $f \sim g$ or $f \sim h$. Since R is présimplifiable $f \sim g$ implies $h \in U(R)$ and $f \sim h$ implies $g \in U(R)$. Hence f has unique factorization. If f is not an irreducible then f is a product of two nonunit element of R. Since $f(X) \neq 0$ we have only one choice for f, namely $f = 4$. So f can not be written as a product of more than two nonunit elements.*

Let R be an integral domain. It is well-known that R is a $UFD \Leftrightarrow R[X]$ is a UFD. But this is not true if R is a commutative ring. In fact $R[X]$ is a UFR if and only if R is a finite direct product of UFD's [5, Theorem 2.7.(1)]. Another way to state this theorem is that if A is an $SPIR$ then $A[X]$ is not a UFR. In fact if A is an $SPIR$ then $A[X]$ is neither a U-HFR nor an HFR. To see this we need to find some irreducible elements in $A[X]$.

Lemma 10 *Let A be an $SPIR$ and $P = (p)$ be the prime ideal of A. Let $q(X) = p + c_1X + ... + c_kX^k + ... + c_mX^m \in A[X]$. If $c_i \in P$ for $i \neq k$ and $c_k = 1$ then $q(X)$ is irreducible in $A[X]$ for every positive integer k.*

Proof. Suppose $q(X) = fg$ in $A[X]$ where $f = a_0 + a_1X + ... + a_sX^s$ and $g = b_0 + b_1X + ... + b_tX^t$. Then $p = a_0b_0 \Rightarrow p \sim a_0$ or $p \sim b_0$, say $p \sim a_0$. Then $a_0 = up$ for some $u \in U(A)$ and $b_0 \in U(A)$. Without loss of generality we may take $b_0 = 1$. It suffices to show that $b_i \in P$ for $i = 1, ..., t$. Suppose $b_j \notin P$ for some $j \in \{1, ..., t\}$. Assume that b_j is the first coefficient of g such that $b_j \notin P$ and b_l is the last coefficient of g such that $b_l \notin P$ $(j, l > 0)$. Now consider the UFD $R = A[X]/P[X] \cong (A/P)[X]$. Then we have $\bar{X}^k = \bar{f}\bar{g} = \bar{f}.(\bar{1} + \bar{b}_j\bar{X}^j + ... + \bar{b}_l\bar{X}^l)$ in R. $\bar{X}$ is an irreducible in R and $\bar{g} = (\bar{1} + \bar{b}_j\bar{X}^j + ... + \bar{b}_l\bar{X}^l) \notin U(R)$. But $\bar{X}$ can not be an associate of any factor of $\bar{g}$. This contradicts unique factorization in R. Hence $1 - g \in P[X]$ and $g \in U(A[X])$. ■

Proposition 11 *If A is an $SPIR$ which is not a field then $A[X]$ is neither a U-HFR nor an HFR.*

Proof. Let $P = (p)$ be the unique prime ideal of A and let n be the least positive integer such that $P^n = (0)$, $n > 1$. Now consider $p^{n-1}X^k$ in $A[X]$. Then

$$p^{n-1}X^k = p^{n-1}.(p + X^k) = p^{n-1}.X^k$$

are two nonassociate irreducible factorizations of $p^{n-1}X^k$. The first has length n while the other has $n - 1 + k$. To complete the proof we need only to check the irreducibility of factors. It is clear that the prime element p remains prime in $A[X]$. Irreducibility of X comes from the fact that X is irreducible if and only if $A[X]$ is indecomposable [4, Theorem 6.4.] and by Lemma 10, $p + X^k$ is also an irreducible in $A[X]$. ■

Hence the following theorem can not be generalized to the case where I need not be proper.

Theorem 12 *Let A be an SPIR and I be a nonzero proper ideal of A . Then $R = A+XI[X]$ is a U-HFR if and only if I is the prime ideal of A.*

Proof. It is enough to show that if P is the prime ideal of A, then $R = A + XP[X]$ is a U-HFR. Let $f(X) = p^{\alpha_0}a_0 + p^{\alpha_1}a_1X + ... + p^{\alpha_s}a_sX^s$ be a nonunit element of R where $\alpha_i \geq 1$, $a_i \in U(A)$, and $t = \min\{\alpha_0, \alpha_1, \ldots, \alpha_s\}$. Suppose $t < n$. Then $f(X) \neq 0$. Since R is présimplifiable, any irreducible factorization of $f(X)$ is the same as a U-decomposition. We show that in any irreducible factorization of $f(X)$, there are exactly t irreducible factors. Suppose $f = f_1f_2 ... f_k$ with each f_i irreducible in R. Let $c(f)$ be the ideal generated by the coefficients of f. Then $c(f) = (p^t)$ and $c(f) = (p)$.

Since $c(f)$ is locally principal $c(f) = c(f_1)c(f_2) \dots c(f_k)$. Hence $k = t$. If $f(X) = 0$ then there are exactly n irreducible factors in any U-decomposition of f. Hence R is a U-HFR. ■

Theorem 13 *Let A be UFR such that* $A = A_1 \times A_2 \times \dots \times A_n$ *finite direct product of UFD's and SPIR's and* $I = I_1 \times I_2 \times \dots \times I_n$ *be an ideal of A. Then* $A + XI[X]$ *is a U-HFR if and only if each* I_i *is either 0 or a prime ideal of* A_i *or may be* A_i *in the case where* A_i *is an integral domain.*

Proof. Take $A + XI[X] = (A_1 \times A_2 \times \dots \times A_n) + X(I_1 \times I_2 \times \dots \times I_n)[X] = (A_1 + XI_1[X]) \times \dots \times (A_n + XI_n[X])$. Then $A + XI[X] = R_1 \times \dots \times R_n$ where $R_i = A_i + XI_i[X]$. If A_i is a UFD then R_i is an HFD if and only if I_i is a prime ideal of A_i or $I_i = A_i$ [11] and if A_i is an $SPIR$ then R_i is a U-HFR if and only if I_i is a prime ideal of A_i (Theorem 12). The result follows from Theorem 2. ■

References

[1] A.G. Ağargün, D.D. Anderson and S. Valdes-Leon, Factorization in commutative rings with zero divisors, III, *Rocky Mountain J. Math.* **31** (2001), 1-21.

[2] D.D. Anderson, D.F. Anderson and M. Zafrullah, Factorization in integral domains, *J. Pure Appl. Algebra* **69** (1990), 1-19.

[3] D.D. Anderson, D.F. Anderson and M. Zafrullah, Factorization in integral domains, II, *J. Algebra* **152** (1992), 78-93.

[4] D.D. Anderson and S. Valdes-Leon, Factorization in commutative rings with zero divisors, *Rocky Mountain J. Math.* **26** (1996), 439-480.

[5] D.D. Anderson and R. Markanda, Unique factorization rings with zero divisors, *Houston J. Math.* **11** (1985), 15-30.

[6] M. Axtell, U-factorizations in commutative rings with zero divisors, *Comm. Algebra* **30** (2002), 1241-1255.

[7] M. Axtell, S. Forman, N. Roersma and J. Stickles, Properties of U-factorizations, *Int. J. Comm. Rings* **2**(2) (2003), 83-99.

[8] C.R. Fletcher, Unique factorization rings, *Proc. Camb. Phil. Soc.* **65** (1969), 579-583.

[9] C.R. Fletcher, The structure of unique factorization rings, *Proc. Camb. Phil. Soc.*, **67** (1970), 535-540.

[10] R. Gilmer, *Multiplicative Ideal Theory*, Dekker, New York (1972).

[11] N. Gonzalez, S. Pellerin and R. Robert, Elasticity of A+XI[X] domains where A is a UFD, *J. Pure Appl. Algebra* **160** (2001), 183-194.

[12] N. Roersma, U-factorizations in commutative rings with zero divisors, *Rose-Hulman Inst. Tech. Undergraduate Math J.* **2**(2) (2001).

[13] A. Zaks, Half-factorial domains, *Bull. Amer. Math. Soc.* **82** (1976), 721-724.

In: Mathematics, Game Theory and Algebra Compendium ISBN 978-1-60456-659-8
Editor: Jacob H. Mathias, pp. 165-181

Chapter 17

Some Diophantine Equations Associated to Seminormal Cohen-Kaplansky Domains

Abdallah Badra [*] ***and Martine Picavet-L'Hermitte*** [†]
Laboratoire de Mathématiques Pures
Université Blaise Pascal, 63177 Aubière-Cedex, France

Communicated by Mohammad Zafrullah

Abstract

A Cohen-Kaplansky domain (CK domain) R is an integral domain where every nonzero nonunit element of R is a finite product of irreducible elements and such that R has only finitely many nonassociate irreducible elements. In this paper, we investigate seminormal CK domains and obtain the form of their irreducible elements. The solutions of a system of diophantine equations allow us to give a formula for the number of distinct factorizations of a nonzero nonunit element of R, with an asymptotic formula for this number.

1991 Mathematics Subject Classification: Primary 11D04, 1315; Secondary 13A05, 20M14

Key Words and phrases: CK domain, half-factorial, seminormal, t-closed, diophantine equations

1. Introduction

Let R be an *atomic* integral domain, that is, each nonzero nonunit element of R can be written as a finite product of irreducible elements (or *atoms*). The simplest situation is when R has only a finite number of (nonassociate) atoms. Such a domain R was called a *Cohen-Kaplansky domain (CK domain)* by D.D. Anderson and J.L. Mott in [2] who obtained many

*E-mail address: Abdallah.Badra@math.univ-bpclermont.fr
†E-mail address: Martine.Picavet@math.univ-bpclermont.fr

conditions equivalent to R being a CK domain, after I.S. Cohen and I. Kaplansky [4] inaugurated the study of CK domains. In Section 2 we recall and give basic results on CK domains.

An atomic domain R is called a *half-factorial* domain (*HFD*) if each factorization of a nonzero nonunit element of R into a product of atoms has the same length (Zaks [15]). A ring R is called *seminormal* if whenever $x, y \in R$ satisfy $x^3 = y^2$, there is $a \in R$ with $x = a^2,\ y = a^3$ [14]. Section 3 is devoted to the study of seminormal CK domains. In particular, we show that a seminormal CK domain is half-factorial and obtain some equivalent conditions for a CK domain to be seminormal. As factorization properties of CK domains and seminormality are preserved by localization, we consider a local seminormal CK domain R. Let $\bar{R}$ be its integral closure. Then $\bar{R}$ is a DVR with maximal ideal $\bar{R}p$, which is also the maximal ideal of R. Moreover the atoms of R are of the form vp, where v is a unit of $\bar{R}$. If $\mathcal{U}(\bar{R})$ (resp. $\mathcal{U}(R)$) is the group of units of $\bar{R}$ (resp. R), the factor group $\mathcal{U}(\bar{R})/\mathcal{U}(R)$ is a finite cyclic group. Let $\bar{u}$ be a generator of $\mathcal{U}(\bar{R})/\mathcal{U}(R)$ and n the order of $\bar{u}$. If $x = vp^k$ is a nonzero nonunit element of R with $\bar{v} = \bar{u}^r,\ r \in \{0, \ldots, n-1\}$, in $\mathcal{U}(\bar{R})/\mathcal{U}(R)$, the distinct factorizations of x in R into atoms are deduced from the system of diophantine equations in $(a_1, \ldots, a_n) \in \mathbb{N}^n$:

$$(S)\quad \begin{cases} \displaystyle\sum_{i=1}^{n} a_i = k \\[2ex] \displaystyle\sum_{i=1}^{n} \overline{ia_i} = \bar{r} \text{ in } \mathbb{Z}/n\mathbb{Z} \end{cases}$$

The calculation of the number of solutions of this system is the object of Section 4. If we denote by $\eta(x)$ the number of non-associated irreducible factorizations of x into atoms, we get that $\eta(x)$ is the number of solutions of the system (S).

Section 5 ends this paper with the asymptotic behaviour of the function η where we use the following result by F. Halter-Koch :

Theorem 1.1 *[6,* Theorem 1*]. Let H be an atomic monoid such that each nonunit x has finitely many non-associated factorizations into irreducibles. Suppose that there are only finitely many irreducible elements of H which divide some power of x. There exists two constants $A \in \mathbb{Q}$ and $d \in \mathbb{N},\ A > 0$ such that $\eta(x^n) = An^d + O(n^{d-1})$.*

An explicit value for A and d is obtained for a local seminormal CK domain.

For a ring R, we denote by $\mathrm{Max}(R)$ the set of maximal ideals of R and by $\mathcal{U}(R)$ its group of units. Let $x, y \in R$. We say that x and y are *associates* ($x \sim y$) if there exists $u \in \mathcal{U}(R)$ such that $x = uy$. For an integral domain R, we denote by $\bar{R}$ its integral closure. The conductor $[R : \bar{R}]$ of an integral domain R in its integral closure is called the *conductor* of R. For a finite set S, we denote by $|S|$ the number of elements of S. For $x \in \mathbb{R}$, we set $[x] = \sup\{n \in \mathbb{Z} \mid n \leq x\}$.

2. Basic Results on CK Domains

We first recall some of useful results concerning CK domains.

Theorem 2.1 *[2,* Theorem 4.3*] For an integral domain R, the following statements are equivalent.*

1. *R is a CK domain.*
2. *$\bar{R}$ is a semilocal PID with $\bar{R}/[R:\bar{R}]$ finite and $|\operatorname{Max}(R)| = |\operatorname{Max}(\bar{R})|$.*
3. *R is a one-dimensional semilocal domain with R/M finite for each nonprincipal maximal ideal M of R, $\bar{R}$ is a finitely generated R-module (equivalently, $[R:\bar{R}] \neq 0$), and $|\operatorname{Max}(R)| = |\operatorname{Max}(\bar{R})|$.*

This theorem implies the following properties.

Proposition 2.2 *[2,* Theorem 4.3, Theorem 3.1, Theorem 2.1 and Corollary 2.5 *] Let R be a CK domain. Then*

1. *R is Noetherian and for each $x \in \bar{R}$, there exists an $n \in \mathbb{N}^*$ with $x^n \in R$.*
2. *$\mathcal{U}(\bar{R})/\mathcal{U}(R)$ is a finite group.*
3. *R_M is a CK domain for each maximal ideal M of R. In particular, $\bar{R}_M$ is a DVR.*
4. *Let T be an overring of R. Then T is also a CK domain.*
5. *The atoms of R are primary.*

D.D. Anderson and J.L. Mott [2] say that a pair of rings $R \subset S$ is a *root extension* if for each $s \in S$, there exists an $n = n(s) \in \mathbb{N}^*$ with $s^n \in R$. For such an extension we have $|\operatorname{Max}(R)| = |\operatorname{Max}(S)|$. Hence $R \subset \bar{R}$ is a root extension when R is a CK domain.

Proposition 2.3 *Let R_1 and R_2 be two CK domains with the same integral closure R'. Then $R = R_1 \cap R_2$ is a CK domain with integral closure R'.*

Proof. Set $R = R_1 \cap R_2$. Define $I_1 = [R_1 : R']$, $I_2 = [R_2 : R']$ and $I = [R : R']$. Then $I_1 \cap I_2$ is a common ideal of R' and R contained in I so that $I \neq 0$. Let $a, b \in R'$ with $b \neq 0$ and i a nonzero element of I. Then ia and ib are in R and hence $a/b = ia/ib$ shows that R has the same quotient field as R'. Moreover, $R \subset R'$ is a root extension. Then R' is obviously the integral closure of R and is a semilocal PID. Since R'/I_1 and R'/I_2 are finite, this gives that $R'/(I_1 \cap I_2)$ is also finite because isomorphic to a subring of $R'/I_1 \times R'/I_2$, so that R'/I is finite.

Moreover, we have $|\operatorname{Max}(R)| = |\operatorname{Max}(R')|$ because $R \subset R'$ is a root extension. Applying Theorem 2.1, (2), we get that R is a CK domain with integral closure R'. ■

Corollary 2.4 *Let D be a DVR with maximal ideal M such that D/M is finite. Let I be a nonzero ideal of D. The set of underrings of D with integral closure D and with conductor I has a least element and all these underrings are CK domains.*

Proof. Set $\mathcal{E} = \{R \text{ underring of } D \mid \bar{R} = D,\ [R : D] = I\}$. Since D/M is finite, so is D/I. Indeed, if $M = Dp$ for some atom $p \in D$, then $I = Dp^n$, for some integer n and an obvious induction shows that $|D/I| = |D/M|^n$. Consider $R \in \mathcal{E}$. Then the finiteness of D/I implies the finiteness of R/I. So D is a finitely generated R-module because D/I is a finitely generated R/I-module. It follows that $|\operatorname{Max}(R)| = 1$ and R is a CK domain by Theorem 2.1, (2).

Since D/I is finite, there are finitely many subrings of D/I, and so finitely many $R \in \mathcal{E}$. Let R and $S \in \mathcal{E}$ and set $T = R \cap S$. By Proposition 2.3, T is a CK domain with conductor $J \supset I$. But $T \subset R$ implies $J \subset I$, so that $J = I$ and $T \in \mathcal{E}$. Therefore the intersection of all elements of $\mathcal{E}$ is a CK domain with conductor I and integral closure D and is the least element of $\mathcal{E}$. ■

3. Characterization of Seminormal CK Domains

Let R be an integral domain with quotient field K. We say that R is *t-closed* if whenever $x \in K$ and $x^2 - rx, x^3 - rx^2 \in R$ for some $r \in R$, then $x \in R$ [9]. A t-closed integral domain is seminormal. Recall that an integral domain R is said to be a *pseudo-valuation domain* (PVD) if there exists a valuation overring V of R such that $\operatorname{Spec}(R) = \operatorname{Spec}(V)$ [8] and an integral domain R is said to be a *locally pseudo-valuation domain* (locally PVD) if each of its localizations at a prime ideal is a PVD [5].

Proposition 3.1 *Let R be a one-dimensional Noetherian integral domain such that its integral closure $\bar{R}$ is a finitely generated R-module. The following conditions are equivalent:*

1. *R is seminormal and the canonical map* $\operatorname{Spec}(\bar{R}) \to \operatorname{Spec}(R)$ *is bijective.*
2. *R is t-closed.*
3. *R is a locally PVD.*
4. *The conductor I of R is a radical ideal in $\bar{R}$ and* $|\operatorname{Max}(R)| = |\operatorname{Max}(\bar{R})|$.

In particular, a CK domain R is seminormal if and only if R is t-closed.

Proof. (1) ⇔ (2) is [9, Proposition 3.7].

(2) ⇔ (3) is [10, Corollary 3.4].

(2) ⇔ (4) comes from [9, Corollary 3.8 and Proposition 2.8]. Indeed, for any $P \in \operatorname{Max}(R)$, the conductor of R_P is I_P. ■

We obtain as a corollary a first characterization of local seminormal (or t-closed) CK domains.

Corollary 3.2 *Let R be a local CK domain with integral closure $\bar{R} \neq R$. Let $\bar{R}p$ be the maximal ideal of $\bar{R}$. Then R is seminormal if and only if $\mathcal{U}(\bar{R})p \subset R$.*

Proof. Assume that R is seminormal. By Proposition 3.1 (4), $\bar{R}p$ is the conductor of R, so that $\mathcal{U}(\bar{R})p \subset \bar{R}p \subset R$.

Conversely, if $\mathcal{U}(\bar{R})p \subset R$, we get that $\mathcal{U}(\bar{R})p^n \subset R$ for any integer n and $\bar{R}p \subset R$ gives that $\bar{R}p$ is the conductor of R so that R is seminormal. ■

In the nonlocal case, this condition is not fulfilled :

Corollary 3.3 *Let R be a CK domain with integral closure $\bar{R} \neq R$. Let $\bar{R}p_i$, $i = 1, \dots, n$, be the maximal ideals of $\bar{R}$.*

Then $\mathcal{U}(\bar{R})p_i \subset R$ for any $i = 1, \dots, n$, implies that R is seminormal and $n = 1$.

Proof. The case $n = 1$ is the previous Corollary. Assume $n > 1$. Any nonunit of $\bar{R}$ is in R. Moreover, $\bar{R}p_1$ and $\bar{R}p_2$ are comaximal ideals of $\bar{R}$. For any $u \in \mathcal{U}(\bar{R})$, there exists $v, w \in \bar{R}$ such that $u = vp_1 + wp_2 \in R$. Then $\bar{R} = R$, a contradiction. ■

Corollary 2.4 has a new formulation in the seminormal case.

Corollary 3.4 *Let D be a DVR with maximal ideal M such that D/M is finite. The set of seminormal underrings of D with integral closure D is linearly ordered.*

Proof. Let R be a seminormal proper underring of D. Since its conductor is a radical ideal of D, it has to be M, a maximal ideal in R so that R/M is a subfield of the finite field D/M. But the set of subfields of D/M is linearly ordered.

Let R_1, R_2 be two seminormal proper underrings of D with integral closure D. Their conductor is M and we have, for instance, $R_1/M \subset R_2/M$, which gives $R_1 \subset R_2$. ■

Here is a fundamental link between seminormal CK domains and factorization.

Proposition 3.5 *A seminormal CK domain is half-factorial.*

Proof. Let R be a seminormal CK domain and $P \in \mathrm{Max}(R)$. Then R_P is a PVD by Proposition 3.1 and a CK domain by Proposition 2.2 (3). So R_P is a HFD for any $P \in \mathrm{Max}(R)$ [2, Theorem 6.2]. The same holds for R [2, Theorem 6.1]. ■

The following theorem gives the additional condition necessary for a CK half-factorial domain to be seminormal.

Theorem 3.6 *Let R be a CK domain with integral closure $\bar{R}$. Let $\bar{R}p_i$, $i = 1, \dots, n$, be the maximal ideals of $\bar{R}$. Then R is seminormal if and only if R is a HFD and $\mathcal{U}(\bar{R})p_1 \cdots p_n \subset R$. Moreover, if these conditions are satisfied, we can choose $p_i \in R$ for each $i = 1, \dots, n$.*

Proof. We can assume $R \neq \bar{R}$ (the case $R = \bar{R}$ is trivial).

Let R be a seminormal CK domain. Then R is a HFD by the previous Proposition and the conductor I of R is a product of some of the $\bar{R}p_i$. It follows that $\mathcal{U}(\bar{R})p_1 \cdots p_n \subset R$.

Conversely, assume that R is a HFD and $\mathcal{U}(\bar{R})p_1 \cdots p_n \subset R$ and let I be the conductor of R. For each $i = 1, \dots, n$, set $P_i = R \cap \bar{R}p_i$, $R_i = R_{P_i}$ and $\overline{R_i} = \overline{R_{P_i}} = \bar{R}_{P_i}$.

First, we show that we may assume $p_i \in R$ for each $i = 1, \dots, n$.

- If P_i is comaximal with I, then $R_i = \overline{R_i}$ and $p_i/1$ is an atom in R_i [2, Theorem 2.1 (2)]. Then there exists a P_i-primary atom $p \in R$ and $s \in R \setminus P_i$ such that $sp_i = p$, which implies $s \in \mathcal{U}(\bar{R})$, so that $\bar{R}p_i = \bar{R}p$.

- Let P_i be non comaximal with I and let x be a P_i-primary atom in R. There exist $u \in \mathcal{U}(\bar{R})$ and an integer k such that $x = up_i^k$ since $x \notin P_j$ for any $j \neq i$. But R_i is a HFD, which implies that $x/1 \simeq p_i/1$ in $\overline{R_i}$ [2, Theorem 6.3] and so $k = 1$. Then $x \simeq p_i$ in $\bar{R}$.

The assumption can be rewritten $\mathcal{U}(\bar{R})p_1 \cdots p_n \subset R$ with $p_i \in R$ for each $i = 1, \dots, n$. This gives finally $\bar{R}p_1 \cdots p_n \subset I \subset R$ and I is a radical ideal in $\bar{R}$. Moreover, R being a CK domain, we get $|\mathrm{Max}(R)| = |\mathrm{Max}(\bar{R})|$ and thus R is seminormal by Proposition 3.1 (4). ■

In the local case, we obtain another characterization for a CK half-factorial domain to be seminormal.

Proposition 3.7 *Let R be a local CK domain with integral closure $\bar{R}$. Then R is seminormal if and only if R is a HFD and has $|\mathcal{U}(\bar{R})/\mathcal{U}(R)|$ nonassociate atoms.*

Proof. We can assume $R \neq \bar{R}$ (the case $R = \bar{R}$ is trivial).

Let R be seminormal. Then R is a HFD by the previous Theorem. Let $\bar{R}p$ be the maximal ideal of $\bar{R}$ and let $a_1, \dots, a_n$ be the nonassociate atoms of R. They are of the form $a_i = u_i p,\ u_i \in \mathcal{U}(\bar{R})$ by [2, Theorem 6.3 (3)]. But since R is seminormal, its conductor is $\bar{R}p$. It follows that $up \in R$ for any $u \in \mathcal{U}(\bar{R})$. Let up, vp be two atoms of R, where $u, v \in \mathcal{U}(\bar{R})$. Then up and vp are associates in R if and only if there exists $w \in \mathcal{U}(R)$ such that $up = wvp$, which is equivalent to $\bar{u} = \bar{v}$ in $\mathcal{U}(\bar{R})/\mathcal{U}(R)$. Hence two atoms up, vp of R, with $u, v \in \mathcal{U}(\bar{R})$, are nonassociates in R if and only if $\bar{u} \neq \bar{v}$ in $\mathcal{U}(\bar{R})/\mathcal{U}(R)$. Then R has $|\mathcal{U}(\bar{R})/\mathcal{U}(R)|$ nonassociate atoms (see also [2, Corollary 5.6]).

Conversely, let R be a HFD with $n = |\mathcal{U}(\bar{R})/\mathcal{U}(R)|$ nonassociate atoms. They are of the form $a_i = u_i p,\ u_i \in \mathcal{U}(\bar{R}),\ i = 1, \dots, n$ and $\{\bar{u}_1, \dots, \bar{u}_n\} = \mathcal{U}(\bar{R})/\mathcal{U}(R)$. It follows that $up \in R$ for any $u \in \mathcal{U}(\bar{R})$. In particular, $p \in R$ so that $p^n \in R$ for any integer $n > 0$ and we get that $\bar{R}p \subset R$. Then $\bar{R}p$ is the conductor of R and R is seminormal. ■

A seminormal CK domain has a property which is not too far from unique factorization. In [3], S.T. Chapman, F. Halter-Koch and U. Krause defined an integral domain R to be *inside factorial* with *Cale basis* $\mathcal{Q}$, if, for every nonzero nonunit $x \in R$, there exists some $n \in \mathbb{N}^*$ such that x^n has a unique factorization, up to units, into elements of $\mathcal{Q}$.

Proposition 3.8 *Let R be a seminormal CK domain with integral closure $\bar{R}$. Then R is inside factorial with Cale basis $\{p_1, \dots, p_n\}$, where the $\bar{R}p_i$ are the maximal ideals of $\bar{R}$ with $p_i \in R$ for $i = 1, \dots, n$.*

Proof. We have seen in Theorem 3.6 that we can choose p_i in R, where the $\bar{R}p_i$ are the maximal ideals of $\bar{R}$.
The atoms of R are of the form $u_{ij}p_i$, with $u_{ij} \in \mathcal{U}(\bar{R}),\ i = 1, \dots, n$ [2, Theorem 2.1 (2)]. Let $r = |\mathcal{U}(\bar{R})/\mathcal{U}(R)|$. Then $u^r \in R$ for any $u \in \mathcal{U}(\bar{R})$. Let x be a nonzero nonunit of R. As an element of $\bar{R}$, it can be written $x = u \prod p_i^{\alpha_i},\ u \in \mathcal{U}(\bar{R})$. Then $x^r = u^r \prod p_i^{r\alpha_i}$ with $u^r \in \mathcal{U}(R)$ and this factorization into the p_i is obviously unique. ■

Remark 3.9 *Under assumptions of the previous Proposition, let e be the exponent of the factor group $\mathcal{U}(\bar{R})/\mathcal{U}(R)$. Then e is the least integer r such that x^r has a unique factorization, up to units, into elements of $\{p_1, \dots, p_n\}$, for every nonzero nonunit $x \in R$. Indeed, e is the least integer r such that $u^r \in \mathcal{U}(R)$ for any $u \in \mathcal{U}(\bar{R})$.*

We can calculate this exponent. D.D. Anderson, D.F Anderson and M. Zafrullah call in [1] an atomic domain with almost all atoms prime a generalized CK domain. A CK domain is obviously a generalized CK domain. We can still assume $R \neq \bar{R}$. Then, if I is the conductor of R, we have the isomorphism $\mathcal{U}(\bar{R})/\mathcal{U}(R) \simeq \mathcal{U}(\bar{R}/I)/\mathcal{U}(R/I)$ by [11, Theorem 2] (the result was obtained for algebraic orders but a generalization to one-dimensional Noetherian domains R with integral closure which are finitely generated R-

modules can be easily made). Since R is seminormal, I is a radical ideal in $\bar{R}$. After a reordering, write $I = \prod_{i=1}^{m} \bar{R}p_i$.

Then $\mathcal{U}(\bar{R})/\mathcal{U}(R) \simeq \prod_{i=1}^{m} [\mathcal{U}(\bar{R}/\bar{R}p_i)/\mathcal{U}(R/P_i)]$, *where* $P_i = R \cap \bar{R}p_i$ *since* $I = \prod_{i=1}^{m} P_i$ *as an ideal of R.*

Set $q_i = |R/P_i|$ *and* $k_i = [\bar{R}/\bar{R}p_i : R/P_i]$. *Then* $e_i = (q_i^{k_i} - 1)/(q_i - 1)$ *is the order (and the exponent) of the finite cyclic group* $\mathcal{U}(\bar{R}/\bar{R}p_i)/\mathcal{U}(R/P_i)$ *and* $e = \mathrm{lcm}(e_1, \ldots, e_m)$.

We are now able to obtain all the factorizations into atoms of a nonzero nonunit element of a seminormal CK domain with the number of distinct factorizations into atoms. We can restrict to the local case by the following proposition.

Proposition 3.10 *Let R be a CK domain with maximal ideals $P_1, \ldots, P_n$. Set $R_i = R_{P_i}$ and define $\eta_i(z)$ to be the number of distinct factorizations into atoms of R_i of a nonzero $z \in R_i$. Then* $\eta(x) = \prod_{i=1}^{n} \eta_i(x/1)$ *for a nonzero $x \in R$.*

Proof. By [2, Theorem 2.1 (2)], the atoms of R are primary and the atoms of R_i are the P_i-primary atoms of R. Moreover, if x is a nonzero nonunit element of R, then x is written in a unique way $x = x_1 \cdots x_n$, where x_i is a P_i-primary element of R for each $i = 1, \ldots, n$ [7, Corollary 1.7]. Indeed, by [1, Corollary 5], a CK domain is *weakly factorial* (such that every nonunit is a product of primary elements), and a weakly factorial domain is a weakly factorial monoid for the multiplicative structure. So, we get $\eta(x) = \prod_{i=1}^{n} \eta(x_i)$ and $\eta(x_i) = \eta_i(x_i/1)$ for each i by [2, Theorem 2.1 (2)] since a factorization of x_i into atoms of R leads to a factorization of $x_i/1$ into atoms of R_i and conversely. ■

To end, we give the form of atoms in a local seminormal CK domain.

Theorem 3.11 *Let R be a local seminormal CK domain with integral closure $\bar{R}$. Let $\bar{R}p$ be the maximal ideal of $\bar{R}$, with $p \in R$. Set $n = |\mathcal{U}(\bar{R})/\mathcal{U}(R)|$ and choose $u \in \mathcal{U}(\bar{R})$ such that $\bar{u}$ is a generator of the cyclic group $\mathcal{U}(\bar{R})/\mathcal{U}(R)$. Then*

1. *A set of all nonassociate atoms of R is $\{u^i p \mid i = 0, \ldots, n-1\}$.*

2. *Let $x = vp^k$, $k \in \mathbb{N}^*$, $v \in \mathcal{U}(\bar{R})$. Let $r \in \{0, \ldots, n-1\}$ be such that $\bar{v} = \bar{u}^r$. The number of nonassociated factorizations of x into atoms of R is equal to the number of solutions $(a_1, \ldots, a_n) \in \mathbb{N}^n$ of the system of diophantine equations :*

$$(S) \quad \begin{cases} \displaystyle\sum_{i=1}^{n} a_i = k \\ \displaystyle\sum_{i=1}^{n} \overline{i a_i} = \bar{r} \text{ in } \mathbb{Z}/n\mathbb{Z} \end{cases}$$

Proof. As above, we can assume $R \neq \bar{R}$. Then $\bar{R}p$ is the conductor of R so that $\bar{R}/\bar{R}p$ is a finite field by Theorem 2.1 (3) and $\mathcal{U}(\bar{R}/\bar{R}p)$ is a finite cyclic group. It follows that $\mathcal{U}(\bar{R})/\mathcal{U}(R) \simeq \mathcal{U}(\bar{R}/\bar{R}p)/\mathcal{U}(R/\bar{R}p)$ (Remark 3.9) is also a finite cyclic group. Let $u \in \mathcal{U}(\bar{R})$ be such that $\bar{u}$ is a generator of $\mathcal{U}(\bar{R})/\mathcal{U}(R)$.

(1) In view of Proposition 3.7, we can choose $\mathcal{A} = \{u^i p\},\ i = 1, \ldots, n$, as a set of nonassociate atoms of R since the u^i are the representatives of the elements of $\mathcal{U}(\bar{R})/\mathcal{U}(R)$ and $u^n p$ is an associate of p in R.

(2) Set $p_i = u^i p,\ i = 1, \ldots, n$, and let x be a nonzero nonunit element of R which is not an atom. Then $x = vp^k,\ k > 1$ with a unique $v \in \mathcal{U}(\bar{R})$. A factorization of x into elements of $\mathcal{A}$ is of the form $x = w\prod_{i=1}^{n} p_i^{a_i},\ w \in \mathcal{U}(R),\ a_i \in \mathbb{N}$. This gives $x = w\prod_{i=1}^{n}(u^i p)^{a_i} = vp^k$ $(*)$, which implies, by identification in $\bar{R}$, the equalities

$$v = w\prod_{i=1}^{n} u^{ia_i} \text{ and } k = \sum_{i=1}^{n} a_i \quad (**)$$

Consider another factorization $x = w'\prod_{i=1}^{n} p_i^{a'_i},\ w' \in \mathcal{U}(R),\ a'_i \in \mathbb{N}$. We get then $k = \sum_{i=1}^{n} a_i = \sum_{i=1}^{n} a'_i$ and $v = w\prod_{i=1}^{n} u^{ia_i} = w'\prod_{i=1}^{n} u^{ia'_i}$. These two factorizations coincide if and only if $a_i = a'_i$ for each i. In this case, we have $w = w'$.

In $\mathcal{U}(\bar{R})/\mathcal{U}(R)$ we have the relation $\bar{v} = \prod_{i=1}^{n} \bar{u}^{ia_i} = \bar{u}^r$ where $r \in \{0, \ldots, n-1\}$ by $(**)$, that is $r \equiv \sum_{i=1}^{n} ia_i \pmod n$, or equivalently, $\bar{r} = \sum_{i=1}^{n} \overline{ia_i}$ in $\mathbb{Z}/n\mathbb{Z}$. Then $(a_1, \ldots, a_n) \in \mathbb{N}^n$ is a solution of the system (S).

Conversely, let $(a'_1, \ldots, a'_n) \in \mathbb{N}^n$ satisfying (S).

Set $x' = \prod_{i=1}^{n} p_i^{a'_i} = \prod_{i=1}^{n}(u^i p)^{a'_i} = u^{a'_1+2a'_2+\cdots+na'_n} p^{a'_1+a'_2+\cdots+a'_n}$.

But $\sum_{i=1}^{n} ia'_i = r + sn,\ s \in \mathbb{Z}$, gives $x' = u^r(u^n)^s p^k$ and $\bar{v} = \bar{u}^r$ implies $u^r = w'v$, where $w' \in \mathcal{U}(R)$. So we get $x' = w'(u^n)^s vp^k = w'(u^n)^s x$, with $w'(u^n)^s \in \mathcal{U}(R)$ and $x \sim x'$ in R. We deduce that two distinct solutions of (S) give two distinct factorizations of x into atoms of R and the number of nonassociated factorizations of x into atoms of R is equal to the number of solutions $(a_1, \ldots, a_n) \in \mathbb{N}^n$ of (S). ∎

We are going to calculate the number of solutions of such a system in the next section.

4. On the Number of Solutions of a System of Two Special Diophantine Equations

In this section, we use the following notation. Let $n, r \in \mathbb{N},\ k, s \in \mathbb{Z}$ with $n > 0$ and $0 \leq r \leq n-1$. We consider the following systems of diophantine equations in $(a_1, \ldots, a_n) \in \mathbb{N}^n$:

$$S(n,k,r) \begin{cases} \displaystyle\sum_{i=1}^{n} a_i = k \\ \displaystyle\sum_{i=1}^{n} \overline{ia_i} = \bar{r} \text{ in } \mathbb{Z}/n\mathbb{Z} \end{cases} \quad \text{and} \quad S'(n,k,s) \begin{cases} \displaystyle\sum_{i=1}^{n} a_i = k \\ \displaystyle\sum_{i=1}^{n} ia_i = s \end{cases}$$

We denote respectively by $N(n,k,r)$ and $p(n,k,s)$ the numbers of solutions $(a_1, \ldots, a_n) \in \mathbb{N}^n$ of $S(n,k,r)$ and $S'(n,k,s)$. Obviously, we have $N(n,k,r) = p(n,k,r) = 0$ for $k < 0$. It is easy to see that

$$N(n,k,r) = \sum_{i \geq 0} p(n,k,r+in) = \sum_{i=[\frac{k-r}{n}]}^{[k-\frac{r}{n}]} p(n,k,r+in)$$

At last, for $n, k \in \mathbb{N},\ k > 0$, we set :

$$F(n,k,x) = \frac{x^k(1-x^{n+k-1})(1-x^{n+k-2})\cdots(1-x^n)}{(1-x)(1-x^2)\cdots(1-x^k)}$$

where x is a variable.

Remark 4.1 *It follows that $p(n,k,s)$ is also the number of partitions of s into k summands $b_j \in \mathbb{N}$ such that $1 \leq b_1 \leq \cdots \leq b_k \leq n$.*

Proposition 4.2 *With the previous notation, for $k > 0$, we have $F(n,k,x) = \sum_{s \geq 0} p(n,k,s)x^s$. Moreover, $F(n,k,x)$ is a polynomial in x.*

Proof. The generating function for the numbers $p(n,k,s)$ is the two-variable series $\varphi(x,y) = \sum_{s,k \geq 0} p(n,k,s)x^s y^k = \dfrac{1}{(1-yx)(1-yx^2)\cdots(1-yx^n)}$ because of

$$\frac{1}{(1-yx)(1-yx^2)\cdots(1-yx^n)} = \prod_{i=1}^{n}\left(\sum_{a_i \geq 0} y^{a_i}x^{ia_i}\right) =$$

$$\sum_{a_1 \geq 0, \ldots, a_n \geq 0} y^{a_1+\cdots+a_n}x^{a_1+2a_2+\cdots+na_n} = \sum_{k \geq 0, s \geq 0} p(n,k,s)y^k x^s$$

We can write $\varphi(x,y) = \sum_{k \geq 0} \varphi_k(x)y^k$ with $\varphi_k(x) = \sum_{s \geq 0} p(n,k,s)x^s$, for all $k \geq 0$.

We can easily check that $(1 - yx^{n+1})\varphi(x, xy) = (1 - yx)\varphi(x, y)$, which implies $(1 - x^k)\varphi_k(x) = (x - x^{n+k})\varphi_{k-1}(x)$ for $k > 0$, so that

$$\varphi_k(x) = \frac{(x - x^{n+k})(x - x^{n+k-1}) \cdots (x - x^{n+1})}{(1 - x^k)(1 - x^{k-1}) \cdots (1 - x)} \varphi_0(x), \text{ for } k > 0.$$

But $\varphi_0(x) = 1$. Hence $\varphi_k(x) = F(n, k, x)$ for $k > 0$.

To end, F is a polynomial in x since $p(n, k, s) = 0$ for large s. ■
We can now calculate $N(n, k, r)$.

Theorem 4.3 *With the previous notation, for $k > 0$, let $F_0, \dots, F_{n-1}$ be the n-components of $F(n, k, x)$, i.e. $F(n, k, x) = \sum_{r=0}^{n-1} x^r F_r(x^n)$. Then $N(n, k, r) = F_r(1)$.*

Proof. Write $F(n, k, x) = \sum_{j \geq 0} f_j x^j,\ f_j \in \mathbb{Q}$. Then $F_r(x^n) = \sum_{i \geq 0} f_{r+in} x^{ni} = \sum_{i \geq 0} p(n, k, r + in) x^{ni}$ and $F_r(1) = \sum_{i \geq 0} p(n, k, r + in) = N(n, k, r)$. ■

The value of $F_r(1)$ gives then the value of $N(n, k, r)$.

Theorem 4.4 *With the previous notation, set $d = \gcd(n, k)$ for $k, n > 0$. Then*

$$N(n, k, r) = \frac{1}{n}\binom{n + k - 1}{k} + \frac{1}{k} \sum_{l=1}^{d-1} \left(\cos\left(\frac{2lr\, pi}{d}\right) \prod_{1 \leq j \leq k-1, d | jl} \binom{n + j}{j} \right)$$

In particular, $N(n, k, r) = \frac{1}{n}\binom{n + k - 1}{k}$ for any $r \in \{0, \dots, n - 1\}$ when $d = 1$.

Proof. We use the relation $F(n, k, x) = \sum_{t=0}^{n-1} x^t F_t(x^n)$. We set $\alpha = e^{\frac{2i\pi}{n}}$. For all $r, m \in \{0, \dots, n - 1\}$, we have $\alpha^{-rm} F(n, k, \alpha^m) = \sum_{t=0}^{n-1} \alpha^{tm-rm} F_t(\alpha^{nm}) = \sum_{t=0}^{n-1} \alpha^{(t-r)m} F_t(1)$.

Summing on m we get

$$\sum_{m=0}^{n-1} \alpha^{-rm} F(n, k, \alpha^m) = \sum_{m=0}^{n-1} \left(\sum_{t=0}^{n-1} \alpha^{(t-r)m} F_t(1) \right) =$$

$$\sum_{t=0}^{n-1} \left(\sum_{m=0}^{n-1} \alpha^{(t-r)m} F_t(1) \right) = \sum_{t=0}^{n-1} F_t(1) \left(\sum_{m=0}^{n-1} \alpha^{(t-r)m} \right) = \sum_{t=0}^{n-1} F_t(1) n \delta_{rt} = nF_r(1)$$

So we obtain $F_r(1) = \frac{1}{n} \sum_{m=0}^{n-1} \alpha^{-rm} F(n, k, \alpha^m)$.

Now, we have to calculate $u_m = F(n,k,\alpha^m)$, where

$$\begin{aligned} F(n,k,x) &= x^k \frac{(1-x^{n+k-1})(1-x^{n+k-2})\cdots(1-x^{n+1})(1-x^n)}{(1-x^{k-1})(1-x^{k-2})\cdots(1-x)(1-x^k)} \\ &= x^k \frac{x^n-1}{x^k-1}\prod_{j=1}^{k-1}\left(\frac{x^{n+j}-1}{x^j-1}\right) \end{aligned}$$

which is a polynomial in x, so that $F(n,k,\alpha^m)$ has a sense.

Using L'Hopital's rule, we are going to calculate the values of $\frac{x^n-1}{x^k-1}$ and $\frac{x^{n+j}-1}{x^j-1}$ for $j=1,\ldots,k-1$, at $x=\alpha^m$, $m=0,1,\ldots,n-1$.

- If $n \not| mk$, then $\frac{\alpha^{mn}-1}{\alpha^{mk}-1}=0$.

  If $n|mk$, then $\left[\frac{x^n-1}{x^k-1}\right]_{x=\alpha^m} = \lim_{x\to\alpha^m}\frac{nx^{n-1}}{kx^{k-1}} = \frac{n}{k}$. Moreover, in this case, $\alpha^{mk}=1$.

Let $j\in\{1,\ldots,k-1\}$.

- If $n \not| mj$, then $\frac{\alpha^{m(n+j)}-1}{\alpha^{mj}-1}=1$.

  If $n|mj$, then $\left[\frac{x^{n+j}-1}{x^j-1}\right]_{x=\alpha^m} = \lim_{x\to\alpha^m}\frac{(n+j)x^{n+j-1}}{jx^{j-1}} = \frac{n+j}{j}$.

To sum up, we obtain $u_m=0$ if $n \not| mk$ and $u_m = \frac{n}{k}\prod_{1\le j\le k-1, n|jm}\frac{n+j}{j}$ if $n|mk$. In particular, $u_0 = \frac{n}{k}\prod_{j=1}^{k-1}\frac{n+j}{j} = \frac{n(n+1)\cdots(n+k-1)}{1\cdots(k-1)k} = \binom{n+k-1}{k}$.

Set $d=\gcd(n,k)$ and $n=n'd,\ k=k'd$ so that $\gcd(n',k')=1$.

Then $n|mk \Leftrightarrow n'|mk' \Leftrightarrow n'|m$.

If $n' \not| m$, then $u_m=0$

If $n'|m$, set $m=ln'$.

Then $n|mj \Leftrightarrow n'd|ln'j \Leftrightarrow d|lj$ so that $u_{ln'} = \frac{n}{k}\prod_{1\le j\le k-1, d|lj}\frac{n+j}{j}$.

This implies

$$\begin{aligned} N(n,k,r) &= \frac{1}{n}\binom{n+k-1}{k} + \frac{1}{n}\sum_{l=1}^{d-1}\alpha^{-rln'}u_{ln'} \\ &= \frac{1}{n}\binom{n+k-1}{k} + \frac{1}{n}\frac{n}{k}\sum_{l=1}^{d-1}\left(\alpha^{-rln'}\prod_{1\le j\le k-1, d|jl}\frac{n+j}{j}\right) \\ &= \frac{1}{n}\binom{n+k-1}{k} + \frac{1}{k}\sum_{l=1}^{d-1}\left(e^{\frac{-2i\pi rln'}{n}}\prod_{1\le j\le k-1, d|jl}\frac{n+j}{j}\right) \end{aligned}$$

which is a real number.

So, we get $N(n,k,r) = \frac{1}{n}\binom{n+k-1}{k} + \frac{1}{k}\sum_{l=1}^{d-1}\left(\cos\left(\frac{2lr\pi}{d}\right)\prod_{1\le j\le k-1, d|jl}\frac{n+j}{j}\right)$.

In particular, if $d = 1$, we get $N(n,k,r) = \frac{1}{n}\binom{n+k-1}{k}$ since we have an empty sum. ■

By the way, keeping the same notation, the following corollary results :

Corollary 4.5 *With the previous notation, we have* $\sum_{r=0}^{n-1} N(n,k,r) = \binom{n+k-1}{k}$.

Proof. It is enough to sum the formula of Theorem 4.4. We can also get it in view of $\sum_{r=0}^{n-1} N(n,k,r) = \sum_{r=0}^{n-1} F_r(1) = F(n,k,1) = \binom{n+k-1}{k}$. ■

Remark 4.6 $N(n,k,r)$ *is a d-periodic function in* r.

Corollary 4.7 *With the previous notation, we have* $N(n,k,r) = N(k,n,r)$.

Proof. We use the formula of Theorem 4.4

$$N(n,k,r) = \frac{1}{n}\binom{n+k-1}{k} + \frac{1}{k}\sum_{l=1}^{d-1}\left(\cos\left(\frac{2lr\pi}{d}\right)\prod_{1\le j\le k-1, d|jl}\binom{n+j}{j}\right)$$

where $d = \gcd(n,k)$. If $n = k$, there is nothing to prove. So, assume $n \neq k$.

• It is easily seen that $\frac{1}{n}\binom{n+k-1}{k} = \frac{1}{k}\binom{k+n-1}{n}$.

• The result is gotten if we prove that

$$\frac{1}{k}\prod_{1\le j\le k-1, d|jl}\left(\frac{n+j}{j}\right) = \frac{1}{n}\prod_{1\le j\le n-1, d|jl}\left(\frac{k+j}{j}\right)$$

for any $l \in \mathbb{N}$ such that $1 \le l \le d-1$.

For such an l and $a, b \in \mathbb{N}$, set $A(a,b) = \{j \in \mathbb{N} \mid a \le j \le b \text{ and } d|jl\}$. We may assume $n > k$. Then

$$\frac{1}{n}\prod_{1\le j\le n-1, d|jl}\left(\frac{k+j}{j}\right) = \frac{1}{n}\prod_{j\in A(1,n-1)}\left(\frac{k+j}{j}\right) = \frac{1}{n}\frac{\prod_{j\in A(1,n-1)}(k+j)}{\prod_{j\in A(1,n-1)} j}$$

But

$$\begin{aligned} A(1,n-1) &= A(1,n-k-1)\cup A(n-k+1,n-1)\cup\{n-k\} \\ &= A(k+1,n-1)\cup A(1,k-1)\cup\{k\} \end{aligned}$$

It follows that

$$\prod_{j\in A(1,n-1)}(k+j) = n\left(\prod_{j\in A(1,n-k-1)}(k+j)\right)\left(\prod_{j\in A(n-k+1,n-1)}(k+j)\right)$$

and

$$\prod_{j\in A(1,n-1)} j = k\left(\prod_{j\in A(k+1,n-1)} j\right)\left(\prod_{j\in A(1,k-1)} j\right)$$

Moreover, $j \in A(1, n-k-1) \Leftrightarrow k+j \in A(k+1, n-1)$ since $d|jl \Leftrightarrow d|(k+j)l$.
So we get $\prod_{j\in A(1,n-k-1)}(k+j) = \prod_{j\in A(k+1,n-1)} j.$
In the same way, we have $j \in A(n-k+1, n-1) \Leftrightarrow t = k+j-n \in A(1, k-1)$ since $d|jl \Leftrightarrow d|(k+j-n)l$.
So we get $\prod_{j\in A(n-k+1,n-1)}(k+j) = \prod_{t\in A(1,k-1)}(n+t) = \prod_{j\in A(1,k-1)}(n+j).$

It follows that

$$\frac{1}{n}\prod_{j\in A(1,n-1)}\left(\frac{k+j}{j}\right) = \frac{n\left(\prod_{j\in A(k+1,n-1)} j\right)\left(\prod_{j\in A(n-k+1,n-1)}(k+j)\right)}{nk\left(\prod_{j\in A(k+1,n-1)} j\right)\left(\prod_{j\in A(1,k-1)} j\right)}$$

$$= \frac{1}{k}\frac{\prod_{j\in A(n-k+1,n-1)}(k+j)}{\prod_{j\in A(1,k-1)} j} = \frac{1}{k}\frac{\prod_{j\in A(1,k-1)}(n+j)}{\prod_{j\in A(1,k-1)} j}$$

$$= \frac{1}{k}\prod_{j\in A(1,k-1)}\left(\frac{n+j}{j}\right)$$

and we are done. ■

When $\gcd(n,k) > 1$, we obtain a simpler evaluation for $N(n,k,r)$.

Theorem 4.8 *With the previous notation, set* $d = \gcd(n,k)$ *for* $k, n > 0$ *and assume* $d > 1$. *Then*

$$N(n,k,r) = \frac{1}{n}\binom{n+k-1}{k} + \frac{1}{k}\sum_{1<\delta\leq d,\delta|d}\frac{\varphi(\delta)\mu(\delta/\gcd(r,d))}{varphi(\delta/\gcd(r,d))}\binom{\frac{n}{\delta}+\frac{k}{\delta}-1}{\frac{n}{\delta}}$$

where φ *and* μ *are respectively the Euler function and the Möbius function.*

In particular, we have

$$N(n,k,0) = \frac{1}{n}\binom{n+k-1}{k} + \frac{1}{k}\sum_{1<\delta\leq d,\delta|d}\varphi(\delta)\binom{\frac{n}{\delta}+\frac{k}{\delta}-1}{\frac{n}{\delta}}$$

and

$$N(n,k,r) = \frac{1}{n}\binom{n+k-1}{n} + \frac{1}{k}\sum_{1<\delta\leq d,\delta|d} \mu(\delta)\binom{\frac{n}{\delta}+\frac{k}{\delta}-1}{\frac{n}{\delta}}$$

when $r > 0$ *and* $\gcd(r,d) = 1$.

Proof. Set $S = \sum_{l=1}^{d-1}\left(\cos\left(\frac{2lr\pi}{d}\right)\prod_{1\leq j\leq k-1,d|jl}\left(\frac{n+j}{j}\right)\right)$ with the notation of Theorem 4.4. We can write

$$\begin{aligned} S &= \sum_{1\leq\delta'\leq d-1,\delta'|d}\left(\sum_{1\leq l\leq d-1,\gcd(l,d)=\delta'}\left(\cos\left(\frac{2lr\pi}{d}\right)\prod_{1\leq j\leq k-1,d|jl}\left(\frac{n+j}{j}\right)\right)\right) \\ &= \sum_{1<\delta\leq d,\delta|d} \sigma_\delta \end{aligned}$$

where $\delta = \frac{d}{\delta'}$ and

$$\sigma_\delta = \sum_{1\leq l\leq d-1,\gcd(l,d)=\delta'}\left(\cos\left(\frac{2lr\pi}{d}\right)\prod_{1\leq j\leq k-1,d|jl}\left(\frac{n+j}{j}\right)\right)$$

For $\delta' = \gcd(l,d)$, we have $d|jl$ and $1 \leq j \leq k \Leftrightarrow \frac{d}{\delta'}$ divides $j\frac{l}{\delta'}$ and $1 \leq j \leq k$ $\Leftrightarrow \delta$ divides j and $1 \leq j \leq k \Leftrightarrow j = i\delta$ and $1 \leq i \leq \frac{k}{\delta}$.

It follows that

$$\prod_{1\leq j\leq k-1,d|jl}\left(\frac{n+j}{j}\right) = \prod_{1\leq i\leq\frac{k}{\delta}-1}\left(\frac{\frac{n}{\delta}+i}{i}\right) = \binom{\frac{n}{\delta}+\frac{k}{\delta}-1}{\frac{n}{\delta}}$$

and

$$\sigma_\delta = \binom{\frac{n}{\delta}+\frac{k}{\delta}-1}{\frac{n}{\delta}}\sum_{1\leq l\leq d-1,\gcd(l,d)=\delta'}\cos\left(\frac{2lr\pi}{d}\right)$$

Consider

$$\begin{aligned} \tau_\delta &= \sum_{1\leq l\leq d-1,\gcd(l,d)=\delta'}\cos\left(\frac{2lr\pi}{d}\right) = sum_{1\leq l\leq d-1,\gcd(l,d)=\delta'}\cos\left(\frac{2r\pi(\frac{l}{\delta'})}{\delta}\right) \\ &= \sum_{1\leq l'\leq\delta-1,\gcd(l',\delta)=1}\cos\left(\frac{2l'r\pi}{\delta}\right) \end{aligned}$$

where $l' = \frac{l}{\delta'}$.

But τ_δ is also the real part of the Ramanujan sum

$$c(r,\delta) = \sum_{1 \le l' \le \delta-1, \gcd(l',\delta)=1} e^{\frac{2il'r\pi}{\delta}}.$$

We have an explicite representation for $c(r,\delta)$ due to Hölder (see [13, Theorem 7.37, chapter 7, page 464]) by $c(r,\delta) = \dfrac{\varphi(\delta)\mu(m)}{\varphi(m)}$, where φ and μ are respectively the Euler function and the Möbius function, and where $m = d/\gcd(d, r\delta') = \delta/\gcd(r,\delta)$. Since $c(r,\delta)$ is a real number, we obtain $\tau_\delta = c(r,\delta)$ and the result is gotten.

In particular, we have the following two special cases

- $r = 0$ gives $\tau_\delta = \varphi(\delta)$

and

- $\gcd(r,d) = 1$ with $r > 0$ gives $\tau_\delta = \mu(\delta)$. ■

Example 4.9 *We are going to find the distinct factorizations into atoms of an element of a local seminormal CK domain.*

Let $\omega = (1+\sqrt{5})/2$ *and consider the PID* $\mathbb{Z}[\omega]$. *Since* 2 *is inert in* $\mathbb{Z}[\omega]$, *the ring* $S = \mathbb{Z}[2\omega]$ *is weakly factorial and t-closed, and so is a generalized CK domain with conductor* $2\mathbb{Z}[\omega]$, *a maximal ideal in* $\mathbb{Z}[\omega]$ *[11, Theorem 2] and [12, Example (2), page 177]. Set* $R = S_{2\mathbb{Z}[\omega]}$, *which is a local seminormal CK domain and* 2 *is an atom in* $\bar{R}$ *and* R. *In view of [12, Theorem 1.2, Proposition 2.1 and Proposition 3.1], we have* $|\mathcal{U}(\bar{R})/\mathcal{U}(R)| = 3$. *Set* $x = 32 = 2^5$. *By Theorems 3.11 and 4.4, we get* $\eta(x) = \dfrac{1}{3}\dbinom{7}{5} = 7$ *since* $\gcd(3,5) = 1$. *As* ω *is the fundamental unit of* $\mathbb{Z}[\omega]$, *its class generates the cyclic group* $\mathcal{U}(\bar{R})/\mathcal{U}(R)$. *We can choose* $p = 2$, $p' = 2\omega$, $p'' = 2\omega^2$ *for the nonassociate atoms of* R. *The different nonassociated factorizations of* x *into atoms of* R *are the following:*
$x = p^5 = \omega^{-3}p^3p'p'' = \omega^{-3}p^2p'^3 = \omega^{-6}p^2p''^3 = \omega^{-6}pp'^2p''^2 = \omega^{-6}p'^4p'' = \omega^{-9}p'p''^4$.

5. On the Asymptotic Behaviour of the Number of Distinct Factorizations into Atoms in a Seminormal CK Domain

As we saw in Section 3, we can restrict to the local case to evaluate the number of distinct factorizations into atoms of an element of a CK domain. To calculate this number for some special elements, we use results of Section 4.

Theorem 5.1 *Let* R *be a local seminormal CK domain with integral closure* $\bar{R}$. *Let* $\bar{R}p$ *be the maximal ideal of* $\bar{R}$, *with* $p \in R$. *Set* $n = |\mathcal{U}(\bar{R})/\mathcal{U}(R)|$.
Let $x = vp^k$, $k \in \mathbb{N}^*$, $v \in \mathcal{U}(\bar{R})$. *The number of nonassociated factorizations of* x^m, $m \in \mathbb{N}^*$ *into atoms of* R *is of the form* $\eta(x^m) = \dfrac{k^{n-1}}{n!}m^{n-1} + O(m^{n-2})$.

In particular, if x *is an atom of* R, *then* $\eta(x^m) = \dfrac{1}{n!}m^{n-1} + O(m^{n-2})$.

Proof. We can use Theorem 1.1 since its assumptions are satisfied by a CK domain. So $\eta(x^m)$ is of the form $\eta(x^m) = Am^d + O(m^{d-1})$ for $m \in \mathbb{N}^*$, where $A \in \mathbb{Q}$,

$d \in \mathbb{N}$, $A > 0$. Then, it is enough to find an equivalent of $\eta(x^m)$. For any $m \in n\mathbb{N}$, we have $v^m \in \mathcal{U}(R)$ and x^m is associated to p^{mk}, so that we can assume that n divides m to get A and d. In view of Theorem 3.11, we are led to calculate the number $N(n, km, 0) = \eta(x^m)$ of solutions $(a_1, \ldots, a_n) \in \mathbb{N}^n$ of the system gotten in Theorem 4.4 :

$$(S) \begin{cases} \sum_{i=1}^{n} a_i = km & (1) \\ \\ \sum_{i=1}^{n} \overline{ia_i} = \bar{0} & (2) \quad \text{in } \mathbb{Z}/n\mathbb{Z} \end{cases}$$

But, by Corollary 4.7, we have, since $n = \gcd(n, mk)$

$$N(n, km, 0) = N(km, n, 0) =$$

$$\frac{1}{mk}\binom{mk+n-1}{n} + \frac{1}{n}\sum_{l=1}^{n-1}\left(\cos\left(\frac{2lr\pi}{n}\right)\prod_{1\leq j\leq n-1, n|jl}\binom{mk+j}{j}\right)$$

where $r = 0$.

First, we have

$$\frac{1}{mk}\binom{mk+n-1}{n} = \frac{(mk+n-1)\cdots(mk+1)}{n!} \sim \frac{(mk)^{n-1}}{n!} = m^{n-1}\frac{k^{n-1}}{n!}.$$

Now, consider $\displaystyle \frac{1}{n}\sum_{l=1}^{n-1}\left(\prod_{1\leq j\leq n-1, n|jl}\binom{mk+j}{j}\right)$ since $r = 0$.

Because of $l \leq n - 1 < n$, we cannot have $n|l$, so that $j \neq 1$ and we have at most $n - 2$ factors in the product.

It follows that $\displaystyle \prod_{1\leq j\leq n-1, n|jl}\binom{mk+j}{j} \leq (mk+n)^{n-2} = O(m^{n-2})$. As we have a sum of $n - 1$ terms, we get that $\eta(x^m) \sim \dfrac{k^{n-1}}{n!}m^{n-1}$. ∎

References

[1] D.D. Anderson, D.F. Anderson and M. Zafrullah, Atomic domains in which almost all atoms are prime, *Comm. Algebra* **20** (1992), 1447-1462.

[2] D.D. Anderson and J.L. Mott, Cohen-Kaplansky domains: Integral domains with a finite number of irreducible elements, *J. Algebra* **148** (1992), 17-41.

[3] S.T. Chapman, F. Halter-Koch and U. Krause, Inside factorial monoids and integral domains, *J. Algebra,* **252** (2002), 350-375.

[4] I.S. Cohen and I. Kaplansky, Rings with a finite number of primes I, *Trans. Amer. Math. Soc.* **60** (1946), 468-477.

[5] D.E. Dobbs and M. Fontana, Locally pseudo-valuation domains, *Annali Mat. Pura Appl.* **134** (1983), 147-168.

[6] F. Halter-Koch, On the asymptotic behaviour of the number of distinct factorizations into irreducibles, *Ark. Mat.* **31** (1993), 297-305.

[7] F. Halter-Koch, Divisor theories with primary elements and weakly Krull domains, *Boll. UMI*, **9-B** (1995), 417-441.

[8] J.R. Hedstrom and E.G. Houston, Pseudo-valuation domains, *Pac. J. Math.* **75** (1) (1978), 137-147.

[9] G. Picavet and M. Picavet-L'Hermitte, Anneaux t-clos, *Comm. Algebra* **23** (1995), 2643-2677.

[10] M. Picavet-L'Hermitte, t-closed pairs, *Lecture Notes in Pure and Appl. Math.*, Marcel Dekker, New York 185 (1997), 401-415.

[11] M. Picavet-L'Hermitte, Factorization in some orders with a PID as integral closure, *Algebraic Number Theory and Diophantine Analysis*, de Gruyter, New York (2000), 365-390.

[12] M. Picavet-L'Hermitte, Weakly factorial quadratic orders, *Arab. J. Sci. Eng.* **26** (2001), 171-186.

[13] D. Redmond, *Number Theory*, Marcel Dekker, New York (1996).

[14] R.G. Swan, On seminormality, *J. Algebra* **67** (1980), 210-229.

[15] A. Zaks, Half-factorial domains, *Bull. Amer. Math. Soc.* **82** (1976), 721-724.

In: Mathematics, Game Theory and Algebra Compendium ISBN 978-1-60456-659-8
Editor: Jacob H. Mathias, pp. 183-194

Chapter 18

ON INTEGRALLY CLOSED GOING-DOWN RINGS

David E. Dobbs
Department of Mathematics, University of Tennessee
Knoxville, Tennessee 37996-1300, U. S. A.

Abstract

If R is a (commutative nonzero unital) ring, then $Z(R)$ and $Nil(R)$ denote the sets of zero-divisors and nilpotent elements of R, respectively. The collection of rings R such that $Nil(R)$ is a divided prime ideal of R is denoted by $\mathcal{H}$, and for such an R, the canonical map from the total quotient ring of R to $R_{Nil(R)}$ is denoted by $\phi = \phi_R$. Idealization examples show that the conditions "R is a ring such that $Z(R) = Nil(R)$" and "$R \in \mathcal{H}$" are logically independent. For rings R such that $Z(R) = Nil(R)$, the following three results are shown: R is a quasilocal going-down ring if and only if $R + PR_P$ is an integral ring extension of R for each $P \in \text{Spec}(R)$; if R is integrally closed, then R is a quasilocal going-down ring if and only if R is a divided ring; and if R is integrally closed, then $R \in \mathcal{H}$. For rings $R \in \mathcal{H}$, the following two results are shown: R is a going-down (resp., a divided) ring $\Leftrightarrow R/Nil(R)$ is a going-down (resp., a divided) domain $\Leftrightarrow \phi(R)$ is a going-down (resp., a divided) ring; and if $\phi(R)$ is integrally closed, then R is a quasilocal going-down ring if and only if R is a divided ring. Examples are given of an integrally closed quasilocal going-down ring $R \in \mathcal{H}$ such that R is not a divided ring; and of an integrally closed quasilocal EGD ring $R \in \mathcal{H}$ such that R is not a going-down ring. The class of ϕ-coherent rings is defined and is used to give new sufficient conditions for ϕ-Prüfer rings.

1. Introduction

All rings considered in this note are commutative with $1 \neq 0$, and all ring extensions are unital. If R is a ring, then $Z(R)$ denotes the set of zero-divisors in R; $Nil(R)$ denotes the set of nilpotent elements of R; $U(R)$ denotes the set of units of R; $\text{Spec}(R)$ denotes the set of prime ideals of R; $\text{Tot}(R)$ denotes the total quotient ring of R; R' denotes the integral closure of R (in $\text{Tot}(R)$); and $R_{red} := R/Nil(R)$ denotes the canonically associated reduced ring of R. A ring R is said to be *integrally closed* in case $R' = R$. By an *overring of* a ring R, we mean any ring T such that $R \subseteq T \subseteq \text{Tot}(R)$.

Recall from [8], [12] that a(n integral) domain R is called a *going-down domain* if $R \subseteq T$ satisfies the going-down property (GD, in the notation of [18, p. 28]) for all domain (resp., overring) extensions T of R; and from [11] that a ring R is called a *going-down ring* if R/P is a going-down domain for each $P \in \text{Spec}(R)$. A domain is a going-down ring if and only if it is a going-down domain [11, p. 2]. The most familiar examples of going-down domains are Prüfer domains and domains of Krull dimension 1. Any going-down domain must be treed [8, Theorem 2.2] (in the sense that none of its prime ideals can contain incomparable prime ideals), though a going-down ring need not be treed [11, Example 1]. Also, recall from [3], [9] that if R is a ring (resp., domain) and $P \in \text{Spec}(R)$, then P is called a *divided prime ideal of/in* R if $P \subseteq Rr$ for all $r \in R \setminus P$; and that R is called a *divided ring* (resp., *divided domain*) if each $P \in \text{Spec}(R)$ is divided in R. Any divided ring is a quasilocal treed going-down ring ([3, Proposition 1 (a)], [11, Remark (c), p. 4]), but the converse is false, even for domains [9, Example 2.9]. The result that motivates much of this paper is [19, Corollary 11]: an integrally closed domain is a quasilocal going-down domain (if and) only if it is a divided domain.

The above-mentioned motivating result was generalized in [9, Corollary 2.8], with "an integrally closed domain" replaced by "a seminormal domain". The latter result was, in turn, generalized in [6, Theorem 3.4] to the context of seminormal (in the sense of [20]) rings. As seminormal rings are necessarily reduced, we focus here on possible generalizations of the motivating result to various classes of rings that are not necessarily reduced. One fruitful such class consists of the rings R such that $Z(R) = Nil(R)$. Indeed, it was shown in [11, Corollary 2.6] that if R is a ring such that $Z(R) = Nil(R)$, then R is a going-down ring if and only if R is an EGD ring (in the sense that $R \subseteq T$ satisfies GD for each overring T of R); without the hypothesis that $Z(R) = Nil(R)$, neither implication is valid [11, Examples 1 and 2]. The tractability of this class of rings is due, in part, to the following fact. If R is a ring such that $Z(R) = Nil(R) =: Q$, then there is a canonical R-algebra isomorphism $R_Q \cong \text{Tot}(R)$ [11, Proposition 2.3 (b)] and, for each $P \in \text{Spec}(R)$, this isomorphism restricts to allow R_P to be viewed (up to R-algebra isomorphism) as an overring of R [6, Proposition 2.5 (a)]. One of our main results, Theorem 2.2 (a), generalizes a result for domains [10, Proposition 2.1], by establishing that a ring R such that $Z(R) = Nil(R)$ is a quasilocal going-down ring if and only if $R + PR_P$ is an integral ring extension of R for each $P \in \text{Spec}(R)$.

Another fruitful generalization of domains is the collection $\mathcal{H} := \{R \mid R$ is a ring such that $Nil(R)$ is a divided prime ideal of $R\}$. For $R \in \mathcal{H}$, the ring homomorphism $\phi = \phi_R$: $\text{Tot}(R) \rightarrow R_{Nil(R)}$ is defined by $\phi(\frac{a}{b}) = \frac{a}{b}$ for all $a \in R, b \in R \setminus Z(R)$. The ϕ-theoretic approach has been used in a series of papers (of which [1], [2], [4] and [5] are typical) to generalize several concepts and results from the setting of domains to that of rings possibly with nontrivial zero-divisors. With this inspiration, we say that a ring $R \in \mathcal{H}$ is a *ϕ-going-down ring* (resp., a *ϕ-divided ring*) if $\phi(R)$ is a going-down ring (resp., a divided ring). Corollary 2.9 and Proposition 2.6 establish that $R \in \mathcal{H}$ is a *ϕ-going-down ring* (resp., a *ϕ-divided ring*) $\Leftrightarrow$ R is a going-down ring (resp., a divided ring) $\Leftrightarrow$ R_{red} is a going-down domain (resp., a divided domain).

We return now to the motivating question of generalizing [19, Corollary 11], [9, Corollary 2.8] beyond the context of domains. It was shown in [6, Example 3.10] that there exists a family of quasilocal integrally closed treed going-down rings that fail to be divided rings. However, each of these rings R is such that $Z(R) \neq Nil(R)$ and $R \notin \mathcal{H}$. Our basic question thus has the following sharpening: if R is an integrally closed ring such that $Z(R) = Nil(R)$ or $R \in \mathcal{H}$, must it be the case that R is a quasilocal going-down ring (if and) only if R is a divided ring? We show that the answer is "Yes and no". More precisely, Theorem 2.4 establishes that if R is an integrally closed ring such that $Z(R) = Nil(R)$, then R is a quasilocal going-down ring (if and) only if R is a divided ring. On the other hand, Example 2.13 constructs an integrally closed ring $R \in \mathcal{H}$ such that R is a quasilocal going-down ring but R is not a divided ring. Nevertheless, the ϕ-theoretic approach leads to the following positive result in Corollary 2.9 (d): if $R \in \mathcal{H}$ is such that $\phi(R)$ is integrally closed, then R is a quasilocal going-down ring (if and) only if R is a divided ring.

It is natural to try to relate the two main classes of non-domains that we have mentioned above. We show in Proposition 2.12 that the conditions "R is a ring such that $Z(R) = Nil(R)$" and "$R \in \mathcal{H}$" are logically independent. However, a logical connection does exist within the class of integrally closed rings, as Proposition 2.7 establishes that if an integrally closed ring R satisfies $Z(R) = Nil(R)$, then $R \in \mathcal{H}$. Also noteworthy are Proposition 2.14 and Example 2.15, which establish a non-reversible implication between "R is a going-down ring" and "R is an EGD ring" for the rings $R \in \mathcal{H}$.

Motivated in part by the nonnil-Noetherian rings introduced in [5], we define the class of ϕ-coherent rings as consisting of those $R \in \mathcal{H}$ such that $\phi(R)$ is a coherent ring. This concept is used in Proposition 2.17, an analogue of the characterization [7, Corollary 4] of Prüfer domains as the integrally closed coherent going-down domains, to give new sufficient conditions for the ϕ-Prüfer rings that were introduced in [1].

The examples in Proposition 2.12, as well as those in Examples 2.13 and 2.15, are constructed as idealizations. Recall that if A is a ring and E is a (unital) A-module, then the *idealization* $R := A(+)E$ is the ring whose additive structure is the abelian group $A \oplus E$ and whose multiplication is defined by $(a_1, e_1)(a_2, e_2) = (a_1a_2, a_1e_2 + a_2e_1)$. A convenient reference for the basic facts about idealizations is [17]. For instance, Spec$(A(+)E) = \{P(+)E \mid P \in \text{Spec}(R)\}$. One useful consequence is that an idealization $A(+)E$ is quasilocal if and only if A is quasilocal.

Any unexplained material is standard, as in [15], [18].

The author wishes to thank Ayman Badawi for some stimulating correspondence.

2. Results

We begin by recalling the following result, for the sake of completeness. As usual, we say that a ring extension $R \subseteq T$ is *unibranched* if the canonical contraction map Spec$(T) \to$ Spec$(R), Q \mapsto Q \cap R$, is a bijection.

Proposition 2.1. *[6, Theorem 3.3] Let $R \in \mathcal{H}$ such that $Z(R) = Nil(R)$. Then R is a quasilocal going-down ring if and only if there exists an integral overring T of R such that T is a divided ring and the ring extension $R \subseteq T$ is unibranched.*

Without assuming *both* "$R \in \mathcal{H}$" and "$Z(R) = Nil(R)$", one still has interesting characterizations of the quasilocal going-down rings R: see Proposition 2.6 (a) and part (a) of the next result. Theorem 2.2 (c) also recalls a useful characterization of the divided rings R such that $Z(R) = Nil(R)$.

Theorem 2.2. *Let R be a ring such that $Z(R) = Nil(R)$. Then:*

(a) R is a quasilocal going-down ring if and only if $R + PR_P$ is an integral ring extension of R for each $P \in Spec(R)$.

(b) Let $P \in Spec(R)$. Then P is a divided prime ideal of R if and only if $PR_P = P$ (when R_P is identified as an overring of R).

(c) R is a divided ring if and only if $PR_P = P$ for each $P \in Spec(R)$ (when each R_P is identified as an overring of R).

Proof. We recalled in the Introduction that the "$Z(R) = Nil(R)$" hypothesis permits R_P to be viewed as an overring of R for each $P \in \text{Spec}(R)$. It follows, under this hypothesis, that $R \subseteq R + PR_P \subseteq \text{Tot}(R)$ for each $P \in \text{Spec}(R)$, and so the assertions in (a)–(c) are meaningful.

(a) Suppose first that $R+PR_P$ is an integral ring extension of R for each $P \in \text{Spec}(R)$. Then, by modifying the domain-theoretic proof that (e) $\Rightarrow$ (a) in [10, Proposition 2.1], we obtain that R is a quasilocal EGD ring. (We leave these details to the reader.) Therefore, since $Z(R) = Nil(R)$, [11, Proposition 2.3 (a) and Theorem 2.5] ensure that R is a going-down ring.

Conversely, suppose that R is a quasilocal going-down ring. Then, since $Z(R) = Nil(R)$, [11, Proposition 2.3 (a) and Theorem 2.4 (b)] ensure that R is an EGD ring. We now proceed to show that if $P \in \text{Spec}(R)$, then the overring $R + PR_P$ is integral over R.

Put $Q := Z(R) = Nil(R)$. As $Q \subseteq P$, we can use [11, Proposition 2.3 (b)] and [6, Proposition 2.5 (b)] to view $R \subseteq R_P \subseteq R_Q = \text{Tot}(R)$. Put $D := R/Q$, and consider $\overline{P} := P/Q \in \text{Spec}(D)$. Note that D is a quasilocal going-down domain, by [11, Proposition 2.1 (b)]. Therefore, by [9, Corollary 2.6], we conclude that $D \subseteq D + \overline{P}D_{\overline{P}} =: E$ is an integral overring extension of domains. In fact, that result shows that for each $v \in \overline{P}D_{\overline{P}} =: N$, there exists a positive integer m such that $v^j \in \overline{P}$ for each integer $j \geq m$.

Now, consider the canonical isomorphism $\theta : D_{\overline{P}} \rightarrow R_P/QR_P$ (cf. [15, Proposition 5.8]). This map is given by

$$\theta(\frac{r+Q}{s+Q}) = \frac{r}{s} + QR_P \text{ for all } r \in R, s \in R \setminus P.$$

It is easy to check that θ induces the following three isomorphisms:

$$D \rightarrow (R + QR_P)/QRP, \quad \overline{P} \rightarrow (P + QR_P)/QRP, \quad N \rightarrow PR_P/QRP,$$

and hence an isomorphism $E = D + N \rightarrow (R + PR_P)/QRP$. By using these isomorphisms to reformulate the above information, we see that for each $w \in PR_P$, there exists a positive integer m such that $w^j \in P + QR_P$ for each integer $j \geq m$. In particular, there exists $t \in P \subseteq R$ such that $u := w^m - t \in QR_P$. However, each element of QR_P is nilpotent (since $Q = Nil(R)$), and so there exists a positive integer n such that $u^n = 0$. By expanding $(w^m - t)^n = 0$, we obtain an integrality equation for w over R (of degree mn). In particular, each element of PR_P is integral over R, and the proof of (a) is complete.

As for the remaining assertions, note that (b) is a special case of [6, Proposition 2.5

(c)], and that (c) follows from (b) and the definition of a divided ring. □

The next result summarizes the basic connection between the "quasilocal going-down ring" and "divided ring" concepts.

Proposition 2.3. *Any divided ring is a quasilocal treed going-down ring.*

Proof. The " quasilocal treed" conclusion was noted in [3, Proposition 1 (a)]. For the "going-down ring" conclusion, see either [11, Remark (c), p. 4] or [6, Proposition 3.1]. □

The converse of Proposition 2.3 is false, even if R is a domain: see [9, Example 2.9] and [8, Theorem 2.2]. As noted in the Introduction, the converse does hold, however, for integrally closed domains R. In our main positive result, we next generalize this equivalence to the "$Z(R) = Nil(R)$" context.

Theorem 2.4. *Let R be an integrally closed ring such that $Z(R) = Nil(R)$. Then R is a quasilocal going-down ring if and only if R is a divided ring.*

Proof. The "if" assertion is immediate from Proposition 2.3 (and does not need the "integrally closed" hypothesis). Conversely, suppose that R is a quasilocal going-down ring. By Theorem 2.2 (c), it suffices to prove that if $P \in \text{Spec}(R)$, then $PR_P = P$. Now, by Theorem 2.2 (a) and [6, Proposition 2.5 (a)], $T := R + PR_P$ is an integral overring of R. As R is integrally closed, it follows that $T = R$, whence $PR_P \subseteq R$ and $PR_P = PR_P \cap R = P$, as desired. □

Remark 2.5. One cannot simply omit the hypothesis that $Z(R) = Nil(R)$ in Theorem 2.4. Indeed, the construction in [6, Example 3.10] supplies a family of integrally closed quasilocal going-down rings R (there called "A") which fail to be divided rings. Necessarily, such rings R satisfy $Z(R) \neq Nil(R)$, by Theorem 2.4. In fact, they also satisfy $R \notin \mathcal{H}$, by the proof of [6, Example 3.10]. This leaves open the question whether an integrally closed quasilocal going-down ring in $\mathcal{H}$ must be a divided ring. We answer this question in Example 2.13 below.

Next, we obtain the basic characterizations of the "quasilocal going-down ring" and "divided ring" concepts for the "$R \in \mathcal{H}$" context.

Proposition 2.6. *Let R be a ring. Then:*

(a) Assume, in addition, that $Nil(R) \in \text{Spec}(R)$ (for instance, assume that either $Z(R) = Nil(R)$ or $R \in \mathcal{H}$). Then R is a going-down (resp., quasilocal going-down) ring if and only if R_{red} is a going-down (resp., quasilocal going-down) domain.

(b) Assume, in addition, that $R \in \mathcal{H}$. Then R is a divided ring if and only if R_{red} is a divided domain.

Proof. (a) In view of the canonical bijection $\text{Spec}(R_{red}) \rightarrow \text{Spec}(R)$ supplied by a standard homomorphism theorem, it is clear that R is quasilocal if and only if R_{red} is quasilocal. Moreover, it has already been noted in [13, p. 563], as an easy consequence of another part of that same homomorphism theorem, that R is a going-down ring if and only if R_{red} is a going-down ring. Accordingly, since the hypothesis that $Nil(R) \in \text{Spec}(R)$ ensures that R_{red} is a domain, (a) now follows from the fact that a domain is a going-down ring if and only if it is a going-down domain [11, p. 2].

(b) The assertion was established as part of [6, Proposition 2.11]. We next offer a somewhat simpler proof. It is a consequence of the definitions that a domain is a divided ring if and only if it is a divided domain. Thus, the "only if" assertion is known (see [11, Remark (c), p. 4] or [3, Corollary 3]) and does not require the "$R \in \mathcal{H}$" hypothesis. Conversely, suppose that $D := R_{red}$ is a divided domain. We will show that each $P \in \text{Spec}(R)$ is a divided prime ideal of R, i.e., that $P \subseteq Rr$ for each $r \in R \setminus P$. Consider $\overline{P} := P + Nil(R) \in \text{Spec}(D)$ and $\overline{r} := r + Nil(R) \in D$. Since $\overline{P}$ is a divided prime ideal of D and $\overline{r} \in D \setminus \overline{P}$, we have that $\overline{P} \subseteq D\overline{r}$, whence $P \subseteq Rr + Nil(R)$. However, since $R \in \mathcal{H}$ and $r \in R \setminus Nil(R)$, we have that $Nil(R) \subseteq Rr$, whence $P \subseteq Rr$, as required. □

We next give a useful point of contact between the "$Z(R) = Nil(R)$" and "$R \in \mathcal{H}$" concepts. It might also be be useful to note (shades of [9, Corollary 2.8]!) that the proof of Proposition 2.7 goes through if "integrally closed" is weakened to "root-closed".

Proposition 2.7. *If an integrally closed ring R satisfies $Z(R) = Nil(R)$, then $R \in \mathcal{H}$.*

Proof. Since $Z(R) = Nil(R)$, it is clear that $Nil(R) \in \text{Spec}(R)$. It only remains to show that $Nil(R)$ is a divided prime ideal of R. We will show that if $r \in R \setminus Nil(R)$, then $Nil(R) \subseteq Rr$. Consider any element $s \in Nil(R)$. Put $t := \frac{s}{r} \in \text{Tot}(R)$. Since $s^n = 0$ for some positive integer n, we have that $t^n = 0$. As R is integrally (and, hence, root-) closed, $t \in R$. In other words, there exists $a \in R$ such that $\frac{s}{r} = \frac{a}{1} \in \text{Tot}(R)$. Thus, $s = ar \in Rr$, as desired. □

Remark 2.8. We pause to indicate how to use Propositions 2.6 and 2.7 in order to produce an alternate proof of Theorem 2.4. As recalled in Proposition 2.3, any divided ring is quasilocal. Therefore, our task is to show that if R is a quasilocal integrally closed ring such that $Z(R) = Nil(R)$, then R is a going-down ring if and only if R is a divided ring. Note that $R \in \mathcal{H}$, by Proposition 2.7. Moreover, $D := R_{red}$ is a quasilocal domain.

It is important to prove that D inherits the "integrally closed" property from R, and there at least two ways to carry out this proof. The first uses an overtly ϕ-theoretic approach (which is trivial here, since the "$Z(R) = Nil(R)$" hypothesis ensures that $R \cong \phi(R)$ canonically). In detail, [2, Lemma 2.8] yields that ($R \cong$) $\phi(R)$ is integrally closed if and only if D is integrally closed. For a proof that is not overtly ϕ-theoretic, let $K := \text{Tot}(R)$, and note as in the proof of Proposition 2.7 that $Nil(K) = Nil(R)$. As it is easy to check that $K_{red} = K/Nil(R) =: F$ is the quotient field of D, we have that R is a pullback, namely, $R = D \times_F K$. If $\pi : K \rightarrow F$ is the canonical surjection, then the lore of pullback theory (cf. [14]) yields that $R' = \pi^{-1}(D')$. In particular, R is integrally closed if and only if D is integrally closed.

Now, the assertions in Proposition 2.6 reduce our task to showing that the quasilocal integrally closed domain D is a going-down domain if and only if it is a divided domain. Accordingly, an application of the motivating result (as in either [19, Corollary 11] or [9, Corollary 2.8]) completes the alternate proof of Theorem 2.4.

Recall from [2] that if $R \in \mathcal{H}$, then R is said to be *ϕ-integrally closed* if $\phi(R)$ is integrally closed; and, by [2, Lemma 2.8], this condition is equivalent (given $R \in \mathcal{H}$) to the domain R_{red} being integrally closed. Similarly, in the spirit of definitions in papers such as [1] and [2], we say that if $R \in \mathcal{H}$, then R is a *ϕ-going-down ring* (resp., a *ϕ-divided ring*) if $\phi(R)$ is a going-down ring (resp., divided ring). A complete characterization of these concepts is given in the next result.

Corollary 2.9. *Let $R \in \mathcal{H}$. Then:*

(a) The following three conditions are equivalent:

(1) R is a ϕ-going-down ring (i.e., $\phi(R)$ is a going-down ring);

(2) $\phi(R)$ is an EGD ring;

(3) R is a going-down ring.

(b) The following four conditions are equivalent:

(1) $\phi(R)$ is a quasilocal going-down ring;

(2) $\phi(R)$ is a quasilocal EGD ring;

(3) R is a quasilocal ϕ-going-down ring;

(4) R is a quasilocal going-down ring.

(c) R is a ϕ-divided ring if and only if R is a divided ring.

(d) Assume also that R is ϕ-integrally closed (i.e., that $\phi(R)$ is integrally closed). Then R is a quasilocal going-down ring if and only if R is a divided ring.

Proof. (a), (b) It is known that $\ker(\phi) \subseteq Nil(R)$. So, by a standard homomorphism theorem, the contraction map $\text{Spec}(\phi(R)) \to \text{Spec}(R)$ is an isomorphism of partially ordered sets (under inclusion). In particular, $\phi(R)$ is quasilocal if and only if R is quasilocal. Moreover, since it is also known that $Z(\phi(R)) = Nil(\phi(R))$, [11, Corollary 2.6] ensures that $\phi(R)$ is a going-down ring if and only if $\phi(R)$ is an EGD ring. Thus, it suffices to show that $\phi(R)$ is a going-down ring if and only if R is a going-down ring. This, however, is easy to see, for [1, Lemma 2.5] ensures that, up to isomorphism, $\phi(R)$ and R have the same factor domains.

(c) It is enough to repeat the final sentence of the proof of (a).

(d) By (b) and (c), an equivalent task is to show that $\phi(R)$ is a quasilocal going-down ring if and only if $\phi(R)$ is a divided ring. Since $\phi(R)$ is integrally closed and $Z(\phi(R)) = Nil(\phi(R))$, an application of Theorem 2.4 (or Remark 2.8) completes the proof. □

The next two results set up the machinery that will produce the desired examples.

Proposition 2.10. *Let D be a domain with quotient field F. Consider the D-module $E := F/D$, and put $R := D(+)E$. Then:*

(a) $Nil(R) = 0(+)E$ is a divided prime ideal of R, and so $R \in \mathcal{H}$.

(b) $Z(R) = R \setminus U(R)$. In particular, $R = Tot(R)$, and so R is trivially both integrally closed and an EGD ring.

Proof. (a) Since D is a reduced ring, an easy calculation shows that $Nil(R) = 0(+)E$; and this is a prime ideal of R since $0 \in \text{Spec}(D)$. As for the "divided" conclusion, it suffices to show that if $0 \neq d \in D$ and $f \in F$, then $Nil(R) \subseteq R(d, f + D)$. For this, note that if $g \in F$, then $(0, g + D) = (0, gd^{-1} + D)(d, f + D)$. (For a module-theoretic point of view on the proof of the "divided" conclusion, see [2, Remark 1].)

(b) Clearly, $U(R) \subseteq U(D)(+)E$. The reverse inclusion also holds. Indeed, if $d \in U(D)$ and $f \in F$, then the element $z := (d^{-1}, -d^{-2}f + D) \in R$ satisfies $(d, f + D)z = (1, 0)$, the identity element of R. Thus, $U(R) = U(D)(+)E$.

Of course, $Z(R) \subseteq R \setminus U(R)$. It suffices to establish the reverse inclusion. Consider any element $x \in R \setminus U(R)$. We will show that $x \in Z(R)$. Without loss of generality, x is not nilpotent. Therefore, by (a) and the above comments, $x = (\delta, g + D)$ for some $\delta \in D \setminus (U(D) \cup \{0\})$ and some $g \in F$. Then $x \in Z(R)$ since the nonzero element $y := (0, \delta^{-1} + D) \in R$ satisfies $xy = (0, 1 + D) = (0, D) = 0 \in R$. □

Proposition 2.11. *Let D be a domain with quotient field F. Put $R := D(+)D$. Then:*

(a) $Z(R) = Nil(R) = 0(+)D$ is a prime ideal of R.

(b) $Tot(R) = F(+)F$.

(c) $R \in \mathcal{H}$ if and only if D is a field.

(d) $R' = D'(+)F$. Thus, R is integrally closed if and only if D is a field.

Proof. (a) Argue as in the first sentence in the proof of Proposition 2.10 (a) to establish the second asserted equality. The first asserted equality follows easily from the fact that D is a domain.

(b) View R as a subring of $F(+)F$ in the usual way. Note that if $d, e \in D$ with $d \neq 0$, then $(d, e)(d^{-1}, -ed^{-2}) = (1, 0)$, the identity element of R. Therefore, by (a), each non-zerodivisor of R is invertible in $F(+)F$, and so $\mathrm{Tot}(R) \subseteq F(+)F$. For the reverse inclusion, consider any element $x \in F(+)F$. Write $x = (a/c, b/c)$ for some $a, b, c \in D$ such that $c \neq 0$. Observe that $x(c, 0) = (a, b)$, whence $x = (a, b)(c, 0)^{-1} \in \mathrm{Tot}(R)$.

(c) From the point of view announced in [2, Remark 1], the assertion follows since D is a divisible D-module if and only if D is a field. We next offer a direct proof of (c).

Suppose first that D is a field (namely, F). Then, by (b), R is its own total quotient ring. In particular, R is integrally closed. Thus, by combining (a) with Proposition 2.7, we conclude that $R \in \mathcal{H}$.

For the (contrapositive of the) converse, we assume that d is a nonzero nonunit of D and we must show that, although $Nil(R)$ is a prime ideal of R, it is not divided in R. In fact, since D is a domain, it is easy to check that $Nil(R) \not\subseteq R(d, 0)$, for there do not exist elements $e, f \in D$ such that $(0, 1) = (e, f)(d, 0)$.

(d) In view of (b), projection onto the first coordinate shows that $R' \subseteq D'(+)F$. Similarly, $D'(+)0 \subseteq R'$. Moreover, $0(+)F = Nil(F(+)F) \subseteq R'$. Thus,

$$D'(+)F = (D'(+)0) + (0(+)F) \subseteq R'.$$

The conclusions are now immediate. □

It is now an easy matter to use Propositions 2.10 and 2.11 to show that neither "R is a ring such that $Z(R) = Nil(R)$" nor "$R \in \mathcal{H}$" implies the other.

Proposition 2.12. *The conditions "R is a ring such that $Z(R) = Nil(R)$" and "$R \in \mathcal{H}$" are logically independent.*

Proof. First, take $R := D(+)D$ as in Proposition 2.11, with D a domain but not a field. Then, by parts (a) and (c) of Proposition 2.11, $Z(R) = Nil(R)$ but $R \notin \mathcal{H}$. Notice, by

Proposition 2.7, that any ring with the properties established in this paragraph must fail to be integrally closed. And, by Proposition 2.11 (d), such is indeed the case for the ring that we have just constructed.

Next, take $R := D(+)E$ as in Proposition 2.10, with D a domain but not a field. By Proposition 2.10 (a) and (the proof of) Proposition 2.10 (b), $R \in \mathcal{H}$ but $Z(R) \neq Nil(R)$. (Indeed, if d is any nonzero nonunit of D, then $(d, e) \in Z(R) \setminus Nil(R)$ for any $e \in E$.) □

We next show that one cannot replace "ϕ-integrally closed" with "integrally closed" in the statement of Corollary 2.9 (d). In other words, a ϕ-theoretic formulation is essential if one wishes to generalize [19, Corollary 11] and [9, Corollary 2.8] to the rings in $\mathcal{H}$.

Example 2.13. There exists an integrally closed ring $R \in \mathcal{H}$ such that R is a quasilocal going-down ring but R is not a divided ring.

Proof. Take D to be a quasilocal going-down domain that is not a divided domain (constructed, for instance, as in [9, Example 2.9]). Let F be the quotient field of D, let $E := F/D$, and take $R := D(+)E$, as in Proposition 2.10. By part (b) of that result, R is integrally closed. Moreover, $R \in \mathcal{H}$ by Proposition 2.10 (a). Thus, by parts (a) and (b) of Proposition 2.6, it suffices to show that R_{red} is a quasilocal going-down domain that is not a divided domain. This, in turn, follows from the choice of D, for Proposition 2.10 (a) ensures that

$$R_{red} = R/Nil(R) = (D(+)E)/(0(+)E) \cong D.$$

Given the roles of the EGD concept in parts (a) and (b) of Corollary 2.9, it is natural to ask about possible implications between "R is a going-down ring" and "R is an EGD ring" for $R \in \mathcal{H}$. The next two results settle this matter.

Proposition 2.14. *Let R be a ring such that $Nil(R) \in Spec(R)$ (for instance, suppose that $R \in \mathcal{H}$). If, in addition, R is a going-down ring, then R is an EGD ring.*

Proof. By [13, Proposition 2.1 (a)], it suffices to show that R is a locally irreducible ring, i.e., that each prime ideal of R contains a unique minimal prime ideal of R. This, however, is evident since $Nil(R) \in \text{Spec}(R)$ (cf. [18, Theorems 10 and 25]). □

Example 2.15. There exists an integrally closed quasilocal EGD ring $R \in \mathcal{H}$ such that R is not a going-down ring.

Proof. Take D to be a quasilocal domain that is not a going-down domain (for instance, the formal power series ring in two analytically independent indeterminates over a field). Let F be the quotient field of D, let $E := F/D$, and take $R := D(+)E$, as in Proposition 2.10. By part (b) of that result, R is an integrally closed EGD ring. Also, R is quasilocal since D is quasilocal. Moreover, $R \in \mathcal{H}$ by Proposition 2.10 (a). Thus, by Proposition 2.6 (a), it suffices to show that R_{red} is not a going-down domain. This, in turn, follows since (as we showed in the proof of Example 2.13), $R_{red} \cong D$. □

Recall from [1] that a ring $R \in \mathcal{H}$ is said to be a *ϕ-Prüfer ring* if $\phi(I)$ is an invertible ideal of $\phi(R)$ for each ideal I of R such that $I \not\subseteq Nil(R)$; equivalently, by [1, Theorem

2.2], if $\phi(R)$ is a Prüfer ring in the sense of [16], i.e., if every finitely generated regular ideal of $\phi(R)$ is invertible; equivalently, by [1, Theorem 2.6], if R_{red} is a Prüfer domain. It seems natural to seek a ϕ-theoretic analogue of the result [7, Corollary 4] that a domain is a Prüfer domain if and only if it is an integrally closed coherent going-down domain. Accordingly, we say that a ring $R \in \mathcal{H}$ is a *ϕ-coherent ring* if $\phi(R)$ is a coherent ring. Our final result uses this concept to give new sufficient conditions for a ring to be a ϕ-Prüfer ring. First, we need a lemma that is of some independent interest. As usual, if A is a ring and n is a positive integer, then A^n denotes the direct sum of n copies of A.

Lemma 2.16. *Let $R \in \mathcal{H}$, and put $D := R_{red}$. If R is a ϕ-coherent ring, then D is a coherent domain.*

Proof. We will show that each nonzero finitely generated ideal J of D is finitely presented. As usual, the map on associated reduced rings that is induced by ϕ allows us to identify D with $\phi(R)_{red}$. Then, by [1, Lemma 2.4], $J = I/Nil(\phi(R))$ for some finitely generated ideal I of $\phi(R)$ such that $I \not\subseteq Nil(\phi(R))$. Pick $i \in I \setminus Nil(\phi(R))$. Since $\phi(R) \in \mathcal{H}$, we have that $Nil(\phi(R)) \subseteq \phi(R)i$. As $Nil(\phi(R))$ is a prime ideal of $\phi(R)$, it follows easily that $Nil(\phi(R)) = Nil(\phi(R))i = Nil(\phi(R))I$. Thus, there is a canonical isomorphism

$$I \otimes_{\phi(R)} D = I \otimes_{\phi(R)} \phi(R)/Nil(\phi(R)) \cong I/Nil(\phi(R))I = I/Nil(\phi(R)) = J.$$

Also, since $\phi(R)$ is a coherent ring by hypothesis, there exists a positive integer n and a finitely generated $\phi(R)$-module N figuring in a short exact sequence

$$0 \to N \to \phi(R)^n \to I \to 0$$

of $\phi(R)$-modules. Applying the (right-exact) functor $\cdot \otimes_{\phi(R)} D$ to this sequence produces the exact sequence

$$N \otimes_{\phi(R)} D \to D^n \to I \otimes_{\phi(R)} D \to 0$$

of D-modules. Observe that $K := N \otimes_{\phi(R)} D$ is a finitely generated D-module, and therefore so is its image in D^n. Hence, in view of the above canonical isomorphism, we see that $J \cong D^n/\mathrm{im}(K)$, which is a finitely presented D-module, as desired. □

Proposition 2.17. *Let $R \in \mathcal{H}$. Consider the following four conditions:*

(1) R is a ϕ-integrally closed, ϕ-coherent, ϕ-going-down ring;

(2) R is a ϕ-integrally closed, ϕ-coherent going-down ring;

(3) $\phi(R)$ is integrally closed and R is a ϕ-coherent going-down ring;

(4) R is a ϕ-Prüfer ring.

Then (1) $\Leftrightarrow$ (2) $\Leftrightarrow$ (3) $\Rightarrow$ (4).

Proof. Note that (1) $\Leftrightarrow$ (2) by Corollary 2.9 (a); and (2) $\Leftrightarrow$ (3) by the definition of "ϕ-integrally closed". It remains to prove that (3) $\Rightarrow$ (4).

As usual, we identify the domain $D := R_{red} \cong \phi(R)_{red}$. Recall that $K := \mathrm{Tot}(\phi(R)) = R_{Nil(R)}$ is such that $Nil(K) = Nil(\phi(R))$ and K_{red} is the quotient field of D. Therefore, by using the pullback-theoretic methods in the second paragraph of Remark 2.8 (cf. also [2, Lemma 2.8]), we see that $D' = \phi(R)'/Nil(\phi(R)) = \phi(R)'_{red}$. In particular, D is integrally closed if and only if $\phi(R)$ is integrally closed.

(3) $\Rightarrow$ (4): Recall, from [1, Theorem 2.6], that (4) is equivalent to D being a Prüfer domain, i.e. by [7, Corollary 4], to D being an integrally closed coherent going-down domain. Assume (3). Then D is integrally closed, by the preceding paragraph; D is coherent by Lemma 2.16; and D is a going-down domain by Proposition 2.6 (a). The proof is complete. □

If $R \in \mathcal{H}$ is a ϕ-Prüfer ring, then R is a ϕ-integrally closed (ϕ-)going-down ring (by combining [1, Theorem 2.6], the second paragraph of the above proof, and Proposition 2.6 (a)), but we doubt that R must be ϕ-coherent.

Given the preliminary work on ϕ-coherence in Lemma 2.16 and Proposition 2.17 and the ϕ-theoretic analogue of "Noetherian" in papers such as [5], we close by encouraging a deeper ϕ-theoretic study of analogues of coherence and related weaker properties such as the finite conductor property and the quasi-coherent property. In this regard, we have two observations. On the positive side, note that the finite conductor property actually sufficed to play the role attributed above to "coherent" in [7, Corollary 4]. Finally, we sound a cautionary note relative to the possibility of finding positive results for Prüfer rings R such that $Z(R) = Nil(R)$. (By Proposition 2.7, any such $R \in \mathcal{H}$.) Note that the ring R constructed in [6, Example 2.18 (a)] illustrates that a Prüfer ring $R \notin \mathcal{H}$ need not be a finite conductor ring (and, hence, need not be coherent).

References

[1] D. F. Anderson and A. Badawi, On ϕ-Prüfer rings and ϕ-Bezout rings, *Houston J. Math.*, to appear.

[2] D. F. Anderson and A. Badawi, *On ϕ-Dedekind and ϕ-Krull rings*, submitted for publication.

[3] A. Badawi, On divided commutative rings, *Comm. Algebra* **27** (1999), 1465–1474.

[4] A. Badawi, On ϕ-chained rings and ϕ-pseudo-valuation rings, *Houston J. Math.* **29** (2001), 725–736.

[5] A. Badawi, On Nonnil-Noetherian rings, *Comm. Algebra* **31** (2003), 1669–1677.

[6] A. Badawi and D. E. Dobbs, On locally divided rings and going-down rings, *Comm. Algebra* **29** (2001), 2805–2825.

[7] D. E. Dobbs, On going down for simple overrings, *Proc. Amer. Math. Soc.* **39** (1973), 515–519.

[8] D. E. Dobbs, On going-down for simple overrings, II, *Comm. Algebra* **1** (1974), 439–458.

[9] D. E. Dobbs, Divided rings and going-down, *Pacific J. Math.* **67** (1976), 353–363.

[10] D. E. Dobbs, Coherence, ascent of going-down, and pseudo-valuation domains, *Houston J. Math.* **4** (1978), 551–567.

[11] D. E. Dobbs, Going-down rings with zero-divisors, *Houston J. Math.* **23** (1997), 1–12.

[12] D. E. Dobbs and I. J. Papick, On going-down for simple overrings, III, *Proc. Amer. Math. Soc.* **54** (1976), 35–38.

[13] D. E. Dobbs and G. Picavet, Weak Baer going-down rings, *Houston J. Math.* **29** (2003), 559–581.

[14] M. Fontana, Topologically defined classes of commutative rings, *Ann. Mat. Pura Appl.* **123** (1980), 331–355.

[15] R. Gilmer, *Multiplicative Ideal Theory*, Marcel Dekker, New York, 1972.

[16] M. Griffin, Prüfer rings with zero divisors, *J. Reine Angew. Math.* **239/240** (1969), 55–67.

[17] J. Huckaba, *Commutative Rings with Zero Divisors*, Marcel Dekker, New York/Basel, 1988.

[18] I. Kaplansky, *Commutative Rings*, rev. ed., Univ. Chicago Press, Chicago, 1974.

[19] S. McAdam, Simple going-down, *J. London Math. Soc.* **13** (1976), 167–173.

[20] R. G. Swan, On seminormality, *J. Algebra* **67** (1980), 210–229.

In: Mathematics, Game Theory and Algebra Compendium
Editor: Jacob H. Mathias, pp. 195-201
ISBN 978-1-60456-659-8

Chapter 19

COMPUTATIONAL OF THE INTEGRAL CLOSURE

L. El Fadil
Department of Mathematics, Faculty of Sciences Dhar-Mahraz,
P.O. Box 1796, Fes-Atlas, Fes, Morocco

Abstract

The Pohst-Zassenhaus process and the Dedekind criterion give an excellent tool to compute the index of the $\mathbb{Z}$-order $\mathbb{Z}[\alpha]$, then to compute the integral closure of $\mathbb{Z}$ in $\mathbb{Q}[\alpha]$. We extend these results, partially, to any extension of a Dedekind domain.

Introduction

Throughout this paper, R is a Dedekind domain with quotient field K, A is a unit commutative K-algebra, $\alpha \in A$ is an integral element over R and $T(X)$ its minimal polynomial over K. (observe that $T(X)$ is a monic polynomial lies in $R[X]$). If $R = \mathbb{Z}$ and $\mathbb{Q}[\alpha]$ is a number field, then the Dedekind criterion, given in [1, pp. 305-307], gives an excellent tool to enlarge the $\mathbb{Z}$-order $\mathbb{Z}[\alpha]$, successively, for every prime integer p such that p^2 divides the discriminant of $T(X)$, until we obtain the integral closure of $\mathbb{Z}$ in $\mathbb{Q}[\alpha]$. In this paper, our aim is to extend this criterion to any extension of an arbitrary Deddekind domain : We enlarge the R-order $R[\alpha]$, successively, for every maximal ideal $\mathcal{P}$ of R such that $\mathcal{P}^2$ contains the discriminant of $T(X)$, until we obtain the integral closure of R in $K[\alpha]$. Also we give a condition, which allows us to test if $R[\alpha]$ is the integral closure of R in $K[\alpha]$.

For every maximal ideal $\mathcal{P}$ of R, denote by $k(\mathcal{P})$ the residual field $R/\mathcal{P}$, $R_{\mathcal{P}}$ the localize of R by $\mathcal{P}$ and $S_{\mathcal{P}}$ the scalar extension of S by $R_{\mathcal{P}}$. For two polynomials P and Q lie in $F[X]$, where F is a field, denote by $M(P,Q)$ (resp. $D(P,Q)$) their least common multiple (resp. their greatest common divisor), i.e., $M(P,Q)$ (resp. $D(P,Q)$) is the monic polynomial generates $PF[X] \cap QF[X]$ (resp. $PF[X] + QF[X]$). $Res(P,Q)$ is the resultant of P and Q. Finally $deg(P)$ is the degree of the polynomial of P.

1. Preliminaries

Let Λ be a R order of $K[\alpha]$, i.e., Λ is a unit subring of $K[\alpha]$, finitely generated as a R-module, which contains a K-basis of $K[\alpha]$. For every $x \in \Lambda$, x induces an endomorphism l_x, of the R-module Λ, defined by $l_x(s) = xs$ for every $s \in \Lambda$. Since R is a domain, the characteristic polynomial $C_x(X)$ of the endomorphism $l_x : \Lambda_{\mathcal{P}} \longrightarrow \Lambda_{\mathcal{P}}$ is a monic polynomial of $R[X]$, which does not depend on the choice of the maximal ideal $\mathcal{P}$. Set $C_x(X) = X^n + a_{n-1} + ... + a_0$. Define $T_{\Lambda/R}(x) = -a_{n-1}$ and $N_{\Lambda/R}(x) = a_0$. These maps generalize the classical notions of the trace and the norm maps, when they are defined from a separable extension of fields L/K by : $T_{L/K}(x) = \sum_{1=1}^{n} \sigma_i(x)$ and $N_{L/K}(x) = \prod_{1=1}^{n} \sigma_i(x)$, where $\sigma_1, \sigma_2, ..., \sigma_n$ are the K-isomorphisms of L into its algebraic closure. The linear form $T_{\Lambda/R}$ induces a bilnear form of Λ defined by $T_{\Lambda/R}(x, y) = T_{\Lambda/R}(xy)$ for every $(x, y) \in \Lambda^2$. If Λ is a free R-module of rank n, then the determinant of the bilnear form $T_{\Lambda/R}$ with respect to a R-basis $(e_1, ..., e_n)$ of Λ is denoted by $D(e_1, ..., e_n)$, and called the discriminant of $(e_1, ..., e_n)$. The discriminant ideal of Λ over R is the ideal of R generated by $D(e_1, ..., e_n)$. Generally, if Λ is a projective R-module of constant rank n, then for every system $(a_1, ..., a_n) \in \Lambda^n$, the discriminant of $(a_1, ..., a_n)$ is the determinant $D(a_1, ..., a_n)$ of the matrix $(T_{\Lambda/R}(a_i a_j))_{1 \le i,j \le n}$. The discriminant ideal, $D_R(\Lambda)$, of Λ over R is the ideal of R generated by all discriminants $D(a_1, ..., a_n)$, where $(a_1, ..., a_n) \in \Lambda^n$.

For two free submodules M and N of A, with the same rank over R, there is a nonsingular R-linear map f with $f(M) = N$. The principal ideal of R, generated by the determinant of f, depends only of M and N, which will denote by : $[M : N]$, i.e., $[M : N] = det(f)R$. If the restriction on M and N to be R-free is removed, then the ideal $[M : N]$ is well defined by : for every maximal ideal $\mathcal{P}$ of R, $[M : N]_{\mathcal{P}} = [M_{\mathcal{P}} : N_{\mathcal{P}}]$. If $N \subset M$, then $[M : N]$ is called the index of N in M. For more details of the discriminant (see [2, 4, 5]).

Let Λ be a R-order of $K[\alpha]$ and $\mathcal{O}$ the integral closure of R in $K[\alpha]$. The index $[\mathcal{O} : \Lambda]$ is called the index of the R-order Λ. The R-order Λ will be called a $\mathcal{P}$-maximal order if $\mathcal{P}$ does not divide the index $[\mathcal{O} : \Lambda]$, i.e., $\Lambda_{\mathcal{P}}$ is the integral closure of $R_{\mathcal{P}}$ in $K[\alpha]$.

Proposition 1.1 *Let Λ be a R-order of $K[\alpha]$ and $I_{\mathcal{P}}$ the $\mathcal{P}$-radical of Λ, i.e., $I_{\mathcal{P}} = \prod_{i=1}^{g} \mathcal{M}_i$, where $\mathcal{M}_1$,...,$\mathcal{M}_g$ are the maximal ideals of Λ above $\mathcal{P}$. Then*

(1) $I_{\mathcal{P}} = \{x \in \Lambda \,,\, \exists m \ge 1 \,,\; x^m \in \mathcal{P}\Lambda\}$.

(2) *There exists an integer $m \ge 1$ such that $I_{\mathcal{P}}^m \subset \mathcal{P}\Lambda$.*

Proof. (1) Let $x \in \Lambda$ and $m \ge 1$ be an integer such that $x^m \in \mathcal{P}\Lambda$. Since $\mathcal{P}\Lambda \subset \mathcal{M}_i$ for every i, $x^m \in \mathcal{M}_i$, and then $x \in \mathcal{M}_i$ for every i. Hence $x \in I_{\mathcal{P}}$.
Conversely, since $\Lambda/\mathcal{P}\Lambda$ is a $K(\mathcal{P})$-algebra of finite dimension, $\Lambda/\mathcal{P}\Lambda$ is an Artinian ring. Hence its radical is nilpotent, i.e., for every $x \in I_{\mathcal{P}}$, there exists an integer $m \ge 1$ such that $x^m \in \mathcal{P}\Lambda$.
(2) Since $I_{\mathcal{P}}$ is an ideal of Λ and Λ is a finitely generated module over the Noetherian ring R, $I_{\mathcal{P}}$ is a finitely generated R-module. Let $(x_1, ..., x_r)$ be a finite set of generators of $I_{\mathcal{P}}$. For every i, there exists $m_i \ge 1$ such that $x_i^{m_i} \in \mathcal{P}\Lambda$. Let $m = \sum_{i=1}^{r} m_i$. Then $I_{\mathcal{P}}^m \subset \mathcal{P}\Lambda$.■

The following Proposition generalizes Theorem 6.1.3, given in [1], page 304.

Proposition 1.2 *Let Λ be a R-order of $K[\alpha]$.*
Set $\Lambda' = \{x \in K[\alpha] \mid xI_{\mathcal{P}} \subset I_{\mathcal{P}}\}$. Then Λ' is a $\mathcal{P}$-maximal R-order of $K[\alpha]$. Furthermore, either Λ is a $\mathcal{P}$-maximal R-order of $K[\alpha]$, or $\mathcal{P}^n \subset [\Lambda' : \Lambda] \subset \mathcal{P}$, where n is the degree of $T(X)$.

Proof. Since $I_{\mathcal{P}}$ is an ideal of Λ, Λ' is a ring containing Λ. Let $x \in \Lambda'$. Since $\mathcal{P} \subset I_{\mathcal{P}}$, $x\mathcal{P} \subset xI_{\mathcal{P}} \subset I_{\mathcal{P}} \subset \Lambda$, and then $\Lambda \subset \Lambda' \subset \mathcal{P}^{-1}\Lambda$. Hence Λ' is a torsion free finitely generated R-module containing Λ. Thus Λ' is a R-order of $K[\alpha]$. The fact that $\Lambda \subset \Lambda' \subset \mathcal{P}^{-1}\Lambda$ implies that $\mathcal{P}^n \subset [\Lambda' : \Lambda]$. Consequently, if $\Lambda' \neq \Lambda$, then $\mathcal{P}^n \subset [\Lambda' : \Lambda] \subset \mathcal{P}$.
To show that Λ' is a $\mathcal{P}$-maximal R-order of $K[\alpha]$, we can assume that R is a discrete valuation domain with maximal ideal $\mathcal{P} = \pi R$.
First, we have $\Lambda' \subset \mathcal{O}$. Let $r \geq 1$ be an integer such that $[\mathcal{O} : \Lambda'] = \pi^r R$. Then $\pi^r\mathcal{O} \subset \Lambda'$. Assume that $\Lambda' \neq \mathcal{O}$. Let n be the largest integer such that $J_{\mathcal{P}}^n\mathcal{O} \not\subset \Lambda'$, where $J_{\mathcal{P}}$ is the $\mathcal{P}$-radical of Λ'. Let $x \in J_{\mathcal{P}}^n\mathcal{O} - \Lambda'$. Then $xJ_{\mathcal{P}} \subset \Lambda'$. Since $J_{\mathcal{P}}^{n+m+1}\Lambda' \subset J_{\mathcal{P}}^m\Lambda' \subset \mathcal{P}\Lambda'$, where m is the integer defined in Proposition 1.1, it follows that if $y \in J_{\mathcal{P}}$, then $xy \in J_{\mathcal{P}}$. So $xJ_{\mathcal{P}} \subset J_{\mathcal{P}}$, and then $x \in \Lambda'$. This is a contradiction. So $\Lambda' = \mathcal{O}$. ■

2. Main Results

In this section our aim is to extend Theorem 6.1.4, given in [1], page 306, to any extension of a Dedekind domain. We start by Proposition 2.3, in which we compute the integral closure of R in $K[\alpha]$ and the index of the R-order $R[\alpha]$, where R is a principal domain.

The following Proposition generalizes Theorem 6.1.4, given in [1], page 306. Indeed, R is an arbitrary principal domain and $K[\alpha]$ is not necessarily a field.

Proposition 2.1 *Let R be a principal domain and $\mathcal{P} = \pi R$ a maximal ideal of R.*
Let $\bar{T}(X) = \prod_{i=1}^r \bar{t}_i(X)^{e_i}$ the factorization of $\bar{T}$ of irreducible polynomials in $K(\mathcal{P})[X]$, $g(X) = \prod_{i=1}^r t_i(X)$, $h(X) = \prod_{i=1}^r t_i(X)^{e_i-1}$ and $f(X) = \frac{gh(X)-T(X)}{\pi} \in R[X]$.
Let U and D be two monic polynomials of $R[X]$ such that $D(\bar{f}, \bar{g}, \bar{h}) = \bar{D}$ and $\bar{U}(X) = \frac{\bar{T}}{D(\bar{f},\bar{g},\bar{h})}$ in $K(\mathcal{P})[X]$. Then
1) The $\mathcal{P}$-radical of $R[\alpha]$ is given by : $I_{\mathcal{P}} = \mathcal{P}R[\alpha] + g(\alpha)R[\alpha]$.
2) The R-order $R[\alpha]'$, defined in Proposition 1.2, is given by :

$$R[\alpha]' = R[\alpha] + \frac{1}{\pi}U(\alpha)R[\alpha].$$

In particular, $[\mathcal{O} : R[\alpha]] = \mathcal{P}^d R$, and then $D_R(\Lambda) = (\mathcal{P})^{f-2d}R$, where $f = v_{\mathcal{P}}(Disc(P))$, $d = deg(D)$ and $v_{\mathcal{P}}$ is the associated $\mathcal{P}$-valuation to the maximal ideal $\mathcal{P}$.

In order to prove this Proposition, we need the following Lemmas :

Lemma 2.2 *Under the hypotheses of Proposition 2.3, let* $x = \frac{A(\alpha)}{\pi} \in \frac{1}{\pi}R[\alpha]$. *Then*
1) $x\pi \in I_{\mathcal{P}}$ *if and only if* $\bar{g}(X)$ *divides* $\bar{A}(X)$ *in* $K(\mathcal{P})[X]$.
2) Let $k(X) \in R[X]$ *be a monic polynomial lifts* $\frac{\bar{g}}{D(\bar{f},\bar{g})}$. *Then* $xg(\alpha) \in I_{\mathcal{P}}$ *if and only if* $\overline{hk}(X)$ *divides* $\bar{A}(X)$ *in* $K(\mathcal{P})[X]$.

The proof is the same as that of Lemma 6.1.5, given in [1], page 307.

Lemma 2.3 *Let* U *be a monic polynomial of* $R[X]$ *of degree* $r \leq n$ *and* $s \in R^*$. *Set* $M = R[\alpha] + \dfrac{U(\alpha)}{s}R[\alpha]$. *Then* M *is a free* R*-module of rank* n *and* $[M : R[\alpha]] = s^{n-r}R$.

Proof. Let $x \in M$. Then there exists two polynomials L and N of $R[X]$ such that $deg(UN) \leq n-1$ and $x = L(\alpha) + \frac{U(\alpha)}{s}N(\alpha)$. By the Euclidean algorithm, we can assume that $deg(L) < deg(U)$. Hence $\mathcal{F} = (1, \alpha, ..., \alpha^{r-1}, \frac{U(\alpha)}{s}, \alpha\frac{U(\alpha)}{s}, .., \alpha^{n-r}\frac{U(\alpha)}{s})$ is a R-basis of M. Let $f : R[\alpha] \longrightarrow M$, which transforms $(1, \alpha, ..., \alpha^{n-1})$ to $\mathcal{F}$. Then $det(f) = s^{n-r}$. So
$[M : R[\alpha]] = s^{n-r}R$. ∎

We show now Proposition 2.3.
1) First, we have $\mathcal{P}R[\alpha] \subset I_{\mathcal{P}}$. Moreovr, since every $e_i \leq n$, $\bar{T}(X)$ divides $\bar{g}^n(X)$ in $K(\mathcal{P})[X]$, and then $\bar{g}^n(\bar{\alpha}) = 0$, i.e., $g^n(\alpha) \in \mathcal{P}R[\alpha]$. Hence $g(\alpha) \in I_{\mathcal{P}}$. Therefore $\mathcal{P}R[\alpha] + g(\alpha)R[\alpha] \subset I_{\mathcal{P}}$.
Conversely, let $x = A(\alpha) \in R[\alpha]$. If $x \in I_{\mathcal{P}}$, then there exists an integer m such that $x^m \equiv 0$ (mod $\mathcal{P}R[\alpha]$), i.e., $\bar{A}^m(\alpha)$ in $R[\alpha]/\mathcal{P}R[\alpha]$. Since $R[\alpha]/\mathcal{P}R[\alpha]$ is a free $K(\mathcal{P})$-algebra of rank n, $\bar{T}(X)$ is the minimal polynomial of $\bar{\alpha}$ over $K(\mathcal{P})$. So $\bar{T}(X)$ divides $\bar{A}^m(X)$, and then for every i, $\bar{t}_i(X)$ divides $\bar{A}(X)$ in $K(\mathcal{P})[X]$. Consequently, $\bar{g}(X)$ divides $\bar{A}(X)$ in $K(\mathcal{P})[X]$. That means that $x \in \mathcal{P}R[\alpha] + g(\alpha)R[\alpha]$.
2) Let $x = \frac{A(\alpha)}{\pi} \in \frac{1}{\pi}R[\alpha]$. Set $J_{\mathcal{P}}$ the $\mathcal{P}$-radical of $R[\alpha]'$. From Proposition 1.2, $x \in R[\alpha]'$ if and only if $xJ_{\mathcal{P}} \subset J_{\mathcal{P}}$, i.e., $x\pi \in J_{\mathcal{P}}$ and $xg(\alpha) \in J_{\mathcal{P}}$. From Lemma 2.4, that means that $\bar{g}(X)$ divides $\bar{A}(X)$ and $\overline{hk}(X)$ divides $\bar{A}(X)$ in $K(\mathcal{P})[X]$. Hence $D(\bar{g}, \overline{kh})$ divides $\bar{A}(X)$ in $K(\mathcal{P})[X]$.

$$M(\bar{g}, \overline{hk}) = \bar{k}M(D(\bar{f}, \bar{g}), \bar{h}) = \frac{\bar{g}}{D(\bar{f}, \bar{g})}\frac{\bar{h}D(\bar{g}, \bar{f})}{D(\bar{f}, \bar{g}, \bar{h})} = \frac{\bar{T}}{D(\bar{f}, \bar{g}, \bar{h})} = \bar{U}.$$

Finally, $R[\alpha]' = R[\alpha] + \frac{1}{\pi}U(\alpha)R[\alpha]$.
To achieve the proof, it suffices to use Lemma 2.5.

Corollary 2.4 *Under the hypotheses of Proposition 2.3,* $R_{\mathcal{P}}[\alpha]$ *is the integral closure of* $R_{\mathcal{P}}$ *in* $K[\alpha]$ *if and only if* $D(\bar{f}, \bar{g}, \bar{h}) = 1$ *in* $K(\mathcal{P})[X]$.
Let $\mathcal{O}$ *be the integral closure of* R *in* $K[\alpha]$ *and* $\mathcal{P}_1, ..., \mathcal{P}_s$ *the maximal ideals of* R *such that* $disc(T) \in \mathcal{P}_i^2$ *for every* i. *For every* i, *let* $\mathcal{P}_i = \pi_i R$, $\bar{T}(X) = \prod_{j=1}^{r} \bar{t}_j(X)^{e_j}$ *be the*

factorization of $\bar{T}$*, of irreducible polynomials, in* $K(\mathcal{P})[X]$*,* $g(X) = \prod_{j=1}^{r} t_j(X)$*,* $h(X) = \prod_{j=1}^{r} t_j(X)^{e_j-1}$ *and* $f(X) = \frac{gh(X)-T(X)}{\pi_i} \in R[X]$*. Let* D_i *be a monic polynomial of* $R_{\mathcal{P}_i}[X]$ *lifts of* $D(\bar{f}, \bar{g}, \bar{h})$*. Then*

$$[\mathcal{O} : R[\alpha]] = \prod_{i=1}^{s} \mathcal{P}_i^{d_i} \text{ and } D_R(\mathcal{O}) = \prod_{i=1}^{s} \mathcal{P}_i^{f_i-2d_i},$$

where $f_i = v_{\mathcal{P}_i}(Disc(T))$ *and* $d_i = deg(D_i)$ *for every* $1 \leq i \leq s$*.*

Now, we are ready to give our main result, Theorem 2.7, which, partially, generalizes Theorem 6.1.4, given in [1], page 306, to any extension over a Dedekind domain.

Theorem 2.5 *Let* $\mathcal{O}$ *be the integral closure of* R *in* $K[\alpha]$ *and* $\mathcal{P}_1, ..., \mathcal{P}_s$ *the maximal ideals of* R *such that* $disc(T) \in \mathcal{P}_i^2$ *for every* i*. For every* i*, let* $\mathcal{P}_i R_{\mathcal{P}_i} = \pi_i R_{\mathcal{P}_i}$*,* $\bar{T}(X) = \prod_{j=1}^{r} \bar{t_j}(X)^{e_j}$ *be the factorization of* $\bar{T}$*, of irreducible polynomials, in* $K(\mathcal{P})[X]$*,* $g(X) = \prod_{j=1}^{r} t_j(X)$*,* $h(X) = \prod_{j=1}^{r} t_j(X)^{e_j-1}$ *and* $f(X) = \frac{gh(X)-T(X)}{\pi_i} \in R_{\mathcal{P}_i}[X]$*. Let* D_i *be a monic polynomial of* $R_{\mathcal{P}_i}[X]$ *lifts of* $D(\bar{f}, \bar{g}, \bar{h}) = \bar{D}_i$*. Then*
1)

$$[\mathcal{O} : R[\alpha]] = \prod_{i=1}^{s} \mathcal{P}_i^{d_i} \text{ and } D_R(\mathcal{O}) = \prod_{i=1}^{s} \mathcal{P}_i^{f_i-2d_i},$$

where $f_i = v_{\mathcal{P}_i}(Disc(T))$ *and* $d_i = deg(D_i)$ *for every* $1 \leq i \leq s$*.*
2) In particular, $R[\alpha]$ *is the integral closure of* R *in* $K[\alpha]$ *if and only if for every maximal ideal* $\mathcal{P}$ *of* R *such that* $disc(T) \in \mathcal{P}^2$*, if* $e_j \geq 2$*, then* $v_{\mathcal{P}}(Res(T, t_j)) = deg(t_j)$ *for every* $1 \leq j \leq r$*, where* $\bar{T}(X) = \prod_{j=1}^{r} \bar{t_j}(X)^{e_j}$ *the factorization of* $\bar{T}$*, of irreducible polynomials, in* $K(\mathcal{P})[X]$*.*

Proof. 1) Since R is a Dedekind domain, every ideal of R is defined by its $\mathcal{P}$-valuation, i.e., $I = \prod_{\mathcal{P}} \mathcal{P}^{v_{\mathcal{P}}(I)}$. Then it suffices to define the $\mathcal{P}$-valuation of $[\mathcal{O} : R[\alpha]]$ for every maximal ideal $\mathcal{P}$ of R. Hence we can assume that R is a discrete valuation domain with maximal ideal $\mathcal{P}_i = \pi_i R$. So we can use Proposition 2.3 and Corollary 2.6 to achieve the Proof of 1).
2) Let $\mathcal{P}$ be a maximal ideal of R. To within a permutation, we can assume that for every $1 \leq i \leq s$, $e_i \geq 2$ and $e_i = 1$ elsewhere. Hence $D(\bar{f}, \bar{g}, \bar{h}) = D(\bar{f}, \bar{g_1})$ in $K(\mathcal{P})[X]$, where $g_1(X) = \prod_{j=1}^{s} t_j(X)$. Consequently, $D(\bar{f}, \bar{g}, \bar{h}) = 1$ if and only if $D(\bar{f}, \bar{t_j}) = 1$ for every $1 \leq j \leq s$. Let $1 \leq j \leq s$. Since $T(X) = g(X)h(X) - \pi f(X)$, $Res(T, t_j)) = (-\pi)^{deg(t_j)} Res(f, t_j))$. Since $D(\bar{f}, \bar{t_j}) = 1$ if and only if $v_{\mathcal{P}}(Res(f, t_j)) = 0$, $v_{\mathcal{P}}(Res(T, t_j)) = deg(t_j)$ if and only if $D(\bar{f}, \bar{g}, \bar{h}) = 1$ in $K(\mathcal{P})[X]$.

3. Applications

1) Let $R = \mathbb{Z}$, d an integer, p a prime integer, which divides d and $T(X) = X^{p^r} - d \in \mathbb{Z}[X]$. Set $\mathbb{Q}[\alpha] = \mathbb{Z}[X]/(T(X))$. Then $Disc(T) = (-p^r)^{p^r} d^{p^r-1}$.

Let q be a prime integer, which divides d. Modulo q, $\bar{T} = X^{p^r}$, $\bar{g}(X) = X$, $\bar{h}(X) = X^{p^r-1}$ and $f(X) = \frac{-d}{q}$. Thus if q^2 does not divide d, then $D(\bar{f}, \bar{g}, \bar{h}) = 1$, else $D(\bar{f}, \bar{g}, \bar{h}) = X$. Consequently, $R[\alpha]$ is the integral closure of R in $\mathbb{Q}[\alpha]$ if and only if d is square free.
For $d = p^2m$, where m is an integer square free and p does not divide m, $\mathcal{O} = \mathbb{Z}[\alpha] + \frac{1}{p}\alpha^{p^r-1}\mathbb{Z}[\alpha]$ is the integral closure of $\mathbb{Z}$ in $\mathbb{Q}[\alpha]$.

2) Let m be an integer square free such that $m \equiv 2$ or $m \equiv 3$ modulo 4. Let $R = \mathbb{Z}[\sqrt{m}]$. Then R is a Dedekind domain.
a) Let p be a prime integer, which divides m and $T(X) = X^{p^r} - \sqrt{m} \in R[X]$. Then $Disc(T) = \mp(p^r)^{p^r}\sqrt{m}^{p^r-1}$. Let $\mathcal{P}$ be a maximal ideal of R such that $Disc(T) \in \mathcal{P}^2$. Then $\mathcal{P} = qR + \sqrt{m}R$, where q is a prime, which divides m. Modulo $\mathcal{P}$, $\bar{T} = X^{p^r}$, $Res(T, X) = T(0) = -\sqrt{m}$ and $v_{\mathcal{P}}(\sqrt{m}) = 1 = deg(X)$. Hence $R[m^{\frac{1}{p^r}}]$ is the integral closure of R in $\mathbb{Q}[m^{\frac{1}{p^r}}]$, where $m^{\frac{1}{p^r}}$ is a root of $T(X)$. So it follows that $\mathbb{Z}[m^{\frac{1}{2p^r}}]$ is the integral closure of $\mathbb{Z}$ in $\mathbb{Q}[m^{\frac{1}{2p^r}}]$.
b) Let $T(X) = X^{p^r} - a \in R[X]$. Then $Disc(T) = \mp(p^r)^{p^r}a^{p^r-1}$.
Let $\mathcal{P}$ be maximal ideal of R such that $Disc(T) \in \mathcal{P}^2$.
1th case, $a \in \mathcal{P}$. Then $\bar{T}(X) = X^{p^r}$ and $Res(T, X) = T(0) = -a$.
2^d case, $a \notin \mathcal{P}$. Then $\mathcal{P} = pR + \sqrt{m}R$, $\bar{T} = (X - a_1)^{p^r}$, where $a = a_1 + a_2\sqrt{m} \in \mathbb{Z}[\sqrt{m}]$ and $Res(T, X - a_1) = T(a_1) = a_1^{p^r} - a_1 - a_2\sqrt{m}$.
Hence $R[\alpha]$ is the integral closure of R in $\mathbb{Q}[\sqrt{m}, \alpha]$ if and only if a is square free in R.

4) Let K be a field of characteristic p, $R = K[X]$ and L the quotient field of R. Let $Q(X)$ be a polynomial of $K[X]$ and $P_n(Y) = Y^{p^n} + Q(X)Y + Q(X) \in K[X][Y]$. Set $R[\alpha_n] = R[Y]/(P_n(Y))$. Then $Disc(P_n) = Q(X)^{p^n}$. Let $P(X)$ be an irreducible polynomial of $R[X]$, which divides $Q(X)$. Modulo $P(X)$, $\bar{P}_n(Y) = Y^{p^n}$ and $Res(P_n, Y) = P_n(0) = Q(X)$. Then $R[\alpha_n]$ is the integral closure of R in $L[\alpha_n]$ if and only if $Q(X)$ is square free.

References

[1] H. Cohen, A course in computational algebraic number theory, *GTM* **138**, Springer, 1996.

[2] M. Charkani and L. El fadil, Generalization of the discriminant and applications. (to appear in the forthcoming issues of the *Arabian Journal of Sciences and Engineering*)

[3] L. El fadil, *Discriminant and separability.* (submitted)

[4] J.W. Cassels and A. Frohlich. *Algebraic number theory*, Academic Press, London and New York, 1967.

[5] A. Fröhlich and M. Taylor, Algebraic number theory, *Combridge Studies in Advenced Mathematics* **27**,CUP, 1992.

[6] G. Janusz, *Algebric Number Theory*, Academic Press, New York , 1973.

[7] S. Lang, *Algebraic number theory*, Springer-Verlag Berlin Heidelberg, New York, Paris, Tokyo, 1986.

[8] O. Zariski and P.Samuel, *Commutative Algebra*, Springer-Verlag, Berlin, New York , *GTM* **28**, 1991.

[9] L.C. Washington. *Introduction to cyclotomic fields*, Springer-Verlag, Berlin, Heidelberg, New York, 1982.

In: Mathematics, Game Theory and Algebra Compendium
Editor: Jacob H. Mathias, pp. 203-209
ISBN 978-1-60456-659-8

Chapter 20

Negacyclic Codes of Length 2^e over Z_4

Taher Abualrub and Marwan Abukhaled
Department of Mathematics and Statistics
American University of Sharjah, Sharjah-UAE

Abstract

In this paper we study negacyclic codes of length $n = 2^e$ for some integer e. We found the set of generators for this type of codes as ideals in the ring $D_4 = Z_4[x] / \langle x^n + 1 \rangle$. It will be shown that any negacyclic code has the form $\langle \alpha(x+1)^m \rangle$, $\alpha = 1, 2$, and $m = 0, 1, ..., n-1$. It will also be shown that the element 2 is in every nonzero code of the form $\langle (x+1)^m \rangle$. As such, it is concluded that negacyclic codes of length $n = 2^e$ for some integer e have a low minimum distance. Dual codes of length $n = 2^e$ over Z_4 will also be studied. It will be shown that $\langle 2 \rangle$ is the only *self-dual* code of length 2^e.

1. Introduction

The study of linear and cyclic codes over Z_4 has provided useful results in coding theory [1-4, 6-8]. It was shown in [6] that important families of binary nonlinear codes are in fact images under the Gray map of linear codes over Z_4. For example, the Gray map [6] was used to show that Kerdock and Preparata codes can be constructed as binary images via the Gray map of linear codes over Z_4.

Wolfmann in [10] studied negacyclic codes of odd length over Z_4. He showed that this is an important class of codes and must be given some attention from the coding community. He defined a (linear) negacyclic code over Z_4 to be an ideal in the ring $D_4 = Z_4[x] / \langle x^n + 1 \rangle$. In this correspondence, we note that negacyclic codes mean linear negacyclic codes. In [10], Wolfmann also showed that over Z_4, the structure of negacyclic codes is similar to that of cyclic codes. Furthermore, he showed that if C is a negacyclic code of odd length n then its polynomial representation I is a principle ideal generated by a constant polynomial or a polynomial of the kind $g(x) = a(x)[b(x) + 2]$

where $x^n + 1 = a(x)b(x)c(x)$ in $Z_4[x]$ and where $a(x)$, $b(x)$, and $c(x)$ are pairwise coprime polynomials. He also studied the Gray image of negacyclic codes and showed that the Gray image of a linear negacyclic code over Z_4 is a binary-distance-invariant (not necessary linear cyclic code). One of the many problems that Wolfmann posted for future studies in [10] was the study of negacyclic codes of even length.

In this paper, we study negacyclic codes of even length over Z_4. In particular, the length $n = 2^e$ for some integer e. It will be shown that the structure of the ring D_4 is different from that of R_4. It will also be shown that the class of negacyclic codes of length $n = 2^e$ is not a big class of codes as in the case of cyclic codes. The fact that the ring D_4 is a principle ideal ring and hence ideals (negacyclic codes) are principle ideals will be established. We will also show the nonexistence of nontrivial self-dual negacyclic codes of length $n = 2^e$. Furthermore, it will be shown that this class of codes is not a good class since most codes have a low minimum distance. The objectives of this study will be reached in the following sequel: section 2 covers a study of the ring D_4 and the structure of negacyclic codes in it whereas section 3 covers a study of the structure of dual negacyclic codes.

We assume throughout this paper that $n = 2^e$.

2. Negacyclic Codes

By $Z_4 = \{0, 1, 2, 3\}$, it is meant the ring of integers modulo 4.

Construct the ring

$$D_4 = Z_4[x] / \langle x^n + 1\rangle = \{f(x) : f(x) = a_0 + a_1x + ...a_{n-1}x^{n-1}\}$$

where $a_i \in Z_4 \ \forall\ i = 0, 1, ..., n - 1$, and $x^n = -1$.

In [1], it was shown that $(x^n + 1)$ is an irreducible polynomial over Z_4.

Since any element c in Z_4 can be written as $c = a + 2b$ where $a,\ b$ are elements in Z_2 then any element in D_4 can be expressed in the form

$$f(x) = \sum_{i=0}^{n-1} a_i(x + 1)^i + 2\sum_{i=0}^{n-1} b_i(x + 1)^i \quad \text{where } a_i,\ b_i = 0, 1, \text{ for all } i = 0, 1, ..., n - 1.$$

As in the case of cyclic codes, it is convenient to represent code words of length n by polynomials modulo $x^n + 1$. We identify $v = (v_0, ..., v_{n-1})$ with the polynomial $v(x) = v_0 + v_1x + ... + v_{n-1}x^{n-1}$ in the ring D_4.

Definition 1 *By a negacyclic code over Z_4, it is meant an ideal in the ring D_4.*

The proof of the following lemma can be found in [1].

Lemma 1 *Let* $m = 2^k$ *where* $k > 0$. *Then* $x^m + 1 = (x+1)^m + 2x^{m/2}$ *in* $Z_4[x]$.

A particular instant of the this lemma is that, in the ring D_4 we have, $0 = x^n + 1 = (x+1)^n + 2x^{n/2}$. This implies that $(x+1)^n = 2x^{n/2} \Rightarrow (x+1)^{2n} = 0$. This proves the following lemma:

Lemma 2 $(x+1)$ *is a nilpotent element in* D_4 *whose nilindex is equal to* $2n$.

Notice that if a and b are nilpotent elements in a ring R then using the Binomial Theorem on the ring D_4 it can easily be shown that $(a \pm b)$ is also a nilpotent element.

Lemma 3 *Let* $f(x) = \sum_{i=0}^{n-1} a_i(x+1)^i + 2\sum_{i=0}^{n-1} b_i(x+1)^i$ *be an element in* D_4 *where* $a_i,\ b_i = 0, 1$. *Then* $f(x)$ *is a unit in* D_4 *iff* $a_o = 1$.

Proof. $\Rightarrow$) Suppose $f(x)$ is a unit. If $a_0 = 0$, then $f(x) = (x+1)^j h(x)$ where $j \geq 1$. But then

$$f(x)^{2n} = \left[(x+1)^j h(x)\right]^{2n} = 0.$$

Therefore, $f(x)$ is not a unit.

$\Leftarrow$) Suppose $f(x) = 1 + \sum_{i=1}^{n-1} a_i(x+1)^i + 2\sum_{i=0}^{n-1} b_i(x+1)^i$. Since $(x+1)$ is nilpotent then

$$\sum_{i=1}^{n-1} a_i(x+1)^i + 2\sum_{i=0}^{n-1} b_i(x+1)^i$$

is also a nilpotent element with

$$\left[\sum_{i=1}^{n-1} a_i(x+1)^i + 2\sum_{i=0}^{n-1} b_i(x+1)^i\right]^{2n} = 0.$$

Hence $f(x) = 1 + a$ where $a^{2n} = 0$. Then,

$$(a+1)(-1(a^{2n-1} - a^{2n-2} + a^{2n-2} - a^{2n-3} + ... - 1) = -(a^{2n} - 1) = 1.$$

Hence $f(x)$ is a unit. ∎

Lemma 4 D_4 *is a local ring with maximal ideal* $M = \langle x+1 \rangle = \langle x-1 \rangle$.

Proof. In D_4, it was established that

$$0 = x^n + 1 = (x+1)(x^{n-1} - x^{n-2} + x^{n-3} - ... + 1) + 2.$$

It follows that $2 \in \langle x+1 \rangle = \langle x-1 \rangle$. Since any element in D_4 has the form

$$f(x) = \sum_{i=0}^{n-1} a_i(x+1)^i + 2\sum_{i=0}^{n-1} b_i(x+1)^i \text{ where } a_i = 0, 1,\ b_i = 0, 1 \text{ for all } i = 0, 1, ..., n-1,$$

then the only elements that are not in M are the ones for which $a_0 = 1$. But these are unit elements by the above lemma. Therefore D_4 is a local ring with maximal ideal $M = \langle x+1 \rangle = \langle x-1 \rangle$. ∎

Since $(x+1)^n = 2x^{n/2}$ and $x^{n/2}$ is a unit in D_4 then $2 = \langle (x+1)^n \rangle$.

Lemma 5 *If R is a finite local ring with maximal ideal $M = \langle a \rangle$. Then any ideal I in R is given by $I = \langle a^k \rangle$.*

Proof. Let $i \in I \subseteq M = \langle a \rangle$. Then $i = ar_1$ for some $r_1 \in R$. If r_1 is a unit then $I = M$, otherwise $r_1 \in M$ since R is a local ring. Hence $r_1 = ar_2$ for some $r_2 \in R$ which implies that $i = a^2 r_2$. Continuing in this process leads to $i = a^k$ for some integer k. Therefore $I = \langle a^k \rangle$ for some integer k. ■

The main result can now be stated:

Theorem 1 *D_4 is a principal ideal ring with nonzero ideals (negacyclic codes) given by $C = \langle \alpha(x+1)^m \rangle$ where $\alpha = 1,\ 2,$ and $m = 0, 1, 2, ..., n-1$.*

Proof. Follows directly from lemmas 4 and 5 in addition to the fact that $\langle 2 \rangle = \langle (x+1)^n \rangle$. ■

Since $\langle 2 \rangle = \langle (x+1)^n \rangle$, then 2 is an element in any nonzero negacyclic code of the form $\langle (x+1)^m \rangle$. This shows that the minimum Hamming and Lee wights of any negacyclic code of length $n = 2^e$ and of the form $\langle (x+1)^m \rangle$ is equal to 1 and 2 respectively. Also, this shows that the minimum Euclidean weight for this type of codes is at most 4. For this reason, this class of codes may be classified as a bad one in terms of the minimum distance.

As a result of this Theorem, the following corollary seems to hold.

Corollary 2 *The number of negacyclic codes of length $n = 2^e$ over Z_4 is equal to $2n + 1$.*

3. Dual Codes

Definition 2 *Let $u = (u_1, \ldots, u_n)$ and $v = (v_1, \ldots, v_n)$ be any two vectors over Z_4. Define an inner product over Z_4 by $u \cdot v = u_1 v_1 + \ldots + u_n v_n$. the vectors u and v are said to be orthogonal if $u \cdot v = 0$.*

Definition 3 *Let C be a negacyclic code of length n. The dual of C is denoted by $C^\perp$ and is defined by $C^\perp = \{u : u.v = 0\ \forall v \in C\}$.*

The following theorem is the same as in [8] except that a field is replaced by a ring. We state it without proof.

Theorem 3 *Let $f(x) = a_0 + a_1 x + ... + a_{n-1} x^{n-1}$ and $b(x) = b_0 + b_1 x + ... + b_{n-1} x^{n-1}$. Then $a(x)b(x) = 0$ iff the vector $(a_0, a_1, ..., a_{n-1})$ is orthogonal to the vector $(b_{n-1}, b_{n-2}, ..., b_0)$ and to all its cyclic shifts.*

Definition 4 *If $f(x) = a_0 + a_1x + ... + a_rx^r$, then the reciprocal of $f(x)$ is the polynomial $f^*(x) = a_r + a_{r-1}x + ... + a_0x^r$. Symbolically $f^*(x)$ can be represented by $f^*(x) = x^r f(\frac{1}{x})$.*

If I is an ideal in D_4, then $I^* = \{f^*(x) : f(x) \in I\}$ is also an ideal.

Definition 5 *Let I be an ideal in D_4. The annihilator of I (denoted by $A(I)$), which is an ideal in D_4 is given by*

$$A(I) = \{g(x) : f(x)g(x) = 0 \text{ for all } f(x) \in I\}$$

It is clear that if C is a negacyclic code whose associated ideal I, then the associated ideal for $C^\perp$ is $A^*(I)$.

The following lemma and its proof can be found in [1].

Lemma 6 *If $f(x) = (x+1)^r$, then $f(x) = f^*(x)$.*

Theorem 4 *Let C be a negacyclic code of length n over Z_4.*

1. If $C = \langle (x+1)^i \rangle$, then $C^\perp = \langle 2(x+1)^{n-i} \rangle$.
2. If $C = \langle 2(x+1)^i \rangle$, then $C^\perp = \langle (x+1)^{n-i} \rangle$.

Proof. Suppose $C = \langle (x+1)^i \rangle$. $2(x+1)^{n-i}(x+1)^i = 2(x+1)^n = 0$. Hence $\langle 2(x+1)^{n-i} \rangle \subseteq A(C)$. Suppose that $A(C) = \langle \alpha(x+1)^j \rangle$ for some $j = 0, 1, ..., n-1$. Since $\alpha(x+1)^j.(x+1)^i = 0$, then α must equal 2. The smallest power j satisfies $2(x+1)^j.(x+1)^i = 0$ is $j = n-i$. Therefore, $\langle 2(x+1)^{n-i} \rangle = A(C)$. From lemma 6, $A(C) = A^*(C) = C^\perp$ and consequently $C^\perp = \langle 2(x+1)^{n-i} \rangle$. This establishes (1). The proof of two is similar. ■

Table 1. The Z_4 Negacyclic Codes of Length 8

Gen. Poly.	Hamming Weight	Lee Weight	Dual	Hamming Weight	Lee Weight
$(x+1)$	1	2	$2(x+1)^7$	8	16
$(x+1)^2$	1	2	$2(x+1)^6$	4	8
$(x+1)^3$	1	2	$2(x+1)^5$	4	8
$(x+1)^4$	1	2	$2(x+1)^4$	2	4
$(x+1)^5$	1	2	$2(x+1)^3$	2	4
$(x+1)^6$	1	2	$2(x+1)^2$	2	4
$(x+1)^7$	1	2	$2(x+1)^1$	2	4
2	1	2	self-dual	1	2

As a result of the this theorem the following corollary seems to hold.

Corollary 5 *$C = \langle 2 \rangle$ is the only self-dual negacyclic code of length $n = 2^e$ over Z_4.*

Example: The following table list all negacyclic codes of length 8 with their duals and their Hamming and Lee weights.

4. Conclusion

The negacyclic codes studied here are taken over Z_4 and have a length of $n = 2^e$. Contrary to the case of cyclic codes of the same length, it was shown that the ring D_4 is a principal ideal ring. We constructed a set of generators for this type of codes and their duals. It was also shown that the element 2 is in every negacyclic code of the form $\langle (x+1)^m \rangle$ which implies that this class of negacyclic codes has a low minimum distance. In particular, these codes are asymptotically bad. It was also shown that there is only one elf-dual code of length 2^e which is the trivial self-dual code $\langle 2 \rangle$. It will be very interesting to see what kind of applications might arise from this type of codes. Open problems include the study of negacyclic codes of other lengths. In particular the study of negacyclic codes of length $n = 2e$ where e is an odd integer.

References

[1] T. Abualrub and R. Oehmke, "On the Generators of Z_4 Cyclic Codes," Submitted *IEEE Tran. Inform. Theory.*

[2] T. Abualrub and R. Oehmke, "Cyclic Codes of length 2^e over Z_m," *Proc. International Workshop on Coding and Cryptography*, WCC 2001, Paris France, 2001, 15-21.

[3] T. Blackford, "Cyclic Codes over Z_4 of Odly Even Length," *Proc. International Workshop on Coding and Cryptography*, WCC 2001, Paris France, 2001, 83-92.

[4] A. R. Calderbank and N. J. A. Sloane, "Modular and $p-$adic Cyclic Codes," Des. Codes Cryptogr., vol. 6, pp. 21-35, 1995.

[5] G. Castagnoli, J. L. Massey, P. A. Schoeller, and N. Von Seeman, "On Repeated Root Cyclic Codes," *IEEE. Tran. Inform Theory*, vol. 37, no. 2, pp. 337-342, 1991.

[6] A. R. Hammons, P. V. Kumar and A. R. Calderbank, N. J. A. Sloane, and P. Sole, "The Z_4-Linearity of Kerdock, Preparata, Goethals, and Related Codes," *IEEE Trans. Inform. Theory*, vol. 40, 301-319, Mar. 1994.

[7] P. Kanwar and S. R. Lopez-Permouth, "Cyclic Codes over the Integers Modulo p," *Finite Fields and their Applications*, vol. 3, number 4 (1997), pp. 334-352.

[8] V. Pless and Z. Qian, "Cyclic Codes and Quadratic Residue Codes over Z_4," *IEEE, Trans. Inform. Theory*, vol. 42, pp1594-1600, 1996.

[9] J. H. Van Lint, "Repeated-Root Cyclic Codes," *IEEE Trans. on Inform. Theory*, vol. 37, no. 2, pp. 343-345, 1991.

[10] J. Wolfmann, "Negacyclic and Cyclic Codes over Z_4," *IEEE Trans. on Inform. Theory*, vol. 45, n0. 7, pp 2527-2532, November 1999.

In: Mathematics, Game Theory and Algebra Compendium
Editor: Jacob H. Mathias, pp. 211-219
ISBN 978-1-60456-659-8

Chapter 21

COOPERATIVE STOCHASTIC GAMES IN STATIONARY STRATEGIES

Elena V. Baranova and Leon A. Petrosyan
Saint-Petersburg State University
Saint-Petersburg, Russia

Consider the infinite graph tree $G = (Z, L)$ where Z is the set of all vertices and L is the point to set mapping determined on the set of vertices Z. Thus $L_z \subset Z$ for every vertex of graph G. The set L_z is finite for every $z \in Z$. Each vertex z is considered as n-person simultaneous game

$$\Gamma(z) = \langle N, X_1^z, \ldots, X_n^z, K_1^z, \ldots, K_n^z \rangle.$$

Suppose that the game $\Gamma(z)$ is one of the t simultaneous n-person games

$$\Gamma^j = \langle N, X_1^j, \ldots, X_n^j, K_1^j, \ldots, K_n^j \rangle, j = 1, \ldots, t,$$

where N is the set of players which is the same for all games Γ^j, $j = 1, \ldots, t$, X_i^j ($i = 1, \ldots, n$) is the set of strategies of player i in the game Γ^j, K_i^j is the payoff function of player i in the game Γ^j (we suppose that $K_j^i\left(x_1^j, \ldots, x_n^j\right) \geqslant 0$). The n-tuple $x^j = \left(x_1^j, \ldots, x_n^j\right)$ is called the situation in the game Γ^j. Suppose that the set of strategies of each player is finite for any one of the games $\Gamma^1, \ldots, \Gamma^t$ and for every vertex $z \in Z$ there are no such vertices $y_1 \in L_z$ and $y_2 \in L_z$, $y_1 \neq y_2$, so that corresponding games Γ^j corresponding to them coinside.

In each vertex $z \in Z$ the transition probabilities to the next vertices $y \in L_z$ depending on the situation x^z realized in game $\Gamma(z)$ are given

$$p(z, y; x^z) \geqslant 0,$$

$$\sum_{y \in L_z} p(z, y; x^z) = 1,$$

where $p(z, y; x^z)$ is the probability that the game $\Gamma(y)$ will be played next after the game $\Gamma(z)$ under condition that the situation x^z has realized in $\Gamma(z)$. Suppose that q is the probability of the game end on each step. Define the first vertex (root) of the graph tree G as $z_0 \in Z$.

Stochastic game with random duration $\overline{G}(z_0)$ develops in the following way. On the first step the simultaneous game $\Gamma(z_0)$ (one of t defined simultaneous games $\Gamma^1, \dots, \Gamma^t$) takes place, suppose that the situation x^{z_0} is realized in this game. Further the game either stops with the probability q, $0 \leqslant q \leqslant 1$, or with the probability $(1-q)$ continues and after moves to the next vertices of the graph tree with the probabilities $p(z_0, y; x^{z_0})$, depending on the situation realized in $\Gamma(z_0)$. On the step k of the stochastic game with random duration $\overline{G}(z_0)$ the simultaneous game $\Gamma(z_k)$ (one of t defined simultaneous games $\Gamma^1, \dots, \Gamma^t$) takes place and the situation x^{z_k} is realized. The game $\overline{G}(z_0)$ either ends with the probability q, $0 \leqslant q \leqslant 1$, on the step k, or with the probability $(1-q)$ continues and after moves to the next vertices of graph tree G with the probabilities $p(z_k, z_{k+1}; x^{z_k})$ depending on the situation x^{z_k} which has realized on the step k. If $\overline{G}(z_0)$ is the stochastic game described above, $\overline{G}(z_k)$ is a subgame of the game $\overline{G}(z_0)$ originating from the simultaneous game $\Gamma(z_k)$ in the vertex $z_k \in Z$ of the graph G.

Because of the stochastic structure of the game we consider the mathematical expectation of a payoff as the payoff. For the mathematical expectation of player i's payoff in game $\overline{G}(z_0)$ the following formula is true

$$E_i(z_0) = qK_i^{z_0}(x^{z_0}) + (1-q)\left[K_i^{z_0}(x^{z_0}) + \sum_{y \in L_{z_0}} p(z_0, y; x^{z_0}) E_i(y)\right]$$
$$= K_i^{z_0}(x^{z_0}) + (1-q) \sum_{y \in L_{z_0}} p(z_0, y; x^{z_0}) E_i(y),$$

where $E_i(y)$ is the mathematical expectation of player i's payoff in the subgame $\overline{G}(y)$ beginning from the vertex $y \in Z$ of the graph G. Suppose that the series in the right part of the above equality converge for each $y \in L_z$, $z_0 \in Z$ and $x^{z_0} \in X^{z_0}$.

Strategy $\eta_i(\cdot)$ of the player i in the game $\overline{G}(z_0)$ is the rule, that determines the choice of a strategy in the game $\Gamma(z)$ for each simultaneous game $\Gamma(z)$. Consider stochastic game $\overline{G}(z_0)$ in stationary strategies, i.e.

$$\eta(z) = x_i^j \in X_i^j$$

for each z, for which $\Gamma(z) = \Gamma^j$, $(j = 1, \dots, t), i \in N$. Let the set of pure stationary strategies of player i in the subgame $\overline{G}^k$ (the subgame beginning from the vertex in which the simultaneous game Γ^k is defined) be denoted as Ξ_i^k $(k = \overline{1, t})$. That is, in the game defined in stationary strategies the players' choice of a strategy in simultaneous games doesn't depend neither on the history (the history of vertex $z \in Z$ is the sequence of vertices, realized in a stochastic game, the last of which is z), nor on the vertex, in which the game is at present, but depends only on the simultaneous game.

Since there is a finite set of t simultaneous games $\Gamma^1, \ldots, \Gamma^t$, it is sufficient to consider t subgames $\overline{G}^j$ ($j = 1, \ldots, t$), i.e. t subgames of the game $\overline{G}(z_0)$, beginning from the simultaneous games $\Gamma^1, \ldots, \Gamma^t$.

Suppose that $\eta(\cdot) = (\eta_1(\cdot), \ldots, \eta_n(\cdot))$ is the situation in stationary strategies, so that $\eta(z) = x_i^j \in X_i^j$ for each z: $\Gamma(z) = \Gamma^j$, $(j = 1, \ldots, t), i \in N$. If $\eta_i(\cdot)$ is the strategy of player i in the game $\overline{G}(z_0)$, the stationary strategy in the subgame $\overline{G}(z)$ of the game $\overline{G}(z_0)$ will be denoted by $\eta_i(\cdot)$. The following formula is true for the mathematical expectation of player i's payoff:

$$E_i(z) = K_i^z(x^z) + (1-q) \sum_{y \in L_z} p(z, y; x^z) E_i(y).$$

Since we consider the game $\overline{G}(z_0)$ in stationary strategies and the set of simultaneous games defined in the vertices of the graph as finite, the transition probabilities $p(z, y; x^z)$ from vertex $z \in Z$ to $y \in L_z$, under condition that the situation x^z has realized in the simultaneous game $\Gamma(z)$, can be written in matrix form. Further we'll denote the transition probabilities $p(z, y; x^z)$ as $p(i, j; x^i)$, under condition that the simultaneous game Γ^i is defined in the vertex $z \in Z$ and the game Γ^j is defined in $y \in L_z$. Form the matrix of transition probabilities:

$$\Pi(\eta(\cdot)) = \begin{pmatrix} p(1,1;x^1) & \ldots & p(1,t;x^1) \\ p(2,1;x^2) & \ldots & p(2,t;x^2) \\ \ldots & \ldots & \ldots \\ p(t,1;x^t) & \ldots & p(t,t;x^t) \end{pmatrix},$$

where $\eta(\cdot) = (\eta_1(\cdot), \ldots, \eta_n(\cdot))$ is a situation in stationary strategies. The formula for the mathematical expectation of the player i's payoff , under condition that the stationary strategy profile (situation) $\eta(\cdot)$ has realized, is as follows:

$$E_i(\eta(\cdot)) = K_i(\eta(\cdot)) + (1-q)\Pi(\eta(\cdot))E_i(\eta(\cdot)), \tag{1}$$

where $E_i(\eta(\cdot)) = \left(E_i^1(x^1), \ldots, E_i^t(x^t)\right)^T$ (the sign ^{T} means transposition). ($E_i^j(\cdot)$ is the mathematical expectation of player i's payoff in the subgame, beginning from the simultaneous game Γ^j), E is the identity matrix of dimension $t \times t$, $K_i(\eta(\cdot)) = \left(K_i^1(x^1), \ldots, K_i^t(x^t)\right)^T$ where $K_i^j(x^j)$ is the player i's payoff in the simultaneous game Γ^j, under condition that the situation x^j has realized in this game. If $\det(E - (1 - q)\Pi(\eta(\cdot))) \neq 0$, from (1) we receive

$$E_i(\eta(\cdot)) = \left(E - (1-q)\Pi(\eta(\cdot))\right)^{-1} K_i(\eta(\cdot)). \tag{2}$$

Denote vector of stationary strategies maximizing the sum of mathematical expectations of players' payoffs (cooperative solution) in the subgame beginning from the vertex $z \in Z$: $\Gamma(z) = \Gamma^j$ as $\overline{\eta}^j(\cdot) = (\overline{\eta}_1^j(\cdot), \ldots, \overline{\eta}_n^j(\cdot))$.

For the maximum of the sum of mathematical expectations of players' payoffs the Bellman equation holds

$$V(N) = \max_{\substack{\eta_i \in \Xi_i^1 \\ \cdots \\ \eta_i \in \Xi_i^t \\ i \in N}} \left[\sum_{i \in N} K_i(\eta(\cdot)) + (1-q)\left(\Pi(\eta(\cdot))V(N)\right) \right],$$

where $V(N) = (V^1(N), \ldots, V^t(N))^T$ and $V^j(N)$ is the maximum of the sum of mathematical expectations of payoffs in the subgame beginning from the simultaneous game Γ^j.

$$V(N) = \sum_{i \in N} K_i(\overline{\eta}(\cdot))(E - (1-q)\left(\Pi(\overline{\eta}(\cdot))\right)), \tag{3}$$

Define an auxiliary stochastic zero-sum subgame G_S^z between coalition $S \subset N$ and coalition $N \backslash S$ for each vertex z of the graph G (since one of the games $\Gamma^1, \ldots, \Gamma^t$ associates with each vertex of the graph and we consider only stationary strategies, it is necessary to consider only t games). Set the value of function $V^j(S)$ as the lower value of zero-sum stochastic game G_S^z, if the simultaneous game Γ^j is defined in vertex z, i.e.

$$V^j(S) = \max_{\eta_S(\cdot)} \min_{\eta_{N \backslash S}(\cdot)} \sum_{i \in S} E_i^j(\eta_S(\cdot), \eta_{N \backslash S}(\cdot)), j = \overline{1,t} \tag{4}$$

$$V^j(\emptyset) = 0.$$

Consider cooperative stochastic subgame $\overline{\overline{G}}(z)$ with the characteristic function defined by formulae (3), (4). Further, for definiteness sake, suppose that the simultaneous game Γ^j is defined in the vertex z of the graph G. Denote the set of imputations in the game $\overline{\overline{G}}(z)$ ($\overline{\overline{G}}^j$) as $I(z)$ (I^j):

$$I(z) = I^j = \left\{ \alpha_i^j : \sum_{i \in N} \alpha_i^j = V^j(N), \alpha_i^z \geqslant V^j(\{i\}) \right\}.$$

By the solution of the game is meant any fixed subset $C(z) \subset I(z)$. It can be one of the classical cooperative solutions (C-core, N-core, NM-solution, Shapley Value). For simplicity sake, suppose that $C(z)$ is Shapley Value:

$$C(z) = Sh(z) = \{Sh_1(z), ..., Sh_n(z)\} \in I(z).$$

Definition 1. *The vector-function $\beta(z) = \beta^j = \left(\beta_1^j, \ldots, \beta_n^j\right)$ is called cooperative payoff distribution procedure (CPDP), if*

$$\sum_{i \in N} \beta_i^j \leqslant \max_{\substack{\eta_i \in \Xi_i^j \\ i \in N}} \sum_{i \in N} K_i^j(x_1^j, \ldots, x_n^j), \tag{5}$$

where the expression in the right part is the maximal total payoff of players in the game Γ^j.

The sequence of situations $x^{z_0}, x^{z_1}, \ldots$, where x^{z_i} is the situation realized in the game $\Gamma(z_i)$, $z_i \in L_{z_{i-1}}$ is called the path in the multistage game. In a stochastic non zero-sum subgame $\overline{\overline{G}}(z_k)$ the sum of β_i^j for a fixed i is a random variable (the sum of values β_i^j, $\ldots$, β_i^t, $\ldots$, calculated along the path $\overline{x}^{z_k,\cdots}$) in the situation $\overline{\eta}^j(\cdot) = (\overline{\eta}_1^j(\cdot), \ldots, \overline{\eta}_n^j(\cdot))$ (if the simultaneous game Γ^j is defined in the vertex). Denote the mathematical expectation of such β_i counted along the path $\overline{x}^{z_k,\cdots}$ in cooperative nonzero-sum subgame $\overline{\overline{G}}(z_k)$ as $B_i(z_k)$. Since the set of simultaneous games is finite $\{\Gamma_1, \ldots, \Gamma_t\}$, it is sufficient to consider t subgames. B_i satisfies the following equation:

$$B_i = \beta_i + (1-q)\Pi(\overline{\eta}(\cdot))B_i, \tag{6}$$

where $B_i = (B_i^1, \ldots, B_i^t)^T$ (B_i^j is the mathematical expectation of β_i counted in subgame $\overline{\overline{G}}(z)$, where $z : \Gamma(z) = \Gamma^j$), E is the identity matrix of dimention $t \times t$, $\beta_i = (\beta_i^1, \ldots, \beta_i^t)^T$. (6) can be rewritten in the form: $B_i = (E - (1-q)\Pi(\overline{\eta}(\cdot)))^{-1}\beta_i$.

Calculate Shapley Value for each subgame $\overline{\overline{G}}(z)$, where $z \in Z$ (i.e. actually t Shapley Values).

$$Sh_i(z) = Sh_i^j = \sum_{\substack{S \subset N \\ S \ni i}} \frac{(|S|-1)!\,(n-|S|)!}{n!} \left(V^j(S) - V^j(S \backslash \{i\})\right), j = 1, \ldots, t,$$

$i \in N$ and $|S|$ is the number of the elements of S (here Sh_i^j is a component of Shapley Value in the subgame $\overline{\overline{G}}(z)$, beginning from the simultaneous game Γ^j).

Define $\gamma_i = (\gamma_i^1, \ldots, \gamma_i^t)^T$ from the system of equations

$$Sh_i = \gamma_i + (1-q)\Pi(\overline{\eta}(\cdot))Sh_i, \tag{7}$$

where $Sh_i = (Sh_i^1, \ldots, Sh_i^t)^T$ and $\Pi(\overline{\eta}(\cdot))$ is a matrix of the transition probabilities, defined above.

Lemma 1. *Vector $\gamma_i = (\gamma_i^1, \ldots, \gamma_i^t)^T$ defined by formula (5) is CPDP.*

Before the beginning of the game the players come to an agreement about the choice of the set of strategies which guarantee the maximal total payoff and expect to receive the components of Shapley Value $Sh(z_0)$. Motion along the path $\overline{x}^{z_0}, \overline{x}^{z_1}, \ldots$ corresponds to the development of the game during the time. After the first step the game moves to the vertex z_1, i.e. actually the players get to the new game $\overline{\overline{G}}(z_1)$, which is the subgame of $\overline{\overline{G}}(z_0)$. For the protection of cooperation on this step the players must expect getting payoffs in accordance with Shapley Value, calculated for the subgame $\overline{\overline{G}}(z_1)$. Unfortunately realizing the payments on each stage of the game in accordance with the payoffs in simultaneous games on these stages, it is impossible to obtain that the residuary payments should be the components of Shapley Value for the subgame beginning from this stage. This shows the time-inconsistency of Shapley Value. It is necessary to redistribute payoffs of players in a simultaneous game, provided the time-inconsistency of Shapley Value is overcome.

Definition 2. *Shapley Value $Sh(z_0)$ is called time-consistent in game $\overline{\overline{G}}(z_0)$, if there exists such nonnegative CPDP $\beta(z)$ (for every $z \in Z$) that*

$$Sh_i = \beta_i + (1-q)\Pi(\overline{\eta}(\cdot))Sh_i \tag{8}$$

for every $i \in N$. Here $Sh_i = (Sh_i^1, \dots, Sh_i^t)^T$, $\beta_i = (\beta_i^1, \dots, \beta_i^t)^T$, ($\beta_i^j$ is the i-th element of CPDP for cooperative stochastic subgame, beginning from the vertex z: $\Gamma(z) = \Gamma^j$) and $\Pi(\overline{\eta}(\cdot))$ is the matrix of the transition probabilities, described above.

If Shapley Value is time-consistent, then realizing payments to the players on each stage of the path in accordance with CPDP one can obtain that Shapley Value calculated in subgame $\overline{\overline{G}}(z)$ coincides with the payoffs, which remain for getting by the players in the subgame $\overline{\overline{G}}(z)$ of the game $\overline{\overline{G}}(z_0)$.

If $\det(E - (1-q)\Pi(\overline{\eta}(\cdot))) \neq 0$, the formula (8) will be as follows:

$$Sh_i = (E - (1-q)\Pi(\overline{\eta}(\cdot)))^{-1}\beta_i. \tag{9}$$

The equation (9) has the only solution in respect to β_i if and only if $\det(E - (1 - q)\Pi(\eta(\cdot))) \neq 0$. In this case we get the following formula:

$$\beta_i = (E - (1-q)\Pi(\overline{\eta}(\cdot)))Sh_i. \tag{10}$$

In a general case we can't guarantee the nonnegativeness of the elements of vector $\beta_i = (\beta_i^1, \dots, \beta_i^t)$. We get identical formulas (6), (9) for the values B_i and Sh_i at β_i. That is, the following lemma takes place.

Lemma 2. $B_i = Sh_i$, $i \in N$.

Lemma 2 says that the mathematical expectation of sums β_i (which is equal to B_i), which are the payments to the players realized along the trajectory of subgame $\overline{\overline{G}}(z)$, equals to the mathematical expectation of player i's payoff in this subgame (i.e. to the i component of Shapley Value Sh_i).

Example. $N = \{1,2\}$ is the set of players. The simultaneous game Γ^1 is defined in the vertex z_0, the game Γ^1 is defined in z_1 and so on.

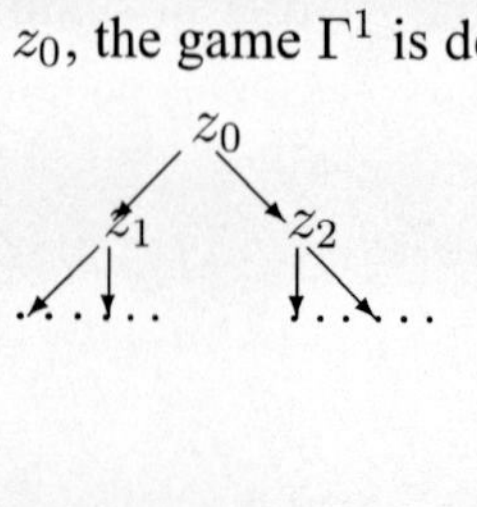

The probability to end the game $\overline{\overline{G}}(z_0)$ in each vertex of the graph G is equal to $q = \frac{3}{4}$.

$$\Gamma^1 : \begin{pmatrix} (5,5) & (0,8) \\ (8,0) & (1,1) \end{pmatrix}, \quad \Gamma^2 : \begin{pmatrix} (3,0) & (6,4) \\ (5,6) & (2,2) \end{pmatrix}.$$

Denote strategies of the first (second) player in the game Γ^1 as x_1, x_2 (y_1, y_2) and strategies of the first (second) player in the game Γ^2 as a_1, a_2 (b_1, b_2). Transition probabilities are as following:

$$\begin{array}{llll}
p_1^1(x_1,y_1) = \frac{1}{2}, & p_1^1(x_1,y_2) = \frac{1}{6}, & p_2^1(a_1,b_1) = \frac{1}{3}, & p_2^1(a_1,b_2) = \frac{1}{2}, \\
p_1^2(x_1,y_1) = \frac{1}{2}, & p_1^2(x_1,y_2) = \frac{5}{6}, & p_2^2(a_1,b_1) = \frac{2}{3}, & p_2^2(a_1,b_2) = \frac{1}{2}, \\
p_1^1(x_2,y_1) = \frac{5}{6}, & p_1^1(x_2,y_2) = \frac{1}{3}, & p_2^1(a_2,b_1) = \frac{1}{2}, & p_2^1(a_2,b_2) = \frac{2}{3}, \\
p_1^2(x_2,y_1) = \frac{1}{6}, & p_1^2(x_2,y_2) = \frac{2}{3}, & p_2^2(a_2,b_1) = \frac{1}{2}, & p_2^2(a_2,b_2) = \frac{1}{3}.
\end{array}$$

Since the set of strategies in simultaneous games Γ^1, Γ^2 of each player consists of two elements, 16 different situations in stationary strategies are possible in multistage cooperative game $\overline{\overline{G}}(z_0)$. Denote them as $\eta^1, \ldots, \eta^{16}$. Moreover, $\eta^j = (\eta_1^j, \eta_2^j)$ $(j = \overline{1,16})$ where $\eta_1^j(\eta_2^j)$ is a stationary strategy of the first (second) player and η_i^j consists of two elements the first of which is a strategy of the player i in the game Γ^1, the second one is a strategy of the player i in the game Γ^2

1)For $\eta^1 = ((x_1, a_1), (y_1, b_1))$:

$$\Pi(\eta^1) = \begin{pmatrix} \frac{1}{2} & \frac{1}{2} \\ \frac{1}{3} & \frac{2}{3} \end{pmatrix}.$$

$E_1(\eta^1)$ and $E_2(\eta^1)$ can be calculated by formula (2):

$$E_1(\eta^1) = \begin{pmatrix} \frac{424}{47} \\ \frac{360}{47} \end{pmatrix}, \qquad E_2(\eta^1) = \begin{pmatrix} \frac{400}{47} \\ \frac{180}{47} \end{pmatrix}.$$

Do the same calculations for the rest of situations in stationary strategies.

Calculate the value of characteristic function by formula (4)

$$V(\emptyset) = \begin{pmatrix} V^1(\emptyset) \\ V^2(\emptyset) \end{pmatrix} = \begin{pmatrix} 0 \\ 0 \end{pmatrix},$$

$$V(\{1\}) = \begin{pmatrix} V^1(\{1\}) \\ V^2(\{1\}) \end{pmatrix} = \begin{pmatrix} \frac{52}{25} \\ \frac{35}{9} \end{pmatrix}, V(\{2\}) = \begin{pmatrix} V^1(\{2\}) \\ V^2(\{2\}) \end{pmatrix} = \begin{pmatrix} \frac{20}{13} \\ \frac{32}{13} \end{pmatrix}.$$

$$V(\{1,2\}) = \begin{pmatrix} V^1(\{1,2\}) \\ V^2(\{1,2\}) \end{pmatrix} = \begin{pmatrix} \frac{40}{3} \\ \frac{29}{2} \end{pmatrix}.$$

$$Sh(\{1\}) = \begin{pmatrix} Sh^1(\{1\}) \\ Sh^2(\{1\}) \end{pmatrix} = \begin{pmatrix} \frac{3727}{468} \\ \frac{6764}{975} \end{pmatrix},$$

$$Sh(\{2\}) = \begin{pmatrix} Sh^1(\{2\}) \\ Sh^2(\{2\}) \end{pmatrix} = \begin{pmatrix} \frac{6236}{975} \\ \frac{3059}{468} \end{pmatrix}.$$

Calculate the CPDP. A stationary strategy $\eta^3 = ((x_1, a_2), (y_1, b_1))$ gives maximum to $V^1(\{1,2\})$ and a strategy $\eta^7 = ((x_1, a_2), (y_2, b_1))$ gives maximum to $V^2(\{1,2\})$.

$$Sh_i^1 = \beta_i^1 + (1-q)\left(p_1^1(\eta_3)Sh_i^1 + p_1^2(\eta_3)Sh_i^2\right)$$

$$Sh_i^2 = \beta_i^2 + (1-q)\left(p_2^1(\eta_7)Sh_i^1 + p_2^2(\eta_7)Sh_i^2\right)$$

$$Sh_i = \beta_i + (1-q)\begin{pmatrix} p_1^1(\eta_3) & p_1^2(\eta_3) \\ p_2^1(\eta_7) & p_2^2(\eta_7) \end{pmatrix} Sh_i$$

Form the system of linear equations for CPDP for the first player:

$$\begin{pmatrix} \frac{7}{6} & \frac{1}{6} \\ \frac{1}{6} & \frac{7}{6} \end{pmatrix} \begin{pmatrix} \beta_1^1 \\ \beta_1^2 \end{pmatrix} = \begin{pmatrix} \frac{6764}{975} \\ \frac{3727}{468} \end{pmatrix}.$$

The solution of this system is a vector

$$\begin{pmatrix} \beta_1^1 \\ \beta_1^2 \end{pmatrix} = \begin{pmatrix} \frac{571057}{93600} \\ \frac{475001}{93600} \end{pmatrix}$$

Similarly,

$$\begin{pmatrix} \beta_2^1 \\ \beta_2^2 \end{pmatrix} = \begin{pmatrix} \frac{460497}{93600} \\ \frac{447321}{93600} \end{pmatrix}$$

Since $\beta_i^j \geqslant 0$ $(i \in N, j = 1,2)$, Shapley Value is time-consistent. The condition taken from lemma 2 is realized and $Sh(\{1\}) = B_1 = \begin{pmatrix} \frac{6764}{975} \\ \frac{3727}{468} \end{pmatrix}$, and $Sh(\{2\}) = B_2 = \begin{pmatrix} \frac{6236}{975} \\ \frac{3059}{468} \end{pmatrix}$.

References

[1] C.W. Clark, *Bioeconomic modelling and fisheries management*, Wiley, NY, 1985.

[2] L.A. Petrosyan, E.M. Baranova, E.V. Shevkoplyas Multistage cooperative games with random duration, *Proc. of Institute of mathimatics and mechanics of the Ural Branch of the Russian Academy of Sciences*, collected articles *"Optimal control and differential games"* **10**, no. 2 (2004), Ekaterinburg, 116–130.

[3] L.S. Shapley, *Stochastic games*, *Proc. Nat. Acad. USA* **39** (1953), 1095–1100.

[4] K. Najim K., A.S. Poznyak, E.Gomez, Adaptive policy for two finite Markov chains zero-sum stochastic games with unknown transition matrices and average payoffs, *Automatica* **37**, no. 7 (2001), 1007–1018.

[5] J. Flesch , F. Thuijsman , S. Vrieze, Optimality in different strategy classes in zero-sum stochastic games, *Math. Methods of Operat. Res.* **54**, no. 2 (2001), 315–322.

[6] L.A. Petrosjan, Cooperative stochastic games, *Proc. "10th Inter. Symposium on Dynamic Games and Appl."* **2** (2002), St. Peterburg, 687–690.

[7] D.W.K. Yeung, L.A. Petrosjan, *Subgame consist coòperative solutions in stochastic differential games*, *J. optimiz. theory and appl.* **120**, no. 3 (2004), 651–666.

In: Mathematics, Game Theory and Algebra Compendium ISBN 978-1-60456-659-8
Editor: Jacob H. Mathias, pp. 221-231

Chapter 22

TWO LEVEL GAMES

Elisa Caprari
Dipartimento di Ricerche Aziendali, Facoltà di Economia
Università degli Studi di Pavia, Pavia, Italy
Roberto E. Lucchetti*
Dipartimento di Matematica, Politecnico di Milano
Milano, Italy

1. Introduction

Consider the following model of a game. There are two players, one is called the leader, the other one the follower. Each one of them has a set of available strategies, say X and Y respectively, from which he must select "a good one". They both have a payoff function, say f and g, depending on the strategies selected by both the players (i.e. defined on $X \times Y$). The development of the game is as follows: the leader announces his choice x, among his available strategies in X. The follower selects an element from his strategy space Y in order to maximize his payoff function $g(x, \cdot)$. Then he announces his optimal strategy $BS(x)$. Thus the leader will get the payoff $f(x, BS(x))$. From this it is clear that the task of the leader is to find a strategy x in order to maximize $f(\cdot, BS(\cdot))$. In the literature there are several results concerning this type of two level problems, called Stackelberg games in the theory of games. We refer for instance to the papers [2], [8], [9], [12], [13], [17] [20] and to the references therein.

Observe that our previous description of the game is rather loose, for the main reason that $BS(x)$, even if non empty, usually it is not a singleton, but rather it contains many elements. Thus a choice must be done, according to the fact whether the leader can influence or not the choice of the follower among the elements of $BS(x)$. In this note most of the results are concerned with the case when the leader can drive the choice of the followers (i.e. he can force the choice of a specific vector y inside $BS(x)$; observe that the followers are indifferent between the various alternatives in $BS(x)$). There is no other reason for our

---

*Research of this author was partially supported by Ministero dell'Istruzione, dell'Università e della Ricerca (COFIN 2001).

choice than the fact that this case is mathematically easier to handle.
Our aim in this note is to go beyond this well studied model, in order to introduce situations which are more complicated and interesting, and to provide some initial existence results. For instance, one can suppose concrete situations where the followers are more than one. This in turn originates two possible models, according to the fact that these players can act cooperatively or not. Here, we briefly analyze both situations. More interesting and more challenging, at least in our opinion, is the case when the leaders are more than one. One can imagine, as an example of such a situation, of several insurance companies having to decide the fees for car insurance, and the followers are in this case the car owners. For this type of model, we give an initial and partial result. Also, we only deal with the case when the two leaders act non cooperatively. It is our opinion that the cooperative case, even if interesting, needs another approach, since the outcome of a cooperative game is simply a vector of distributions, and not announcements of strategies, which seems to be at the basis of the two level problem mechanism, as far as the leader party is concerned.

The note is organized as follows. In Section 2 we analyze the case when there are one leader and several followers acting non cooperatively. In Section 3 there is always one leader, and the followers act cooperatively. In the final section, we describe the situation where there are two leaders acting non cooperatively, and several followers acting cooperatively. In all cases we provide existence results for a solution of the game. We do not consider the case in which there are two leaders and many followers, all of them acting non cooperatively, since in this case it does not seem to be easy to produce an existence result of an equilibrium of the game.

Although we now give the essentials on game theory we shall need in the sequel, we assume the reader to be familiar with the classical non-cooperative and cooperative game theory. Thus we do not describe in details all the solution concepts we mention, and we just quote the well known results we shall need about them. The reader interested in more details, or in the proofs of the mentioned properties, can consult any book dealing with game theory, for instance the well known book by G. Owen [18] .

A *non-cooperative game* between two players is usually defined as a quadruple $(X, Y, f : X \times Y \to \mathbb{R}, g : X \times Y \to \mathbb{R})$. The meaning: X and Y are the strategy spaces of the two players. Every choice of a pair $(x, y) \in X \times Y$ determines an outcome of the game. The two players have preferences on these outcomes, expressed by their utility functions f, g. A Nash equilibrium for such a game is a pair $(\overline{x}, \overline{y}) \in X \times Y$ such that

$$f(\overline{x}, \overline{y}) \geq f(x, \overline{y}) \qquad \forall x \in X, \tag{1}$$

and

$$g(\overline{x}, \overline{y}) \geq g(\overline{x}, y) \qquad \forall y \in Y \tag{2}$$

(see [14] , [15] and [16]). It is clear why a Nash equilibrium can be considered as a satisfactory outcome of a game: if a pair of strategies is to be suggested to the players, and if this is a Nash equilibrium, no one of the players has an incentive to deviate from the strategy proposed to him. At the end of the section we provide an existence result for a Nash equilibrium.

A *side payment cooperative game* between n players is a pair $(N, v : 2^N \to \mathbb{R})$ such that N is a set of n elements (the most natural choice being $N = \{1, 2, \ldots, n\}$), representing the set of the players, and such that $v(\emptyset) = 0$. An element $S \subset N$ is called a coalition. The function v is thus defined on the set of the coalitions, and $v(S)$ represents the amount of utility the coalition S is able to guarantee to itself, once it is formed. Side payment means that $v(S)$ can be distributed (at least in principle) in any possible way among the members of the coalition, and how this will take place is not matter of the model. Let us denote by G^n the 2^{n-1}-dimensional linear space of the side payment games. G^n can be endowed with the following distance: for $u, v \in G^n$,

$$d(u, v) = \max_{S \in 2^n} |u(S) - v(S)|.$$

In general, a solution for a side payment game is a subset of $\mathbb{R}^n$. A vector x in this subset represents a utility distribution for the players: x_1 is the amount given to player one etc.

The first concept of solution for cooperative games we present is the so called *nucleolus* (see [5] and [19]). To define this concept we introduce the set of imputations of the game v:

$$I(v) = \{x \in \mathbb{R}^n : x(N) = v(N),\ \ x_i \geq v(\{i\})\}, \tag{3}$$

where, for each coalition S, we use the notation:

$$\sum_{i \in S} x_i = x(S). \tag{4}$$

For each $x \in I(v)$ we consider the 2^n dimensional vector $q(x)$ whose components are the numbers $v(S) - x(S)$, arranged in decreasing order, and S runs over all the coalitions of N. We shall say that x is at least as acceptable as y (with respect to v), and write $x \succsim y$, if $q(x)$ is not greater than $q(y)$ according to the lexicographical order. If $q(x)$ is smaller than $q(y)$, we shall say that x is more acceptable than y and write $x \succ y$. The nucleolus $N(v)$ of a game v is the set

$$N(v) = \{x \in I(v) : x \succsim y \text{ for all } y \in I(v)\}. \tag{5}$$

The underlying idea in the definition of nucleolus is that a vector in it minimizes the maximum complaint of the coalitions. It is well known that, for every cooperative game v, the nucleolus $N(v)$ consists of a single vector, and that $N : G^n \to \mathbb{R}^n$ is a continuous function.

Another well known and very important concept of solution for a side payment game, is given by the *core* $C(v)$, defined as follows: $C(v) \subset \mathbb{R}^n$ is the set

$$C(v) = \{x \in I(v) : x(S) \geq v(S)\}, \tag{6}$$

where $I(v)$ is the set of the imputations of the game v, defined in (3), and $x(S)$ is defined in (4).

Also the underlying idea in the definition of the core is clear: a vector is in the core if and only if it is an admissible distribution of utilities ($x(N) = v(N)$) and no coalition can argue against it, in the sense that it could do better by acting alone. An important feature of the core is that it can be defined for general cooperative games, not only for side payment games. Here we just remember that the core is a (closed) subset of $\mathbb{R}^n$, which can contain several vectors, and which can also be empty. If the game is *balanced*, then the core is nonempty. We shall denote by B^n the family of the balanced games. Finally, it is easy to show that $N(v) \in C(v)$ whenever $C(v) \neq \emptyset$.

To conclude with this overview of solutions for side payment games, let us finally define the *Shapley value* $S(v)$ of the game v. $S(v) \in R^n$ is the vector $S = (s_1, \ldots, s_n)$ defined as:

$$s_i = \sum_{T \in 2^N, i \in T} \frac{(n - |T|)!\,(|T| - 1)!}{n!} \left[v(T) - v(T \setminus \{i\}) \right]. \tag{7}$$

The Shapley value is also called *power index* for the players, since the term $v(T) - v(T \setminus \{i\})$ represents the marginal contribution of the player i to the coalition T, and so s_i is a probabilistic combination of the marginal contribution of the player i to all coalitions, thus representing his power within the game. The properties of the Shapley value we use in this note are that it is a continuous function from G^n to R^n (actually, it is Lipschitz with Lipschitz constant 2) and that it acts linearly on the space G^n.

We finish this introduction by recalling two known facts about multifunctions that will be used later.

Let us be given X, Y topological spaces, and a multifunction (i.e. a multivalued function) $T : X \rightrightarrows Y$.

Definition 1 *$T : X \rightrightarrows U$ is lower semicontinuous at a point $x_0 \in X$ if $T(x_0) = \emptyset$ or if for every open set A in U such that $T(x_0) \cap A \neq \emptyset$, there exists $V(x_0)$ such that for each $x \in V$, $T(x) \cap A \neq \emptyset$.*

Definition 2 *$T : X \rightrightarrows U$ is upper semicontinuous at a point $x_0 \in X$ if for every open set A in U such that $T(x_0) \subset A$, there exists $V(x_0)$ such that $T(V) \subset A$.*

A function $f : X \to Y$ which is either lower or upper semicontinuous at a point x as a multifunction, is actually continuous as a function at the point. We shall consider only multifunctions T such that $T(x)$ is a closed subset of Y for all x: in this case it is easy to see that an upper semicontinuous multifunction has closed graph. Under the compactness assumptions we shall make in this paper, actually the two properties are equivalent.

Next, we establish the following important propositions, due to Berge (see [1] and [10]), that will be largely used in the sequel. The setting is the following.

We have X, Y topological spaces, $f : X \times Y \to R$ is a given function, and $T : X \rightrightarrows Y$ a given multifunction. Define:

$$m(x) = \sup\{f(x, y) : \ y \in T(x)\}, \qquad i(x) = \inf\{f(x, y) : \ y \in T(x)\}.$$

The following results deal with upper semicontinuity of m and i.

Proposition 1 *Let X be a compact set, let $T(x) : X \rightrightarrows Y$ be compact valued. Suppose moreover T has closed graph. Let $f : X \times Y \longrightarrow R$ be upper semicontinuous on $X \times T(X)$. Then m is upper semicontinuous.*

Observe that under the above assumptions (which could be a little weakened) m assumes maximum on X.

Proposition 2 *Let $T : X \rightrightarrows Y$ be lower semicontinuous on X and let $f : X \times Y \longrightarrow R$ be upper semicontinuous on $X \times T(X)$. Then i is upper semicontinuous.*

Observe that in the second proposition no compactness assumption is needed.

The final result we need to remind is the Nash existence theorem for a Nash equilibrium in a non-cooperative game [14].

Theorem 1 *Under the following assumptions:*

1. *X and Y are compact and convex subsets of an Euclidean space;*
2. *f, g are continuous on $X \times Y$;*
3. *for each $y \in Y$, $f(\cdot, y)$ is quasi concave on Y;*
4. *for each $x \in X$ $g(x, \cdot)$ is quasi concave on Y;*

the game $(X, Y, f, g : X \times Y \to R)$ has an equilibrium.

We remember that a function $u : C \to R$, where C is a convex set, is said to be quasi concave provided its level sets: $\{x \in C : u(x) \geq a\}$, are convex sets for all $a \in R$. The proof of the Nash theorem is a simple application of a fixed point theorem, such as, for instance, the Kakutani theorem.

2. One Leader, n Non-cooperative Followers

In this section we consider the case where there are one leader and n followers acting non cooperatively. First of all let us settle the necessary notation. In order to avoid heavy writing, we shall suppose $n = 2$, but all our results can be extended, with no effort, to the general case.
We are given subsets X, Y, Z of Euclidean spaces, and functions $f, g, h : X \times Y \times Z \to R$. In this context, the game can be modelled as $(X, Y, Z, f, g, h : X \times Y \times Z \to R)$.
The lower level problem consists in finding, for each leader's strategy $x \in X$, $(\overline{y}, \overline{z}) \in Y \times Z$ such that

$$g(x, \overline{y}, \overline{z}) \geq g(x, y, \overline{z}) \tag{8}$$

for each $y \in Y$, and

$$h(x, \overline{y}, \overline{z}) \geq h(x, \overline{y}, z) \tag{9}$$

for each $z \in Z$. I.e., the followers look for the set of the Nash equilibria of the game, parameterized by $x \in X$, $(Y, Z, g(x, \cdot, \cdot), h(x, \cdot, \cdot) : Y \times Z \to R)$. We denote by $NE(x)$

the set of the Nash equilibria for the two followers, when x is announced by the leader, i.e. the set of the pairs $(\overline{y}, \overline{z})$ verifying (8) and (9) above. Then the problem for the leader is:

$$\max_{x \in X} \max_{(y,z) \in NE(x)} f(x, y, z) \,. \tag{10}$$

Thus a solution for the game is a triple $(\bar{x}, \bar{y}, \bar{z})$ such that:

- $(\bar{y}, \bar{z})$ is a Nash equilibrium for the game with payoff functions $g(\bar{x}, \cdot, \cdot)$ and $h(\bar{x}, \cdot, \cdot)$;
- $f(\bar{x}, \bar{y}, \bar{z}) = \max_{x \in X} \max_{(y,z) \in NE(x)} f(x, y, z)$.

We assume:

1. X is compact;
2. Y and Z are compact and convex;
3. g, h are continuous on $X \times Y \times Z$;
4. for each $x \in X$ and $z \in Z$ $g(x, \cdot, z)$ is quasi concave on Y;
5. for each $x \in X$ and $y \in Y$ $h(x, y, \cdot)$ is quasi concave on Z;
6. f is upper semicontinuous on $X \times Y \times Z$.

From the Nash Theorem 1, it follows that, under 1.,2.,3.,4. and 5. above, for every $x \in X$ the multifunction $NE(x)$ of the Nash equilibria of the followers is nonempty. Also, it is clear that $NE(x)$ is a compact set for every x.

It is now possible to prove the following existence theorem.

Theorem 1 *Consider the two level game introduced in this section. Under the above assumptions, the problem P has solutions.*

Proof. Let $v(x) = \max_{(y,z) \in NE(x)} f(x, y, z)$. Let us prove that v assumes maximum on X. To do the job, we appeal to Proposition 1. The unique fact to be checked is that NE is closed graph, i. e. if

$$\begin{aligned} g(x_n, \overline{y}_n, \overline{z}_n) &\geq g(x_n, y, \overline{z}_n) \qquad \forall y \in Y, \\ h(x_n, \overline{y}_n, \overline{z}_n) &\geq h(x_n, \overline{y}_n, z) \qquad \forall z \in Z, \\ \text{and} \qquad (x_n, \overline{y}_n, \overline{z}_n) &\longrightarrow (x, \overline{y}, \overline{z}) \,, \end{aligned}$$

then

$$\begin{aligned} g(x, \overline{y}, \overline{z}) &\geq g(x, y, \overline{z}) \qquad \forall y \in Y, \\ h(x, \overline{y}, \overline{z}) &\geq h(x, \overline{y}, z) \qquad \forall z \in Z. \end{aligned}$$

But this is clear, from the continuity assumptions made on g and h. Thus there exists $\bar{x} \in X$ such that

$$\max_{(y,z)\in NE(\bar{x})} f(\bar{x}, y, z) = \max_{x\in X} \max_{(y,z)\in NE(x)} f(x, y, z).$$

Now a solution of the problem exists since $NE(\bar{x})$ is a compact set and $f(\bar{x}, \cdot, \cdot)$ is upper semicontinuous. ■

To end this section, we remark that the above problem was already considered in [6] and [7], and that the above theorem can clearly be applied to the case when at the lower level there is a zero sum game.

3. One Leader, n Cooperative Followers

In this section we consider the case where there are one leader and n followers acting cooperatively. From the mathematical point of view, the problem we consider here will be very similar to that one described in the previous section. The problem for the leader is formalized by the same equation (10), while the followers still take x as one of the data of the problem, and play a (classical) game, parameterized by x. The difference, from the point of view of Game Theory, lies in the fact that the followers play cooperatively.

Let us describe the setting. Suppose we are given a subset X of an Euclidean space, a function $f : X\times R^n \to R$, and a function $v : X \times 2^N \to R$, where $N = \{1, \dots, n\}$, with $v(x, \emptyset) = 0$ for all $x \in X$. The game is then modelled by $(X, N, f : X \times R^n \to R, v : X \times 2^N \to R)$.

Having in mind that the followers play a cooperative game, we can suppose that their choice will be one of the known solutions for side payment games: the Core, the Bargaining set, the Kernel, the Equal division core, the Least core, the Least tax core, the Nash Bargaining set, the Nucleolus. Let us call $SOL(x)$ one of these solution sets of the game $v(x, \cdot)$, played by the followers once the leader announces his strategy x. Then a solution for Leader-Followers problem P we are considering is a pair $(\bar{x}, \bar{y})$ such that:

- $\bar{y} \in SOL(\bar{x})$;
- $f(\bar{x}, \bar{y}) = \max_{x\in X} \max_{u\in SOL(x)} f(x, u)$.

At this point it is quite clear that the problem, from a mathematical point of view, is the same as that one of the previous section. The formal difference is only in the fact that there the multifunction NE was generated by the Nash equilibria of the game played by the followers, here the multifunction SOL is generated by a cooperative game. But mathematically the same theorem holds, once one knows that the (relevant) properties of the two multifunctions NE and SOL are the same. The key point to apply the same pattern is to know under which assumptions SOL is closed graph. The paper [11] is dedicated to show that all these solution concepts give raise to upper semicontinuous multifunctions, defined on appropriate classes of games. Thus we can state:

Theorem 1 *Consider the two level game introduced in this section. Let X be compact, let $f : X\times R^n \to R$ be upper semicontinuous at $X\times R^n$ and let SOL be one of the solution concepts for a side payment game recalled before. Then the problem P has solutions.*

But actually in the cooperative case we can do something more, because some of these solution concepts enjoy more regularity properties than the Nash equilibria multifunction. We concentrate our attention on the nucleolus, the core and the Shapley value. In the case of the nucleolus and of the Shapley value, the first great simplification is due to the fact that, for every strategy x announced by the leader, the reply of the leader is unique (i.e. $SOL(x)$ is a singleton). In this case then the problem of the leader is simply to maximize, over X, the function $m(\cdot) = f(\cdot, SOL(\cdot))$.

Theorem 2 *Consider the two level game introduced in this section. Let X be a closed subset of an Euclidean space, and suppose to each $x \in X$ it is associated a game $v(\cdot, x) \in G^n$. Let SOL be either the nucleolus or the Shapley value multifunction associated to the game played by the followers. Let f be upper semicontinuous on $X\times R^n$ and suppose $\lim_{\|(x,y)\|\to\infty} f(x, y) = -\infty$. Then P admits a solution.*

Proof. The function m is upper semicontinuous, due to continuity of SOL. Moreover it is coercive, so that it assumes maximum. ■

So far, we have assumed that the leader has (or believes to have) some control on the behavior of the followers, since it is interested in maximizing $\sup_y f(x, y)$, where y ranges on all possible best choices of the followers among the optimal ones (given x): in the opposite case he would try to maximize $\inf_y f(x, y)$. The reason of our choice should be clear: since most of the solution concepts for games enjoy upper (and not lower) semicontinuity properties, in order to apply the suitable proposition by Berge, we need to work with a sup function. Only lower semicontinuity results for multifunctions allow to deal with the case of an inf function. There is however a solution (multifunction) concept, the core, which is lower and upper semicontinuous at the same time. Thus it is possible to establish a result for the "pessimistic" case too. Using Proposition 2, the following result easily follows:

Theorem 3 *Let X be a compact subset of an Euclidean space. Suppose to each $x \in X$ it is associated a game $v(\cdot, x) \in B^n$. Let f be upper semicontinuous on $X\times R^n$. Suppose the problem for the leader is:*

$$\max_{x\in X} \min_{u\in C(x)} f(x, u), \tag{11}$$

where $C(x)$ is the core of the game $v(\cdot, x)$. Then the problem admits a solution.

Observe that compactness of X could be substituted by an ad hoc coercivity assumption, since Proposition 2 does not require compactness in order to show lower semicontinuity of i.

4. Two Leaders Play a Non-cooperative Game

In this section we want to say some words about the situation where there are two leaders playing a non-cooperative game, and several followers playing another game, parameterized by the choices of the leaders. It is not difficult to imagine which type of situation could be modelled by this type of game: however tackling it mathematically does not seem to be easy at all. Thus, here we give only a first result of this type, concerning a particular case where at the lower level a cooperative game is played.

Let us describe the setting. There are two players, the leaders, and several players, say l, the followers. The data are an $m \times n$ bimatrix A, with coefficients (a_{ij}, b_{ij}), an $m \times n$ matrix V, where each entry v_{ij} represents a cooperative game, and two concave functions $h, k : R^l \to R$. The pure strategies of the two leaders are the choice of a row/column respectively. These choices determine a pair ij, and this picks out coefficients (a_{ij} and b_{ij}) from the bimatrix A, and a game v_{ij} from V. The followers play in such a case the game v_{ij}. The payoff functions of the two leaders are $f(ij) = a_{ij} + h(S_{ij})$ and $g(ij) = b_{ij} + k(S_{ij})$, where S_{ij} represents the Shapley value of the game v_{ij}. It is quite natural in this setting to try to prove existence of a solution in mixed strategies. Let us formalize it: as it is well known, a mixed strategy for the first player, which must select a row, is a vector of the simplex S^m (remember, m is the number of the rows). And the (extended) strategy space of the second leader is the simplex S^n. Once elements $\lambda \in S^m$ and $\mu \in S^n$ are selected by the two players, it is natural to suppose that the followers will play the game $v_{\lambda\mu} = \sum_{ij} \lambda_i \mu_j v_{ij}$. And the payoff function of the first (second) player will be: $f(\lambda, \mu) = \sum_{ij} \lambda_i \mu_j a_{ij} + h(S(v_{\lambda\mu}))$ $(g(\lambda, \mu) = \sum_{ij} \lambda_i \mu_j b_{ij} + k(S(v_{\lambda\mu})))$. A solution of this game is then a triple $(\bar{\lambda}, \bar{\mu}, \bar{S})$, where:

1. $(\bar{\lambda}, \bar{\mu})$ is a Nash equilibrium for the two leaders, with the above payment functions f, g;

2. $\bar{S}$ is the Shapley value of the game $v_{\bar{\lambda}\bar{\mu}}$.

We can now state our result.

Theorem 1 *Under the assumptions described above, the problem has a solution.*

Proof. We quite naturally use the Nash Theorem 1. What is needed is continuity of f and g and their quasi concavity with respect to the variable of one player, for each fixed value of the variable of the other player: in formula $f(\cdot, \mu)$ must be a quasi concave function and $g(\lambda, \cdot)$ must be a quasi concave function. In our setting, the crucial point, which easily provides concavity of the above functions, is the fact that the Shapley value, as already mentioned, acts linearly on the space of the side payment cooperative games. ■

Let us make some final comments.

Showing existence of an equilibrium point when there are more than one leader seems to be intimately connected with the use of a fixed point argument. Beyond the case of having a contraction, fixed points theorems usually require some convexity or generalized convexity assumptions. In our setting this means that we need to have some form of convexity in the payment functions of the leaders, with respect to their own variables, and thus some convexity with respect to the solution of the lower level problem. This seems to be a very strong assumption to do.
Another problem we did not consider here is the case when at the upper level a cooperative

game is played. We believe that all this subject is worth of more thought and research: our note is just a try of opening to this type of study, also to see the reaction of other scholars. In line with what we were saying in the previous sentence, we did not search for the most general formulation and the most general assumptions in order to establish existence results. If the subject will be considered of some interest, further research and generalizations will be provided.

References

[1] C. Berge, *Topological Spaces*, New York, 1963.

[2] D. Flam Sjur, L. Mallozzi, J. Morgan, A new look for Stackelberg-Cournot equilibria in oligopolistic markets, *Economic Theory* **20**, no. 1 (2002), 183–188.

[3] W. Hildenbrand, Core and equilibria of a cooperative game, *Naval Research Logistic Quarterly* **12**, 223–259.

[4] S. Kakutani, A Generalization of Brouwer's Fixed Point Theorem, *Duke Mathematical Journal* **8** (1941), 457–459.

[5] E. Kohlberg, On the nucleolus of a Charaterietic function game, *SIAM J. Appl. Math.* **20**, no. 1 (1971), 62–67.

[6] M.B. Lignola, J. Morgan, *Existence of solutions to Bilevel Variational Problems in Banach Spaces*, on "Equilibrium problems and Variational Models" Eds. F. Giannessi, A. Maugeri, P. Pardalos, Kluwer Academic Publishers, 2001.

[7] M.B. Lignola, J. Morgan, Existence of solutions to Generalized Bilevel Programming Problem, A. Migdalas et al (eds.), *Multilevel Optimization Algorithms and Applications*, 333–359.

[8] M.B. Lignola, J. Morgan, Stability of Regularized Bilevel Programming Problems, *Journal of Optimization Theory and Applications*, **93**, no. 3 (1997), 575–596.

[9] M.B. Lignola, J. Morgan, Well-posedness for Optimization Problems with Constraints defined by Variational Inequalities having a unique solution, *Journal of Global Optimization* **16** (2000), 57–67.

[10] R.E. Lucchetti, Stability in Optimization, *Serdica Math. J.* **15** (1989), 34–48.

[11] R.E. Lucchetti, F. Patrone, S. Tjis, A. Torre, Continuity properties of solutions concepts for cooperative games, *O.R. Spektrum* **9** (1987), 101–107.

[12] L. Mallozzi, J. Morgan, Weak Stackelberg problem and mixed solutions under data perturbations, *Optimization* **32**, no. 3 (1995), 269–290.

[13] L. Mallozzi, J. Morgan, Mixed Strategies for Hierarchical Zero-sum Games, *Annals of the international Society of Dynamic Games* **6** (2002), 65–77.

[14] J. F. Nash, Equilibrium points in N-person games, *Proc. Nat. Acad. Sci., U.S.A.*, **36** (1950a), p. 48-49.

[15] J. F. Nash, The bargaining problem, *Econometrica* **18** (1950b), 155–162.

[16] J. F. Nash, Non-cooperative games, *Annals of Mathematics* **54** (1951), 286–295.

[17] I. Nishizaki, M. Sakawa, Stackelberg solutions to multiobjective two level linear programming problems, *J. Optim. Theory and Appl.* **103**, no. 1 (1999), 161–12.

[18] G. Owen, *Game theory*, Second edition, Academic Press, Orlando, 1982.

[19] D. Schmeidler, The nucleolus of a characteristic function game, *SIAM J. Appl. Math.* **17**, 1163–1170.

[20] M. Simaan, J. B. Jr. Cruz, A Stackelberg solution for games with many players, *IEEE Trans. Automatic Control* **AC-18**, no. 3 (1973), 322–324.

In: Mathematics, Game Theory and Algebra Compendium ISBN 978-1-60456-659-8
Editor: Jacob H. Mathias, pp. 233-240

Chapter 23

CONJUGATE–SET GAME FOR A NONLINEAR PROGRAMMING PROBLEM*

H. Kawasaki
Graduate School of Mathematics, Kyushu University Hakozaki
Fukuoka, Japan

1. Introduction

The conjugate point was introduced by Jacobi to give a sufficient optimality condition for the simplest problem in the calculus of variations

$$(SP)\quad \text{Minimize} \quad \int_0^T f(t, x(t), \dot{x}(t))dt \quad \text{subject to} \quad x(0) = A,\ x(T) = B,$$

where A and B are given points in R^n, $T > 0$ fixed, and f is a smooth function. Conjugate points are defined as zero points of a non-trivial solution $y(t)$ of the Jacobi equation with the initial condition $y(0) = 0$. Jacobi proved that if a feasible solution $x(t)$ satisfies the Euler equation and the strengthened Legendre condition, and if there are no points conjugate to $t = 0$ on $[0, T]$, then $x(t)$ is a weak minimum for (SP), see e.g. [1].

Recently, a conjugate point theory was proposed for a minimization problem of a smooth function $f(x)$ with n variables

$$(P_0) \qquad \text{Minimize} \quad f(x), \qquad x \in R^n$$

in [3]. The Jacobi equation for (P_0) is a difference equation that the descending principal minors of the Hesse matrix $f''(x)$ satisfy. According to Sylvester's criterion, an $n \times n$-symmetric matrix $A = (a_{ij})$ is positive-definite if and only if its descending principal minors $|A_k|$ $(k = 1, \ldots, n)$ are positive, where $A_k := (a_{ij})_{1\le i,\ j\le k}$, see e.g. [5]. Furthermore, $y_k := |A_k|$ satisfies the following recursion relation.

$$y_k = \sum_{i=0}^{k-1} \sum_{\rho\in S(i+1,k)} \varepsilon(\rho) a_{i+1\rho(i+1)} a_{i+2\rho(i+2)} \cdots a_{k\rho(k)} y_i, \quad k = 1, \ldots, n, \qquad (1)$$

*This research was partially supported by the Grant-in Aid for General Scientific Research from the Japan Society for the Promotion of Science 14340037.

where $y_0 := 1$, $\varepsilon(\rho)$ denotes the sign of ρ, and $S(k+1, n)$ denotes the set of all permutations ρ on $\{k+1, \ldots, n\}$ satisfying that there is no $\ell > k$ such that ρ is closed on $\{\ell+1, \ldots, n\}$. We call the recursion relation (1) the Jacobi equation for A. We say that k is (strictly) conjugate to 1 if the solution $\{y_i\}$ of the Jacobi equation with the initial condition $y_0 = 1$ changes the sign for the first time from positive to non-positive (negative) at k. Namely,

$$y_1 > 0, \; \ldots, \; y_{k-1} > 0, \text{ and } y_k \leq 0 \quad (y_k < 0), \tag{2}$$

see [2]. Then, it is evident that A is positive definite if and only if there are no points conjugate to 1. Similarly, if there exists a number k such that k is strictly conjugate to 1, then A is not nonnegative definite.

By the way, since each variable x_k in (P_0) plays the same role in general, there are no reasons to start with $k = 1$ in order to define conjugate points. So it is natural to deal with $\{1, \ldots, k\}$ or $\{x_1, \ldots, x_k\}$ rather than the endpoints 1 and k. This idea leads us to a conjugate set and a cooperative game. We call the game the conjugate-set game.

This paper is organized as follows. In Section 2, we first define a conjugate set, and next present a fundamental theorem on conjugate sets. In Section 3, we introduce the conjugate-set game. We give several examples and compute their Shapley values.

2. Strict Conjugate Sets

In this section, we define (minimal strict) conjugate sets and give a fundamental theorem on conjugate sets.

Definition 2.1. *Let $A = (a_{ij})$ be an $n \times n$ symmetric matrix, and $I = \{i_1, \ldots, i_k\}$ a subset of $\{1, \ldots, n\}$. If a submatrix $(a_{ij})_{i,j \in I}$ of A has a non-positive (negative) principal minor, then we call I a (strict) conjugate set. For the sake of convenience, we call the corresponding set of variables $\{x_k\}_{k \in I}$ a (strict) conjugate set. When any proper subset J of a (strict) conjugate set I is not a (strict) conjugate set, we call I (or $\{x_k\}_{k \in I}$) a minimal (strict) conjugate set.*

Theorem 2.1. *Let $A = (a_{ij})$ be a $2m+1$-multi diagonal matrix, that is,*

$$a_{ij} = 0 \quad \textit{if} \quad |i-j| \geq m+1. \tag{3}$$

Then any number of any minimal (strict) conjugate set jumps at most m. In particular, when A is a tridiagonal matrix, any minimal (strict) conjugate set consists of sequential numbers.

Proof. Let I be an arbitrary minimal (strict) conjugate set. Assume that I is divided into non-empty sets I_1 and I_2 such that

$$j - i \geq m+1 \quad \forall i \in I_1, \; \forall j \in I_2. \tag{4}$$

Denote by $S(I)$, $S(I_1)$, and $S(I_2)$ the set of permutations on I, I_1, and I_2, respectively. Denote by $A(I)$ the submatrix $(a_{ij})_{i,j \in I}$. Similarly, we define $A(I_1)$ and $A(I_2)$. Then, we get from (3) and (4) that $a_{i\sigma(i)} = 0$ for any $\sigma \in S(I)$ satisfying

$$\exists i \in I_1, \quad \sigma(i) \in I_2. \tag{5}$$

Hence, we may omit $\sigma \in S(I)$ satisfying (5) in the definition of $|A(I)|$. So,

$$|A(I)| = \sum_{\sigma \in S(I)} \varepsilon(\sigma) \prod_{i \in I} a_{i\sigma(i)} = \sum_{\sigma \in S(I),\ \sigma(I_1)=I_1} \varepsilon(\sigma) \prod_{i \in I} a_{i\sigma(i)}.$$

For such a $\sigma \in S_I$, we denote by σ_1 and σ_2 its restriction on I_1 and I_2, respectively. Then,

$$\begin{aligned} |A(I)| &= \sum_{\sigma \in S(I),\ \sigma(I_1)=I_1} \varepsilon(\sigma_1)\varepsilon(\sigma_2) \prod_{i \in I_1} a_{i\sigma_1(i)} \prod_{i \in I_2} a_{i\sigma_2(i)} \\ &= \left(\sum_{\sigma_1 \in S(I_1)} \varepsilon(\sigma_1) \prod_{i \in I_1} a_{i\sigma_1(i)} \right) \left(\sum_{\sigma_2 \in S(I_2)} \varepsilon(\sigma_2) \prod_{i \in I_2} a_{i\sigma_2(i)} \right) \\ &= |A(I_1)||A(I_2)|. \end{aligned}$$

Since I is a (strict) conjugate set, $|A(I)|$ is nonpositive (negative). Hence either $|A(I_1)|$ or $|A(I_2)|$ is nonpositive (negative), which contracts the minimality of I. This completes the proof.

Example 2.1. *Let S_1 denote the sphere with center $(3/2, 0, 3/2)$ and radius $1/\sqrt{2}$, S_2 the sphere with center $(-3/2, 0, 3/2)$ and radius $1/\sqrt{2}$, and C the circle in yz-plane with center $(0, 0, R)$ and radius $R > 0$. Then our problem is to find $(X_1, X_2, X_3) \in S_1 \times S_2 \times C$ that minimizes the area of the triangle $X_1X_2X_3$. We may take as the objective function*

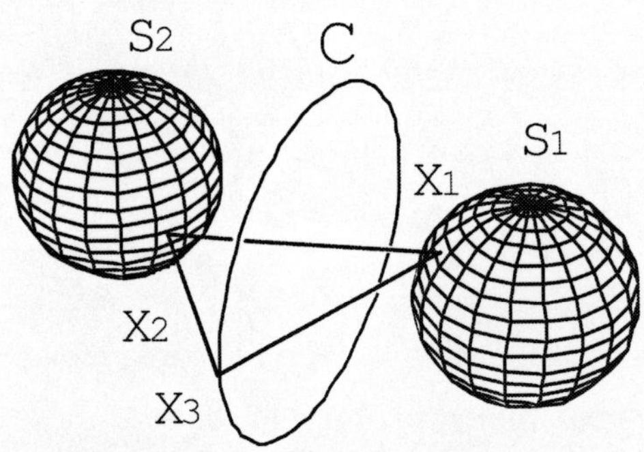

Figure 1. Minimal triangle 1.

$$f(\theta_1, \phi_1, \theta_2, \phi_2, \theta_3) := \| (X_1 - X_3) \times (X_2 - X_3) \|^2 .$$

where

$$X_1 = \left(\frac{1}{\sqrt{2}} \sin\theta_1 \cos\phi_1 + \frac{3}{2}, \frac{1}{\sqrt{2}} \sin\theta_1 \sin\phi_1, \frac{1}{\sqrt{2}} \cos\theta_1 + \frac{3}{2} \right),$$

$$X_2 = \left(\frac{1}{\sqrt{2}} \sin\theta_2 \cos\phi_2 - \frac{3}{2}, \frac{1}{\sqrt{2}} \sin\theta_2 \sin\phi_2, \frac{1}{\sqrt{2}} \cos\theta_2 + \frac{3}{2} \right),$$

and

$$X_3 = (0, R\sin\theta_3, R\cos\theta_3 + R)$$

for some $0 \le \theta_1 \le \pi$, $0 \le \phi_1 < 2\pi$, $0 \le \theta_2 \le \pi$, $0 \le \phi_2 < 2\pi$, *and* $0 \le \theta_3 < 2\pi$. *We*

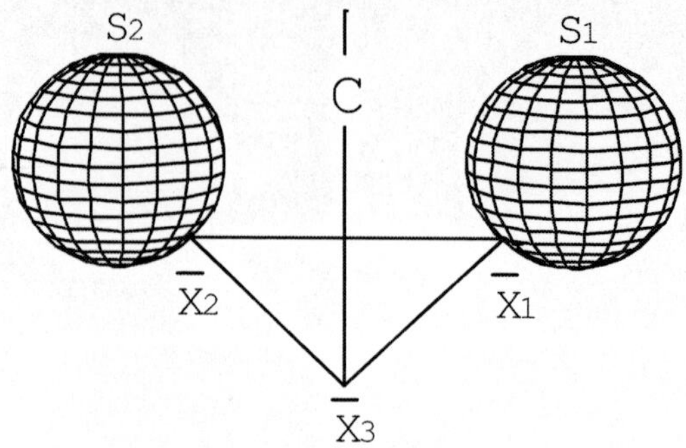

Figure 2. Minimal triangle 2.

test whether $(\bar{X}_1, \bar{X}_2, \bar{X}_3) := (1,0,1,-1,0,1,0,0,0)$ *gives a minimal area or not, that is, whether* $\bar{\theta} := (\bar{\theta}_1, \bar{\phi}_1, \bar{\theta}_2, \bar{\phi}_2, \bar{\theta}_3) := (3\pi/4, \pi, 3\pi/4, 0, \pi)$ *is a local minimum of* f *or not. It is easily seen that* $\bar{\theta}$ *is a stationary point of* f *and that the Hesse matrix* $f''(\bar{\theta})$ *is given by*

$$\begin{pmatrix} 4 & 0 & -2 & 0 & 0 \\ 0 & 3 & 0 & 0 & -2R \\ -2 & 0 & 4 & 0 & 0 \\ 0 & 0 & 0 & 3 & 2R \\ 0 & -2R & 0 & 2R & 8R(R-1) \end{pmatrix}. \tag{6}$$

By exchanging the second row (column) with the third row (column), the Hesse matrix becomes

$$A := \begin{pmatrix} 4 & -2 & 0 & 0 & 0 \\ -2 & 4 & 0 & 0 & 0 \\ 0 & 0 & 3 & 0 & -2R \\ 0 & 0 & 0 & 3 & 2R \\ 0 & 0 & -2R & 2R & 8R(R-1) \end{pmatrix}. \tag{7}$$

Since the descending principal minors of A are given by

$$|A_1| = 4, \ |A_2| = 12, \ |A_3| = 36, \ |A_4| = 108, \ |A_5| = 24R(2R-3),$$

$\bar{\theta}$ *is (not) minimal when* $R > 3/2$ *(*$R < 3/2$*). In particular, when* $R < 1$*, since the last diagonal element* $8R(R-1)$ *is negative,* $\{\theta_3\}$ *is a minimal strict conjugate set, which implies that the area of the triangle gets bigger by changing only* θ_3*. On the other hand, when* $1 < R < 3/2$*,* $\{\phi_1, \phi_2, \theta_3\}$ *is a minimal strict conjugate set, which implies that a certain increment*

$$\Delta\theta := (\Delta\theta_1, \Delta\phi_1, \Delta\theta_2, \Delta\phi_2, \Delta\theta_3) = (0, \Delta\phi_1, 0, \Delta\phi_2, \Delta\theta_3)$$

decreases the objective function. Indeed, when $R = 5/4$ *for example, by taking* $\Delta\theta = (0, -\Delta\phi_1, 0, \Delta\phi_1, -\Delta\phi_1)$*, we have*

$$\begin{aligned} f(\bar{\theta} + \Delta\theta) &= (\cos\Delta\phi_1 - 3)^2 (8\cos^2\Delta\phi_1 - 5\cos\Delta\phi_1 + 5) \\ &\simeq 32 - 6\Delta\phi_1^2 - \frac{\Delta\phi_1^4}{4} + O(\Delta\phi_1^5). \end{aligned}$$

Hence $f(\bar{\theta} + \Delta\theta) < f(\bar{\theta})$ *for any sufficiently small* $\Delta\phi_1$. *Finally, we add a note from the view point of Theorem 2.1. If we number the variables* $(\theta_1, \phi_1, \theta_2, \phi_2, \theta_3)$ *as* $(x_1, \ldots, x_5)$, *then the indices corresponding to the minimal set* $\{\phi_1, \phi_2, \theta_3\}$ *are 2, 4 and 5, which jumps 1.*

3. Conjugate-set Game

In Example 2.1, we have just seen that if there exists a strict conjugate set, we can improve a solution by moving the variables of the strict conjugate set simultaneously. In this section, we give an answer to the question that how much does each variable of the strict conjugate set contribute to decrease the objective function. For this purpose, we define a cooperative game based on conjugate sets and compute its Shapley value.

Definition 3.1. *For any subset* S *of* $\{1, \ldots, n\}$, *we define a characteristic function* $v(S)$ *by the maximum number* $0 \leq k \leq n$ *of disjoint strict conjugate sets contained in* S. *Let* X *denote the set of all variables* $\{x_1, \ldots, x_n\}$. *For any subset* $X_S := \{x_i;\ i \in S\}$ *of* X, *we defnie* $v(X_S) := v(S)$. *We call this cooperative game the conjugate-set game.*

The following lemma is obvious from the definition of $v(S)$.

Lemma 3.1. (a) $0 \leq v(S) \leq n$ *for any* S. (b) $v(\phi) = 0$. (c) *If* $S \cap T = \phi$, *then* $v(S) + v(T) \leq v(S \cup T)$.

As is well-known, the Shapley value is defined by

$$\phi_i(v) = \sum_{i \in S} \frac{v(S) - v(S - \{i\})}{n\binom{n-1}{s-1}}, \tag{8}$$

where s denotes the cardinal number of S. It is regarded as a measure to evaluate how much does player i contribute in the cooperative game with the characteristic function v.

Example 3.1. *In Example 2.1,* $\{\phi_1, \phi_2, \theta_3\}$ *was the unique minimal strict conjugate set when* $1 < R < 3/2$. *So,*

$$v(Y) = \begin{cases} 1 & \text{if} \quad \{\phi_1, \phi_2, \theta_3\} \subset Y, \\ 0 & \text{if} \quad \{\phi_1, \phi_2, \theta_3\} \not\subset Y. \end{cases} \tag{9}$$

Then the Shapley value is given by

$$\phi_1(v) = \phi_3(v) = 0, \quad \phi_2(v) = \phi_4(v) = \phi_5(v) = \frac{1}{3}. \tag{10}$$

(3.3) is so expected. In order to present a non-trivial example, we consider the following extremal problem.

$$(P_1) \qquad \text{Minimize} \quad f(x) := \sum_{k=0}^{n} f_k(x_k, x_{k+1}), \quad x := (x_1, \ldots, x_n) \in R^n,$$

where x_0 and x_{n+1} are given. For example, the shortest polygonal path problem on a surface S is formulated as (P_1), see Example 3.2 below. Problem (P_1) has a nice property that the Hesse matrix is tridiagonal. So, the Jacobi equation (1) reduces to a recursion relation of three adjacent principal minors y_{k-2}, y_{k-1} and y_k. In [4], we analyzed conjugate points for constant tridiagonal Hesse matrices

$$A := \begin{pmatrix} a & b & & \\ b & a & \ddots & \\ & \ddots & \ddots & b \\ & & b & a \end{pmatrix}, \tag{11}$$

where a, $b \in R$. Without loss of generality, we may assume that $a > 0$ and $b = \pm 1$. Combining Theorem 2.1 and the main result of [4, Theorem 5.1], we get the following theorem.

Theorem 3.1. *(a) When $a \geq 2$, there are no strict conjugate sets. (b) When $0 < a < 2$, let k denote the first number satisfying $(k+1)\varphi > \pi$, where φ $(0 < \varphi < \pi)$ is the argument of the solution of the characteristic equation*

$$y^2 - ay + 1 = 0. \tag{12}$$

If $k \leq n$, then $\{1, 2, \ldots, k\}$, $\{2, 3, \ldots, k+1\}$,..., $\{n-k+1, n-k+2, \ldots, n\}$ are minimal strict conjugate sets.

Proof. (a) is a direct consequence of [4, Theorem 5.1]. (b): By [4, Theorem 5.1], when $0 < a < 2$, k is strictly conjugate to 1, which implies that $\{1, 2, \ldots, k\}$ is a strict conjugate set. On the other hand, by Theorem 2.1, any minimal strict conjugate set consists of a sequential numbers. Suppose that $\{1, 2, \ldots, k\}$ is not minimal. Then there exists a pair of numbers $(\ell, m) \neq (1, k)$ such that $\{\ell, \ell+1, \ldots, m\}(\subset \{1, 2, \ldots, k\})$ is a minimal strict conjugate set. So, we see from the form of A that $m - \ell + 1(< k)$ is strictly conjugate to 1. This contradicts that k is strictly conjugate to 1.

Example 3.2. *Let S be an ellipsoid defined by*

$$\frac{x^2}{a^2} + \frac{y^2}{a^2} + \frac{z^2}{c^2} = 1.$$

Let us find the shortest polygonal path joining two given points $A = (a, 0, 0)$ and $B = (a\cos T, a\sin T, 0)$, where $T > 0$ and each knot X_k is chosen from a longitude $\ell_k := \{(a\sin\theta\cos k\Delta t, a\sin\theta\sin k\Delta t, c\cos\theta) : 0 < \theta < \pi\}$ and $\Delta t := T/(n+1)$. Since each knot is expressed as

$$X_k = (a\sin\theta_k\cos k\Delta t,\ a\sin\theta_k\sin k\Delta t,\ c\cos\theta_k)$$

by some $0 < \theta_k < \pi$, the length of X_kX_{k+1}, say f_k, is a function of (θ_k, θ_{k+1}). Hence the total length is given by $f(\theta_1, \ldots, \theta_n) := \sum_{k=0}^{n} f_k(\theta_k, \theta_{k+1})$, where $\theta_0 = \theta_{n+1} := \pi/2$. Next, let $\bar{X}_k$ denote the intersection of the equator and ℓ_k. Then $\bar{\theta} := (\pi/2, \ldots, \pi/2)$

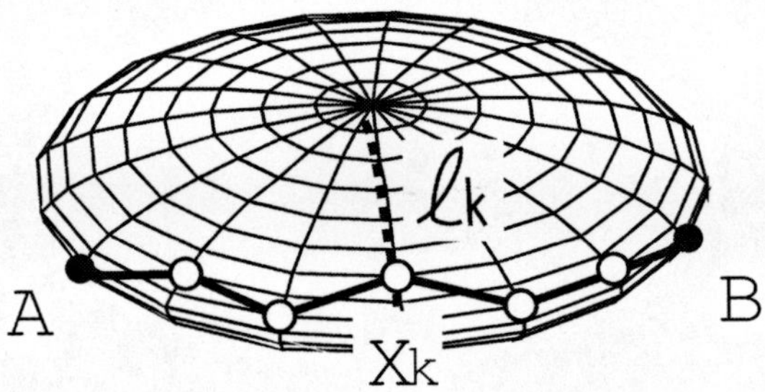

Figure 3. Shortest polygonal path problem.

corresponds to the equatorial polygonal path $A\bar{X}_1 \cdots \bar{X}_n B$. *It is easily seen that* $\bar{\theta}$ *is a stationary point of* $f(\theta_1, \ldots, \theta_n)$. *Furthermore, since*

$$f_k''\left(\frac{\pi}{2}, \frac{\pi}{2}\right) = \frac{c^2}{2a\sin\frac{\Delta t}{2}}\begin{pmatrix} d & -1 \\ -1 & d \end{pmatrix}, \quad d := 1 - 2\frac{a^2}{c^2}\sin^2\frac{\Delta t}{2},$$

the Hesse matrix $f''\left(\frac{\pi}{2}, \ldots, \frac{\pi}{2}\right)$ *is equal to*

$$\frac{c^2}{2a\sin\frac{\Delta t}{2}}\begin{pmatrix} 2d & -1 & & \\ -1 & 2d & \ddots & \\ & \ddots & \ddots & -1 \\ & & -1 & 2d \end{pmatrix}. \tag{13}$$

Since $0 < 2d < 2$ *for any sufficiently small* Δt, *this example reduces to case (b) of Theorem 3.1. So, denoting by* k *the first number satisfying* $(k+1)\varphi > \pi$, *where* $0 < \varphi < \pi$ *the argument of the solution of the characteristic equation* $y^2 - 2d + 1 = 0$, *we conclude that* $\{1, 2, \ldots, k\}$, $\{2, 3, \ldots, k+1\}$,..., $\{n-k+1, n-k+2, \ldots, n\}$ *are minimal strict conjugate sets. As a special case, we consider the following two cases: (1)* $n = 5,\ k = 4$ *and (2)* $n = 9,\ k = 4$. *In the case of* $n = 5,\ k = 4$,

$$v(S) = \begin{cases} 1 & \textit{if S contains } \{1,2,3,4\} \textit{ or } \{2,3,4,5\} \\ 0 & \textit{otherwise.} \end{cases}$$

Hence

$$(\phi_1(v),\ \phi_2(v),\ \phi_3(v),\ \phi_4(v),\ \phi_5(v)) = \left(\frac{1}{20}, \frac{6}{20}, \frac{6}{20}, \frac{6}{20}, \frac{1}{20}\right).$$

In the case of $n = 9,\ k = 4$, *there are two disjoint minimal strict conjugate sets as Fig. 5. So,*

$$v(S) = \begin{cases} 2 & \textit{if } \ Y \supset \{1,2,3,4\} \cup \{6,7,8,9\} \\ 1 & \textit{if } \ Y \textit{ contains just one minimal strict conjugate set} \\ 0 & \textit{otherwise.} \end{cases}$$

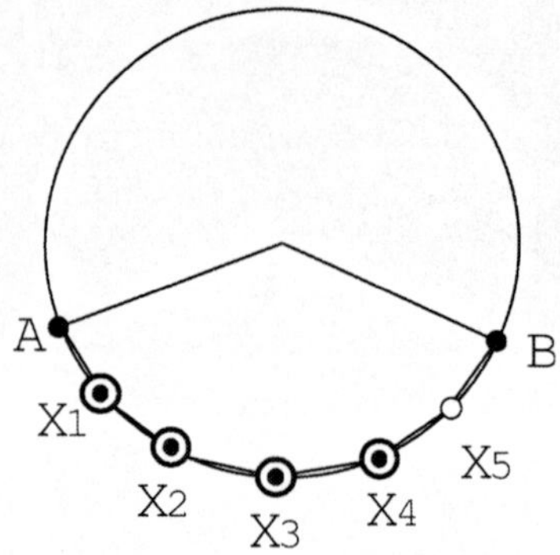

Figure 4. $n = 5,\ k = 4$.

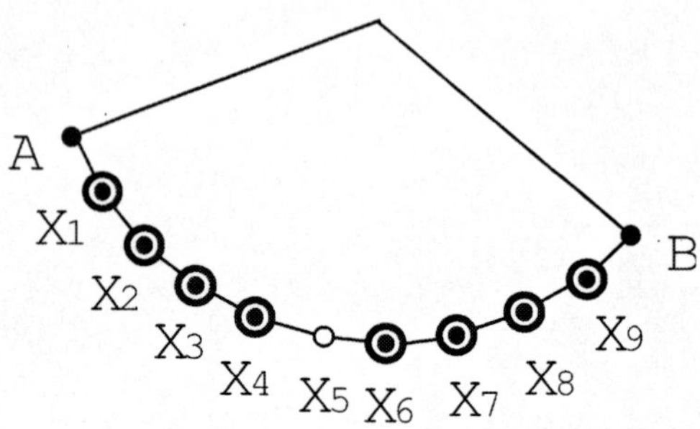

Figure 5. $n = 9,\ k = 4$.

Therefore

$$(\phi_1(v), \ldots, \phi_9(v)) = \frac{1}{360}\,(23,\ 86,\ 104,\ 122,\ 50,\ 122,\ 104,\ 86,\ 23)\,,$$

which implies that X_5 *contributes less than* X_2, X_3, X_4, X_6, X_7 *and* X_8.

References

[1] I.M. Gelfand, S.V. Fomin, *Calculus of variations*, Prentice Hall, 1963.

[2] H. Kawasaki, Conjugate points for a nonlinear programming problem with constraints, *J. Nonlinear Convex Anal.* **1** (2000), 287–293.

[3] H. Kawasaki, A conjugate points theory for a nonlinear programming problem, *SIAM J. Control Optim.* **40** (2001), 54–63.

[4] H. Kawasaki, Analysis of conjugate points for constant tridiagonal Hesse matrices of a class of extremal problems, *Optim. Methods and Software* **18** (2003), 197–205.

[5] G. Strang, *Linear Algebra and its Applications*, Academic Press, New York, 1976.

In: Mathematics, Game Theory and Algebra Compendium ISBN 978-1-60456-659-8
Editor: Jacob H. Mathias, pp. 241-257

Chapter 24

DYNKIN'S STOPPING GAMES WITH ZERO PAYOFFS FOR SEPARATE STOPPING

Victor Domansky
St. Petersburg Institute for Economics and Mathematics
Russian Academy of Sciences, Saint - Petersburg, Russia

1. Introduction

Stopping games are generalizations of optimal stopping problems for stochastic processes (see e.g. Shiryaev (1977)) for the case when several decision makers with different goals are involved.

There are various game setting of stopping problems. According to one of them Player 1 selects a stopping time and Player 2 picks a distribution of the process (see Sakaguchi M., (1992), Irle A. (1995)). Another approach assumes that players observe independently different stochastic processes. The players' payoffs depend on the whole combination of states where the processes are stopped (see e.g. Domansky V. (1976), Presman E., Sonin I. (1977), Mazalov V. (1987), Szajowski K. (1995)).

We consider the game setting of stopping problem for a Markov chain going back to Dynkin (1969). Two players observe the Markov sequence $x_n, n = 0, 1, \dots$ with the state space X. Four functions

$$a_{11},\ a_{12},\ a_{21},\ c : X \to R$$

are given over X.

Each player may stop the chain at any stage n. When the chain is stopped the game is finished and Player 1 wins from Player 2 the sum, depending on the player who stopped the chain and on its current state x. The payoff $a_{11}(x)$ corresponds to simultaneous stopping at the state x. The payoffs $a_{12}(x)$ and $a_{21}(x)$ correspond to stopping by Player 1 and Player 2 only. Stopping games in this setting were introduced by Neveu (1975), who called them "Dynkin's games".

The function c determines the "payoff at infinity". If the game continues infinitely Player 1 gets its limit along the trajectory. Thus, the "payoff at infinity" depends on the limiting behavior of the chain. "Payoffs at infinity" for "Dynkin's games" in this form were introduced by Elbakidze (1976).

We denote $\Gamma_c(x)$ the game, depending on the initial state x and on the limit payoff c, distinguishing the latter parameter. Dynamic programming considerations imply that the value function $v_c(x)$ of the game $\Gamma_c(x)$ should satisfy the optimality equation that is a game analogue of Wald-Bellman equation. Thus, the value of stopping game is a fixed point of operator T, functioning over the set of functions over the state space X. In general, such fixed point is not unique and its choice depends on the function c, determining the "payoff at infinity". Generally, resolving of optimality equation requires employing mixed strategies. Mixed behavior strategies determine randomized stopping rules.

In most cases the authors studying Dynkin's games (Kiefer (1971), Neveu (1975), Bismut (1979)), Alario-Lazaret et al.(1982), Lepeltier et al.(1984)) make two assumptions: 1) the "the payoff at infinity" is zero; 2) payoffs satisfy the inequalities $a_{12} \leq a_{12} \leq a_{21}$ resulting in solvability of optimality equations with use of pure strategies only. Elbakidze (1976) was first who dropped the assumption 1) but held the assumption 2). In the works of Yasuda (1985) and Rosenberg, Solan, Vielle (2000) the assumption 2) is dropped, but the assumption 1) is not. It allows to obtain the solutions for these games as the limits of solutions for discounted games.

As in my previous papers (Domansky (1978), (2001), (2002), (2004)), here we drop both these assumptions. We investigate solvability of optimality equations by means of mixed strategies and the impact of "the payoff at infinity" c on the choice of solution.

My previous investigations on the topics include the following result: Suppose that payoffs a_{12} and a_{21} belong to the class L of functions with bounded expectation of modulus supremum along the chain trajectory. Then we may construct the harmonic functions $\overline{h}_{12}(x)$ and $\underline{h}_{21}(x)$ representing an upper limit of the payoff a_{12} and a lower limit of the payoff a_{21} respectively. Let $\overline{h}_{12}(b)$ and $\underline{h}_{21}(b)$ be corresponding functions on the Martin exit boundary.

Under assumption that payoff a_{11} also belongs to the class L the value $v_c(x)$ of the game $\Gamma_c(x)$ depends on the value of "payoff at infinity" at points b of Martin exit boundary where the inequalities $\overline{h}_{12}(b) \leq c(b) \leq \underline{h}_{21}(b)$ hold. In other words, to be taken into account "payoff at infinity" should be a "saddle point" at the boundary point b.

Here we confine our attention to subclasse of Dynkin games with zero payoffs a_{12} and a_{21}, corresponding to separate stopping by only one of Players. The payoff $a = a_{11}$, corresponding to simultaneous stopping by both Players and the "payoff at infinity" c are positive. We show that the value $v_c(x)$ of such game is a non-negative subharmonic function satisfying the inequality $v_c(x) \leq c(x)$. The main problem is, whether it is more than zero.

It follows from the above-mentioned result that such games have zero values under assumption that the payoff a belongs to the class L. It is not the case if we allow the payoff $a(x)$ to be unbounded. It makes the situation more complicated as the growth rate of the payoff a is included into the set of factors determining the impact of "the payoff at infinity" c at the value of the stopping games $\Gamma_c(x)$.

Solving these games requires indispensable employing randomized strategies. A stationary randomized strategy of stopping is given by a function $\tau : X \to R^1$ indicating the probability to stop the chain if it hits the state x.

The crucial role for the character of solution for these games play properties of stationary strategies given by functions reversely proportional to the payoff a. The necessary and sufficient condition for the value of the game to be more than zero is that these strategies do not stop the chain for sure.

It becomes possible under the following conditions:

a) all states of the chain are transient;

b) the potential of the function $c \cdot a^{-1}$ with respect to a specially modified Markov chain is finite.

We give the bounds for values and for optimal randomized stopping times by means of these potentials.

For Markov chains with a trivial Martin boundary, i.e. for the case when the only harmonic functions belonging to the class L are the constants, the necessary and sufficient condition for the value of the game to be more than zero is finiteness of potential of the function a^{-1} with respect to the initial Markov chain. This is just the case considered in my paper.

2. Superharmonic and Subharmonic Functions. Martin Boundaries

In this section we cite some well known results for Markov chains with a countable state space. These results allow us to describe required properties of randomized stopping rules for Markov chains. The related topics include the structure of superharmonic and subharmonic functions, Martin exit boundaries and potentials for these chains. We follow the book of Dynkin and Yushkevich (1969).

Let x_n, $n = 0, 1, 2, \ldots$ be a Markov chain with a countable state space X and with a matrix of transition probabilities $P(x, y)$. Probabilities and mathematical expectations corresponding to initial state $x = x_0 \in X$ are denoted $\mathbf{P}_x$ and $\mathbf{E}_x$.

Studies on optimal stopping problems for Markov chains substantially use means of theory of superharmonic and subharmonic functions with respect to the operator P, defined with the formula

$$Pg(x) = \mathbf{E}_x g = \mathbf{E}[g(x_1)|x_0 = x].$$

A function f is called superharmonic (subharmonic) if $Pf \leq f$ ($Pf \geq f$). A function f is called harmonic if $Pf = f$.

Let L be the set of functions $g : X \to R^1$, such that

$$\mathbf{E}_x \sup_n |g(x_n)| < \infty \quad \forall x \in X.$$

Let $\mathcal{R}_\alpha$ be the recurrent classes of the chain and let $\mathcal{T}$ be the set of transient states of the chain.

Let $P_{0,\alpha}$ be restriction of matrix P to $\mathcal{T} \times \mathcal{R}_\alpha$, $P_{0,0}$ restriction of P to $\mathcal{T} \times \mathcal{T}$. Then the series

$$G_0 = \sum_{n=0}^{\infty} P_{0,0}^n$$

converges. Let f_α (respectively f_0) be the restriction of f to $\mathcal{R}_\alpha$ (respectively to $\mathcal{T}$).

Proposition 2.1. *A function $f \in L$ is superharmonic (subharmonic) iff $f_\alpha = \lambda_\alpha \mathbf{I}_\alpha$, where λ_α is a constant, and the function*

$$f_0 - G_0 \sum_\alpha P_{0,\alpha} \lambda_\alpha \mathbf{I}_\alpha$$

is superharmonic (subharmonic).

This proposition reduces studying superharmonic and subharmonic functions to the case when all states are transient (excluding the set of absorbing states where the function is equal to zero).

Consider the potential matrix

$$G = \sum_{n=0}^{\infty} P^n.$$

where P^n is the n-step transition matrix and P^0 is the identity matrix. The term $G(x, y)$ of the potential matrix gives the expected sojourn time at the state y for the initial state x. This time is infinite if y is a recurrent state attainable from x. A function $u \geq 0$ can have finite potential

$$Gu = \lim_{n \to \infty} \sum_0^n P^m u,$$

only if its support contains in the set T of transient states. The potential Gu of a function $u \geq 0$ is a superharmonic function.

Proposition 2.2. *A superharmonic function $f \in L$ has a unique representation as a sum (Riesz decomposition)*

$$f(x) = h_f(x) + \pi_f(x),$$

where $h_f(x)$ is a harmonic function

$$h_f(x) = \lim_{n \to \infty} P^n f(x) = \mathbf{E}_x \lim_{n \to \infty} f(x_n),$$

and a superharmonic function π_f is a potential of the non-negative function $f(x) - Pf(x)$. The following limit equalities hold:

$$\lim_{n \to \infty} \pi_f(x_n) = 0, \lim_{n \to \infty} f(x_n) = \lim_{n \to \infty} h_f(x_n).$$

We obtain the analogous representation for a subharmonic function $g \in L$, examining a superharmonic function $f = -g \in L$,

Corollary. *A subharmonic function $g \in L$ has a unique representation as a difference*

$$g(x) = h_g(x) - \pi_g(x),$$

where $h_g(x)$ is a harmonic function

$$h_g(x) = \lim_{n \to \infty} P^n g(x) = \mathbf{E}_x \lim_{n \to \infty} g(x_n),$$

and a superharmonic function π_g is a potential of the non-negative function $Pg(x) - g(x)$. The following limit equalities hold:

$$\lim_{n\to\infty} \pi_g(x_n) = 0, \lim_{n\to\infty} g(x_n) = \lim_{n\to\infty} h_g(x_n).$$

Constructing of Martin exit boundary for a transient Markov chain allows to describe the asymptotic behavior of this random process, to reveal the structure of the set of harmonic functions, and to obtain the representation of Poisson integral type for harmonic functions.

Let H_0 be the set of non-negative harmonic functions $h \in L$, satisfying $h(0) = 1$, where 0 is a certain chosen point in X. H_0 is a convex compact set in pointwise convergence topology. Martin exit boundary B is defined as a set of extreem points of H_0.

For $x \in X, b \in B$, let the function $K(x, b)$ (Martin kernel) be the value at the point $x \in X$ of the harmonic function corresponding to the point $b \in B$.

Proposition 2.3. *For any harmonic function $h \in L$ there exists a bounded measure ν_h on B providing the integral representation*

$$h(x) = \int_B K(x, b)\nu_h(db), \quad \forall x \in X.$$

Inversely, for any bounded measure ν on B the integral $\int_B K(x, b)\nu(db)$ represents the harmonic function $h_\nu \in L$ on X.

Let ν_1 be the measure on B defining the harmonic function identically equal to one: $1 = \int_B K(x, b)\nu_1(db)$ for all $x \in X$.

Proposition 2.4. *Any harmonic function $h \in L$ P_x-a.s. has a limit $\lim_{n\to\infty} h(x_n)$. This limit defines an integrable function $\hat{h} \in L^1(B, \nu_1)$ at Martin exit boundary B. The function h is equal to the expectation of its boundary values.*

$$h(x) = \mathbf{E}_x \lim_{n\to\infty} h(x_n) = \mathbf{E}_x h(x_\infty) = \int_B \hat{h}(b)K(x, b)\nu_1(db).$$

3. Properties of Randomized Stopping Strategies

The results of this subsection belong to the author.

Let τ be a stationary randomized strategy of stopping given by a function $\tau : X \to R^1$. Here $\tau(x)$ is the probability to stop the chain at state x if it was not stopped earlier:

$$\tau(x) = \mathbf{P}(\tau = n | \tau \geq n; x_n = x).$$

We put

$$P_\tau(x, y) = (1 - \tau(x))P(x, y),$$

i.e. $P_\tau = (I - [\tau]) \cdot P$, where $[\tau]$ is a diagonal matrix with $[\tau](x, x) = \tau(x)$, $[\tau](x, y) = 0$ for $x \neq y$.

Consider the distribution of states where the chain is stopped with the strategy τ:

$$H_\tau(x, y) = \mathbf{P}\{x_\tau = y|x_0 = x\} = \sum_{n=0}^{\infty} \mathbf{P}(x_n = y|\tau \geq n; x_0 = x)\mathbf{P}(\tau = n|\tau \geq n; x_n = y).$$

The matrix H_τ has the form

$$H_\tau = \sum_{n=0}^{\infty} P_\tau^n[\tau].$$

The function

$$S_\tau(x) = \sum_{y \in X} H_\tau(x, y)$$

gives the probability that the strategy τ stops the chain for the initial state $x_0 = x$.

The function $P^n S_\tau$ gives the probability that this strategy stops the chain after the step $n-1$. Hence $\pi_\tau = (I - P)S_\tau$ is the probability that the chain will be stopped at initial step only. Therefore S_τ is a superharmonic function. Its Riesz decomposition has the form

$$S_\tau = h_\tau + G\pi_\tau,$$

where the harmonic function

$$h_\tau = \lim_{n \to \infty} P^n H_\tau \cdot 1$$

gives the probability that the strategy τ stops the chain infinitely many times, for the initial state $x_0 = x$. This event is an invariant event, i.e. it has the form $x_\infty \in B_\tau$ for a certain stopping subset $B_\tau \subset B$ of the Martin exit boundary. Therefore

$$h_\tau(x) = \mathbf{P}_x\{x_\infty \in B_\tau\} = \int_{B_\tau} K(x, b)d\nu_1.$$

We say that the strategy τ stops the chain when it converges to the boundary point $b \in B_\tau$.

Definition 3.1. We say that a strategy τ is *completely stopping* if $B_\tau \equiv B$ and, consequently, $S_\tau = h_\tau = \mathbf{1}$.

Definition 3.2. We say that a strategy τ is *transitive* if $B_\tau \equiv \emptyset$ and, consequently, $h_\tau = 0$, $S_\tau = G\pi_\tau$.

Proposition 3.1. *A strategy τ is transitive iff the potential $G\tau$ of the function τ is finite.*

Proposition 3.2. *For a strategy τ, let $h_\tau \neq 0, 1$. Then a strategy τ is completely stopping for the chain with transition probabilities $p(x, y)h_\tau(y)/h_\tau(x)$. A strategy τ is transitive for the chain with transition probabilities $p(x, y)(1 - h_\tau(y))/(1 - h_\tau(x))$.*

Corollary. *For any function $e : X \to (\varepsilon, 1]$, where $\varepsilon > 0$, the stopping sets B_τ and $B_{e\tau}$ coinside.*

Consider the positive harmonic function

$$\hat{c} = \mathbf{E}_x \underline{\lim}_{n \to \infty} c(x_n),$$

and let $\hat{c}(b)$ be the corresponding function on the Martin exit boundary.

Observe that the limit $\lim_{n\to\infty}\hat{c}(x_n)$ exists a.s. and the equality

$$\lim_{n\to\infty}\hat{c}(x_n) = \underline{\lim}_{n\to\infty}c(x_n) = c(b)$$

holds. Thus, $\underline{\lim}_{n\to\infty}c(x_n)$ may be replaced with $\lim_{n\to\infty}\hat{c}(x_n)$. As this limit only is essential for the game, further we put $c(x) = \hat{c}(x)$, and write $\lim_{n\to\infty}c(x_n)$ instead of $\underline{\lim}_{n\to\infty}c(x_n)$, presupposing c to be a harmonic function a priori.

4. Optimality Equations and Games with Zero Values

Here we assume that payoffs satisfy the following conditions:

$$a_{21}(x) = a_{12}(x) = 0, \quad a_{11}(x) = a(x) > 0, \quad c(x) > 0, \quad \forall x \in X.$$

The state space X is countable, the transition probabilities are given with the infinite matix $P = [P(x,y)], x, y \in X$. The "payoff at infinity" $c(x)$ is a positive harmonic function.

Strategies of players for the game $\Gamma_c(x)$ are randomized stopping times τ and σ for the chain. Let T and Σ be the sets of strategies of Players 1 and 2 correspondingly.

Denote the expected gain of Player 1 for the game $\Gamma_c(x)$ under strategies $\tau \in T$ and $\sigma \in \Sigma$ by $K_c(\tau, \sigma|x)$:

$$K_c(\tau,\sigma|x) = \mathbf{E}_x(\mathbf{I}_{[\tau=\sigma<\infty]}a_{11}(x_\tau) + \mathbf{I}_{[\tau=\sigma=\infty]}\lim_{n\to\infty}c(x_n)).$$

The optimality equation for the value $v_c(x)$ of the game $\Gamma_c(x)$ takes the form

$$v_c(x) = \frac{a(x)\cdot\sum_{y\in X}p(x,y)\cdot v_c(y)}{a(x)+\sum_{y\in X}p(x,y)\cdot v_c(y)}. \tag{4.1}$$

Thus, the value of stopping game is a fixed point of operator T, functioning over the set of integrable functions according to formula

$$Tu(x) = \frac{a(x)\cdot\sum_{y\in X}p(x,y)\cdot u(y)}{a(x)+\sum_{y\in X}p(x,y)\cdot u(y)}. \tag{4.2}$$

In general, such fixed point is not unique and its choice depends on the function c, determining "the payoff at infinity". Notice that the function c is not taken into account with the equation (1.1) by itself.

The function $v_0(x) = 0$ for all $x \in X$ is a solution of the equation (4.1), although the strategy $\sigma_0(x) = 0, \forall x \in X$ provides Player 2 the loss $c(x)$. The best answer of Player 1 to the strategy of Player 2 "not to stop at all" is the strategy "not to stop at all" too, resulting in $c(x)$.

On the other hand, the best answer of Player 2 to the strategy of Player 1 "not to stop at all" is the strategy "to stop immediately" resulting in payoff zero.

Thus, resolving of equation (4.1) requires employing mixed strategies. Mixed behavior strategies determine randomized stopping rules.

Proposition 4.1. *For any state x such that the set of recurrent states is accessible from x a.s. and for any "payoff at infinity" c the value $v_c(x)$ is equal to zero.*

Proof. For any $0 < \varepsilon \leq A$ the strategy of Player 2 $\sigma(y) = \varepsilon \cdot a(y)^{-1} \wedge 1$, for all recurrent states y stops the chain. It provides Player 2 the loss not exceeding ε. Therefore, it is an ε-optimal strategy. This proves Proposition 4.1.

Remark. *In this game Player 2 has no optimal strategy.*

Further we assume that all states of the chain are transient. This means that the potential matrix

$$G(x, y) = \sum_{n=0}^{\infty} P^n(x, y)$$

is finite for all $x, y \in X$.

Proposition 4.2. *The upper value $\overline{v}_c(x)$ and the lower value $\underline{v}_c(x)$ of the game $\Gamma_c(x)$ are non-negative subharmonic functions satisfying the inequalities*

$$\underline{v}_c(x) \leq \overline{v}_c(x) \leq c(x).$$

Proof. The upper value $\overline{v}_c(x)$ and the lower value $\underline{v}_c(x)$ of the game $\Gamma_c(x)$ are solutions of the equation (4.1). Any solution v of the equation (4.1) satisfies the inequalities $0 \leq v \leq P \cdot v$. This means that v is a non-negative subharmonic function. As it was said above the strategy $\sigma_0(x) = 0, \forall x \in X$, provides Player 2 the loss $c(x)$. Hence $\overline{v}_c(x) \leq c(x)$. The identically valid inequality $\underline{v}_c(x) \leq \overline{v}_c(x)$ completes the proof of Proposition 4.2.

Let

$$\overline{v}_c = h_{\overline{v}_c} - G \cdot \pi_{\overline{v}_c}; \qquad \underline{v}_c = h_{\underline{v}_c} - G \cdot \pi_{\underline{v}_c}$$

be the Riesz decompositions of subharmonic functions $\overline{v}_c$ and $\underline{v}_c$.

Proposition 4.3. *The kernel $\pi_{\overline{v}_c}$ of potential in the Riesz decompositions for $\overline{v}_c$ satisfies the relations*

$$\pi_{\overline{v}_c}(x) = \frac{\overline{v}_c(x) \cdot P\overline{v}_c(x)}{a(x)} < \frac{(c(x))^2}{a(x)}.$$

Analogous relations are fulfilled for the kernel $\pi_{\underline{v}_c}(x)$.

Proof. The kernel $\pi_{\overline{v}}$ of potential for subharmonic function $\overline{v}_c$ is given by

$$\pi_{\overline{v}_c}(x) = P\overline{v}_c(x) - \overline{v}_c(x),$$

It follows from equation (3.1) that

$$(P\overline{v}_c(x) - \overline{v}_c(x)) \cdot a(x) = \overline{v}_c(x) \cdot P\overline{v}_c(x).$$

Since $\overline{v}_c(x) \leq P(x, y)\overline{v}_c(y) \leq c(x)$, we get the second inequality of Proposition 4.3. The proof for $\pi_{\underline{v}_c}$ is analogous.

Theorem 4.1. *If the stationary strategy $\sigma_{1/a}$, given by the function $a(y)^{-1} \wedge 1$, is completely stopping; then for any "payoff at infinity" c the value $v_c(x)$ of the game $\Gamma_c(x)$ is equal to zero. Player 2 has no optimal strategy.*

Proof. Consider the stationary strategy $\sigma_{\varepsilon/a}$ of Player 2 given by the function $\varepsilon \cdot a(y)^{-1} \wedge 1$. This strategy is completely stopping according to Corollary of Proposition 3.2. and, consequently, the equality

$$K_c(\tau_\infty, \sigma_\varepsilon) = 0.$$

holds for any "payoff at infinity". On the other hand, the equality

$$K_c(\tau, \sigma_\varepsilon) = \varepsilon$$

holds for any stopping strategy τ of Player 1. Hence, the strategy σ_ε is ε-optimal strategy. It proves Proposition 4.4.

5. Games with Transitive Strategies $\sigma_{1/a}$

In this section we consider the games $\Gamma_c(x)$ such that the stationary strategy $\sigma_{1/a}$, given by the function $a(y)^{-1} \wedge 1$, is transitive.

According to Proposition 3.1. the stationary strategy $\sigma_{1/a}$ is transitive iff the potential of function a^{-1} is finite, i.e.

$$S_{1/a}(x) = \sum_{y \in X} G(x, y) \cdot a(y)^{-1} < \infty, \quad \forall x \in X. \tag{5.1}$$

Proposition 5.1. *If the strategy $\sigma_{1/a}$ is transitive, then for the Riesz decompositions of subharmonic functions $\overline{v}_c$ and $\underline{v}_c$ the equalities*

$$h_{\overline{v}_c}(x) = h_{\underline{v}_c}(x) = c(x)$$

hold.

Proof. Consider the stationary strategies $\tau_{c/a}$ and $\sigma_{c/a}$ of Player 1 and Player 2 respectively given by the same function

$$\tau_{c/a}(x) = \sigma_{c/a}(x) = c(x) \cdot a(x)^{-1} \wedge 1, \quad \forall x \in X.$$

The strategies $\tau_{c/a}$ and $\sigma_{c/a}$ are transitive. It follows from (5.1) that

$$S_{c/a}(x) = \sum_{y \in X} G(x, y) c(y) \cdot a(y)^{-1} < \infty, \quad \forall x \in X.$$

The probability that the strategy $\tau_{c/a}$ stops the chain with initial state x does not exceed $S_{c/a}(x) \wedge 1$.

Consider breaking Markov chains $x_n^{\tau_{c/a}}$ and $x_n^{\sigma_{c/a}}$ with substochastic transition probabilities

$$P^{\tau_{c/a}}(x, y) = P^{\sigma_{c/a}}(x, y) = (1 - \tau_{c/a}(x)) P(x, y),$$

resulting from initial chain under strategies $\tau_{c/a}$ and $\sigma_{c/a}$.

The function $c(x)$ is a superharmonic function with respect to the chains $x_n^{\tau_{c/a}}$ and $x_n^{\sigma_{c/a}}$. Consequently, $c(x)$ is the value of maximizing stopping problem for the chain $x_n^{\sigma_{c/a}}$

with the current payoff $c(x)$ and the "payoff at infinity" $\lim_{n\to\infty} c(x_n)$. The best answer of Player 1 to the strategy $\sigma_{c/a}$ is to stop immediately.

The best answer of Player 2 to the strategy $\tau_{c/a}$ is not to stop at all, i.e. $\sigma_\infty(x) = 0, \forall x \in X$. This strategy ensure Player 2 the loss

$$w_c(x) = \min_{\sigma\in\Sigma} K_c(\tau_{c/a}, \sigma|x) = K_c(\tau_{c/a}, \sigma_\infty|x). \tag{5.2}$$

The function $w_c(x)$ is a harmonic function with respect to the chains $x_n^{\tau_{c/a}}$ and $x_n^{\sigma_{c/a}}$, i.e.

$$w_c(x) = (1 - \tau_{c/a}(x)) \sum_{y\in X} p(x,y)w_c(y), \forall x \in X. \tag{5.3}$$

The function $w_c(x)$ satisfies the inequalities

$$w_c(x) \geq c(x) \cdot (1 - (S_{c/a}(x) \wedge 1)), \forall x \in X.$$

Hence, the lower value $\underline{v}_c(x)$ of the game $\Gamma_c(x)$ satisfies the inequalities

$$\underline{v}_c(x) \geq w_c(x) \geq c(x) \cdot (1 - (S_{c/a}(x) \wedge 1)).$$

Since

$$\lim_{n\to\infty} S_{c/a}(x_n) = 0,$$

we obtain

$$\lim_{n\to\infty} \underline{v}_c(x_n) \geq \lim_{n\to\infty} w_c(x_n) = \lim_{n\to\infty} c(x).$$

Identically true inequality $\underline{v}_c(x) \leq \overline{v}_c(x)$ in combination with previous inequality proves that

$$\lim_{n\to\infty} \underline{v}_c(x_n) = \overline{v}_c(x_n) = \lim_{n\to\infty} c(x).$$

It follows that

$$h_{\overline{v}_c}(x) = h_{\underline{v}_c}(x) = c(x).$$

It completes the proof of Proposition 5.1.

Theorem 5.1. *If the strategy $\sigma_{1/a}$ is transitive, then a) for any non-negative harmonic functions $c(x) \in L$ the game $\Gamma_c(x)$ has a value; the value $v_c(x)$ is the unique solution of equation (3.1) satisfying the "boundary condition at infinity":*

$$\lim_{n\to\infty} v_c(x_n) = \lim_{n\to\infty} c(x_n); \tag{5.4}$$

b) the stationary strategies τ_c^* and σ_c^* given by

$$\tau_c^*(x) = \sigma_c^*(x) = v_c(x) \cdot a(x)^{-1} \qquad \forall x \in X,$$

are optimal.

Proof. a) According to Proposition 5.1 functions $\underline{v}_c(x)$ and $\overline{v}_c(x)$ are fixed points of operator T satisfying the same "boundary condition at infinity" (5.4).

It was shown in the paper [ ] that if the function $R(x) \in L$ is a solution of optimality equation for stopping game, then the game $\Gamma_R(x)$ with the "payoff at infinity" defined by

R has a value $v_R(x)$ and $v_R(x) = R(x)$. Hence, both functions are equal to the value $v_c(x)$ of the game $\Gamma_c(x)$:

$$\underline{v}_c(x) = \overline{v}_c(x) = v_c(x).$$

This proves the assertion a) of Theorem 5.1.
b) It is easy to verify that $v_c(x) \in L$ is a harmonic function with respect to the breaking chain $x_n^{\tau_c^*}$ with transition probabilities

$$P^{\tau_c^*}(x, y) = (1 - \tau_c^*(x))P(x, y).$$

It follows that

$$v_c(x) = \min_{\sigma \in \Sigma} K_c(\tau_c^*, \sigma | x).$$

Analogously, $v_c(x) \in L$ is a harmonic function with respect to the breaking chain $x_n^{\sigma_c^*}$ and, consequently,

$$v_c(x) = \max_{\tau \in T} K_c(\tau, \sigma_c^* | x).$$

This proves the assertion b) of Theorem 5.1.

The next Proposition defines iterative procedure allowing to construct the value $v_c(x)$ of the game $\Gamma_c(x)$.

Proposition 5.2. *The function v_c satisfies inequalities*

$$T^n w_c(x) < v_c(x) < T^n c(x),$$

where $w_c(x)$ is defined with formula (5.2);
the sequence $T^n w_c(x)$ is nondecreasing, the sequence $T^n c(x)$ is nonincreasing, both sequences converge to $v_c(x)$.

Proof. By definition the function $w_c(x)$ is the result of best reply of Player 2 to the strategy $\tau_{c/a}$. The sequence $T^n w_c(x)$ is nondecreasing because it represents the results of consecutive improvemnts of the strategy $\tau_{c/a}$:

$$T^n w_c(x) = \sup_{\tau \in T_n(\tau_{c/a})} \inf_{\sigma \in \Sigma} K_c(\tau, \sigma | x),$$

where

$$T_n(\tau_{c/a}) = \{\tau : \tau = \tau_{c/a} \quad \text{for} \quad t \geq n\}.$$

Anlogously, the sequence $T^n w_c(x)$ is nonincreasing because it represents the results of consecutive improvemnts of the strategy $\sigma(x) = 0, \forall x$ of Player 2, i.e. the strategy "not to stop at all".

Therefore, the limits $\lim_{n\to\infty} T^n w_c(x)$ and $\lim_{n\to\infty} T^n c(x)$ exist. Both functions $\lim_{n\to\infty} T^n w_c(x)$ and $\lim_{n\to\infty} T^n c(x)$ are fixed points of operator T satisfying the boundary condition (5.4) at infinity. According to Theorem 5.1.

$$\lim_{n\to\infty} T^n w_c(x) = \lim_{n\to\infty} T^n c(x) = v_c(x),$$

where the function $v_c(x)$ is the unique solution of equation ( ) satisfying the boundary condition ( ) at infinity. This proves Proposition 5.2.

6. Games with Nontransitive and Nonstopping Strategies $\sigma_{1/a}$

In this section we consider the games $\Gamma_c(x)$ such that the stationary strategy $\sigma_{1/a}$, given by the function $a(y)^{-1} \wedge 1$, is not completely stopping and not transitive. In this case the stopping set B^- and the non-stopping set B^+ are nonempty complementary subsets of Martin exit boundary.

Define the harmonic function c_{B^+} by the formula

$$c_{B^+}(x) = \int_{B^+} K(x,b)c(b)d\nu_1.$$

Here $K(x,b)$ is the Martin kernel for the chain, $c(b)$ are boundary values of function c, and ν_1 is the measure on B defining the harmonic function identically equal to one.
In particular, let $P^+(x)$ be the probaility that a trajectory starting at the state x finishes at the set B^+ of Martin boundary. Then $P^+(x) = \mathbf{1_{B^+}(x)}$, where $\mathbf{1(x)} \equiv \mathbf{1}$ is a harmonic function identically equal to one.

Proposition 6.1. *Harmonic components of Risz decompositions of subharmonic functions $\overline{v}_c$ and $\underline{v}_c$ coinside and are equal to $c_{B^+}(x)$.*

Proof. First let us show that for any $x \in X$ the in equality $\overline{v}_c(x) \le c_{B^+}(x)$. In fact

$$\overline{v}_c(x) = \inf_{\sigma\in\Sigma}\sup_{\tau\in T} K_c(\tau,\sigma|x) \le \inf_{\varepsilon}\sup_{\tau\in T} K_c(\tau,\sigma_\varepsilon|x) \le \inf_{\varepsilon} K_c(\tau_\infty,\sigma_\varepsilon|x) = c_{B^+}(x).$$

As $c_{B^+} \le c$, we have

$$\underline{v}_{c_{B^+}}(x) \le \underline{v}_c(x),$$

and it is sufficient to show that

$$h_{\underline{v}_{c_{B^+}}(x)}(x) = c_{B^+}(x).$$

Consider stationary strategy $\tau_{c_{B^+}}$ of Player 1 given by function

$$\tau_{c_{B^+}}(x) = c_{B^+}(x)\cdot a(x)^{-1}, \quad \forall x \in X.$$

Consider modified Markov chain $\tilde{x}_n$ with the same state space X and transition probabilities

$$[\tilde{P}(x,y)] = [(c_{B^+}(x))^{-1}P(x,y)c_{B^+}(y)], \quad x,y \in X.$$

The Martin boundary $\tilde{B}$ for this chain coinside with the subset B^+ of the Martin boundary of initial chain. The stopping set $\tilde{B}^-$ for stationary strategies σ_ε is empty.

It follows that the potential of the function $a(x)^{-1}$ with respect to the chain $\tilde{x}_n$ is finite, i.e.

$$\tilde{S}_{1/a}(x) = \sum_{y\in X} \tilde{G}(x,y)a(y)^{-1} < \infty, \quad \forall x \in X.$$

Potential matrices $\tilde{G}(x,y)$ and $G(x,y)$ are connected with the relations

$$\tilde{G}(x,y) = (c_{B^+}(x))^{-1}G(x,y)c_{B^+}(y).$$

It follows that for the potential $S_{\tau_{c_{B+}}}$ of the function

$$c_{B^+}(x) \cdot a(x)^{-1}$$

the following equality holds:

$$S_{\tau_{c_{B+}}}(x) = c_{B^+}(y) \cdot \tilde{S}_{1/a}(x) < \infty, \quad \forall x \in X.$$

Since the potential of the function $c_{B^+}a(x)^{-1}$ with respect to the initial chain x_n is finite for any initial state x, the strategy $\tau_{c_{B+}}$ is transitive. The probability that this strategy stops the chain for initial state x does not exceed $S_{\tau_{c_{B+}}}(x) \wedge 1$.

The best answer of Player 2 to the strategy $\tau_{c_{B+}}$ in the game $\Gamma_{c_{B+}}(x)$ is the strategy $\sigma_\infty(x) = 0$ for all $x \in X$. This strategy guarantees him the loss

$$w_{c_{B+}}(x) = \min_{\sigma \in \Sigma} K_{c_{B+}}(\tau_c, \sigma|x) = K_{c_{B+}}(\tau_c, \sigma_\infty|x).$$

The function $w_{c_{B+}}(x)$ satisfies inequalities

$$w_{c_{B+}}(x) \geq c_{B^+}(x) \cdot (1 - (S_{\tau_{c_{B+}}}(x) \wedge 1).$$

Hence, the lower value $\underline{v}_{c_{B+}}(x)$ of the game $G_{c_{B+}}(x)$ satisfies inequalities

$$\underline{v}_c(x) \geq w_c(x) \geq c_{B^+}(x) \cdot (1 - (S_{\tau_{c_{B+}}}(x) \wedge 1).$$

Since

$$\lim_{n\to\infty} S_{\tau_{c_{B+}}}(x_n) = 0,$$

we get

$$\lim_{n\to\infty} \underline{v}_{c_{B+}}(x_n) \geq \lim_{n\to\infty} w_{c_{B+}}(x_n) = \lim_{n\to\infty} c_{B^+}(x).$$

It follows that

$$\lim_{n\to\infty} \underline{v}_{c_{B+}}(x_n) = \lim_{n\to\infty} c_{B^+}(x).$$

Consequently

$$h_{\overline{v}_c}(x) = h_{\underline{v}_c}(x) = c_{B^+}(x).$$

This finishes the proof of Proposition 6.1.

The main result of this section contains in the following Theorem:

Theorem 6.1. *a)For any positive harmonic function $c \in L$ the game $\Gamma_c(x)$ has a value; the value $v_c(x)$ is a unique solution of optimality equation (5.1), satisfying the "boundary condition at infinity"*

$$\lim_{n\to\infty} v_c(x_n) = \lim_{n\to\infty} c_{B^+}(x_n); \tag{6.1}$$

b) the stationary strategy τ_c^* of Player 1 given with

$$\tau_c^*(x) = v_c(x) \cdot a(x)^{-1} \qquad \forall x \in X,$$

is optimal; in general, Player 2 has ε-optimal strategies only.

The proof of Theorem 6.1. is based on Proposition 6.1. and is analogous to the proof of Theorem 5.1.

Corollary. *The value $v_c(x)$ of the game $\Gamma_c(x)$ depends on values $c(b)$ of "the payoff at infinity" only at Martin boundary points belonging to the non-stopping set B^+. If a trajectory converges to a boundary point $b \in B^-$, then ε-optimal strategy of Player 2 stops the chain almost surely.*

The following construction allows to reduce studying stopping games $\Gamma_c(x)$ with non-stopping set $B^+ \neq B$ to the games $\Gamma_c(x)$ with $B^+ = B$.

Consider the modified stopping game $\tilde{\Gamma}_1(x)$ for modified Markov chain $\tilde{x}_n$ with transition probabilities

$$[\tilde{P}(x,y)] = [(c_{B^+}(x))^{-1} P(x,y) c_{B^+}(y)], \quad x, y \in X.$$

and with payoffs $\tilde{a} = a/c_{B^+}$ and $c = 1$.

According to Theorem 5.1, the game $\tilde{\Gamma}_1(x)$ has the value $\tilde{v}_1(x)$ satisfying the "boundary condition at infinity"

$$\lim_{n\to\infty} \tilde{v}_c(\tilde{x}_n) = 1. \tag{6.2}$$

Proposition 6.2. *The value v_c of the game $\Gamma_c(x)$ is given with the equality*

$$v_c(x) = c_{B^+}(x) \cdot \tilde{v}_1(x).$$

Proof. The function $\tilde{v}_1(x)$ satisfies the optimality equation (5.1) with respect to the modified Markov chain $\tilde{x}_n$

$$\tilde{v}_1(x) = \frac{a(x) \cdot (c_{B^+}(x))^{-1} \cdot (c_{B^+}(x))^{-1} \sum_{y\in X} P(x,y) c_{B^+}(y) \cdot \tilde{v}_1(y)}{a(x) \cdot (c_{B^+}(x))^{-1} + (c_{B^+}(x))^{-1} \sum_{y\in X} P(x,y) c_{B^+}(y) \cdot \tilde{v}_1(y)}.$$

It follows from this equality that the function $c_{B^+}(x) \cdot \tilde{v}_c(x)$ satisfies the optimality equation (5.1) with respect to the initial Markov chain x_n. By (6.2), it follows that this function satisfies the "boundary condition at infinity" (6.1). According to Theorem 6.1, this function is equal to the value $v_c(x)$ of the game $\Gamma_c(x)$. It proves Proposition 6.2.

7. Illustrative Examples

Consider the chain x_n with state space

$$X = Z^1 = \{\ldots, -1, 0, 1, \ldots\},$$

and with transition probabilities

$$P(x, x+1) = p, \quad P(x, x-1) = q, \quad \text{for} \quad x > 0,$$

$$P(x, x-1) = p, \quad P(x, x+1) = q, \quad \text{for} \quad x < 0,$$

where $p > q, p + q = 1$. For $x = 0, P(0, -1) = P(0, 1) = 1/2$.

The chain x_n can be treated as a symmetric random walk over the integers with repulsion from the central point 0. The exit Martin boundary B contains two points $+\infty$ and $-\infty$.

Probabilities $P(x, +\infty)$ and $P(x, -\infty)$ of absorption at points $+\infty$ and $-\infty$ are given with

$$P(x, +\infty) = 1 - 1/2(q/p)^x, \quad P(x, -\infty) = 1/2(q/p)^x, \quad \text{for} \quad x \geq 0,$$

$$P(x, +\infty) = 1/2(q/p)^x, \quad P(x, -\infty) = 1 - 1/2(q/p)^x, \quad \text{for} \quad x < 0.$$

Any harmonic function c can be represented as a linear combination of two functions $P(x, +\infty)$ and $P(x, -\infty)$:

$$c(x) = c(+\infty) \cdot P(x, +\infty) + c(+\infty) \cdot P(x, -\infty).$$

The potential matrix $G(x, y) = \sum_{n=0}^{\infty} P^n(x, y)$ is given with formulas

$$G(x, y) = P(x, +\infty)(p - q)^{-1} \quad \text{for} \quad 0 < x < y,$$

$$G(x, y) = P(x, -\infty)(p - q)^{-1} \quad \text{for} \quad y < 0 < x,$$

and with symmetric formulas for $x < 0$.

For a function $a : Z \to R_+$ put

$$S_{1/a}^{+} = \sum_{x=1}^{\infty} (a(x))^{-1},$$

$$S_{1/a}^{-} = \sum_{x=-1}^{-\infty} (a(x))^{-1}.$$

We obtain the following result:

a) If both quantities $S_{1/a}^{+}$ and $S_{1/a}^{-}$ are finite, then the non-stopping set B^+ coinsides with the whole Martin boundary $B = (+\infty, -\infty)$.

In this case the harmonic component of the value $v_c(x)$ of the game $\Gamma_c(x)$ is equal to

$$c(x) = c(+\infty) \cdot P(x, +\infty) + c(-\infty) \cdot P(x, -\infty).$$

b) If only one of two quantities $S_{1/a}^{+}$ or $S_{1/a}^{-}$ is finite, then the non-stopping set B^+ contains only one point: $B^+ = +\infty$ or $B^+ = -\infty$ correspondingly.

In this case the harmonic component of the value $v_c(x)$ of the game $\Gamma_c(x)$ is equal to

$$c^+(x) = c(+\infty) \cdot P(x, +\infty),$$

or

$$c^-(x) = c(-\infty) \cdot P(x, -\infty),$$

correspondingly.

c) If both quantities $S_{1/a}^{+}$ and $S_{1/a}^{-}$ are infinite, then the stopping set B^- coinsides with the whole Martin boundary $B = (+\infty, -\infty)$.

In this case, for any "payoff at infinity" c the value $v_c(x)$ of the game $\Gamma_c(x)$ is equal to zero.

References

[1] M. Alario-Lazaret, J.P. Lepeltier, B. Marchal, Dynkin Games, Stochastic Differential Systems (Bad Honnef) 23-32 (1982), *Lecture notes in Control and Information Sciences*, **43**, Springer-Verlag.

[2] J.M. Bismuth, Sur un problem de Dynkin, *Z.Warsch. V. Geb.* **39** (1977), 31–53.

[3] V. Domansky, *On some game problems of optimal stopping for a sequence of sums of random variables*, In: Modern directions in game theory, Vilnius: Mokslas, 1976, 86–93.

[4] V. Domansky , Game problems of optimal stopping for Markov chains, Survey of seminars on prob.theory and math. statistics, LOMI Academy of Sciences URSS, 1977, *Probability theory and its appl.* **23**, no. 4 (1978), 863–865.

[5] V. Domansky, Randomized optimal stopping rules for a class of stopping games, *Game Theory and Applications*, Huntington (N.Y.), 7 (2001), p.33–43.

[6] V. Domansky, Randomized optimal stopping times for a class of stopping games, *Probability theory and its appl.* **46**, no. 6 (2002), 1–11.

[7] V. Domansky, Dynkin games with randomized optimal stopping rules, *Annals of the International Society of Dynamic Games*, Birkhauser (Boston, MA,USA), 7 (2004), 247–262.

[8] E.B. Dynkin, A game variant of optimal stopping problem, *Dokl. Acad, nauk,* **185**, no. 1 (1969), 16–19.

[9] E.B. Dynkin, A.A. Yushkevich, *Theorems and Problems on Markov Processes*, Plenum Press, N.Y., 1969.

[10] N. Elbakidze, Constructing value and optimal policies for game stopping problem for Markov process, *Probability theory and its appl.* **21**, no. 1 (1976), 164–169.

[11] A. Irle, Games of stopping with infinite horizon, ZOR, *Mathematical Methods of Operations Research* **42** (1995), 345–359.

[12] J.P. Lepeltier, M.A. Maingueneau, Le jeu de Dynkin en theorie generale sans l'hypothese de Mokobodski, *Stochastics* **13** (1984), 25–44.

[13] V. Mazalov, *Game stopping times*, Novosibirsk, Nauka, Siberian branch, 1987.

[14] J. Neveu, *Discrete-Parameter Martingales*, North-Holland, Amsterdam, 1975.

[15] E. Presman, I. Sonin, *Game problems of optimal stopping. Existence and uniqueness of equilibrium points*, In: Probabilistic problems for economics control, Moscow: Nauka, 1977, 115–144.

[16] D. Rosenberg, E. Solan and V. Vieille, *Stopping games with randomized strategies*, Preprint. Laboratoire d'Econometrie de l'ecole polytechnique, Paris, France, 2000.

[17] M. Sakaguchi, Sequential deception games, *Math. Japonica* **37**, no. 5 (1992), 813–826.

[18] A. Shirjaev, *Statistical sequential analysis*, Moscow, Nauka, 1976.

[19] K. Szajowski, Optimal stopping of a discrete Markov process by two decision makers, *SIAM J. Control and Optimization* **33**, no. 5 (1995), 1392–1410.

[20] M. Yasuda, On a randomized strategy in Neveu's stopping problem, *Stochastic Processes Appl.* **21** (1985), 159–166.

In: Mathematics, Game Theory and Algebra Compendium ISBN 978-1-60456-659-8
Editor: Jacob H. Mathias, pp. 259-267

Chapter 25

AN INVESTMENT ALLOCATION GAME WITH A COST

Andrey Garnaev and Alexey Solovyev
Department of Computer Modelling and Multiprocessor Systems
Faculty of Applied Mathematics and Control Processes
St. Petersburg State University, Saint - Petersburg, Russia

1. Introduction

Market share models describe the manner in which a fixed market is partitioned among a number of competitors, given the effort allocations of each of the competitors. In this models, market shares have the form

$$\frac{\text{your allocation}}{\text{your allocation} + \text{opponent's allocation}}$$

Models that specify payoffs as a function of competitive effort allocations of this form can be often met in the economics, game theory and operation research literature (Schmalensee [6], Shapley and Shubik [8], Shakun [7], Monahan [4]).

In this paper we consider the model where there is initial allocation of the resource from a third party. Nash equilibrium are derived. The case of one and two firms game are investigated. Also the Stackelberg behaviour of the players is considered.

2. One Firm Game

In this section we consider one firm game. Namely we assume that a firm is planning to allocate some financial resources in n independent markets. Let V_i denote the sales potential of the ith market. We assume that already A_i and D_i dollars of effort to market i are located by the firm and some third part. If the firm allocates x_i dollars of effort to market i then it gets the share of the market i equals $V_i(x_i + A_i)/(x_i + A_i + D_i)$. We assume that budget of the firm is X and if it allocates $\sum_{i=1}^{n} x_i$ resources then the investment cost is $C\sum_{i=1}^{n} x_i$ where $C > 0$. The aim of each firm is to plan its budget so that to maximize total firm profit minus the investment cost. So, the payoffs to the firm if it employs the strategy $x = (x_1, \dots, x_n)$ is given by

$$M(x) = \sum_{i=1}^{n} \frac{V_i(x_i + A_i)}{x_i + A_i + D_i} - C \sum_{i=1}^{n} x_i,$$

Since M is concave in x, the Kuhn - Tucker Theorem (Mangasaria [3]) implies the following theorem.

Theorem 1. *x^* is the optimal strategy if and only if there is a non-negative ω such that*

$$\frac{\partial}{\partial x_i} M(x^*, y^*) = \frac{V_i D_i}{(x_i^* + A_i + D_i)^2} - C \begin{cases} = \omega & \text{for } x_i^* > 0, \\ \leq \omega & \text{for } x_i^* = 0, \end{cases} \tag{1}$$

where

$$\omega \begin{cases} \geq 0 & \text{for } \sum_{i=1}^{n} x_i^* = X, \\ = 0 & \text{otherwise,} \end{cases} \tag{2}$$

If $V_i D_i/(A_i + D_i)^2 \leq C$ for $i \in [1, n]$ then by Theorem 1. the strategy $x^* = (0, \dots, 0)$ is optimal. So, we assume that there is an i such that $V_i D_i/(A_i + D_i)^2 > C$. For non-negative ω let

$$Q(\omega) = \{i : V_i D_i/(A_i + D_i)^2 > C + \omega\},$$

$$x_i(\omega) = \begin{cases} \sqrt{\dfrac{V_i D_i}{C + \omega}} - A_i - D_i & \text{if } i \in Q(\omega), \\ 0 & \text{otherwise} \end{cases}$$

and

$$K(\omega) = \sum_{i=1}^{n} x_i(\omega).$$

Clearly K is a continuous decreasing function in $[0, \infty)$ and $K(\omega) \to 0$ for $\omega \to \infty$ and $K(\omega) \to 0$. Thus, if $K(0) > X$ there is a positive $\bar{\omega}$ such that $K(\bar{\omega}) = X$. Thus, the following theorem holds.

Theorem 2. *The strategy $x(\omega_*)$ is optimal where $\omega_* = 0$ if $K(0) \leq X$ and $\omega_* = \bar{\omega}$ if $K(0) > X$.*

3. Two Firm Game

In this section we consider two firm allocation game. Two firms, say 1 and 2, compete against each other in n independent markets. The total sales potential in each market is fixed and is shared between the competitors on the basis of both the quality and the magnitude of the marketing efforts expended by each competitor. Let V_i denote the sales potential of the ith market. We assume that already A_i, B_i and D_i dollars of effort to market i are located by firm 1, firm 2 and some third part. These value are fixed and known to both players. If firm 1 and firm 2 allocates x_i and y_i dollars of effort to market i, respectively. Then the share of market i getting by firm 1 and 2 is $V_i(x_i + A_i)/(x_i + A_i + y_i + B_i + D_i)$ and $V_i(y_i + B_i)/(x_i + A_i + y_i + B_i + D_i)$. We assume that budget of firm 1 and 2 is X and Y, respectively. If firm 1 (2) allocates $\sum_{i=1}^n x_i$ ($\sum_{i=1}^n y_i$) resources then the investment cost is $C_1 \sum_{i=1}^n x_i$ ($C_2 \sum_{i=1}^n y_i$) where $C_j > 0$ for $j = 1, 2$. The aim of each firm is to plan its budget so that to maximize total firm profit minus the investment cost.So, the payoffs to firms if firm 1 and 2 employ strategy $x = (x_1, \ldots, x_n)$ and $y = (y_1, \ldots, y_n)$, respectively, are given by

$$M_1(x, y) = \sum_{i=1}^n \frac{V_i(x_i + A_i)}{(x_i + A_i) + (y_i + B_i) + D_i} - C_1 \sum_{i=1}^n x_i,$$
$$M_2(x, y) = \sum_{i=1}^n \frac{V_i(y_i + B_i)}{(x_i + A_i) + (y_i + B_i) + D_i} - C_2 \sum_{i=1}^n y_i,$$

where

$$\sum_{i=1}^n x_i \le X, \quad \sum_{i=1}^n y_i \le Y$$

and $x_i \ge 0$, $y_i \ge 0$ for $i \in [1, n]$.
We will look for a Nash equilibrium, i.e. we want to find $(x^*, y^*) \in \Delta_1 \times \Delta_2$ such that

$$M_1(x, y^*) \le M_1(x^*, y^*) \text{ for any } x \in \Delta_1,$$
$$M_2(x^*, y) \le M_2(x^*, y^*) \text{ for any } y \in \Delta_2,$$

where Δ_1 and Δ_2 are the sets of all the strategies of firm 1 and 2 respectively.

4. Solution of the Two Firms Game

Since M_1 and M_2 are concave in x and y, the Kuhn - Tucker Theorem (Mangasaria [3]) implies the following theorem.

Theorem 3. *(x^*, y^*) is a Nash equilibrium if and only if there are non-negative ω and ν such that*

$$\frac{\partial}{\partial x_i} M(x^*, y^*) = \frac{V_i(y_i^* + B_i + D_i)}{(x_i^* + A_i + y_i^* + B_i + D_i)^2} - C_1 \begin{cases} = \omega & \text{for } x_i^* > 0, \\ \le \omega & \text{for } x_i^* = 0, \end{cases} \tag{3}$$

$$\frac{\partial}{\partial y_i} M(x^*, y^*) = \frac{V_i(x_i^* + A_i + D_i)}{(x_i^* + A_i + y_i^* + B_i + D_i)^2} - C_2 \begin{cases} = \nu & \text{for } y_i^* > 0, \\ \le \nu & \text{for } y_i^* = 0, \end{cases} \tag{4}$$

where

$$\omega \begin{cases} \ge 0 & \text{for } \sum_{i=1}^n x_i^* = X, \\ = 0 & \text{otherwise}, \end{cases} \tag{5}$$

$$\nu \begin{cases} \ge 0 & \text{for } \sum_{i=1}^n y_i^* = Y, \\ = 0 & \text{otherwise}. \end{cases} \tag{6}$$

Lemma 1. *Let (x^*, y^*) be a Nash equilibrium and $A_i = B_i = D_i = 0$ then $x_i^* > 0$, $y_i^* > 0$ and*

$$x_i^* = V_i \frac{\nu + C_2}{(\nu + \omega + C_1 + C_2)^2}, \quad y_i^* = V_i \frac{\omega + C_1}{(\nu + \omega + C_1 + C_2)^2}.$$

For simplification let us introduce the following notation

$$\Omega = \omega + C_1 \quad \text{and} \quad N = \nu + C_2.$$

Lemma 2. *Let (x^*, y^*) be a Nash equilibrium and $D_i > 0$, then*

(i) *if $x_i^* = 0$ and $y_i^* = 0$ then*

$$X_i \leq \Omega, \quad Y_i \leq N,$$

where

$$X_i = \frac{V_i(B_i + D_i)}{(A_i + B_i + D_i)^2}, \quad Y_i = \frac{V_i(A_i + D_i)}{(A_i + B_i + D_i)^2},$$

(ii) *if $x_i^* > 0$ and $y_i^* = 0$ then*

$$x_i^* = \sqrt{\frac{V_i(B_i + D_i)}{\Omega}} - A_i - B_i - D_i \tag{7}$$

and

$$X_i > \Omega, \quad N \geq \Xi(B_i, D_i, V_i, \Omega)$$

where

$$\Xi(A, D, V, t) = \frac{1}{A + D}\left(\sqrt{V(A + D)t} - At\right),$$

(iii) *if $x_i^* = 0$ and $y_i^* > 0$ then*

$$y_i^* = \sqrt{\frac{V_i(A_i + D_i)}{N}} - A_i - B_i - D_i$$

and

$$Y_i > N, \quad \Omega \geq \Xi(A_i, D_i, V_i, N),$$

(iv) *if $x_i^* > 0$ and $y_i^* > 0$ then*

$$x_i^* = \frac{V_i N - 2D_i \Omega N - 2D_i \Omega^2 + \sqrt{V_i^2 N^2 + 4V_i D_i \Omega N^2 + 4V_i D_i N^3}}{2(\Omega + N)^2} - A_i,$$

$$y_i^* = \frac{V_i \Omega - 2D_i \Omega N - 2D_i N^2 + \sqrt{V_i^2 \Omega^2 + 4V_i D_i N \Omega^2 + 4V_i D_i \Omega^3}}{2(\Omega + N)^2} - B_i$$

and

$$\Omega < \Xi(A_i, D_i, V_i, N), \quad N < \Xi(B_i, D_i, V_i, \Omega).$$

Proof. (i) follows immediately from (3) and (4) where $x_i^* = y_i^* = 0$.

(ii) Let $x_i^* > 0$ and $y_i^* = 0$. Then, by (3) we have that

$$\frac{V_i(B_i + D_i)}{(x_i^* + A_i + B_i + D_i)^2} = \Omega.$$

Thus,

$$\frac{V_i(B_i + D_i)}{(A_i + B_i + D_i)^2} > \Omega$$

and x_i^* is given by (7). Then, by (4) we have that

$$\begin{aligned} N &\geq \frac{V_i(x_i^* + A_i + D_i)}{(x_i^* + A_i + B_i + D_i)^2} \\ &= \frac{1}{(B_i + D_i)}\left(\sqrt{V_i(B_i + D_i)\Omega} - B_i\Omega\right) \end{aligned}$$

and the result follows.

(iii) can be proved similarly to (ii).

(iv) Let $x_i^* > 0$ and $y_i^* > 0$. Then, by (3) and (4) we have that

$$\frac{V_i(y_i^* + B_i + D_i)}{(x_i^* + A_i + y_i^* + B_i + D_i)^2} = \Omega, \qquad \frac{V_i(x_i^* + A_i + D_i)}{(x_i^* + A_i + y_i^* + A_i + D_i)^2} = N.$$

and the result follows.

Define for each pair (Ω, N) of positive real numbers the following four subsets of the integer interval $[1, n]$:

$$\begin{aligned} I_{00} &= \{i \in [1, n] : X_i \leq \Omega,\ Y_i \leq N\}, \\ I_{10} &= \{i \in [1, n] : X_i > \Omega,\ N \geq \Xi(B_i, D_i, V_i, \Omega)\}, \\ I_{01} &= \{i \in [1, n] : Y_i > N,\ \Omega \geq \Xi(A_i, D_i, V_i, N)\}, \\ I_{11} &= \{i \in [1, n] : \Omega < \Xi(A_i, D_i, V_i, N),\ N < \Xi(B_i, D_i, V_i, \Omega)\}. \end{aligned}$$

By Theorem 3. a Nash equilibria (x^*, y^*) can be considered as a pair $(x^*(\Omega, N), y^*(\Omega, N))$ depending on two parameters $\Omega = \omega + C_1$ and $N = \nu + C_2$. Then, by Lemma 2. we have the following lemma.

Lemma 3. *Let $(x^*(\Omega, N), y^*(\Omega, N))$ be a Nash equilibrium, then*

(i) *if $i \in I_{00}$ then $x_i^* = y_i^* = 0$,*

(ii) *if $i \in I_{10}$ then $x_i^* > 0 = y_i^*$,*

(iii) *if $i \in I_{01}$ then $y_i^* > 0 = x_i^*$,*

(iv) *if $i \in I_{11}$ then $x_i^* > 0, y_i^* > 0$.*

Let

$$K(\Omega,N)=\sum_{k=1}^{n}x_i(\Omega,N) \quad \text{and} \quad H(\Omega,N)=\sum_{k=1}^{n}y_i(\Omega,N)$$

where $(x(\Omega,N),y(\Omega,N))$ is given by Lemma 2., then by Lemma 3.

$$K(\Omega,N)=\sum_{i\in I_{10}}\left(\sqrt{\frac{V_i(B_i+D_i)}{\Omega}}-A_i-B_i-D_i\right)$$
$$+\sum_{i\in I_{11}}\left(\frac{V_iN-2D_i\Omega N-2D_i\Omega^2+\sqrt{V_i^2N^2+4V_iD_i\Omega N^2+4V_iD_iN^3}}{2(\Omega+N)^2}-A_i\right)$$

$$H(\Omega,N)=\sum_{i\in I_{01}}\left(\sqrt{\frac{V_i(A_i+D_i)}{N}}-A_i-B_i-D_i\right)$$
$$+\sum_{i\in I_{11}}\left(\frac{V_i\Omega-2D_i\Omega N-2D_iN^2+\sqrt{V_i^2\Omega^2+4V_iD_iN\Omega_i^2+4V_iD_i\Omega^3}}{2(\Omega+N)^2}-B_i\right)$$

Then Lemma 3. and Theorem 3. imply the following theorem.

Theorem 4. *A Nash equilibrium $(x^*(\Omega,N),y^*(\Omega,N))$ is given by Lemmas 2. and 3. where (Ω,N) are determined by the following systems of non-linear equations and inequations:*

(i) *if the following inequations hold true then $\Omega=C_1$ and $N=C_2$:*

$$K(C_1,C_2)\le X, \quad H(C_1,C_2)\le Y,$$

(ii) *if the following equation have positive solution that satisfy the next inequation then $\Omega=C_1$ and $N=C_2+\nu$:*

$$H(C_1,C_2+\nu)=Y, \quad K(C_1,C_2+\nu)\le X,$$

(iii) *if the following equation have positive solution that satisfy the next inequation then $N=C_2$ and $\Omega=C_1+\omega$:*

$$K(C_1+\omega,C_2)=X, \quad H(C_1+\omega,C_2)\le Y,$$

(iv) *if the following system of non-linear equations have positive solution then $\Omega=C_1+\omega$ and $N=C_2+\nu$:*

$$K(C_1+\omega,C_2+\nu)=X, \quad H(C_1+\omega,C_2+\nu)=Y.$$

5. A Numerical Example

Let us consider a numerical example where $V=(1000,2000,4000)$, $D=(100,20,20)$, and $X=1000, Y=700$. Thus, $V/D=(10,100,200)$. For this case there are unique Nash equilibrium given in Table 1 for different values of C_1 and C_2.

Table 1. Nash equilibrium, payoffs, and ω and ν

C_1, C_2	x^*	y^*	M_1	M_2	ω	ν
(0,0)	(143.40, 286.33, 570.27)	(79.35, 205.72, 414.94)	3832	2701	1.72	2.34
(1,1)	(143.40, 286.33, 570.27)	(79.35, 205.72, 414.94)	3832	2701	0.72	1.34
(1,2)	(143.40, 286.33, 570.27)	(79.35, 205.72, 414.94)	3832	2701	0.72	0.34
(1,3)	(152.25, 284.26, 563.45)	(37.72, 146.12, 298.54)	3343	686	0.64	0.00
(2,1)	(119.46, 249.32, 497.51)	(83.18, 204.81, 411.98)	1855	2211	0.00	1.40
(3,1)	(60.07, 162.68, 327.98)	(94.66, 202.16, 403.18)	1176	2869	0.00	1.47
(2,2)	(119.46, 249.33, 497.52)	(83.19, 204.82, 411.99)	1855	1512	0.00	0.40
(3,3)	(68.50, 166.13, 333.05)	(68.50, 166.13, 333.05)	1471	1471	0.00	0.00
(10, 100)	(0.00, 43.25, 69.44)	(0.00, 0.00, 0.00)	3346	0	0.00	0.00
(100, 10)	(0.00, 0.00, 0.00)	(0.00, 43.25, 69.44)	0	3346	0.00	0.00
(100, 100)	(0, 0, 6.18)	(0, 0, 6.18)	146	146	0.00	0.00

6. Stackelberg Equilibrium

In the previous sections players act simultaneously. Stackelberg proposed an approach in which one player, called the dominant player or Leader, moves first and the outcome of that player's choice is made known to the other player called Follower before the other player makes his choice. We consider the Stackelberg in a particular case of our game. Two firms start to compete for a given market of total value, V, by investing a certain amount of resources into the project through advertising. Let D recourses already be invested by other participants (besides these two firms) in the market and this value is fixed and known to both firms. It is reasonable to assume that $D \leq V$ since there is no sense to invest more resources than the expected feedback. If firm 1 allocates $x > 0$ and firm 2 allocates $y > 0$, then the proportion of the market that firm 1 and firm 2 get are $x/(x+y+D)$ and $y/(x+y+D)$. The pure strategy spaces for this game are the sets $X = Y = [0, \infty)$. It would not hurt to restrict the strategy spaces to $[0, V]$; no firm would like to invest more resources than the expected return. The payoffs for the two players are the profits minus expenses, i.e. pure profit:

$$M_1(x,y) = \frac{Vx}{x+y+D} - x,$$
$$M_2(x,y) = \frac{Vy}{x+y+D} - y,$$

First, let us find out what happens in the case of one firm game i.e. when there is only one firm (say, firm 1) which is going to enter the market. Then the return to Firm 1, if it invests x resources, is $M(x) = Vx/(x+D) - x$. The firm will choose x to maximize this quantity. We may find the point at which the maximum occurs by taking a derivative with respect to x, setting it to zero and solving relative to x. The resulting equation is $VD - (x+D)^2 = 0$, whose positive solution is $x = \sqrt{VD} - D$. Thus, the maximal pure profit is $(\sqrt{V} - \sqrt{D})^2$.

To find Nash equilibrium (x_*, y_*) in the two firm game, we look for a pure strategy for each player that is the best response to the other's strategy. We find simultaneously the value of x_* that maximizes $M_1(x, y_*)$ and the value of y_* that maximizes $M_1(x_*, y)$ by setting the partial derivatives to zero. Thus,

$$\frac{V}{x + y_* + D} - \frac{Vx}{(x + y_* + D)^2} = 1, \tag{8}$$

$$\frac{V}{x_* + y + D} - \frac{Vy}{(x_* + y + D)^2} = 1. \tag{9}$$

Solving these equations simultaneously and denoting the result by x_* and y_*, we find the positive solution

$$x_* = y_* = \frac{\sqrt{V^2 + 8VD}}{8} + \frac{V - 4D}{8}.$$

The payoff vector corresponding to the Nash equilibrium (x_*, y_*) is (v, v) where

$$v = \frac{V^2 - 10VD + (V + 2D)\sqrt{V(V + 8D)}}{4(V + \sqrt{V(V + 8D)})}.$$

It is interesting that if $D = 0$ then in the one firm game (when there is no opponent on the market) the optimal solution is not to invest money since the whole market belongs to the monopolist anyway. In the case of the two firms game, $x_* = y_* = V/4$ and $v = V/4$, and so the competition makes the firms spend one forth their money on investments.

In the case Stackelberg equilibrium firm 1 (Leader) chooses an amount x to invest. This amount is then made known to firm 2 (Follower) who then chooses an amount y for investments to maximize his profit. So, by (9)

$$y(x) = \sqrt{V(x + D)} - x - D.$$

Since firm 1 now knows that firm 2 will choose this best response, firm 1 prefers to choose x to maximize

$$M_1(x, y(x)) = \frac{\sqrt{V}x}{\sqrt{V + x}} - x$$

This function is maximized at x which is the unique positive root of the equation

$$\sqrt{V}(2D + x) - 2\sqrt{(D + x)^3} = 0.$$

Let us have a look at a numerical example. Let $V = 100$ then the payoff to Leader and to Follower and to players employing NE strategies for $D \in [1, 5]$ are given by Table 2.

Table 2. The payoffs for the Stackelberg and Nash equilibrium

D	Leader payoff	x_*	Follower payoff	y_*	NE payoff	x_*
1	24.037	25.894	23.175	24.965	24.029	24.999
2	23.139	26.617	21.628	24.878	23.111	24.963
3	22.295	27.212	20.282	24.753	22.242	24.919
4	21.498	27.706	19.09	24.502	21.416	24.861
5	20.742	28.118	18.021	24.43	20.629	24.790

7. Conclusions

In this paper we consider the investment allocation game where there is an initial allocation of the resource from a third party. Nash equilibrium are derived. The case of one and two firms game are investigated. Also the Stackelberg behaviour of the players is considered. It is interesting to note that compared to the Nash equilibrium case Leader invests more resources than in the Nash equilibrium while Follower invests almost the same resources as in the Nash equilibrium but it leads to a different outcome, namely to a small increasing Leader profit compared to the Nash equilibrium and an essential decrease in Follower's profit.

References

[1] J.M. Danskin, *The theory of max-min*, Springer-Verlag, Berlin, 1967.

[2] A.Y. Garnaev, *Search Games and Other Applications of Game Theory*, Springer, Heidelberg, New York, 2000.

[3] O.L. Mangasarian, *Nonlinear Programming*, McGraw-Hill, New York, 1969.

[4] G.E. Monahan, The Structure of Equilibria in Market Share Attraction Models, *Managment Sciences* **18** (1987), 146–153.

[5] L.A. Petrosjan, N.A. Zenkevich, *Game Theory*, World Scientific, London, 1996.

[6] R. Schmalensee, A Model of Promotional Competition in Ologopoly, *Rev. Econ.Stud.* **43** (1976), 493–507.

[7] M.F. Shakun, Advertising Expenditures in Coupled Markets - a Game theory Approach, *Managment Sciences* **11** (1965), B42–B47.

[8] L. Shapley, M. Shubik, Stochastic Games, Oligopoly Theory and Competitive Resource Allocation, In *Dynamic Optimization and Mathematical Economics* (P.T. Lu, Ed.) New York, Plenum, 1979.

In: Mathematics, Game Theory and Algebra Compendium ISBN 978-1-60456-659-8
Editor: Jacob H. Mathias, pp. 269-282

Chapter 26

AN OPTIMAL INSURANCE POLICY IN THE INDIVIDUAL RISK MODEL SEEN AS A BARGAINING GAME

Alexey Y. Golubin
Department of Operations Research
Moscow Institute of Electronics and Mathematics, Russia

1. Introduction

This paper is devoted to a problem of optimal risk-sharing in an insurance scheme called the individual risk model (see, e.g., Borch (1990) and Lemaire (1995) for a detailed discussion of the model). Consider an insurance company (insurer) an a group of n potential insurance purchasers (insureds), each of them facing a risk represented by a stochastic variable X_j, $j = 1, \ldots, n$. Denote by $Y_i = Y_i(X_1, \ldots, X_n)$ the risk of the i-th agent after the exchange. The problem is to find functions Y_i, $i = 0, \ldots, n$ in a sertain class so that they will be "optimal" from view-points of all $n + 1$ agents.

The game theory concepts have been implemented in analysis of risk-sharing in insurance by Borch's works (1960a, 1960b, 1990). His theorem characterizing Pareto-optimal risk exchanges in a reinsurance market was extended to a constrained case in Gerber (1978). The Borch's risk exchange model of reinsurance, which is a special kind of n-person cooperative game, has been developed by Lemaire (1979, 1990), Lemaire and Baton (1981) where such characteristics as the Pareto-optimal payoffs, core, bargaining set, value of the game are investigated. Aase (2002) considered the model in a more general setting, including the competitive equilibrium notion, risk allocation problems in incomplete financial markets. Borm et al. (1998) investigated risk exchange in an insurance market, regarding it as a cooperative game with restrictions on forming coalitions and assuming only exponentially distributed losses. A parametrical optimization of insurance policies in an individual risk model was studied in Golubin (2002a) under various assumptions on distribution of the summary risk.

The problem we study in this paper differs from the classical model of risk exchange by two points. First, in the individual risk model each insured is independent of the others so that the insured shares his risk with the insurer, not with the other insureds. Second, a premium to be paid by the insured is defined by the mean value principle which is commonly used in insurance practice. This means that the premium enters the payoffs as a functional of a risk-sharing rule, not as an independent variable chosen joinly with the risk-sharing. We show that the model can be treated as a bargaining game and suggest ways of determining the Nash's solution and the Kalai-Smorodinsky's solution to the game.

The paper is organized as follows. In section 2 the notation and description of the insurance game to be studied are given. Section 3 presents a way of constructing the set of Pareto-optimal policies in the game. It is shown that the Pareto-optimal policy is of the form either a deductible or upper limit policy, depending to the value of a Pareto multiplier (an insurer's weight). Section 4 applies the Nash concept for a game solution to the considered model. We reduce finding the Nash's solution to solving an equation with respect to the insurer's weight in Pareto-optimal policy. In section 5 we similarly use the Kalai-Smorodinsky game solution concept to determine the optimal insurance policy in our game. We also compare the obtained policy with that generated as the Nash's solution. The results are illustrated by a numerical example.

2. The Individual Risk Model as a Bargaining Game

Consider an insurer and a group of n insureds. The group is assumed to be *homogeneous* in the sense that the individual losses X_j, $j = 1, \dots, n$ are independent identically distributed stochastic variables with a distribution function $F(x) = P\{X_j \le x\}$.

Denote by $supp\, F$ the support of $F(x)$, i.e., the least closed set S on the real line such that $P\{X_1 \in S\} = 1$. It is assumed that $supp\, F = [0, T]$, the case of the infinite interval $supp\, F = [0, \infty)$ is also admissible. (Note that this class of distributions includes not only ones with a positive density on $[0, T]$ but also distributions with several point masses on $[0, T]$ and a "defect" density $f(x) > 0$ on this interval, which is quite common in insurance applications (see Lemaire (1995).) Each of n insureds has the same initial capital w_1 and the same utility function $u_1(x)$, meaning that the insured prefers a stochastic variable X to Y if and only if $E\, u_1(w_1 + X) > E\, u_1(w_1 + Y)$. The insurer's utility function and his/her initial capital are denoted, respectively, by $u_0(x)$ and w_0. The functions $u_0(x)$ and $u_1(x)$ are supposed to be twice continuously differentiable on the relevant domains, $u_i'(x) > 0\; i = 0, 1$, $u_0''(x) \le 0$, and $u_1''(x) < 0$, so that all the insureds are supposed to be risk-averse, while the insurer may also be risk-neutral: $u_0(x) = x$.

Thus, we assume the insurance market to consist of $n + 1$ players, the insurer and the n identical insureds. The insurer and an insured are negotiating to conclude a treaty on an insurance policy to be used. Any coalitions within the group of insureds are not allowed, so each insured stands as a separate player in bargaining with the insurer. An *insurance policy* is identified with a measurable function $I(x)$ on $[0, T]$ satisfying the constraint $0 \le I(x) \le x$ which means that a reimbursement is always nonnegative and not greater than the loss size. The insurance payment to the j-th insured is $I(X_j)$, and the residual loss $X_j - I(X_j)$ is covered by the insured himself. Then the summary risk taken by the insurer is $X = \sum_{j=1}^{n} I(X_j)$.

The individual premium, i.e., the price paid by every insured, is calculated by a widely used principle of the expected value (see, e.g., Borch (1990)): $P = (1+\alpha)E\, I(X_1)$, where

$\alpha > 0$ is a given loading coefficient.

Once a policy $I(x)$ is chosen, the expected utilities of final capitals of the insurer and the j-th insured are $J_0[I] \stackrel{def}{=} E\,u_0(w_0 + nP - \sum_{j=1}^{n} I(X_j))$ and $J_j[I] \stackrel{def}{=} E\,u_1(w_1 - P - X_j + I(X_j))$, $j = 1, \ldots, n$. In view of these $n+1$ criteria that the corresponding players wish to maximize we come to an $(n+1)$-person bargaining game

$$\begin{cases} J_0[I] \to \max \\ J_j[I] \to \max,\ j = 1, \ldots, n \\ I \in \Pi = \{I(\cdot) : 0 \le I(x) \le x\} \end{cases}$$

This notion was introduced by Nash in 1950 (see also Luce and Raiffa (1957) for detailed discussion). In our case the role of pay-offs is played by the expected utilities $J_i[I]$, $i = 0, \ldots, n$ in which the latter n criteria are identical, $J_i[\cdot] \equiv J_1[\cdot]$ for $i \ge 1$. The *disagreement point* of the game that corresponds to the situation where no treaty is concluded is the point of the utilities under the policy $I(x) \equiv 0$:

$$(J_0[0], J_1[0], \ldots, J_1[0]) \text{ with } J_0[0] = u_0(w_0) \text{ and } J_1[0] = E\,u_1(w_1 - X_1).$$

Two concepts of the solution to the bargaining game are known by now. They are Nash's and Kalai-Smorodinsky's solutions. Both assume that the players act rationally, which means that an optimal policy I should be individually rational, $J_i[I] \ge J_i[0]$ (player will not agree to lessen his initial utility), and Pareto-optimal (I cannot be dominated in the sense of all $n+1$ criteria). The Nash's and Kalai-Smorodinsky's principles defined the solutions to the game amongst this set of individually rational Pareto-optimal policies.

3. Pareto-optimal Policies

In this section we construct the set of Pareto-optimal policies for the above-introduced game. Recall that a policy $\hat{I}(x)$ is called Pareto-optimal, if there is no other insurance policy $I(x)$ such that $J_i[I(\cdot)] \ge J_i[\hat{I}(\cdot)]$ for $i = 0, \ldots, n$ and at least one of the inequalities is strict. As is known (see, e.g., Gerber (1978), Zeleny (1982)), a family of Pareto-optimal solutions can be obtained by maximization of a weighted sum of the criteria. In our case, this method results in the following: fix a constant $\rho \in (0, 1)$ and then maximize the functional

$$\rho J_0[I] + (1-\rho) n J_1[I]$$

over the set of insurance policies $I(x)$. Here the Pareto multiplier ρ is associated with a "weight" of the insurer in making the contract, while $1 - \rho$ is a "weight" of the insured. For covenience we rewrite the problem above as

$$J[I] \equiv \delta J_0[I] + n J_1[I] \to max \qquad \text{subject to } I \in \Pi, \tag{1}$$

where $\delta \stackrel{def}{=} \rho/(1-\rho) \in (0, \infty)$ will in the sequel be referred to as an insurer's relative weight. Due to concavity of J_0 and J_1, the one-parameter family of maximizers $\{\hat{I}_\delta\}_{\delta > 0}$ consists of all Pareto-optimal policies (except degenerated cases related to $\rho = 0 (= 1)$).

The theorem below gives a characterization of the Pareto-optimal policy $\hat{I}(x) = \hat{I}_\delta(x)$. Briefly, it states that $\hat{I}(x)$ is either a *policy with an upper limit* of full coverage (see (2) below) or a *policy with a deductible* (see (3)), involving in both cases a coinsurance above some level $\hat{x}$ of individual loss. To formulate the theorem, we need some notation: denote $B(x) = w_1 - \hat{P} - x + \hat{I}(x)$, $\hat{P} = (1+\alpha)E\hat{I}(X_1)$, and a stochastic-valued function $A(x) = w_0 + n\hat{P} - \hat{I}(x) - \sum_1^{n-1}\hat{I}(X_j)$.

Theorem 1. *A policy* $\hat{I}(x)$ *solves* (1) *if and only if conditions* (2)-(6) *hold:*

$$\text{either } \hat{I}(x) = x \text{ on } x \le \hat{x} \text{ and } 0 < \hat{I}(x) < x \text{ on } x > \hat{x}, \tag{2}$$

$$\text{or } \hat{I}(x) = 0 \text{ on } x \le \hat{x} \text{ and } 0 < \hat{I}(x) < x \text{ on } x > \hat{x}, \tag{3}$$

for some $0 \le \hat{x} \le T$. *In both cases,* $\hat{I}(x)$ *on* $[\hat{x}, T]$ *satisfies the equation*

$$\hat{I}'(x) = \frac{u_1''(B(x))}{\delta E\{u_0''(A(x))\} + u_1''(B(x))}, \tag{4}$$

with initial condition, respectively, either $\hat{I}(\hat{x}) = \hat{x}$ *or* $\hat{I}(\hat{x}) = 0$. *Moreover, the level* $\hat{x}$ *satisfies the equation*

$$\hat{U}(\hat{x}) - (1+\alpha)E\hat{U}(X_1) = 0, \tag{5}$$

$$\text{otherwise } \hat{I}(x) \equiv x (\equiv 0) \text{ if } \hat{U}(0) - (1+\alpha)E\hat{U}(X_1) < 0 (> 0), \tag{6}$$

where $\hat{U}(x) \equiv \delta E\{u_0'(A(x))\} - u_1'(B(x))$.

Proof. Let $\hat{I}(x)$ be an optimal policy in (1). Fix any policy $I \in \Pi$ and consider the policy $\lambda\hat{I}(x) + (1-\lambda)I(x)$, where $\lambda \in [0,1]$ is a parameter. As $\hat{I}(x)$ is optimal, the derivative of $J[\lambda\hat{I} + (1-\lambda)I]$ with respect to λ at the point $\lambda = 1$ is nonnegative. Note that this inequality is also sufficient for optimality $\hat{I}$ in (1) since J is a concave functional.

Recall that the functional

$$J_1[I] = \int_0^T u_1(w_1 - P - x + I(x))dF(x) \text{ and}$$

$$J_0[I] = \int_0^T E\{u_0(w_0 + nP - I(x) - \sum_{j=1}^{n-1} I(X_j))\}dF(x).$$

Then, after differentiating $J(\lambda) = J[\lambda\hat{I} + (1-\lambda)I]$ at $\lambda = 1$, we obtain that the inequality $J'(\lambda)_{\lambda=1} \ge 0$ is equivalent to

$$\int_0^T (\hat{U}(x) - \hat{C})I(x)dF(x) \ge \int_0^T (\hat{U}(x) - \hat{C})\hat{I}(x)dF(x)$$

$$\text{for any } I : 0 \le I(x) \le x,$$

where the function

$$\hat{U}(x) = \delta E\{u_0'(A(x))\} - u_1'(B(x)) \text{ and } \hat{C} = (1+\alpha)E\hat{U}(X_1). \tag{7}$$

Thus, $\hat{I}(x)$ is a solution to (1) if and only if it minimizes the integral

$$\int_0^T (\hat{U}(x) - \hat{C})I(x)dF(x) \tag{8}$$

subject to $0 \leq I(x) \leq x$.

An optimality condition for problem (8) is given by the generalized Neyman-Pearson lemma (see Karlin and Studden (1966, Chap. VIII, XIV)). According to the lemma, an admissible function $\hat{I}(x)$ is optimal in problem (8) if and only if on the interval $[0, T]$ it satisfies

$$\hat{I}(x) = \begin{cases} 0, & \text{if } \hat{U}(x) - \hat{C} > 0 \\ x, & \text{if } \hat{U}(x) - \hat{C} < 0 \end{cases} \quad \text{up to a set of zero } F\text{-measure.} \tag{9}$$

Suppose first that $\hat{U}(0) < \hat{C}$. Then, as x increases from $x = 0$, we have by (9) that $\hat{I}(x)$ remains equal to x up to a point $\hat{x}$ at which $\hat{U}(x)$ becomes equal to $\hat{C}$ — note that $\hat{U}(x)$ is increasing (see (7)) with $\hat{I}(x) = x$. On the remaining interval $[\hat{x}, T]$, $\hat{U}(x)$ can neither decrease ($\hat{I}(x) = x$ for $\hat{U}(x) < \hat{C}$, so $\hat{U}(x)$ would have increased for such x) nor increase — by (9) we would have $\hat{I}(x) = 0$ for $\hat{U}(x) > \hat{C}$ and, according to (7), $\hat{U}(x)$ must decrease. Thus, the function $\hat{U}(x)$ is increasing on the interval $[0, \hat{x}]$, where $\hat{I}(x) = x$, and identically equals $\hat{C}$ on $[\hat{x}, T]$. The latter may be empty, which means $\hat{U}(x) - \hat{C} < 0 (> 0)$ on $[0, T]$ and in this case $\hat{I} \equiv x (\equiv 0)$ respectively. Differentiating the equation $\hat{U}(x) = \hat{C}$ on $[\hat{x}, T]$, from (7) we have

$$-\delta E\{u_0''(A(x))\}\hat{I}'(x) - u_1''(B(x))(\hat{I}'(x) - 1) = 0$$

and, hence, $\hat{I}(x)$ satisfies equation (4) with initial condition $\hat{I}(\hat{x}) = \hat{x}$.

Suppose that $\hat{U}(0) > \hat{C}$. Similar to the reasonings above, we have that for some $\hat{x}$ the function $\hat{U}(x)$ is decreasing on $[0, \hat{x}]$, i.e., $\hat{I}(x) = 0$ for $x \in [0, \hat{x}]$, and $\hat{U}(x) = \hat{C}$ on the interval (possibly empty) $[\hat{x}, T]$. Differentiating the equation $\hat{U}(x) = \hat{C}$ on $[\hat{x}, T]$ gives the required result. Finally, if $\hat{U}(0) = \hat{C}$ then, evidently, $\hat{U}(x) \equiv \hat{C}$ and $\hat{I}(x)$ satisfies (4) on the whole interval $[0, T] = [\hat{x}, T]$ with initial condition $\hat{I}(0) = 0$. □

In a particular case of $n = 1$ (i.e. there are only two players in the game, the insurer and one insured), $A(x)$ in (4) becomes a nonstochastic function, $A(x) = w_0 + \hat{P} - \hat{I}(x)$. Then the Theorem 1's statement coincides with the result obtained in Golubin (2002b) for a two-agent risk exchange problem.

Remark 1. In the limiting case $\delta \to \infty$, (1) converts into a problem of maximization of the insurer's utility $J_0[I] \equiv Eu_0(w_0 + nP - \sum_{j=1}^{n} I(X_j))$ only. Then Theorem 1 (see (4)) gives that the Pareto-optimal $\hat{I}(x)$ is a stop-loss policy $\hat{I}(x) = \min\{x, \hat{x}\}$. Oppositely, if $\delta \to 0$ then (1) converts into maximizing the insured's utility $J_1[I]$. The right-hand side of (4) becomes equal to 1 in the limit and, by (4), the optimal $\hat{I}(x)$ is a deductible policy $\hat{I}(x) = (x - \hat{x})_+ \stackrel{def}{=} \max\{x - \hat{x}, 0\}$ without coinsurance. An analogous result holds if the insurer is supposed to be risk-neutral. Indeed, $u_0''(x) \equiv 0$ hence $\hat{I}'(x) = 1$ on $[\hat{x}, T]$, so that $\hat{I}(x) = (x - \hat{x})_+$ where the deductible level $\hat{x}$ is determined by conditions (5)-(6) in Theorem 1.

Proposition 1 below answers the question: which of the two kinds (the deductible or upper limit policy) does $\hat{I}(x)$ take under a fixed weight $\delta > 0$? Let $I_0(\delta, x)$ be a policy that satisfies equation (4) on $[0, T]$ with initial condition $I_0(\delta, 0) = 0$ (that is, a policy with neither deductible nor upper limit), where $0 < \delta < \infty$ is viewed now as a parameter.

Proposition 1. *Let δ^* be a unique solution to the equation*

$$U(0) - C = 0, \tag{10}$$

where expressions for $U(\cdot)$ and C are the same as that in (7) for $\hat{U}(\cdot)$ and $\hat{C}$ but with the policy $I_0(\delta, x)$ instead of $\hat{I}(x)$. Then the Pareto-optimal $\hat{I}(x)$ is a policy with an upper limit (with a deductible) $\hat{x} > 0$ if and only if $\delta > (<)\ \delta^*$.

Proof. Since δ^* is a root of (10) then, as was shown in Theorem 1, $\hat{U}(x) - \hat{C} \equiv 0$ on $[0, T]$ under $\hat{I}(x) = I_0(\delta^*, x)$ and, hence, $I_0(\delta^*, x)$ is optimal in problem (1) where δ is taken equal to δ^*.

Let, first, the prescribed weight δ be such that $\delta < \delta^*$. Suppose the contrary to the statement proved, that is, the optimal $\hat{I}(x)$ is a policy with an upper limit $\hat{x} > 0$. As was shown in the Theorem's 1 proof, this means that $\hat{U}(0) - \hat{C} < 0$. If we continuously decrease δ up to zero, the corresponding optimal policy $\hat{I}(\delta, x)$ continuously changes to $\hat{I}(0, x)$, a solution to (1) with $\delta = 0$. By the definition (see (7)), an expression for $\hat{U}(0) - \hat{C}$ under $\delta = 0$ is

$$(1 + \alpha) E\, u_1'(w_1 - P - X_1 + \hat{I}(0, X_1)) - u_1'(w_1 - P) \tag{11}$$

with $P = (1 + \alpha) E\, \hat{I}(0, X_1))$. Observing that $-x + \hat{I}(0, x)$ is a nonincreasing function, and $u_1''(x) < 0$, we have that expression (11) is positive. Therefore there exists a point δ' such that $\hat{U}(0) - \hat{C} = 0$ for some $\delta = \delta'$ and, hence, the optimal $\hat{I}(\delta', x)$ is a policy with neither upper limit nor deductible. As was proved above, it implies that $\delta' \in (0, \delta^*)$ is a solution to (10), but this contradicts the assumption that δ^* is the only solution to (10). The obtained contradiction proves that the policy $\hat{I}(x)$ we have started from is a policy with a deductible.

Let $\delta > \delta^*$. By the same reasonings as above, we obtain that in this situation $\hat{I}(x)$ is a policy with an upper limit $\hat{x} > 0$. □

Example 1. Consider the case of quadratic utility functions $u_i(x) = -\frac{1}{2}\beta_i x^2 + x$, $i = 0, 1$. (Note that $u_i'(x) > 0$ on the interval $(-\infty, 1/\beta_i)$ so it is natural to assume that the initial capitals $w_i < 1/\beta_i$, $i = 0, 1$.)

Since $u_i''(x) \equiv -\beta_i$, the right-hand side of (4) is a constant $\theta = \beta_1/(\delta\beta_0 + \beta_1)$. Then the Pareto-optimal $\hat{I}(x)$ is a piece-wise linear function, it is either an upper limit policy of the form $I_1(\hat{x}, x) = x - (1-\theta)(x-\hat{x})_+$ or a deductible policy of the form $I_2(\hat{x}, x) = \theta(x-\hat{x})_+$. To determine the kind of $\hat{I}(x)$, observe that now the policy $I_0(\delta, x) = \theta x$, therefore equation (10) in Proposition 1: $U(0) - (1 + \alpha) E\, U(X_1) = 0$, where

$$U(x) = \delta E\{u_0'(w_0 - nP + \hat{I}_0(\delta, x) - \sum_1^{n-1} I_0(\delta, X_j))\} - u_1'(w_1 - P - x + I_0(\delta, x)),$$

takes the form

$$\delta(1 - \beta_0 w_0) - \beta_1 E\, X_1 \left(1 + \alpha \frac{\delta \beta_0 n + \beta_1}{\delta \beta_0 + \beta_1}\right) = 1 - \beta_1 w_1. \tag{12}$$

As is easily seen, (12) has a unique solution, the "critical weight" $\delta^* > 0$, that depends on the loss distribution $F(x)$ only through its mean $E\, X_1$ and increases when n goes up. According to Proposition 1, if the given insurer's relative weight $\delta > (=)\, \delta^*$ then the optimal $\hat{I}(x)$ is a policy $I_1(\hat{x}, x)$ with an upper limit $\hat{x} > (=)\, 0$. If $\delta < \delta^*$ then $\hat{I}(x)$ is a policy $I_2(\hat{x}, x)$ with a deductible $\hat{x} > 0$.

Suppose $E\, X_1 = 1, \beta_0 = 0.1, \beta_1 = 1, w_0 = 1, w_1 = 0$, and $\alpha = 1$, regarding n as a parameter. From (12) we obtain the values of the "critical" weight $\delta^* = \delta^*(n)$

n	1	5	50	200
δ^*	3.402	4.846	48.484	214.604

Suppose further that the individual loss X_1 is uniformly distributed on $[0, 2] = [0, T]$. A level $\hat{x}$ in a deductible policy (if $\delta < \delta^*$) or in an upper limit policy (if $\delta > \delta^*$) is determined by conditions (5)-(6) in Theorem 1. In our case we get the following results:

δ	1.000	2.000	10.00	100.0	250.0
		$n = 5$			
$\hat{x}$	1.936	1.244 ‖	1.601	2.000	2.000
		$n = 50$			
$\hat{x}$	1.509	1.315	0.955 ‖	0.123	0.193
		$n = 200$			
$\hat{x}$	1.386	1.328	1.236	0.687 ‖	0.008

Here deductible values are separated from upper limit values by ‖. For instance, if $n = 5$ then the deductible $\hat{x}$ decreases from 1.936 to 1.244 as δ increases. After passing through zero (at $\delta = \delta^* = 4.846$) deductible converts into upper limit that increases from zero, passing through 1.601 at $\delta = 10$ and achiving then the extreme value $2 (= T)$.

4. Constructing a Nash's Solution

In 1950 Nash suggested a concept of bargaining game solution based on four axioms the reasonable players should obey. In particular, it is supposed that the solution should belong to the set of individually rational Pareto-optimal decisions, and that in a symmetric situation the players should maximize the joint gain and then divide it equally among themselves. The axioms result in that the decision the players agree upon is a maximizer of the product of their pay-off gains. Applied to our game, this means the following. Denote by $J^0[I] \stackrel{def}{=}$

$J_0[I] - J_0[0]$ and $J^1[I] \stackrel{def}{=} J_1[I] - J_1[0]$ the shifted utilities of the insurer and insured so that the initial utilities are now $(J^0[0], J^1[0]) = (0,0)$. Then the Nash's solution is a maximizer in the problem

$$\max_I J^0[I](J^1[I])^n \quad \text{s.t. } J^0[I] \geq 0,\ J^1[I] \geq 0. \tag{13}$$

Throughout the sequel of the paper we will assume the existence of a policy I such that

$$J^0[I] > 0,\ J^1[I] > 0. \tag{14}$$

This condition is a bit stronger than just nonemptyness of the set of individually rational Pareto-optimal policies. From the other hand it is the Slater condition in mathematical programming, related to problem (13).

In order to construct the Nash's solution, recall that in the previous section we obtain a one-parameter set $\{\hat{I}_\delta\}_{\delta>0}$ of Pareto-optimal policies in the game. Denote by $\Delta \stackrel{def}{=} \{\delta > 0 :\ J^k[\hat{I}_\delta] > 0,\ k = 0,1\}$ the set (nonempty) corresponding Pareto-optimal policies that satisfy (14). The next statement reduces the problem of finding the Nash's solution to solving an equation with respect to δ.

Proposition 2. *A policy I^N is the Nash's solution to the game if and only if $I^N = \hat{I}_{\delta'}$, where δ' solves equation* (15) *on* Δ

$$\frac{J^1[\hat{I}_\delta]}{J^0[\hat{I}_\delta]} = \delta. \tag{15}$$

Proof. Let I^N be the Nash's solution, or, equivalently, a maximizer in the problem

$$\max_I \ln J^0[I] + n \ln J^1[I] \quad \text{s.t. (14).} \tag{16}$$

Due to concavity of the goal functional, the necessary and sufficient optimality condition in (16) takes the form

$$\frac{d}{d\lambda}\{\ln J^0[\lambda I^N + (1-\lambda)I] + n \ln J^1[\lambda I^N + (1-\lambda)I]\}_{\lambda=1} \geq 0, \text{ or}$$

$$\frac{J^1[I^N]}{J^0[I^N]}\frac{d}{d\lambda}J^0[\lambda I^N + (1-\lambda)I]_{\lambda=1} + n\frac{d}{d\lambda}J^1[\lambda I^N + (1-\lambda)I]_{\lambda=1} \geq 0 \tag{17}$$

for any I satisfying (14).

Hence I^N maximizes the functional

$$\delta' J^0[I] + nJ^1[I] \quad \text{s.t. } J^0[I] \geq 0,\ J^1[I] \geq 0$$

(we can admit the nonstrict inequalities because of continuity of J^0 and J^1). Applying a known theorem from mathematical programming on Lagrange function for a concave problem under the Slater condition (see, e.g., Bazaraa and Shetty (1979, Chap. 6.2)) and noting that the inequalities are not binding at I^N, we have that I^N maximizes $\delta' J^0[I] + nJ^1[I]$ over the set Π of all policies. Therefore I^N is a Pareto-optimal policy $\hat{I}_{\delta'}$, with δ' solving (15).

Let $\delta' \in \Delta$ is a solution to (15). By construction, $\hat{I}_{\delta'}$ maximizes $\delta' J^0[I] + nJ^1[I]$ over Π and, hence, condition (17) holds for any $I \in \Pi$. This implies that $\hat{I}_{\delta'}$ solves problem (16), i.e., $\hat{I}_{\delta'}$ is the Nash's solution. □

Now we answer the following question: which of the two kinds, the deductible or upper limit policy, the Nash's solution belongs to. Recall that a kind of a given Pareto-optimal policy is determined in Proposition 1 through a comparison of δ and the "critical weight" δ^* defined in (10). Inequality $\delta < (>)\delta^*$ means that $\hat{I}_\delta$ is a deductible (upper limit) policy. Proposition 3 below gives a kind of the Nash's solution I^N, dealing with the value of δ^* only, without calculation of the Nash's weight δ' in $I^N = \hat{I}_{\delta'}$.

Proposition 3. *The Nash's solution I^N is a deductible (upper limit) policy if and only if*

$$\frac{J^1[\hat{I}_{\delta^*}]}{J^0[\hat{I}_{\delta^*}]} < (>)\delta^*. \tag{18}$$

Proof. Show first that the left-hand side of (15) is a nonincreasing function in δ. Evidently, it is sufficient to prove the following: If $\delta_1 < \delta_2$ and I_k is a maximizer in the problem $\max_{I\in\Pi} \delta_k J^0[I] + nJ^1[I]$, $k = 1, 2$, then $J^0[I_1] \leq J^0[I_2]$ and $J^1[I_1] \geq J^1[I_2]$. Suppose the contrary, and let first $J^0[I_1] > J^0[I_2]$. By the definition of I_2,

$$\delta_2 J^0[I] + nJ^1[I] \leq \delta_2 J^0[I_2] + nJ^1[I_2] \text{ for any } I \in \Pi. \tag{19}$$

From the other hand, $\delta_1 J^0[I_1] + nJ^1[I_1] \geq \delta_1 J^0[I_2] + nJ^1[I_2]$ and, hence, $\delta_2 J^0[I_1] + nJ^1[I_1] \geq \delta_1 J^0[I_2] + nJ^1[I_2] + (\delta_2 - \delta_1)J^0[I_1] > \delta_2 J^0[I_2] + nJ^1[I_2]$ as $J^0[I_1] > J^0[I_2]$ by supposition. Thus we get a contradiction with (19). The second case $J^1[I_1] < J^1[I_2]$ is treated in an analogous way.

By the proved monotonicity of $\frac{J^1[\hat{I}_\delta]}{J^0[\hat{I}_\delta]}$ in δ, inequality (18) means exactly that a solution to (15), i.e. the Nash's weight δ', is less (more) than δ^* correspondingly. According to Proposition 1, the Nash's solution $I^N = \hat{I}_{\delta'}$ is a deductible (upper limit) policy if and only if (18) holds. □

5. Constructing a Kalai-Smorodinsky's Solution

Another concept for a solution to the bargaining game was introduced in Kalai and Smorodinsky (1975). They criticized the Nash's solution as it does not satisfy a monotonicity condition. Their modification of Nash's axioms leads to that the new solution is determined as an intersection of the Pareto-optimal curve on the pay-off space and the straight line linking the disagreement point of initial pay-offs and the "ideal" point formed by maximum possible pay-offs for each of the players. In our $(n+1)$-person bargaining game, the n insureds have identical expected utilities $J_k[I] \equiv J_1[I]$, $k = 1, \ldots, n$ therefore Kalai-Smorodinsky's solution (KS-solution) can be determined as an intersection of the Pareto-optimal curve on the two-dimensional space $M = \{(J^0[I], J^1[I]) : I \in \Pi\}$ of shifted utilities and the straight line L linking the origin and the "ideal" point (J^0_*, J^1_*) where $J^k_* \stackrel{def}{=} \sup_{I\in\Pi} J^k[I]$, $k = 0, 1$.

Proposition 4. *A policy I^{KS} is the KS-solution to the game if and only if $I^{KS} = \hat{I}_{\delta''}$, where δ'' solves equation* (20) *on* Δ

$$\frac{J^1[\hat{I}_\delta]}{J^0[\hat{I}_\delta]} = \frac{J^1_*}{J^0_*}. \tag{20}$$

Proof. Denote by γ an angle betwen the line L linking the point $(0,0)$ with the ideal point (J^0_*, J^1_*) and the abscissa axis. Note that $0 < \gamma < \pi/2$ by condition (14). The definition of KS-solution given above is equivalent to the following: a Pareto-optimal point $(J^0[\hat{I}_\delta], J^1[\hat{I}_\delta])$ lies in the intersection if and only if

$$\frac{J^0[\hat{I}_\delta]}{J^1[\hat{I}_\delta]} = \tan\gamma,$$

which coincides with (20). □

Similar to the case of Nash's solution, we can establish a kind of the KS-solution in terms of the critical weight value δ^*.

Proposition 5. *The KS-solution I^{KS} is a deductible (upper limit) policy if and only if*

$$\frac{J^1[\hat{I}_{\delta^*}]}{J^0[\hat{I}_{\delta^*}]} < (>) \frac{J^1_*}{J^0_*}. \tag{21}$$

The proof is conducted along the same lines as in the Proposition 3's proof.

Example 2. Let the insurer be risk-neutral, that is, $u_0(x) \equiv x$. According to Theorem 1 (see also Remark 1), any Pareto-optimal policy is a deductible policy of the form $\hat{I}(x) = (x - \hat{x})_+$, where the deductible level $\hat{x}$ depends on the weight δ. It is required to determine 1) the Nash's solution and 2) the KS-solution to our game (i.e. the corresponding levels $\hat{x}$) if the insured has a quadratic utility function $u_1(x) = -\frac{1}{2}\beta x^2 + x$.

Since $J_0[\hat{I}] = E\,u_0(w_0 + n\hat{P} - \sum_1^n \hat{I}(X_i)) = w_0 + \alpha n \int_{\hat{x}}^T \bar{F}(x)dx$ and $J_0[0] = w_0$, we get $J^0[\hat{I}] = J_0[\hat{I}] - J_0[0] = \alpha n \hat{m}_1$ where $\hat{m}_1 \stackrel{def}{=} \int_{\hat{x}}^T \bar{F}(x)dx$. Similarly, $J^1[\hat{I}] = -\frac{1}{2}\beta[\hat{m}_1((\alpha^2 - 1)\hat{m}_1 - 2w_1\alpha + 2(1+\alpha)E\,X_1) - 2\hat{m}_2] - \alpha\hat{m}_1$, where $\hat{m}_2 \stackrel{def}{=} \int_{\hat{x}}^T x\bar{F}(x)dx$. By condition (5) in Theorem 1 that determines $\hat{x}$ under a fixed δ, we have an equation

$$1 - \beta w_1 - \alpha^{-1}\beta(\hat{x} - (1+\alpha)E\,X_1 + (1-\alpha^2)\hat{m}_1 = \delta. \tag{22}$$

Remark that if $\delta = 0$ then (22) becomes an equation for determining a deductible level $\tilde{x}$ which maximizes the insured's expected utility only. Note also that the left-hand side of (22) is decreasing in $\hat{x}$ therefore its root $\hat{x}$ remains less than $\tilde{x}$ as long as $\delta > 0$.

1) To find the Nash's solution, we make use of equation (15)

$$\frac{J^1[\hat{I}_\delta]}{J^0[\hat{I}_\delta]} = \delta.$$

As is easily seen, if $\delta' > 0$ is a solution to it then necessarily δ' belongs to the set Δ related to "strictly" rational Pareto-optimal policies because $J^0[\hat{I}_{\delta'}] = \alpha n \hat{m}_1 > 0$ and, hence, $J^1[\hat{I}_{\delta'}] > 0$. Inserting expression (22) for δ, we rewrite (15) as an equation with respect to $\hat{x}$

$$\frac{(\alpha^2-1)(n+1/2)}{n+1}\hat{m}_1 - \hat{x}\frac{n}{n+1} - \frac{\hat{m}_2}{(n+1)\hat{m}_1} + A = 0, \tag{23}$$

where $A = \alpha(\beta^{-1} - w_1) + (1+\alpha)E\,X_1$. Its solution x^N gives the Nash's optimal policy $I^N(x) = (x - x^N)_+$.

Consider, for numerical implementation, a case of exponentially distributed individual losses $X_i \sim 1 - e^{-x/\mu}$, $i = 1, \ldots, n$. Then (23) converts into

$$(1+\alpha)\mu\left[\frac{(\alpha-1)(n+1/2)}{n+1}e^{-\hat{x}/\mu} + 1\right] - \hat{x} - \frac{\mu}{n+1} + \alpha(\beta^{-1} - w_1) = 0.$$

Let $\mu = 1, \beta = 0.5, w_1 = 0$. Solving the equation above yields for the loading $\alpha = 0.5$:

n	1	5	25	100	225	400	625
x^N	1.917	2.261	2.394	2.424	2.429	2.431	2.432

for $\alpha = 1$:

x^N	3.500	3.833	3.961	3.990	3.995	3.997	3.998

for $\alpha = 1.5$:

x^N	5.006	5.338	5.466	5.495	5.500	5.502	5.503

Compare the obtained Nash's solutions and a deductible level $\tilde{x}$ maximizing the insured's utility only, which is determined by equation (22) with $\delta = 0$ in the right-hand side. As follows from the table below, the Nash's solution x^N quite rapidly tends to $\tilde{x}$ as the number of insureds $n \to \infty$

α	0.5	1	1.5
$\tilde{x}$	2.434	4.0	5.505

2) In order to construct the KS-solution, we rewrite equation (20) in Proposition 4, taking into account that now the "ideal" point $(J^0_*, J^1_*) = (J^0[I^*], J^1[I^{**}])$ is defined by the "ideal" policies: $I^*(x) = x$ and $I^{**}(x) = (x - \tilde{x})_+$ that maximize, correspondingly, the insurer's and insured's expected utilities. After rearranging the terms, (20) takes the form of an equation with respect to $\hat{x}$:

$$\psi(\hat{x}) = \psi(\tilde{x}) \int_{\tilde{x}}^{T} \bar{F}(x)dx / E\,X_1, \tag{24}$$

where the function

$$\psi(x) \stackrel{def}{=} \frac{(\alpha^2-1)}{2}m_1 - \frac{m_2}{m_1} + \alpha(\beta^{-1} - w_1) + (1+\alpha)E\,X_1$$

$$\text{with } m_1 = \int_x^T \bar{F}(t)dt \text{ and } m_2 = \int_x^T t\bar{F}(t)dt.$$

We see that for our case the KS-solution does not depend on the number of insureds n.

Suppose we have exponentially distributed individual losses $X_i \sim 1 - e^{-x/\mu}$, then (24) becomes

$$\psi(\hat{x}) = \psi(\tilde{x})e^{-\tilde{x}/\mu} \qquad (25)$$

with $\psi(x) = (1+\alpha)\mu[\frac{1}{2}(\alpha-1)e^{-x/\mu} + 1] - x - \mu + \alpha(\beta^{-1} - w_1)$.
Let $\mu = 1, \beta = 0.5, w_1 = 0$, and the loading coefficient α be a varying parameter. From equation (25) we obtain the following KS-solutions

α	0.5	0.75	1	1.25	1.5	1.75	2
x^{KS}	1.501	2.266	3.018	3.765	4.511	5.257	6.005

For convenience we also present the insured's optimal levels:

$\tilde{x}$	2.434	3.233	4.000	4.755	5.505	6.254	7.003

Like the Nash's solution x^N, the KS-optimal deductible x^{KS} does not exceed the level $\tilde{x}$ and increases as α increases (which reflects insured's desire to lessen the insurer's share of his risk as insurance becomes more expensive). Under a fixed α, x^N always exceeds x^{KS}, that is, the KS-solution transfers a greater share of risk to the insurer than that in the Nash's solution.

Remark 2. As is demonstrated by Example 2, the concept of Nash's solution in our model with a risk-neutral insurer is "unjust" for the insurer since an increasing number of insureds makes him less and less influential in making the contract. In the limit, as $n \to \infty$, the insurer agrees on the policy chosen by the insured only, as if insurer were excluded from the players ($\delta = 0$ in problem (1)). It is not surprising because (see (16)) I^N is a maximizer in the problem

$$\max_I \ln(\alpha n E\,I(X_1)) + n\ln(J^1[I])$$

and, as n increases, the weight n on the second term $\ln(J^1[I])$ related to the insured's utility increases faster than the weight in the first term. We can conclude that the concept of KS-solution is preferable to the Nash's one, at least when the number of insureds n is large enough.

References

[1] K.K. Aase, Perspectives of Risk Sharing, *Scand. Actuarial J.* **2** (2002), 73–128.

[2] B. Baton, J. Lemaire, The Core of a Reinsurance Market, *ASTIN Bulletin* **12** (1981), 57–71.

[3] M. Bazaraa, C. Shetty, *Nonlinear Programming. Theory and Algorithms*, John Wiley, New York, 1979.

[4] K. Borch, Reciprocal Reinsurance Treaties, *ASTIN Bulletin* **1** (1960), 171–191.

[5] K. Borch, Reciprocal Reinsurance Treaties Seen as a Two-person Cooperative Game, *Skand. Aktuarietidskrift* **43** (1960), 29–58.

[6] K. Borch, *Economics of Insurance*, Edited by Aase K.K. and Sandmo A., Amsterdam, North-Holland, 1990.

[7] P. Borm, J. Suijs, A.D. Waeggenaere, Stochastic Cooperative Games in Insurance, *Insurance: Mathematics and Economics* **22** (1998), 209–228.

[8] H.U. Gerber, Pareto-optimal Risk Exchanges and Related Decision Problems, *ASTIN Bulletin* **10** (1978), 25–33.

[9] A.Y. Golubin, Franchise Optimization in the Static Insurance Model, *Journal of Mathematical Sciences* **112** (2002), 4126–4140.

[10] A,Y. Golubin, A.Y. Pareto-optimal risk sharing between the insurer and insured, Abstracts of communications, Tenth All-Russian School-Colloqium on Stochastic Methods, *Rewiew of Applied and Industrial Mathematics* **9**, no. 3 (2002), 602– 603, (in Russian).

[11] E. Kalai, M. Smorodinsky, Other Solutions to the Nash Bargaining Problem, *Econometrica* **43** (1975), 513–518.

[12] S. Karlin, W. Studden, *Tchebycheff Systems: with Applications in Analysis and Statistics*, Interscience Publishers, 1966.

[13] J. Lemaire, A Non Symmetrical Value For Games Without Transferable Utilities. Application to Reinsurance, *ASTIN Bulletin* **10** (1979), 195–214.

[14] J. Lemaire, Cooperative Game Theory and Its Insurance Applications, *ASTIN Bulletin* **21** (1990), 17–40.

[15] J. Lemaire, *Bonus-Malus Systems in Automobile Insurance*, Kluwer Academic Publishers, 1995.

[16] R. Luce, H. Raiffa, *Games and Decisions*. John Wiley, New York, 1957.

[17] J. Nash, The Bargaining Problem, *Econometrica* **18** (1950), 155–162.

[18] A. Raviv, The Design of an Optimal Insurance Policy, *The American Economic Review* **69** (1979), 84–96.

[19] M. Zeleny, *Multiple Criteria Decision Making*, McGrow-Hill, New York, 1982.

In: Mathematics, Game Theory and Algebra Compendium ISBN 978-1-60456-659-8
Editor: Jacob H. Mathias, pp. 283-299

Chapter 27

A Fishery Game Model with Migration: Reserved Territory Approach

Vladimir V. Mazalov and Anna N. Rettieva
Institute of Applied Mathematical Research
Karelian Research Center of RAS
Russia

1. Introduction

A dynamic game model of a bioresource management problem (fisheries) is considered. The center (state) which determines the reserved portion of the reservoir (where fishing is prohibited), and the players (fishing farms) which harvest fish are the participants of the game. Each player is an independent decision maker, guided by the considerations of maximizing the profit from fish sale. We consider finite and infinite planning horizon. Pontryagin's maximal principle and Hamilton-Jacobi-Bellman equation were applied to determine Nash and Stakelberg equlibriums.

2. Game Model

Let us consider the center, which determines the reserved area of the reservoir denoted by s, $0 \le s \le 1$. We consider the strategies of the two players which exploit the fish stock during T time periods. Let us divide the water area into two parts: S_1 and S_2, where fishing is prohibited and allowed, respectively. Denote by x_1 and x_2 the size of the population per unit area of S_1 and S_2, respectively. Then $s = S_1/S$ is the reserved area. There is a migratory exchange between the two parts of the reservoir with the exchange coefficient $\gamma = q/s$, where q is the exchange rate.

The dynamics of the fishery is described by the system of equations:

$$\begin{cases} x_1'(t) = \varepsilon x_1(t) + \gamma_1(x_2(t) - x_1(t)), \\ x_2'(t) = \varepsilon x_2(t) + \gamma_2(x_1(t) - x_2(t)) - u(t) - v(t), \; x_i(0) = x_i^0, \end{cases} \tag{1}$$

where $x_1(t) \geq 0$ – size of the population at time t in the reserved area; $x_2(t) \geq 0$ – size of the population at time t in the area where fishing is allowed; ε – natural growth rate of the population; $u(t) \geq 0$ – first farm's fishing efforts at time t; $v(t) \geq 0$ – second farm's fishing efforts at time t; $s(t)$ – reserved portion of the reservoir and $\gamma_i = q/s$ – coefficients of the migratory exchange, $i = 1, 2$.

Then the payoffs of the two players over a fixed time period $[0, T]$ are

$$\begin{array}{l} J_1 = \int_0^T e^{-rt}[m_1((x_1(t) - \bar{x}_1)^2 + (x_2(t) - \bar{x}_2)^2) + c_1 u(t)^2 - p_1 u(t)]\, dt\,, \\ J_2 = \int_0^T e^{-rt}[m_2((x_1(t) - \bar{x}_1)^2 + (x_2(t) - \bar{x}_2)^2) + c_2 v(t)^2 - p_2 v(t)]\, dt\,, \end{array} \tag{2}$$

where $\bar{x}_i(t)$ – size of the population which is optimal for reproduction, m_i – penalty for deviation from the state $(\bar{x}_1, \bar{x}_2)$, c_i – catching costs of the i-th player and p_i – market price for each player, $i = 1, 2$.

Let's denote

$$c_{ir} = c_i e^{-rt}\,, \quad m_{ir} = m_i e^{-rt}\,, \quad p_{ir} = c_i e^{-rt}\,, \quad i = 1, 2.$$

We consider different optimality principles.

2.1. Nash Optimal Solution

We are interested in the optimal solution of the following problem:

$$\begin{cases} J_1(u^*, v^*) \leq J_1(u, v^*)\,, \ \forall u\,, \\ J_1(u^*, v^*) \leq J_1(u^*, v)\,, \ \forall v\,. \end{cases}$$

The Hamilton function for player I is

$$\begin{aligned} H_1 &= m_{1r}((x_1(t) - \bar{x}_1)^2 + (x_2(t) - \bar{x}_2)^2) + c_{1r} u(t)^2 - p_{1r} u(t) + \\ &+ \lambda_{11}(t)(\varepsilon x_1(t) + \gamma_1(x_2(t) - x_1(t))) + \\ &+ \lambda_{12}(t)(\varepsilon x_2(t) + \gamma_2(x_1(t) - x_2(t)) - u(t) - v(t))\,. \end{aligned}$$

Let's find the maximum of H_1:

$$\frac{\partial H_1}{\partial u} = 2c_{1r} u(t) - p_{1r} - \lambda_{12}(t) = 0\,.$$

Then the maximum is achieved at the point

$$u(t) = \frac{\lambda_{12}(t) + p_{1r}}{2c_{1r}}\,.$$

According to the maximum principle [4]

$$\begin{aligned} \lambda_{11}'(t) &= -\tfrac{\partial H_1}{\partial x_1} = -2m_{1r}(x_1(t) - \bar{x}_1) - \lambda_{11}(t)(\varepsilon - \gamma_1) - \lambda_{12}(t)\gamma_2\,, \\ \lambda_{12}'(t) &= -\tfrac{\partial H_1}{\partial x_2} = -2m_{1r}(x_2(t) - \bar{x}_2) - \lambda_{12}(t)(\varepsilon - \gamma_2) - \lambda_{11}(t)\gamma_1\,, \end{aligned}$$

and the transversability conditions are

$$\lambda_{1i}(T) = 0\,,\ i = 1, 2\,.$$

In a similar manner, for player II

$$\begin{aligned} H_2 &= m_{2r}((x_1(t) - \bar{x}_1)^2 + (x_2(t) - \bar{x}_2)^2) + c_{2r}v(t)^2 - p_{2r}v(t) + \\ &+ \lambda_{21}(t)(\varepsilon x_1(t) + \gamma_1(x_2(t) - x_1(t))) + \\ &+ \lambda_{22}(t)(\varepsilon x_2(t) + \gamma_2(x_1(t) - x_2(t)) - u(t) - v(t))\,. \end{aligned}$$

It yields

$$v(t) = \frac{\lambda_{22}(t) + p_{2r}}{2c_{2r}}\,,$$

and equations for additional variables are

$$\begin{aligned} \lambda'_{21}(t) &= -\frac{\partial H_2}{\partial x_1} = -2m_{2r}(x_1(t) - \bar{x}_1) - \lambda_{21}(t)(\varepsilon - \gamma_1) - \lambda_{22}(t)\gamma_2\,, \\ \lambda'_{22}(t) &= -\frac{\partial H_2}{\partial x_2} = -2m_{2r}(x_2(t) - \bar{x}_2) - \lambda_{22}(t)(\varepsilon - \gamma_2) - \lambda_{21}(t)\gamma_1\,, \end{aligned}$$

with the transversability conditions

$$\lambda_{2i}(T) = 0\,,\ i = 1, 2\,.$$

Finally, to find the Nash equilibrium we have to solve the following system:

$$\begin{cases} x'_1(t) = \varepsilon x_1(t) + \gamma_1(x_2(t) - x_1(t))\,, \\ x'_2(t) = \varepsilon x_2(t) + \gamma_2(x_1(t) - x_2(t)) - \frac{\lambda_{12}(t) + p_{1r}}{2c_{r1}} - \frac{\lambda_{22}(t) + p_{2r}}{2c_{2r}}\,, \\ \lambda'_{11}(t) = -2m_{1r}(x_1(t) - \bar{x}_1) - \lambda_{11}(t)(\varepsilon - \gamma_1) - \lambda_{12}(t)\gamma_2\,, \\ \lambda'_{12}(t) = -2m_{1r}(x_2(t) - \bar{x}_2) - \lambda_{12}(t)(\varepsilon - \gamma_2) - \lambda_{11}(t)\gamma_1\,, \\ \lambda'_{21}(t) = -2m_{2r}(x_1(t) - \bar{x}_1) - \lambda_{21}(t)(\varepsilon - \gamma_1) - \lambda_{22}(t)\gamma_2\,, \\ \lambda'_{22}(t) = -2m_{2r}(x_2(t) - \bar{x}_2) - \lambda_{22}(t)(\varepsilon - \gamma_2) - \lambda_{21}(t)\gamma_1\,, \\ \lambda_{i1}(T) = \lambda_{i2}(T) = 0\,,\ x_i(0) = x_i^0\,. \end{cases}$$

Let's recall our notations and introduce new variables:

$$\bar{\lambda}ij = \lambda ije^{rt}\,,\ i, j = 1, 2\,.$$

Then for these variables we get the system:

$$\begin{cases} x'_1(t) = \varepsilon x_1(t) + \gamma_1(x_2(t) - x_1(t))\,, \\ x'_2(t) = \varepsilon x_2(t) + \gamma_2(x_1(t) - x_2(t)) - \frac{\bar{\lambda}_{12}(t) + p_1}{2c_1} - \frac{\bar{\lambda}_{22}(t) + p_2}{2c_2}\,, \\ \bar{\lambda}'_{11}(t) = -2m_1(x_1(t) - \bar{x}_1) - \bar{\lambda}_{11}(t)(\varepsilon - \gamma_1 - r) - \bar{\lambda}_{12}(t)\gamma_2\,, \\ \bar{\lambda}'_{12}(t) = -2m_1(x_2(t) - \bar{x}_2) - \bar{\lambda}_{12}(t)(\varepsilon - \gamma_2 - r) - \bar{\lambda}_{11}(t)\gamma_1\,, \\ \bar{\lambda}'_{21}(t) = -2m_2(x_1(t) - \bar{x}_1) - \bar{\lambda}_{21}(t)(\varepsilon - \gamma_1 - r) - \bar{\lambda}_{22}(t)\gamma_2\,, \\ \bar{\lambda}'_{22}(t) = -2m_2(x_2(t) - \bar{x}_2) - \bar{\lambda}_{22}(t)(\varepsilon - \gamma_2 - r) - \bar{\lambda}_{21}(t)\gamma_1\,, \\ \bar{\lambda}_{i1}(T) = \bar{\lambda}_{i2}(T) = 0\,,\ x_i(0) = x_i^0\,. \end{cases} \tag{3}$$

Theorem 1.

$$u^*(t) = \frac{\bar{\lambda}_{12}(t) + p_1}{2c_1}$$

and

$$v^*(t) = \frac{\bar{\lambda}_{22}(t) + p_2}{2c_2},$$

with $\bar{\lambda}_{i2}$, $i = 1, 2$ *satisfing (3), form the Nash optimal solution of the problem (1)-(2).*

Proof. See [8].

Example.

Modelling was carried out for the following values: $q = 0.2$, $\gamma_1 = \gamma_2 = q/s$, $\varepsilon = 0.08$, $m_1 = m_2 = 0.09$, $c_1 = c_2 = 10$, $p_1 = p_2 = 100$, $T = 200$, $r = 0.1$.

Let the initial size of the population be $x_1(0) = 50$, $x_2(0) = 50$. And the optimal for reproduction population sizes are $\bar{x}_1 = 100$ and $\bar{x}_2 = 100$.

You can see the optimal values of $u^*(t)$ and $v^*(t)$ (equal for both players) in Fig.3. The figure indicates that the player's strategies should equal 8 almost all the time. The size of the population in the reserved area grows from 50 to 160 individuals (Fig.1). The size of the population in the area where fishing is allowed grows from 50 to 140 individuals (Fig.2).

The players' profits given that they use the optimal strategies, are $J_1 = J_2 = 103.6457652$.

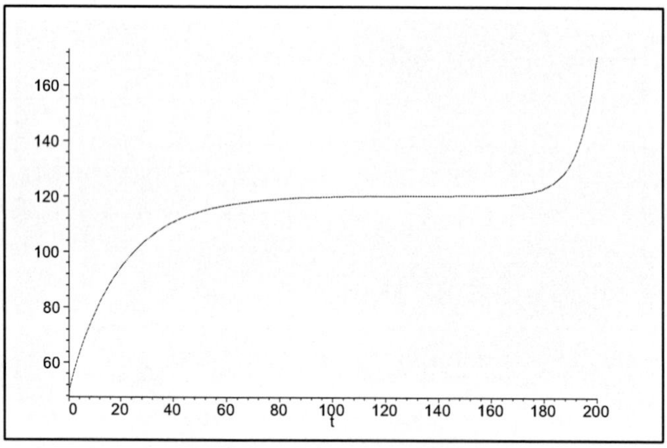

Figure 1. Values of $x_1^*(t)$.

2.2. Stackelberg Optimal Solution

We are interested in the optimal solution of the following problem:

$$\begin{cases} \max\limits_{v \in R(u^*)} J_1(u^*, v) \le \max\limits_{v \in R(u)} J_1(u, v), \\ \text{where } R(u) = \{v \,|\, J_2(u, v) \le J_2(u, v')\}, \\ \text{and } v^* \in R(u^*). \end{cases}$$

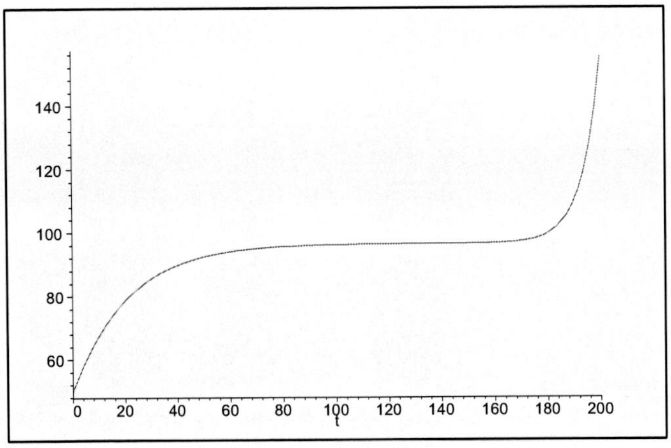

Figure 2. Values of $x_2^*(t)$.

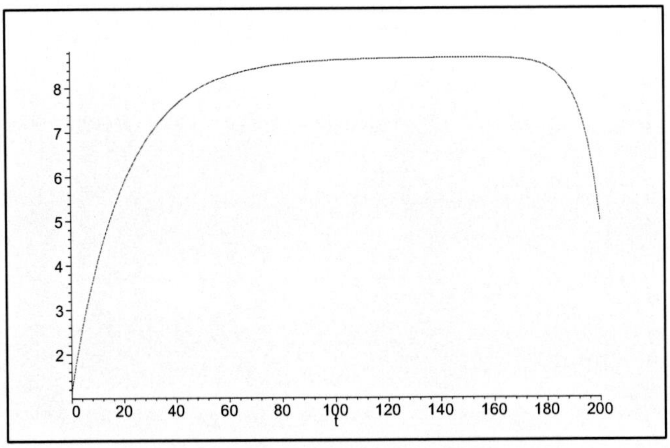

Figure 3. Values of $u^*(t)$.

For solving this problem we use Pontryagin's maximum principle modified for a two level game.

We fix some strategy $u(t)$, and the Hamilton function for player II is

$$\begin{aligned} H_2 &= m_{2r}((x_1(t)-\bar{x}_1)^2+(x_2(t)-\bar{x}_2)^2)+c_{2r}v(t)^2-p_{2r}v(t)+\\ &+ \lambda_{21}(t)(\varepsilon x_1(t)+\gamma_1(x_2(t)-x_1(t)))+\\ &+ \lambda_{22}(t)(\varepsilon x_2(t)+\gamma_2(x_1(t)-x_2(t))-u(t)-v(t))\,. \end{aligned}$$

Wherefore

$$v(t)=\frac{\lambda_{22}(t)+p_{2r}}{2c_{2r}},$$

and the equations for additional variables are

$$\begin{aligned} \lambda_{21}'(t) &= -\tfrac{\partial H_2}{\partial x_1} = -2m_{2r}(x_1(t)-\bar{x}_1)-\lambda_{21}(t)(\varepsilon-\gamma_1)-\lambda_{22}(t)\gamma_2\,,\\ \lambda_{22}'(t) &= -\tfrac{\partial H_2}{\partial x_2} = -2m_{2r}(x_2(t)-\bar{x}_2)-\lambda_{22}(t)(\varepsilon-\gamma_2)-\lambda_{21}(t)\gamma_1\,, \end{aligned}$$

with the transversability conditions

$$\lambda_{2i}(T) = 0\,, \;\; i = 1, 2\,.$$

We substitute this strategy of player II into the system (1) and combine it with the equations for additional variables

$$\begin{cases} x_1'(t) = \varepsilon x_1(t) + \gamma_1(x_2(t) - x_1(t))\,, \\ x_2'(t) = \varepsilon x_2(t) + \gamma_2(x_1(t) - x_2(t)) - u - \frac{\lambda_{22}(t)+p_{2r}}{2c_{2r}}\,, \\ \lambda_{21}'(t) = -2m_{2r}(x_1(t) - \bar{x}_1) - \lambda_{21}(t)(\varepsilon - \gamma_1) - \lambda_{22}(t)\gamma_2\,, \\ \lambda_{22}'(t) = -2m_{2r}(x_2(t) - \bar{x}_2) - \lambda_{22}(t)(\varepsilon - \gamma_2) - \lambda_{21}(t)\gamma_1\,, \\ \lambda_{21}(T) = \lambda_{22}(T) = 0\,, \;\; x_i(0) = x_i^0\,. \end{cases}$$

Then we repeat all operations using the Pontryagin's maximum principle. The Hamilton function for player I is

$$\begin{aligned} H_1 \;\; = \;\; & m_{1r}((x_1(t) - \bar{x}_1)^2 + (x_2(t) - \bar{x}_2)^2) + c_{1r}u(t)^2 - p_{1r}u(t) + \\ + \;\; & \lambda_{11}(t)(\varepsilon x_1(t) + \gamma_1(x_2(t) - x_1(t))) + \\ + \;\; & \lambda_{12}(t)(\varepsilon x_2(t) + \gamma_2(x_1(t) - x_2(t)) - u(t) - \tfrac{\lambda_{22}(t)+p_{2r}}{2c_{2r}}) + \\ + \;\; & \mu_1(t)(-2m_{2r}(x_1(t) - \bar{x}_1) - \lambda_{21}(t)(\varepsilon - \gamma_1) - \lambda_{22}(t)\gamma_2) + \\ + \;\; & \mu_2(t)(-2m_{2r}(x_2(t) - \bar{x}_2) - \lambda_{22}(t)(\varepsilon - \gamma_2) - \lambda_{21}(t)\gamma_1) \end{aligned}$$

Let's find the maximum of H_1:

$$\frac{\partial H_1}{\partial u} = 2c_{1r}u(t) - p_{1r} - \lambda_{12}(t) = 0\,.$$

Then the maximum is achieved at the point

$$u(t) = \frac{\lambda_{12}(t) + p_{1r}}{2c_{1r}}\,.$$

According to the maximal principle [4]

$$\begin{cases} \lambda_{11}'(t) = -2m_{1r}(x_1(t) - \bar{x}_1) - \lambda_{11}(t)(\varepsilon - \gamma_1) - \lambda_{12}(t)\gamma_2 + 2m_{2r}\mu_1(t)\,, \\ \lambda_{12}'(t) = -2m_{1r}(x_2(t) - \bar{x}_2) - \lambda_{12}(t)(\varepsilon - \gamma_2) - \lambda_{11}(t)\gamma_1 + 2m_{2r}\mu_2(t)\,, \\ \mu_1'(t) = -\frac{\partial H_1}{\partial \lambda_{21}} = \mu_1(t)(\varepsilon - \gamma_1) + \mu_2(t)\gamma_1\,, \\ \mu_2'(t) = -\frac{\partial H_1}{\partial \lambda_{22}} = \frac{\lambda_{12}(t)}{2c_{2r}} + \mu_2(t)(\varepsilon - \gamma_2) + \mu_1(t)\gamma_2\,, \end{cases}$$

and the transversability conditions are

$$\lambda_{2i}(T) = 0\,, \;\; \mu_i(0) = 0\,.$$

Finally, to find the Stackelberg equilibrium we have to solve the following system:

$$\begin{cases} x_1'(t) = \varepsilon x_1(t) + \gamma_1(x_2(t) - x_1(t)), \\ x_2'(t) = \varepsilon x_2(t) + \gamma_2(x_1(t) - x_2(t)) - \frac{\lambda_{12}(t)+p_{1r}}{2c_{1r}} - \frac{\lambda_{22}(t)+p_{2r}}{2c_{2r}}, \\ \lambda_{11}'(t) = -2m_{1r}(x_1(t) - \bar{x}_1) - \lambda_{11}(t)(\varepsilon - \gamma_1) - \lambda_{12}(t)\gamma_2 + 2m_{2r}\mu_1(t), \\ \lambda_{12}'(t) = -2m_{1r}(x_2(t) - \bar{x}_2) - \lambda_{12}(t)(\varepsilon - \gamma_2) - \lambda_{11}(t)\gamma_1 + 2m_{2r}\mu_2(t), \\ \lambda_{21}'(t) = -2m_{2r}(x_1(t) - \bar{x}_1) - \lambda_{21}(t)(\varepsilon - \gamma_1) - \lambda_{22}(t)\gamma_2, \\ \lambda_{22}'(t) = -2m_{2r}(x_2(t) - \bar{x}_2) - \lambda_{22}(t)(\varepsilon - \gamma_2) - \lambda_{21}(t)\gamma_1, \\ \mu_1'(t) = \mu_1(t)(\varepsilon - \gamma_1) + \mu_2(t)\gamma_1, \\ \mu_2'(t) = \frac{\lambda_{12}(t)}{2c_{2r}} + \mu_2(t)(\varepsilon - \gamma_2) + \mu_1(t)\gamma_2, \\ \lambda_{i1}(T) = \lambda_{i2}(T) = 0, \; x_i(0) = x_i^0, \; \mu_i(0) = 0. \end{cases}$$

Let's recall our notations and introduce new variables:

$$\bar{\lambda}ij = \lambda ije^{rt},$$

Then for these variables we get the system:

$$\begin{cases} x_1'(t) = \varepsilon x_1(t) + \gamma_1(x_2(t) - x_1(t)), \\ x_2'(t) = \varepsilon x_2(t) + \gamma_2(x_1(t) - x_2(t)) - \frac{\bar{\lambda}_{12}(t)+p_1}{2c_1} - \frac{\bar{\lambda}_{22}(t)+p_2}{2c_2}, \\ \bar{\lambda}_{11}'(t) = -2m_1(x_1(t) - \bar{x}_1) - \bar{\lambda}_{11}(t)(\varepsilon - \gamma_1 - r) - \bar{\lambda}_{12}(t)\gamma_2 + 2m_2\mu_1(t), \\ \bar{\lambda}_{12}'(t) = -2m_1(x_2(t) - \bar{x}_2) - \bar{\lambda}_{12}(t)(\varepsilon - \gamma_2 - r) - \bar{\lambda}_{11}(t)\gamma_1 + 2m_2\mu_2(t), \\ \bar{\lambda}_{21}'(t) = -2m_2(x_1(t) - \bar{x}_1) - \bar{\lambda}_{21}(t)(\varepsilon - \gamma_1 - r) - \bar{\lambda}_{22}(t)\gamma_2, \\ \bar{\lambda}_{22}'(t) = -2m_2(x_2(t) - \bar{x}_2) - \bar{\lambda}_{22}(t)(\varepsilon - \gamma_2 - r) - \bar{\lambda}_{21}(t)\gamma_1, \\ \mu_1'(t) = \mu_1(t)(\varepsilon - \gamma_1) + \mu_2(t)\gamma_1, \\ \mu_2'(t) = \frac{\bar{\lambda}_{12}(t)}{2c_2} + \mu_2(t)(\varepsilon - \gamma_2) + \mu_1(t)\gamma_2, \\ \bar{\lambda}_{i1}(T) = \bar{\lambda}_{i2}(T) = 0, \; x_i(0) = x_i^0, \; \mu_i(0) = 0. \end{cases} \tag{4}$$

Theorem 2. *The strategies*

$$u^*(t) = \frac{\bar{\lambda}_{12}(t) + p_1}{2c_1}$$

and

$$v^*(t) = \frac{\bar{\lambda}_{22}(t) + p_2}{2c_2},$$

with $\bar{\lambda}_{i2}$, $i = 1, 2$ *satisfing (4), form the Stackelberg optimal solution of the problem (1)-(2).*

Example.

Modelling was carried out for the same values as in section 1.1.

You can see the optimal values of $u^*(t)$ in Fig.6 and those of $v^*(t)$ in Fig.7. The figures indicate that player's I startegy should equal 6 almost all the time, and player's II strategy – equal 12. The size of the population in the reserved area grows from 50 to 200 individuals (Fig.4). The size of the population in the area where fishing is allowed grows from 50 to 180 individuals (Fig.5),

The players' profits given that they use the optimal strategies, are $J_1 = -253.7144427$, $J_2 = 943.6131678$.

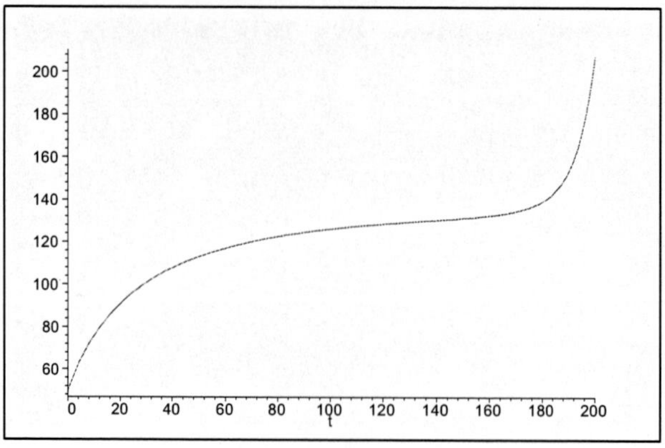

Figure 4. Values of $x_1^*(t)$.

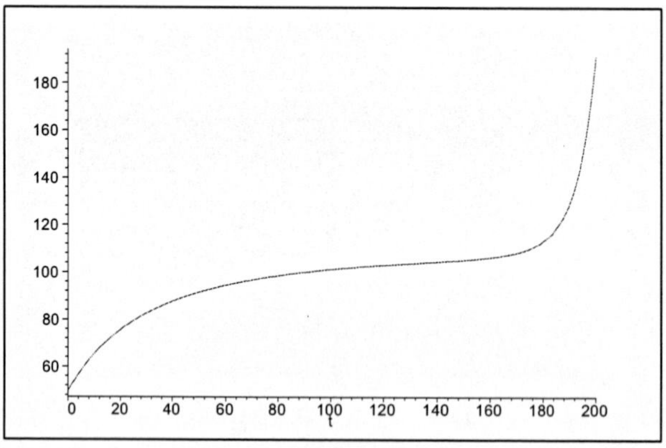

Figure 5. Values of $x_2^*(t)$.

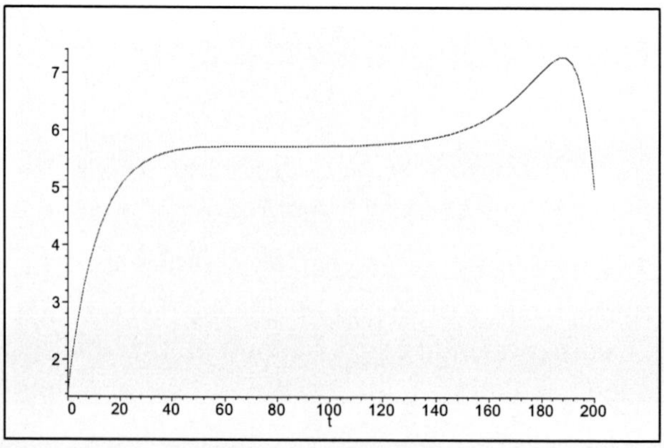

Figure 6. Values of $u^*(t)$.

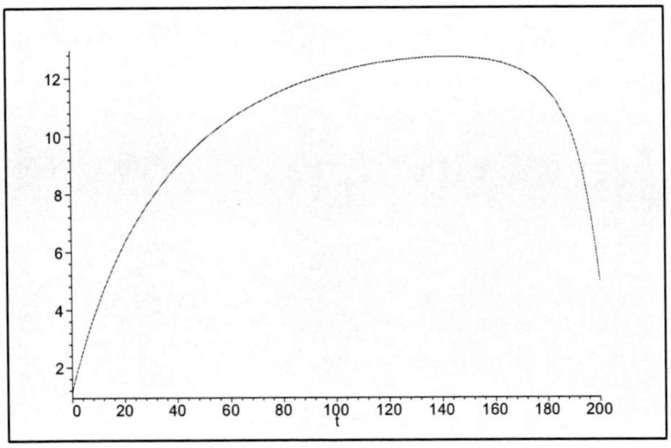

Figure 7. Values of $v^*(t)$.

Table 1. Player I profit

$s(t)$	0.1	0.2	0.3	0.4	0.5	0.6	0.7	0.8	0.9
Nash	11591	6670	3177	1017	103	353	1689	4040	7341
Stakelberg	11240	6329	2837	673	-253	-26	1277	3582	6820

Table 2. Player II profit

$s(t)$	0.1	0.2	0.3	0.4	0.5	0.6	0.7	0.8	0.9
Nash	11591	6670	3177	1017	103	353	1689	4040	7341
Stakelberg	12426	7498	4002	1846	943	1210	2571	4955	8294

Let's compare the players' payoffs when we use different optimality principles.

Profit of player I, which corresponds to different sizes of the reserved area $s(t)$, are shown in Table 1, and profit of player II – in Table 2.

In the case of the Nash equilibrium both players are in the same conditions, so their strategies and profits are equal.

In the case of the Stackelberg equilibrium player I is the leader and, as Tables 1, 2 show, this equilibrium is better for player I, but worse for player II. This solution gives the gain to player I, whereas player II carries all the expenses of maintaining a stable population development.

3. Model Over an Infinite Horizon

The dynamics of the fishery, as before, is described by the system of equations:

$$\begin{cases} x_1'(t) = \varepsilon x_1(t) + \gamma_1(x_2(t) - x_1(t)), \\ x_2'(t) = \varepsilon x_2(t) + \gamma_2(x_1(t) - x_2(t)) - u(t) - v(t), \ x_i(0) = x_i^0, \end{cases} \tag{5}$$

where $x_1(t) \geq 0$ – size of the population at time t in the reserved area; $x_2(t) \geq 0$ – size of the population at time t in the area where fishing is allowed; ε – natural growth rate of the population; $u(t) \geq 0$ – first farm's fishing efforts at time t; $v(t) \geq 0$ – second farm's fishing efforts at time t; $s(t)$ – reserved portion of the reservoir and $\gamma_i = q/s$ – coefficients of the migratory exchange, $i = 1, 2$.

Then the discounted payoffs of the two players over an infinite horizon at a rate r are

$$\begin{aligned} J_1 &= \int_0^\infty e^{-rt}[m_1((x_1(t)-\bar{x}_1)^2 + (x_2(t)-\bar{x}_2)^2) + c_1 u(t)^2 - p_1 u(t)]\,dt\,, \\ J_2 &= \int_0^\infty e^{-rt}[m_2((x_1(t)-\bar{x}_1)^2 + (x_2(t)-\bar{x}_2)^2) + c_2 v(t)^2 - p_2 v(t)]\,dt\,, \end{aligned} \tag{6}$$

where $\bar{x}_i(t)$ – population size optimal for reproduction, m_i – penalty for deviation from the state $(\bar{x}_1, \bar{x}_2)$, c_i – catching costs of the i-th player and p_i – market price for each player, $i = 1, 2$.

When we investigated the model with a finite planning horizon, we used Pontryagin's maximum principle. The infinite horizon gives us an opportunity to use Hamilton–Jacobi–Bellman equation.

Let's consider different optimality principles for infinite horizon model.

4. Nash Optimal Solution

Let's fix the second player's strategy and consider the problem of determining the optimal strategy of player I.

Define the value function $V(x)$ for our problem

$$V(x_1, x_2) = \min_u \{ \int_0^\infty e^{-rt}[m_1((x_1(t)-\bar{x}_1)^2 + (x_2(t)-\bar{x}_2)^2) + c_1 u(t)^2 - p_1 u(t)]\}\,.$$

The Hamilton–Jacobi–Bellman equation is

$$\begin{aligned} rV(x_1, x_2) &= \min_u \{ m_1((x_1-\bar{x}_1)^2 + (x_2-\bar{x}_2)^2) + c_1 u^2 - p_1 u + \tfrac{\partial V}{\partial x_1}(\varepsilon x_1 + \gamma_1(x_2 - x_1)) + \\ &+ \tfrac{\partial V}{\partial x_2}(\varepsilon x_2 + \gamma_2(x_1 - x_2) - u - v)\} \end{aligned}$$

Let's find the maximum over u:

$$2c_1 u - \frac{\partial V}{\partial x_2} - p_1 = 0\,.$$

Then the maximum is achieved at the point

$$u = (\frac{\partial V}{\partial x_2} + p_1)/2c_1\,.$$

Substituting it into the equation, we get

$$rV(x_1, x_2) = m_1((x_1 - \bar{x}_1)^2 + (x_2 - \bar{x}_2)^2) - \frac{(\frac{\partial V}{\partial x_2} + p_1)^2}{4c_1} \begin{array}{l} + \frac{\partial V}{\partial x_1}(\varepsilon x_1 + \gamma_1(x_2 - x_1)) + \\ + \frac{\partial V}{\partial x_2}(\varepsilon x_2 + \gamma_2(x_1 - x_2) - v) \end{array}$$

It is easy to verify that the quadratic form provides a solution to this equation.
Let $V(x_1, x_2) = a_1 x_1^2 + b_1 x_1 + a_2 x_2^2 + b_2 x_2 + k x_1 x_2 + l$.
Then the player's I strategy is

$$u(x) = \frac{2a_2 x_2 + b_2 + k x_1 + p_1}{2c_1},$$

where the coefficients satisfy the system of equations

$$\begin{cases} ra_1 &= m_1 - \frac{k^2}{4c_1} + 2a_1(\varepsilon - \gamma_1) + k\gamma_2 \\ rb_1 &= -2m_1\bar{x}_1 - \frac{kb_2}{2c_1} - \frac{kp_1}{2c_1} + b_1(\varepsilon - \gamma_1) + b_2\gamma_2 - kv \\ ra_2 &= m1 - \frac{a_2^2}{c_1} + 2a_2(\varepsilon - \gamma_2) + k\gamma_1 \\ rb_2 &= -2m_1\bar{x}_2 - \frac{a_2 b_2}{c_1} - \frac{a_2 p_1}{c_1} + b_2(\varepsilon - \gamma_2) + b_1\gamma_1 - 2a_2 v \\ rk &= -\frac{a_2 k}{c_1} + k(\varepsilon - \gamma_1) + 2a_1\gamma_1 + 2a_2\gamma_2 + k(\varepsilon - \gamma_2) \\ rl &= m_1\bar{x}_1^2 + m_1\bar{x}_2^2 - \frac{b_2^2}{4c_1} - \frac{b_2 p_1}{2c_1} - \frac{p_1^2}{4c_1} - b_2 v \end{cases} \tag{7}$$

Analogously for player II

$$v(x) = \frac{2\alpha_2 x_2 + \beta_2 + k_2 x_1 + p_2}{2c_2},$$

where the coefficients satisfy the system of equations

$$\begin{cases} r\alpha_1 &= m_2 - \frac{k_2^2}{4c_1} + 2\alpha_1(\varepsilon - \gamma_1) + k_2\gamma_2 \\ r\beta_1 &= -2m_2\bar{x}_1 - \frac{k_2\beta_2}{2c_1} - \frac{k_2 p_1}{2c_1} + \beta_1(\varepsilon - \gamma_1) + \beta_2\gamma_2 - k_2 u \\ r\alpha_2 &= m2 - \frac{\alpha_2^2}{c_1} + 2\alpha_2(\varepsilon - \gamma_2) + k_2\gamma_1 \\ r\beta_2 &= -2m_2\bar{x}_2 - \frac{\alpha_2\beta_2}{c_1} - \frac{\alpha_2 p_1}{c_1} + \beta_2(\varepsilon - \gamma_2) + \beta_1\gamma_1 - 2\alpha_2 u \\ rk_2 &= -\frac{\alpha_2 k_2}{c_1} + k_2(\varepsilon - \gamma_1) + 2\alpha_1\gamma_1 + 2\alpha_2\gamma_2 + k_2(\varepsilon - \gamma_2) \\ rl_2 &= m_2\bar{x}_1^2 + m_2\bar{x}_2^2 - \frac{\beta_2^2}{4c_1} - \frac{\beta_2 p_1}{2c_1} - \frac{p_1^2}{4c_1} - \beta_2 u \end{cases} \tag{8}$$

So, we proved the following theorem.

Theorem 3.

$$u^*(x) = \frac{2a_2x_2 + b_2 + kx_1 + p_1}{2c_1}$$

and

$$v^*(x) = \frac{2\alpha_2x_2 + \beta_2 + k_2x_1 + p_2}{2c_2},$$

form the Nash optimal solution of the problem (5)-(6), where the coefficients are defined from (7) and (8).

Example.

Modelling was carried out for the following values: $q = 0.2$, $\gamma_1 = \gamma_2 = q/s$, $\varepsilon = 0.08$, $m_1 = m_2 = 0.09$, $c_1 = c_2 = 10$, $p_1 = p_2 = 100$, $T = 200$, $r = 0.1$.

Let the initial size of the population be $x_1(0) = 50$, $x_2(0) = 50$. And the optimal for reproduction population sizes are $\bar{x}_1 = 100$ and $\bar{x}_2 = 100$.

You can see the optimal values of $u^*(t)$ and $v^*(t)$ (equal for both players) in Fig.10. The figure indicates that players' strategies should increase from 2 to 8. The size of the population in the reserved area grows from 50 to 120 individuals (Fig.8). The size of the population in the area where fishing is allowed grows from 50 to 90 individuals (Fig.9).

The players' payoffs given that they use the optimal strategies, are $J_1 = J_2 = 388.0627019$.

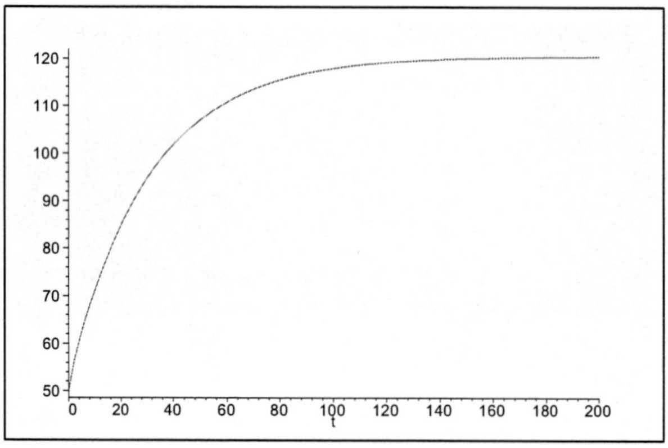

Figure 8. Values of $x_1^*(t)$.

4.1. Stackelberg Optimal Solution

Using Hamilton–Jacobi–Bellman equation for player II we get

$$v(x) = \frac{2\alpha_2x_2 + \beta_2 + k_2x_1 + p_2 + \sigma u}{2c_2},$$

where the coefficients are defined from (8).

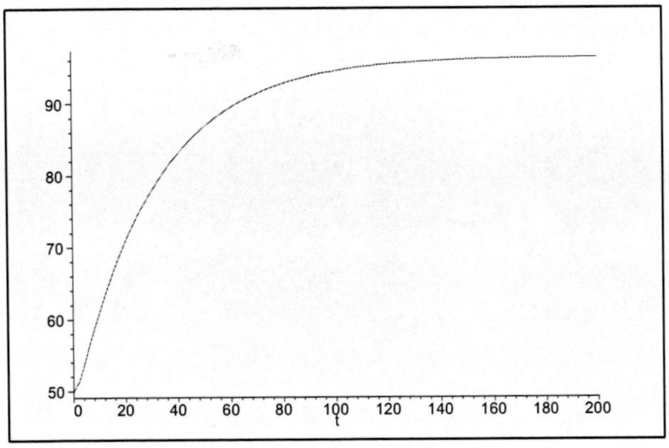

Figure 9. Values of $x_2^*(t)$.

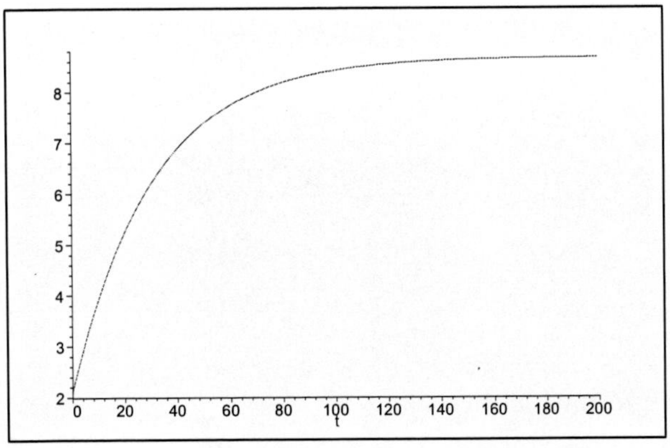

Figure 10. Values of $u^*(t)$.

Define the value function $V(x)$ for player's I problem

$$V(x_1, x_2) = \min_u \{ \int_0^\infty e^{-rt} [m_1((x_1(t) - \bar{x}_1)^2 + (x_2(t) - \bar{x}_2)^2) + c_1 u(t)^2 - p_1 u(t)] \} .$$

The Hamilton–Jacobi–Bellman equation is

$$\begin{aligned} rV(x_1, x_2) &= \min_u \{ m_1((x_1 - \bar{x}_1)^2 + (x_2 - \bar{x}_2)^2) + c_1 u^2 - p_1 u + \tfrac{\partial V}{\partial x_1}(\varepsilon x_1 + \gamma_1(x_2 - x_1)) + \\ &+ \tfrac{\partial V}{\partial x_2}(\varepsilon x_2 + \gamma_2(x_1 - x_2) - u - \tfrac{2\alpha_2 x_2 + \beta_2 + k_2 x_1 + p_2 + \sigma u}{2c_2}) \} \end{aligned}$$

Let's find the maximum over u:

$$2c_1 u - \frac{\partial V}{\partial x_2}(1 + \frac{\sigma}{2c_2}) - p_1 = 0 ,$$

wherefore

$$u = (\frac{\partial V}{\partial x_2}(2c_2 + \sigma) + 2p_1 c_2)/4c_1 c_2 .$$

Substituting it into the equation, we get

$$\begin{aligned} rV(x_1,x_2) &= m_1((x_1-\bar{x}_1)^2+(x_2-\bar{x}_2)^2)+(\tfrac{\partial V}{\partial x_2})^2\tfrac{(2c_2+\sigma)^2}{8c_1c_2^2}+\tfrac{p_1^2}{2c_1}+\tfrac{\partial V}{\partial x_1}(\varepsilon x_1+\gamma_1(x_2-x_1))+\\ &+ \tfrac{\partial V}{\partial x_2}(x_1(\gamma_2-\tfrac{k_2}{2c_2})+x_2(\varepsilon-\gamma_2-\tfrac{\alpha_2}{c_2})+\tfrac{2c_2+\sigma}{2c_1c_2}-\tfrac{\beta_2+p_2}{2c_2}) \end{aligned}$$

It is easy to verify that the quadratic form provides a solution to this equation.
Let $V(x_1,x_2)=a_1x_1^2+b_1x_1+a_2x_2^2+b_2x_2+gx_1x_2+l$.
Then the player's I strategy is

$$u(x)=((2a_2x_2+b_2+gx_1)(2c_2+\sigma)+2p_1c_2)/4c_1c_2\,,$$

where the coefficients satisfy the system of equations

$$\begin{cases} ra_1 = m_1+g^2\frac{(2c_2+\sigma)^2}{8c_1c_2^2}+2a_1(\varepsilon-\gamma_1)+g(\gamma_2-\frac{k_2}{2c_2}) \\ rb_1 = -2m_1\bar{x}_1+2b_2g\frac{(2c_2+\sigma)^2}{8c_1c_2^2}+g\frac{(2c_2+\sigma)}{2c_1c_2}b_1(\varepsilon-\gamma_1)+b_2(\gamma_2-\frac{k_2}{2c_2})-\frac{g(\beta_2+p_2)}{2c_2} \\ ra_2 = m1+a_2\frac{(2c_2+\sigma)^2}{2c_1c_2^2}+g\gamma_1+2a_2(\varepsilon-\gamma_2-\frac{\alpha_2}{c_2}) \\ rb_2 = -2m_1\bar{x}_2+a_2b_2\frac{(2c_2+\sigma)^2}{2c_1c_2^2}+a_2\frac{(2c_2+\sigma)}{c_1c_2}+b_2(\varepsilon-\gamma_2-\frac{\alpha_2}{c_2})+b_1\gamma_1-a_2\frac{\beta_2+p_2}{c_2} \\ rg = a_2g\frac{(2c_2+\sigma)^2}{2c_1c_2^2}+g(\varepsilon-\gamma_1)+2a_1\gamma_1+2a_2(\gamma_2-\frac{k_2}{2c_2})+g(\varepsilon-\gamma_2-\frac{\alpha_2}{c_2}) \\ rl = m_1\bar{x}_1^2+m_1\bar{x}_2^2+b_2^2\frac{(2c_2+\sigma)^2}{8c_1c_2^2}+b_2\frac{(2c_2+\sigma)}{2c_1c_2}+\frac{p_1^2}{2c_1}-\frac{b_2(\beta_2+p_2)}{2c_2} \end{cases} \tag{9}$$

So, we proved the following theorem.

Theorem 4. *The strategies*

$$u^*(x)=((2a_2x_2+b_2+gx_1)(2c_2+\sigma)+2p_1c_2)/4c_1c_2$$

and

$$v^*(x)=(\sigma(2a_2x_2+b_2+gx_1)(2c_2+\sigma)+4c_1c_2(2\alpha_2x_2+\beta_2+k_2x_1)+2c_2(\sigma p_1+2c_1p_2))/8c_1c_2^2\,,$$

form the Stackelberg optimal solution of the problem (5)-(6), where the coefficients are defined from (8) and (9).

Example.

Modelling was carried out for the same values as in section 2.1.

You can see the optimal values of $u^*(t)$ in Fig.13 and those of $v^*(t)$ in Fig.14. The figure indicates that player's I strategy should increase from 4.6 to 5.2, and player's II strategy – from 0 to 14. The size of the population in the reserved area grows from 50 to 120 individuals (Fig.11). The size of the population in the area where fishing is allowed grows from 50 to 100 individuals (Fig.12).

The players' profits given that they use the optimal strategies, are $J_1=163.4798825$ and $J_2=1153.667808$.

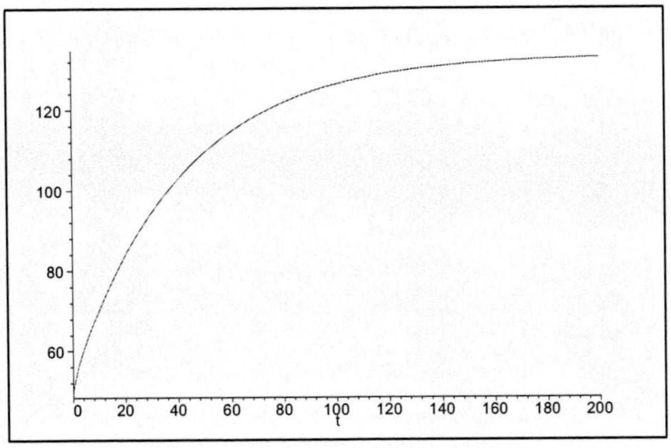

Figure 11. Values of $x_1^*(t)$.

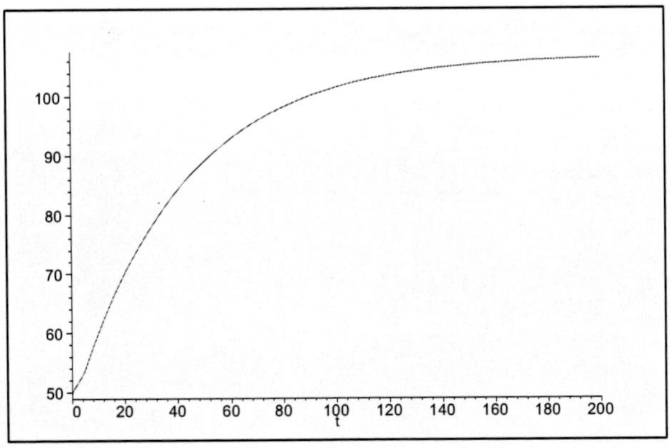

Figure 12. Values of $x_2^*(t)$.

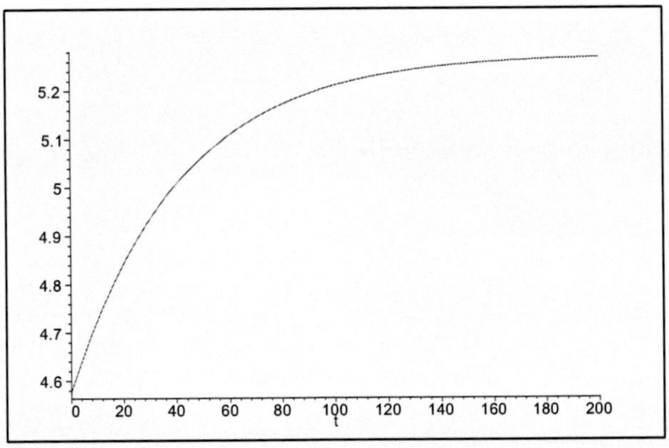

Figure 13. Values of $u^*(t)$.

Let's compare the players' profits when we use different optimal principls.

Profit of player I, which corresponds to different sizes of the reserved area $s(t)$, are shown in Table 3, and profit of player II – in Table 4.

As in the case of the finite planning horizon, the Stackelberg equilibrium is better for

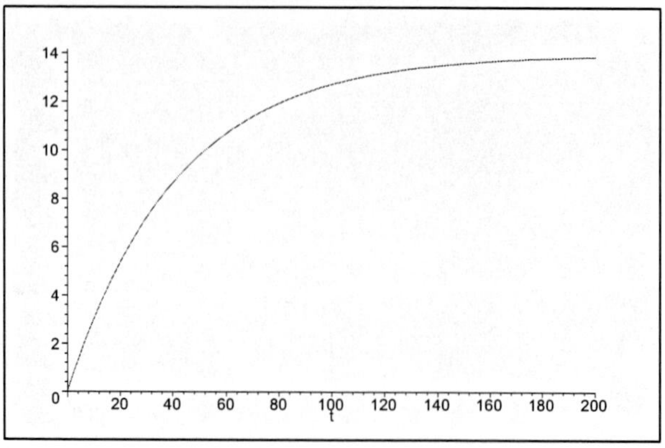

Figure 14. Values of $v^*(t)$.

Table 3. Player I profit

$s(t)$	0.1	0.2	0.3	0.4	0.5	0.6	0.7	0.8	0.9
Nash	11868	6930	3431	1280	388	668	2037	4414	7720
Stakelberg	11641	6702	3203	1053	163	446	1818	4199	7508

Table 4. Player II profit

$s(t)$	0.1	0.2	0.3	0.4	0.5	0.6	0.7	0.8	0.9
Nash	11868	6930	3431	1280	388	668	2037	4414	7720
Stakelberg	12642	7706	4207	2052	1153	1425	2784	5149	8442

player I, but worse for player II than the Nash equilibrium.

In practice, the Hamilton–Jacobi–Bellman equation is more convenient for long–term planning.

References

[1] C.W. Clark, *Bioeconomic modelling and fisheries management*, Wiley, NY, 1985.

[2] H. Ehtamo, R.P. Hamalainen, A cooperative incentive equilibrium for a resource management problem, *J. of Economic Dynamics and Control* **17** (1993), 659–678.

[3] A.A. Petrosjan, V.V. Zakharov, *Mathematical models in environmental policy analysis*, Nova Scince Publ., New York, 1998.

[4] L.S. Pontryagin, V.G. Boltyanski, R.V. Gamrelidze, E.F. Mishenko, *Mathematical theory of optimal processes*, Nauka, 1969, (in Russian).

[5] V.I. Gurman, G.N. Konstantinov, *Planning and prediction of natural–economic systems*, Nauka, Novosibirsk, 1984, (in Russian).

[6] V.V. Mazalov, A.N. Rettieva, About one resource management problem, *Survey in Applied and Industrial Mathematics* **9**, no. 2 (2002), TVP, Moscow, 293–306, (in Russian).

[7] V.V. Mazalov, A.N. Rettieva,A fishery game model with age–distributed population: reserved territory approach, *Game Theory and Applications* **9** (2003), Nova Science Publishers, Inc., 125–139.

[8] T. Basar, G.J. Olsder, *Dynamic noncooperative game theory*, Academic Press, New York, 1982.

In: Mathematics, Game Theory and Algebra Compendium ISBN 978-1-60456-659-8
Editor: Jacob H. Mathias, pp. 301-318

Chapter 28

NON-HIERARCHICAL SIGNALLING: TWO-STAGE FINANCING GAME

Anton Miglo
University of Guelph,
Department of Economics Ontario,Canada
Nikolay Zenkevich
Saint-Petersburg State University, Faculty of Applied
Mathematics and Control Processes, Saint-Petersburg, Russia

1. Introduction

The literature analyzing games where some players have private information about their "types" is usually based on the duality of "good" and "bad" types (GB approach), where "good" type denotes the type with better quality. Depending on the context, the quality could mean the quality of produced good, the ability to work etc. Typically in such a game, the "good" type tries to signal its type to uninformed players by sending the messages which cannot be mimicked by the "bad" type.[1] In contrast, this paper analyzes a signalling game without types hierarchy. Different types have the same average qualities but different performances profiles over time which are their private information. Hence a "good" type in the beginning may become "bad" in the end or "bad" in the beginning may become "good" in the end. We apply this idea to analyze a financing-investment game where firms' insiders have private information about the firm's profit profile over time.[2]

More specifically, we analyze a situation where a firm's initial shareholders have to raise funds for financing an investment project. There is no internal funds available and therefore the financing should be external. The cost of investment is known to the shareholders and to the potential investors while the expected profit is the shareholders' private information.

---

[1] For a review of signalling games see, for instance, Fudenberg and Tirole (1990) or Petrosjan and Zenkevich (1996).

[2] In the similar spirit, some researchers assume that firms have the same average profit but different parameteres of risk which are their private information. For example, in the second part of Brennan and Kraus (1986) cash flows are ordered by mean-preserving spread condition. It is shown that optimal securities are neither convex nor concave in this case. In Brick, Frierman and Kim (1998) firms' profits have the same average value but different variances. The authors do obtain some results about firm dividend policy.

Such a situation in a static context (one period) was well studied in the literature. The equilibrium is typically pooling where all firms issue debt which survive usual equilibrium refinements and which minimizes mispricing (undervaluation) for a "good" type, i. e. for a firm with high expected profit.[3] We thus consider a two-period situation. In each period there is an investment and a profit. As was noted previously we suppose that different types have the same average profit but different profit profiles over time which are their private information. Also, we assume that managers have the choice between issuing debt or equity. The solution of the game we obtained shares with the standard models the existence of pooling equilibrium with debt. However, in our game a separating equilibrium (which is efficient by definition) may exist as well. Which equilibrium prevails depends on the initial distribution of types in the economy.

To provide basic ideas about how the private information about firms profit profile over time can affect financing choice let us suppose that there are only two types of firms. One is "performance-improving" (I) and have an increasing expected profit, while others are "stagnant" (S) and have a flatter or decreasing expected profit. In such an environment, prices can be affected by the "lemon" effect in both periods. [4] Intuitively, I would seem to have an informational advantage in the first period: because of lower profits in this period, this type of firm can capitalize on the adverse selection problem. On the other hand, in the second period the informational advantage passes to S. We show that I and S face very different incentives regarding financial decisions. The point is that, generally speaking, debt has a shorter maturity than equity, which has by definition infinite maturity. Thus, the price of first-period equity is type-independent due to the two-period maturity of equity (contrary to the one-period of debt) and to the fact that both types face the same total profit over the two periods. As a result, if I were to issue equity in the first period, they would always be mimicked by S, who stand to gain in the second period by being perceived as growing and, therefore, as expecting high profits in the second period. The implication is that I are at a disadvantage for equity issues in the first period. This is the main engine driving the results of this article.

While I would definitely prefer debt to equity, incentives for S depend on the macroeconomic situation or on the initial distribution of types in the economy. The idea here is that if the economy is growing—there are on average more performance-improving than stagnating firms—interest rates tend to be more suitable for I. In particular, first-period interest rates would be relatively high compared to those of the second period, because I is considered "bad" in the first period and "good" in the second. Given such an interest rate profile, we show that if S plays debt, it would be beneficial to creditors, but not to the firm. This is because the creditors benefit in the first period due to the high interest rates and to the fact that S does well at that period.

The rest of this paper is organized as follows. The basic model and some preliminary results are presented in Section 2. Sections 3 and 4 provide the analysis of two-type and multiple-type economies respectively. The conclusion is drawn in Section 5.

---

[3] See, for instance, Nachman and Noe (1994).

[4] We use the term "lemon" problem to describe a situation where private information leads to underpricing for a "good" type. See Akerloff (1970) for a classical example.

2. Model

Consider a firm with two-stage investment project. In each stage $t = 1, 2$ an amount b has to be invested. In each stage the project can either be successful (with probability ρ_t) or unsuccessful (with probability $1 - \rho_t$). If the former is the case the revenue $\widetilde{R}_t$ equals 1 and if the latter is the case the revenue equals 0. Total expected revenue over both periods is then $\rho_1 + \rho_2$. Since all firms have the same total discounted profit, total revenue can be normalized to unity without loss of generality. Hence $\rho_1 + \rho_2 = 1$ and we write $\rho_1 = \theta$ and $\rho_2 = 1 - \theta$. Firms differ only through the parameter θ. We assume $b < 1/2$ with the θ's restricted to the interval $[b, 1 - b]$, which implies that the investment has non-negative profitability in each period, i. e. the expected profit is not smaller than the amount of investments in period one ($b \leq \theta$) and in period two ($b \leq 1 - \theta$). A firm has increasing expected revenues and profits if $\theta < 1/2$, the profit profile is flat or declining if $\theta = 1/2$ or $\theta > 1/2$ respectively. If we let θ be distributed according to the density $f(\theta)$, then the total (average) first-period revenue is:

$$y = \int_b^{1-b} \theta f(\theta)\, d\theta$$

Clearly, total second-period revenue is then $1 - y$. This means that the economy is growing (revenues and profits are increasing) if $y < 1/2$, and it is stagnant or declining if $y = 1/2$ or $y > 1/2$.

The firm's shareholders are responsible for capital structure choice, investments and profit distribution. The initial capital structure is 100% equity, with n shares outstanding. The firm maximizes wealth of initial shareholders, who we will call the entrepreneur. Let α_0 and β_0 denote the initial proportions of equity owned respectively by outsiders and entrepreneur and let α_t and β_t denote their proportions immediately after the financing and investment for stage t were done. Clearly, $\alpha_0 = 1$, $\beta_0 = 0$ and $\alpha_t + \beta_t = 1$ for all t.

There exists universal risk-neutrality in this economy. In addition, the competition among investors is perfect. Insider shareholders know the firm's type, but the investors do not. The distribution of types is common knowledge.

To finance the first stage, the firm may issue either debt (d) or equity (e).[5] In both cases, the firm gets amount b from the market, which is immediately invested. Holding free cash flow is costly. The reasons are well-known: "empire building" or inefficient investments and acquisitions, which spread the resources under the manager's control; increasing manager compensations or direct entrenchment; etc.[6] More specifically, we assume that any available free cash flow disappears immediately, producing useless loss for the shareholders.[7] Knowing this the shareholders will never keep the free cash that implies that any available first-period profit will be distributed as dividend. Thus, in the second period only debt fi-

---

[5] More complicated securities are not considered here since the model's implications are all about equity-debt choice. Also, for the simplicity of exposition, we assume that only pure strategies can be played, although this is not crucial to the results.

[6] Easterbrook (1984), Jensen (1986). See also [7] and [14] for empirical analysis about the significance of manager's agency cost in holding cash

[7] We assume free cash to be any available cash at the end of a period, which means any resources that were not invested during the period or any received profits that were not used for interests or dividends.

nancing is possible.[8,9] As in the standard literature in this field, we assume that the contract of debt is enforceable at no cost.

The sequence of events is illustrated in figure 1. We assume that the firm's type is revealed to initial insiders in the period 0. The investors are identical and we will call them simply the market. The market determines the prices of issued securities. Also the market observes the first-period capital structure choice. However, it does not observe the profit previously realized by the firm.

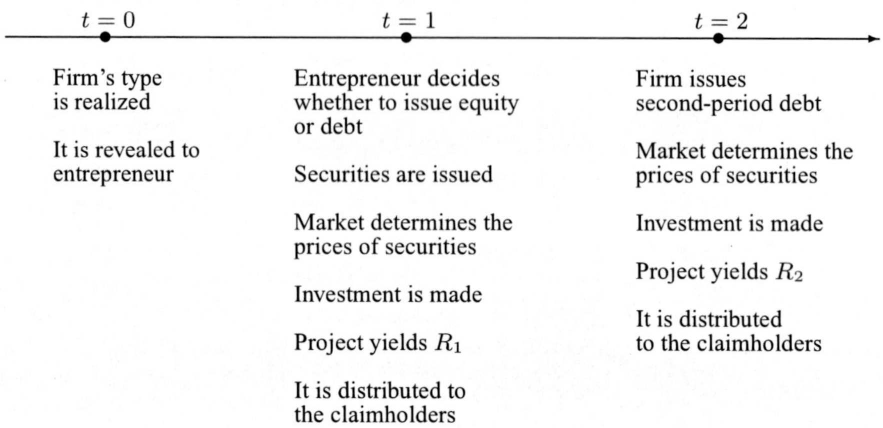

Figure 1. The sequence of events.

Throughout this article, we use the concept of Perfect-Bayesian equilibria (PBE) and also verify that off-equilibrium beliefs survive usual refinements like Cho and Kreps' (1987) intuitive criterion and consistency (Kreps and Wilson, 1982). The intuitive criterion seems to be not very powerful in games where pooling equilibrium is Pareto-efficient (see Cadsby, Frank, and Maksimovic, 1998). Fortunately, this is not the case in the present paper. In addition, note that perfect competition between outsiders implies zero market profit and risk-neutral valuation for any security issued. More specifically, we assume that there are at least two investors and the competition among them is in the Bertrand style (see Cho and Kreps (1987) or Nachman and Noe (1994)). Their pricing strategies are identical and equal to the expected value of the offered securities. Competition in the capital market therefore results in the price that yields zero net profit to investors.

From definition of debt and equity it follows that if debt is issued in period t with face value D then the debtholders expected payoff equals $E\left[\min(\widetilde{R}_t, D)\right]$. The shareholders are residual claimants. If new equity was issued the shareholders share the profit according to the number of shares owned.

---

[8] This is based on Myers and Majluf's (1984) idea that in one-period setting under asymmetric information, equity is never issued. Although our environments are quite different, one can show that the introduction of the possibility of equity issue in the second period does not alter any results.

[9] To simplify, we assume that mixed financing (debt/equity in the first period or cash/debt in the second) is not possible. The basic intuitions developed within this paper are not affected by introducing these possibilities. It is also important to note that the model can be extended by allowing mixed strategies (in game-theoretic terms), which can be interpreted to some extent as real mixed financing.

2.1. Perfect Market

This subsection provides some useful information about benchmark pricing when the market knows the firm's type. Consider strategy e. Denote the issue of shares in period 1 by Δn, the price of issued shares by p_e^1 and the second period debt face value by D_e^2. The relations describing the pricing and the payoffs are:

1) first-period budget constraint:

$$b = p_e^1 \Delta n \tag{1}$$

2) market valuation of second-period debt:

$$b = E\left[\min(\widetilde{R}_2, D_e^2)\right] \tag{2}$$

3) market valuation of equity issued in the first period (recall that n denotes the initial number of shares):

$$p_e^1 = \frac{\overline{R}_1}{n+\Delta n} + \frac{E\left[\max(0, \widetilde{R}_2 - D_e^2)\right]}{n+\Delta n} \tag{3}$$

where $E[\widetilde{R}] = \overline{R}$. Given the identity:

$$\min(R, D) + \max(0, R - D) = R \tag{4}$$

and using equations (1) and (2), we can transform (3) to:

$$p_e^1 = \frac{\overline{R}_1 + \overline{R}_2 - 2b}{n}$$

Since $\overline{R}_1 + \overline{R}_2 = 1$, we get

$$p_e^1 = \frac{1-2b}{n} \tag{5}$$

Remark 1. *p_e^1 depends only on the firm's total profit and not on its profit profile over time.*

Using equation (2) and conducting a similar exercise for strategy d (for type θ), one can obtain the efficient (symmetric information) face values of debt (for the first and second period respectively):

$$D_d^1 = b/\theta, D_d^2 = b/(1-\theta) \tag{6}$$

If $\theta < 1/2$, the interest rate profiles in the case of d corresponding to type θ is downward sloping (and upward sloping if $\theta > 1/2$, respectively). Finally, note that regardless of how the investment is financed, the value of the firm for the entrepreneur is:

$$V = 1 - 2b \tag{7}$$

For example if e is played then the entrepreneur's expected payoff equals

$$\frac{n\overline{R}_1}{n+\Delta n} + \frac{nE\left[\max(0, \widetilde{R}_2 - D_e^2)\right]}{n+\Delta n}$$

Taken into account (3) and (5) this equals $1 - 2b$. As usual, in perfect market, the choice of financing does not matter.

2.2. Asymmetric Information

Now consider the situation where the firm's type is its private information. Let us introduce the payoff-functions. Denote by $V_j(\theta, \widehat{\theta})$ the entrepreneur's final payoff if the firm is of type θ but is perceived as type $\widehat{\theta}$, given the first-period action $j = e, d$. The following explains why the analysis of these functions is useful. Suppose that the market beliefs observing strategy j are characterized by a density function $\mu^j(\theta)$ with support $[b, 1-b]$.

Lemma 1. *Let the market beliefs observing strategy* $j = e, d$ *be* μ^j. *The pricing is then as if the market believes with probability* 1 *that the firm is type* $\widehat{\theta}_j$*, where*

$$\widehat{\theta}_j = \int \theta \mu^j(\theta) d\theta$$

Proof. Consider $j = d$. Let first-period debt face value equals D^1_d. The first-period lenders' expected first-period payoff is then: $\int D^1_d \theta \mu^j(\theta) d\theta$. Risk-neutral valuation implies that it should be equal to b. Thus

$$D^1_d = \frac{b}{\int \theta \mu^j(\theta) d\theta}$$

Analogously for second-period debt:

$$D^2_d = \frac{b}{\int (1-\theta) \mu^j(\theta) d\theta}$$

Lemma 1 follows from (6). Now consider $j = e$. For second-period debt face value the reasoning is exactly as for D^2_d. Now consider the first-period share price. Since the firm's total expected profit equals 1 and since the second-period lenders' expected payoff equals b, the profit of insiders plus the profit of first-period outsiders of firm j equals $1 - b$. Also, in the case of e, α^2_o (that also shows the fraction of equity held by the first-period outsiders in the moment of first-period profit distribution) equals $\frac{\Delta n}{n+\Delta n}$. Thus, expected profit of first-period outsider shareholders is:

$$\frac{\Delta n}{n + \Delta n}(1-b)$$

Risk-neutral valuation implies that the expected revenue of first-period outsiders is b. Thus: $\frac{\Delta n}{n+\Delta n}(1-b) = b$ and $\Delta n = \frac{bn}{1-2b}$, which implies (because we know the budget constraint $p^1_e \Delta n = b$) that

$$p^1_e = \frac{1-2b}{n} \tag{8}$$

End proof.[10]

Note that under perfect information, the first-period share price equals $\frac{1-2b}{n}$, regardless of the issuer's type (equation (5)). The same result holds true under asymmetric information. It provides the intuition as to why growing firms prefer debt to equity—they cannot use

[10] Note that the same result holds true if one introduces the possibility for second-period outsiders to observe first-period profit realization, given that market beliefs are Bayesian.

their informational advantage in the first-period playing equity because the price is always the same.

Consider the features of functions $V_j(\theta, \widehat{\theta})$ for $j = e, d$. If the entrepreneur plays e

$$\alpha_c^2 = \frac{n}{n + \Delta n} = \frac{n}{n + b/p_e^1} = \frac{1 - 2b}{1 - b}$$

as implied by (8). Thus

$$V_e(\theta, \widehat{\theta}) = \frac{1 - 2b}{1 - b}(\theta + (1 - \theta)(1 - \frac{b}{1 - \widehat{\theta}})) \tag{9}$$

Also:

$$V_d(\theta, \widehat{\theta}) = \theta(1 - b/\widehat{\theta}) + (1 - \theta)(1 - b/(1 - \widehat{\theta})) \tag{10}$$

Obviously, $V_j(\theta, \theta) = 1 - 2b, j = e, d$, since this corresponds to complete information valuation. Observe also that $V_d(\theta, 1/2) = 1 - 2b$. The following properties are obvious:

Lemma 2. $\frac{\partial V_e(\theta,\widehat{\theta})}{\partial \theta} > 0$ *and* $\frac{\partial V_e(\theta,\widehat{\theta})}{\partial \widehat{\theta}} < 0.$

The idea behind Lemma 2 is that since the first-best share price in the first period is the same for all types, the types with high θ benefit from their informational advantages in the second period (when they are really "lemons"). On the other hand, a larger $\widehat{\theta}$ means a larger second-period interest rate, which is unprofitable.

Lemma 3. *$V_d(\theta, \widehat{\theta}) \geq 1 - 2b$ if and only if $\theta \leq \widehat{\theta} \leq 1/2$ or $1/2 \leq \widehat{\theta} \leq \theta$. Furthermore*

$$sign(\frac{\partial V_d(\theta, \widehat{\theta})}{\partial \theta}) = sign(\widehat{\theta} - 1/2) \tag{11}$$

$$\widehat{\theta} < \min(\theta, 1/2) \Rightarrow \frac{\partial V_d(\theta, \widehat{\theta})}{\partial \widehat{\theta}} > 0 \tag{12}$$

$$\widehat{\theta} > \max(\theta, 1/2) \Rightarrow \frac{\partial V_d(\theta, \widehat{\theta})}{\partial \widehat{\theta}} < 0 \tag{13}$$

Proof.

$$V_d(\theta, \widehat{\theta}) \geq 1 - 2b \Leftrightarrow \frac{\theta}{\widehat{\theta}} + \frac{1 - \theta}{1 - \widehat{\theta}} \leq 2 \Leftrightarrow$$

$$\Phi(\widehat{\theta}) \equiv \theta(1 - \widehat{\theta}) + (1 - \theta)\widehat{\theta} - 2\widehat{\theta}(1 - \widehat{\theta}) \leq 0,$$

where Φ is convex with roots $\widehat{\theta} = 1/2$ and $\widehat{\theta} = \theta$. This proves the first statement. The proof of (11) follows from:

$$\frac{\partial V_d(\theta, \widehat{\theta})}{\partial \theta} = -\frac{(1 - 2\widehat{\theta})b}{\widehat{\theta}(1 - \widehat{\theta})}$$

To prove (12) and (13) one can check that

$$sign(\frac{\partial V_d(\theta, \widehat{\theta})}{\partial \widehat{\theta}}) = sign(\frac{\theta}{1 - \theta} - (\frac{\widehat{\theta}}{1 - \widehat{\theta}})^2)$$

Now $\widehat{\theta} < \min(\theta, 1/2)$ implies $\frac{\theta}{1-\theta} > (\frac{\widehat{\theta}}{1-\widehat{\theta}})^2$ while $\widehat{\theta} > \max(\theta, 1/2)$ implies $\frac{\theta}{1-\theta} < (\frac{\widehat{\theta}}{1-\widehat{\theta}})^2$. *End proof.*

Intuitively, by analogy with perfect information case, a downward sloping interest rates profile ($\widehat{\theta} \leq 1/2$) is suitable for growing firms, i. e. for firms with $\theta \leq \widehat{\theta}$ and not for firms with lower than average rate of growth ($\theta > \widehat{\theta}$), which are better off with upward sloping interest rate profile. Conversely for the case of stagnating economy ($1/2 \leq \widehat{\theta}$). The intuition behind condition (11) is the same. Now consider equation (12). If the interest rate profile is downward sloping then, for a firm that has lower than average rate of growth, making interest rate profile less upward sloping is profitable. On the other hand, if the interest rate profile is upward sloping then, for a firm with higher than average rate of growth, making the interest rate profile deeper is unprofitable (equation (13)).

Lemma 4. $sign(V_d(\theta, \widehat{\theta}) - V_e(\theta, \widehat{\theta})) = sign(\widehat{\theta} - \theta)$.

Proof. Consider $V_d(\theta, \widehat{\theta}) - V_e(\theta, \widehat{\theta})$. That is:

$$1 - b(\frac{\theta}{\widehat{\theta}} + \frac{1-\theta}{1-\widehat{\theta}}) - \frac{1-2b}{1-b}(\theta + (1-\theta)(1 - \frac{b}{1-\widehat{\theta}})) = \frac{(\widehat{\theta}-\theta)(1-b-\widehat{\theta})b}{(1-b)(1-\widehat{\theta})\widehat{\theta}} \tag{14}$$

The sign of the last expression depend obviously on the sign of $\widehat{\theta} - \theta$. *End proof.*

Figure 2 illustrates the first parts of Lemmas 2 and 3, condition (11) and Lemma 4.

In both cases $V_e(\theta, \widehat{\theta})$ is increasing in θ. When $\widehat{\theta} < 1/2$ (Figure 2a), $V_d(\theta, \widehat{\theta})$ is downward sloping in θ and is upward sloping if $\widehat{\theta} > 1/2$ (Figure 2b). If the latter is the case the slope of $V_e(\theta, \widehat{\theta})$ is greater than that of $V_d(\theta, \widehat{\theta})$ meaning that the payoff from the strategy debt is less sensitive to adverse selection problem as compared to equity. This is in keeping with most of the literature in this field. Intuitively, if a firm is perceived as a less growing type than it is in reality then it will prefer debt to equity. This is because the second-period interest rate is the same in either case, but by playing debt, the firm gains in the first period by being a bad firm-type. If, in contrast, a firm is perceived by the market as a less growing than in reality type, it would prefer equity.

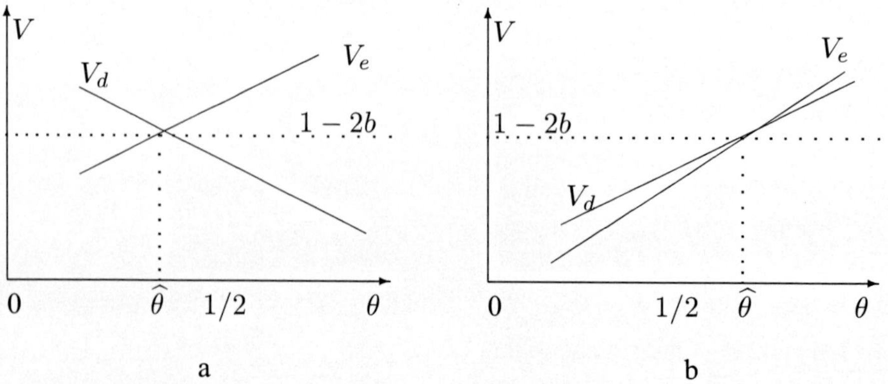

Figure 2. $V_d(\theta, \widehat{\theta})$ **and** $V_e(\theta, \widehat{\theta})$ **when: a)** $\widehat{\theta} < 1/2$**; b)** $\widehat{\theta} > 1/2$.

3. Two-Type Economy

To generate the basic ideas, we first consider a two-type economy. Firm I is characterized by the parameter θ_I, firm S has parameter θ_S where $\theta_I < \theta_S$. By definition, S has better performance in the first period while I in the second (note that both may actually be declining, but S then declines faster). Let μ_0 be the proportion of type I firms, $0 < \mu_0 < 1$. Hence $y = \theta_I\mu_0 + \theta_S(1 - \mu_0)$. Since each firm may play two types of strategy (d or e), there are 4 potential candidates for equilibrium: two separating and two pooling. Given the concepts described in Section 2 a separating equilibrium is defined as follows:

1) type I plays j_I and type S plays $j_S \neq j_I, j_T \in \{e, d\}, T \in \{I, S\}$.

2) $V_{j_I}(\theta_I, \theta_I) \geq V_{j_S}(\theta_I, \theta_S)$ and $V_{j_S}(\theta_S, \theta_S) \geq V_{j_I}(\theta_S, \theta_I)$.

A pooling two-type equilibrium is defined as follows:

1) both type play j.

2) observing the strategy $j_{off} \neq j$ (off-equilibrium path) the market believes that the type is I with probability μ_I and the type is S with probability $\mu_S = 1 - \mu_I$ such that $V_j(\theta_I, y) \geq V_{j_{off}}(\theta_I, \theta_{\mu_p})$ and $V_j(\theta_S, y) \geq V_{j_{off}}(\theta_S, \theta_{\mu_p})$, where $\theta_{\mu_p} = \mu_I\theta_I + \mu_S\theta_S$.

3) If for type T $\max_\theta V_{j_{off}}(\theta_T, \theta) < V_j(\theta_T, y)$ then $\mu_T = 0, T \in \{I, S\}$.

The first condition means that different types play different strategies under separating equilibrium and the same strategy under pooling. The second condition represents the non-deviation condition for each type (individual rationality). Finally, the third condition in the case of pooling equilibrium assures that the equilibrium survives the intuitive criterion of Cho and Kreps (1987). This condition means that the market off-equilibrium beliefs are reasonable in the sense that if for any type T its maximal payoff from deviation is not greater than its equilibrium payoff then the market should place the probability 0 on possible deviations of this type. The definitions above are consistent with standard PBE definition (see, for instance, Fudenberg and Tirole, 1991) with an addition of intuitive criterion which is quite common in such kind of games (see, for instance, Nachman and Noe, 1994). Finally note that Lemma 1 insures that in described above equilibria the market makes zero-profit (competitive rationality).[11]

Proposition 1. *1) The situation where I plays e and S plays d is not an equilibrium; 2) if and only if $\theta_I \leq 1/2$, there exists a separating equilibrium where I plays d and S plays e.*

Proof. (i) *Part 1.* Suppose, in opposite, that such equilibrium exists. Of course, each type would have $1 - 2b$ in a separating equilibrium. From Lemma 2 $V_e(\theta_S, \theta_I) > 1 - 2b$ because $\theta_I < \theta_S$. Thus S would deviate from its equilibrium strategy to e and such equilibrium is impossible.

---

[11] Also note that by definition of pooling, the off-equilibrium beliefs are consistent (Kreps and Willson, 1982). If out of equilibrium path the market believes that the type is θ then it keeps the same beliefs in the second period (it follows from the definition of the payoff functions V). Otherwise the market off-equilibrium beliefs would be inconsistent.

(ii) *Part 2.* Let $\theta_I \leq 1/2$. I does not mimic S. From Lemma 2 we have $V_e(\theta_I, \theta_S) \leq 1-2b$ because $\theta_I < \theta_S$. S does not mimic I. From Lemma 3, $V_d(\theta_S, \theta_I) \leq 1-2b$ because $\theta_I < 1/2$ and $\theta_I < \theta_S$. Now, if $\theta_I > 1/2$ then from Lemma 3 $V_d(\theta_S, \theta_I) > 1-2b$. Thus S would mimic I. *End proof.*

Intuitively, in the equilibrium described in Part 2, I does not deviate because by playing e it is not able to capitalize on its first-period informational advantage. The share's price in the first period does not depend on the firm's type (Lemma 1), while the interest rate in the second period will be unfavorable. S does not deviate because the interest rates profile is downward sloping or flat when $\theta_I \leq 1/2$, making d unprofitable for S (which performs better with upward sloping interest rates profile).

Proposition 2. *If and only if $y \geq 1/2$, pooling with d is an equilibrium.*

Proof. (i) *Part 1.* 1) Existence. Let $y \geq 1/2$. Consider pooling equilibrium where both types play d, which is supported by off-equilibrium market beliefs that the firm is S.[12] First of all, let us verify non-deviation for each type. Since $1/2 \leq y < \theta_S$, we gets from Lemma 3 $V_d(\theta_S, y) \geq 1-2b$. Thus the type S does not deviate. From Lemma 2, we have

$$V_e(\theta_I, y) \geq V_e(\theta_I, \theta_S) \tag{15}$$

The condition of non-deviation for the type I is obviously $V_d(\theta_I, y) \geq V_e(\theta_I, \theta_S)$. The latter follows from the condition (15) and Lemma 4:

$$V_d(\theta_I, y) \geq V_e(\theta_I, y) \geq V_e(\theta_I, \theta_S) \tag{16}$$

Let us now verify that off-equilibrium beliefs survive the intuitive criterion of Cho and Kreps (1987). To show this, let us calculate the maximal payoff of type S in the case that it plays e. Its payoff is evidently maximized if the market's belief places the probability 1 on type I observing equity, i. e. $V_e(\theta_S, \theta_I)$. If off-equilibrium beliefs survive intuitive criterion, this expression must be greater than $V_d(\theta_S, y)$.[13] It follows immediately from Lemmas 2 and 4:

$$V_e(\theta_S, \theta_I) \geq V_e(\theta_S, y) \geq V_d(\theta_S, y)$$

This completes the proof of sufficiency.

2) As for the necessary condition, if $y < 1/2$ then pooling with d is impossible because type S would deviate in e (from Lemma 3 its equilibrium payoff would be less than $1-2b$). *End proof.*

The idea behind the Proposition 2 is simple. Only if growing firms dominate the credit market ($y < 1/2$) will the interest rates profile be downward sloping, creating incentives for stagnating firms to play e.

---

[12] Note that in terms of the definition of pooling given above, we have here $j_{off} = e$, $\mu_S = 1$ and $\theta_p = \theta_S$.

[13] Otherwise μ_S should be equal to 0.

Proposition 3. *1) if $y < 1/2$, pooling with e is not an equilibrium; 2) if $y \geq 1/2$ and if pooling with e exists, then mispricing is greater under that than under pooling with d.*

Proof. see Appendix 1.[14]

Intuitively, if y is low, then in the case of pooling with e second-period interest rate is low, making high profit for the type S. In some cases this profit is even greater than maximal possible profit under the strategy d. This situation is not an equilibrium because the market should set the probability 0 on the possibility for S to play d, making off-equilibrium interest rates suitable for the type I, that would deviate to d.[15] Secondly, if pooling with e exists, then mispricing is greater than it is under pooling with d. Intuitively by analogy with Lemma 4, type I (undervalued under pooling equilibrium, because S can always achieve at least first-best using e as a last resort) prefers pooling with debt over pooling with equity. Propositions 1, 2 and 3 are at the root of two major insights of this paper; they provide clues about the link between initial distribution of types in the economy and individual firm capital structure policy, and they show why debt is a signal of a firm's increasing performance while equity is a signal of decreasing performance.

The main conclusion of the above analysis is that performance-improving firms definitely prefer debt while stagnating firms base their strategy on the macroeconomic situation—if the economy is growing, they will issue equity, and if the economy is stagnating, both strategies can lead to equilibrium. Also note that in a two-type economy, separating equilibrium always dominates pooling by minimal mispricing. However, the intuition about the existence of pooling equilibria is useful and it will be further applied in Sections 4 and 5.

4. Multiple Type Economy

To provide more ideas about the role of macroeconomic situation in this game, consider a multiple type economy, and suppose that $f(\theta) > 0, \forall\theta$ on the support $[b, 1-b]$. Here again we consider two types of equilibria: pooling and semi-separating (or two-strategy equilibrium). A two-strategy equilibrium is defined as follows:

1) There are two sets Θ_e and Θ_d such that any $\theta \in \Theta_e$ plays e and any $\theta \in \Theta_d$ plays d, $\Theta_e \cap \Theta_d = O$, $\Theta_e \cup \Theta_d = [b, 1-b]$ and the measure of probability of O equals 0.

2) $V_e(\theta, \theta^e) \geq V_d(\theta, \theta^d)$, $\forall\theta \in \Theta_e$ and $V_d(\theta, \theta^d) \geq V_d(\theta, \theta^e)$, where $\theta^e = \frac{1}{M(\Theta_e)} \int_{\theta\in\Theta_e} \theta f(\theta) d\theta$ and $\theta^d = \frac{1}{M(\Theta_d)} \int_{\theta\in\Theta_d} \theta f(\theta) d\theta$.

Note that $M(\Theta_j)$ is the measure of probability of Θ_j and θ^j denotes the expected type of firms playing $j \in \{e, d\}$. A pooling equilibrium is defined as follows:

1) all types play j.

---

[14] We use the standard concept of mispricing that can be found, for example, in Nachman and Noe (1994). The magnitude of mispricing in a given equilibrium equals to that of undervalued types. For instance, if the strategy j is played by undervalued type T (the undervaluation is only possible in pooling) then the mispricng equals $1 - 2b - V_j(T, y)$. The overvaluation of overvalued type does not matter. Note that the total profit in the economy is always the same across different equilibriums.

[15] Remark that, since $y < 1/2$ implies $\theta_I < 1/2$, a separating equilibrium exists in this case.

2) observing the strategy $j_{off} \neq j$ (off-equilibrium path) the market beliefs are $\mu^{j_{off}}$ such that $V_j(\theta, y) \geq V_{j_{off}}(\theta, \theta_{\mu^{j_{off}}}), \forall\theta$, where $\theta_{\mu^{j_{off}}} = \int \theta\mu^{j_{off}}(\theta)d\theta$.

3) if for type θ $\max_{\mu} V_{j_{off}}(\theta, \theta_{\mu}) < V_j(\theta, y)$ then $\mu^{j_{off}}(\theta) = 0$, where $\theta_{\mu} = \int \theta\mu(\theta)d\theta$.

In each case the first condition means that both strategies are played in two-strategy equilibrium and that all types play the same strategy under pooling. The second condition represents the non-deviation condition for each type (individual rationality). Finally, the third condition in the case of pooling equilibrium assures that the equilibrium survives the intuitive criterion of Cho and Kreps (1987).

Proposition 4. *1) In any two-strategy equilibrium, there exists $\theta^* \in (b, 1-b)$ such that firms with $\theta > \theta^*$ play e and firms with $\theta < \theta^*$ play d; 2) If $y \leq 1/2$ (non-declining economy), this equilibrium exists.*

(i) *Proof of part 1.* Since both strategies are played, the $V_e(\theta, \theta^e)$ and $V_d(\theta, \theta^d)$ must intersect (otherwise we would have a pooling equilibrium like in Figure 3b) and the intersection is unique since the payoffs are linear in θ, as shown in the foregoing discussion.[16]

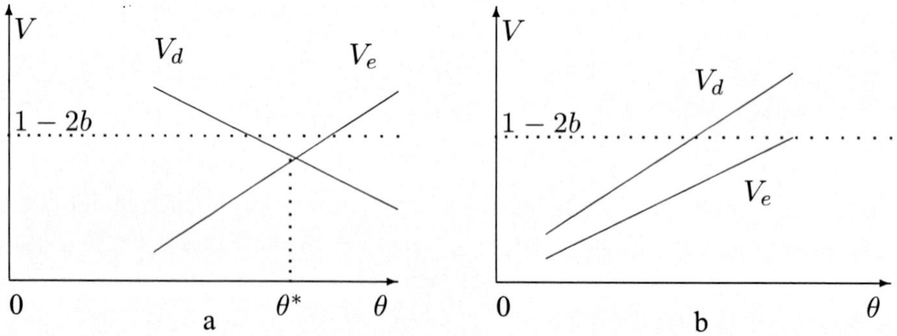

Figure 3. "Separating" and pooling equilibrium in multiple type economy.

Thus, the only candidates for a two-strategies equilibrium are either the one described in the Proposition or one where firms with $\theta > \theta^*$ issue debt and firms with $\theta < \theta^*$ issue equity. If we suppose that the latter is the case, contrary to the Proposition, it is only possible if V_d is increasing and cuts V_e from below. But the interest rates combination supporting this situation is impossible. First of all, it must be that $1 - 2b > V_e(\theta^*, \theta^e) = V_d(\theta^*, \theta^d)$. If it does not, all firms playing debt ($\theta > \theta^*$) have more than first-best, which contradicts risk-neutral valuation. But in this case equity market makes positive profit (all issuing equity firms have less than $1 - 2b$) which contradicts equilibrium's concept. The only possible equilibrium is in Figure 3a where firms with $\theta > \theta^*$ issue equity and firms with $\theta < \theta^*$ issue debt.

[16] Recall that θ^j is the expected type of firms playing $j = e, d$.

(ii) *Proof of part 2.* Let

$$\theta^d(\theta^*) = \frac{1}{F(\theta^*)} \int_b^{\theta^*} \theta f(\theta) d\theta$$

and

$$\theta^e(\theta^*) = \frac{1}{(1 - F(\theta^*))} \int_{\theta^*}^{1-b} \theta f(\theta) d\theta$$

If $y \leq 1/2$, then for any $\theta^* \in (b, 1 - b)$, we have $b < \theta^d(\theta^*) < 1/2$ and:

$$V_d(b, \theta^d(\theta^*)) > 1 - 2b > V_e(b, \theta^e(\theta^*))$$

Similarly, since $\theta^d(\theta^*) < 1/2 < 1 - b$ and $\theta^e(\theta^*) < 1 - b$

$$V_d(1 - b, \theta^d(\theta^*)) < 1 - 2b < V_e(1 - b, \theta^e(\theta^*))$$

Now define $\psi(\theta^*) = V_d(\theta^*, \theta^d(\theta^*)) - V_e(\theta^*, \theta^e(\theta^*))$. From the previous results $\psi(b) > 0 > \psi(1 - b)$, hence from the intermediate value theorem there exists $\theta^* \in (b, 1 - b)$ such that $\psi(\theta^*) = 0$. This proves existence because $V_e(\theta, \theta^e(\theta^*))$ is increasing in θ, while $V_d(\theta, \theta^d(\theta^*))$ is decreasing in θ when $\theta^d(\theta^*) < 1/2$. *End proof.*

Because the first-best share price in the first period is the same for all types, the types with high θ benefit from their informational advantages in the second period (when in actuality they are really "lemons"). Hence, intuitively the separation equilibrium described in Proposition 4 should exist if the payoff in the given debt strategy is decreasing in θ. A sufficient condition for existence of this equilibrium is that $y \leq 1/2$, which means that the economy is not declining on average. Also, in this case, pooling with debt is not an equilibrium. The intuition for this condition is the following: pooling with debt is unprofitable for stagnating firms because, since there are more growing firms, interest rate profile corresponds more with them and, as we know, stagnating firms would lose by playing debt. Thus, stagnating firms tend to signal their type by playing equity. This leads to the following result.

Proposition 5. *1) If and only if $y \geq 1/2$, pooling with d is an equilibrium; 2) if $y < 1/2$, pooling with e is not an equilibrium; 3) if $y \geq 1/2$ and if pooling with e exists, then mispricing is greater under that than under pooling with d.*

Proof. see Appendix 2.

As we can see, the only qualitative difference with the basic model is the fact that here the existence of "separation" equilibrium, outlined in Proposition 4, is subject to the condition that the economy is non-declining. This is not surprising because in a two-type model, like in the one described in Section 3, the grower firm is also growing absolutely, which implies that the interest rate profile is necessarily downward sloping, making it unprofitable for the stagnating firm to mimic this type by playing debt. In the multiple type case, the condition $y \leq 1/2$ insures that in the case of "separation", the equilibrium interest rate profile would necessarily be downward sloping.

5. Conclusion

Let us summarize the analysis of the model. Two equilibriums may exist: separating and pooling. A separating equilibrium exists if and only if the economy is growing on average. In separating equilibrium firms issuing equity have higher performance in the first period and lower performance in the second period. Also equity may be issued only in the case of separating equilibrium because pooling with debt dominates pooling with equity by minimal mispricing when economy is stagnating. One can see that these results contrast the standard GB approach ("good" and "bad" firms) where the solution is typically pooling with debt. We believe that the idea of signalling the performance profile over time (in contrast or in addition) to the signalling about total performance provides an interesting insight which can be applied in other fields of research as well.

Also, the results of the paper are consistent with several financial phenomena:

(i) Firms issuing equity underperform in the long-run as compared to non-issuing firms (measured as a decline of profit, profit to assets ratio or profit per share). This is implied by Propositions 1 and 4: in any equilibrium, where both debt and equity are issued, only the types with low second period profit issue equity in the first period.[17] At the same time the performance of firms issuing equity exceeds the performance of the non-issuing firms at the time of issue (or in the near future after issue). Clearly, this also follows from Propositions 1 and 4.[18] Similarly, the model predicts that leverage is negatively correlated with profitability. In separating equilibrium, first-period low-profitable firms issue debt.[19]

(ii) This paper suggests a new motive for issuing equity (Propositions 1 and 4) that has not been explored in existing literature. When the firm knows that it will be high-profitable in the near future and low-profitable in the long-term, the entrepreneur may want to issue equity.

(iii) This model provides a rationale for the link between debt-equity choice and business cycle. The analysis of the basic model reveals the following ideas. Growing firms prefer debt. The incentives for stagnating firms depend on the macroeconomic situation. If the economy is in contraction, stagnating firms may prefer debt, but will necessarily issue equity when the economy is growing (Propositions 1, 2, 3 and 4). Thus, equity issues seem to be procyclical.

Acknowledgments

Claude Fluet, Thomas Noe, Leon Petrosjan and P. Viswanath have given us especially useful comments on earlier versions of this paper. We are also grateful to seminar participants at the University of Guelph, ESADE, University of Bonn, UQAM, the Bank of Canada, the 2003 MWF, the 2002 CIRPEE and the 2003 SCSE meetings for their suggestions and comments. The financial support awarded by the Social Sciences and Humanities Research Council of Canada and the Institut de finance mathématique de Montréal was instrumental

[17] This conclusion is confirmed by empirical findings (see for example [5], [11], [15] or [16]).

[18] While this point was not the main focus of the empirical research cited above, some authors did stress the point that issuing firms outperform non-issuing firms just before issue, and others documented that issuing firms outperform non-issuing firms in the year of issue and in the first year after issue (see [11] and [16]).

[19] This is consistent, for example, with [9], [20] and [21].

in enabling our continued research. We also appreciate the editing assistance of Kaarla Sundström.

Appendix 1

(i) *Proof of part 1.* We will show that pooling with e does not survive intuitive criterion. Consider pooling with e and assume that if a type deviates to d, it is perceived as the type $\widehat{\theta}_d$. It is enough to show that

$$\min_y V_e(\theta_S, y) \geq \max_{\widehat{\theta}} V_d(\theta_S, \widehat{\theta}_d) \tag{17}$$

where $y < 1/2$ and $\theta_I \leq \widehat{\theta}_d \leq \theta_S$. 1) $\theta_S < 1/2$. From Lemma 3, $V_d(\theta_S, \widehat{\theta}_d) < 1 - 2b$. Since by Lemma 2 $V_e(\theta_S, y) > 1 - 2b$, this proves (17). 2) $\theta_S \geq 1/2$. From (9) minimal value of $V_e(\theta_S, y)$ on feasible support is obtained when $y = 1/2$ (and equals $\frac{1-2b}{1-b}(1 - 2b + 2b\theta_S)$). From (10)

$$V_d(\theta_S, \widehat{\theta}_d) = 1 - b(\frac{\theta_S}{\widehat{\theta}_d} + \frac{1 - \theta_S}{1 - \widehat{\theta}_d}) \tag{18}$$

Local minimum of expression in brackets under positive $\widehat{\theta}_d$ is $\widehat{\theta}_{d\min} = \frac{\theta_S - \sqrt{\theta_S(1-\theta_S)}}{2\theta_S - 1}$. Thus, the maximal payoff of S, if it were playing d, would be:

$$1 - b(2\theta_S - 1)(\frac{\theta_S}{\theta_S - \sqrt{\theta_S(1 - \theta_S)}} + \frac{1 - \theta_S}{\sqrt{\theta_S(1 - \theta_S)} - 1 + \theta_S})$$

Hence (17) is equivalent to the following:

$$\frac{1 - 2b}{1 - b}(1 - 2b + 2b\theta_S) > 1 - b(2\theta_S - 1)(\frac{\theta_S}{\theta_S - \sqrt{\theta_S(1 - \theta_S)}} + \frac{1 - \theta_S}{\sqrt{\theta_S(1 - \theta_S)} - 1 + \theta_S}) \Leftrightarrow$$

$$(1 - b)(2\theta_S - 1)^2\sqrt{\theta_S(1 - \theta_S)} - (2\theta_S^2 - 2\theta_S + \sqrt{\theta_S(1 - \theta_S)})(3 - 4b + 4b\theta_S - 2\theta_S) > 0$$

Since $\theta_S \geq 1/2$, the left side is decreasing in b. Thus it is enough to show that this inequality holds under $b = 1 - \theta_S$ (because $1 - b \geq \theta_S$). It can be verified that $\theta_S(2\theta_S - 1)^2\sqrt{\theta_S(1 - \theta_S)} - (2\theta_S^2 - 2\theta_S + \sqrt{\theta_S(1 - \theta_S)})(6\theta_S - 4\theta_S^2 - 1) > 0$ on feasible support of θ_S. Thus, the market would attribute the probability 0 to the possibility of playing d by S. The market would believe that firm is type I after observing debt. In this case, type I would certainly deviate from pooling with e to d.

(ii) *Part 2* follows from Lemma 4. *End proof.*

Appendix 2

(i) *Proof of part 1.* 1) Sufficiency. Consider pooling equilibrium where all types play d, which is supported by off-equilibrium market beliefs that the firm is the type $1 - b$. First of all, let us verify non-deviation for each type. Consider the case $\theta \geq y$. Since $1/2 \leq y \leq \theta$, we gets from Lemma 3 $V_d(\theta, y) \geq 1 - 2b$. From Lemma 2 we have

$V_e(\theta, 1-b) < V_e(\theta, \theta) = 1 - 2b$. Thus the type θ does not deviate. Now, consider the case $\theta < y$. From Lemma 2, we have

$$V_e(\theta, y) \geq V_e(\theta, 1-b) \tag{19}$$

The condition of non-deviation for the type θ is obviously $V_d(\theta, y) \geq V_e(\theta, 1-b)$. Given the condition (19), is is enough to show that $V_d(\theta, y) \geq V_e(\theta, y)$. This obviously follows from Lemma 4.

Let us now verify that off-equilibrium beliefs survive the intuitive criterion of Cho and Kreps (1987). To show this, let us calculate the maximal payoff of type $1-b$ in the case that it plays e. Its payoff is evidently maximized if the market's belief places the probability 1 on type b observing equity, i. e. $V_e(1-b, b)$. If off-equilibrium beliefs survive intuitive criterion, this expression must be greater than $V_d(1-b, y)$. From Lemmas 2 and 4 we get immediately:

$$V_e(1-b, b) \geq V_e(1-b, y) > V_d(1-b, y)$$

This completes the proof of sufficiency.

2) As for the necessary condition, if $y < 1/2$ then pooling with d is impossible because type $1-b$ would deviate in e (its equilibrium payoff would be less than $1-2b$).

(ii) *Proof of part 2.* We will show that pooling with e does not survive intuitive criterion when $y < 1/2$. First, for θ such that $\theta > y$ it can be shown (analogously to Proposition 4) that

$$\min_y V_e(\theta, y) \geq \max_{\widehat{\theta}} V_d(\theta, \widehat{\theta}) \tag{20}$$

if $y < 1/2$. Thus, the market would attribute the probability 0 to the possibility of playing d by θ if $\theta > y$. Evidently in this case $\theta^d < y$. Thus $V_d(\theta^d, \theta^d) = 1 - 2b$. Also, $V_e(\theta^d, y) < 1 - 2b$. Thus θ^d would deviate and pooling with e is impossible.

(iii) *Proof of part 3.* Let $y \geq 1/2$. According to part 1, pooling with d exists. Assume that pooling with e exists. From Lemma 2 $V_e(\theta, y) < 1 - 2b$ for and only for $\theta < y$. As implied by Lemma 4 for such types $V_d(\theta, y) \geq V_e(\theta, y)$. Thus mispricing under pooling with e is larger that under pooling with d. *End proof.*

References

[1] G. Akerloff, The Market for Lemons: Quality Incertainty and the Market Mechanism, *Quarterly Review of Economics* **74**, no.3 (1970), 488–500.

[2] M. Brennan, A. Kraus, Efficient Financing under Asymmetric Information, *The Journal of Finance XLII*, no. **5** (1987) december, 1225–1243.

[3] I. Brick, M. Frierman, Y.K. Kim, Asymmetric information Concerning the Variance of Cash Flows: The Capital Structure Choice, *International Economic Review* **39**, no. 3 (1998) (Aug.), 745–761.

[4] C.B. Cadsby, M. Frank, V. Maksimovic, Equilibrium Dominance in Experimental Financial Markets.*The Review of Financial Studies*, **11**, no. 1 (1998)(Spring), 189–232.

[5] J. Cai, K. Wei, The Investment and Operating Performance of Japanese Initial Public Offerings, *Pacific-Basin Finance Journal* **5** (1997), 389–417.

[6] I.K. Cho, D. Kreps, Signalling Games and Stable Equilibria, *Quarterly Journal of Economics* **102 (2)** (1987), 179–221.

[7] A. Dittmar, J. Mahrt-Smith, H. Servaes, International Corporate Governance and Corporate Cash Holdings, *Journal of Financial and Quantitative Analysis* **38**, no. 1 (2003)(march), 111–133.

[8] F. Easterbrook, Two Agency-cost Explanation of Dividends, *American Economic Review* **74 (4)** (1984), 650–659.

[9] E. Fama, K. French, Testing Trade-off and Pecking Order Predictions about Dividends and Debt, *The Review of Financial Studies* **15**, no.1 (1001) ,1–33.

[10] D. Fudenberg, J. Tirole, *Game Theory*, Cambridge, Mass. : MIT Press, 1991.

[11] B. Jain, O. Kini, The Post-Issue Operating Performance of IPO Firms,*The Journal of Finance XLIX*, no. **5** (1994), December, 1699–1726.

[12] M.C. Jensen, Agency Cost of Free Cash-flow, Corporate Finance and Takeovers, *American Economic Review* **76**, no. 2 (1986), 323–329.

[13] D.M. Kreps, R. Wilson, Sequential Equilibria, *Econometrica* **50**, no.4 (1982), July, 863–894.

[14] L. Lang, R. Litzenberger, Dividend Announcements: Cash Flow Signalling vs. Free Cash Flow Hypothesis?, *Journal of Financial Economics* **24** (1989), North-Holland, 181–191.

[15] T. Loughran, J. Ritter, The Operating Performance of Firms Conducting Seasoned Equity Offerings, *Journal of Finance* **52**, no.5 (1997) Dec., 1823–1850.

[16] W. Mickelson, M. Partch, K. Shah, Ownership and Operating Performance of Companies That Go Public, *Journal of Financial Economics* **44** (1997), 281–307.

[17] S.C. Myers, N.S. Majluf, Corporate Financing and Investment Decisions When Firms Have Information That Investors Do not have,*Journal of Financial Economics* **13(2)** (1984), June, 187–221.

[18] D.C. Nachman, T.H. Noe, Optimal Design of Securities Under Asymmetric Information, *The Review of Financial Studies, Spring* **7**, no.1 (1994), 1–44.

[19] L.A. Petrosian, N. A. Zenkevich, Game Theory, Singapore; River Edge, N. J.; *World Scientific, Series on optimization* **3** (1996).

[20] R.G. Rajan, L. Zingales, What Do We Know About Capital Structure? Some evidence from international data, *Journal of Finance* **50** (1995), 1421–1460.

[21] S. Titman, R. Wessels, The Determinants of Capital Structure Choice, *Journal of Finance* **43** (1988), 1–19.

[22] N.A. Zenkevich, S. N. Voznyuk, A Game-Theoretic Model of Divisible Goods Bargaining,*Game Theory and Applications* **1** (1996), New York, USA. Nova Science Publishers Inc., 201–208.

[23] N.A. Zenkevich, S. Huang, *Comparison of Two Economic Models for a Business-to-Business Exchange*, ICM Millenium Lectures on Games, Springer-Verlag, Berlin, Heidelberg, 2003, 335–345.

In: Mathematics, Game Theory and Algebra Compendium
Editor: Jacob H. Mathias, pp. 319-333
ISBN 978-1-60456-659-8

Chapter 29

THREE-PLAYER GAME OF 'KEEP-OR-EXCHANGE'

Minoru Sakaguchi
Osaka University
Osaka, Japan

AMS Subject Classification: 60G40, 90C39, 90D10.

1. Three-Player Games of ' Score Showdown'

Consider the three players I, II and III (sometimes they are denoted by 1,2 and 3). Let $X_{ij}(i = 1, 2, 3; j = 1, 2)$ be the random variable (r.v.) observed by player i at the j-th observation. We assume that X_{ij}'s are i.i.d., each with uniform distribution in $[0, 1]$. The game is played in the three stages.

In the first stage, I observes that $X_{11} = x$ and chooses one of either A_1 (*i.e.*, I accepts x) or R_1 (*i.e.*, I rejects x and resamples a new r.v. X_{12}). The observed value x and I's choice of either A_1 or R_1 are informed to II and III. But the observed value of X_{12} is not informed to II and III.

In the second stage, after knowing I's choice of $x\&(A_1 \cup R_1)$, II observes that $X_{21} = y$, and chooses either one of A_2 (*i.e.*, II accepts y) or R_2 (*i.e.*, II rejects y and resamples a new r.v. X_{22}). The observed value y and II's choice of either A_2 or R_2 are informed to III. But the observed value of X_{22} is not informed to III.

In the third stage, after knowing I's choice of $x\&(A_1 \cup R_1)$ and II's choice of $y\&(A_2 \cup R_2)$, III observes that $X_{31} = z$ and chooses either one of A_3 (*i.e.*, III accepts z) or R_3 (*i.e.*, III rejects z and resamples a new r.v. X_{32}).

Let, for $i = 1, 2, 3$,

$$S_i(X_{i1}, X_{i2}) = \left\{ \begin{array}{c} X_{i1} \\ X_{i2} \end{array} \right. , \text{ if } X_{i1} \text{ is} \left\{ \begin{array}{c} \text{accepted} \\ \text{rejected} \end{array} \right. \text{by player } i, \qquad (1)$$

which we call the *score* for player i.

After the third stage is over, showdown is made, the scores are compared, and the player with the highest score among the players becomes the *winner*. Each player aims to maximize the probability of his (or her) winning. We assume that all players are intelligent, and each player should prepare for that any subsequent player must use their optimal strategies.

The three-player game of ‘ Keep-or-Exchange’ (*i.e.*, the score is defined by (1.1)) is solved in Section 3. The solution is found to be very complicate far more than expected. It is compared with that of the two-player case, given in Section 2. In [8] the other two-player games of ‘ Competing Average’, where the score is

$$S_i(X_{i1}, X_{i2}) = \begin{cases} X_{i1} \\ \frac{1}{2}(X_{i1} + X_{i2}) \end{cases}, \text{ if } \begin{cases} X_{i1} \text{ is accepted} \\ X_{i2} \text{ is resampled,} \end{cases} \tag{2}$$

and ‘ Showcase Showdown’ where the score is

$$S_i(X_{i1}, X_{i2}) = \begin{cases} X_{i1} \\ (X_{i1} + X_{i2})I(X_{i1} + X_{i2} \le 1) \end{cases}, \text{ if } \begin{cases} X_{i1} \text{ is accepted} \\ X_{i2} \text{ is resampled,} \end{cases} \tag{3}$$

are solved. The three-player game versions of these two-player games remain to be solved as yet. See also [1∼7].

Intuitively it would seem in the three-player games, that the last-mover has an advantage over the middle-mover, and the middle-mover, in turn, has an advantage over the first-mover. Theorem 2 in the present paper shows that this intuition is correct in ‘ Keep-or-Exchange’, where the score is (1.1). It is an interesting work to investigate whether the counter examples do exist or not.

Three-player simultaneous-move game is solved in Section 4. A conjecture for the solution to the four-player game is given. We observe that player behaves more cautious as he has more competitors.

2. Keep-or-Exchange — Two-Player Game

First we solve the two-player game. We will find, in the next section, that the three-player game is surprisingly complicate to solve, compared with in the two-player case.

Let $W_i (i = 1, 2)$ be the event that player i wins. To find the players' optimal strategies we must derive them in reverse order. Define state $\begin{Bmatrix} (y|x, A_1) \\ (y|x, R_1) \end{Bmatrix}$ for II, to mean that I $\begin{Bmatrix} \text{accepted} \\ \text{rejected} \end{Bmatrix} X_{11} = x$ in the first stage and II has just observed $X_{21} = y$ in the second stage. Then we have

$$p_{2A}(y|x, A_1) = P\{W_2|\text{II accepts } X_{21} = y \text{ in state } (y|x, A_1)\} = I(y > x), \tag{4}$$

$$\begin{aligned} p_{2R}(y|x, A_1) &= P\{W_2|\text{II rejects } X_{21} = y \text{ in state } (y|x, A_1)\} \\ &= P(X_{22} > x) = \bar{x} \equiv 1 - x, \quad \text{indep.of } y, \end{aligned} \tag{5}$$

$$\begin{aligned} p_{2A}(y|x, R_1) &= P\{W_2|\text{II accepts } X_{21} = y \text{ in state } (y|x, R_1)\} \\ &= P(X_{12} < y) = y, \quad \text{indep.of } x, \end{aligned} \tag{6}$$

and

$$\begin{aligned} p_{2R}(y|x, R_1) &= P\{W_2|\text{II rejects } X_{21} = y \text{ in state } (y|x, R_1)\} \\ &= P(X_{12} < X_{22}) = \frac{1}{2}, \quad \text{indep.of } x \text{ and } y, \end{aligned} \tag{7}$$

THEOREM 2.2 *The solution to the two-player game with the score function* (1.1) *is as follows. The optimal strategy for* I *in the first stage is given by:*

$$\textit{Accept (Reject)} X_{11} = x, \quad \textit{if } x > (<) \sqrt{3/8} \approx 0.6124. \tag{8}$$

The optimal strategy for II *in the second stage is given by*

$$\textit{Accept (Reject)} \ X_{21} = y, \ \textit{if } y > (<) \left\{ \begin{array}{c} x \\ 1/2 \end{array} \right\} \ \textit{in state} \ \left\{ \begin{array}{c} (y|x, A_1) \\ (y|x, R_1) \end{array} \right\}. \tag{9}$$

The optimal values are

$$\begin{aligned} P(W_1) &= \frac{1}{3}\left\{1 + 2\,(3/8)^{3/2}\right\} \approx 0.4864 \\ P(W_2) &= 1 - P(W_1) = \frac{2}{3}\left\{1 - (3/8)^{3/2}\right\} \approx 0.5136. \end{aligned} \tag{10}$$

Proof is given in [8].

3. Keep-or-Exchange — Three-Player Game

Let W_i be the event that player i wins. To find the players' optimal strategies, we must derive them in reverse order. Define state $\left\{ \begin{array}{c} (z|xA_1, yA_2) \\ (z|xA_1, yR_2) \end{array} \right\}$ for III, to mean that I accepted $X_{11} = x$ in the first stage, II $\left\{ \begin{array}{c} \text{accepted} \\ \text{rejected} \end{array} \right\}$ $X_{21} = y$ in the second stage, and III has just observed $X_{31} = z$ in the third stage. Also define the other two states $(z|xR_1, yR_2)$ and $(z|xR_1, yA_2)$ similarly. Then we easily find that

$$\begin{aligned} p_{3A}(z|xA_1, yA_2) &\equiv P\{W_3|\text{III accepts } X_{31} = z \text{ in state } (z|xA_1, yA_2)\} \\ &= I(z > y), \quad \text{(since II behaves optimally)} \end{aligned} \tag{11}$$

$$\begin{aligned} p_{3R}(z|xA_1, yA_2) &\equiv P\{W_3|\text{III rejects } X_{31} = z \text{ in state } (z|xA_1, yA_2)\} \\ &= P(X_{32} > y) = \bar{y}, \end{aligned} \tag{12}$$

$$\begin{aligned} p_{3A}(z|xA_1, yR_2) &\equiv P\{W_3|\text{III accepts } X_{31} = z \text{ in state } (z|xA_1, yR_2)\} \\ &= I(z > x)P(z > X_{22}) = zI(z > x), \end{aligned} \tag{13}$$

$$\begin{aligned} p_{3R}(z|xA_1, yR_2) &\equiv P\{W_3|\text{III rejects } X_{31} = z \text{ in state } (z|xA_1, yR_2)\} \\ &= P\{X_{32} > (x \vee X_{22})\} = \frac{1}{2}(1 - x^2), \end{aligned} \tag{14}$$

$$p_{3A}(z|xR_1, yR_2) \equiv P\{W_3|\text{III accepts } X_{31} = z \text{ in state } (z|xR_1, yR_2)\} \quad (15)$$
$$= P(z > X_{12} \vee X_{22}) = z^2,$$

$$p_{3R}(z|xR_1, yR_2) \equiv P\{W_3|\text{III rejects } X_{31} = z \text{ in state } (z|xR_1, yR_2)\} \quad (16)$$
$$= 1/3, \quad \forall(x, y, z),$$

and

$$p_{3A}(z|xR_1, yA_2) \equiv P\{W_3|\text{III accepts } X_{31} = z \text{ in state } (z|xR_1, yA_2)\} \quad (17)$$
$$= I(z > y)P(z > X_{12}) = zI(z > y),$$

$$p_{3R}(z|xR_1, yA_2) \equiv P\{W_3|\text{III rejects } X_{31} = z \text{ in state } (z|xR_1, yA_2)\} \quad (18)$$
$$= P\{X_{32} > (y \vee X_{12})\} = \frac{1}{2}(1 - y^2).$$

THEOREM 3.3 *The solution to the three-player game with the score function* (1.1) *is as follows. The optimal strategy for* I *in the first stage is given by:*

$$\textit{Accept (Reject) } X_{11} = x, \textit{ if } x > (<) x_0 = c^{1/4} \approx 0.68774, \quad (19)$$

where $c \approx 0.22372$ *is a four-order polynomial of* $k^{1/3} \equiv \left(\frac{9-\sqrt{3}}{27}\right)^{1/3} \approx 0.64568$, *given by* (3.24). *The optimal strategy for* II *in the second stage is:*

$$\textit{Accept (Reject) } X_{21} = y, \quad (20)$$

$$\textit{if } y > (<) \begin{cases} y_0(x) \\ k^{1/3} \approx 0.64568 \end{cases}, \textit{ in state } \begin{cases} (y|xA_1) \\ (y|xR_1) \end{cases}$$

where $y_0(x) \equiv \sqrt{h^-(x)}I(0 < x \leq \sqrt{2}-1) + \sqrt{h^+(x)}I(\sqrt{2}-1 < x \leq \xi) + xI(\xi < x \leq 1)$, $h^-(x) \equiv \frac{1}{8}(3 - 2x^2 + 3x^4)$, $h^+(x) \equiv \frac{1}{2}(1 - x + x^2 - x^3)$, $k \equiv \frac{9-\sqrt{3}}{27} \approx 0.26918$ *and* $\xi \approx 0.54368$ *is in a unique root in* $(0, 1)$ *of the equation* $x^3 + x^2 + x - 1 = 0$ *(See Figure 2). Note that* $y_0(x) \geq x, \forall x \in (0, 1)$. *The optimal strategy for* III *in the third stage is given by:*

$$\textit{Accept (Reject) } X_{31} = z, \quad (21)$$

$$\textit{if } z > (<) \begin{cases} y \\ \frac{1}{2}(1 - x^2) \vee x \\ 1/\sqrt{3} \approx 0.57735 \\ \frac{1}{2}(1 - y^2) \vee y \end{cases}, \textit{ in state } \begin{cases} (z|xA_1, yA_2) \\ (z|xA_1, yR_2) \\ (z|xR_1, yR_2) \\ (z|xR_1, yA_2). \end{cases}$$

The optimal value for the three players are

$$P(W_1) \approx 0.32309, \quad P(W_2) \approx 0.33270 \quad \textit{and} \quad P(W_3) \approx 0.34421. \tag{22}$$

Proof. The theorem is proven in the four steps.

Step 1. First we derive the optimal strategy for III in the third stage. Let $P(W_3|xA_1, yA_2)$ be the conditional probability of III's winning given that I accepted $X_{11} = x$, and II accepted $X_{21} = y$. Also let $P(W_3|xA_1, yR_2)$ and the two others denote the analogous conditional probabilities. Then, from the four pairs of (3.1)-(3.2), (3.3)-(3.4), (3.5)-(3.6) and (3.7)-(3.8), we find respectively that

$$\begin{aligned} P(W_3|xA_1, yA_2) &= \int_0^1 \{p_{3A}(z|xA_1, yA_2) \vee p_{3R}(z|xA_1, yA_2)\}\, dz \\ &= \int_0^1 \{I(z > y) \vee \bar{y}\}\, dy = y\bar{y} + \bar{y} = 1 - y^2. \end{aligned} \tag{23}$$

$$\begin{aligned} P(W_3|xA_1, yR_2) &= \int_0^1 \{p_{3A}(z|xA_1, yR_2) \vee p_{3R}(z|xA_1, yR_2)\}\, dz \\ &= \int_0^1 \left\{zI(z > x) \vee \frac{1}{2}(1 - x^2)\right\} dz \end{aligned} \tag{24}$$

$$= \frac{1}{8}(5 - 2x^2 + x^4)I(x \le \sqrt{2} - 1) + \frac{1}{2}(1 + x)(1 - x^2)I(x > \sqrt{2} - 1)$$

(*cf.*, $\frac{1}{2}(1 - x^2) < x \Longleftrightarrow x > \sqrt{2} - 1$, and see Figure 1.)

$$\begin{aligned} P(W_3|xR_1, yR_2) &= \int_0^1 \{p_{3A}(z|xR_1, yR_2) \vee p_{3R}(z|xR_1, yR_2)\}\, dz \\ &= \int_0^1 \left(z^2 \vee \frac{1}{3}\right) dz = (9 + 2\sqrt{3})/27 \approx 0.46163, \end{aligned} \tag{25}$$

$$\begin{aligned} P(W_3|xR_1, yA_2) &= \int_0^1 \{p_{3A}(z|xR_1, yA_2) \vee p_{3R}(z|xR_1, yA_2)\}\, dz \\ &= \int_0^1 \left\{zI(z > y) \vee \frac{1}{2}(1 - y^2)\right\} dz, \end{aligned} \tag{26}$$

which is the same as (3.14) with x replaced by y.

When we compute $p_{3A}(\cdot) \vee p_{3R}(\cdot)$ for each of the four states, we find that the optimal strategy for III in the third stage is given by (3.11). Figure 1 shows its derivation for the case of (3.14).

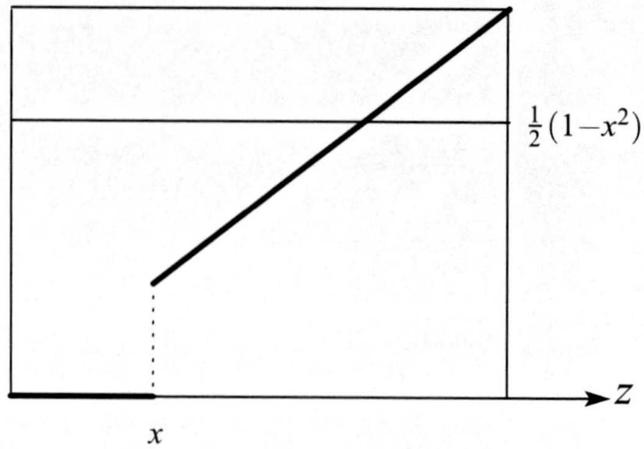

Figure 1. a. Eq.(3.14), where $x \le \sqrt{2}-1$.

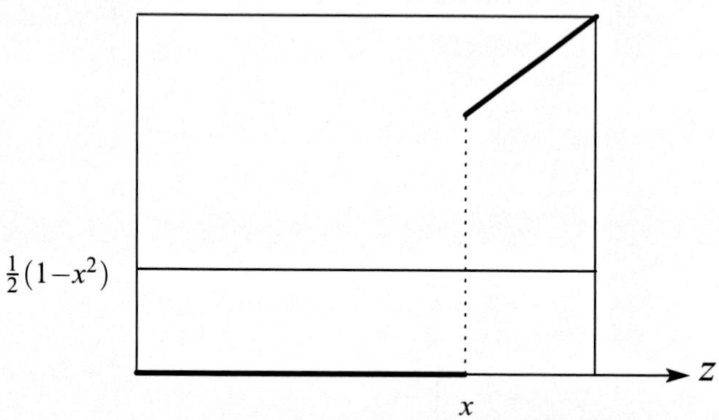

Figure 1. b. Eq.(3.14), where $x > \sqrt{2}-1$.
(Bold line corresponds to $p_{3A}(\cdot)$)

Step 2. By referring to III's optimal strategy given by (3.11), we can obtain the conditional probabilities of II's winning as bellow. Define state $(y|xA_1)$ for II to mean that I accepted $X_{11}=x$ in the first stage, and II has just observed $X_{21}=y$ in the second stage. The other state $(y|xR_1)$ is defined similarly. Then

$$\begin{aligned} p_{2A}(y|xA_1) &\equiv P\left[W_2|\,\{\text{II accepts } X_{21}=y \text{ in state } (y|xA_1)\}\right. \\ &\quad \left.\cap\{\text{III will behave optimally in the third stage}\}\right] \\ &= I(y>x)P(y>X_{31}\vee X_{32}) = I(y>x)y^2, \end{aligned} \tag{27}$$

$$\begin{aligned} p_{2R}(y|xA_1) &\equiv P\left[W_2|\,\{\text{II rejects } X_{21}=y \text{ in state } (y|xA_1)\}\right. \\ &\quad \left.\cap\{\text{III will behave optimally in the third stage}\}\right] \end{aligned} \tag{28}$$

$$= \begin{cases} P\{X_{22}>X_{31}>\frac{1}{2}(1-x^2)\}+P\{X_{31}<\frac{1}{2}(1-x^2), X_{22}>x\vee X_{32}\}, & \text{if } x\le\sqrt{2}-1 \\ P\{X_{22}>x\vee X_{32}, X_{31}<x\}+P\{X_{22}>X_{31}>x\}, & \text{if } x>\sqrt{2}-1 \end{cases}$$

$$= \frac{1}{8}(3-2x^2+3x^4)I(x\le\sqrt{2}-1)+\frac{1}{2}(1-x+x^2-x^3)I(x>\sqrt{2}-1) \equiv h(x), \text{ say.}$$

(Figure 2 shows the function $h(x)$) and also, for the other two conditional probabilities, we have from (3.11) again

$$p_{2A}(y|xR_1) = \begin{cases} P\{X_{31} < \frac{1}{2}(1-y^2), y > X_{12} \vee X_{32}\}, & \text{if } y \le \sqrt{2}-1 \\ P\{y > X_{12} \vee X_{31} \vee X_{32}\}, & \text{if } y > \sqrt{2}-1 \end{cases} \quad (29)$$

$$= \frac{1}{2}y^2(1-y^2)I(y \le \sqrt{2}-1) + y^3 I(y > \sqrt{2}-1).$$

$$p_{2R}(y|xR_1) = P\left\{X_{22} > X_{12}, X_{22} > X_{31} > \frac{1}{\sqrt{3}}\right\} \quad (30)$$

$$+P\left\{X_{31} < \frac{1}{\sqrt{3}}, X_{22} > X_{12} \vee X_{32}\right\}$$

$$= \frac{9-4\sqrt{3}}{27} + \frac{1}{3\sqrt{3}} = \frac{9-\sqrt{3}}{27} \approx 0.26918$$

in which the first probability is obtained by computing $\int_{1/\sqrt{3}}^{1} t\left(t - \frac{1}{\sqrt{3}}\right) dt.$

Let $P(W_2|xA_1)$ be the conditional probability of II's winning given that I accepted $X_{11} = x$, and II and III behave optimally subsequently. Also let $P(W_2|xR_1)$ denote another one similarly. We notice, by Figure 2, that

$$h(x) - x^2 > (=,<)\, 0, \text{ if } x < (=,>)\, \xi \approx 0.54368$$

where ξ is a unique root of the equation $h(x) = x^2$, *i.e.*, $1 - x - x^2 - x^3 = 0$. Therefore, from (3.17)-(3.18),

(3.21)

$$P(W_2|xA_1) = \begin{cases} \int_0^1 (y^2 \vee h^-(x))dy = \frac{1}{3}\left\{1 + 2(h^-(x))^{3/2}\right\}, & \text{if } x < \sqrt{2}-1 \\ \int_0^1 (y^2 \vee h^+(x))dy = \frac{1}{3}\left\{1 + 2(h^+(x))^{3/2}\right\}, & \text{if } \sqrt{2}-1 < x < \xi \\ xh^+(x) + \frac{1}{3}(1-x^3), & \text{if } x > \xi \end{cases}$$

where $h^-(x) \equiv \frac{1}{8}(3 - 2x^2 + 3x^4)$ and $h^+(x) \equiv \frac{1}{2}(1 - x + x^2 - x^3)$. See Figures 2 and 3.

Also, from (3.19)-(3.20), we have

$$P(W_2|xR_1) = \int_0^1 \{p_{2A}(y|xR_1) \vee p_{2R}(y|xR_1)\}\, dy \quad (22)$$

$$= \int_0^1 \left[\left\{\frac{1}{2}y^2(1-y^2)I(y \le \sqrt{2}-1) + y^3 I(y > \sqrt{2}-1)\right\} \vee k\right] dy$$

$$= \frac{1}{4}\left(1 + 3k^{4/3}\right) \approx 0.38035,$$

by using $\int_0^1 (y^3 \vee k)dy = \frac{1}{4}\left(1 + 3k^{4/3}\right)$, and $k \equiv \frac{9-\sqrt{3}}{27} \approx 0.26918$. See Figure 4.

Computations in (3.21) and (3.22) show that the optimal strategy for II in the second stage is given by (3.10) stated in the theorem.

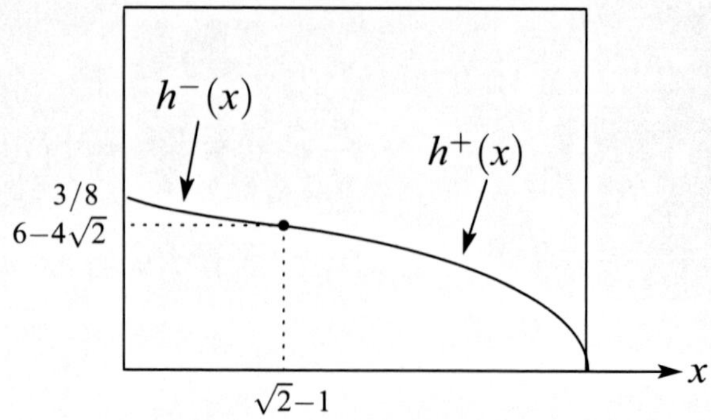

Figure 2. The function $h(x)$.

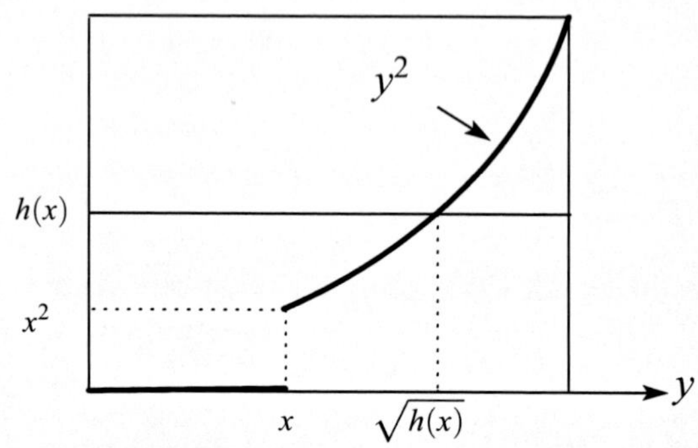

Figure 3. a. Eq.(3.21) where $x \leq \xi$

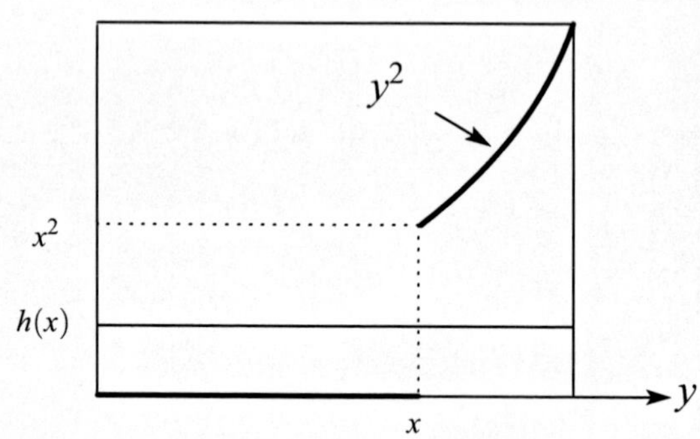

Figure 3. b. Eq.(3.21) where $x > \xi$.

Step 3. We want to derive the optimal strategy for I in the first stage. By referring to II's optimal behavior given by (3.10) and III's optimal behavior given by (3.11), the conditional probability

$$p_{1A}(x) \equiv P[W_1|\{\text{I accepts } X_{11} = x\} \cap\{\text{II and III behave optimally in the subsequent stages}\}]$$

of II's winning is found to be;

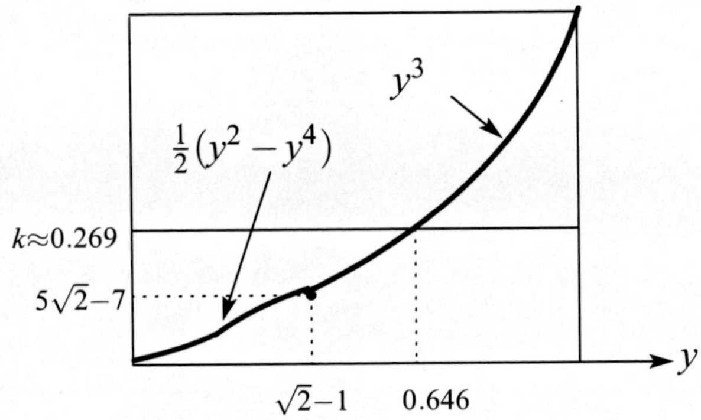

Figure 4. Computation in Eq.(3.22).

For $x \le \sqrt{2}-1$,

$$P\{x > X_{31} > X_{21} > \sqrt{h^-(x)}\} + P\{x > X_{21} > \sqrt{h^-(x)}, X_{31} < X_{21}, x > X_{32}\}$$
$$+P\{X_{21} < \sqrt{h^-(x)}, x > X_{22}, x > X_{31} > \frac{1}{2}(1-x^2)\}$$
$$+P\{X_{21} < \sqrt{h^-(x)}, X_{31} < \frac{1}{2}(1-x^2), x > X_{22} \vee X_{32}\}$$

each term of which corresponds to the play $A_1 \to A_2 \to A_3, A_1 \to A_2 \to R_3, A_1 \to R_2 \to A_3$ and $A_1 \to R_2 \to R_3$, respectively. Since $\sqrt{h^-(x)} > x, \forall x \le \sqrt{2}-1$, the first two terms are zero, and the third term is also zero, because of $x < \frac{1}{2}(1-x^2)$. Hence we have $p_{1A}(x) = \frac{1}{2}x^2(1-x^2)\sqrt{h^-(x)}$.

For $\sqrt{2}-1 < x < \xi$, $p_{1A}(x)$ is, similarly as above,

$$P\{x > X_{31} > X_{21} > \sqrt{h^+(x)}\} + P\{x > X_{21} > \sqrt{h^+(x)}, X_{31} < X_{21}, x > X_{32}\}$$
$$+P\{X_{21} < \sqrt{h^+(x)}, X_{31} > x > X_{22}\} + P\{X_{21} < \sqrt{h^+(x)}, X_{31} < x, x > X_{22} \vee X_{32}\},$$

and since $\sqrt{h^+(x)} > x, \forall x \in (\sqrt{2}-1, \xi)$, the first two terms are zero. Therefore

$$p_{1A}(x) = x\bar{x}\sqrt{h^+(x)} + x^3\sqrt{h^+(x)} = (x - x^2 + x^3)\sqrt{h^+(x)}.$$

And, for $\xi < x < 1$, we have

$$p_{1A}(x) = P\{x > X_{31} > X_{21} > x\} + P\{x > X_{21} > x, X_{31} < X_{21}, x > X_{32}\}$$
$$+P\{X_{21} < x, x > X_{31}, x > X_{22}\} + P\{X_{21} < x, X_{31} < x, x > X_{22} \vee X_{32}\},$$

each term of which corresponds to the play $A_1 \to A_2 \to A_3, A_1 \to A_2 \to R_3, A_1 \to R_2 \to A_3$ and $A_1 \to R_2 \to R_3$, just as directed by (3.10)-(3.11). The first three terms are zero, and so $p_{1A}(x) = x^4$. Hence by summarizing we get

$$p_{1A}(x) = \frac{1}{2}x^2(1-x^2)\sqrt{h^-(x)}I(x \le \sqrt{2}-1) + (x - x^2 + x^3)\sqrt{h^+(x)}I(\sqrt{2}-1 < x \le \xi) + x^4 I(\xi < x \le 1). \quad (23)$$

We find that $p_{1A}(x)$ is increasing with values $0, \sqrt{2}(\sqrt{2}-1)^4 \approx 0.0416, \xi^2(1-\xi+\xi^2) \approx 0.2223$ and 1, at $x = 0, \sqrt{2}-1, \xi$ and 1, respectively.

On the other hand,

$$p_{1R}(x) = P\left[W_1 | \{\text{I rejects } X_{11} = x\} \cap \{\text{II and III behave optimally in the subsequent stages}\}\right]$$

is, by referring to (3.10)-(3.11), found to be:

For $\forall x \in (0,1)$, it is

$$P\{X_{12} > X_{31} > X_{21} > k^{1/3}\} + P\{X_{12} > X_{21} > k^{1/3} \vee X_{31}, X_{12} > X_{32}\}$$

$$+P\{X_{21} < k^{1/3}, X_{31} > \frac{1}{\sqrt{3}}, X_{12} > X_{22} \vee X_{31}\} + P\{X_{21} < k^{1/3}, X_{31} < \frac{1}{\sqrt{3}}, X_{12} > X_{22} \vee X_{32}\},$$

each term of which corresponds to the play $R_1 \to A_2 \to A_3, R_1 \to A_2 \to R_3, R_1 \to R_2 \to A_3$ and $R_1 \to R_2 \to R_3$, respectively. Each term sums up in this order, and hence

$$p_{1R}(x) = \int_{k^{1/3}}^{1} (t - k^{1/3})(1-t)dt + \int_{k^{1/3}}^{1} \frac{1}{2} t(t^2 - k^{2/3})dt \tag{24}$$

$$+ \int_{1/\sqrt{3}}^{1} k^{1/3} \cdot \frac{1}{2}(1-t^2)dt + \frac{1}{3\sqrt{3}} k^{1/3}$$

$$= \left[\frac{1}{6}(1-k) - \frac{1}{2}k^{1/3} + \frac{1}{2}k^{2/3}\right] + \frac{1}{8}(1-k^{2/3})^2 + \left(\frac{9-4\sqrt{3}}{27}\right) k^{1/3} + \frac{\sqrt{3}}{9} k^{1/3}$$

$$= \frac{1}{6}(1-k) + \frac{1}{8}(1+k^{2/3})^2 - \left(\frac{9+2\sqrt{3}}{54}\right) k^{1/3} \approx 0.22372 \equiv c, \text{ say.}$$

Therefore, from (3.23)-(3.24), we obtain

$$P(W_1) = \int_0^1 \{p_{1A}(x) \vee p_{1R}(x)\}dx = \int_0^1 (x^4 \vee c)dx = \frac{1}{5}(1+4c^{5/4}) \approx 0.32309,$$

and the optimal strategy for I in the first stage is given by (3.9).

Step 4. Finally we have to find $P(W_2)$. Applying I's strategy (3.9) to Eq. (3.21)-(3.22), we obtain

$$P(W_2) = \int_0^1 \{P(W_2|xR_1)I(x \le x_0) + P(W_2|xA_1)I(x > x_0)\}dx$$

$$= \frac{1}{4}(1+3k^{4/3})x_0 + \int_{x_0}^1 \left\{xh^+(x) + \frac{1}{3}(1-x^3)\right\} dx \approx 0.26158 + 0.07112 \approx 0.33270,$$

since $x_0 \approx 0.68774 > \xi \approx 0.54368$, and the integral in the r.h.s. is

$$\int_{x_0}^1 \left(\frac{1}{3} + \frac{1}{2}x - \frac{1}{2}x^2 + \frac{1}{6}x^3 - \frac{1}{2}x^4\right) dx \approx 0.07112.$$

Therefore

$$P(W_3) = 1 - P(W_1) - P(W_2) \approx 1 - 0.32309 - 0.33270 = 0.34421,$$

as is given by (3.12).

This completes the proof of Theorem 2. □

Remark 1. *We observe, by Theorems 1 and 2, that the difference between the players' winning probabilities is diminished in the three-player case than in the two-player case. Let $W_{n,i}$ be the winning probability for player i in the n-player game. We showed that*

$$P(W_{2,2}) - P(W_{2,1}) = 0.0272$$

and

$$P(W_{3,3}) - P(W_{3,1}) = \{P(W_{3,3}) - P(W_{3,2})\} + \{P(W_{3,2}) - P(W_{3,1})\} = 0.0211.$$

A conjecture that $P(W_{n,n}) - P(W_{n,1}) \downarrow 0$, as $n \to \infty$, is almost prohibitive to prove.

Remark 2. *We give a numerical example which shows how Theorems 1 and 2 work.*

	Two-player game	**Three-player game**
1st stage	If I observes $X_{11} = x = 0.482$, then he announces 0.482 & R_1 (since $x < \sqrt{3/8} \approx 0.6124$) and exchange x to X_{12}	If I observes $X_{11} = x = 0.482$, then he announces 0.482 & R_1 (since $x < x_0 = 0.6877$) and exchange x to X_{12}
2nd stage	If II obs. $X_{21} = y = 0.644$, then he accepts it (since $y > 1/2$).	If II obs. $X_{21} = y = 0.644$, then he announces 0.644 & R_2 (since $y < y_0 = 0.6457$) and exchange it to X_{22}
3rd Stage		If III obs. $X_{31} = z = 0.581$, he accepts it (since $z > 1/\sqrt{3} \approx 0.5774$)
Showdown	I(II) wins if $X_{12} > (<) 0.644$.	Players' scores are X_{12}, X_{22}, and 0.581, resp. Player with the highest score wins.

Remark 3. *It seems to us that the sequential game discussed in the present paper doesn't belong to the area of dynamic programming. The result obtained in the two-player game is not applicable to the three-player game.*

4. Simultaneous-Move Game

In the simultaneous-move version of the game, the unfair information acquisition by the players disappears. Each player $i, i = 1, 2, 3$, privately observes X_{i1} and chooses either one of A_i or R_i. The observed value and choice by each player are unknown to his (or her) opponents. Suppose that players' strategies have the form of :

$$\text{I accepts (rejects) } X_{11} = x, \text{ if } x > (<) a,$$
$$\text{II accepts (rejects) } X_{21} = y, \text{ if } y > (<) b,$$
$$\text{III accepts (rejects) } X_{31} = z, \text{ if } z > (<) c.$$

Let $M_i(a, b, c) \equiv P\{W_i|\text{I, II, and III choose } a, b \text{ and } c, \text{ respectively}\}, i = 1, 2, 3$. Evidently $\sum_{i=1}^{3} M_i(a, b, c) = 1, \forall(a, b, c) \in [0, 1]^3$, and, by symmetry, $M_i(a, a, a) = 1/3, \forall i, \forall a \in [0, 1]$.

Let $p_{AAA}, p_{RRR}, p_{AAR}$, etc., denote the winning probability for I when the players' choice-triple is $A - A - A, R - R - R, A - A - R$, etc. Also let $q_{AAA}, q_{RRR}, q_{AAR}$ ($r_{AAA}, r_{RRR}, r_{AAR}$) etc, similarly denote the winning probability for II (III). Then we find that

$$M_1(a, b, c) = p_{AAA} + p_{RRR} + (\text{other six probabilities}), \tag{25}$$
$$M_2(a, b, c) = q_{AAA} + q_{RRR} + (\text{other six probabilities}), \tag{26}$$
$$M_3(a, b, c) = r_{AAA} + r_{RRR} + (\text{other six probabilities}), \tag{27}$$

where

$$\begin{aligned}
p_{AAA} &= P\{X_{11} > a, X_{21} > b, X_{31} > c, X_{11} > X_{21} \vee X_{31}\} = \int_{a\vee(b\vee c)}^{1} (t-b)(t-c)dt, \\
p_{RRR} &= P\{X_{11} < a, X_{21} < b, X_{31} < c, X_{12} > X_{22} \vee X_{32}\} = \frac{1}{3}abc, \\
p_{ARR} &= P\{X_{11} > a, X_{21} < b, X_{31} < c, X_{11} > X_{22} \vee X_{32}\} = bc\int_{a}^{1} t^2 dt, \\
p_{AAR} &= P\{X_{11} > a, X_{21} > b, X_{31} < c, X_{11} > X_{21} \vee X_{32}\} = c\int_{a\vee b}^{1} t(t-b)dt, \\
p_{ARA} &= P\{X_{11} > a, X_{21} < b, X_{31} > c, X_{11} > X_{22} \vee X_{31}\} = b\int_{a\vee c}^{1} t(t-c)dt, \\
p_{RAA} &= P\{X_{11} < a, X_{21} > b, X_{31} > c, X_{12} > X_{21} \vee X_{31}\} = a\int_{b\vee c}^{1} (t-b)(t-c)dt, \\
p_{RRA} &= P\{X_{11} < a, X_{21} < b, X_{31} > c, X_{12} > X_{22} \vee X_{31}\} = ab\int_{c}^{1} t(t-c)dt, \\
p_{RAR} &= P\{X_{11} < a, X_{21} > b, X_{31} < c, X_{12} > X_{21} \vee X_{32}\} = ac\int_{b}^{1} t(t-b)dt, \\
q_{AAA} &= P\{X_{11} > a, X_{21} > b, X_{31} > c, X_{21} > X_{11} \vee X_{31}\} = \int_{b\vee(a\vee c)}^{1} (t-a)(t-c)dt, \\
q_{RRR} &= P\{X_{11} < a, X_{21} < b, X_{31} < c, X_{22} > X_{12} \vee X_{32}\} = \frac{1}{3}abc, \\
q_{ARR} &= P\{X_{11} > a, X_{21} < b, X_{31} < c, X_{22} > X_{11} \vee X_{32}\} = bc\int_{a}^{1} t(t-a)dt,
\end{aligned}$$

$$
\begin{aligned}
q_{AAR} &= P\{X_{11} > a, X_{21} > b, X_{31} < c, X_{21} > X_{11} \vee X_{32}\} = c\int_{a\vee b}^{1} t(t-a)dt,\\
q_{ARA} &= P\{X_{11} > a, X_{21} < b, X_{31} > c, X_{22} > X_{11} \vee X_{31}\} = b\int_{a\vee c}^{1} (t-a)(t-c)dt,\\
q_{RAA} &= P\{X_{11} < a, X_{21} > b, X_{31} > c, X_{21} > X_{12} \vee X_{31}\} = a\int_{b\vee c}^{1} t(t-c)dt,\\
q_{RRA} &= P\{X_{11} < a, X_{21} < b, X_{31} > c, X_{22} > X_{12} \vee X_{31}\} = ab\int_{c}^{1} t(t-c)dt,\\
q_{RAR} &= P\{X_{11} < a, X_{21} > b, X_{31} < c, X_{21} > X_{12} \vee X_{32}\} = ac\int_{b}^{1} t^2dt,
\end{aligned}
$$

and

$$
\begin{aligned}
r_{AAA} &= P\{X_{11} > a, X_{21} > b, X_{31} > c, X_{31} > X_{11} \vee X_{21}\} = \int_{c\vee(a\vee b)}^{1} (t-a)(t-b)dt,\\
r_{RRR} &= P\{X_{11} < a, X_{21} < b, X_{31} < c, X_{32} > X_{12} \vee X_{22}\} = \frac{1}{3}abc,\\
r_{ARR} &= P\{X_{11} > a, X_{21} < b, X_{31} < c, X_{32} > X_{11} \vee X_{22}\} = bc\int_{a}^{1} t(t-a)dt,\\
r_{AAR} &= P\{X_{11} > a, X_{21} > b, X_{31} < c, X_{32} > X_{11} \vee X_{21}\} = c\int_{a\vee b}^{1} (t-a)(t-b)dt,\\
r_{ARA} &= P\{X_{11} > a, X_{21} < b, X_{31} > c, X_{31} > X_{11} \vee X_{22}\} = b\int_{a\vee c}^{1} t(t-a)dt,\\
r_{RAA} &= P\{X_{11} < a, X_{21} > b, X_{31} > c, X_{31} > X_{12} \vee X_{21}\} = a\int_{b\vee c}^{1} t(t-b)dt,\\
r_{RRA} &= P\{X_{11} < a, X_{21} < b, X_{31} > c, X_{31} > X_{12} \vee X_{22}\} = ab\int_{c}^{1} t^2dt,\\
r_{RAR} &= P\{X_{11} < a, X_{21} > b, X_{31} < c, X_{32} > X_{12} \vee X_{21}\} = ac\int_{b}^{1} t(t-b)dt,
\end{aligned}
$$

First we have to noice that

$$M_i(a,a,a) = \frac{1}{3}, i = 1,2,3, \forall a \in [0,1].$$

We prove this for $i = 1$ only. Proof is the same for $i = 2, 3$. From (4.1) we have

$$
\begin{aligned}
M_1(a,a,a) &= [p_{AAA} + p_{RRR} + (\text{other six probabilities})]_{a=b=c}\\
&= (1+a)\int_a^1 (t-a)^2dt + \frac{1}{3}a^3 + a^2\int_a^1 t^2dt + 2(a+a^2)\int_a^1 t(t-a)dt\\
&= \frac{1}{3}(1+a)(1-a)^3 + \frac{1}{3}a^3 + \frac{1}{3}a^2(1-a^3) + 2(a+a^2)\cdot\frac{1}{6}(2-3a+a^3),
\end{aligned}
$$

which is easily shown to be equal to $1/3, \forall a \in [0,1]$.

THEOREM 4.4 *Solution to the simultaneous-move three-player game. The game has a unique equilibrium point* (a^*, a^*, a^*), *and the common equilibrium value* $1/3$, *where* $a^* \approx 0.691$ *is a unique root in* $[0,1]$ *of the equation*

$$2a^4 = 1 - a + a^2 - a^3. \tag{28}$$

Proof. Eq. (4.1) becomes, if $b = c$,

$$\begin{aligned} M_1(a, b.b) &= [p_{AAA} + p_{RRR} + (\text{other six probabilities})]_{b=c} \qquad (29)\\ &= \int_{a\vee b}^{1} (t-b)^2 dt + \frac{1}{3}ab^2 + b^2 \int_a^1 t^2 dt \\ &\quad +2b \int_{a\vee b}^{1} t(t-b)dt + a \int_b^1 (t-b)^2 dt + 2ab \int_b^1 t(t-b)dt, \end{aligned}$$

and so

$$\begin{aligned} \frac{\partial M_1(a,b,b)}{\partial a} &= b^2(\frac{1}{3} - a^2) + \int_b^1 (t-b)^2 dt \qquad (30)\\ &\quad +2b \int_b^1 t(t-b)dt - I(a \geq b)(a-b)(a-b+2ab) \\ &= -a^2b^2 + \frac{1}{3}(1 - b + b^2 - b^3 + b^4) - I(a \geq b)(a-b)(a-b+2ab). \end{aligned}$$

Therefore

$$\left.\frac{\partial M_1(a,b,b)}{\partial a}\right|_{a=0} = \frac{1}{3}(1 - b + b^2 - b^3 + b^4) > 0,$$

and

$$\left.\frac{\partial M_1(a,b,b)}{\partial a}\right|_{a=b-0} = -b^4 + \frac{1}{3}(1 - b + b^2 - b^3 + b^4) = \frac{1}{3}(1 - b + b^2 - b^3 - 2b^4) = 0,$$

if b satisfies $2b^4 = 1 - b + b^2 - b^3$. Hence, from (4.5)-(4.6), we find that $M_1(a, b, b)$ is a concave function of a with the value $\frac{1}{3}(1 - b + b^2 - b^3 + b^4)$ at $a = 0, 1$, and we have $\max_b M_1(a^*, b, b) = M_1(a^*, a^*, a^*)$ where a^* is a unique root of the equation (4.4).

The arguments about $M_i(a, b, c), i = 2, 3$ are made similarly as about $M_1(a, b, c)$. We can find, from (4.2)-(4.3) and probability groups $q \ldots$ and $r \ldots$, that

$$\begin{aligned} M_2(a, b, a) &= [q_{AAA} + q_{RRR} + (\text{other six probabilities})]_{a=c} \\ &= \text{the same as } M_1(a, b, b) \text{ with } a \text{ and } b \text{ interchanged.} \end{aligned}$$

$$\frac{\partial M_2(a,b,a)}{\partial b} = -a^2b^2 + \frac{1}{3}(1 - a + a^2 - a^3 + a^4) - I(b \geq a)(b-a)(b-a+2ab). \qquad (31)$$

$$\begin{aligned} M_3(a, a, c) &= [r_{AAA} + r_{RRR} + (\text{other six probabilities})]_{a=b} \\ &= \text{the same as } M_2(a, b, a) \text{ with } b \text{ and } c \text{ interchanged.} \end{aligned}$$

$$\frac{\partial M_3(a,a,c)}{\partial c} = -a^2c^2 + \frac{1}{3}(1 - a + a^2 - a^3 + a^4) - I(c \geq a)(c-a)(c-a+2ca). \qquad (32)$$

Therefore the conditions in equilibrium

$$\frac{\partial M_1(a,b,b)}{\partial a} = 0, \quad \frac{\partial M_2(a,b,a)}{\partial b} = 0, \quad \text{and} \quad \frac{\partial M_3(a,a,c)}{\partial c} = 0$$

give the eq.point (a^*, a^*, a^*) just as mentioned in the theorem. This completes the proof of the theorem. □

The two-player game is solved in [8]. Let

$$M_1(a,b) \equiv P\{W_1|\text{I (II) chooses } a(b)\} = 1 - M_2(a,b).$$

Then it is shown that

$$M_1(a,b) = \frac{1}{2}\{-a^2b + (a+1)(1-b+b^2)\} - I(a \geq b)\frac{1}{2}(a-b)^2,$$

$$\frac{\partial M_1(a,b)}{\partial a} = -ab + \frac{1}{2}(1-b+b^2) - I(a \geq b)(a-b).$$

And we have the following

THEOREM 4.4 *Solution to the simultaneous-move two-player game. The game has a unique saddle point* (g, g) *and the saddle value* $\frac{1}{2}$, *where* $g = \frac{1}{2}(\sqrt{5}-1) \approx 0.61803$ (For the proof, see [8]).

Remark 4. *The optimal threshould number is g (golden bisection number) in the two-player game and it increases to* $a^* \approx 0.691$ *in the three-player game. Furthermore, by considering Theorems 3 and 4, where* $g^2 = 1 - g$ *and* $2a^4 = 1 - a + a^2 - a^3$ *(Eq.(4.4)), we have a conjecture that the simultaneous-move four-player game has a unique eq.point* (a^*, a^*, a^*, a^*) *and the common eq.value* $1/4$, *where* $a^* \approx 0.738$ *is a unique root in* $[0,1]$ *of the equation* $3a^6 = 1 - a + a^2 - a^3 + a^4 - a^5$. *Player behaves more cautious as he (or her) has more competitors.*

References

[1] P.R. Coe, W. Butterworth,Optimal stopping in the ‘ ‘ Showcase Showdown”, *The American Statistician* **49** (1995), 271–275.

[2] S. Even, ‘ The price is Right’ game, *American Math. Monthly* **73** (1966), 180–182.

[3] T. Feder, Toetjes, *American Math. Monthly* **97** (1990), 785–794.

[4] C. Ferguson, T.S. Ferguson, On the Borel and von Neumann poker models, *Game Th. Appl.* **9** (2003), 19–35.

[5] T.S. Ferguson, C. Genest, *Toetjes na*, to appear.

[6] L.A. Petrosjan, N. A. Zenkevich, *Game Theory*, World Scientific, Singapore, 1996.

[7] M. Sakaguchi, Best-choice games where arbitration comes in, *Game Th. Appl.* **9** (2003), 141–149.

[8] M. Sakaguchi, *Two-player games of ‘ ‘Score Showdown”*, to appear.

In: Mathematics, Game Theory and Algebra Compendium ISBN 978-1-60456-659-8
Editor: Jacob H. Mathias, pp. 335-359

Chapter 30

DYNAMIC NONCOOPERATIVE R&D IN DUOPOLY WITH SPILLOVERS AND TECHNOLOGY GAP*

Shravan Luckraz
Department of Economics, Faculty of Arts and Social Sciences
National University of Singapore, Singapore

1. Introduction

It has been well established that when one firm independently develops a cost reducing innovation, the firm's competitors benefit in the sense that they can use the innovation to reduce their own costs. When such spillover effects are significant, noncooperative firms might be expected to research too little from the standpoint of the industry since each firm tends to ignore the positive externality which its research generates on the cost of its rival firm (see D'Aspremont and Jacquemin (1988), Henriques (1990) and Simpson and Vonortas (1994)). However, when spillovers are endogenous it is also observed that the firm's disincentive to engage in R&D activity is partially offset because its own R&D can potentially enhance its capacity to absorb its rival's technology (see Katsoulacos and Ulph (1998), Kultti and Takalo (1998), Kamien and Zang (2000) and Grunfeld (2003)). Moreover, reduced costs of rival firms due to spillovers will lead all firms to compete more intensively in the product market. Empirical findings by Cohen and Levinthal (1989) reinforce the fact that spillovers have two opposing effects on R&D investment in strategic games: firstly, they increase the firm's incentive to raise its own R&D and, secondly, they create a disincentive for the rival firm to invest in R&D as free riding becomes a better strategy. A possible explanation for

*This paper is a substantially revised version of a chapter of the author's Ph.D. thesis. I thank Julian Wright for his useful comments and suggestions which helped to improve this paper. I am also grateful to Ivan Png, Mark Donoghue, Sougata Poddar, Albert Hu and the participants of the NUS Industrial Organization Workshop for their suggestions. I am indebted to Rabah Amir, Steffen Jorgensen, Vladimar.V. Mazalov, Tamer Basar, Rodney Beard and other participants of the Eleventh International Symposium of Dynamic Games and Applications for their suggestions. I remain responsible for any errors.

this behavior is that there exists a threshold level of spillovers beyond which the firm has no incentive to increase its R&D activities.

The purpose of this paper is to show how the dynamics of the technology gap between firms helps demarcate the opposing effects of spillovers on R&D incentives. We explore the theoretical link between spillovers as pioneered by D'Aspremont and Jacquemin (1988) (henceforth AJ) and technology gap by allowing the rate of spillovers to depend on the latter[1]. Intuitively, when the follower lags far behind the leader it enjoys larger spillovers and has fewer incentives to conduct its own R&D, but as it moves closer to the frontier[2] it is "forced" to innovate as its free riding possibility set becomes smaller. Thus, if there exists a relationship between spillovers and R&D incentives an analogous link must also exist between the latter and the level of technology gap.

In order to demonstrate the relationship between technology gap and R&D incentives, we develop a two stage game of process R&D and output competition for an ex-ante asymmetric duopoly with one-way spillovers[3]. In the model, at the first stage the two firms conduct process R&D and in the second stage, they compete in Cournot fashion in the product market. We go one step further than Katsoulacos and Ulph (1998) and Kultti and Takalo (1998) who were the first[4] to endogenize spillovers in an AJ framework. Our R&D spillover function does not depend solely on the absorptive capacity effect as in the latter studies since it also takes into account the size of the technology gap between the two firms. We seek to extend existing theoretical framework by incorporating the impact of such endogenous spillovers on the benefits and the costs of R&D. The effect of the spillovers on the cost of undertaking R&D[5] has the following interpretation. Assuming that the spillover rate is endogenous (positively related to the size of the technology gap), then the further away the firm is from the frontier, the less technologically efficient it is, that is; it finds it more costly to undertake R&D when the technology gap is large. Hence, firms operating well within the frontier incur greater costs of doing research since the size of the technology gap (or endogenized spillover) is large.

We also consider the dynamic version of a two stage R&D game since we cannot observe changes in the magnitude of the gap over time in a static model. We derive our results based on the steady state values of R&D as well as on their transitional dynamic paths. We present three different (though non-mutually exclusive) sets of results. First, we present a variant of the static AJ model with one-way endogenous[6] spillovers. We show that the existence of a subgame perfect Nash equilibrium (SPNE) requires that the level of spillovers

---

[1] While more recent studies have attempted to endogenize spillovers in an AJ framework (see Amir, Estignev and Wooders (2003)), this is the first attempt to show that the nature of such endogenity turns on the level of technology gap between the two firms.

[2] The frontier is defined as the level of technological efficiency of the leader firm.

[3] In contrast to the traditional AJ framework in which both firms benefit from spillovers, we consider the case in which only the follower can free ride off the leader. Amir and Wooders (2000) also consider one-way spillovers in a two stage game of process R&D.

[4] Subsequent attempts to endogenize spillovers in an AJ framework have been made by Kamien and Zang (2000), Amir, Evstignev and Wooders (2003) and Grunfeld (2003).

[5] In the current literature, while spillovers increase the benefits of the firm's R&D by reducing its costs of production by an amount proportional to its rival's investments, they do not affect the cost of undertaking R&D.

[6] We assume that the marginal cost of production of the follower also depends on the technology gap between the leader and itself. The AJ model will be the special case where the technology gap reduces to zero.

to be low and the initial marginal cost to be high. We show that the relationship between the free-riding behavior of the laggard and the level of spillovers is non-monotonic. We observe that they are positively related as long as the size of spillovers is small.

Secondly, we develop a dynamic version of the latter model in a differential game setting. It is shown that if each firm in the industry takes into account the dynamic strategic response of its rival, results can be derived by looking at the transitional dynamics of the firms' reaction functions in the neighborhood of the steady state. While in low cost industries[7] , we find that there exists no steady state with complete catch-up[8] , in high cost industries we observe that there exists a unique and stable steady state with complete catch-up (ex-post symmetry).Lastly, we provide a general framework for analyzing dynamic AJ models with one-way endogenous spillovers. We derive some general conditions that would guarantee the existence of a steady state in a more general class of two stage R&D games with spillovers. In doing so, we also outline the cases when R&D spillovers can act as a deterrent to future research.

Our contribution to the literature might be described as follows. While recent attempts to endogenize spillovers in the AJ paradigm take into account the firm's absorptive capacity only, we show, by introducing the concept of technology gap, that the follower firm's incremental R&D effort does not only enhance its capacity to learn (by reducing its own R&D costs) but also, after some point in time, begins to reduce its marginal benefits too [9] . This is due to decreasing returns to scale to imitation or reverse engineering activities. Thus, the follower firm benefits more from the spillovers when the technology gap is large than when it is small; in other words, the laggard's marginal benefits from the spillovers decrease in the level of the technology gap. Also, its cost of R&D falls as the gap becomes smaller due to the absorptive capacity effect. In view of the key role that spillovers play in the current two stage R&D game literature, we believe that it necessary to ascertain whether the existing results remain robust to a more general version of one-way endogenous spillovers. Moreover, our framework will nicely capture the notion that both the R&D benefits and costs of a lagging firm change with its research expenditure when the technology gap is endogenized.

The remainder of the paper is organized as follows. Section 2 presents a static version of the AJ model with one way spillovers. In Section 3 we study the dynamic version of the AJ model. A general framework is proposed in Section 4. We conclude in Section 5.

2. D'Aspremont and Jacquemin (AJ) Revisited - The Static Case

In this section we look at the augmented version of the AJ model. Consider an industry with a duopolistic market structure in which two firms (firm 1 and firm 2) engage in a two stage R&D game. At the first stage, firms 1 and 2 conduct process R&D by choosing their research intensity (the amount by which they reduce their costs of production) X_1 and X_2 respectively.n the second stage, the firms compete in Cournot fashion in the product market.

---

[7] Here we refer to the marginal cost of production.

[8] By complete catch-up we mean that the technology gap equals zero.

[9] Kamien and Zang (2000) emphasize that the followers themselves must invest in R&D in order to take advantage of the R&D innovations of others (the absorptive capacity effects).

As in AJ we assume that the demand faced by the two rivals is linear with the slope -1[10]. The demand schedule is given by

$$P = A - Q\,, \tag{1}$$

where $Q = q_i + q_j$, $i \neq j, i = 1, 2$ and q_i s the output of firm i.

We impose an ex-ante asymmetry between the two firms both on the marginal benefit and on the cost of their R&D. In particular, on the marginal benefit (or marginal cost reduction) side of R&D, we assume a one-way spillover structure in which only firm 1, the "follower"[11] can benefit from spillovers from firm 2, the "leader", but not vice-versa. Moreover, the spillover rate depends positively on the technology gap between the two firms. On the cost side of R&D, we assume that the firm benefiting from spillovers (the follower) incurs a higher R&D cost than the leader. Also, the larger the technology gap between the two firms, the higher the R&D cost of the follower. In other words, our assumption states that while free-riding opportunities reduce as the technology gap becomes smaller, so does the R&D cost for a follower firm. The per unit production marginal costs for the follower and the leader are given respectively by the following equations.

$$C_1 = C - X_1 - \beta X_2 \tag{2}$$

$$C_2 = C - X_2 \tag{3}$$

where $X_1, X_2 \in [0, C)$.

We shall assume β to be given by[12]

$$\beta \equiv \frac{b\,(X_2 - X_1)}{1 + bX_2} \tag{4}$$

where $b \in (0, 1)$ is the spillover parameter. We also impose the following parameter restriction

$$C > A \tag{5}$$

β is the level of spillovers, due to the leader's(firm 2) R&D, which accrues to the follower (firm 1). A crucial departure from the standard AJ framework is that there exists an asymmetry characterized by one-way spillovers. The leader's marginal cost (3) is decreasing in the level of research that it undertakes and is ex-ante unaffected by the R&D of the follower. The latter's per unit cost (2) is not only reduced by its own R&D effort but also by a proportion of that of the leader's. The spillover parameter b, can account for involuntary leakage or voluntary exchange of technological information. One can also interpret b as a parameter that is inversely related to the degree of patent protection or appropriability. The parameter restriction given by (5) implies the existence of large per unit marginal cost in the industry. It can be easily verified that if the R&D levels for both firms are zero, they will choose not to produce in the second stage. Thus, a minimum R&D level, at least by firm 2, is necessary to ensure a positive output in the product market. While our assumptions restrict our study only to industries where costs are large enough so that (5) is satisfied, they do not rule

[10] This does not lead to a loss of generality.

[11] Note that "leader" and "follower" are not used in the Stackelberg sense here.

[12] In the next section, we provide a formal way of deriving this expression.

the possibility in which the optimal R&D levels for both firms are zero, so that they both do not produce in the second stage. Thus, the infinimum (zero) of the interval containing X_i for $i = 1, 2$ is included in the interval. However, we assume that the initial marginal cost (without R&D) is so large that it is not feasible for firm$i$ to undertake any R&D level X_i such that $X_i = C$. In other words, the supremum (C) of the interval containing for $i = 1, 2$ is not included in the interval. As a corollary, we also have for all and also that if . Therefore, it is never possible for firm i to reduce its marginal cost to zero in our setting and hence, unlike other AJ models in which the additive marginal benefit functions are unable to exclude the possibility of maximal costs reductions (See Amir, 2000), our assumptions of large costs rule out this possibility. The firms R&D costs functions are assumed to be as follow,

$$C_1^{R\&D} = (1 + \beta) C(X_1) + bX_2 \quad \text{for firm 1} \tag{6}$$

$$C_2^{R\&D} = C(X_2) \quad \text{for firm 2} \tag{7}$$

where $C(.)$ is monotonically increasing in its argument

The R&D cost function of the follower (6) is new and a justification is warranted. The reason for having the term $(1 + \beta)$ n the follower's R&D cost function is owing to the fall in its research costs when it approaches the technological frontier. It is not difficult to imagine β as representing the distance between the technology levels of the two firms, in which case a higher (lower) β (smaller) technology gap. Thus, initially, when the follower is a new entrant, β is high and the marginal benefits of free riding are large. However, when its technological efficiency approaches that of the leader, becomes lower and due to decreasing returns to imitation the marginal benefits begin to fall. Now, when the follower's technology level approaches that of the leader, it must be the case that its R&D costs fall and gradually converge on that of the leader. Hence, while technology catch-up implies reductions in the benefits of free riding, it also lowers research costs. If $\beta = 0$ it means that catch-up is complete as the follower's and the leader's marginal benefits and marginal costs of R&D become identical. If $\beta < 0$, the follower would have leapfrogged the leader and the marginal benefit of free riding would be negative (the last term in the RHS of (2) becomes greater or equal to zero so the total benefit due to spillovers is declining).Also, the cost of R&D of the follower also becomes smaller as becomes β more negative. The last term on the RHS of (6) shows that some learning or reverse engineering costs are incurred by the follower in order to benefit from the leader's research. Thus we assume that the externalities accrued by the follower do not take the form of a pure public good. This assumption also implies that if it is optimal for the follower to free-ride completely on the leader, by choosing $X_1 = 0$, then it also has to bear some costs given by bX_2 . Models which employ the additive marginal benefit function as in the AJ framework are often criticized for they cannot exclude the possibility that a firm can fully free-ride costlessly on its rival (see Amir, 2000). The cost function given by (6) allows our model to survive the so-called Amir's critique. Moreover, for simplicity, we shall assume that $C(X_i) = X_i$[13] for $i = 1, 2$ is linear and is given by $C(X_i) = X_i$. This assumption which

[13] While a strictly convex cost function is one of the necessary ingredients to guarantee the existence of a maximum point in the AJ class of games, the endogenous spillover rate of our model makes this assumption redundant. In other, words, due to (4), usual parameter restrictions can be used to ensure the quasi -concavity

does not violate the quasi-concavity of the first stage profit function will be relaxed in the next sections. It is also to be noteworthy that the underlying asymmetry (leader/ follower configuration) in our model is only observed through the marginal cost reduction and the R&D cost functions and that we do not impose the condition $X_2 > X_1$. Hence, even when the follower leapfrogs the leader so that $X_2 \leq X_1$, the forms of the benefit and cost functions do not change. Thus, while the follower, being a new entrant with low R&D, can free-ride on the leader's existing technology by incurring some costs, the leader in turn cannot learn or benefit from the laggard's technology.

The timing of the game is as follows. In the first stage the firms choose their respective R&D levels X_i and in the second stage they choose their respective R&D quantities q_i . As is customary in the analysis of staged games, we work backwards from the last stage to the first. Thus the firm's problem in the second stage is given by

$$\underset{q_i}{Max}\,(A-(q_i+q_j))\,q_i - C_i q_i, i = 1,2, i \neq j \tag{8}$$

It is well known that the Cournot second stage quantity and profit function for firm i is given by

$$q_i = \frac{A - 2C_i + C_j}{3} \tag{9}$$

$$\pi_i = \frac{(A - 2C_i + C_j)^2}{9} \tag{10}$$

(2), (3), (6), (7) and (10) imply that the first stage objective functions for the follower and the leader are given respectively by

$$\underset{X_1}{Max}\frac{(A-2\,(C-X_1-\beta X_2)+C-X_2)^2}{9} - (1+\beta)\,C\,(X_1) \quad \text{for firm 1} \tag{11}$$

$$\underset{X_2}{Max}\frac{(A-2\,(C-X_2)+C-\beta X_2)^2}{9} - C\,(X_2) \quad \text{for firm 2} \tag{12}$$

where β is given by (4).

The firms' R&D reaction functions are given respectively by[14]

$$-\,(A-C+2X_1-X_2)+\frac{9}{4}\,(1+2bX_2-2bX_1) = \frac{2bX_2\,(X_2-X_1)}{1+bX_2} \quad \text{for firm 1} \tag{13}$$

$$\frac{2}{9}\left(A-C+\left(2-\frac{b\,(X_2-X_1)}{1+bX_2}\right)X_2-X_1\right)\frac{2+2bX_2+bX_1}{(1+bX_2)^2} = 1 \quad \text{for firm 2} \tag{14}$$

Proof. See Appendix ■

of (11) and (12).Moreover, the Hessian matrix of (11) and (12) are found to be negative definite for some region of the parameter space. Poof can be provided upon request.

[14] The second order conditions are satisfied for small values of b and $C > A$. Proof can be provided upon request

It can be shown that (13) simplifies to

$$\frac{9}{4}-(A-C)+\left(\left(\frac{27}{4}-(A-C)\right)b+1\right)X_2-\left(\frac{9}{2}b+2\right)X_1-\frac{9}{2}b^2X_2X_1+\left(\frac{9}{2}b^2-b\right)X_2^2=0 \tag{15}$$

We can now demonstrate our results via the following propositions.

Proposition 0.1. *Assume that $C > A$ and b is small, then (i) there exists a level of X_2^M beyond which X_1 starts to fall, that is, there exists a level of the leader's investment beyond which the laggard will start to free-ride.(ii) there exists an inverted U-shaped relationship between b and $-\Delta X_2^M$[15], that is, while free-riding increases with the spillover rate initially, it becomes inversely related to the spillover rate as the latter gets gets larger.*

Proof. See Appendix ■

By providing a key relationship between the free-riding and the spillover rate, Proposition 2.1 provides some theoretical support to the findings of Aghion et al. (2001) that there exists an inverted U-shaped relationship between imitation and innovation. As shown in the proof, there exists a cut-off value of the spillover rate b beyond which free-riding starts to fall. Thus, as pointed out by Cameron (1999) there are diminishing returns to scale to imitation. Intuitively, this happens when the follower has extracted all possible imitative possibilities from the leader.

Proposition 0.2. *Assume that $C > A$ and b is small, then the Nash equilibrium of the game exists and is non-unique.*

Proof. See Appendix ■

As shown in the appendix, the derivation of Proposition 2.2 is based on the graphs of reaction functions (13) and (14). Since we have already assumed $C > A$ from (5) and we know from the existence proof (see Appendix) that both firms choose some positive X_i , the parameter value C can be chosen ex-post so that X_i for $i = 1, 2$are strictly less than C[16] . We also show that more than one (nonzero) fixed point can exist. Therefore in a leader/follower market configuration both firms will optimally undertake some minimum R&D as long as the spillover rate parameter is not too high and the initial marginal cost is not too low. This is not surprising as excess externalities might lead to a very low level of R&D effort by the follower who would rather free ride on the leader. As a result, the latter's R&D effort might be reduced to zero.

---

[15] Note that $-\Delta X_2^M$(change in the maximum value of firm 2's R&D) captures free-riding.

[16] Note that this is possible since the upper bound (C) of X_i is not in the interior of the interval containing X_i .

3. The Dynamic Case

In this section we shall exploit the fact that spillovers are endogenous, together with the dynamic nature of technology gap to extend the existing AJ framework. Subtracting (2) from (3), we can define the gap at time t as

$$G_t = (1 - \beta) X_{2t} - X_{1t} \tag{16}$$

We shall assume[17] that the spillover takes the following form

$$\beta_t \equiv bG_t \tag{17}$$

where $b \in (0, 1)$

(17) implies a positive relationship between the spillover rate and the technology gap. As mentioned earlier, we emphasize the idea that when the gap is large, the benefits to free riding are greater than when the gap is small.

Since the firms will be playing a differential game, some amendments to the features of the model in the previous section are made without resulting in a loss of generality and without changing the essence of the previous game. First, in the dynamic game, the firms will be choosing their R&D investments I_{it} (rate of change of R&D over time) rather than the R&D level X_{it} . We do not restrict I_{it} to be positive. A negative I_{it} is realized when an asset is sold or disposed. Thus, we have

$$\dot{X}_{it} = I_{it} \tag{18}$$

for $i = 1, 2$

For simplicity it is assumed the depreciation rate of zero. Secondly, the cost of such investments for the follower and the leader are given respectively by

$$C_{1t}^{R\&D} = \frac{I_{1t}^2 (1 + \beta_{1t})}{2X_{1t}} \quad \text{for firm 1} \tag{19}$$

$$C_{2t}^{R\&D} = \frac{I_{2t}^2}{2X_{2t}} \quad \text{for firm 2} \tag{20}$$

We observe from (19) that if firm 1 decides to fully free-ride on the leader by choosing a very low X_{1t} , then its cost of R&D would become very large (close to ∞). Such a formulation is robust to the Amir's critique.

The timing of the game is such that there are two stages in each period. Now, since in the second stage of each period the firms choose their output, which are ex-post functions of X_{it} and X_{jt}only, the reduced forms of the firms' first stage objective functions can be derived so that the only choice variables of the dynamic game are R&D investment levels. Thus, firms compete to find their optimal R&D time path. Moreover, we shall use the concept of open-loop Nash equilibrium [18] to solve the differential game. It is also important to note

[17] Linearity is assumed for simplicity and to make the solution more tractable.

[18] An open-loop Nash equilibrium is found when a competitor takes his rival's reaction function solely as a function of time in his dynamic optimization problem. For more details see Dockner et al (2000). We assume an open-loop equilibrium as in Peretto (1996) since we are unable to find a closed-form solution to analyze the properties of the model for a closed-loop or Markov perfect equilibrium. In principle, if the objective function is of linear quadratic form, the closed-loop equilibrium can be found by setting the Hamilton Jacobi Bellman (HJB) equation.

that unlike the case described by Vencatachellum (1998), the open loop Nash equilibrium in our model does not coincide with the myopic strategy whereby the firm does not take into account the R&D of its rival while choosing its optimal path[19] . We shall now use (16) and (17) to endogenize the spillover rate in the model. (17) and (19) can be rewritten as

$$\beta_t \equiv \frac{b\left(X_{2t} - X_{1t}\right)}{1 + bX_{2t}} \tag{21}$$

$$C_{1t}^{R\&D} = \frac{I_{1t}^2}{2X_{1t}} \left(\frac{1 + 2bX_{2t} - bX_{1t}}{1 + bX_{2t}} \right) \tag{22}$$

This special case arises since the Hamiltonian function of his model is linear and separable in its rival's stock of human capital and hence, the latter term vanishes at the first order condition.

Definition 0.1. *The pair* (θ_1, θ_2) *is called an open-loop Nash equilibrium with function* θ_j *mapping* $t \in [0, T)$ *to a real number if for each* $j = 1, 2$ *, an optimal control path* $I_j\,(.)$ *of the problem below exists and is given by* $I_j\,(t) = \theta_j\,(t)$.

Using (10), (18), (21) and (22), we can write firm 1 dynamic problem as

$$V_1 = \underset{I_{1t}}{Max} \int_0^\infty e^{-\rho t} \left\{ \frac{1}{9} \left(A - 2\left(C - X_{1t} - \frac{b\left(X_{2t} - X_{1t}\right) X_{2t}}{\left(1 + bX_{2t}\right)} \right) + C - X_{2t} \right)^2 - C_{1t}^{R\&D} \right\} dt^{20} \tag{23}$$

$s.t\ \dot{X}_{it} = I_{it}, i = 1, 2,\ X_{20} > X_{10}$ [21] is given and $\underset{Lim\,T \to \infty}{X_{iT}} \geq 0$, ρ is the rate of interest

Using (11), (18), (20) and (21), we can write firm 2's dynamic problem as

$$V_2 = \underset{I_{2t}}{Max} \int_0^\infty e^{-\rho t} \left\{ \frac{1}{9} \left(A - 2\left(C - X_{2t}\right) + C - X_{1t} - \frac{b\left(X_{2t} - X_{1t}\right) X_{2t}}{\left(1 + bX_{2t}\right)} \right)^2 - C_{2t}^{R\&D} \right\} dt$$

$$s.t\ \dot{X}_{it} = I_{it}, i = 1, 2,\ X_{20} > X_{10} \text{ is given and } \underset{Lim\,T \to \infty}{X_{iT}} \geq 0 \tag{24}$$

The current value Hamiltonian function[22] and first order conditions for firm 1 are given by

---

[19] This special case arises since the Hamiltonian function of his model is linear and separable in its rival's stock of human capital and hence, the latter term vanishes at the first order condition.

[20] We assume that the price of investment (P_t) is equal to zero so that $P_t I_{it} + C_{it}^{R\&D} = C_{it}^{R\&D}$. If investment is negative in some period ($I_{it} < 0$), then the firm incurs a loss of $C_{it}^{R\&D}$.

[21] Here we assume that the leader's R&D level is higher, that is, the technology gap is positive at least initially.

[22] We assume that the exogenous variables A, b and C are chosen so that they belong to the subset of real numbers which would satisfy the Mangasarian second order sufficient conditions for the dynamic problems of the two firms.

$$\begin{aligned} H_1 &= \frac{1}{9}\left(A-2\left(C-X_{1t}-\frac{b\left(X_{2t}-X_{1t}\right)X_{2t}}{\left(1+bX_{2t}\right)}\right)+C-X_{2t}\right)^2 \\ &- \frac{I_{1t}^2}{2X_{1t}}\left(\frac{1+2bX_{2t}-bX_{1t}}{1+bX_{2t}}\right)+\lambda_{1t}I_{1t}+\lambda_{2t}I_{2t} \end{aligned} \tag{25}$$

$$\frac{dH_1}{dI_{1t}}=0\Rightarrow\frac{I_{1t}}{X_{1t}}\left(\frac{1+2bX_{2t}-bX_{1t}}{1+bX_{2t}}\right)=\lambda_{1t} \tag{26}$$

$$\dot{\lambda}_{1t}=\rho\lambda_{1t}-\frac{4}{9\left(1+bX_{2t}\right)}\left(A-C+2X_{1t}+\left(\frac{2b\left(X_{2t}-X_{1t}\right)}{1+bX_{2t}}-1\right)X_{2t}\right)+\frac{I_{1t}^2}{2X_{1t}^2}\left(\frac{-1-2bX_{2t}}{1+bX_{2t}}\right) \tag{27}$$

$$\begin{aligned} \dot{\lambda}_{2t}=\rho\lambda_{2t} &- \frac{2\left(2bX_{2t}-2bX_{1t}-1\right)}{9\left(1+bX_{2t}\right)^2}\left(A-C+2X_{1t}+\left(\frac{2b\left(X_{2t}-X_{1t}\right)}{1+bX_{2t}}-1\right)X_{2t}\right) \\ &+ \frac{I_{2t}^2bX_{1t}}{2X_{1t}^2}\left(\frac{1+bX_{1t}}{\left(1+bX_{2t}\right)^2}\right) \end{aligned} \tag{28}$$

$$\underset{Lim\ T\to\infty}{e^{-\rho T}\lambda_{1T}X_{1T}}=0 \tag{29}$$

$$\underset{Lim\ T\to\infty}{e^{-\rho T}\lambda_{2T}X_{2T}}=0 \tag{30}$$

The current value Hamiltonian function and first order conditions for firm 2 are given by

$$H_2=\frac{1}{9}\left(A-C+\left(2-\frac{b\left(X_{2t}-X_{1t}\right)}{\left(1+bX_{2t}\right)}\right)X_{2t}-X_{1t}\right)^2-\frac{I_{2t}^2}{2X_{2t}}+\mu_{1t}I_{2t}+\mu_{2t}I_{1t} \tag{31}$$

$$\frac{dH_2}{dI_{2t}}=0\Rightarrow\frac{I_{2t}}{X_{2t}}=\mu_{1t} \tag{32}$$

$$\dot{\mu}_{1t}=\rho\mu_{1t}-\frac{2\left(2+2bX_{2t}+2bX_{1t}\right)}{9\left(1+bX_{2t}\right)^2}\left(A-C+\left(2-\frac{b\left(X_{2t}-X_{1t}\right)}{1+bX_{2t}}\right)X_{2t}-X_{1t}\right)-\frac{I_{2t}^2}{2X_{2t}^2} \tag{33}$$

$$\dot{\mu}_{2t}=\rho\mu_{2t}-\frac{2}{9\left(1+bX_{2t}\right)}\left(A-C+\left(2-\frac{b\left(X_{2t}-X_{1t}\right)}{1+bX_{2t}}\right)X_{2t}-X_{1t}\right) \tag{34}$$

$$\underset{Lim\ T\to\infty}{e^{-\rho T}\mu_{1T}X_{2T}}=0 \tag{35}$$

$$\underset{Lim\ T\to\infty}{e^{-\rho T}\mu_{2T}X_{1T}}=0 \tag{36}$$

The above equations enable us to derive the main result of this section. In particular we are now in a position to look at the dynamic equilibrium of the R&D two stage game in which the technology gap evolves over time.

Proposition 0.1. *Assume that at the steady state $I_{1t}=I_{2t}=0$. If the conditions given in the above open loop differential game hold, then (i) there exists a steady state subgame perfect Nash equilibrium given by $X_{1t}^*=X_{2t}^*=C-A$ for all$C>A$,(ii) the equilibrium is unique,(iii) there exists a stable path converging to it and (iv) the path is unique as it is the Saddle Path, (v) at the steady state equilibrium of the full game,$q_{it}\to 0$ as $X_{it}\to C-A$ for $i=1,2$ and there exists a neighborhood around the steady state in which q_{it} both and X_{it} are falling.*

Proof. See Appendix. ■

Proposition 3.1 states that if the R&D of both firms were to stop growing at a so-called "steady state" then it must be the case that technology catch-up has taken place in the industry as long as the marginal costs are large enough. Therefore, when the basic AJ model is augmented to allow for an endogenous spillover rate which depends on the dynamics of the technology gap between the two firms, ex-post symmetry occurs. Thus, the follower, benefiting from the externalities, builds up more absorptive capacity and the leader responds to this by reducing its own R&D. As a result, the technology gap shrinks over time.

The intuition behind this proposition is best described in terms of two opposite driving forces affecting R&D incentives. The first one originates from the competitive pressure which the follower puts on the leader so that the latter has no other alternative than to increase its own R&D effort in order to maintain its market share advantage. The second effect arises when the leader reduces its R&D effort in anticipation that the follower might free ride excessively from its research activities. Our findings corroborate those of Grunfeld (2003) who shows that the relative strengths of the two effects are determined by the follower's level of absorptive capacity. Thus when the size of the technology gap is small (high absorptive capacity), there are weaker incentives for R&D efforts. It is also noteworthy that the model predicts a unique and stable Nash equilibrium in contrast to the previous section.[23]

We also find that when the R&D levels of both firms converge to their steady state values, the second period quantity produced converge to zero. We also show in the appendix that this can happen only if the investment levels of both firms are negative in the neighborhood of the steady state. Thus, this is a clear example of creative destruction or what is also referred to as the Schumpeterian effect at work. Intuitively, the laggard's free-riding behavior causes the leader to invest too little. As a result, inadequate R&D levels fail to reduce the firms' marginal cost of production by an amount which would guarantee a positive quantity of goods produced at all times.

4. A General Model of Dynamic R&D with Endogenous Spillovers

The aim of this section is to provide a more general class of results for dynamic R&D games with uni-directional spillovers. In particular we ask which criteria (if any) are required to determine whether a Schumpeterian effect (as in Section 3) or a Non-Schumpeterian effect will prevail in a differential two stage R&D game with endogenous spillovers.

In what follows, we will develop a theory of dynamic optimal investment in an economy where the representative industry is a duopoly with one-way endogenous spillovers and in which firms play a two stage R&D game in each period. Among the few necessary amendments to the model of Section 3 are: An economy in which savers are the shareholders of the firms, a more general specification of the R&D cost function, a generic distance func-

[23] Note that $I_{1t} = I_{2t} = 0$ not the only way of characterizing steady state. $\frac{I_{1t}}{X_{1t}} = \frac{I_{2t}}{X_{2t}}$ is another possible steady state configuration in which all variables grow at a constant rate in equilibrium. Also note that the requirement that$C > A$ would imply that it is possible that the optimal (product) output for one firm is zero in some periods.

tion to characterize the technology gap and some basic assumptions on the smoothness of the profit function. The requirement of a consumer-side economy is necessary not so much for the sake of completing the model in the general equilibrium sense but most importantly to allow for a platform (stock market) on which the firm's R&D investment can be valued.

4.1. The Model

We shall consider an economy as in Aghion, Harris, Howitt and Vickers (2001) in which labor supply is perfectly elastic and each consumer consumes a constant proportion of their income in each period. This setting has two important properties. First, the discount rate is equal to the rate of interest, that is ρ [24] $= r_t$ and secondly, labor supply decisions are exogenously given. Since our main concern is to look at the dynamics of the duopolists' R&D investments, such an extreme simplification of the consumer economy allows us to "transfer" all the dynamics from consumption to production. The production side of the economy consists of n identical duopolistic industries in which the firms play a two stage R&D game. Here one should note that owing to the assumption that the n industries are identical, the growth rate of the whole economy can be extrapolated from the growth rate of the representative industry. Now, since the only variables growing in any industry are the two rival firms' R&D spending, it must be the case that the dynamic two stage R&D game fully characterizes the economy's behavior. Thus, this "reduced form" of the dynamic general equilibrium model of the current section, is similar to the starting points of the R&D games discussed in Sections 3 and 4.

The duopolists will take the demand derived from the consumer problem as given and compete in Cournot fashion in the output market in the second stage of the game. We shall make the following assumptions:

Assumption 0.1. *The first stage reduced form benefit function is given by*

$$\pi_i \equiv \pi_i \left(X_i, X_j, \beta_t\right), \ i \neq j, i = 1, 2 \tag{37}$$

$$\beta_t \equiv \beta_t \left(X_i, X_j, b\right), b \in (0, 1) \tag{38}$$

Also $\pi_i(\cdot)$ is twice continuously differentiable, increasing and strictly concave in X_i. Moreover it is decreasing in X_j and $\pi_{iX_i}(\cdot)$ is bounded from above.
And $\beta_t(\cdot) \in (0,1), \beta_{tX_1}(\cdot) \leq 0, \beta_{tX_2}(\cdot) \geq 0$

We ought to impose a few more qualifications on (37) and (38). First, the demand schedule derived in the consumer problem and the marginal effective costs of production (2) and (3) are chosen such that the conditions imposed on the profit function in assumption 4.1 are satisfied[25] .

Secondly, given the dependence of (37) on $\beta_t(X_i, X_j, b)$, one can observe that it is possible that $\pi_{1X_2}(\cdot) \geq 0$ due to the free riding effect. It is, therefore, plausible that at first glance our assumption that $\pi_{1X_2}(\cdot) \leq 0$ may seem contradicting. However, if we recall

[24] Abuse of notation: ρ has been used as the interest rate in the previous section. However, due to the above property they will be treated as equal henceforth.

[25] It is not difficult to verify that there exist a class of functions for the demand schedule and marginal cost curve that will satisfy those conditions.

that the first stage profit function of firm 1 (the follower) is positively related to the effective cost of firm 2 (leader), we would understand that x_2 increases in (apart from the free riding effect) also reduces the leader's cost and hence the follower's profit. Hence the correct interpretation of this negative cross partial derivative would be that the negative effect (due to the leader's own cost reductions) of the leader's incremental research on the follower's profits exceeds the latter's benefits from free riding on the leader's research. Another feature of assumption 4.1 is that $\beta_t(\cdot)$ can be regarded as a distance function that corresponds to the technology gap and it can be shown that in any metric space there exists a general class of functions satisfying the property:$\beta_t(X_i, X_j, b) = 0$ if and only if $X_i = X_j$.

Assumption 0.2. *The R&D costs functions of firm 1 is given by*

$$C_{1t}^{R\&D} = I_1 + \phi\left(\frac{I_1}{X_1}\right) I_1 \left(1 + \beta_t\left(X_i, X_j, b\right)\right) \tag{39}$$

$$C_{1tI_{1t}}^{R\&D} = 1 + \left(\phi(\cdot) + \left(\frac{I_1}{X_1}\right)\phi'(\cdot)\right)\left(1 + \beta_t(\cdot)\right) \geq 0$$

$$C_{1tI_{1t}I_{1t}}^{R\&D} = \left(\left(\frac{I_1}{X_1^2}\right)\phi''(\cdot) + \left(\frac{2}{X_1}\right)\phi'(\cdot)\right)\left(1 + \beta_t(\cdot)\right) \geq 0 \text{ and } C_{1tX_{1t}X_{1t}}^{R\&D} \geq 0 \tag{40}$$

The R&D cost functions of firm 2 is given by

$$C_{2t}^{R\&D} = I_2 + \phi\left(\frac{I_2}{X_2}\right) I_2 \tag{41}$$

$$C_{2tI_{2t}}^{R\&D} = 1 + \left(\phi(\cdot) + \left(\frac{I_2}{X_2}\right)\phi'(\cdot)\right) \geq 0$$

$$C_{2tI_{2t}I_{2t}}^{R\&D} = \left(\left(\frac{I_2}{X_2^2}\right)\phi''(\cdot) + \left(\frac{2}{X_1}\right)\phi'(\cdot)\right) \geq 0 \text{ and } C_{2tX_{2t}X_{2t}}^{R\&D} \geq 0 \tag{42}$$

where I_i is the rate of change of X_i over time [26] *while $\phi'(\cdot)$ and $\phi''(\cdot)$ are the first and second derivatives of $\phi(\cdot)$ with respect to I_i .*

Assumption 4.2 simply implies a cost function that is monotonically increasing and convex in investment and is decreasing and convex in R&D stock. (39) and (41) also imply that the price of investment is equal to 1 for both firms. This completes the model.

4.2. Solving the Model

By using the principle of backward induction and bearing in mind that the second stage output level has already been chosen in terms of X_{it} and X_{jt} , we solve for the optimal I_{it} andI_{jt} in the first stage of the game by solving for the subgame perfect open loop Nash equilibrium. Firm 1 objective function is given by

$$\underset{I_{1t}}{Max} \int_0^\infty e^{-\rho t}\left\{\pi_{1t}(\cdot) - I_{1t} - \left(1 + \beta(\cdot)\right) I_{1t}\phi(\cdot)\right\} dt$$

[26] Time subscripts have been omitted in the above functions for convenience

$$s.t\ \dot{X}_{it} = I_{it} - \delta X_{it}, i = 1, 2,\ X_{20} > X_{10} \text{ is given and } \underset{Lim\ T\to\infty}{X_{iT}} \geq 0\ ,\delta \text{ is the depreciation rate} \tag{43}$$

Firm 2 objective function is given by

$$\underset{I_{2t}}{Max} \int_0^\infty e^{-\rho t} \left\{\pi_{2t}(\cdot) - I_{2t} - I_{2t}\phi(\cdot)\right\} dt$$

$$s.t\dot{X}_{it} = I_{it} - \delta X_{it}, i = 1, 2, X_{20} > X_{10} \text{ is given and } \underset{Lim\ T\to\infty}{X_{iT}} \geq 0 \tag{44}$$

The Hamiltonian function and its first order conditions for firm 1 can be given as

$$H_1 = \pi_{1t}(\cdot) - I_{1t} - (1 + \beta(\cdot)) I_{1t}\phi(\cdot) + \sum_{i=1}^{2} \lambda_{it}(I_{it} - \delta X_{it}) \tag{45}$$

$$\frac{dH_1}{dI_{1t}} = 0 \Rightarrow \lambda_{1t} = 1 + (1 + \beta(\cdot)) \left(\phi(\cdot) + \left(\frac{I_{1t}}{X_{1t}}\right)\phi'(\cdot)\right) \tag{46}$$

$$\dot{\lambda}_{1t} = (\rho + \delta)\lambda_{1t} - \left(\pi_{1tX_{1t}}(\cdot) + (1 + \beta(\cdot))\left(\frac{I_{1t}^2}{X_{1t}^2}\right)\phi'(\cdot) - I_{1t}\phi(\cdot)\beta_{X_{1t}}(\cdot)\right) \tag{47}$$

$$\dot{\lambda}_{2t} = (\rho + \delta)\lambda_{2t} - \left(\pi_{1tX_{2t}}(\cdot) - I_{1t}\phi(\cdot)\beta_{X_{2t}}(\cdot)\right) \tag{48}$$

$$\dot{X}_{it} = I_{it} - \delta X_{it}, i = 1, 2 \tag{49}$$

$$\underset{Lim\ T\to\infty}{e^{-\rho T}\lambda_{iT}X_{iT}} = 0 \tag{50}$$

The Hamiltonian function and its first order conditions for firm 2 can be given as

$$H_2 = \pi_{2t}(\cdot) - I_{2t} - I_{2t}\phi(\cdot) + \sum_{i=1}^{2} \mu_{it}(I_{it} - \delta X_{it}) \tag{51}$$

$$\frac{dH_2}{dI_{2t}} = 0 \Rightarrow \mu_{2t} = 1 + \left(\phi(\cdot) + \left(\frac{I_{2t}}{X_{2t}}\right)\phi'(\cdot)\right) \tag{52}$$

$$\dot{\mu}_{2t} = (\rho + \delta)\mu_{2t} - \left(\pi_{2tX_{2t}}(\cdot) + \left(\frac{I_{2t}^2}{X_{2t}^2}\right)\phi'(\cdot)\right) \tag{53}$$

$$\dot{\mu}_{1t} = (\rho + \delta)\mu_{1t} - \left(\pi_{2tX_{1t}}(\cdot)\right) \tag{54}$$

$$\dot{X}_{it} = I_{it} - \delta X_{it}, i = 1, 2 \tag{55}$$

$$\underset{Lim\ T\to\infty}{e^{-\rho T}\mu_{iT}X_{iT}} = 0 \tag{56}$$

Given the above first order conditions[27] we can now derive the main propositions of this paper. We start by deriving an important lemma which constitutes the basis for our propositions.

[27] Note that though $\phi(\cdot)$ has been used interchangeably in both firms' problems, they are not equal in general since while for firm 1 the argument of $\phi(\cdot)$ is $\frac{I_1}{X_1}$, the argument for firm 2 will be $\frac{I_2}{X_2}$.

Lemma 0.1. *If assumptions 4.1 and 4.2 hold, then λ_{1t} and μ_{2t} are the net marginal values of the follower's and the leader's R&D respectively.*

Proof. For the follower (firm 1), rewriting (47), yields

$$\dot{\lambda}_{1t} - (\rho + \delta)\lambda_{1t} = -\left(\pi_{1tX_{1t}}(\cdot) + (1 + \beta(\cdot))\left(\frac{I_{1t}^2}{X_{1t}^2}\right)\phi'(\cdot) - I_{1t}\phi(\cdot)\beta_{X_{1t}}(\cdot)\right) \quad (57)$$

Multiplying both sides by $e^{-(\rho+\delta)t}$ and taking the integral with respect to time from zero to infinity on both sides and rewriting we have

$$\int_0^\infty \frac{d}{dt}\left(e^{-(\rho+\delta)t}\lambda_{1t}\right)dt = -\int_0^\infty e^{-(\rho+\delta)t}\left(\pi_{1tX_{1t}}(\cdot) + (1 + \beta(\cdot))\left(\frac{I_{1t}^2}{X_{1t}^2}\right)\phi'(\cdot) - I_{1t}\phi(\cdot)\beta_{X_{1t}}(\cdot)\right)dt \quad (58)$$

From the First Fundamental Theorem of Calculus we have,

$$\frac{d}{dt}\left(e^{-(\rho+\delta)t}\lambda_{1t}\right)|_0^\infty = -\int_0^\infty e^{-(\rho+\delta)t}\left(\pi_{1tX_{1t}}(\cdot) + (1 + \beta(\cdot))\left(\frac{I_{1t}^2}{X_{1t}^2}\right)\phi'(\cdot) - I_{1t}\phi(\cdot)\beta_{X_{1t}}(\cdot)\right)dt \quad (59)$$

Using the transversality condition (50) we have

$$\lambda_{10} = \int_0^\infty e^{-(\rho+\delta)t}\left(\pi_{1tX_{1t}}(\cdot) + (1 + \beta(\cdot))\left(\frac{I_{1t}^2}{X_{1t}^2}\right)\phi'(\cdot) - I_{1t}\phi(\cdot)\beta_{X_{1t}}(\cdot)\right)dt \quad (60)$$

Thus at any time t we have,

$$\lambda_{1t} = \int_0^\infty e^{-(\rho+\delta)t}\left(\pi_{1tX_{1t}}(\cdot) + (1 + \beta(\cdot))\left(\frac{I_{1t}^2}{X_{1t}^2}\right)\phi'(\cdot) - I_{1t}\phi(\cdot)\beta_{X_{1t}}(\cdot)\right)dt \quad (61)$$

w from (61), we observe that the bracketed expression consists of three terms. The first one is the marginal increase in profits due to an incremental unit of R&D (X_{it}). The sum of the second and third terms (both being positive since $\beta_{X_{1t}}(\cdot)$) gives us the marginal reduction in the cost of investment due to the incremental unit of R&D (X_{it}). Thus the RHS of (61) should give us the marginal value of the follower's R&D (X_{it}) at any time t discounted by the time preference parameter and the depreciation rate. Hence it gives us the net marginal value of the follower's R&D at any point in time. By analogous arguments one can find the net marginal value of the leader's R&D at any point in time and it will be given by the following expression;

$$\mu_{2t} = \int_0^\infty e^{-(\rho+\delta)t}\left(\pi_{2tX_{2t}}(\cdot) + \left(\frac{I_{1t}^2}{X_{1t}^2}\right)\phi'(\cdot)\right)dt \quad (62)$$

■

Lemma 4.1 gives an important result about the shadow price of the two firms' R&D levels. Savers in this economy would use λ_{1t} and μ_{2t} as indicators[28] for the value of their investments in the stock market.

---

[28] Although in practice the shadow price of capital, which is also referred to as Tobin's q in the investment literature, cannot be empirically observed, an equivalent average measure can be used. For more details see Hayashi (1982).

Proposition 0.1. *If assumptions 4.1 and 4.2 hold, then* $\lambda_{1t} = \mu_{2t}$ *in equilibrium, that is the marginal values of the two firms' R&D should be equal.*

Proof. From Lemma 4.1, we know that $\lambda_{1t} > 0$ and $\mu_{2t} > 0$ are the marginal values of the follower's and the leader's respectively. We claim that $\lambda_{1t} = \mu_{2t}$ in equilibrium. Suppose not, then we have two possible cases (i)$\lambda_{1t} > \mu_{2t}$or (ii) $\lambda_{1t} < \mu_{2t}$. Case (i) implies that the valuation of firm 1 at time t is greater than that of firm 2. A frictionless stock market would imply that savers would transfer theirs funds from the firm with lower valuation to the firm with the higher valuation such that the marginal value of the latter will start to fall (recall that from assumption 4.1 that the firm's benefits from R&D increase at a decreasing rate and from assumption 4.2, the reductions in the cost of investment due to an additional unit of R&D also decrease at a decreasing rate) while that of the firm with the lower valuation will start to rise. Thus the no-arbitrage condition would imply that case (i) cannot be equilibrium. By analogous arguments one can show that case (ii) cannot hold[29] ■

Proposition 4.1 states that if investors (identical savers of this economy) were to allocate their funds in the representative industry comprising of the leader and the follower firms, then the valuations (in terms of the marginal benefits of shareholders) of these two firms on the stock market must equalize in equilibrium. The economic intuition behind this result is given as follows. In a dynamic economy in which agents invest in the two firms of the representative industry, the shadow price of such investments can be observed in terms of the valuation of these firms on the stock market. Now since shareholders are always seeking higher returns they would increase (decrease) their investment in the firm with higher (lower) valuation. But as this process goes on the marginal values of the firms would fall (rise) until they become equal in equilibrium. Proposition 4.1 also clearly justifies the importance of having a consumer side in our model. Thus although the R&D activities of the two firms are not a priori comparable, when their respective valuations are translated to the investor's portfolio, we can find a relationship between their marginal values.

Proposition 0.2. *If assumptions 4.1 and 4.2 hold, then the two firms' R&D investment strategies are strategic complements; that is, an increase in the leader's investment rate will lead to an increase in the follower's investment rate and vice versa.*

Proof. Lemma 4.1 gives us two expressions for λ_{1t} and μ_{2t} respectively. Equating them as per Proposition 4.1, and using (46) and (52), we have

$$1 + (1 + \beta_t (X_{1t}, X_{2t}, b)) \left(\phi \left(\frac{I_{1t}}{X_{1t}} \right) + \left(\frac{I_{1t}}{X_{1t}} \right) \phi' \left(\frac{I_{1t}}{X_{1t}} \right) \right) = \left(\phi \left(\frac{I_{2t}}{X_{2t}} \right) + \left(\frac{I_{2t}}{X_{2t}} \right) \phi' \left(\frac{I_{2t}}{X_{2t}} \right) \right) \tag{63}$$

Now treating X_{1t} , X_{2t} and b as exogenous and by taking total derivatives on both sides of (63), we can make use of the convexity assumption (in 5.2.2) to establish a positive relationship between X_{1t} and X_{2t}. ■

Proposition 4.2 gives us a very important result in the theory of differential two stage dynamic R&D games with one-way endogenous spillovers. It states that an increase in the

[29] At $\lambda_{1t} = \mu_{2t}$, the consumer is indifferent between investing in firm 1 and investing in firm 2.

leader's R&D will initially encourage the follower to increase its R&D. Subsequently, the leader will respond to the latter increase to by raising its own R&D level even higher. In this way the technological frontier is always shifting outwards. Intuitively, proposition 4.2 can be better described along these lines. If both firms in the duopoly face rising marginal costs of R&D, then an incremental increase in the level of R&D by one firm will increase its marginal cost as well. Now, since in equilibrium the marginal costs of the firms must be equal, the best response of the rival firm must be to increase its level of R&D. The rationale for the result is that when the follower undertakes some research, the leader's technological advantage starts to fall and in order to maintain its lead it has to invest further. Thus the laggard always pushes the leader to do more research and as a result a competitive market structure prevails at all times.

Proposition 0.3. *If assumptions 4.1 and 4.2 hold, then (i) in the absence of spillovers both technology catch-up and symmetric investment are sufficient conditions for the existence of a steady state in the dynamic two stage R&D game (Schumpeterian economy) (ii) in the presence of spillovers, only technology catch-up is a sufficient condition for the existence of a steady state in the dynamic two stage R&D game (Non-Schumpeterian economy)*

Proof. Define the steady state as the balanced growth path on which ${}_gQ_t = 0$ where ${}_gQ_t$ is the growth rate of total output in the representative industry. The latter fully characterizes the behavior of the whole economy. Now, since we know that the only two variables growing in the industry are the two firms R&D stock, we assume that ${}_gQ_t = 0$ is equivalent to ${}_gX_{1t} ={}_g X_{2t}$.[30] Thus, at the steady state, we shall have $\frac{I_{1t}}{X_{1t}} = \frac{I_{2t}}{X_{2t}}$(Note that the steady state $I_{1t} = I_{2t} = 0$ of the previous section can be seen as a special case of this.)

(i) Technology catch-up (sufficiency) .Consider equation (63) for the case where $\beta = 0$, that is there are no spillovers.

$$\left(\phi\left(\frac{I_{1t}}{X_{1t}}\right) + \left(\frac{I_{1t}}{X_{1t}}\right)\phi'\left(\frac{I_{1t}}{X_{1t}}\right)\right) = \left(\phi\left(\frac{I_{2t}}{X_{2t}}\right) + \left(\frac{I_{2t}}{X_{2t}}\right)\phi'\left(\frac{I_{2t}}{X_{2t}}\right)\right) \tag{64}$$

Now suppose catch up takes place at some time t, then $X_{1t} = X_{2t}$. (64) becomes

$$\left(\phi\left(\frac{I_{1t}}{X_{1t}}\right) + \left(\frac{I_{1t}}{X_{1t}}\right)\phi'\left(\frac{I_{1t}}{X_{1t}}\right)\right) = \left(\phi\left(\frac{I_{2t}}{X_{1t}}\right) + \left(\frac{I_{2t}}{X_{1t}}\right)\phi'\left(\frac{I_{2t}}{X_{1t}}\right)\right) \tag{65}$$

Since from assumption 4.2 we know that $\left(\phi\left(\frac{I_{it}}{X_{1t}}\right) + \left(\frac{I_{it}}{X_{1t}}\right)\phi'\left(\frac{I_{it}}{X_{1t}}\right)\right)$is monotonically increasing in I_{it}, it must be the case that $I_{1t} = I_{2t}$ (symmetric investment). Thus, since $X_{1t} = X_{2t}$, it must be that $\frac{I_{1t}}{X_{1t}} = \frac{I_{2t}}{X_{2t}}$ and hence technology catch-up implies the existence of a steady state.

---

[30] In general, if total output Q_t can be expressed as a function of the technology gap only , then for the class of distance functions assumed in our framework, ${}_gX_{1t} ={}_g X_{2t}$ if and only if that ${}_gQ_t = 0$. An example of a differential game in which ${}_gX_{1t} ={}_g X_{2t}$ at the steady state can be found in Traca and Reis (2003). The latter prove that ${}_gX_{1t} ={}_g X_{2t}$ should hold at the steady state for a Cobb-Douglas distance function. We also note that the spillover distance function used in the previous section also satisfies this property. More specifically, it can be shown that ${}_gQ_t = 0$ if the growth rate of the spillover distance function converges to zero and this occurs if and only if $\frac{bX_{1t}}{1+bX_{1t}} = \frac{bX_{2t}}{1+bX_{2t}}$or ${}_gX_{1t} ={}_g X_{2t}$ for large bX_{1t} and bX_{2t} .

Symmetric Investment (sufficiency)Now suppose that we are at the state where $I_{1t} = I_{2t}$, then (64) becomes

$$\left(\phi\left(\frac{I_{1t}}{X_{1t}}\right)+\left(\frac{I_{1t}}{X_{1t}}\right)\phi'\left(\frac{I_{1t}}{X_{1t}}\right)\right)=\left(\phi\left(\frac{I_{1t}}{X_{2t}}\right)+\left(\frac{I_{1t}}{X_{2t}}\right)\phi'\left(\frac{I_{1t}}{X_{2t}}\right)\right) \tag{66}$$

Since from assumption 4.2 we know that $\left(\phi\left(\frac{I_{it}}{X_{1t}}\right)+\left(\frac{I_{it}}{X_{1t}}\right)\phi'\left(\frac{I_{it}}{X_{1t}}\right)\right)$is monotonically increasing in X_{it}, it must be the case that $X_{1t} = X_{2t}$. But $X_{1t} = X_{2t}$ implies catch-up if the technology gap belongs to the special class of functions satisfying $T(X_{1t}, X_{2t}) = 0$ if and only if $X_{1t} = X_{2t}$. Again here it must be that $\frac{I_{1t}}{X_{1t}} = \frac{I_{2t}}{X_{2t}}$ since $I_{1t} = I_{2t}$.

(ii) Technology catch-up (sufficiency) Consider equation (63) again

$$1+(1+\beta_t(X_{1t},X_{2t},b))\left(\phi\left(\frac{I_{1t}}{X_{1t}}\right)+\left(\frac{I_{1t}}{X_{1t}}\right)\phi'\left(\frac{I_{1t}}{X_{1t}}\right)\right)=\left(\phi\left(\frac{I_{2t}}{X_{2t}}\right)+\left(\frac{I_{2t}}{X_{2t}}\right)\phi'\left(\frac{I_{2t}}{X_{2t}}\right)\right) \tag{67}$$

Now suppose catch up takes place at some time t, then $X_{1t} = X_{2t}$.Given the condition from assumption 4.1 that $\beta_t(X_i, X_j, b) = 0$ if and only if $X_{1t} = X_{2t}$, (63) reduces to

$$\left(\phi\left(\frac{I_{1t}}{X_{1t}}\right)+\left(\frac{I_{1t}}{X_{1t}}\right)\phi'\left(\frac{I_{1t}}{X_{1t}}\right)\right)=\left(\phi\left(\frac{I_{2t}}{X_{1t}}\right)+\left(\frac{I_{2t}}{X_{1t}}\right)\phi'\left(\frac{I_{2t}}{X_{1t}}\right)\right)$$

Since from assumption 4.2 we know that $\left(\phi\left(\frac{I_{it}}{X_{1t}}\right)+\left(\frac{I_{it}}{X_{1t}}\right)\phi'\left(\frac{I_{it}}{X_{1t}}\right)\right)$is monotonically increasing in I_{it}, it must be the case that $I_{1t} = I_{2t}$ and hence $\frac{I_{1t}}{X_{1t}} = \frac{I_{2t}}{X_{2t}}$. This completes the proof.[31] ■

Proposition 4.3 gives us an important relationship between the existence of a steady state and the level of technology gap between firms. Before any further description of this result one important general observation between the technology gap and R&D incentives ought to be made. If the technology gap becomes zero at some point in time, then around its neighborhood the leader's R&D level has to be increasing (decreasing) at a lower (higher) rate than that of the follower since $X_{10} > X_{20}$ were assumed to be the initial conditions. Moreover, if symmetric investment always implies technology catch-up, then it must be the case that at some point in time before the catch-up state the leader's investment level rose (fell) at a lower (higher) rate than that of the follower. Intuitively, the follower's incremental investment reduces the leader incentives to undertake high levels of research due to the free riding effect. Hence, this is an example of a Schumpeterian effect due to the possibility of creative destruction. As a corollary to this observation, we have the fact that in a state in which symmetric investment does not imply that the gap shrinks to zero, it is possible that the follower's incremental increase in R&D effort leads to an even larger incremental increase in the leader's effort (excluding of course the case where $I_{1t} = I_{2t} = 0$). Thus in this scenario the leader would always innovate further to maintain its technological lead. This is an example of a Non-Schumpeterian effect whereby the leader always innovates further for fear of losing its technological advantage. Now, since part (i) of proposition 4.3 describes

[31] We leave it as an exercise to the reader to disprove that symmetric investment is a sufficient condition for the existence of a steady state in the presence of spillovers.

a situation in which states with symmetric investment without catch-up cannot exist, it relates better to the first scenario (Schumpeterian) and since in part (ii) of proposition 4.3 it is possible to have symmetric investment in states with no catch-up, it is closer to the second scenario (Non-Schumpeterian).

Proposition 4.3 states that if there is no technological leakage between the leader and the follower in the dynamic two stage R&D game, then convergence (catch-up or ex-post symmetry) both in R&D stock and at the R&D investment level are sufficient conditions for the existence of a steady state. However, if we introduce endogenous externalities into the model, then it is possible to have symmetric investment with R&D stock asymmetry in which the leader remains the leader and thrives to maintain its lead. While this result primarily helps demarcate between the Schumpeterian and the Non-Schumpeterian effects in a general class of dynamic R&D games, it also leads to an interesting policy implication; with less appropriation (less strict rules against the act of reverse engineering or fewer rules hindering the diffusion of technology) we can always have an economy that enjoys Non-Schumpeterian growth in the sense that competition always takes place in a competitive rather than in a monopolistic environment. This emerges directly from poposition 4.3 which postulates that the presence of spillovers can lead to a Non-Schumpeterian economy. Our results, together with the empirical findings of Cohen and Levinthal (1989) provide a strong case in favor of lower appropriations.

5. Summary and Concluding Remarks

We have proposed a representation of an asymmetric two stage R&D game in which technology diffuses from the leader to the follower if and only if the latter undertakes some minimum R&D of its own. Moreover, while such externalities increase the benefits of producing goods, it also increases the cost of doing research. This paper has extended the basic AJ framework by including uni-directional endogenous spillovers and allowing such externalities to depend on the size of the technology gap. In addition, it accommodates these additional features by posting a dynamic two stage game in which the dynamics of the gap can be captured. These two features (endogenous spillovers and R&D costs increasing in the spillover rate) allow us to derive some important relationships between the technology gap and research incentives, as well as the dynamic equilibrium of a differential game.

Our analysis of a static game reveals primarily (in an extended AJ framework) that there exists a non-monotonic relationship between free-riding behavior and the spillover rate. However, the SPNE of the game was not unique. From this observation and the fact that technology gap evolves over time, we formulate an R&D differential game with one-way endogenous spillovers. We found that the technology gap would shrink to zero if there were to exist steady state equilibrium in our dynamic example. Furthermore, the equilibrium would be stable if the industry costs were large enough.

In a more general setting, we found, under some reasonable assumptions, that catch-up and symmetric investment are sufficient conditions for the existence of a steady state in a framework with no spillovers and that steady states with no catch-up exist only in the presence of spillovers. Thus, we infer that the latter can promote welfare (growth) by enhancing dynamic competition in an economy. Promising directions for further investigation include the extension of our analysis to a cooperative setting with research joint ventures (as has been done traditionally for the static and exogenous spillover case). Also, proper characterization of the stability conditions that would guarantee a Saddle Path to the different steady states discussed in Section 4 is an avenue for further inquiry.

6. Appendix

6.1. Derivations of (13) and (14)

For firm 1, FOC of (11) (by substituting β by (10) when appropriate) with respect to X_1 results in

$$\frac{4}{9(1+bX_2)}\left(A-C+2X_1+\left(\frac{2b(X_2-X_1)}{1+bX_2}-1\right)X_2\right)=\left(\frac{1+2bX_2-2bX_1}{1+bX_2}\right) \quad \text{(A1)}$$

Simplifying (A1) gives (13).

For firm 2, FOC of (12) with respect to X_2 results in

$$\frac{2(2+2bX_2+bX_1)}{9(1+bX_2)^2}\left(A-C+\left(2-\frac{b(X_2-X_1)}{1+bX_2}\right)X_2-X_1\right)=1 \quad \text{(A2)}$$

Simplifying (A2) gives (14).

6.2. Proof for Proposition 2.1

We shall derive this result from the reaction function of the laggard given by (15). By making X_1 the subject of formula, we find the roots of the graph, that is, we find the values of X_2 when $X_1 = 0$ Thus, we have at $X_1 = 0$

$$X_2 = \frac{-b\left(\frac{27}{4}-f\right)-1\pm\sqrt{\left(\left(\frac{27}{4}-f\right)b+1\right)^2-4\left(\frac{9}{4}-f\right)\left(\frac{9}{2}b^2-b\right)}}{9b^2-2b} \quad \text{(A3)}$$

where $f = A - C$.

Now, it can be shown that for $f < 0$ and $b < \frac{2}{9}$, there will be one positive root and one negative root which lies to the right of the vertical asymptote. The vertical intercept is given by

$$X_1 = \frac{\frac{9}{4}-f}{\frac{9}{4}b-f} \quad \text{(A4)}$$

Moreover the part of the graph that is relevant for the parameter space $f < 0$, $b < \frac{2}{9}$ and $X_i > 0$ is an inverted u-shaped parabola which attains a maximum in the positive orthant. This implies that the laggard's investment will increase up to the maximum point after which it starts to free-ride. The maximum point is where

$$X_2^M = \frac{b\left(\frac{27}{4}+f\right)+1}{2b-9b^2} \quad \text{(A5)}$$

We now try to determine the sign of X_2^M w.r.t b.Using (A4), we find that the sign is negative if $b < \frac{1}{9}$ and positive if $\frac{1}{9} < b < \frac{2}{9}$. Thus, since $-\Delta X_2^M$ is a proxy to free-riding, we conclude that free-riding increases with as long as $b < \frac{1}{9}$, and decreases with it for $\frac{1}{9} < b < \frac{2}{9}$. Hence, the relationship between (free-riding) is non-monotonic and is an inverted U-shaped.

6.3. Proof for Proposition 2.2

(A2) divided by (A1) gives

$$\frac{(A-C+2X_2-X_1)(2+2bX_2+bX_1)-\frac{9}{2}(1+bX_2)^2}{\frac{9}{4}(1+2bX_2-2bX_1)-(A-C+2X_1-X_2)} = \left(\frac{1+2bX_2+bX_1}{2}\right) \quad \text{(A6)}$$

(A5) can be simplified as

$$\frac{27}{2} - 6(A-C) + \left(\left(\frac{63}{2}-6(A-C)\right)b-6\right)X_2 - b\left(\frac{27}{4}+3(A-C)\right)X_1 - \left(\frac{9}{2}b^2+3b\right)X_2X_1 + \left(18b^2-6b\right)X_2^2 - \frac{9}{2}b^2X_1^2 = 0 \text{ (A7)} \quad \text{(A7)}$$

Now it can be shown that (A7) is the graph of a non-degenerate conic. Thus, it can be shown to be either a hyperbola or ellipse with some rotation and/or translation from the origin. To prove for existence of a Nash equilibrium we find the X_1 and X_2 intercepts of this conic (which is a reduced-form of the reaction function of the leader firm) and compare their values with the X_1 and X_2 intercepts of the parabola (the reaction function of the laggard firm), the roots of which are given in (A3). From (A7), we find the values of X_2 when $X_1 = 0$. Thus,

$$X_2 = \frac{\left(6-\left(\frac{63}{2}-6f\right)b\right)-1 \pm \sqrt{\left(\left(\frac{63}{2}-6f\right)b-6\right)^2-(54-24f)\left(18b^2-6b\right)}}{2\left(18b^2-6b\right)} \quad \text{(A8)}$$

and at $X_2 = 0$, X_1will be

$$X_1 = \frac{b\left(\frac{27}{4}+3f\right) \pm \sqrt{\left(\left(\frac{27}{4}+3f\right)b\right)^2+\frac{9}{2}b^2(54-24f)}}{-9b^2} \quad \text{(A9)}$$

Now it can be shown that if b is low enough and C is large enough, the X_1 intercept of the conic is always greater than that of the parabola. Moreover, if C is large enough, is low enough so that terms in b^2can be ignored, the X_2 intercept of the conic is always less than that of the parabola. If we denote the difference between the parabola and the conic by some function $\Psi(X_1, X_2)$, then as is well-known from the Weierstrass Intermediate Value Theorem that if a continuous function on an interval is sometimes positive and sometimes negative, it must be zero at some point. This proves for existence of Nash equilibrium. However, we cannot guarantee uniqueness since both the parabola and the conic are non-monotonic functions. This completes the proof.

6.4. Proof for Proposition 3.1

Assume that at the steady state $I_{1t} = I_{2t} = 0$, then replacing $I_{1t} = 0$ in (26), we have

$$\lambda_{1t} = 0 \quad \text{(A10)}$$

Replacing (A10) in (27) and using the fact that (A10) implies $\dot{\lambda}_{1t} = 0$, we have (after re arranging)

$$A - C + 2X_{1t} + \left(\frac{2b\,(X_{2t} - X_{1t})}{1 + bX_{2t}} - 1 \right) X_{2t} = 0 \tag{A11}$$

Replacing $I_{2t} = 0$ in (32), we have

$$\mu_{1t} = 0 \tag{A12}$$

Replacing (A12) in (33) and using the fact that (A12) implies $\dot{\mu_{1t}} = 0$, we have (after re arranging)

$$A - C + \left(2 - \frac{b\,(X_{2t} - X_{1t})}{1 + bX_{2t}} \right) X_{2t} - X_{1t} = 0 \tag{A13}$$

Solving (A11) and (A13) simultaneously gives

$$X_{1t}^* = X_{2t}^* = C - A = -\frac{1}{b} \tag{A14}$$

Now since $b \in (0,1)$ and $X_{1t}^*, X_{2t}^* \geq 0$ we choose our parameter values such that $C > A$.[32]

he above proves part (i) and (ii) of Proposition 4.1.1. To check the stability we have to linearize the system of differential equations given by (26), (27), (32) and (33) in the neighborhood of the steady state. After some manipulations of these four equations we manage to reduce them the following two equations

$$\left(\frac{1 + 2bX_{2t} - bX_{1t}}{X_{1t}\,(1 + bX_{2t})} \right) \dot{I}_{1t} + \frac{b\,(1 + bX_{1t})}{X_{1t}\,(1 + bX_{2t})^2} I_{1t} I_{2t} - \frac{(1 + 2bX_{1t})}{2X_{1t}^2\,(1 + bX_{2t})} I_{1t}^2 -$$
$$\left(\frac{\rho\,(1 + 2bX_{2t} - bX_{1t})}{X_{1t}\,(1 + bX_{2t})} \right) I_{1t} + \frac{4}{9\,(1 + bX_{2t})} \left(A - C + 2X_{1t} + \left(\frac{2b\,(X_{2t} - X_{1t})}{1 + bX_{2t}} - 1 \right) X_{2t} \right) = 0 \tag{A15}$$

$$\frac{\dot{I}_{2t}}{X_{2t}} - \frac{I_{2t}^2}{2X_{2t}^2} - \frac{\rho I_{2t}}{X_{2t}} + \frac{2\,(2 + 2bX_{2t} + 2bX_{1t})}{9\,(1 + bX_{2t})^2} \left(A - C + \left(2 - \frac{b\,(X_{2t} - X_{1t})}{1 + bX_{2t}} \right) X_{2t} - X_{1t} \right) = 0 \tag{A16}$$

We first compute the Jacobian determinant of the above evaluated at the steady state value $X_{1t}^* = X_{2t}^* = C - A$. Note that the calculation is made simpler since we know that the steady state values of $I_{1t}^*, I_{2t}^*, \dot{I}_{1t}^*, \dot{I}_{2t}^*$are equal to zero. Bearing this fact in mind and after taking the total derivatives of the above expressions with respect to X_{1t} and X_{2t} with we arrive (after some algebraic manipulations) to the following Jacobian matrix:

$$\underset{2\times 2}{J} = \begin{matrix} \frac{8}{9(1-b(A-C))^2} & \frac{-4(1+b(A-C))}{9(1-b(A-C))^2} \\ \frac{2(3b(A-C)-2)}{9(1-b(A-C))^3} & \frac{2(2-3b(A-C))(2-b(A-C))}{9(1-b(A-C))^3} \end{matrix} \tag{A17}$$

Now it can be shown that

$$|J| = \frac{8\,(3b\,(A - C))}{b\,((A - C) - 1)^4} \tag{A18}$$

[32] Note that this "high cost industries" assumption might imply that the second stage output for this example is negative in some periods.

Since $|J| \leq 0$ holds by definition for the parameter restrictions[33] of our model, we conclude that the steady state equilibrium is stable. Hence there exists a stable path converging to the unique steady state of our model. However, it is possible that there is more than one path converging to such steady state. We next prove that this path is unique; that is it is a Saddle Path.

In control theory it is well known that if the set of eigenvalues given by the matrix in (A17) contains a positive real part and a negative real part, then the path to the steady state of the system described above is a Saddle path. It can be verified[34] that (A17) has two eigenvalues and given the parameter restrictions of our model, the fact that one is positive and the other is negative holds by definition.

For part (v), we consider four different cases. (1) The investment levels of both firms are negative in the neighborhood of the steady state. (2) The investment level of firm 1 is negative while that of firm 2 is positive. (3) The investment level of firm 1 is positive while that of firm 2 is negative. (4) The investment levels of both firms are positive. For case (1), we let the R&D level of firm 1 and firm 2 be $C - A + \epsilon_1$ and $C - A + \epsilon_2$ for small $\epsilon_1, \epsilon_2 > 0$ respectively. We then compute $q_{it} = \frac{A - 2C_i + C_j}{3}$ evaluated at $C - A + \epsilon_1$ and $C - A + \epsilon_2$ for $i = 1, 2$. We find that $q_{it} > 0$ for $i = 1, 2$ as long as $|\epsilon_1 - \epsilon_2|$ is not too large. By analogous reasoning, we find that for each of the cases (2) - (4), at least one firm does not have an incentive to produce a positive output. Thus $q_{it} > 0$ for $i = 1, 2$ only for case (1). Hence, we have a positive mass in the neighborhood of the steady state equilibrium of the full game only if the R&D stocks of both firms are falling in its neighborhood. This completes the proof.

References

[1] P. Aghion, C. Harris, P. Howitt and J. Vickers, Competition, Imitation and Growth with Step by Step Innovation, *Review of Economic Studies* **68** (2001), 467–492.

[2] R. Amir, Modelling Imperfectly Appropriable R&D via Spillovers, *International Journal of Industrial Organization* **18** (2000), 1013–1032.

[3] R. Amir, J. Wooders, Effects of One-Way Spillovers over Market Shares, Industry Price, Welfare and R&D Cooperation, *Journal of Economics and Management Strategy* **18** (1999), 223–249.

[4] R. Amir, J. Wooders, One-Way, Spillovers, Endogenous Innovator/Imitator Roles and Research Joint Ventures, *Games and Economic Behavior* **31** (2000), 1–25.

[5] R. Amir, I. Evstignev, J. Wooders, Non Cooperative versus Cooperative R&D with Endogenous Spillover rates, *Games and Economic Behavior* **42** (2003), 183–207.

[6] G. Cameron, The Sun Also Rises. Productivity Convergence between Japan and the USA, *Working Paper*, Nuffield College, Oxford, 1999.

[33] Note we use once more the "high cost" condition that $C > A$.

[34] The derivations can be provided upon request.

[7] D. Campisi, P. Mancuso, A. Nastasi, R&D Competition, Absorptive Capacity and Market Shares, *Journal of Economics* **73** (2001), 57–80.

[8] W. Cohen, D. Levinthal, Innovation and Learning: The Two Faces of R&D, *Economic Journal* **99** (1989), 569–596.

[9] C. D'Aspremont, A. Jacquemin, Cooperative and Non cooperative R&D in Duopoly with Spillovers, *American Economic Review* **78** (1988), 1133–1137.

[10] E.J. Dockner, V.N. Long, S. Gerhard, S. Jorgensen, *Differential Games in Economics and Management Science*, Cambridge University Press, 2000.

[11] L. Grunfeld, Meet me Halfway but Don't Rush: Absorptive Capacity and Strategic Investment Revisited, *Working Paper* (NIIA), 2003.

[12] F. Hayashi, Tobin's Marginal q and Average q: A Neoclassical Interpretation *Econometrica* **50** (1982), 213–224.

[13] I. Henriques, Cooperative and Non cooperative R&D in Duopoly with Spillovers: Comment, *American Economic Review* **80** (1990), 638–640.

[14] M. Kamien, E. Muller, I. Zang, Research Joint Ventures and R&D Cartels, *American Economic Review* **82** (1992), 1293–1306.

[15] M. Kamien, I. Zang, Meet me Halfway: Research Joint Ventures and Absorptive Capacity, *International Journal of Industrial Organization* **18** (2000), 995–1012.

[16] Y. Katsoulacos, D. Ulph, Endogenous Spillovers and Research Joint Ventures, *The Journal of Industrial Economics* **XLVI** (1998), 333–357.

[17] K. Kultti, T. Takalo, R&D Spillovers and Information Exchange, *Economics Letters* **61** (1998), 21–23.

[18] S. Martin, Spillovers, Appropriablity and R&D, *Working Paper*, 2000.

[19] P. Peretto, Sunk Costs, Market Structure and Growth, *International Economic Review* **37** (1996), 895–923.

[20] J. Reinganum, A Dynamic Game of R&D: Patent Protection and Competitive Behavior, *Econometrica* **50** (1982), 671–688.

[21] L.E. Ruff, Research and Technological Progress in a Cournot Economy, *Journal of Economic Theory* **1** (1969), 397–415.

[22] J.A. Schumpeter, *The Theory of Economic Development*, Oxford University Press, 1934.

[23] D. Simpson, N. Vonortas, Cournot Equilibrium with Imperfectly Appropriable R&D, *The Journal of Industrial Economics* **42** (1994), 79–92.

[24] K. Suzumura, Cooperative and Non cooperative R&D in an Oligopoly with Spillovers, *American Economic Review* **82** (1992), 1307–1320.

[25] D. Traca, A. Reis, *Spillovers and the Competitive Pressure for Long-Run Innovation*, Unpublished, 2003.

[26] D. Vencatachellum, Endogenous Growth with Strategic Innovations, *Journal of Economic Dynamics and Control* **23** (1998), 233–254.

[27] D. Vencatachellum, A Differential R&D Game: Implications for Knowledge- Based Growth Models, *Journal of Optimization Theory and Applications* **96** (1998), 175–189.

In: Mathematics, Game Theory and Algebra Compendium ISBN 978-1-60456-659-8
Editor: Jacob H. Mathias, pp. 361-371

Chapter 31

Mathematics of the JIPTO and Theory of the Pursuit

Grigoriy Tomski
UNESCO

1. Introduction

We know that the Game Theory initially was the study of the parlour games. The mathematical theory of the JIP (Jeux Intellectuels de Poursuite - Intellectual Games of Pursuit) draws the attention of the mathematicians since 1989 [1, 2]. The success of the JIPTO (Jeux Intellectuel de Poursuite de Tomski), popularized on an international scale thanks to the support of UNESCO and used like teaching support and creative stimulant [2-5], reinforces this interest. The whole of the geometrical proposals on the JIPTO and the other JIP presents an interesting extension of the traditional geometry.

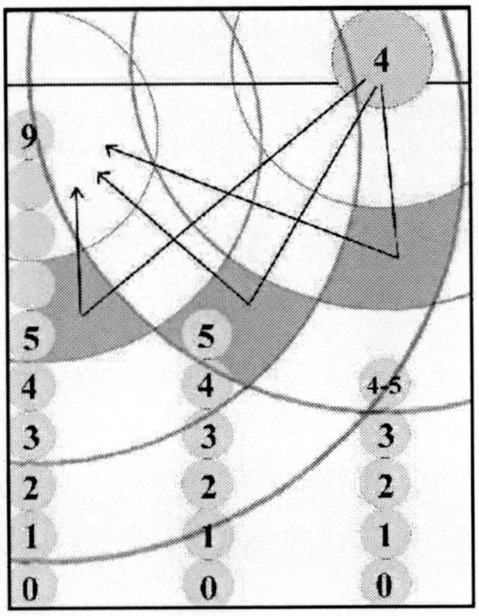

2. The Particularity of the JIPTO

The success of the JIPTO and the rich possibility of his use as teaching support is explained by the diversity of its versions and the fact why its board has the form of ”vertical triptych” with the decorative aspect which enables to create a rich collection of the artistic and decorative compositions of the JIPTO.

Basic versions of this game: **JIPTO-B1-A1, JIPTO-B1-A2, JIPTO-B2-A1, JIPTO-B2-A2** model the pursuit of the five ”fugitives” by a ”pursuer”. The ”fugitives” try to reach the other side of the board avoiding be touched by the ”pursuer” ; this one tries to capture the ”fugitives” further possible from its side. The board of the JIPTO is divided into three principal zones by lines I, II, III. Each ”fugitive” which touches one of the these lines, marks a point by reached line.

In fact, the JIPTO is a family of parlour games, on the same board, which has been developed starting from some initial versions. The process of construction of the various versions of the JIPTO is founded on the progressive complication of the rules: each created game generates several games corresponding to the modifications and complications of the rules. Certain modifications create more or less perceptible complication of preceding version, while the others are real transformations which allow, for example, on the basis of intellectual version of the JIPTO to create versions of the JIPTO of chance or cooperative versions.

The board of the JIPTO does not have boxes (squares). One can use a hexagonal grid, but in this case a figure can only move in six directions. These simplified games, derived from the JIPTO, we call them the ”Jiptoïdes”. It is interesting that several versions of Jiptoïde, for example, with ”castlings” of the personalized pieces are as or even more complicated as the Chess.

Since 1993, thanks to the support of UNESCO, we began the popularization of the JIPTO on an international scale within the framework of the FIDJIP (International Federation of the JIPTO, http://www.jipto.com).

3. Classification of the Versions

Let us indicate by **JIPTO-B1** the basic JIPTO where the ”pursuer” is twice taller than the ”fugitives” and by **JIPTO-B2** the basic version where all the pieces have same dimensions (in this case a ”pursuer” must carry out two steps when the ”fugitives” advance of a step).

Let us designate by **JIPTO-A1** the versions of JIPTO with first method of calculation of the points when we keep with a ”fugitive” a marked point by reached line. Let us designate by **JIPTO-A2** the versions of JIPTO with second method of calculation of points when we cancel the gained point by a ”fugitive” on a line if it moves back then behind this line. In this case we just analyze the positions at the moment of capture.

The four principal basic versions of the JIPTO are:

1. **JIPTO-B1-A1** is at the same time **JIPTO-B1** and **JIPTO-A1**.

2. **JIPTO-B1-A2** is at the same time **JIPTO-B1** and **JIPTO-A2**.

3. **JIPTO-B2-A1**

4. **JIPTO-B2-A2**

So we can give to each "fugitive" a value. In this case:

- the "fugitive" of value 1 which touches one of the lines I, II or III marks one point by reached line;
- the "fugitive" of value 2 which touches one of the lines I, II or III marks two points by reached line;
- the "fugitive" of value **N** which touches one of the lines I, II or III marks **N** points by reached line.

Let us designate by **JIPTO-P(a, b, c, d, e)** the versions of the JIPTO with the personalized pieces of "fugitives" with the values has **a, b, c, d, e.** We play generally the following versions:

5. **JIPTO-B1-A2-P(1, 2, 3, 4, 5)**

6. **JIPTO-B2-A2-P(1, 2, 3, 4, 5)**

7. **JIPTO-B1-A2-P(1, 1, 1, 1, 5)**

8. **JIPTO-B2-A2-P(1, 1, 1, 1, 5)**

Let us designate by **JIPTO-R(n)** the family of JIPTO with **n** castlings of the pieces of the "fugitives". We often play **JIPTO-R(3)** and **JIPTO-R(5).** The following versions have a great strategic richness:

9. **JIPTO-B1-A2-P(1, 2, 3, 4, 5)-R(3)**

10. **JIPTO-B2-A2-P(1, 2, 3, 4, 5)-R(3)**

11. **JIPTO-B1-A2-P(1, 1, 1, 1, 5)-R(3)**

12. **JIPTO-B2-A2-P(1, 1, 1, 1, 5)-R(3)**

13. **JIPTO-B1-A2-P(1, 2, 3, 4, 5)-R(5)**

14. **JIPTO-B2-A2-P(1, 2, 3, 4, 5)-R(5)**

15. **JIPTO-B1-A2-P(1, 1, 1, 1, 5)-R(5)**

16. **JIPTO-B2-A2-P(1, 1, 1, 1, 5)-R(5)**

After the first round, the players exchange pieces and play the second round. By comparing the profits of the "fugitives" in these two rounds, we determine who has gained.

Three persons must play three rounds:

- In the first round the first player handles the piece of the "pursuer", the second player advances two "fugitives" and the third advances three others "fugitives". We play normally but the second and third players count their points separately, the first player does not gain any point during this round.

- In the second round the first player advances two "fugitives", the second player advances three "fugitives" and the third cheek with the piece of the "pursuer".

- During the third round the first player advances three "fugitives", the second plays with the piece of the "pursuer" and the third handles two pieces of the "fugitives".

In each round each one counts its points (when it moves the "fugitifs"). The profit of the player is equal to the sum of its points.

Four participants play in a similar way but they must play four handles. So the number of handles is equal to the number of players.

Let us designate by **JIPTO-N** the versions of the JIPTO with **N** > **2** players, for example:

17. **JIPTO-B1-A1-3**
18. **JIPTO-B1-A2-3**
19. **JIPTO-B2-A1-3**
20. **JIPTO-B2-A2-3**
21. **JIPTO-B1-A2-P(1, 2, 3, 4, 5)-3**
22. **JIPTO-B2-A2-P(1, 2, 3, 4, 5)-3**
23. **JIPTO-B1-A2-P(1, 1, 1, 1, 5)-3**
24. **JIPTO-B2-A2-P(1, 1, 1, 1, 5)-3**
25. **JIPTO-B1-A2-P(1, 2, 3, 4, 5)-R(3)-3**
26. **JIPTO-B2-A2-P(1, 2, 3, 4, 5)-R(3)-3**
27. **JIPTO-B1-A2-P(1, 1, 1, 1, 5)-R(3)-3**
28. **JIPTO-B2-A2-P(1, 1, 1, 1, 5)-R(3)-3**
29. **JIPTO-B1-A2-P(1, 2, 3, 4, 5)-R(5)-3**
30. **JIPTO-B2-A2-P(1, 2, 3, 4, 5)-R(5)-3**
31. **JIPTO-B1-A2-P(1, 1, 1, 1, 5)-R(5)-3**
32. **JIPTO-B2-A2-P(1, 1, 1, 1, 5)-R(5)-3**

We can then transform each one of these games concerned of chance **JIPTO-P(1. 2, 3, 4, 5)-HR(d)** with an ordinary die (six numbered faces from 1 to 6) by always advancing the corresponding piece (when we obtain 6 then one advances the "pursuer") :

33. **JIPTO-B1-A2-P(1. 2, 3, 4, 5)-HR(d)**
34. **JIPTO-B2-A2-P(1. 2, 3, 4, 5)-HR(d)**
35. **JIPTO-B1-A2-P(1. 2, 3, 4, 5)-R(3)-HR(d)**
36. **JIPTO-B2-A2-P(1. 2, 3, 4, 5)-R(3)-HR(d)**
37. **JIPTO-B1-A2-P(1. 2, 3, 4, 5)-R(5)-HR(d)**
38. **JIPTO-B2-A2-P(1. 2, 3, 4, 5)-R(5)-HR(d)**

and corresponding games with several participants.

We often play the **JIPTO-P(a, b, c, d, e)** with symbolic values of the pieces of "fugitives".

Let us describe some classes of the most practised JIPTO :

JIPTO-ZS

The family of JIPTO with a zone of safety for the "fugitives" where they cannot be captured.

JIPTO-O(f)

The family of JIPTO with impassable obstacle for the "fugitives".

JIPTO-O(p)

The family of JIPTO with impassable obstacle for a "pursuer".

JIPTO-S(1)

The family of JIPTO with **n** pieces of "assistants (first-aid workers)". A captured "fugitive" which is touched by a "assistant" start again its movement. The version **JIPTO-S(1)** is most widespread.

During 1980-1992 we tested thousands of the versions of the JIPTO, for example

JIPTO-B1-A2-P(1, 2, 3, 4, 5)-R(3)-ZS-O(f)-O(p)-S(1)-3-HR(d)

In fact, the number of the versions of the JIPTO is infinite and each one easily finds several versions with its taste.

4. Modèles Mathematical

Let us describe, for example, the analytical model of basic version **JIPTO-B1-A2.**

<u>Positions</u>: Let us consider the displacement of the centers of the pieces. Let us designate by

$$\mathbf{P(n) = (x(n), y(n)),\ 2 \leq x(n) \leq 28,\ 2 \leq y(n) \leq 38}\ ,$$

the center of the "pursuer" (its position) and by

$$\mathbf{F(n, i) = (x(n, i), y(n, i)), 1 \leq x(n, i) \leq 29, 1 \leq y(n, i) \leq 39} ,$$

the position of the "fugitive" of number **i = 1, 2, 3, 4, 5,** where **n** is the number of the step (**n = 0, 1, 2...**). Thus **P(0), F(0, 1), F(0, 2), F(0, 3), F(0, 4), F(0, 5)** are the initial positions. We describe here the play with the "fugitives" of diameter 2 and a "pursuer" of diameter 4.

Trajectory of the "pursuer" : A line

$$\mathbf{X = P(0)\, P(1) \ldots P(n) \ldots} \quad (1)$$

where **P(n -1) P(n) = 4 (N = 1, 2...)** is called a trajectory of the P ("pursuer").

Capture: That is to say

$$\mathbf{M(1) = \{(x, y) : 0 < x < 30, 20 \leq y < 29 \}},$$
$$\mathbf{M(2) = \{(x, y) : 0 < x < 30, 29 \leq y < 38\}},$$
$$\mathbf{M(3) = \{(x, y) : 0 < x < 30, 38 \leq y \leq 39\}}.$$

The "fugitive" number **i** is captured with l'instant **N** if

$$\mathbf{P(N)\, F(n, i) < 3} \text{ and } \mathbf{F(N,i)} \text{ do not belong to } \mathbf{M(3)} ;$$

in this case, **F(N+1, i) = A,** where **A = (0,0).** It is said that **F(N, i)** is the position of capture.

Trajectories of the "fugitives" : A line

$$\mathbf{X(i) = F(0, i)\, F(1, i) \ldots F(n, i) \ldots F(N, i)} \quad (2)$$

where **F(N, i)** is the position of capture, that is to say **F(N, i)** belongs to **M(3)** and **F(N-1, i)** belongs to **M(2);**

$$\mathbf{F(n\text{-}1, i)\, F(n, i) = 2 \ (n = 1, 2, \ldots , N)}$$

a trajectory of F_i ("fugitive" is called; number i).

Number of points: Let us define the **K(i)** number of points corresponding to the trajectory (2) in a following way:

K(i) = j if **F(N, i)** belongs to appartient à **M(j) (j = 1, 2, 3)**, if not sinon **K(i) = 0.**

Principal criterion: The number

$$\mathbf{K = K(1) + K(2) + K(3) + K(4) + K(5)} \quad (3)$$

is the number of points of the "fugitives" in a round of the JIPTO which is made up by a whole of 6 trajectories (1), (2).

Programs and strategies: One can describe a trajectory (1) by the equation

$$\mathbf{x(n+1) = x(n) + 4 \cos u(n), y(n+1) = y(n) + 4 \sin u(n),} \quad (4)$$

where a number **u(n)** is called the program of P from the instant **n** to the instant **n+1 (n = 0,1...)**. A trajectory (2) is described by the equation

$$\mathbf{x(n+1, i) = x(n, i) + 2 \cos v(n, i), y(n+1, i) = y(n, i) + 2 \sin v(n, i),} \quad (5)$$

where a number **v(n, i)** is called the program of F_i from the instant **n** to the instant **n+1.**

A sequence

$$\mathbf{U(n) = (u(0), u(1), ..., u(n-1))} \quad (6)$$

is the program of P on **n** not. A program (6) and one initial position **P(0)=(x(0), y(0))** define by the equation (4) a piece **X(P(0), U(n))** of trajectory of P. A sequence

$$\mathbf{V(n, i) = (v(0, i), v(1, i), ..., v(n-1, i))} \quad (7)$$

is the program of F_i on **n** not. A program (7) and one initial position **F(0,i)** define by the equation (5) a piece **X(F(0, i), V(n, i))** of trajectory of $F_i$to a position of capture.

The players choose their programs (6) and (7) according to the available information and of the rules of the game. One calls these actions the strategies of the players. The trajectories (1) and (2) depend on the initial positions, of the strategies of the players (of the opponent also).

In the **JIPTO-B2** the trajectories are made up of circles of diameter **d = 2.** One can study the games with **d ¡ 2**. With the limit when **d** tends towards 0, one obtains the "ideal mathematical JIPTO". In this mathematical model the positions of the "fugitive" and the "pursuer" are geometrical points and the trajectories are lines. The ideal mathematical JIPTO is easier to analyze, it is useful for the analysis of the basic real JIPTO with **d** rather small.

Let us designate by **P(t)** a position of the "pursuer" (P) and by **F(t)** a position of the "fugitive" (F) at the instant **t**. A round starts at the instant **t = 0.** One does not count the pauses for the reflexions.

A line, composed of the points **P(t)** is called a trajectory P. A trajectory of F is a line, composed of the points **F(t)**. We have:

$$\mathbf{P(s)\,P(t) \leq a(t - s), F(s)\,F(t) \leq b(t - s)} \text{ for } \mathbf{t > s},$$

where **a** is a maximum speed of P and **b** is a maximum speed of F (**a > b**, for example **a = 2** and **b = 1**). Let us note that in the ideal mathematical JIPTO the "pursuer" and the "fugitives" can change the direction of their displacements at any time. In order to facilitate the analysis, one can study the simpler ideal mathematical JIPTO with less number of the "fugitives" (one, two or three), less lines (one or two), but also with a small number of steps (one, two, three, etc).

5. Modeling of the Strategies

Let us give some examples of the strategies of the ”pursuer” the most used.

Pure pursuit: The ”pursuer” follows a ”fugitive”. He always takes his step towards a position of this ”fugitive”.

Cut way: If a ”fugitive” goes in a direction, the ”pursuer” defines a position of corresponding capture and moves towards this position. When the ”fugitive” changes its direction, the pursuer defines a new position of capture and goes towards this position.

Stopping of road: Initially the ”pursuer” arrives to a position right between ”fugitive” and one of the lines I, II, III (which cut out the zones M(1), M(2), M(3)) and then, after each step of this ”fugitive”, the ”pursuer” restores such a situation (when a line which passes by the centers of the positions of the ”pursuer” and of ”fugitive” is perpendicular to line I) while approaching this one.

The various combinations of these strategies can give effective ”superstratégies”. Now let us describe some strategies of ”fugitive”.

Go towards the aim: One advances straight towards line III. One can allow the ”fugitive” to move back if the ”pursuer” approaches him.

Escape: A ”fugitive” further takes its step to be possible from the ”pursuer”.

One can combine these two strategies in the various ”revolving movements” when ”fugitive” tries to pass towards line III and flees the ”pursuer” when it approaches too much him.

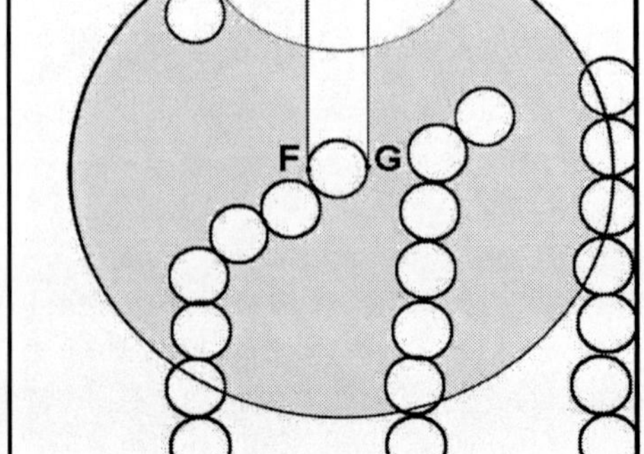

6. Sources of the Irresolute Problems

The whole of the geometrical proposals on the various versions of the JIPTO and other JIP presents an interesting extension of the traditional geometry. The elementary geometry represents a scientific system which, starting from the clear and simple bases, leads gradually to a series of noncommonplace consequences. The development of the geometrical theory of the continuation [6] widens its field of application.

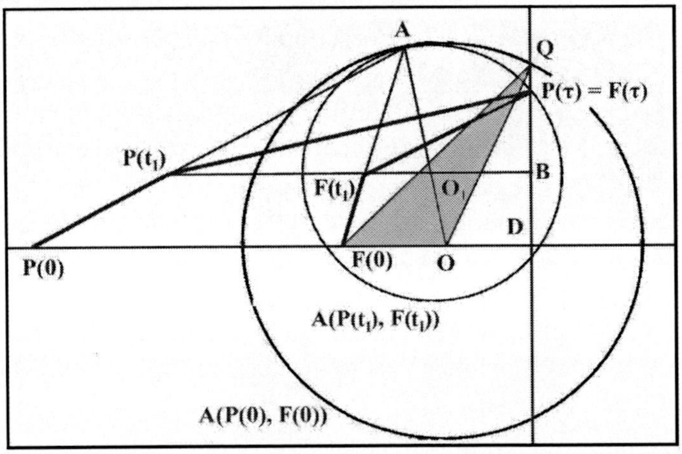

The initiation to the geometry gives the easiest access to the acquisition of the mathematical culture. We evaluate the mathematical culture at an individual in the following way:

Initial level: one starts to understand the concept of mathematisation;

Mean level: one acquires a mathematical knowledge which can go from the very elementary knowledge until knowledge of the complex mathematical theories;

Higher level: one is able to create new mathematical knowledge.

Our criteria classify among the people with the mathematical culture of the higher level the great mathematicians of Antiquity.

One can divide the mean level into some levels according to the additional criteria, for example, the criterion of ingeniousness:

The ordinary mean level: one can solve mathematical problems which do not claim ingeniousness;

The advanced mean level: one is able easily to reproduce the demonstrations of the studied theorems and to propose clever solutions with problems already solved by the others.

One can subdivide each one of these two levels according to the criterion of the acquired mathematical knowledge: the advanced mean level of the elementary school (of the high school, the college, the university, or by class).

The study more elementary geometry makes it possible to reach the mathematical culture of the mean level. The mathematical modeling of the versions of the JIPTO and initiation with the geometrical theory of the pursuit give multiple possibilities of learning how to create new mathematical knowledge and of thus reaching the mathematical culture of the higher level.

Each principal version of the JIPTO can be object of a thesis entitled " The mathematical theory of game **JIPTO-V**", where **V = JIPTO-B1-A1, JIPTO-B1-A2-P(1,2,3,4,5)...**

But the hope to find in a foreseeable future the optimal strategies even in the initial versions of the JIPTO is very mean. Their descriptions must be founded on the geometrical intuition.

Compared to the first criterion, the basic JIPTO is a game with infinite sets of strategies and a discontinuous criterion. It does not exist in this case a method of calculation numerical of the optimal strategies.

After our research, carried out during 1990-2000, we think that the optimal values two versions **JIPTO-B1-A1, JIPTO-B1-A2, JIPTO-B2-A1, JIPTO-B2-A1** compared to the first criterion are equal to 8 or 9.

In the case of a null game of JIPTO, to decide between the players, one uses very natural additional criteria. The JIPTO belongs thus to the class of the dynamic plays with perfect l information but with several criteria.

These plays are very difficult for l'analyse theoretical. In this case, one seeks the optimal strategies compared to the second criterion among the optimal strategies compared to the first criterion, etc. The existence of the optimal strategies compared to the additional criteria is improbable.

Compared to the fourth additional criterion the JIPTO is a fast play of capture with five "fugitives", indescribable strategic restrictions and geometrical restrictions on displacements. But the fast problem of pursuit with the strategic and geometrical restrictions and with only two fugitives is not solved yet.

7. Mathematics of the JIPTO and Differential Games

The theories of the various versions of the JIPTO will not be separated in a tight way. On the contrary an optimal or sufficiently effective strategy in a version of the JIPTO will be often rather effective in more several other versions. Thus the construction of an effective strategy presents in fact the discovery of a method which can be useful for the study of the mathematical models, for example, within the framework of the differential game theory.

In the opposite direction, research in mathematical theory of the pursuit, in particular, the results on the games with the "life line" and on the games of fast capture with several "fugitive" is interesting for the theories of the versions of the JIPTO.

However the most adequate models of the versions of the JIPTO are not differential games. Even in the cases of the versions of the "ideal mathematical JIPTO", where the circles (representing the pieces) are replaced by the geometrical points, there are several characteristics which do not exist in the other classes of the differential games, for example, the possibility in certain cases of limiting the number of correction of directions of the displacement of the players which can sometimes facilitate research. In addition, the existence of the several criteria in the majority of the versions of the JIPTO complicates the task of the analysis supplements even for the simplest versions and generates the quantity of the new very difficult mathematical problems.

References

[1] L. Petrosyan, G. Tomski, *Elementarnye zadachi presledovaniya i ubeganiya (Elementary problems of pursuit and evasion)*, University of Yakutsk, 1989.

[2] L. Petrosyan, G. Tomski, *Cherez igri k tvorchestvu (From games to creativity)*, Novosibirsk: Nauka, 1991.

[3] G .Tomski, T.Tomski, *JIPTO : Jeux de réflexion pour tous*, Paris : ACL Editions, 1996.

[4] G.Tomski, *JIPTO*, Paris : JIPTO international, 2000.

[5] G.Tomski, *Mathématiques du JIPTO*, Paris : JIPTO international, 2002.

[6] L. Petrosyan, G. Tomski, *Geometriya prostogo presledovaniya (Geometry of simple pursuit)*, Novosibirsk: Nauka, 1983.

In: Mathematics, Game Theory and Algebra Compendium ISBN 978-1-60456-659-8
Editor: Jacob H. Mathias, pp. 373-387

Chapter 32

Investment Decisions under Uncertainty and Evaluation of American Options

Alexander A. Vasin and Vladimir V. Morozov*
Faculty of Computational Mathematics and Cybernetics
Moscow State University, Russia

1. Introduction

Capital projects evaluation and selection is an important problem in investment finance, and it is also of interest for actuarial science. A traditional approach to such investment is based on the net present value (NPV) concept, see Jorgensen (1963), Tobin (1969) and Kellison (1991). Nickell (1978) and Abel (1990) provide detailed surveys of development of this approach. In order to make an investment decision about a single project, one should compute the NPV which is the difference between the present values of proceeds and expenditures related to the project and invest if this difference is positive. If there are several possible projects, an investor should consider all variants of investment and choose the one with the maximum NPV. Under this approach, there are issues that arise in calculating the NPV, in particular, how to take into account risks related to the projects, and what discount rate (or rates) should be used. The survey by Lewin et al. (1994) discusses these issues as well as an alternative method based on the internal rate of return.

However, the data on investment decisions shows that business practice does not correspond to the mentioned rule, see the same survey, and also Dixit and Pindyck (1994), D&P below. Firms typically invest only if the rates of return are essentially (2-3 times) higher than those which correspond to the zero NPV. D&P stress that the traditional theory does not take into account the combination of two features typical for such projects: irreversibility of investment and a possibility to delay it. Under these conditions, a firm with an opportunity to invest is holding an option similar to American call option: it has the right

*The authors gratefully acknowledge the support from Society of Actuaries, the Committee on Knowledge Extension Research, research project *Mathematical Models and Software for Financial Organizations at Risky Markets.* We thank referees of SOA for useful comments.

but not the obligation to buy an asset at any time. When a firm makes an irreversible expenditure, it gives up the possibility to use new information that might affect the desirability of the investment. Thus, a firm should compare the NPV's of two variants: investment now or delaying the decision (while the traditional theory corresponds to a now or never proposition). The basic investment model of this sort was developed by McDonald and Sigel (1986), McD&S below.

McKean(1965) derived the optimal stopping rule for the perpetual (without expiration date) American option. This paper provides an optimal investment rule for McD&S model. D&P developed many extensions of McD&S model where a firm faces one investment opportunity. However, in many practical situations there exist several alternative investment opportunities. The present paper aims to examine the corresponding option evaluation problem. The difficulty is that in general case such evaluation cannot be reduced to separate evaluation of every project. This difficulty arises if a firm has to choose only one project because of limited resources of some kind. Usually, it's impossible to solve the problem in explicit formulas. That's why suboptimal investment strategies and bounds for the option value are very useful. For example, if some upper bound looks unsatisfactory for the investor he may reject the purchase of the option. We restrict our study to the case of two alternative projects with equal sunk costs. In this case the optimal investment problem reduces to evaluation of the perpetual American call option on the maximum of two assets (PACOMTA for short). The holder of the PACOMTA can buy at any time one of the two assets paying the fixed strike price.

Broadie and Detemple (1997) (B&D below) characterized the optional exercise regions for a number of American option contracts on two or more assets. Below we give more precise characterization of the exercise region of the PACOMTA. First, the boundary of the exercise region forms right angles with axes. Second, it approaches asymptotically the border lines of a similar region for the PACOMTA with zero strike prices (see Gerber and Shiu, 1996). Third, the exercise region belongs to the area corresponding to the lower bound on the price of the PACOMTA. This bound is the maximum of the prices of the perpetual options on exclusive assets. Proceeding from these properties of the boundary we derive the upper bound for the option value. The survey of bounds for the standard American option see in Broadie and Detemple (1996).

2. Formulation of the Problem

Consider a firm which meets two investment opportunities. Every possible project i ($i = 1, 2$) requires a fixed sunk cost I_i at the start of its realization and then brings profit $P_i(t)$ that depends on time t and satisfies geometrical Brownian motion equation

$$dP_i(t) = \alpha_i P_i(t)dt + \sigma_i P_i(t)dz_i(t) \tag{2.1}$$

with drift α_i, $0 < \alpha_i < r$, and volatility σ_i where $z_1(t)$ and $z_2(t)$ are standard Wiener processes with a constant correlation ρ, and r is the constant rate of interest. The firm observes the current profits $P_1(t), P_2(t)$, and chooses the proper project and the moment for investment, proceeding from this information.

Let a function $f(P)$ be twice differentiable. Then the Ito's formula for process (2.1) gives the differential $df(P_i(t))$ as

$$df(P_i(t)) = f'(P_i(t))(\alpha_i dt + \sigma_i dz_i(t)) + \frac{1}{2} f''(P_i(t))\sigma_i^2 dt.$$

Hence, for $f(P) = \log(P)$ the process $f(P_i(t))$ follows the equation

$$d\log(P_i(t)) = (\alpha_i - \sigma_i^2/2)dt + \sigma_i dz_i(t). \tag{2.2}$$

Vector value $(X_1(t), X_2(t)) = (z_1(t)/\sqrt{t}, z_2(t)/\sqrt{t})$ is random with the bivariate normal probability density function

$$n_2(x_1, x_2) = \frac{1}{2\pi\sqrt{1-\rho^2}} \exp\left(- \frac{x_1^2 - 2\rho x_1 x_2 + x_2^2}{2(1-\rho^2)} \right).$$

Denote $P_i = P_i(0),\ \tilde{\alpha}_i = \alpha_i - \sigma_i^2/2,\ i = 1,2$. The equation (2.2) implies $P_i(t) = P_i \exp(X_i(t)\sigma_i\sqrt{t} + \tilde{\alpha}_i t)$.

If the firm invests to the project i at time t then

$$V(P_i(t)) = E \int_0^\infty \exp(-r\tau)P_i(t)\exp(X_i(\tau)\sigma_i\sqrt{\tau} + \tilde{\alpha}_i\tau)d\tau = \frac{P_i(t)}{r - \alpha_i}$$

is the expected present value of the profit flow. The price of the option to invest is

$$F(P_1, P_2) = \max_T E[\exp(-rT)(\max_{i=1,2}(P_i(T)/(r - \alpha_i) - I_i)],$$

where $a_+ = \max(a, 0)$ and the maximum is taken over all stopping times T. A definition of the stopping time see in Duffie (2001). The option evaluation problem is to calculate $F(P_1, P_2)$ and find the corresponding optimal stopping time T^*.

In the case $I_1 = I_2 = K$ this problem is equivalent to evaluation of the PACOMTA with strike price K, asset prices $S_i(t) = P_i(t)/(r - \alpha_i)$, and (in the absence of arbitrage) dividend rates $\delta_i = r - \alpha_i$, $i = 1, 2$.

3. The Option Value for Projects with Zero Volatilities

Consider a portfolio with m investment projects such that $\sigma_i = 0,\ i = 1, ..., m$. Then the future profit flow is given by $P_i(t) = P_i \exp(\alpha_i t)$ for every i. If the firm invests to project i at the time t then the NPV is $\exp(-rt)(P_i(t)/(r - \alpha_i) - I_i)$. It is easy to see that the optimal time t_i of investment to the project i is when the profit reaches rI_i. More precisely, $t_i = 0$ for $P_i \geq rI_i$, otherwise $t_i = \log(rI_i/P_i)/\alpha_i$. The corresponding NPV is

$$F_i(P_i) = \begin{cases} (\alpha_i I_i/(r - \alpha_i))(P_i/(rI_i))^{r/\alpha_i} & \text{for } P_i < rI_i, \\ P_i/(r - \alpha_i) - I_i & \text{for } P_i \geq rI_i. \end{cases}$$

For incompatible projects, the optimal investment rule is to choose the optimal project $j \in \text{Arg}\max_i F_i(P_i)$ and to invest at the time t_j. Hence, the option value is $F(P_1, ..., P_m) = \max_i F_i(P_i)$.

Consider for illustration the optimal investment rule in case of two projects. Let $\alpha_1 > \alpha_2$ and $\alpha_2 I_2(r - \alpha_1) > \alpha_1 I_1(r - \alpha_2)$. Then the curve $\mathcal{C} = \{(P_1, P_2) \in \mathbb{R}^2_+ \ : \ F_1(P_1) = F_2(P_2)\}$ and lines $P_i = rI_i, \ i = 1, 2$, divide the quadrant $\mathbb{R}^2_+$ into four parts $D_i, \ i = 1, 2, 3, 4$ (see Figure 3.1). The arrows show directions of vectors $(dP_1(t), dP_2(t))$.

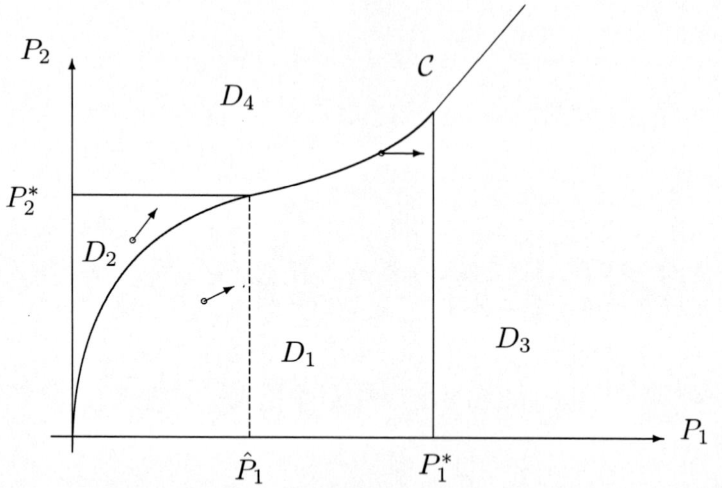

Figure 3.1. The optimal investment rule under zero volatilities.

The optimal investment rule is as follows: For initial (P_1, P_2) lying in D_1 the firm should wait until $P_1(t)$ reaches $P_1^* = rI_1$ and then invest to the first project. For (P_1, P_2) lying in D_2 the firm should invest to the second project as soon as $P_2(t) = P_2^* = rI_2$. For (P_1, P_2) lying on the curve $\mathcal{C}$ and $P_1 \le \hat{P}_1$ the firm should act in any of specified two ways. For $P_1 \in (\hat{P}_1, rI_1)$ the firm should invest to the second project immediately or wait until $P_1(t) = P_1^*$ and then invest to the first project. For (P_1, P_2) lying in D_3 or in D_4 the firm should invest immediately to the first project or, correspondingly, to the second one. So, $D_3 \cup D_4$ represents immediate exercise region for the option to invest. In the next section we study a boundary of the exercise region for a PACOMTA.

4. Properties of the Exercise Region Boundary

The process $S_i(t)$ satisfies the equation (2.1). Denote

$$S_i = S_i(0) = P_i/(r - \alpha_i), a_i = \log S_i + \tilde{\alpha}_i t, \ b_i = \sigma_i \sqrt{t}, \ i = 1, 2.$$

The pair

$$(S_1(t), S_2(t)) = (\exp(X_1(t)b_1 + a_1), \exp(X_2(t)b_2 + a_2))$$

is a random vector value with the bivariate lognormal probability density function

$$\psi(s,t) = \psi(s_1, s_2, t) = \frac{1}{s_1 s_2 b_1 b_2} n_2\left(\frac{\log s_1 - a_1}{b_1}, \frac{\log s_2 - a_2}{b_2}\right),$$

where $s_1 > 0, s_2 > 0$. Then

$$C(S_1, S_2) \overset{\text{def}}{=} F(S_1(r - \alpha_1), S_2(r - \alpha_2))$$
$$= \max_T E[\exp(-rT)(\max_{i=1,2} S_i(T) - K)] = \max_T W(T)$$

is the value of the PACOMTA. The stopping time

$$T^* = \inf\{t \geq 0 \ : \ C(S_1(t), S_2(t)) = (\max_{i=1,2} S_i(t) - K)_+\}$$

is an optimal option-exercise strategy (see Duffie, 2001 and Shiryaev, 1978). So, $\mathcal{E} = \{(S_1, S_2) \in \mathbb{R}_+^2 \ : \ C(S_1, S_2) = (\max_{i=1,2} S_i - K)_+\}$ is an immediate exercise region. For the perpetual American call on a single asset i, the optimal stopping time and the value are

$$T_i^* = \inf\{t \geq 0 \ : \ S_i(t) \geq S_i^*\},$$

$$C_i(S_i) = \begin{cases} \dfrac{K}{\beta_i - 1}\left(\dfrac{S_i(\beta_i - 1)}{\beta_i K}\right)^{\beta_i} & \text{for } S_i < S_i^*, \\ S_i - K & \text{for } S_i \geq S_i^*, \end{cases}$$

where

$$S_i^* = \frac{K\beta_i}{\beta_i - 1}, \quad \beta_i = \frac{-\tilde{\alpha}_i + \sqrt{\tilde{\alpha}_i^2 + 2r\sigma_i^2}}{\sigma_i^2} > 1, \quad i = 1, 2.$$

This result is attributed to McKean (1965). Note that function $C_i(S_i)$ is differentiable at the point S_i^* and $C_i(S_i) > S_i - K$ for all $S_i < S_i^*$.

B&D proved the following result formulated here for the perpetual option.

Proposition 4.1. *Let $\mathcal{E}$ represent the immediate exercise region for a PACOMPTA. Define the subregions $\mathcal{E}_i = \{(S_1, S_2) \in \mathcal{E} \ : \ S_i > S_{3-i}\}$ for $i = 1, 2$. Then $\mathcal{E}$ satisfies the following properties.*

(i) $\mathcal{E} \cap \{(S_1, S_2) \in \mathbb{R}_+^2 \ : \ S_1 \neq S_2\} = \emptyset$.

(ii) *Subregion $\mathcal{E}_1$ is convex.*

(iii) $(S_1, S_2) \in \mathcal{E}_1$ *implies* $(\lambda S_1, \lambda S_2) \in \mathcal{E}_1$ *for all* $\lambda \geq 1$.

(iv) $(S_1, S_2) \in \mathcal{E}_1$ *implies* $(S_1, \lambda S_2) \in \mathcal{E}_1$ *for all* $0 \leq \lambda \leq 1$.
The symmetric properties to (ii),(iii),*and* (iv)*hold for the subregion* $\mathcal{E}_2$.

So, the border of the subregion $\mathcal{E}_i$ includes a graph of some nondecreasing convex function $S_i = G_i(S_{3-i})$ with $G_i(0) = S_i^*$ for $i = 1, 2$.

The early exercise premium representation expresses the value of the American option as the value of the corresponding European option plus the gains from early exercise. Kim (1990), Jacka (1991), and Carr, Jarrow, and Myneni (1992) provide such a representation for the standard American option when the underlying asset price follows a geometric Brownian motion process. B&D gave the following analogous formula for an American max-option. Taking into account that the value of the perpetual European call option is equal to zero, we obtain

$$C(S_1, S_2) = \int_0^\infty \int_{s_1 \geq G_1(s_2)} \exp(-rt)(\delta_1 s_1 - rK)\psi(s,t)dsdt \tag{4.1}$$

$$+ \int_0^\infty \int_{s_2 \geq G_2(s_1)} \exp(-rt)(\delta_2 s_2 - rK)\psi(s,t)dsdt.$$

Note that the differences $\delta_i s_i - rK, \ i = 1, 2$, are positive since $s_i \geq G_i(s_{3-i}) \geq G_i(0) = S_i^* > rK/\delta_i$. Hence, we have

$$\int_0^\infty \int_{s_i \geq G_i(s_{3-i})} \exp(-rt)(\delta_i s_i - rK)\psi(s,t)dsdt \tag{4.2}$$

$$\leq \int_0^\infty \int_{s_i \geq S_i^*} \exp(-rt)(\delta_i s_i - rK)\psi(s,t)dsdt = C_i(S_i), \quad i = 1, 2.$$

Using standard technique, we can rewrite (4.1) as

$$C(S_1, S_2) = \int_0^\infty \int_{-\infty}^\infty [\delta_1 S_1 \exp(-\delta_1 t)N(d_{11}) - rK\exp(-rt)N(d_{12})]n(x_2)dx_2dt$$

$$+ \int_0^\infty \int_{-\infty}^\infty [\delta_2 S_2 \exp(-\delta_2 t)N(d_{21}) - rK\exp(-rt)N(d_{22})]n(x_1)dx_1dt, \tag{4.3}$$

where

$$d_{ij} = \frac{-\log(G_i(\exp(b_{3-i}x_{3-i} + (2-j)\rho b_1 b_2 + a_{3-i}))) + b_i \rho x_{3-i} + (2-j)b_i^2 + a_i}{b_i\sqrt{1-\rho^2}}$$

for $i, j = 1, 2$. (See Kim 1990, page 570, for an analogous formula for the perpetual American option on a single asset.) We use formula (4.3) to obtain the following result. Denote

$$\alpha = \alpha_1 - \alpha_2, \ \sigma^2 = \sigma_1^2 - 2\rho\sigma_1\sigma_2 + \sigma_2^2, \ \tilde{\alpha} = \alpha - \frac{\sigma^2}{2}, \ \theta_{1,2} = \frac{-\tilde{\alpha} \pm \sqrt{\tilde{\alpha}^2 + 2\delta_2\sigma^2}}{\sigma^2},$$

$$c_1 = \left(\frac{\theta_1}{\theta_1 - 1}\right)^{\frac{\theta_1}{\theta_1-\theta_2}} \left(\frac{-\theta_2}{1-\theta_2}\right)^{\frac{-\theta_2}{\theta_1-\theta_2}}, \; c_2 = \left(\frac{\theta_1}{\theta_1 - 1}\right)^{\frac{1-\theta_1}{\theta_1-\theta_2}} \left(\frac{-\theta_2}{1-\theta_2}\right)^{\frac{\theta_2-1}{\theta_1-\theta_2}},$$

$$\underline{C}(S_1, S_2) = \max_{i=1,2} C_i(S_i), \;\; \underline{\mathcal{E}} = \{(S_1, S_2) \in \mathbb{R}^2_+ : \underline{C}(S_1, S_2) = (\max_{i=1,2} S_i - K)_+\}.$$

Proposition 4.2. *Let $\mathcal{E}$ be the immediate exercise region for a PACOMTA, and graph of function $G_i(S_{3-i})$ represent the border of the subregion $\mathcal{E}_i = \{(S_1, S_2) \in \mathcal{E} \; : \; S_i > S_{3-i}\}$ for $i = 1, 2$. Then $\mathcal{E}$ satisfies the following properties.*

(i) $G_i(0) = S_i^*, \; G_i'(0) = 0.$

(ii) *The graph of $G_i(S_{3-i})$ approaches asymptotically the line $S_i = c_i S_{3-i}$.*

(iii) *The inequality $\underline{C}(S_1, S_2) \le C(S_1, S_2)$ holds for all $(S_1, S_2) \in \mathbb{R}^2_+$, and $\mathcal{E} \subset \underline{\mathcal{E}}$.*

See the proof in Appendix and Figure 4.1 for the illustration. Denote

$$\underline{\mathcal{E}}^c = \mathbb{R}^2_+ \backslash \underline{\mathcal{E}}, \;\text{ and if }\; \beta_{3-i} > \beta_i \;\text{ then }\; \hat{S}_i = C_i^{-1}(S_{3-i}^* - I_i) = S_i^* \left(\frac{\beta_i - 1}{\beta_{3-i} - 1}\right)^{1/\beta_i}.$$

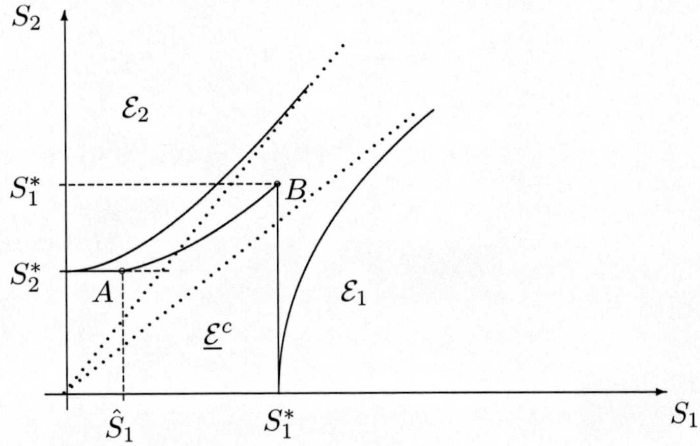

Figure 4.1. The immediate exercise region $\mathcal{E}$ for the PACOMTA.

Figure 4.1 shows sets $\mathcal{E} = \mathcal{E}_1 \cup \mathcal{E}_2$ and $\underline{\mathcal{E}}^c$ in the case $\beta_1 < \beta_2, \; c_2\hat{S}_1 < S_2^*$. The border of the set $\underline{\mathcal{E}}^c$ between points $A = (\hat{S}_1, S_2^*)$ and $B = (S_1^*, S_1^*)$ is the curve $S_2 - I_2 = C_1(S_1)$. The other part of this border is the broken line connecting consequently the points $A, (0, S_2^*), (0, 0), (S_1^*, 0)$ and B.

From the assertion (iii) of Proposition 4.2 it follows

$$G_i(S_{3-i}) \ge C_{3-i}(S_{3-i}) + I_i \;\text{ for all }\; S_{3-i} \le S_{3-i}^* \;\text{ and }\; i = 1, 2. \tag{4.4}$$

5. Upper Bound for the Option Value

Let $\mathcal{G}_i$ be the set of all nondecreasing convex functions of variable S_{3-i} satisfying the properties (i),(ii) of Proposition 4.2 and (4.4). Clearly, $G_i \in \mathcal{G}_i$. Function

$$\overline{G}_i(S_{3-i}) = \begin{cases} \max(c_i S_{3-i}, S_i^*, C_{3-i}(S_{3-i}) + K) & \text{for } S_{3-i} \le S_{3-i}^*, \\ \max(c_i S_{3-i}, S_i^*) & \text{for } S_{3-i} > S_{3-i}^*, \end{cases}$$

also belongs to $\mathcal{G}_i$. Note that $\mathcal{G}_i$ is convex, and $\tilde{G}_i, \hat{G}_i \in \mathcal{G}_i$ implies $\max(\tilde{G}_i, \hat{G}_i) \in \mathcal{G}_i$.

Denote $\overline{C}(S_1, S_2)$, $\overline{d}_{ij}$ the expressions of $C(S_1, S_2), d_{ij}$ with $\overline{G}_i$ substituted for G_i, $i = 1, 2$. Inequalities $\overline{G}_i(S_{3-i}) \le G_i(S_{3-i})$ take place for all $S_{3-i} \ge 0$. Hence, (4.1) implies $C(S_1, S_2) \le \overline{C}(S_1, S_2)$. So, $\overline{C}(S_1, S_2)$ is an upper bound for $C(S_1, S_2)$.

Proposition 5.1. *If $\beta_{3-i} > \beta_i$ let the following inequality hold*

$$c_{3-i}\hat{S}_i \ge S_{3-i}^*. \tag{5.1}$$

Then

$$\overline{C}(S_1, S_2) = \int_0^\infty (S_1 \exp(-\delta_1 y^2) H_{11}(y) + S_2 \exp(-\delta_2 y^2) H_{21}(y)) dy$$

$$-K \int_0^\infty \exp(-r y^2)(H_{12}(y) + H_{22}(y)) dy,$$

for all $(S_1, S_2) \in \mathbb{R}_+^2$ satisfying $S_i < \overline{G}_i(S_{3-i})$, $i = 1, 2$, where for $i, j = 1, 2$,

$$H_{ij}(y) = N\Big(\frac{1}{\sqrt{1-\rho^2}}(x_{3-ij}(y) + \rho d_{ij1}(y))\Big)\frac{dN(d_{ij1}(y))}{dy}$$

$$+N\Big(-\frac{\sigma}{\sigma_i\sqrt{1-\rho^2}}\Big(x_{3-ij}(y) + \Big(\frac{\sigma_i\rho - \sigma_{3-i}}{\sigma}\Big)d_{ij2}(y)\Big)\Big)\frac{dN(d_{ij2}(y))}{dy},$$

$$x_{3-ij}(y) = \frac{\log(S_i^*/(c_i S_{3-i})) - (\tilde{\alpha}_{3-i} + \rho\sigma_1\sigma_2(2-j))y^2}{\sigma_{3-i} y},$$

$$d_{ijk}(y) = \begin{cases} \dfrac{\log(S_i/S_i^*) + (\tilde{\alpha}_i + \sigma_i^2(2-j))y^2}{\sigma_i y} & \text{for } k = 1, \\ \dfrac{\log(S_i/(c_i S_{3-i})) + (\tilde{\alpha}_i - \tilde{\alpha}_{3-i} + (\sigma_i^2 - \rho\sigma_1\sigma_2)(2-j))y^2}{\sigma y} & \text{for } k = 2. \end{cases}$$

See the proof in Appendix. Note that (5.1) is true if $\beta_1 = \beta_2$.

Calculations show that $\overline{C}(S_1, S_2)$ is an increasing function and the absolute error of it's approximation to $C(S_1, S_2)$ decreases if the larger price of the assets is fixed and the

Table 5.1. Bounds $\overline{C}(S_1, S_2)$ and $\underline{C}(S_1, S_2)$.

$\overline{C}(S_1, S_2)$	$\sigma_1 = 0.005$	$\sigma_1 = 0.01$	$\sigma_1 = 0.05$	$\sigma_1 = 0.1$
$S_1 = S_2 = 8$	4.15	4.26	5.13	6.28
$S_1 = 8, S_2 = 6$	4.048399	4.0513	4.51	5.45
$\underline{C}(S_1, S_2) = C_1(S_1)$	4.048395	4.0506	4.12	4.31

smaller one is decreasing. Table 5.1 gives bounds $\overline{C}(S_1, S_2)$ and $\underline{C}(S_1, S_2)$ for the example with symmetrical values of parameters:

$$r = 0.05,\ \rho = 0.2,\ \sigma_1 = \sigma_2,\ \delta_1 = \delta_2 = 0.01,\ K = 10.$$

Observing last two entries of the third column, one can conclude that for all $S_2 \leq 6$ and $\sigma_1 = 0.01$ the following inequalities hold

$$\underline{C}(8, S_2) = 4.0506 \leq C(8, S_2) \leq \overline{C}(8, S_2) \leq \overline{C}(8, 6) = 4.0513.$$

So, $\overline{C}(8, S_2)$ gives a good approximation to $C(8, S_2)$ in this case.

6. Discussion

In section 4 we examined the exercise boundary for the American call option written on the maximum of two assets without expiration date. In section 5 we obtained the upper bound for the option value in the integral form. This bound may be calculated with help of the comprehensive computer system (e.g. Maple). Our approach can be applied to other perpetual American options on two assets, such as exchange options (see Margrabe, 1978), spread options, and so on (see B&D). We think these results may be generalized on American dual strike option or the option evaluation of alternative projects with nonequal sunk costs.

Now pay attention to investment decision rules. In section 3 the optimal decision rule was formulated for projects with zero volatilities. This rule may serve also as an approximation to the analogous rule for projects with very small volatilities. In section 4 the lower bound $\underline{C}(S_1, S_2)$ for a PACOMTA value corresponds to the optimal decision rule T_j^* for the perpetual American call written on a single asset $j = \arg\max_{i=1,2} C_i(S_i)$. This rule may be improved as follows: to buy at the time t some asset i as soon as

$$S_i(t) = S_i^* \text{ and } C_i(S_i(t)) \geq C_{3-i}(S_{3-i}(t)); \tag{6.1}$$

to exercise the option denote by T^0 the minimal time t satisfying (6.1) for some asset i. This decision rule is similar to the optimal investment rule in section 3. Let

$$(S_1, S_2) \in D_1 = \{(S_1, S_2) \in \mathbb{R}_+^2\ :\ C_1(S_1) > C_2(S_2),\ S_1 < S_1^*\}.$$

If the point $(S_1(t), S_2(t))$ remains in D_1 then both rules, corresponding to T^0 and T_j^*, recommend to buy the first asset as soon as $S_1(t)$ reaches S_1^*. But if the point $(S_1(t), S_2(t))$ crosses the curve $\mathcal{C} = \{(S_1, S_2) \in \mathbb{R}_+^2\ :\ C_1(S_1) = C_2(S_2)\}$ and enters into region

$D_2 = \{(S_1, S_2) \in \mathbb{R}^2_+ \,:\, C_1(S_1) < C_2(S_2),\ S_2 < S_2^*\}$ then the rule corresponding to T^0 uses this information and prescribes a better decision: to buy the second asset as soon as $S_2(t)$ reaches S_2^*. In the game-theoretical terminology, T^0 dominates T_j^*.

The computation of the value $W(T^0)$ is a rather tiresome task. But if $(S_1^*, S_2^*) \in \mathcal{C}$ one can solve it by using a technique offered by Broadie and Detemple (1996) for the standard American option. Moreover, their technique is applicable to decision rules $T(L_1, L_2)$ dependent on positive parameters L_1, L_2 and defined as

$$T(L_1, L_2) = \min\{t \,:\, \min_{i=1,2}(L_i - S_i(t)) \geq 0\}.$$

Note that if $(S_1^*, S_2^*) \in \mathcal{C}$ then $T^0 = T(S_1^*, S_2^*)$. A challenging problem is to find a lower bound $\max\limits_{L_1 \geq S_1, L_2 \geq S_2} W(T(L_1, L_2))$ for the PACOMTA and for more general dual strike American option under fixed S_1, S_2.

Appendix

The next lemma finds the known integrals (see, e.g., Besson and Partrat, 1992).

Lemma A.1. *For* $\nu = 1, 2$ *and positive* u, a, b, *the following formulas take place*

$$\frac{1}{\sqrt{2\pi}} \int_0^\infty x^{-\nu+1/2} \exp\left(- \frac{u(x + 1/x)}{2} \right) dx = \frac{1}{\sqrt{u}} \exp(-u), \tag{A.1}$$

$$B_\nu(a, b) = \frac{1}{\sqrt{2\pi}} \int_0^\infty t^{-\nu+1/2} \exp\left(- \frac{at + b/t}{2} \right) dt = a^{(\nu-2)/2} b^{(1-\nu)/2} \exp(-\sqrt{ab}). \tag{A.2}$$

Proof. The change of variable $y = 1/x$ proves equity of integrals in the left-hand side of $(A.1)$ for $\nu = 1, 2$. The other change $y = x^{1/2} - x^{-1/2}$ gives

$$\frac{1}{\sqrt{2\pi}} \int_0^\infty x^{-1/2} \exp\left(- \frac{u(x + 1/x)}{2} \right) dx = \frac{1}{\sqrt{2\pi}} \int_{-\infty}^\infty \exp\left(- \frac{u(y^2 + 2)}{2} \right) dy$$

$$+ \frac{1}{\sqrt{2\pi}} \int_{-\infty}^\infty \frac{y}{\sqrt{y^2 + 4}} \exp\left(- \frac{u(y^2 + 2)}{2} \right) dy = \frac{1}{\sqrt{u}} \exp(-u).$$

The change of variable $x = t\sqrt{a/b}$ proves $(A.2)$. □

Let $n(x)$ and $N(x)$ be standard density and cumulative probability normal distribution functions.

Lemma A.2. *For all* $c \geq 0,\ \delta > 0$ *and all* a, b, *the following formula take place*

$$\delta \int_0^\infty \exp(-\delta t) \int_{-\infty}^\infty N(ax + b\sqrt{t} + c/\sqrt{t}) n(x) dx dt \tag{A.3}$$

$$= 1 + \frac{1}{2}\left(\frac{b}{\sqrt{b^2 + 2(a^2+1)\delta}} - 1\right) \exp\left(-\frac{c(b + \sqrt{b^2 + 2(a^2+1)\delta})}{a^2+1}\right).$$

If, besides, $(b + ac)^2 + (-c^2 + 2\delta)(a^2 + 1) > 0$ *then*

$$\int_0^\infty \exp(-\delta t) t^{-1/2} \int_{-\infty}^\infty n(ax + b\sqrt{t}) \exp(c\sqrt{t}x) n(x) dx dt \tag{A.4}$$

$$= ((b + ac)^2 + (-c^2 + 2\delta)(a^2 + 1))^{-1/2}.$$

Proof. It is easy to verify that

$$\int_{-\infty}^{x_0} n(ax + b) n(x) dx = \frac{1}{\sqrt{a^2+1}} n\Big(\frac{b}{\sqrt{a^2+1}}\Big) N\Big(\sqrt{a^2+1}\Big(x_0 + \frac{ab}{a^2+1}\Big)\Big) \tag{A.5}$$

for all a, b, x_0. Take at first $c > 0$ and denote $z = ax + b\sqrt{t} + c/\sqrt{t},\ A = a^2 + 1$. Integrating by parts and using $(A.5)$, we have

$$\delta \int_0^\infty \exp(-\delta t) \int_{-\infty}^\infty N(z) n(x) dx dt = -\exp(-\delta t) \int_{-\infty}^\infty N(z) n(x) dx \Big|_0^\infty$$

$$+ \frac{1}{2} \int_0^\infty \exp(-\delta t) \left(\frac{b}{t^{1/2}} - \frac{c}{t^{3/2}}\right) \int_{-\infty}^\infty n(z) n(x) dx dt$$

$$= 1 + \frac{1}{2} \int_0^\infty \exp(-\delta t) \left(\frac{b}{t^{1/2}} - \frac{c}{t^{3/2}}\right) \frac{1}{\sqrt{A}} n\left(\frac{b\sqrt{t} + c/\sqrt{t}}{\sqrt{A}}\right) dt$$

$$= 1 + \frac{1}{2\sqrt{A}} \exp\Big(-\frac{bc}{A}\Big) \left(bB_1\left(\frac{b^2 + 2A\delta}{A}, \frac{c^2}{A}\right) - cB_2\left(\frac{b^2 + 2A\delta}{A}, \frac{c^2}{A}\right)\right).$$

$(A.2)$ implies $(A.3)$. Tending $c \to 0+$ we get $(A.3)$ with $c = 0$. $(A.5)$ implies $(A.4)$. □

Lemma A.3. *Let function* $G_i(S_{3-i}, K)$ *correspond to the boundary of subregion* $\mathcal{E}_i(K)$ *of the immediate exercise region* $\mathcal{E}(K)$ *for a PACOMTA with the strike price* K. *Then* $\mathcal{E}$ *satisfies the following properties.*

(i) *For all* $\lambda > 0$, $\mathcal{E}_i(\lambda K) = \lambda \mathcal{E}_i(K)$, *and* $G_i(\lambda S_{3-i}, \lambda K) = \lambda G_i(S_{3-i}, K)$, *i.e.,* G_i *is a homogeneous function.*

(ii) $G_i(S_{3-i}, K)$ *has asymptote* $S_i = c_i S_{3-i}$, *where coefficient* $c_i \geq 1$ *doesn't depend on* K.

Proof. (i) Let $C(S_1, S_2, K)$ be the value of the PACOMTA with strike price K. It follows from the definition of the option value that C is a homogeneous function. Points (S_1, S_2) of $\mathcal{E}(K)$ satisfy equation $C(S_1, S_2, K) = \max_{i=1,2} S_i - K$, where the both sides are homogeneous functions. If the point (S_1, S_2) satisfies this equation then the point $(\lambda S_1, \lambda S_2)$ does it with strike price λK. This proves (i).

(ii) Consider subregion $\mathcal{E}_i(K)$. From the assertions (i),(iii) of Proposition 4.1 it follows the existence of asymptote $S_i = c_i S_{3-i}$ of function $G_i(S_{3-i})$ with coefficient $c_i = \inf\{S_i/S_{3-i} : (S_1, S_2) \in \mathcal{E}_i(K)\} \geq 1$. The proved part (i) of this Lemma implies independence of c_i on K. □

Proof of Proposition 4.2. Choose the point (S_1, S_2) on the boundary of subregion $\mathcal{E}_1(K)$, i.e., $S_1 = G_1(S_2)$.

(i) Using inequality (4.2) with $i = 2$ and equities $C(S_1, S_2) = G_1(S_2) - K$, $N(d_{1j}) = 1 - N(-d_{1j})$, one can write (4.3) as

$$\int_0^\infty \int_{-\infty}^\infty [\delta_1 G_1(S_2) \exp(-\delta_1 t) N(-d_{11}) - rK \exp(-rt) N(-d_{12})] n(x_2) dx_2 dt + o(S_2) = 0.$$

Differentiation of this equation by S_2 at the point $S_2 = 0$ yields

$$G_1'(0) \int_0^\infty \int_{-\infty}^\infty \Big[\delta_1 \exp(-\delta_1 t)\Big(N(-h_1) + \frac{n(-h_1)(\exp(\sigma_2\sqrt{t}x_2 + (\rho\sigma_1\sigma_2 + \tilde{\alpha}_2)t) - 1)}{\sigma_1\sqrt{t(1-\rho^2)}}\Big)$$

$$-\frac{rK}{G_1(0)} \exp(-rt) \frac{n(-h_2)(\exp(\sigma_2\sqrt{t}x_2 + \tilde{\alpha}_2 t) - 1)}{\sigma_1\sqrt{t(1-\rho^2)}}\Big] n(x_2) dx_2 dt = 0,$$

where

$$h_j = \frac{\sigma_1\rho\sqrt{t}x_2 + (\tilde{\alpha}_1 + (2-j)\sigma_1^2)t}{\sigma_1\sqrt{t(1-\rho^2)}}, \quad j = 1, 2, \quad G_1(0) = S_1^* = \frac{K\beta_1}{\beta_1 - 1}.$$

Using Lemma A.2, we obtain the expression for double integral

$$\frac{0.5(\beta_1 - 1)\sigma_1^2}{\sqrt{\tilde{\alpha}_1^2 + 2r\sigma_1^2}} + \left(\delta_1 - \frac{(\beta_1 - 1)r}{\beta_1}\right)\left(\frac{1}{\sqrt{(\tilde{\alpha}_1 + \rho\sigma_1\sigma_2)^2 + 2\delta_2\sigma_1^2}} - \frac{1}{\sqrt{\tilde{\alpha}_1^2 + 2r\sigma_1^2}}\right)$$

$$= \frac{0.5\sigma_1^2(\beta_1 - 1)\beta_1 + \alpha_1\beta_1 - r}{\beta_1\sqrt{\tilde{\alpha}_1^2 + 2r\sigma_1^2}} + \frac{r - \alpha_1\beta_1}{\beta_1\sqrt{(\tilde{\alpha}_1 + \rho\sigma_1\sigma_2)^2 + 2\delta_2\sigma_1^2}}$$

$$= \frac{r - \alpha_1\beta_1}{\beta_1\sqrt{(\tilde{\alpha}_1 + \rho\sigma_1\sigma_2)^2 + 2\delta_2\sigma_1^2}} > 0$$

which yields $G_1'(0) = 0$.

(ii) Note that $\lim_{S_{3-i}\to\infty} G_i(S_{3-i})/S_{3-i} = c_i,\ i = 1, 2$. It implies

$$\lim_{S_2\to\infty} \log\left(\frac{G_1(S_2\exp(b_2x_2 + \rho b_1b_2 + \tilde{\alpha}_2 t))}{G_1(S_2)}\right) = \sigma_2\sqrt{t}x_2 + (\rho\sigma_1\sigma_2 + \tilde{\alpha}_2)t,$$

$$\lim_{S_2\to\infty} \log\left(\frac{G_2(G_1(S_2)\exp(b_1x_1 + \rho b_1b_2 + \tilde{\alpha}_1 t))}{S_2}\right) = \log(c_1c_2) + \sigma_1\sqrt{t}x_1 + (\rho\sigma_1\sigma_2 + \tilde{\alpha}_1)t.$$

Dividing equation (4.3) on S_2 and using last two equities, we find limits of both sides of (4.3), as $S_2 \to \infty$

$$c_1 = c_1\int_0^\infty \delta_1\exp(-\delta_1 t)\int_{-\infty}^{\infty} N\left(\frac{(\sigma_1\rho - \sigma_2)\sqrt{t}x_2 + (\tilde{\alpha} + \sigma^2)t}{\sigma_1\sqrt{t(1-\rho^2)}}\right) n(x_2)dx_2dt$$

$$+\int_0^\infty \delta_2\exp(-\delta_2 t)\int_{-\infty}^{\infty}\left(1 - N\left(\frac{\log(c_1c_2) + (\sigma_1 - \sigma_2\rho)\sqrt{t}x_1 + \tilde{\alpha}t}{\sigma_2\sqrt{t(1-\rho^2)}}\right)\right)n(x_1)dx_1dt.$$

(A.3) implies

$$c_1(\theta_1 - 1) = \theta_1(c_1c_2)^{\theta_2}. \qquad (A.6)$$

Analogous considerations of the boundary of the subregion $\mathcal{E}_2$ give

$$c_2(-\theta_2) = (1 - \theta_2)(c_1c_2)^{1-\theta_1}. \qquad (A.7)$$

Solution of the system (A.6),(A.7) yields the above formulas for c_1, c_2.

(iii) The inequality $\underline{C}(S_1, S_2) \le C(S_1, S_2)$ follows from the definition of the value for the PACOMTA. Hence, we obtain

$$(\max_{i=1,2} S_i - K)_+ \le \underline{C}(S_1, S_2) \le C(S_1, S_2).$$

So, $(\max_{i=1,2} S_i - K)_+ = C(S_1, S_2)$ implies $(\max_{i=1,2} S_i - K)_+ = \underline{C}(S_1, S_2)$, and $\mathcal{E} \subset \underline{\mathcal{E}}$. □

Proof of Proposition 5.1. The inequality $c_i\hat{S}_{3-i} \ge S_i^*$ implies $\max(c_iS_{3-i}, S_i^*) \ge C_{3-i}(S_{3-i}) + K$ for all $S_{3-i} \le S_{3-i}^*$.

Hence, $\overline{G}_i(S_{3-i}) = \max(c_iS_{3-i}, S_i^*)$ and

$$\overline{d}_{ij} = \begin{cases} \dfrac{\log(S_i/S_i^*) + \sigma_i\rho\sqrt{t}x_{3-i} + (\sigma_i^2(2-j) + \tilde{\alpha}_i)t}{\sigma_i\sqrt{t(1-\rho^2)}} & \text{for } x_{3-i} \le x_{3-i,j}(t), \\ \dfrac{\log(S_i/(c_iS_{3-i})) + (\sigma_i\rho - \sigma_{3-i})\sqrt{t}x_{3-i}}{\sigma_i\sqrt{t(1-\rho^2)}} & \\ +\dfrac{(\tilde{\alpha}_i - \tilde{\alpha}_{3-i} + (\sigma_i^2 - \rho\sigma_1\sigma_2)(2-j))t}{\sigma_i\sqrt{t(1-\rho^2)}} & \text{for } x_{3-i} > x_{3-i,j}(t), \end{cases}$$

where

$$x_{3-i,j}(t) = \frac{\log(S_i^*/(c_iS_{3-i})) - (\tilde{\alpha}_{3-i} + \rho\sigma_1\sigma_2(2-j))t}{\sigma_{3-i}\sqrt{t}}$$

is a solution of the equation

$$c_i S_{3-i} \exp(\sigma_{3-i}\sqrt{t}x_{3-i} + (\tilde{\alpha}_{3-i} + \rho\sigma_1\sigma_2(2-j))t) = S_i^*.$$

Let us simplify the expression for $\overline{C}(S_1, S_2)$. At first show that for fixed x_{3-i}

$$\lim_{t\downarrow 0} \overline{d}_{ij} = -\infty, \quad i = 1, 2, \; j = 1, 2. \tag{A.8}$$

If $c_i S_{3-i} < S_i^*$ then

$$S_i < \overline{G}_i(S_{3-i}) = S_i^*, \quad \log(S_i/S_i^*) < 0, \quad \lim_{t\downarrow 0} x_{3-i,j}(t) = \infty.$$

If $c_i S_{3-i} = S_i^*$ then

$$S_i < \overline{G}_i(S_{3-i}) = c_i S_{3-i} = S_i^*, \quad \log(S_i/(c_i S_{3-i})) = \log(S_i/S_i^*) < 0.$$

If $c_i S_{3-i} > S_i^*$ then

$$S_i < \overline{G}_i(S_{3-i}) = c_i S_{3-i}, \quad \log(S_i/(c_i S_{3-i})) < 0, \quad \lim_{t\downarrow 0} x_{3-i,j}(t) = -\infty.$$

In all cases (A.8) is true. Taking into account (A.8) and integrating by parts, we obtain

$$\overline{C}(S_1, S_2) = \int_0^\infty \int_{-\infty}^\infty [S_1 \exp(-\delta_1 t) n(\overline{d}_{11})\overline{d}_{11}' - K \exp(-rt) n(\overline{d}_{12})\overline{d}_{12}'] n(x_2) dx_2 dt$$

$$+ \int_0^\infty \int_{-\infty}^\infty [S_2 \exp(-\delta_2 t) n(\overline{d}_{21})\overline{d}_{21}' - K \exp(-rt) n(\overline{d}_{22})\overline{d}_{22}'] n(x_1) dx_1 dt, \tag{A.9}$$

where $\overline{d}_{ij}'$ is a derivative of $\overline{d}_{ij}$ by t. Taking inner integrals of (A.9) with help of (A.5) and changing variable $t = y^2$, we conclude the proof. □

References

[1] A.B. Abel, *Consumption and Investment*, in Handbook of Monetary Economics, eds. Benjamin Friedman and Frank Hahn, New York: North-Holland, 1990, 725–778.

[2] J.-L. Besson, C. Partrat, Trend et Systèmes de Bonus-Malus, *ASTIN Bulletin* **22** (1992), 12–30.

[3] M. Broadie, J. Detemple, The American Option Valuation: New Bounds, Approximations, and a Comparison of Existing Methods, *Review of Financial Studies* **9** (1996), 1211–1250.

[4] M. Broadie, J. Detemple, The Valuation of American Options on Multiple Assets, *Mathematical Finance* **7** (1997), 241–286.

[5] P. Carr, R. Jarrow, R. Mineni, Alternative Characterization of American Put Options, *Mathematical Finance* **2** (1992), 87–106.

[6] A.K. Dixit, R.S. Pindyck *Investment under Uncertainty*, Princeton: Princeton University Press, 1994.

[7] S.D. Jacka, Optimal Stopping and the American Put, *Mathematical Finance* **1** (1991), 1–14.

[8] D. Jorgensen, Capital Theory and Investment Behavior, *American Economic Review* **53** (1963), 247–259.

[9] D. Duffie, *Dynamic Asset Pricing Theory*, Princeton: Princeton University Press, 2001.

[10] H.U. Gerber, E.S.W. Shiu, Martingale Approach to Pricing Perpetual American Options on Two Stocks, *Mathematical Finance* **6** (1996), 303–322.

[11] S.G. Kellison, *The Theory of Interest*, Homewood: Irwin, 1991.

[12] I.J. Kim, The Analytical Valuation of American Options, *Review of Financial Studies* **3** (1990), 547–572.

[13] W. Margrabe, The Value of an Option to Exchange One Asset to Another, *Journal of Finance* **33** (1978), 177–186.

[14] C.G. Lewin, et al., *Capital Projects*, Institute of Actuaries and Faculty of Actuaries: London, 1994.

[15] R.L. McDonald, D.R. Siegel, The Value of Waiting to Invest, *Quarterly Journal of Economics* **101** (1986, November), 707–728.

[16] H.P. McKean, A Free Boundary Problem for the Heat Equation Arising from a Problem in Mathematical Economics, an Appendix to Samuelson (1965), *Industrial Management Review* **6** (1965), 32–39.

[17] S.J. Nickell, *The Investment Decisions of Firms*, New York: Cambridge University Press, 1978.

[18] A.N. Shiryaev, *Optimal Stopping Rules*, New-York: Springer-Verlag, 1978.

[19] Tobin, General Equilibrium Approach to Monetary Theory, *Journal of Money, Credit and Banking* **1** (1969), 15–29.

In: Mathematics, Game Theory and Algebra Compendium ISBN 978-1-60456-659-8
Editor: Jacob H. Mathias, pp. 389-397

Chapter 33

On a Continuous Dynamic Strategic Market Game

Piotr Więcek and Tadeusz Radzik
Institute of Mathematics, Wrocław University of Technology, Wrocław, Poland

1. Introduction

We consider the game as follows: Two players hold some amounts of money. On every stage of the game one unit of some nondurable commodity is brought to market and players bid parts of their money for the good. The one who proposes more pays for the good and gets right to consume it (if the bids are equal we assume that with probability half each of the players gets possibility of consuming the good). On the next day the money paid by the players is redistributed among them according to some probability distribution. Then another unit of the commodity is brought to market and bids are made, and so on. The objective of each player is to maximize discounted sum of his own consumption over infinitely many stages.

Model of two-person stochastic game of this type was proposed by Secchi and Sudderth [3,4,5]. In first two papers constant-sum versions of this game were studied, in [3] with money finite but continuous, in [4] money having some nondivisible units. In [5], a generalization of the results for the discrete model from [4] to games with arbitrary utility functions (non-constant-sum game) was made. In all of the above papers, some results about the existence and the form of optimal (in [3] ε-optimal) pure strategies, when discount factor is small or approaches 1, were presented. In our previous paper [8], results for the discrete model were extended for all the values of the discount factor.

In the present paper we modify the continuous model of Secchi and Sudderth from [3], and show that for discount factor not greater than $\frac{2}{3}$ and every $\varepsilon > 0$ it possesses an ε-equilibrium in pure stationary strategies of some particular type. We also prove that nothing more can be proved, i.e. that it does not have a stationary equilibrium (0-equilibrium) for any value of the discount factor.

The organization of the paper is as follows: The next section contains the model and all of the new results, while in the last section all the proofs are presented.

2. Model and the Main Result

The game described in first section is a two-person constant-sum stochastic game described with by the following four items:

1. The state of the game in some moment of time is the fraction of money kept by Player 1 at this moment. Thereby state space for the game is the interval $S = [0, 1]$.

2. Actions of the players represent their bids, and thus, the sets of actions available to player 1 and player 2 in state x of the game are $[0, x]$ and $[0, 1 - x]$ respectively.

3. Daily reward to player 1 in state x, when players use actions a and b respectively is
$$r(x, a, b) = \begin{cases} 1 & \text{when } a > b, \\ 1/2 & \text{when } a = b, \\ 0 & \text{when } a < b. \end{cases}$$
Daily reward of player 2 is equal to $\tilde{r} = 1 - r(x, a, b)$.

4. The law of motion between states q is defined for every triple (x, a, b) as follows:
$$q(x, a, b) = \begin{cases} \frac{1}{2}\delta[x] + \frac{1}{2}\delta[x + b] & \text{when } a < b, \\ \frac{1}{4}\delta[x - a] + \frac{1}{2}\delta[x] + \frac{1}{4}\delta[x + a] & \text{when } a = b, \\ \frac{1}{2}\delta[x] + \frac{1}{2}\delta[x - a] & \text{when } a > b, \end{cases}$$
where $\delta[y]$ denotes a degenarate probability distribution concentrated in point y.
q defined as above reflects the fact that on every stage of the game two decisions can be made using "fair coin toss". One is done in every case and it decides what will be the distribution of the money paid by the players for the portion of good in the next stage of the game. The other decision is made only when bids of both players are equal. Then only one of them is chosen to consume the good and pay for it with a help of coin toss.

A *pure strategy* of player 1 is a sequence $\bar{a} = (a_1, a_2, \ldots)$, where $a_n \in [0, x_n]$ is an action available to player 1 on the n-th stage of the game in state x_n, depending on partial history of the game up to its n-th stage $h_n = (x_1, a_1, b_1, \ldots, x_{n-1}, a_{n-1}, b_{n-1}, x_n)$. We say that pure strategy $\bar{f} = (a_1, a_2, \ldots)$ of player 1 is *stationary*, when for every moment n, $a_n = f(x_n)$ for some $f : S \to S$ (recall that action spaces are subsets of S). In the sequel we shall identify stationary strategy $\bar{f}$ with f. The strategies of the other player are defined similarly.

Strategies of the players together with the law of motion determine the unique probability distribution $P_{x,f,g}$ on the histories of the game $(x, a_1, a_2, x_1, a_2, a_3, \ldots)$. Then the expected *discounted rewards* to players 1 and 2, respectively, are defined by the formulas:
$$J_\beta(x, f, g) = E_{x,f,g}\left(\sum_{n=1}^{\infty} \beta^{n-1} r(x_n, a_n, b_n)\right),$$
$$\tilde{J}_\beta(x, f, g) = E_{x,f,g}\left(\sum_{n=1}^{\infty} \beta^{n-1} \tilde{r}(x_n, a_n, b_n)\right),$$

where $\beta \in (0, 1)$ is fixed discount factor.

We say that the (constant-sum) game has a *value* V_β if

$$V_\beta(x) = \sup_f \inf_g J_\beta(x, f, g) = \inf_g \sup_f J_\beta(x, f, g).$$

For an $\varepsilon \geq 0$, a pair of (stationary) strategies (f, g) is called the *(stationary) ε-equilibrium* for the discounted stochastic game, iff for every $x \in S$,

$$J_\beta(x, f, g) \geq J_\beta(x, \pi, g) - \varepsilon \quad \text{and} \quad \tilde{J}_\beta(x, f, g) \geq \tilde{J}_\beta(x, f, \gamma) - \varepsilon,$$

for every strategies π and γ of players 1 and 2 respectively. For $\varepsilon = 0$ a pair satisfying the above inequalities is called simply *(stationary) equilibrium*.

In the sequel, The stochastic game defined above will be denoted by Γ_β.

Our main result is the following.

Theorem 1 *For every $\beta \in (0, \frac{2}{3}]$ game Γ_β*

1. *has a value $V_\beta(x)$ which is nondecreasing in x, nonnegative and bounded above by $\frac{1}{1-\beta}$;*
2. *for every $\varepsilon > 0$ has a pure strategy stationary ε-equilibrium (f_1, f_2) such that for every $x \in [0, \frac{1}{2}]$*

$$f_1(x) = x \leq f_2(x) < x + \varepsilon$$

and

$$f_1(x) = f_2(1 - x), \quad f_2(x) = f_1(1 - x)$$

for $x \in [\frac{1}{2}, 1]$;

3. *does not have a stationary equilibrium.*

Remark 1 *As it will be shortly seen, the proof of the third part of the above theorem does not require the assumption that $\beta \in (0, \frac{2}{3}]$. We only use fact that game Γ_β has a value V_β. Of course, if for some β game does not possess a value, then it does not have a stationary strategy equilibrium. Therefore, the third part of Theorem 1 can be written in a stronger form:*
For each $\beta \in (0, 1)$, game $\overline{\Gamma}_\beta$ does not have a stationary equilibrium.

Remark 2 *All the other results are proved for $\beta \leq \frac{2}{3}$. It is an interesting open problem, what happens when discount factor is greater. The results proved for a discrete counterpart of this game in our previous paper [8] suggest that also in this situation there are ε-pure equilibria in game $\overline{\Gamma}_\beta$. On the other hand, the proof from our case cannot be repeated for this more general situation.*

3. Proof of Theorem 1

Throughout whole of this section we assume that β is a fixed number from $(0, \frac{2}{3}]$.

We start with introducing some notation. Let

$$M_\beta = \left\{ v : [0,1] \to [0, \frac{1}{1-\beta}] \mid v \text{ is nondecreasing} \right.$$
$$\left. \text{and } v(x) + v(1-x) = \frac{1}{1-\beta} \text{ for } x \in [0,1] \right\}.$$

We shall proceed the proof through a couple of lemmata. In the first four of them instead of game Γ_β we shall consider a class of one-stage games $\{\Gamma_\beta(v,x) : v \in M_\beta, x \in [0,1]\}$. The sets of actions available to players 1 and 2 in game $\Gamma_\beta(v,x)$ are respectively intervals $[0,x]$ and $[0,1-x]$. The payoff of player 1 is defined as follows:

$$K_{xv}(a,b) = \begin{cases} \frac{\beta}{2}[v(x) + v(x+b)] & \text{for } a < b \\ \frac{1}{2} + \frac{\beta}{4}[v(x-b) + 2v(x) + v(x+b)] & \text{for } a = b \\ 1 + \frac{\beta}{2}[v(x) + v(x-a)] & \text{for } a > b. \end{cases} \quad (1)$$

The payoff of player 2 is $\widetilde{K_{xv}}(a,b) = \frac{1}{1-\beta} - K_{xv}(a,b)$. Game $\Gamma_\beta(v,x)$ can be interpreted in the following way: Game Γ_β (starting in state x) is stopped during the second stage and on that stage players are paid value of function v in current state of the game.

Part of the proof is based on results concerning discrete counterparts of games $\Gamma_\beta(v,x)$, denoted $\Gamma^k_\beta(v,x)$. Let, for $v \in M_\beta$, $k \in \mathbb{N}$ and $x \in [0,1]$, $\Gamma^k_\beta(v,x)$ denote game defined in the same way as $\Gamma_\beta(v,x)$, but with sets of actions available to players 1 and 2 replaced by $\{0, \frac{x}{k}, \frac{2x}{k}, \ldots, x\}$ and $\{0, \frac{x}{k}, \frac{2x}{k}, \ldots, \frac{[k(1-x)]}{k}\}$ respectively.

We are ready to write first lemma.

Lemma 1 *For every $v \in M_\beta$ and every $x \in [0,1]$, function $K_{xv}(a,b)$ is*

(a)constant in a on the set $a < b$;

(b)nonincreasing in a on the set $a > b$;

(c)constant in b on the set $b < a$;

(d)nondecreasing in b on the set $b > a$.

Proof. It is a simple consequence of monotonicity of v and definition of K_{xv}. ∎

Lemma 2 *Let $k \in \mathbb{N}$, $v \in M_\beta$ and $x \in [0, \frac{1}{2}]$. If*

$$v(2x) - v(0) \le \frac{2}{\beta} \quad (2)$$

then either (x,x) or $(x, x+\frac{x}{k})$ is an equilibrium in game $\Gamma^k_\beta(v,x)$.

Proof. By (1), we directly verify that inequality (2) is equivalent to the following

$$K_{xv}(x - \frac{x}{k}, x) \leq K_{xv}(x, x) \leq K_{xv}(x, x - \frac{x}{k}). \tag{3}$$

If $2x + \frac{x}{k} > 1$ then, by definition, both players have the same set of their pure strategies $\{0, \frac{x}{k}, \ldots, x\}$ in game $\Gamma^k_\beta(v, x)$. Hence, we easily state by (3) and Lemma 1 that (x, x) is an equilibrium in game $\Gamma^k_\beta(v, x)$.

Assume therefore for the rest of the proof that $2x + \frac{x}{k} \leq 1$ and suppose first that $K_{xv}(x, x) \leq K_{xv}(x, x + \frac{x}{k})$. Then, similarly as before, (3) and Lemma 1 clearly imply that (x, x) is an equilibrium in game $\Gamma^k_\beta(v, x)$.

The other possibility is that $K_{xv}(x, x) > K_{xv}(x, x + \frac{x}{k})$. Then we easily see with the help of (c) and (d) of Lemma 1 that pure strategy $x + \frac{x}{k}$ is the best response of player 2 for strategy x of player 1. On the other hand, by (a) of Lemma 1, $K_{xv}(x - \frac{ix}{k}, x + \frac{x}{k}) = \text{const}$ for $i = 0, 1, \ldots, k$. Consequently, $(x, x + \frac{x}{k})$ is an equilibrium in game $\Gamma^k_\beta(v, x)$ in the considered case, completing the proof. ■

The above lemma will be essential to prove the following.

Lemma 3 *For every $v \in M_\beta$ and every $x \in [0, 1]$, game $\Gamma_\beta(v, x)$ has a value $V_v(x)$, such that $V_v \in M_\beta$. Moreover, for every $\varepsilon, \delta > 0$ it has an ε-equilibrium $(a_0(x), b_0(x))$ such that $a_0(x) = x \leq b_0(x) < x + \delta$ for $x \leq \frac{1}{2}$ and $a_0(x) = b_0(1 - x)$, $b_0(x) = a_0(1 - x)$ for $x \geq \frac{1}{2}$.*

Proof. Since $v \in M_\beta$, $v(2x) - v(0) \leq \frac{1}{1-\beta}$ for all x in $[0, \frac{1}{2}]$, but $\frac{1}{1-\beta} \leq \frac{2}{\beta}$ for $\beta \leq \frac{2}{3}$ and consequently (2) also holds.

First we consider the case $x = \frac{1}{2}$. Then by Lemma 2, (x, x) is an equilibrium in game $\Gamma^k_\beta(v, x)$ for any $k \in \mathbb{N}$, and consequently, $K_{xv}(x - \frac{x}{k}, x) \leq K_{xv}(x, x) \leq K_{xv}(x, x - \frac{x}{k})$. Hence, using (a) and (c) of Lemma 1, we immediately get, $K_{xv}(a, x) \leq K_{xv}(x, x) \leq K_{xv}(x, b)$ for all $0 \leq a, b \leq \frac{1}{2}$, which proves that $(\frac{1}{2}, \frac{1}{2})$ is an equilibrium in game $\Gamma^k_\beta(v, x)$ when $x = \frac{1}{2}$.

Now consider the case $x < \frac{1}{2}$. Fix $\varepsilon, \delta > 0$, and v and take $k \in \mathbb{N}$ such that

$$K_{xv}(x, x + \frac{x}{k}) - K_{xv}(x, x^+) < \varepsilon \text{ and } \frac{x}{k} \leq \delta. \tag{4}$$

Again, by Lemma 2, either (x, x) or $(x, x + \frac{x}{k})$ is an equilibrium in game $\Gamma^k_\beta(v, x)$. Hence, $K_{xv}(x, x) \leq K_{xv}(x, x - \frac{x}{k})$ or $K_{xv}(x, x + \frac{x}{k}) \leq (x, x - \frac{x}{k})$. But (c) and (d) of Lemma 1 imply that $K_{xv}(x, x^-) = K_{xv}(x, x - \frac{x}{k})$ and $K_{xv}(x, x + \frac{x}{k}) \geq K_{xv}(x, x^+)$. Therefore

$$K_{xv}(x, x) \leq K_{xv}(x, x^-) \quad \text{or} \quad K_{xv}(x, x^+) \leq K(x, x^-). \tag{5}$$

Now, notice that by (5) and Lemma 1, $\inf_{b \in [0,1-x]} K_{xv}(x, b)$ is either equal to $K_{xv}(x, x)$ or to $K_{xv}(x, x^+)$.

Therefore, if $\inf_{b \in [0,1-x]} K_{xv}(x, b) < K_{xv}(x, x^+)$ then by (d) of Lemma 1, $K_{xv}(x, x) < K_{xv}(x, x + \frac{x}{k}$. It implies that $(x, \frac{x}{k})$ cannot be an equilibrium in game $\Gamma^k_\beta(v, x)$, and consequently (x, x) must be the equilibrium in $\Gamma^k_\beta(v, x)$. Hence, by (a) of Lemma 1

$$K_{xv}(x, x) \geq K_{xv}(x - \frac{x}{k}, x) = K_{xv}(x^-, x)$$

and (x, x) is also an equilibrium in game $\Gamma_\beta(v, x)$.

When $\inf_{b\in[0,1-x]} K_{xv}(x, b) = K_{xv}(x, x^+)$, then (4) guarantees that $K_{xv}(x, x + \frac{x}{k}) - \inf_{b\in[0,1-x]} K_{xv}(x, b) < \varepsilon$. But again (a) of Lemma 1 implies that also $K_{xv}(x, x + \frac{x}{k}) = \sup_{a\in[0,x]} K_{xv}(a, x + \frac{x}{k})$ and so $(x, x + \frac{x}{k})$ is an ε-equilibrium in game $\Gamma_\beta(v, x)$.

Now, take $x > \frac{1}{2}$. One may easily show that for every $a \le x$ and $b \le 1 - x$,

$$K_{xv}(a, b) + K_{1-x,v}(b, a) = \frac{1}{1-\beta}. \tag{6}$$

It follows from the above inequality that, whenever (a^*, b^*) is an ε-equilibrium in game $\Gamma_\beta(v, 1 - x)$, (b^*, a^*) is an ε-equilibrium in game $\Gamma_\beta(v, x)$. Therefore we may put $(a_0(x), b_0(x)) = (b_0(1 - x), a_0(1 - x))$.

Now, let $x \in [0, 1]$. It follows from the above considerations that for any $\varepsilon > 0$ and $v \in M_\beta$ there is an ε-equilibrium in game $\Gamma^k_\beta(v, x)$. This clearly implies that the game has a value $V_v(x)$. What we need to show now is that $V_v \in M_\beta$.

First notice that it follows immediately from definition of K_{xv} that for all $(a, b) \in [0, x] \times [0, 1 - x]$

$$0 \le K_{xv}(a, b) \le 1 + \beta \sup_{y\in[0,1]} v(y) \le 1 + \frac{\beta}{1-\beta} = \frac{1}{1-\beta},$$

where third inequality is because $v \in M_\beta$.

Now take two $x_1, x_2 \in [0, 1]$, $x_1 < x_2$. We will show that $V_v(x_1) \le V_v(x_2)$. In x_1 the set of pure strategies available to player 1 is $[0, x_1]$, while pure strategies available to player 2 are $[0, 1 - x_1]$. When we increase the state to x_2, the set of strategies available to player 1 increases to $[0, x_2]$, while the set of strategies available to player 2 decreases to $[0, 1 - x_2]$. Since the player 1 is the one who maximizes his payoff, to show $V_v(x_1) \le V_v(x_2)$, it is enough to show that

$$K_{x_1v}(a, b) \le K_{x_2v}(a, b)$$

for every $(a, b) \in [0, x_1] \times [0, 1 - x_2]$. But this is true by definition of K_{xv} and fact that $v \in M_\beta$ and thus is nondecreasing.

The last thing we have to show is that $V_v(x) + V_v(1 - x) = \frac{1}{1-\beta}$ for every $x \in [0, 1]$. Notice that since for every $\varepsilon > 0$ game $\Gamma_\beta(v, x)$ has an ε-equilibrium (a_0, b_0) such that $a_0(x) = b_0(1 - x)$, $b_0(x) - a_0(1 - x)$ for all $x \in [0, 1]$, the following inequalities must hold:

$$\begin{aligned} K_{xv}(a_0(x), b_0(x) - \varepsilon &\le V_v(x) \le K_{xv}(a_0(x), b_0(x)) + \varepsilon, \\ K_{1-x,v}(b_0(x), a_0(x) - \varepsilon &\le V_v(1 - x) \le K_{1-x,v}(b_0(x), a_0(x)) + \varepsilon. \end{aligned}$$

If we add these inequalities side by side, using (6) we get

$$\frac{1}{1-\beta} - \varepsilon \le V_v(x) + V_v(1 - x) \le \frac{1}{1-\beta} + \varepsilon.$$

But since this is true for all $\varepsilon > 0$, we get the desired equality, ending the proof that $V_v \in M_\beta$. ∎

In the next lemma, two parts of Theorem 1 are established.

Lemma 4 *The game Γ_β has a value $V \in M_\beta$ and for every $\varepsilon > 0$ it has an ε-pure stationary equilibrium $(f_1^\varepsilon, f_2^\varepsilon)$ such that for every $x \in [0, \frac{1}{2}]$ $f_1^\varepsilon(x) = x \leq f_2^\varepsilon(x) < x + \varepsilon$ and $f_1^\varepsilon(x) = f_2^\varepsilon(1-x)$, $f_2^\varepsilon(x) = f_1^\varepsilon(1-x)$ when $x \in [\frac{1}{2}, 1]$.*

Proof. Define the map $T : M_\beta \to M_\beta$ putting

$$T(v)(x) = V_v(x).$$

Take two $v_1, v_2 \in M_\beta$. Note that by (1) for every $x \in [0,1]$ and for all $(a,b) \in [0,x] \times [0, 1-x]$

$$|K_{xv_1}(a,b) - K_{xv_2}(a,b)| \leq \beta \sup_{y \in [0,1]} |v_1(y) - v_2(y)|,$$

which clearly implies that also

$$|T(v_1)(x) - T(v_2)(x)| = |V_{v_1}(x) - V_{v_2}(x)| \leq \beta \sup_{y \in [0,1]} |v_1(y) - v_2(y)|.$$

But this means that T is a contraction in M_β endowed with uniform distance. Since this is a closed subset of Banach space of all bounded functions from $[0,1]$ to $\mathbb{R}^+$, we can use Banach's fixed point theorem to show that T has a unique fixed point V. Now, for any fixed $\varepsilon > 0$ and every state x, take a pure $(1-\beta)\varepsilon$-equilibrium in game $\Gamma_\beta(V, x)$, $(f_1^\varepsilon(x), f_2^\varepsilon(x))$, from Lemma 3 (with $\delta = \varepsilon$). Clearly, by the definition of K_{xV} and the fact that V is a fixed point of map T, $(f_1^\varepsilon(x), f_2^\varepsilon(x))$, $\varepsilon > 0$, and V satisfy all the assumptions of Theorem 2.2 in [3] and thereby are ε-optimal stationary families for players 1 and 2, and the value for the game Γ_β, respectively. ∎

The last couple of lemmata will be used in the proof of the fact that the game Γ_β does not have a pure strategy stationary equilibrium (0-equilibrium).

Lemma 5 *For every $x > 0$, $V(x) > 0$.*

Proof. Suppose on the contrary that there exists $x_0 > 0$ such that $V(x_0) = 0$. Since $V \in M_\beta$ (by Lemma 3.5), $V(\frac{1}{2}) = \frac{1}{2(1-\beta)} > 0$, and consequently x_0 can be chosen in such a way that $0 < x_0 < 1/2$, $V(x_0) = 0$ and $V(2x_0) > 0$.

Now let $(f_1^\varepsilon, f_2^\varepsilon)$ be an ε-pure stationary equilibrium in game Γ_β described by Lemma 4. From the proof of Lemma 4 one can easily see that the value of the game $\Gamma_\beta(V, x_0)$ is equal to $V(x_0)$ and so by the definition of ε-equilibrium, $V(x_0) = \lim_{\varepsilon \to 0^+} K_{x_0 V}(f_1^\varepsilon(x_0), f_2^\varepsilon(x_0))$. Hence, by a simple analysis of the payoff function $K_{x_0 V}$ for the game $(\Gamma_\beta(V, x_0)$ (see formula 1), we can deduce

$$\begin{aligned} V(x_0) &= \lim_{\varepsilon \to 0^+} K_{x_0 V}(f_1^\varepsilon(x_0), f_2^\varepsilon(x_0)) \\ &= \min\{\frac{\beta}{2}[V(x_0) + V(2x_0^+)], \frac{1}{2} + \frac{\beta}{4}[V(0) + 2V(x_0) + V(2x_0)]\}. \end{aligned}$$

But since $V(x_0) = 0$,

$$0 = \min\{\frac{\beta}{2}V(2x_0^+), \frac{1}{2} + \frac{\beta}{4}[V(0) + V(2x_0)]\},$$

which is impossible, because of the assumed inequality $V(2x_0) > 0$ and the fact that V is nondecreasing. Therefore our supposition is false, which proves the lemma. ∎

Lemma 6 *Game* $\Gamma_\beta(V,0)$ *does not have an equilibrium.*

Proof. Suppose that game $\Gamma_\beta(V,0)$ has an equilibrium. Since player 1 has only one action available, his optimal strategy has to be pure. On the other hand, if player 2 has an optimal answer to it in mixed strategies he also must have an optimal answer in pure strategies. Hence, our supposition implies that $\Gamma_\beta(V,0)$ possesses a pure-strategy equilibrium. From definition of K_{0V} and Lemma 1 it is not difficult to see that

$$V(0) = \min\{\frac{1}{2} + \beta V(0), \frac{\beta}{2}[V(0) + V(0^+)]\}. \tag{7}$$

Suppose $\frac{1}{2} + \beta V(0) < \frac{\beta}{2}[V(0) + V(0^+)]$. It follows that

$$V(0^+) - V(0) > \frac{1}{\beta}. \tag{8}$$

Of course in this case (7) is of the form

$$V(0) = \frac{1}{2} + \beta V(0),$$

or equivalently

$$V(0) = \frac{1}{2(1-\beta)}.$$

But this is impossible, since by (8)

$$V(0^+) > \frac{1}{2(1-\beta)} + \frac{1}{\beta} > \frac{1}{2(1-\beta)} = \frac{1}{1-\beta} - V(0) = V(1).$$

($V \in M_\beta$ and thereby is nondecreasing). Hence, since $K_{0V}(0,0) = \frac{1}{2} + \beta V(0)$, we conclude that $(0,0)$ cannot be an equilibrium in game $\Gamma_\beta(V,0)$ and $V(0) = \frac{\beta}{2}[V(0) + V(0^+)]$ or equivalently $V(0^+) = \frac{2-\beta}{\beta}V(0)$.

In view of what we've already shown, our supposition that $\Gamma_\beta(V,0)$ has a pure strategy equilibrium implies that it is of the form $(0,\gamma)$, where $\gamma > 0$.

$$\frac{\beta}{2}[V(0) + V(\gamma)] = K_{0V}(0,\gamma) = V(0) = \frac{\beta}{2}[V(0) + V(0^+)],$$

and thereby, for $0 < x < \gamma$, $V(x) = V(0^+) = \frac{2-\beta}{\beta}V(0)$. Let $x = \frac{\gamma}{3}$. Then clearly

$$K_{xV}(0,b) = (2-\beta)V(0) \tag{9}$$

for every $0 < b < \frac{2}{3}\gamma$ and

$$K_{xV}(0,0) = \frac{1}{2} + (2-\beta)V(0). \tag{10}$$

Now, using (d) of Lemma 1 we get that

$$K_{xV}(0,\frac{\gamma}{2}) = K_{xV}(a,\frac{\gamma}{2}) \quad \text{for } a \le \frac{\gamma}{3}. \tag{11}$$

On the other hand, combining (9), (10) and (a) of Lemma 1, we can show that

$$K_{xV}(0, \frac{\gamma}{2}) = \min_{b \in [0, 1-\frac{\gamma}{3}]} K_{xV}(0, b).$$

But this together with (11) implies that $(0, \frac{\gamma}{2})$ is an equilibrium in game $\Gamma_\beta(V, x)$ and $V(x) = (2-\beta)V(0)$, which is a contradiction, since $V(x) = \frac{2-\beta}{\beta}V(0)$ and $V(x) > 0$ by Lemma 5. ■

Proof of Theorem 1: First two parts of the theorem were shown in Lemma 4. The third part is a direct consequence of Lemma 6, since game Γ_β has a pure stationary strategy equilibrium iff all of the games $\Gamma_\beta(V, x)$, $x \in [0, 1]$ have pure strategy equilibria. ■

References

[1] I. Karatzas, M. Shubik, W.D. Sudderth, Construction of Stationary Markov Equilibria in a Strategic Market Game, *Mathematics of Operations Research* **19(4)** (1992), 975–1006.

[2] I. Karatzas, M. Shubik, W.D. Sudderth, A Strategic Market with Secured Lending, *Journal of Mathematical Economics* **28** (1997), 207–247.

[3] P. Secchi, W.D. Sudderth, How to Bid for a Pizza, *Quaderni di Dipartamento* **42(5-96)** (1996), Dipartamento di Economia Politica e Metodi Quantitivi, Università di Pavia.

[4] P. Secchi, W.D. Sudderth, A Two-person Strategic Market Game, *Quaderni di Dipartamento* **83(6-98)** (1998), Dipartamento di Economia Politica e Metodi Quantitivi, Università di Pavia.

[5] P. Secchi, W.D. Sudderth, A Simple Two-Person Stochastic Game with Money, *Annals of Dynamic Games* **7** (2003) (Nowak Ed.), Birkhauser, Boston (to appear).

[6] L.S. Shapley, Stochastic Games, *Proceedings of the National Academy of Sciences* **39** (1953), 1095–1100.

[7] M. Shubik, W. Whitt, *Fiat Money in an Economy with One Nondurable Good and No Credit. A Noncooperative Sequential Game*, In *Topics in Differential Games* (Blaquiere Ed.), North Holland, Amsterdam, 401–449, 1973.

[8] P. Więcek, *Pure Equilibria in a Simple Dynamic Model of Strategic Market Game*, 2004, (submitted)

In: Mathematics, Game Theory and Algebra Compendium ISBN 978-1-60456-659-8
Editor: Jacob H. Mathias, pp. 399-405

Chapter 34

DISTORTION OF LENGTHS UNDER FUNCTIONS STARLIKE WITH RESPECT TO A BOUNDARY POINT

***V. Karunakaran** and K. Bhuvaneswari*†**
School of Mathematics, Madurai Kamaraj University
Madurai - 625 021, India

Abstract

An analytic function f in the open unit disc D of the complex plane is said to be starlike with respect to the boundary point origin if $0 \in \partial\Omega$ where $\Omega = f(D)$ and for every point $w \in \Omega$, the line segment $(0, w] = \{tw : 0 < t \leq 1\}$ lies in Ω. Let S denote the class of analytic univalent functions in the open unit disc D satisfying $f(0) = 1$, $\lim\limits_{r \to 1^-} f(r) = 0$, $\lim\limits_{r \to 1^-} f'(r) = 0$ and $f(D)$ is starlike with respect to the origin. Let $C(r)$ be the image in Ω of the semi-open ray joining $z = r$ to $z = 1$ and let $l(r) = \int\limits_r^1 |f'(u)|du$ be its length. We prove that $l(r) < 3|f(r)|$ for all r with $0 < r < 1$.

AMS Subject Classification: 30C45

Key Words and Phrases: Starlike function with respect to boundary point,ray image, Starlike.

1. Introduction

Let $w = f(z)$ be regular and univalent in $|z| < 1$ with $f(0) = 0$. Suppose that f maps the open unit disc D of the complex plane onto a starlike region. The problem of determining bounds on $\frac{L(r,\theta)}{|f(re^{i\theta})|}$ where $L(r, \theta) = \int\limits_0^r |f'(ue^{i\theta})|du$ (the length of the image of the ray joining $z = 0$ and $z = re^{i\theta}$ under f) has a long history (see [1],[2],[3],[4],[5],[9]). On the other hand the concept of a univalent function starlike with respect to a boundary point

*E-mail address: vkarun_mku@yahoo.co.in
†E-mail address: bhuvanamku@yahoo.co.in

was first introduced and characterised in ([6],[7],[8]). The object of this paper is to show that $\frac{l(r)}{|f(r)|}$ is bounded as f varies over a certain subclass of univalent functions starlike with respect to the boundary point origin. Indeed we shall show that $L(f) = \sup_{0<r<1} \frac{l(r)}{|f(r)|}$ is bounded by 3. Some open problems are also discussed.

2. Preliminaries

Let $\mathbb{C}$ be the complex plane and let D be the open unit disc $\{z : |z| < 1\}$. A complex region Ω with $0 \in \partial\Omega$ is called starlike with respect to the origin if for every point $w \in \Omega$ the line segment $(0, w] = \{tw : 0 < t \leq 1\}$ lies in Ω. Also we call a univalent function f of D onto a region Ω starlike with respect to the boundary point origin if Ω is starlike with respect to the boundary point origin. Denote by S_0^* the class of all such functions.

Let $\mathcal{G}$ be the class of all univalent functions f of D that satisfy

$$f(0) = 1 \quad \text{and} \quad \text{Re}\{2z\frac{f'(z)}{f(z)} + \frac{1+z}{1-z}\} > 0 \quad (z \in D).$$

It is known that if $f \in \mathcal{G}$ then $f^2 \in S_0^*$ and conversely $f(0) = 1$, $\lim_{r\to 1^-} f(r) = 0$ and $f^2 \in S_0^*$ then $f \in \mathcal{G}$. However if $f^2 \in S_0^*$, and $f(0) \neq 1$ (or) $\lim_{r\to 1^-} f(r) \neq 0$ then there exists a real β such that $\lim_{t\to 1^-} f(te^{i\beta}) = 0$ and consequently, $\frac{f(e^{i\beta}z)}{f(0)} \in \mathcal{G}$. This gives at once a complete analytic definition of the functions $f \in S_0^*$ (see [7]).

Let B be the class of all analytic functions from D to itself. For $\alpha > 0$, let $B(\alpha)$ be the sub class of B consisting of all functions w whose angular limits of $\frac{(1-w(z))}{1-z}$ at 1 is α

If $f \in S_0^*$ from the representation theorem (Thm.1 of [7]) it looks as though that if the angular limit of $f'(z)$ at $z = 1$ exists then it should be equal to zero atleast when $\alpha \neq 1/4$ (Note that to each $f \in S_0^*$ there corresponds a $w \in B(\alpha)$ such that $-(1-z)^2\frac{f'(z)}{f(z)} = 4\frac{1-w(z)}{1+w(z)}$ $z \in D$). Hence in this paper we explicitly assume that the angular limit of $f'(z)$ at $z = 1$ exists and is equal to zero.

Now let S denote the class of analytic functions in the open unit disc D satisfying $f(0) = 1$, $\lim_{r\to 1^-} f(r) = 0$, $\lim_{r\to 1^-} f'(r) = 0$ and $f(D) = \Omega$ is starlike with respect to the boundary point origin. Let $C(r)$ be the image in Ω of the semi-open ray joining $z = r$ to $z = 1$ and let $l(r) = \int_r^1 |f'(u)|du$ be its length. We prove that $l(r) < 3|f(r)|$ for all r, with $0 < r < 1$.

Theorem 2.1 *Let $f \in S_0^*$ and $L(f) = \sup_{0<r<1} \frac{l(r)}{|f(r)|}$ where $l(r) = \int_r^1 |f'(u)|du$. Then $L(f) \leq 3$.*

Proof : Since $f \in S_0^*$, $f(z) \neq 0$ for $z \in D$ and we can define an analytic branch of $\sqrt{f}$ which belongs to $\mathcal{G}$. Hence using the well known Hergoltz's representation formula we have

$$H(u) = \frac{uf'(u)}{f(u)} = \frac{1}{2\pi}\int_{-\pi}^{\pi}\left(\frac{1+ue^{it}}{1-ue^{-it}} - \frac{1+u}{1-u}\right)dV(t) \tag{1}$$

where $V(t)$ is increasing, $V(\pi)+\pi = V(\pi)-\pi$. Now we can write

$$\frac{1}{u}\mathrm{Re}H(u) = \frac{\partial}{\partial u}\log|f(u)| = \int_0^{\pi}\frac{2(1+u)(\cos t-1)}{(1-u)(1-2u\cos t+u^2)}dW(t) \tag{2}$$

where $dW(t) = \frac{dV(t)-dV(-t)}{2\pi}$ is non-negative and has total mass 1. Now we have to show that

$$\int_r^1 |f'(u)|du \leq 3|f(r)|. \tag{3}$$

We write

$$\int_r^1 |f'(u)|du = \int_r^1 |H(u)||f(u)|u^{-1}du$$

$$= -\int_r^1 \mathrm{Re}H(u)|f(u)|u^{-1}du + \int_r^1 (|H(u)|+\mathrm{Re}H(u))|f(u)|u^{-1}du \tag{4}$$

and note that the first integral on the right side is $-|f(r)|$ by (2).
From (1) we have

$$|H(u)| \leq \int_0^{\pi}\frac{2u\sqrt{2-2\cos t}}{(1-u)\sqrt{1+u^2-2u\cos t}}dW(t).$$

From (1) and (2) we also have
$|H(u)| + \mathrm{Re}H(u)$

$$\leq \int_0^{\pi}\frac{2u\sqrt{2-2\cos t}}{(1-u)\sqrt{1+u^2-2u\cos t}}dW(t) + \int_0^{\pi}\frac{2u(1+u)(\cos t-1)}{(1-u)(1-2u\cos t+u^2)}dW(t)$$

$$\leq \int_0^{\pi}\left\{\frac{2u\sqrt{2-2\cos t}}{(1-u)\sqrt{1+u^2-2u\cos t}} - \frac{2u(1+u)(1-\cos t)}{(1-u)(1-2u\cos t+u^2)}\right\}dW(t).$$

From (2) we also have

$$\log\left|\frac{f(u)}{f(r)}\right| = \int_r^u \frac{\partial}{\partial u}\log|f(u)|du$$

$$= \int_0^{\pi}\log\frac{(1-u)^2(1-2r\cos t+r^2)}{(1-r)^2(1-2u\cos t+u^2)}dW(t).$$

By using Jensen's inequality [10] we get

$$\frac{|f(u)|}{u} \le |f(r)| \int_0^{\pi} \log \frac{(1-u)^2(1-2r\cos t + r^2)}{u(1-r)^2(1-2u\cos t+u^2)} dW(t).$$

Hence using Fubini's theorem

$$\int_r^1 (|H(u)| + \mathrm{Re} H(u))|f(u)|u^{-1} du \le |f(r)| \int_0^{\pi}\int_0^{\pi} I(t,s) dW(s) dW(t)$$

$$\le \frac{1}{2}|f(r)| \int_0^{\pi}\int_0^{\pi} (I(t,s) + I(s,t)) dW(s) dW(t) \tag{5}$$

where

$$I(t,s) = \int_r^1 \left\{ \frac{2u\sqrt{2-2\cos t}}{(1-u)\sqrt{1+u^2-2u\cos t}} - \frac{2u(1+u)(1-\cos t)}{(1-u)(1-2u\cos t + u^2)} \right\} \left\{ \frac{(1-u)^2(1-2r\cos s + r^2)}{u(1-r)^2(1-2u\cos s + u^2)} \right\} du.$$

In order to prove (3) it suffices to show that

$$\int_0^{\pi}\int_0^{\pi} (I(t,s) + I(s,t)) dW(s) dW(t) \le 4.$$

Since dW is a probability measure, more over I is even in both variables it will be sufficient for our result to show that

$$I(s,t) + I(t,s) \le 4 \quad \text{for} \quad 0 < s < t < \pi.$$

From the above integral we have

$$I(t,s) = \frac{1-2r\cos s + r^2}{(1-r)^2}[J(t,s) - K(t,s)]$$

where

$$J(t,s) = \int_r^1 \frac{2(1-u)\sqrt{2-2\cos t}}{\sqrt{1+u^2-2u\cos t}(1+u^2-2u\cos s)} du$$

$$K(t,s) = \int_r^1 \frac{2(1-u^2)(1-\cos t)}{(1-2u\cos t+u^2)(1-2u\cos s+u^2)} du.$$

Now we evaluate these integrals. In the above integrals we change the variable u to $y = \frac{1+u}{\sqrt{1+u^2-2u\cos t}}$ and use $S = \sin s/2$ and $T = \sin t/2$ to get (after some computations)

$$J(t,s) = \left(\frac{1}{T^2-S^2}\right)\left(\frac{T\sqrt{T^2-S^2}}{\sqrt{1-S^2}}\right) \cdot \left[\log\frac{\frac{\sqrt{1-S^2}}{\sqrt{T^2-S^2}}+\frac{1}{T}}{\frac{\sqrt{1-S^2}}{\sqrt{T^2-S^2}}-\frac{1}{T}} - \log\frac{\frac{\sqrt{1-S^2}}{\sqrt{T^2-S^2}}+\frac{1+r}{\sqrt{1+r^2-2r\cos t}}}{\frac{\sqrt{1-S^2}}{\sqrt{T^2-S^2}}-\frac{1+r}{\sqrt{1+r^2-2r\cos t}}}\right]$$

$$J(s,t) = \left(\frac{2S}{T^2-S^2}\right)\left(\frac{\sqrt{T^2-S^2}}{\sqrt{1-T^2}}\right) \cdot \left[\tan^{-1}\frac{\sqrt{T^2-S^2}}{S\sqrt{1-T^2}} - \tan^{-1}\left(\frac{\sqrt{T^2-S^2}}{\sqrt{1-T^2}}\right)\left(\frac{1+r}{1+r^2-2r\cos s}\right)\right]$$

$$K(t,s) = \frac{T^2}{T^2-S^2}\log\frac{T^2[(1-r)^2+4rS^2]}{S^2[(1-r)^2+4rT^2]}$$

$$K(s,t) = \frac{S^2}{T^2-S^2}\log\frac{T^2[(1-r)^2+4rS^2]}{S^2[(1-r)^2+4rT^2]}.$$

Now for simplification purposes we write

$$k = \frac{\sqrt{T^2-S^2}}{T\sqrt{1-S^2}}, \frac{k}{\sqrt{1-k^2}} = \frac{\sqrt{T^2-S^2}}{S\sqrt{1-T^2}},$$

$$x = \frac{(1+r)T}{\sqrt{(1-r)^2+4rT^2}}, y = \frac{(1+r)S}{\sqrt{(1-r)^2+4rS^2}}$$

$$\frac{x^2-y^2}{x^2y^2} = \left(\frac{(1-r)^2}{(1+r)^2}\right)\left(\frac{T^2-S^2}{S^2T^2}\right).$$

Also we put $y = tx, \quad 0 \le t \le 1$, so that

$$k^2 = \frac{1-t^2}{1-t^2x^2}, \frac{k^2}{1-k^2} = \frac{1-t^2}{t^2(1-x^2)}$$

$$\begin{aligned}
I(t,s)+I(s,t) &= \frac{x^2}{x^2-y^2}\left[k\log\frac{1+k}{1-k} - \log\frac{1-kx}{1+kx}\right] - \frac{x^2+y^2}{x^2-y^2}\log\frac{x^2}{y^2} \\
&\quad + \frac{y^2}{x^2-y^2}\left[\frac{2k}{\sqrt{1-k^2}}\left\{\tan^{-1}\frac{k}{\sqrt{1-k^2}} - \tan^{-1}\frac{ky}{\sqrt{1-k^2}}\right\}\right] \\
&= \frac{\sqrt{1-t^2}}{\sqrt{1-t^2(1-t^2)}}\left\{\log\frac{1+\frac{\sqrt{1-t^2}}{\sqrt{1-t^2}}}{1-\frac{\sqrt{1-t^2}}{\sqrt{1-t^2}}} - \log\frac{1+x\frac{\sqrt{1-t^2}}{\sqrt{1-t^2}}}{1-x\frac{\sqrt{1-t^2}}{\sqrt{1-t^2}}}\right\} \\
&\quad + \frac{(2t^2)\sqrt{1-t^2}}{(1-t^2)(t\sqrt{1-x^2})}\left\{\tan^{-1}\frac{\sqrt{1-t^2}}{t\sqrt{1-x^2}} - \tan^{-1}\frac{tx\sqrt{1-t^2}}{t\sqrt{1-x^2}}\right\} \\
&\quad - \frac{1+t^2}{1-t^2}\log\frac{1}{t^2} \\
&= \frac{1}{\sqrt{1-t^2}\sqrt{1-t^2x^2}}\left\{\log\frac{\sqrt{1-t^2x^2}+\sqrt{1-t^2}}{\sqrt{1-t^2x^2}-\sqrt{1-t^2}} - \log\frac{\sqrt{1-t^2x^2}+x\sqrt{1-t^2}}{\sqrt{1-t^2x^2}-x\sqrt{1-t^2}}\right\}
\end{aligned}$$

$$+\frac{2t}{\sqrt{1-t^2}\sqrt{1-x^2}}\left\{\tan^{-1}\frac{\sqrt{1-t^2}}{t\sqrt{1-x^2}}-\tan^{-1}\frac{x\sqrt{1-t^2}}{\sqrt{1-x^2}}\right\}$$
$$+\frac{2(1+t^2)}{1-t^2}\log t$$

$$\begin{aligned}
&= \frac{2}{\sqrt{1-t^2}\sqrt{1-t^2x^2}}\log\frac{\sqrt{1+t^2x^2}+\sqrt{1-t^2}}{\sqrt{1-t^2}+x\sqrt{1-t^2}}-\frac{2}{\sqrt{1-t^2}\sqrt{1-t^2x^2}}\log t\\
&\quad+\frac{2t}{\sqrt{1-t^2}\sqrt{1-x^2}}\left\{\tan^{-1}\frac{\sqrt{1-t^2}}{t\sqrt{1-x^2}}-\tan^{-1}\frac{x\sqrt{1-t^2}}{\sqrt{1-x^2}}\right\}\\
&\quad+\frac{2(1+t^2)}{1-t^2}\log t\\
&= \frac{2}{\sqrt{1-t^2}\sqrt{1-t^2x^2}}\log\frac{1+\sqrt{1-t^2}\sqrt{1-t^2x^2}}{1+x}\\
&\quad+\frac{2t}{\sqrt{1-t^2}\sqrt{1-x^2}}\tan^{-1}\frac{\sqrt{1-t^2}\sqrt{1-x^2}}{t+x}\\
&\quad+2\left[\frac{1+t^2}{1-t^2}-\frac{1}{\sqrt{1-t^2}\sqrt{1-t^2x^2}}\right]\log t.
\end{aligned}$$

Now we have

$$\frac{2}{\sqrt{1-t^2}\sqrt{1-t^2x^2}}\log\left(\frac{1}{1+x}+\frac{\sqrt{1-t^2}\sqrt{1-t^2x^2}}{1+x}\right)\leq\frac{2}{1+x},$$

$$\frac{-2\log t}{\sqrt{1-t^2}\sqrt{1-t^2x^2}}\leq\frac{-2\log t}{1-t^2}$$

and

$$\frac{2t}{\sqrt{1-t^2}\sqrt{1-x^2}}\left\{\tan^{-1}\frac{\sqrt{1-t^2}\sqrt{1-x^2}}{t+x}\right\}\leq\frac{2t}{t+x}.$$

so that

$$I(t,s)+I(s,t)\leq\frac{2}{1+x}+\frac{2t}{t+x}\leq\frac{4}{1+x}\leq 4. \tag{6}$$

Using (6) we have from (5) that

$$\int_r^1 (|H(u)|+\mathrm{Re}H(u))|f(u)|u^{-1}du\leq 2|f(r)| \tag{7}$$

and as already observed

$$\int_r^1 \mathrm{Re}H(u)|f(u)|u^{-1}du=|f(r)|. \tag{8}$$

Now (7),(8) and (4) give (3). Hence the result.

Note 2.2

1. It will be interesting to find the exact upper bound for $\frac{l(r)}{|f(r)|}$ as f varies over the entire class S_0^* and $0 < r < 1$ is fixed. With the estimates we have obtained in the above theorem it is difficult to guess what this upper bound can be.

2. Eventhough S_0^* is rotationaly invariant (in the sense that $f \in S_0^* \Rightarrow e^{-i\theta} f(ze^{i\theta}) \in S_0^*$), the problem of determining $\sup_{f \in S_0^*} \frac{l(r,\theta)}{|f(r,\theta)|}$ where $l(r,\theta) = \int_r^1 |f'(ue^{i\theta})| du$ will be of interest in this context.

References

[1] R. Balasubramanian,V. Karunakaran and S.Ponnusamy, A proof of Hall's conjecture on starlike mappings,*J. London. Math. Soc.*(2)**48**(1993), 278-288.

[2] F.W. Gehring and W.K. Hayman, An inequality in the theory of conformal mapping,*J. Math. Pures Appl.***41**(1963),353-361.

[3] R.R. Hall, The length of ray images under starlike mappings , *Mathematika* **23**(1976),147-150.

[4] R.R. Hall, A conformal mapping inequality for starlike functions of order $\frac{1}{2}$, *Bull. London Math. Soc.* **12**,(1980),119-126.

[5] V. Karunakaran, Length of ray images under conformal maps, *Proc. Amer. Math. Soc.* **87**, No.2,(1983),289-294.

[6] A.Lecko, On the class of functions starlike with respect to a boundary point, *J. Math. Anal. Appl.***261**(2001),649-664.

[7] A.Lecko and A.Lyzzaik, A note on univalent functions starlike with respect to a boundary point,*J. Math. Anal. Appl.* **282** (2003), 846-851.

[8] M.S.Robertson, Univalent functions starlike with respect to a boundary point,*J. Math. Anal. Appl.* **81** (1981) 327-345.

[9] T. Sheil-small, Some conformal mappings inequalities for starlike and convex functions, *J. London Math. Soc.*(2) **1**(1969), 577-587.

[10] A. Zygmund, *Trignometric series* (University press, Cambridge, 1968)

In: Mathematics, Game Theory and Algebra Compendium ISBN 978-1-60456-659-8
Editor: Jacob H. Mathias, pp. 407-421

Chapter 35

SERVICE FACILITY INVENTORY SYSTEM WITH IMPATIENT CUSTOMERS*

Paul Manuel
Department of Information Science, Kuwait University, Kuwait
B. Sivakumar and G.Arivarignan
Department of Applied Mathematics and Statistics,
Madurai Kamaraj University, Madurai, India

Abstract

In this article we analyze a continuous review (s, S) inventory system at a service facility, wherein an item is demanded by a customer is issued after performing service on the item. The waiting hall size is assumed to be finite. The arrival time points of customers form a Poisson process. The service time, life of item in the stock and the lead time of orders are all assumed to be independently distributed as exponential. The waiting customer independently reneges the system after an exponentially distributed amount of time. The joint probability distribution of the number of customers in the system and the inventory level is obtained in both transient and steady state cases. The measures of system performance in the steady state are derived and long run total expected cost rate is derived. The results are illustrated with numerical examples.

AMS Subject Classification: 90B05, 60J27

Key Words and Phrases: Stochastic Inventory, Random life time, Positive lead time, reneging, Service facility, Poisson demands.

1. Introduction

In most of the inventory models considered in the literature, the demanded items are directly issued from the stock, if available. The demands that occurred during stock - out period are

*The research was supported by National Board for Higher Mathematics, INDIA, research award 48/3/2004/R&D - II/2114.

either not satisfied (lost sales case) or satisfied only after the receipt of the ordered items (backlog case). In the latter, it is assumed either all (full backlog case) or any prefixed number of demands (partial backlogging) that occurred during stock out period are satisfied. For review of these works see Nahmias[10], Raafat[13], Kalpakam and Arivarignan[8], Elango and Arivarignan[6], Liu and Yang[9].

But in the case of inventories maintained at service facilities, the demanded items are issued to the customers only after some service is performed on it. In this situation the items are issued not at the time of demand but after a random time of service. This forces the formation of queues in these models. This necessitates the study of both inventory level and queue length joint distributions. Study of such models is beneficial to organizations who

i) provide service to customers by using items from a stock.

ii) maintain stock of items each of which needs service such as assembly or initialization or installation etc., before it is delivered to the customers.

Examples for the first type include firms engaged in servicing consumer products such as Television sets, Computers etc., and for the second type include firms that supply bicycles which need assembly of its parts, that supply food items which need heating or garnishing and that computers which need installation of basic services. Recently Berman, Kaplan and Shimshak [2] have considered an inventory management system at a service facility which uses one item of inventory for each service provided. They assumed that both demand and service rates are deterministic and constant as such queues can form only during stock outs. They determined optimal order quantity that minimizes the total cost rate.

Berman and Kim [3] analyzed a problem in stochastic environment where customers arrive at service facilities according to a Poisson process and the service times are exponentially distributed with mean inter-arrival time (assumed to be greater than the mean service time), and each service requires one item from inventory. The main result of their work is that under both the discounted cost case and the average cost case, the optimal policy of both the finite and infinite time horizon problem is a threshold ordering policy. The optimal policy in Berman and Kim [3] is derived given that the order quantity is known. A logically related model was studied by Q-M. He, E.M. Jewkes and Buzacott [7], who analyzed a Markovian inventory - production system, where customer demands arrive at a workshop and are processed by a single machine in batch sizes of one. Berman and Sapna [4] studied extensively an inventory control problem at a service facility which uses one item of inventory for the service provided. They assumed Poisson arrivals, arbitrarily distributed service times and zero lead times. They analyzed the system with the restriction that the waiting space being finite. Under a specified cost structure, they derived the optimal ordering quantity that minimizes the long-run expected cost rate.

Elango [5] has considered a Markovian inventory system with instantaneous supply of orders at a service facility. The service time is assumed to have exponential distribution with parameter depending on the number of waiting customers. Perumal and Arivarignan [12] have considered a Markovian inventory system in which the size of the space for the waiting customers is assumed to be infinite. Arivarignan and Sivakumar [1] have considered an inventory system with arbitrarily distributed demand, exponential service time and exponential lead time.

It has long been recognized that reneging is an important feature in many real-world queueing contexts. In fact, Palm [11] introduced reneging as a means of modelling the behaviour of telephone switchboard customers more than 65 years ago. However, due to the explosive growth of the call center industry, there has been renewed interest in such models in recent years. In the call center setting, customer impatience (amplified by large customers loads) leads naturally to large amounts of reneging. Ignoring the presence of reneging can lead to inappropriate sizing of the system and poor staffing allocation. Models in which reneging is present are also potentially valuable in problem contexts (Ward and Glynn [15]).

In this paper we consider a continuous review perishable inventory system with service facility and reneging of waiting customers. The transient and steady state analysis of the joint probability distribution for the inventory levels and the number of customers in the system are derived. The various measures of system performance are computed in the steady state case and the results are illustrated with numerical examples.

2. Model Description

Consider a service facility in which perishable items are stocked and the items are delivered to the demanding customers. The demand is for single item per customer. The maximum capacity of the inventory is S and the maximum number of customers permitted in the waiting hall, including the one in the service, is N. The following assumptions are made:

- The arrival time points of customers form a Poisson process with parameter λ.
- The service time for each customer follows a negative exponential distribution with parameter $\mu(> 0)$.
- Life time of each item has negative exponential distribution with parameter $\gamma(> 0)$. Note that in this model we have assumed that the servicing item also perish.
- An impatient customer leaves the system independently after a random time which is distributed as negative exponential with parameter $\alpha(> 0)$. We assume that the customers can abandon the system even after service has been initiated.
- An order for $Q(= S - s)$ items is placed whenever the inventory level drops to s and the items are received only after a random time which is distributed as exponential with parameter $\beta(> 0)$.

Notations

$[A]_{ij}$: (i, j) element of the matrix A.

$$\delta_{ij} = \begin{cases} 1 & \text{if } i = j, \\ 0 & \text{Otherwise.} \end{cases}$$

A_α^* : Laplace transform of any arbitrary matrix $A(t)$, for $Re(\alpha) > 0$.
$\mathbf{0}$: zero matrix.
$\mathbf{e}^T$: $(1, 1, \dots, 1)$.
I : an identity matrix.
E_1 = $\{0, 1, \dots, S\}$.
E_2 = $\{0, 1, \dots, N\}$.
E = $E_1 \times E_2$.

3. Analyses

Let $L(t)$ denote the inventory level and $X(t)$ denote the number of customers (waiting and being served) in the system, at time t. From the assumptions made on the input and output processes, it can be shown that $(L, X) = \{(L(t), X(t));$ $t \geq 0\}$, on the state space E, is a Markov process. The infinitesimal generator of this process,

$$A = ((\ a((i, q), (j, r))\ )), \quad (i, q), (j, r) \in E$$

can be obtained by using the following arguments:

- The arrival of a customer makes a transition from (i, q) to $(i, q + 1)$ with intensity of transition λ, $i = 0, 1, 2, \dots, S,\ q = 0, 1, 2, \dots, N - 1$.

- The completion of service makes one customer leaves the system and decreases the inventory level by 1. Thus a transition takes place from (i, q) to $(i - 1, q - 1)$ with intensity of transition μ, $i = 1, 2, \dots, S,\ q = 1, 2, \dots, N$.

- A transition from (i, q) to $(i - 1, q)$ will take place when any one of i items perish at a rate of γ; thus intensity of transition is $i\gamma$, $i = 1, 2, \dots, S,\ q = 0, 1, 2, \dots, N$.

- A transition from (i, q) to $(i, q - 1)$ will take place when any one of q customers reneges the system at a rate of α; thus intensity of transition is $q\alpha$, $i = 0, 1, 2, \dots, S,\ q = 1, 2, \dots, N$.

- A transition from (i, q) to $(i + Q, q)$, for $i = 0, 1, 2, \dots, s; q = 0, 1, 2, \dots, N$, will take place with the intensity of transition β when a replenishment occurs.

Hence we get

$$a((i,q),(j,r)) = \begin{cases} \lambda & j=i; & r=q+1; \\ & i=0,1,2,\dots,S; & q=0,1,2,\dots,N-1; \\ i\gamma & j=i-1; & r=q; \\ & i=1,2,\dots,S; & q=0,1,2,\dots,N; \\ k\alpha & j=i; & r=q-1; \\ & i=0,1,2,\dots,S; & q=1,2,\dots,N; \\ \mu & j=i-1; & r=q-1; \\ & i=1,2,\dots,S; & q=1,2,\dots,N, \\ -(\lambda+i\gamma) & j=i; & r=q; \\ & i=s+1,s+2,\dots,S; & q=0; \\ -(\mu+i\gamma+q\alpha) & j=i; & r=q; \\ & i=s+1,s+2,\dots,S; & q=N; \end{cases}$$

$$\begin{cases} -(\beta+\lambda+i\gamma) & j=i; & r=q; \\ & i=1,2,\dots,s; & q=0; \\ -(\beta+\mu+i\gamma+q\alpha) & j=i; & r=q; \\ & i=1,2,\dots,s; & q=N; \\ -(\lambda+\mu+i\gamma+q\alpha) & j=i; & r=q; \\ & i=s+1,s+2,\dots,S; & q=1,2,\dots,N-1; \\ -(\beta+\lambda+\mu+i\gamma+q\alpha) & j=i; & r=q; \\ & i=1,2,\dots,s; & q=1,2,\dots,N-1; \\ -(\lambda+\beta) & j=i; & r=q; \\ & i=0; & q=0; \\ -(\lambda+\beta++q\alpha) & j=i; & r=q; \\ & i=0; & q=1,2,\dots,N-1; \\ -(\beta+q\alpha) & j=i; & r=q; \\ & i=0; & q=N; \\ \beta & j=i+Q;; & r=q; \\ & i=0,1,2,\dots,s; & q=0,1,2,\dots,N; \\ 0 & \text{Otherwise.} & \end{cases}$$

Define

$$A_{ij} = ((\ a((i,q),(j,r))\ ))_{q,r\in E_2},\ i,j\in E_1.$$

The infinitesimal generator A can be written in terms of sub matrices A_{ij}, namely,

$$A = ((\ A_{ij}\ ))$$

where

$$A_{ij} = \begin{cases} A_i, & j=i, & i=0,1,2,\dots,S, \\ B_i, & j=i-1, & i=1,2,\dots,S-1,S, \\ C, & j=i+Q, & i=0,1,\dots,s, \\ \mathbf{0}, & \text{otherwise.} & \end{cases}$$

The sub matrices are given by,

For $i = s+1, s+2, \ldots, S$,

$$[A_i]_{kl} = \begin{cases} \lambda, & l = k+1, \quad k = 0, 1, \ldots, N-1, \\ k\alpha, & l = k-1, \quad k = 1, 2, \ldots, N, \\ -(\lambda + i\gamma), & l = k, \quad k = 0, \\ -(\lambda + i\gamma + k\alpha + \mu), & l = k, \quad k = 1, 2, \ldots, N-1, \\ -(i\gamma + k\alpha + \mu), & l = k, \quad k = N, \\ 0, & \text{otherwise}, \end{cases}$$

For $i = 1, 2, \ldots, s$,

$$[A_i]_{kl} = \begin{cases} \lambda, & l = k+1, \quad k = 0, 1, \ldots, N-1, \\ k\alpha, & l = k-1, \quad k = 1, 2, \ldots, N, \\ -(\lambda + i\gamma + \beta), & l = k, \quad k = 0, \\ -(\lambda + i\gamma + k\alpha + \mu + \beta), & l = k, \quad k = 1, 2, \ldots, N-1, \\ -(i\gamma + k\alpha + \mu + \beta), & l = k, \quad k = N, \\ 0, & \text{otherwise}, \end{cases}$$

$$[A_0]_{kl} = \begin{cases} \lambda, & l = k+1, \quad k = 0, 1, \ldots, N-1, \\ k\alpha, & l = k-1, \quad k = 1, 2, \ldots, N, \\ -(\lambda + \beta), & l = k, \quad k = 0, \\ -(\lambda + k\alpha + \beta), & l = k \quad k = 1, 2, \ldots, N-1, \\ -(\beta + k\alpha), & l = k, \quad k = N, \\ 0, & \text{otherwise}, \end{cases}$$

For $i = 1, 2, \ldots, S$,

$$[B_i]_{kl} = \begin{cases} i\gamma, & l = k \quad k = 0, 1, \ldots, N, \\ \mu, & l = k-1, \quad k = 1, \ldots, N, \\ 0, & \text{otherwise}, \end{cases}$$

$$[C]_{kl} = \begin{cases} \beta, & l = k \quad k = 0, 1, \ldots, N, \\ 0, & \text{otherwise}. \end{cases}$$

It may be noted that all the matrices are square matrices of size $(N+1)$.

3.1. Transient Analysis

Define

$$\psi((i,k),(j,l);t) = Pr\{L(t) = j, X(t) = l | L(0) = i, X(0) = k\},$$
$$(i,k),(j,l) \in E.$$

Consider the matrix

$$\Psi(t) = (\, (\, \psi((i,k),(j,l);t)\,)\,).$$

The Kolmogorov backward differential equation satisfied by $\Psi(t)$, is given by

$$\Psi'(t) = \Psi(t)A$$

and the solution of the above differential equation is given by

$$\Psi(t) = e^{At}$$

where e^{At} represents $I + At + \frac{A^2t^2}{2!} + \cdots$.

Alternatively,

$$\Psi^*(\alpha) = (\alpha I - A)^{-1}.$$

The matrix $(\alpha I - A)$ has the following block partitioned form

$$[(\alpha I - A)]_{ij} = \begin{cases} D_i, & j = i, \quad i = 0, 1, 2, \ldots, S \\ -B_i, & j = i - 1, \quad i = 1, 2, \ldots, S - 1, S \\ -C, & j = i + Q, \quad i = 0, 1, \ldots, s, \\ \mathbf{0}, & \text{Otherwise.} \end{cases}$$

where

$$D_i = \alpha I - A_i, \quad i = 0, 1, 2, \ldots, S.$$

It may be observed that $(\alpha I - A)$ is an almost upper triangular matrix in block partitioned form. That is, if we denote the (i, j)-th sub matrix of $\Psi(= \alpha I - A)$ by Ψ_{ij}, then we have

$$\Psi_{ij} = \mathbf{0}, \quad j = 0, 1, \ldots, i - 2, \quad i = 2, 3, \ldots, S.$$

To compute $\Psi^{-1} = (\alpha I - A)^{-1}$ we proceed as described below : Consider an upper triangular matrix U and an almost upper triangular matrix R having the form

$$U = \begin{pmatrix} U_{00} & U_{01} & U_{02} & \cdots & U_{0S} \\ \mathbf{0} & U_{11} & U_{12} & \cdots & U_{1S} \\ \mathbf{0} & \mathbf{0} & U_{22} & \cdots & U_{2S} \\ \cdots & \cdots & \cdots & \cdots & \cdots \\ \mathbf{0} & \mathbf{0} & \mathbf{0} & \cdots & U_{SS} \end{pmatrix}$$

with $U_{ii} = I, \quad i = 1, 2, \ldots, S$, and

$$R = \begin{pmatrix} R_{00} & R_{01} & R_{02} & \cdots & R_{0(S-1)} & R_{0S} \\ -B_1 & \mathbf{0} & \mathbf{0} & \cdots & \mathbf{0} & \mathbf{0} \\ \mathbf{0} & -B_2 & \mathbf{0} & \cdots & \mathbf{0} & \mathbf{0} \\ \mathbf{0} & \mathbf{0} & -B_3 & \cdots & \mathbf{0} & \mathbf{0} \\ \cdots & \cdots & \cdots & \cdots & \cdots & \cdots \\ \mathbf{0} & \mathbf{0} & \cdots & \cdots & -B_S & \mathbf{0} \end{pmatrix}.$$

such that $\Psi U = R$. We find the sub matrices U_{ij} and R_{0j} by computing the product ΨU and equating it to R. The (i, j)-th sub matrix of ΨU, denoted by $[\Psi U]_{ij}$ is given by

$$\begin{cases} -B_iU_{jj}, & i=1,2,\ldots,S \quad j=i-1 \\ D_iU_{ij}-B_iU_{(i-1)j}, & i=s+1,\ldots,S \quad j=i,i+1,\ldots,S \\ & \text{or} \\ & i=1,2,\ldots,s \quad j=i,i+1,\ldots,Q-1 \\ D_0U_{0j}, & i=0 \quad j=0,1,2,\ldots,Q-1 \\ -CU_{(Q+1)j}+D_0U_{0j}, & i=0 \quad j=Q,Q+1,\ldots,S \\ -CU_{(Q+1)j}+D_iU_{ij}-B_iU_{(i-1)j}, & i=1,2,\ldots,s \quad j=Q+i,\ldots,S. \end{cases}$$

By equating the sub matrices of ΨU to the corresponding elements of R, we get

$$U_{ij}=\begin{cases} (B_{i+1}^{-1}D_{i+1})(B_{i+2}^{-1}D_{i+2})\cdots(B_j^{-1}D_j), & i=0,1,2,\ldots,j+1 \\ & j=1,2,\ldots,Q \\ & \text{or} \\ & i=j-Q,j-Q+1,\ldots,S \\ & j=Q+1,Q+2,\ldots,S \\ -B_{i+1}^{-1}CU_{(i+Q+1)j}+B_{i+1}^{-1}D_{i+1}U_{(i+1)j}, & i=0,1,\ldots,s \\ & j=Q+i,Q+2,\ldots,S \end{cases}$$

and

$$R_{0j}=\begin{cases} D_0, & j=0 \\ D_0U_{0j}, & j=1,2,\ldots,Q-1 \\ -CU_{Qj}+D_0U_{0j}, & j=Q,Q+1,\ldots,S. \end{cases}$$

The equation $\Psi U=R$ implies

$$\begin{aligned} (\Psi U)^{-1} &= R^{-1} \\ U^{-1}\Psi^{-1} &= R^{-1} \\ \Psi^{-1} &= UR^{-1}. \end{aligned}$$

It can be verified that the inverse of R is given by,

$$R^{-1} = \begin{pmatrix} \mathbf{0} & -B_1^{-1} & \mathbf{0} & \cdots & \mathbf{0} & \mathbf{0} \\ \mathbf{0} & \mathbf{0} & -B_2^{-1} & \cdots & \mathbf{0} & \mathbf{0} \\ \cdots & \cdots & \cdots & \cdots & \cdots & \cdots \\ \cdots & \cdots & \cdots & \cdots & \cdots & \cdots \\ \mathbf{0} & \mathbf{0} & \mathbf{0} & \cdots & \mathbf{0} & -B_S^{-1} \\ R_{0S}^{-1} & R_{0S}^{-1}R_{00}B_1^{-1} & R_{0S}^{-1}R_{01}B_2^{-1} & \cdots & R_{0S}^{-1}R_{0(S-2)}B_{S-1}^{-1} & R_{0S}^{-1}R_{0(S-1)}B_S^{-1} \end{pmatrix}$$

Since the expression for R^{-1} involves the term R_{0S}^{-1}, we shall show that the latter exists. From $\Psi U=R$, we get

$$\begin{aligned} \det(\Psi U) &= \det(R) \\ \det(\Psi)\det(U) &= \det(R_{0S})\det(-B_1)\det(-B_2)\cdots\det(-B_S). \end{aligned}$$

Since U is an upper triangular matrix and B_i's are lower triangular matrices, their determinant values are not equal to zero. Hence $\det(R_{0S})$ is not equal to zero. This proves the existence of R_{0S}^{-1}.

From $\Psi^{-1} = UR^{-1}$, we can compute the (i,j) – th sub matrix (denoted by Ψ^{ij}) of $\Psi^{-1} = (\alpha I - A)^{-1}$ and it is given by,

$$\Psi^{ij} = \begin{cases} U_{iS}R_{0S}^{-1}R_{0(j-1)}B_j^{-1} & i = j, j+1, \ldots, S; \quad j = 1, 2, \ldots, S, \\ U_{iS}R_{0S}^{-1} & i = 0, 1, 2, \ldots, S; \quad j = 0, \\ U_{iS}R_{0S}^{-1}R_{0(j-1)} - U_{i(j-1)}B_j^{-1} & i = 0, 1, \ldots, S-1; \quad j = i+1, \ldots, S. \end{cases}$$

3.2. Steady State Analysis

It can be seen from the structure of A that the homogeneous Markov process $\{(L(t), X(t)), t \geq 0\}$ on the finite state space E is irreducible. Hence the limiting distribution

$$\mathbf{\Pi} = (\pi^{(0)}, \pi^{(1)}, \ldots, \pi^{(S)})$$

with $\pi^{(m)} = \left(\pi^{(m,0)}, \pi^{(m,1)}, \ldots, \pi^{(m,N)}\right)$, $m = 0, 1, \ldots, S$, where $\pi^{(i,k)}$ denotes the steady state probability for the state (i,k) of the process, exists and is given by

$$\mathbf{\Pi} A = \mathbf{0} \quad \text{and} \quad \sum_{(i,k)}\sum \pi^{(i,k)} = 1. \tag{1}$$

The first equation of the above yields the following set of equations:

$$\begin{aligned} \pi^{(i)}A_i + \pi^{(i-Q)}C &= \mathbf{0}, \quad i = S \\ \pi^{(i+1)}B_{i+1} + \pi^{(i)}A_i + \pi^{(i-Q)}C &= \mathbf{0}, \quad i = Q, Q+1, \ldots, S-1 \\ \pi^{(i+1)}B_{i+1} + \pi^{(i)}A_i &= \mathbf{0}, \quad i = 0, 1, 2, \ldots, Q-1 \end{aligned}$$

After long simplifications, the above equations, except the first one, yields

$$\begin{aligned} \pi^{(i)} &= (-1)^i \pi^{(0)} A_0 B_1^{-1} A_1 B_2^{-1} \cdots A_{i-1} B_i^{-1} & i = 1, 2, \ldots, Q, \\ &= (-1)^i \pi^{(0)} A_0 B_1^{-1} A_1 B_2^{-1} \cdots A_{i-1} B_i^{-1} - \pi^{(0)} C B_i^{-1} & i = Q+1, \\ &= (-1)^i \pi^{(0)} A_0 B_1^{-1} A_1 B_2^{-1} \cdots A_{i-1} B_i^{-1} & \\ &\quad + (-1)^{i-Q} \pi^{(0)} C B_{Q+1}^{-1} A_{Q+1} \cdots A_{i-1} B_i^{-1} & \\ &\quad + (-1)^{i-Q} \pi^{(0)} \left\{ \sum_{l=1}^{i-Q-1} \left(A_0 B_1^{-1} A_1 B_2^{-1} \cdots A_{l-1} B_l^{-1}\right) C \right. & \\ &\quad \left. \left(B_{Q+l+1}^{-1} A_{Q+l+1} \cdots B_i^{-1}\right) \right\} & i = Q+2, Q+3, \ldots, S, \end{aligned}$$

where $\pi^{(0)}$ can be obtained by solving,

$$\pi^{(S)} A_S + \pi^{(s)} C = \mathbf{0} \text{ and } \sum_{i=0}^{S} \pi^{(i)} \mathbf{e} = 1,$$

that is

$$\begin{aligned} \pi^{(0)} \Bigg[(-1)^S \left(\prod_{i=0}^{S-1} \psi_i \right) A_S + (-1)^S C B_{Q+1}^{-1} \left(\prod_{i=0}^{S-1} \psi_i \right) A_S & \\ + (-1)^S \left\{ \sum_{l=1}^{s_1 - 1} \left(\prod_{i=0}^{l-1} \psi_i \right) C \left(B_{Q+l+1}^{-1} \prod_{i=Q+l+1}^{S-1} \psi_i \right) \right\} A_S & \\ + (-1)^{s_1} \left(\prod_{i=0}^{S-1} \psi_i \right) C \Bigg] &= \mathbf{0}, \end{aligned}$$

and

$$\pi^{(0)}\left[I+\sum_{i=1}^{Q+1}(-1)^i\prod_{i=0}^{i-1}\psi_i - CB_{Q+1}^{-1}\right.$$
$$+\sum_{i=Q+2}^{S1}\left\{(-1)^i\left(\prod_{k=0}^{i-1}\psi_k\right)+(-1)^{i-Q}CB_{Q+1}^{-1}\left(\prod_{k=Q+1}^{i-1}\psi_k\right)\right.$$
$$\left.\left.+(-1)^{i-Q}\left\{\sum_{l=1}^{i-Q-1}\left(\prod_{i=0}^{l-1}\psi_i\right)CB_{Q+l+1}^{-1}\left(\prod_{i=Q+l+1}^{i-1}\psi_i\right)\right\}\right\}\right]\mathbf{e} = 1,$$

where $\psi_i = A_i B_{i+1}$.

4. System Performance Measures

In this section we derive some performance measures of the system under consideration.

4.1. Mean Inventory Level

Let β_1 denote the average inventory level in the steady state. Then we have

$$\beta_1 = \sum_{i=1}^{S} i \sum_{m=0}^{N} \pi^{(i,m)}. \tag{2}$$

4.2. Mean Reorder Rate

Let β_2 denote the mean reorder rate then we have

$$\beta_2 = \mu \sum_{m=1}^{N} \pi^{(s+1,m)} + (s+1)\gamma \sum_{m=0}^{N} \pi^{(s+1,m)}. \tag{3}$$

4.3. Mean Balking Rate

Let β_3 denote the mean balking rate of customers in the steady state which is given by

$$\beta_3 = \lambda \sum_{i=0}^{S} \pi^{(i,N)}. \tag{4}$$

4.4. Mean Reneging Rate

Let β_4 denote the mean reneging rate of customers in the steady state which is given by

$$\beta_4 = \sum_{i=0}^{S} \sum_{m=1}^{N} m\alpha\pi^{(i,m)}. \tag{5}$$

4.5. Mean Failure Rate

Let β_5 denote the mean failure rate of the items in the steady state which is given by

$$\beta_5 = \sum_{i=1}^{S}\sum_{m=0}^{N} i\gamma\pi^{(i,m)}. \tag{6}$$

4.6. Mean Waiting Time

Let $\overline{W}$ denote the mean waiting time of the customers in the steady state which is given by

$$\overline{W} = \frac{\Gamma}{\lambda_a}, \tag{7}$$

where

$$\Gamma = \sum_{m=1}^{N} m \sum_{i=0}^{S} \pi^{(i,m)}, \tag{8}$$

and the effective arrival rate (Ross [14]), λ_a is given by

$$\lambda_a = \lambda \sum_{i=0}^{S}\sum_{m=0}^{N-1} \pi^{(i,m)}. \tag{9}$$

5. Cost Analysis

To compute the total expected cost per unit time, we consider the following costs.

h : The inventory carrying cost per unit item per unit time.
c_1 : setup cost per order.
c_2 : balking cost per customer per unit time.
c_3 : reneging cost per customer per unit time.
c_4 : failure cost per unit item per unit time.
c_5 : waiting time cost of a customer per unit time.

The long run expected cost rate is given by

$$TC(S, s, N) = h\beta_1 + c_1\beta_2 + c_2\beta_3 + c_3\beta_4 + c_4\beta_5 + c_5\overline{W},$$

where β_i's and $\overline{W}$ are as given in (2) - (7).
That is

$$\begin{aligned} TC(S,s,N) &= h\sum_{i=1}^{S} i\sum_{m=0}^{N}\pi^{(i,m)} + c_1\left(\mu\sum_{m=1}^{N}\pi^{(s+1,m)} + (s+1)\gamma\sum_{m=0}^{N}\pi^{(s+1,m)}\right) \\ &+c_2\lambda\sum_{i=0}^{S}\pi^{(i,N)} + c_3\sum_{i=0}^{S}\sum_{m=1}^{N} m\alpha\pi^{(i,m)} \\ &+c_4\sum_{i=1}^{S}\sum_{m=0}^{N} i\gamma\pi^{(i,m)} + c_5\frac{\sum_{m=1}^{N} m\sum_{i=0}^{S}\pi^{(i,m)}}{\lambda\sum_{i=0}^{S}\sum_{m=0}^{N-1}\pi^{(i,m)}}. \end{aligned}$$

Due to the complex form of the limiting distribution, it is difficult to discuss the properties of the cost function analytically. However we present, in the next section some numerical examples to illustrate the results of this work.

6. Numerical Illustrations

We have computed the limiting probability distribution of inventory level for specific values of parameters. For the first table we have assumed
$S = 12, s = 4, N = 5, \lambda = 1.5, \mu = 1.2, \gamma = 0.8, \beta = 1,$
$\alpha = 0.7, h = 0.2, c_1 = 100, c_2 = 0.7, c_3 = 0.6, c_4 = 1, c_5 = 1.2$

Inventory	Number of Customers					
Level	0	1	2	3	4	5
0	.035895	.042939	.030679	.016117	.006775	.002415
1	.034014	.027056	.015807	.007288	.002767	.000885
2	.036387	.029240	.017298	.008077	.003102	.001002
3	.038140	.030987	.018563	.008770	.003405	.001111
4	.039450	.032461	.019697	.009409	.003690	.001216
5	.040408	.033761	.020756	.010012	.003960	.001321
6	.034155	.028886	.017921	.008696	.003455	.001158
7	.029398	.025277	.015829	.007719	.003079	.001039
8	.025625	.022525	.014227	.006964	.002785	.000950
9	.018427	.015082	.009051	.004271	.001658	.000549
10	.012877	.010602	.006404	.003040	.001187	.000395
11	.008017	.006647	.004043	.001931	.000758	.000254
12	.003751	.003137	.001923	.000924	.000364	.000123

Mean Reorder Rate = 0.524656
Mean Inventory Level = 4.429783
Mean Balking Rate = 0.018634
Mean Waiting time of a Customer = 0.798432
Mean Failure Rate = 4.429783
Mean Reneging Rate = 4.429783
Expected Total Cost Rate = 59.249330

Inventory	Number of Customers						
Level	0	1	2	3	4	5	6
0	.048477	.067582	.054908	.032241	.014926	.005733	.001906
1	.032861	.028033	.017488	.008581	.003464	.001185	.000350
2	.030308	.026018	.016379	.008117	.003309	.001142	.000339
3	.028394	.024584	.015650	.007845	.003232	.001126	.000337
4	.026835	.023482	.015138	.007678	.003197	.001124	.000339
5	.025491	.022593	.014765	.007579	.003188	.001131	.000343
6	.024284	.021858	.014493	.007524	.003194	.001142	.000349
7	.020858	.019065	.012810	.006715	.002871	.001033	.000318
8	.018150	.016910	.011520	.006093	.002623	.000950	.000294
9	.015940	.015211	.010504	.005601	.002424	.000883	.000276
10	.014088	.013850	.009686	.005200	.002262	.000827	.000263
11	.009572	.008344	.005321	.002663	.001094	.000381	.000115
12	.006807	.005967	.003831	.001930	.000798	.000279	.000084
13	.004585	.004046	.002618	.001329	.000553	.000194	.000059
14	.002766	.002461	.001605	.000821	.000344	.000121	.000037
15	.001259	.001130	.000744	.000383	.000161	.000057	.000017

Mean Reorder Rate = 0.509933
Mean Inventory Level = 4.331311
Mean Balking Rate = 0.010867
Mean Waiting time of a Customer = 0.682097
Mean Failure Rate = 4.331311
Mean Reneging Rate = 4.331311
Expected Total Cost Rate = 83.044328

As a second example we have considered the values and the calculated joint probability distribution of the inventory level and the number of customers is given below :

$$S = 15, s = 5, N = 6, \lambda = 2, \mu = 1.5, \gamma = 1,$$
$$\beta = 0.8, \alpha = 0.9, h = 0.3, c_1 = 150, c_2 = 0.8, c_3 = 0.9, c_4 = 0.8, c_5 = 1$$

References

[1] Arivarignan, G. and Sivakumar, B. (2003). Inventory System with Renewal Demands at Service Facilities, *Stochastic Point Processes*, in : S.K. Srinivasan and A. Vijayakumar (Eds.), Narosa Publishing House, New Delhi, India, 108 - 123.

[2] Berman, O., Kaplan E. H. and Shimshak D.G. (1993). Deterministic Approximations for Inventory Management at Service Facilities, *IIE Transactions*, **25**, 98-104.

[3] Berman, O. and Kim, E. (1999). Stochastic Inventory Policies for Inventory Management of Service Facilities, *Stochastic Models*, **15**, 695-718.

[4] Berman, O. and Sapna, K. P. (2000). Inventory Management at Service Facilities for Systems with Arbitrarily Distributed Service Times, *Stochastic Models,* **16**, 343 - 360.

[5] Elango, C. (2001). *A Continuous Review Perishable inventory System at Service Facilities*, Ph. D., Thesis, Madurai Kamaraj University, Madurai.

[6] Elango, C. and Arivarignan, G. (2003). A Continuous Review Perishable Inventory Systems with Poisson Demand and Partial Backlogging, *Statistical Methods and Practice : Recent Advances,* in : Balakrishnan, N. etc., (Eds.), Narosa Publishing House, New Delhi, 343 - 356.

[7] He,Q-M, Jewkes, E. M. and Buzacott, J. (1998). An efficient algorithm for computing the optimal replenishment policy for an inventory-production system, *Advances in Matrix Analytic Methods for Stochastic Models,* in : Alfa, A. and Chakravarthy, S. (Eds.), Notable Publications, New Jersey, USA, 381 - 402.

[8] Kalpakam, S. and Arivarignan, G. (1990). Inventory System with Random Supply Quantity, *OR Spektrum*, **12**, 139 - 145.

[9] Liu, L. and Yang, T. (1999). An (s, S) random lifetime inventory model with a positive lead time, *European Journal of Operational Research*, **113**, 52-63.

[10] Nahmias, S. (1982). Perishable Inventory Theory: A Review, *Operations Research*, **30**, 680 - 708.

[11] Palm, C. (1937). Etude des delais d'attente, *Ericsson Technicx,* **5,** 37 - 56.

[12] Perumal, V. and Arivarignan, G. (2002). A Continuous Review Perishable inventory System at Infinite Capacity Service Facilities, *ANJAC Journal of Science,* **1**, 37 - 45.

[13] Raafat, F. (1991). A Survey of Literature on Continuously Deteriorating Inventory Models, *Journal of Operational Research Society*, **42**, 27 - 37.

[14] Ross, S. M. (2000). *Introduction to Probability Models*, Harcourt Asia PTE Ltd, Singapore.

[15] Ward, A. R., and Glynn, P. W. (2003). A Diffusion Approximation for a Markovian Queue with Renegging, *Queueing Systems,* **43**, 103 - 128.

In: Mathematics, Game Theory and Algebra Compendium ISBN 978-1-60456-659-8
Editor: Jacob H. Mathias, pp. 423-435

Chapter 36

CERTAIN WEAKER FORMS OF FUZZY PAIRWISE CONTINUOUS FUNCTIONS AND BIEXTENSION OF FUZZY BITOPOLOGICAL SPACES

P.V. Ramakrishnan and A. Regina Mary*
School of Mathematics,
Madurai Kamaraj University,
Madurai- 625021, Tamil Nadu India

Abstract

Kandil introduced and studied the notion of fuzzy bitopological spaces as a natural generalization of fuzzy topological spaces. The concepts of fuzzy pairwise α-continuity, fuzzy pairwise semi-continuity and fuzzy pairwise pre-continuity were introduced by S.S.Sampath Kumar[7]. The notion of simple extension of a fuzzy topological space was introduced by Sunil[12]. In this paper the concepts of fuzzy pairwise faintly α-continuity, fuzzy pairwise faintly semi-continuity, fuzzy pairwise faintly pre-continuity and the biextension of fuzzy bitopological spaces are introduced. Also the connection between the above new mappings, their characterizations and their behaviour with the biextension of fuzzy bitopological spaces have been investigated. Further, biextensions of fuzzy bitopological spaces and their separation properties are studied.

2000 Mathematics Subject classification: 54A40

Key Words and Phrases: (τ_i, τ_j)Fuzzy semi-open sets, (τ_i, τ_j) fyzzy α-open sets, (τ_i, τ_j)fuzzy pre-open sets, (τ_i, τ_j) fuzzy θ cluster point, (τ_i, τ_j) fuzzy θ open sets, fuzzy pairwise regular and fuzzy pairwise connected fuzzy bitopological spaces.

*E-mail address: regidurai@yahoo.co.in. Corresponding author: Dept. of Mathematics, Fatima college(Autonomous), Madurai, Tamil Nadu, India

1. Preliminaries

Definition 1.1 *[4] Let (X, τ) be a fuzzy topological space[2]. A fuzzy point x_t[4] is said to be quasi-coincident(q-coincident, for short) with a fuzzy set A, denoted by x_t q A, iff $A^c(x) < t$ where $A^c(x) = 1 - A(x)$. A fuzzy set A is said to be q-coincident with B, denoted by A q B, iff there exists $x \in X$ such that $A(x) + B(x) > 1$; We also say that A and B are q-coincident. $A \leq B$ iff A and B^c are not q-coincident(denoted by A $\bar{q}$ B^c).*

Definition 1.2 *[4] Let (X, τ_1) be a fts. A fuzzy set A is said to be a τ_1- fuzzy q-neighbourhood (τ_1-f q-nbhd, for short) of a fuzzy point x_t iff there exists a τ_1-fuzzy open set B such that x_t q $B \leq A$.*

Definition 1.3 *[3] A system (X, τ_1, τ_2) consisting of a set X with two fuzzy topologies[2] τ_1 and τ_2 on X is called a fuzzy bitopological space(fbts, for short).*

Throughout this paper the indices i, j take values in $\{1, 2\}$ and $i \neq j$. For definitions and results not explained in this paper, we refer to the papers [1, 2, 4]. As to the notations and terminologies we shall write τ_i-int(A) and τ_j-cl(A) to mean, respectively, the interior and closure of a fuzzy set A with respect to the fuzzy topologies τ_i and τ_j in a fbts (X, τ_1, τ_2).

Definition 1.4 *[8] Let λ be a fuzzy subset of a fbts(X, τ_1, τ_1). λ is called a (τ_i, τ_j)- fuzzy semi-open((τ_i, τ_j)-fso) set of X if there exists a $\nu \in \tau_i$ such that $\nu \leq \lambda \leq \tau_j$-cl($\nu$). The complement of a (τ_i, τ_j)-fso set is called a (τ_i, τ_j)-fuzzy semi-closed((τ_i, τ_j)-fsc)set*

Theorem 1.1 *[8] Let λ be a fuzzy set of a fbts(X, τ_1, τ_2). Then the following are equivalent.*

1. *λ is a (τ_i, τ_j)-fsc set.*
2. *λ^c is a (τ_i, τ_j)-fso set.*
3. *τ_j-int(τ_i-cl(λ)) $\leq \lambda$*
4. *τ_j-cl(τ_i-int(λ^c)) $\geq \lambda^c$*

Definition 1.5 *[7] Let λ be a fuzzy set of a fbts (X, τ_1, τ_2). λ is called*

1. *a (τ_i, τ_j)-fuzzy α-open (briefly, (τ_i, τ_j)-$f\alpha o$)set of X if $\lambda \leq \tau_i$-int(τ_j-cl(τ_i-intλ))*
2. *a (τ_i, τ_j)-fuzzy α-closed (briefly, (τ_i, τ_j)-$f\alpha c$)set of X if $\lambda \geq \tau_i$-cl(τ_j-int(τ_i-clλ))*
3. *a (τ_i, τ_j)-fuzzy pre-open(briefly,(τ_i, τ_j)-fpo)set of X if $\lambda \leq \tau_i$-int(τ_j-clλ)*
4. *a (τ_i, τ_j)-fuzzy pre-closed(briefly,(τ_i, τ_j)-fpc)set of X if $\lambda \geq \tau_i$-cl(τ_j-intλ)*

Theorem 1.2 *[7] Let λ be a fuzzy set in a fbts (X, τ_1, τ_2)*

1. *λ is (τ_i, τ_j)-fαc if and only if λ^c is (τ_i, τ_j)-$f\alpha o$.*

2. *λ is (τ_i, τ_j)-fpc if and only if λ^c is (τ_i, τ_j)-fpo.*

Note 1.1 *[7] In a fbts (X, τ_1, τ_2), any τ_i-fuzzy open(τ_i-fuzzy closed) set is a (τ_i, τ_j)-$fuzzy\ \alpha$ open ((τ_i, τ_j)-$fuzzy\ \alpha$ closed)set.*

Theorem 1.3 *[7, 8]*

1. *Any union of (τ_i, τ_j)-fαo (resp. (τ_i, τ_j)-fpo, (τ_i, τ_j)-fso) sets is a (τ_i, τ_j)-$f\alpha o$(resp. (τ_i, τ_j)-fpo, (τ_i, τ_j)-fso) set.*

2. *Any intersection of (τ_i, τ_j)-$f\alpha c$ (resp. (τ_i, τ_j)-fpc, (τ_i, τ_j)-fsc) sets is a (τ_i, τ_j)-$f\alpha c$(resp. (τ_i, τ_j)-fpc, (τ_i, τ_j)-fsc) set.*

Theorem 1.4 *[7] A fuzzy set λ in a fbts (X, τ_1, τ_2) is a (τ_i, τ_j)-$f\alpha o$ set if and only if it is (τ_i, τ_j)-fso and (τ_i, τ_j)-fpo.*

Definition 1.6 *[7, 8] Let $f : (X, \tau_1, \tau_2) \to (Y, \eta_1, \eta_2)$ be a mapping from a fbts X to another fbts Y. f is called*

1. *a fuzzy pairwise α-continuous (fpαc, in short) mapping if $f^{-1}(\lambda)$ is a (τ_i, τ_j)-$f\alpha o$ set in X for each $\lambda \in \eta_i$*

2. *a fuzzy pairwise pre-continuous(fppc, in short) mapping if $f^{-1}(\lambda)$ is a (τ_i, τ_j)-fpo set in X for each $\lambda \in \eta_i$*

3. *a fuzzy pairwise semi-continuous(fpsc, in short) mapping if $f^{-1}(\lambda)$ is a (τ_i, τ_j)-fso set in X for each $\lambda \in \eta_i$*

Theorem 1.5 *[7, 8] A map $f : (X, \tau_1, \tau_2) \to (Y, \eta_1, \eta_2)$ from a fbts X to a fbts Y is fpαc(resp, fppc,fpsc) iff the inverse image of each η_i-fc set of Y is a (τ_i, τ_j)-$f\alpha c$ (resp, (τ_i, τ_j)-fpc, (τ_i, τ_j)-fsc) set in X.*

Definition 1.7 *[9] A fuzzy point x_t in a fbts (X, τ_1, τ_2) is called a (τ_1, τ_2)-$fuzzy$ θ-cluster point of a fuzzy set λ in X iff for every τ_1-fuzzy open q-neighbourhood μ of x_t, τ_2-$cl(\mu)$ is q-coincident with λ.*

Interchanging the role of τ_1 *and* τ_2 in the above definition (τ_2, τ_1)- fuzzy θ cluster point can be similarly defined.

Definition 1.8 *[9] A fuzzy point x_t in a fbts (X, τ_1, τ_2) is called a fuzzy pairwise-θ- cluster point of a fuzzy set λ if it is both (τ_1, τ_2)-$fuzzy$ θ- cluster point and (τ_2, τ_1)-$fuzzy$ θ-cluster point of λ. The set of all (τ_i, τ_j)-$fuzzy$ θ-cluster points of a fuzzy set λ is called the (τ_i, τ_j)-fuzzy θ-closure of λ and is denoted by (τ_i, τ_j)- $cl_\theta(\lambda)$.*

Definition 1.9 *[9] A fuzzy set λ of a fbts (X, τ_1, τ_2) is said to be (τ_i, τ_j)-$fuzzy$ θ-closed((τ_i, τ_j)-$f\theta c$, in short) iff $\lambda = (\tau_i, \tau_j)$-$cl_\theta(\lambda)$. The complement of a (τ_i, τ_j)-$fuzzy$ θ-closed set is said to be a (τ_i, τ_j)-$fuzzy$ θ-open set ((τ_i, τ_j)-$f\theta o$, in short).*

Note 1.2 *[9] In a fbts* (X, τ_1, τ_2) *every* (τ_i, τ_j)-$fuzzy$ θ*-open set is* τ_i*-fuzzy open*

Definition 1.10 *[9] A fbts* (X, τ_1, τ_2) *is said to be fuzzy pairwise regular (briefly, FP-regular) iff for each fuzzy point* x_t *in* X *and each* τ_i*- fuzzy open q-neighbourhood* μ *of* x_t*, there exists a* τ_i*- fuzzy open q-neighbourhood* ν *of* x_t *such that* τ_j*-*$cl(\nu) \leq \mu$.

Theorem 1.6 *[9] A fbts* (X, τ_1, τ_2) *is FP-regular iff for any fuzzy set* λ *in* X*,* τ_i*-*$cl(\lambda) = (\tau_i, \tau_j)$*-*$cl_\theta(\lambda)$.

Definition 1.11 *[10] A mapping* $f : (X, \tau_1, \tau_2) \to (Y, \eta_1, \eta_2)$ *is said to be fuzzy pairwise -* θ*-continuous (shortly, fp*θ*c) iff for each fuzzy point* x_t *in* X *and each* η_i*- fuzzy open q-neighbourhood* μ *of* $f(x_t)$*, there exists a* τ_i *- fuzzy open q-neighbourhood* λ *of* x_t *such that* $f[\tau_j\text{-}cl(\lambda)] \leq \eta_j\text{-}cl(\mu)$

Theorem 1.7 *[10] If* $f : (X, \tau_1, \tau_2) \to (Y, \eta_1, \eta_2)$ *from a fbts* X *to a fbts* Y *is a fp*θ*c mapping then* $f^{-1}(\lambda)$ *is* (τ_i, τ_j)-$fuzzy$ θ*-open in* X*, for each* (η_i, η_j)-$fuzzy$ θ*- open set in* Y.

Definition 1.12 *[12] Let* (X, F) *be a fuzzy topological space and suppose that* $g \in I^X$ *and* $g \notin F$*. Then the collection* $F(g) = \{g_1 \vee (g_2 \wedge g)/g_1, g_2 \in F\}$ *is called the simple extension of F determined by g.*

Theorem 1.8 *[12] Let* (X, F) *be a fts and suppose that F(g) be the simple extension of F determined by* g*. Then (a)* $F(g)$ *is a fuzzy topology on* X *and (b)* $F(g) = F \vee G$ *where* $G = \{0, 1, g\}$ *and* $F \vee G$ *is the fuzzy topology generated by* F *and* G

Definition 1.13 *[11] A pair* (λ, μ) *of non-null fuzzy sets in a fbts* (X, τ_1, τ_2) *is said to be a fuzzy pairwise separation (fp-separation, for short) relative to X iff* $\lambda\, \bar{q}\, \mu$, $\lambda\, \bar{q}\, [\tau_1\text{-}cl(\mu)]$ *and* $\tau_2\text{-}cl(\lambda)]\, \bar{q}\, \mu$*. A fuzzy set* γ *in a fbts* X *is said to be fuzzy pairwise connected (fp-connected) iff there do not exist two fuzzy sets* λ *and* μ *in* X *such that* (λ, μ) *is a fp-separation relative to* X *and* $\gamma = \lambda \cup \mu$.

Definition 1.14 *[5] A fbts* (X, τ_1, τ_2) *is called fuzzy pairwise-*T_0 *(FP-*T_0*,for short) if for all* $x, y \in X, x \neq y$*, there exists* $U \in \tau_1 \cup \tau_2$ *such that*

$$U(x) \neq U(y).$$

Definition 1.15 *[5] A fbts* (X, τ_1, τ_2) *is called fuzzy pairwise-*T_1 *(FP-*T_1*,for short) if* $\forall x, y \in X, x \neq y$*, there exists* $U \in \tau_1$ *and* $V \in \tau_2$ *such that*

$$U(x) = 1, U(y) = 0 \text{ and } V(x) = 0, V(y) = 1.$$

Based on the definition of fuzzy pairwise T_2[6] for smooth fuzzy topological spaces we define the following.

Definition 1.16 *A fbts* (X, τ_1, τ_2) *is FP-*T_2 *iff for any two fuzzy points*

$$x_t \neq y_s$$

case 1: *When $x \neq y$, there exists $\mu_1, \mu_2, V_1, V_2 \in I^X$ where μ_1 is τ_i-fuzzy open and μ_2 is τ_i-fuzzy open q-neighbourhood of x_t, V_1 is a τ_j-fuzzy open q-neighbourhood of y_s and V_2 is a τ_j-fuzzy open set such that $x_t \in \mu_1, \mu_1 \bar{q} V_1$ and $y_s \in V_2$, $\mu_2 \bar{q} V_2$.*

case 2: *When $x = y$ and $t < s$(say), there exists $\mu, V \in I^X$ where μ is a τ_i-fuzzy open set and V is a τ_j-fuzzy open q-neighbourhood of y_s such that $x_t \in \mu$ and $\mu \bar{q} V$.*

2. Main Results

Definition 2.1 *Let $f : (X, \tau_1, \tau_2) \to (Y, \eta_1, \eta_2)$ be a mapping from a fbts X to another fbts Y. f is called*

1. *a fuzzy pairwise faintly α-continuous (fpfαc, in short) mapping if $f^{-1}(\lambda)$ is a (τ_i, τ_j)-$f\alpha o$ set in X for each (η_i, η_j)-$f\theta o$ set λ of Y.*
2. *a fuzzy pairwise faintly pre-continuous(fpfpc) mapping if $f^{-1}(\lambda)$ is a (τ_i, τ_j)-fpo set in X for each (η_i, η_j)-$f\theta o$ set λ of Y.*
3. *a fuzzy pairwise faintly semi-continuous(fpfsc) mapping if $f^{-1}(\lambda)$ is a (τ_i, τ_j)-fso set in X for each (η_i, η_j)-$f\theta o$ set λ of Y.*

Theorem 2.1 *Let $f : (X, \tau_1, \tau_2) \to (Y, \eta_1, \eta_2)$ be a mapping from a fbts X to another fbts Y. The following statements are equivalent.*

1. *f is fpfαc*
2. *For each fuzzy point x_t of X and each (η_i, η_j)-$f\theta o$ set μ of Y containing $f(x_t)$, there exists a (τ_i, τ_j)-$f\alpha o$ set λ of X containing x_t such that $f(\lambda) \leq \mu$*
3. *The inverse image of each (η_i, η_j)-$f\theta c$ set of Y is a (τ_i, τ_j)-$f\alpha c$ set in X.*

Proof:

(1)⇒ (2)
Since μ is a (η_i, η_j)-$f\theta$o set of Y containing $f(x_t)$ then by (1), $f^{-1}(\mu)$ is a (τ_i, τ_j)-$f\alpha$o set of X. Let $\lambda = f^{-1}(\mu)$. Then λ is a (τ_i, τ_j)-$f\alpha$o set of X containing x_t and $f(\lambda) \leq \mu$.

(2)⇒ (1)
Let μ be any (η_i, η_j)-$f\theta$o set of Y. Then $f^{-1}(\mu)$ is a fuzzy set of X. Suppose $f^{-1}(\mu) = 0$ then clearly $f^{-1}(\mu)$ is a (τ_i, τ_j)-$f\alpha$o set of X. Otherwise there exists a fuzzy point say $x_t \in f^{-1}(\mu)$. So $f(x_t) \in \mu$. Hence by (2), there exists a (τ_i, τ_j)-$f\alpha$o set say λ_{x_t} of X such that $x_t \in \lambda_{x_t}$ and $f(\lambda_{x_t}) \leq \mu$. Thus we have $x_t \in \lambda_{x_t} \leq f^{-1}(\mu)$. Consequently, we get $f^{-1}(\mu) = \cup\{x_t : x_t \in f^{-1}(\mu)\} \leq \cup\{\lambda_{x_t} : x_t \in f^{-1}(\mu)\} \leq f^{-1}(\mu)$

Hence $f^{-1}(\mu) = \cup\{\lambda_{x_t} : x_t \in f^{-1}(\mu)\}$ which is a (τ_i, τ_j)-$f\alpha$o set in X by the Theorem(1.3)

(1)$\Rightarrow$ (3)
Let μ be a (η_i, η_j)-fuzzy θ closed set of Y. Then μ^c is a (η_i, η_j)-fuzzy θ open set of Y. Thus $f^{-1}(\mu^c)$ is a (τ_i, τ_j)- fuzzy α open set of X. But $f^{-1}(\mu^c) = (f^{-1}(\mu))^c$ So, $f^{-1}(\mu)$ is a (τ_i, τ_j)-fuzzy α closed set of X. Similarly (3) $\Rightarrow$ (1) can be proved.

Theorem 2.2 *Let $f : (X, \tau_1, \tau_2) \to (Y, \eta_1, \eta_2)$ be a mapping from a fbts X to another fbts Y. The following statements are equivalent.*

1. *f is fpfpc(resp. fpfsc)*
2. *For each fuzzy point x_t of X and each (η_i, η_j)-$f\theta o$ set μ of Y containing $f(x_t)$, there exists a (τ_i, τ_j)-fpo(resp. fso) set λ of X containing x_t such that $f(\lambda) \leq \mu$*
3. *The inverse image of each (η_i, η_j)-$f\theta c$ set of Y is a (τ_i, τ_j)-fpc(resp. fsc) set in X.*

Proof: Similar to the proof of the Theorem(2.1)

Remark 2.1 *A fuzzy pairwise α-continuous (resp. fppc, fpsc) function is a fuzzy pairwise faintly α-continuous (resp. fpfpc, fpfsc) function*

Proof: Proof follows from Note(1.2)

Theorem 2.3 *A mapping $f : (X, \tau_1, \tau_2) \to (Y, \eta_1, \eta_2)$ from a fbts X to a fbts Y is fpfαc iff it is fpfpc and fpfsc*

Proof: Let f be a fpfαc function. So $f^{-1}(any\ (\eta_i, \eta_j)$-$f\theta$o set of $Y)$ is a (τ_i, τ_j)-$f\alpha$o set of X. So,by Theorem(1.4), it follows that f is also fpfpc and fpfsc. Conversely let f be fpfpc and fpfsc and let λ be any (η_i, η_j)-$f\theta$o set of Y. Then $f^{-1}(\lambda)$ is both (τ_i, τ_j)-fso and (τ_i, τ_j)-fpo and hence (τ_i, τ_j)-$f\alpha$o. So f is fpfαc.

Remark 2.2 *A fpfαc function need not be a fpαc function. This can be seen from the following example.*

Example 2.1

Let $X = \{x, y\}$. Define fuzzy topologies on X as follows.
$\tau_1 = \{0, 1, x_{\frac{1}{2}}\}, \tau_2 = \{0, 1, x_{\frac{1}{8}}\}$ $\eta_1 = \{0, 1, x_{\frac{3}{4}}\}$ *and* $\eta_2 = \{0, 1, x_{\frac{1}{4}}\}$.Clearly (X, τ_1, τ_2) and (X, η_1, η_2) are fuzzy bitopological spaces.

We claim that $f : (X, \tau_1, \tau_2) \to (X, \eta_1, \eta_2)$ defined by $f(x) = x, \forall x \in X$ is a fpfαc function but not a fpαc function.

For, consider $x_{\frac{3}{4}} \in \eta_1$. $f^{-1}(x_{\frac{3}{4}}) = x_{\frac{3}{4}}$. τ_1-$int(\tau_2$-$cl(\tau_1$-$int\ x_{\frac{3}{4}})) = x_{\frac{1}{2}}$. So we get $f^{-1}(x_{\frac{3}{4}}) \not\leq \tau_1 - int(\tau_2$-$cl(\tau_1$-$int\ x_{\frac{3}{4}}))$. Therefore $f^{-1}(x_{\frac{3}{4}})$ is not a (τ_1, τ_2) fαo set of X. So f is not a fpαc function.

We note that (η_1, η_2)-$cl_\theta(x_{\frac{3}{4}})^c = 1$. For, $x_{\frac{3}{4}}$ *and* 1 are the only η_1- fuzzy open q-neighbourhoods of x_t where $t > \frac{1}{4}$. $\eta_2 - cl(x_{\frac{3}{4}}) = (x_{\frac{1}{4}})^c$ *and* $\eta_2 - cl(1) = 1$. Clearly $1\ q\ (x_{\frac{3}{4}})^c$ *and* $(x_{\frac{1}{4}})^c\ q\ (x_{\frac{3}{4}})^c$. Similarly one can prove that (η_2, η_1)-$cl_\theta(x_{\frac{1}{4}})^c = 1$. So the only (η_i, η_j)-$f\theta c$ sets are the fuzzy sets $0, 1$. Clearly $f^{-1}(0)$ *and* $f^{-1}(1)$ are (τ_i, τ_j)-$f\alpha c$ sets of X. Hence it follows that f is a $fpf\alpha c$ function.

Remark 2.3 *A fpfpc function need not be a fppc function.*

For in the Example(2.1), we note that $f : (X, \tau_1, \tau_2) \to (X, \eta_1, \eta_2)$ is a fpfpc function as a fpfαc function is also a fpfpc function. Also f is not a fppc function. For, τ_1-$int(\tau_2$-$clx_{\frac{3}{4}}) = x_{\frac{1}{2}}$. So for a η_1-fuzzy open set $x_{\frac{3}{4}}$, $f^{-1}(x_{\frac{3}{4}}) \not\leq \tau_1$-$int(\tau_2$-$clf^{-1}(x_{\frac{3}{4}})$. Therefore, $f^{-1}(x_{\frac{3}{4}})$ is not a (τ_1, τ_2)-fpo set of X. So f is not a fppc function.

Remark 2.4 *A fpfsc function need not be a fpsc function as can be seen from the following example.*

Example 2.2

Let $X = \{x, y\}$. Define fuzzy topologies on X as follows.

$$\tau_1 = \{0, 1, A_{x,y}^{\frac{1}{4},\frac{3}{4}}\}, \tau_2 = \{0, 1, B_{x,y}^{\frac{1}{2},\frac{1}{2}}\}$$
$$\eta_1 = \{0, 1, C_{x,y}^{\frac{3}{4},\frac{1}{4}}\}, \eta_2 = \{0, 1, D_{x,y}^{\frac{1}{4},\frac{1}{2}}\}.$$

Define $f : (X, \tau_1, \tau_2) \to (X, \eta_1, \eta_2)$ by $f(x) = x, \forall x \in X$. We claim that f is fpfsc but not fpsc.
For, we note that (η_1, η_2)-$cl_\theta(C^c) = 1$ *and* (η_2, η_1)-$cl_\theta(D^c) = 1$. So the only (η_i, η_j)-$f\theta c$ sets are $0, 1$. So clearly f is a fpfsc function.

Note that $C \in \eta_1, f^{-1}(C) = C$ and $C \not\leq \tau_2$-$cl(\tau_1$-$intC)$. So C is not a $(\tau_1, \tau_2) - fso$ set. Therefore f is not a fpsc function.

Theorem 2.4 *Let $f : (X, \tau_1, \tau_2) \to (Y, \eta_1, \eta_2)$be a fpf$\alpha$c (resp, fpfpc,fpfsc) function from a fbts X to a FP-regular fbts Y. Then f is a fpαc (resp, fppc,fpsc)function.*

Proof: Let λ be a η_i -fuzzy closed set of Y. Since (Y, η_1, η_2) is a Fp-regular fbts by Theorem(1.6), it follows that (η_i, η_j)-$cl_\theta(\lambda) = \eta_i$-$cl(\lambda)$. Since $\eta_i - cl(\lambda) = \lambda$, we get (η_i, η_j)-$cl_\theta(\lambda) = \lambda$ and so λ is a (η_i, η_j)-$f\theta c$ set of Y. Since f is fpfαc (resp, fpfpc,fpfsc), $f^{-1}(\lambda)$ is a (τ_i, τ_j)-$f\alpha c$ (resp,(τ_i, τ_j)-fpc, (τ_i, τ_j)-fsc) set of X. So f is a fpαc (resp, fppc, fpsc) function.

Remark 2.5 *If a fpfαc function $f : (X, \tau_1, \tau_2) \to (Y, \eta_1, \eta_2)$ from a fbts X to a fbts Y is a fpαc function, then Y need not be a FP-regular fbts. This can be seen from the following example.*

Example 2.3

Let $X = \{x, y\}$. Define fuzzy topologies on X as follows.
$\tau_1 = \{0, 1, A_{x,y}^{\frac{1}{2},\frac{1}{4}}\}, \tau_2 = \{0, 1, D_{x,y}^{\frac{1}{4},1}\}$, $\eta_1 = \{0, 1, B_{x,y}^{\frac{1}{2},\frac{1}{2}}\}, \eta_2 = \{0, 1, C_{x,y}^{\frac{1}{2},1}\}$ Define $f : (X, \tau_1, \tau_2) \to (X, \eta_1, \eta_2)$ by $f(x) = x$. Clearly f is a fpαc function and so also a fpfαc function. But, we note that (X, η_1, η_2) is not FP-regular. Since, for a fuzzy point $x_{\frac{3}{4}}$ and a η_1- fuzzy open q-neighbourhood B of $x_{\frac{3}{4}}$, there does not exist any η_1-fuzzy open q-neighbourhood μ of $x_{\frac{3}{4}}$ such that η_2-$cl(\mu) \leq B$.

Theorem 2.5 *Let $(X, \tau_1, \tau_2), (Y, \eta_1, \eta_2)$ and (Z, δ_1, δ_2) be fuzzy bitopological spaces. If $f : (X, \tau_1, \tau_2) \to (Y, \eta_1, \eta_2)$ is a fpfαc function and $g : (Y, \eta_1, \eta_2) \to (Z, \delta_1, \delta_2)$ is a fpθc function then $g \circ f : (X, \tau_1, \tau_2) \to (Z, \delta_1, \delta_2)$ is a fpfαc function.*

Proof: Proof follows from the definition of fpfαc function and Theorem(1.7).

Theorem 2.6 *Let $(X, \tau_1, \tau_2), (Y, \eta_1, \eta_2)$ and (Z, δ_1, δ_2) be fuzzy bitopological spaces. If $f : (X, \tau_1, \tau_2) \to (Y, \eta_1, \eta_2)$ is a fpαc function and $g : (Y, \eta_1, \eta_2) \to (Z, \delta_1, \delta_2)$ is a fpθc function then $g \circ f : (X, \tau_1, \tau_2) \to (Z, \delta_1, \delta_2)$ is a fpfαc function.*

Proof: Proof follows from the definition of fpαc function, Theorem(1.7) and the Note(1.2)

Lemma 2.1 *Let (X, τ_1, τ_2) and (Y, η_1, η_2) be fuzzy bitopological spaces. Let θ_k be the fuzzy product topology[13] generated by τ_k and η_k where k=1,2. If μ is a (η_i, η_j)-$f\theta c$ set of Y then $1 \times \mu$ is a (θ_i, θ_j)-$f\theta c$ set of $X \times Y$.*

Proof: Since μ is a (η_i, η_j)-$f\theta$c set, (η_i, η_j)-$cl_\theta(\mu) = \mu$.

We claim (θ_i, θ_j)-$cl_\theta(1 \times \mu) = 1 \times \mu$. Clearly (θ_i, θ_j)-$cl_\theta(1 \times \mu) \geq 1 \times \mu$. Suppose (θ_i, θ_j)-$cl_\theta(1 \times \mu) > 1 \times \mu$. Then there exists $(x, y)_t \in (\theta_i, \theta_j)$-$cl_\theta(1 \times \mu)$ but $(x, y)_t \notin 1 \times \mu$. $(x, y)_t \notin 1 \times \mu \Rightarrow (1 \times \mu)(x, y) < t \Rightarrow min(1(x), \mu(y)) < t \Rightarrow \mu(y) < t \to y_t \notin \mu$.

We claim that $y_t \in (\eta_i, \eta_j)$-$cl_\theta(\mu)$. Let λ be any η_i-fuzzy open q-neighbourhood of y_t. So $1 \times \lambda$ is a θ_i-fuzzy open q-neighbourhood of $(x, y)_t$, since $(1 \times \lambda)^c = 1 \times \lambda^c$[1] and so $(1 \times \lambda)^c(x, y) = (1 \times \lambda^c)(x, y) = \lambda^c(y) < t$. Since $(x, y)_t \in (\theta_i, \theta_j)$-$cl_\theta(1 \times \mu)$, θ_j-$cl(1 \times \lambda)$ q $(1 \times \mu)$.

So θ_j-$cl(1 \times \lambda)(x_1, x_2) + (1 \times \mu)(x_1, x_2) > 1$ for some $(x_1, x_2) \in X \times Y$. Since θ_j-$cl(1 \times \lambda) \leq \tau_j - cl(1) \times \eta_j$-$cl(\lambda) = 1 \times \eta_j$-$cl(\lambda)$

Hence $(1 \times \eta_j$-$cl(\lambda))(x_1, x_2) + (1 \times \mu)(x_1, x_2) > 1 \Rightarrow (\eta_j$-$cl(\lambda)(x_2) + \mu(x_2) > 1 \Rightarrow \eta_j$-$cl(\lambda)$ q μ.

So, $y_t \in (\eta_i, \eta_j)$-$cl_\theta(\mu)$, but $y_t \notin \mu$ which is a contradiction to the fact that $(\eta_i, \eta_j) - cl_\theta(\mu) = \mu$. Therefore (η_i, η_j)-$cl_\theta(1 \times \mu) = 1 \times \mu$ and so $1 \times \mu$ is a (θ_i, θ_j)-fuzzy θ closed set of $X \times Y$.

Lemma 2.2 *[1] Let $g : X \to X \times Y$ be the graph of a mapping $f : X \to Y$. Then, if λ is a fuzzy set of X and μ is a fuzzy set of Y, $g^{-1}(\lambda \times \mu) = \lambda \cap f^{-1}(\mu)$*

Theorem 2.7 *Let $f : (X, \tau_1, \tau_2) \to (Y, \eta_1, \eta_2)$ be a mapping from a fbts X to another fbts Y. Let θ_k be the fuzzy product topology generated by τ_k and η_k where k=1,2. Then if the graph mapping $g : (X, \tau_1, \tau_2) \to (X \times Y, \theta_1, \theta_2)$ of f defined by $g(x) = (x, f(x)) is fpf\alpha c$ (resp. fpfpc,fpfsc) then f is also fpfαc (resp. fpfpc,fpfsc).*

Proof: Let μ be a (η_i, η_j)-$f\theta$c set of Y. Then by Lemma 2.2 we have $f^{-1}(\mu) = 1 \cap f^{-1}(\mu) = g^{-1}(1 \times \mu)$. Since g is fpfαc (resp. fpfpc,fpfsc) and by Lemma(2.1) $1 \times \mu$ is a (θ_i, θ_j)-$f\theta$c set of $X \times Y$, $f^{-1}(\mu)$ is a (τ_i, τ_j)-$f\alpha c$ (resp. (τ_i, τ_j)-fpc, (τ_i, τ_j)-fsc) set in X. Hence f is fpfαc (resp. fpfpc,fpfsc).

3. Biextension of Fuzzy Bitopological Spaces

Definition 3.1 *Let (X, τ_1, τ_2) be a fbts. Suppose $g_i \in I^X$, $g_i \notin \tau_i$ and $\tau_i(g_i)$ is the simple extension of τ_i determined by g_i ($i \in \{1, 2\}$). Then the fbts $(X, \tau_1(g_1), \tau_2(g_2))$ is called the biextension of (X, τ_1, τ_2).*

Remark 3.1 *Let (X, τ_1, τ_2) and (Y, η_1, η_2) be fbts's.*
Suppose $f : (X, \tau_1, \tau_2) \rightarrow (Y, \eta_1, \eta_2)$ is fpfαc then $f : (X, \tau_1(g), \tau_2(h)) \rightarrow (Y, \eta_1(g_1), \eta_2(h_1))$ need not be a fpfαc function where $g, h, g_1, h_1 \in I^X$ and $g \notin \tau_1, h \notin \tau_2, g_1 \notin \eta_1, h_1 \notin \eta_2$. This can be seen from the following example.

Example 3.1

We have already proved that the map $f : (X, \tau_1, \tau_2) \rightarrow (X, \eta_1, \eta_2)$ defined by $f(x) = x, \forall x \in X$ in the Example(2.1) is a fpfαc map. Let $g = x_1, h = y_1$. Now consider $\tau_1(g) = \{0, 1, x_{\frac{1}{2}}, x_1\} = \tau_1^*(say)$,

$$\tau_2(h) = \{0, 1, x_{\frac{1}{8}}, y_1, A_{x,y}^{\frac{1}{8},1}\} = \tau_2^*(say)$$
$$\eta_1(g) = \{0, 1, x_{\frac{3}{4}}, x_1\} = \eta_1^*(say),$$
$$\eta_2(h) = \{0, 1, x_{\frac{1}{4}}, y_1, B_{x,y}^{\frac{1}{4},1}\} = \eta_2^*(say).$$

We claim that $f : (X, \tau_1^*, \tau_2^*) \rightarrow (X, \eta_1^*, \eta_2^*)$ defined by $f(x) = x, \forall x \in X$ is not a fpfαc function. We shall prove that $x_{\frac{3}{4}}$ is a $(\eta_1^*, \eta_2^*) - f\ \theta$ o set , but $f^{-1}(x_{\frac{3}{4}})$ is not a $(\tau_1^*, \tau_2^*) - f\alpha$o set of X. We note that $x_{\frac{3}{4}}$ is a η_1^*-fuzzy open q-neighbourhood of x_t where $t > \frac{1}{4}$, η_2^*-$cl(x_{\frac{3}{4}}) = B^c = x_{\frac{3}{4}}$ and $x_{\frac{3}{4}}\ \bar{q}\ (x_{\frac{3}{4}})^c$. So, no $x_t (t > \frac{1}{4})$, can be a $(\eta_1^*, \eta_2^*)-$ fuzzy θ cluster point of $(x_{\frac{3}{4}})^c$. Consequently $(\eta_1^*, \eta_2^*) - cl_\theta((x_{\frac{3}{4}})^c = (x_{\frac{3}{4}})^c$. So, $(x_{\frac{3}{4}})^c$ is a $(\eta_1^*, \eta_2^*) - f\theta c$ set and so $x_{\frac{3}{4}}$ is a $(\eta_1^*, \eta_2^*) - f\theta$o set. Now consider $f^{-1}(x_{\frac{3}{4}}) = x_{\frac{3}{4}}$. We note that $\tau_1^* - int(\tau_2^* - cl(\tau_1^* - int\ x_{\frac{3}{4}})) = x_{\frac{1}{2}}$. So $f^{-1}(x_{\frac{3}{4}}) \not\leq \tau_1^* - int(\tau_2^* - cl(\tau_1^* - int\ f^{-1}(x_{\frac{3}{4}})))$. So, $f^{-1}(x_{\frac{3}{4}})$ is not a $(\tau_1^*, \tau_2^*) - f\alpha$o set of X.
Therefore $f : (X, \tau_1^*, \tau_2^*) \rightarrow (X, \eta_1^*, \eta_2^*)$ is not a fpfαc function.

Remark 3.2 *In the Example(3.1) we note that $f^{-1}(x_{\frac{3}{4}})$ is not a (τ_1^*, τ_2^*)-fuzzy pre open set. For,*

$$x_{\frac{3}{4}} \not\leq \tau_1^* - int(\tau_2^* - cl\ x_{\frac{3}{4}}).$$

So, $f : (X, \tau_1^*, \tau_2^*) \rightarrow (X, \eta_1^*, \eta_2^*)$ is not a fpfpc function.
But, $f : (X, \tau_1, \tau_2) \rightarrow (X, \eta_1, \eta_2)$ is a fpfpc function by Remark(2.3).

Remark 3.3 *If a function $f : (X, \tau_1, \tau_2) \rightarrow (Y, \eta_1, \eta_2)$ is fpfαc(resp. fpfpc) then $f : (X, \tau_1^*, \tau_2^*) \rightarrow (Y, \eta_1^*, \eta_2^*)$ need not be fpfαc(resp. fpfpc) where $\tau_i^*(resp.\ \eta_i^*)$ is finer than $\tau_i(resp.\ \eta_i)$ where $i \in \{1, 2\}$.*

Remark 3.4 *If $(X, \tau_1(g), \tau_2(h))$ is the biextension of a fbts (X, τ_1, τ_2), $(Y, \eta_1(g_1), \eta_2(h_2))$ is that of a fbts (Y, η_1, η_2) and if $f : (X, \tau_1(g), \tau_2(h)) \to (Y, \eta_1(g_1), \eta_2(h_1))$ is a fpfαc function then $f : (X, \tau_1, \tau_2) \to (Y, \eta_1, \eta_2)$ need not be a fpfαc function. This can be seen from the following example.*

Example 3.2

Let $X = \{x, y\}$. Define fuzzy topologies on X as follows.

$$\tau_1 = \{0, 1, x_{\frac{1}{2}}\}, \tau_2 = \{0, 1, x_{\frac{1}{4}}\}$$
$$\eta_1 = \{0, 1, A_{x,y}^{\frac{3}{4},\frac{3}{4}}\}, \eta_2 = \{0, 1, B_{x,y}^{\frac{1}{4},\frac{1}{4}}\}.$$

Define $f : (X, \tau_1, \tau_2) \to (X, \eta_1, \eta_2)$ by $f(x) = x, \forall x \in X$.
Let us find $(\eta_1, \eta_2) - cl_\theta(A^c)$. We note that A is a η_1-fo q-neighbourhood of x_t, y_t where $t > \frac{1}{4}, \eta_2 - cl(A) = A$ and $A \bar{q} A^c$. So, no $x_t, y_t (t > \frac{1}{4})$ can be a $(\eta_1, \eta_2)-$ fuzzy θ cluster point of A^c. Hence $(\eta_1, \eta_2) - cl_\theta(A^c) = A^c$. So, A is a $(\eta_1, \eta_2) - f\theta o$ set of X. Note that $f^{-1}(A) = A, \tau_1 - int(\tau_2 - cl(\tau_1 - intA)) = x_{\frac{1}{2}}$ and $A \not\leq \tau_1 - int(\tau_2 - cl(\tau_1 - intA))$. So, A is not a $(\tau_1, \tau_2) - f\alpha o$ set. Therefore $f : (X, \tau_1, \tau_2) \to (X, \eta_1, \eta_2)$ is not a fpfαc function.

$$\tau_1(A) = \{0, 1, x_{\frac{1}{2}}, A\} = \tau_1^*(say), \tau_2(B) = \{0, 1, x_{\frac{1}{4}}, B\} = \tau_2^*(say)$$
$$\eta_1(x_{\frac{1}{4}}) = \{0, 1, A, x_{\frac{1}{4}}\} = \eta_1^*(say), \eta_2(x_{\frac{1}{8}}) = \{0, 1, B, x_{\frac{1}{8}}\} = \eta_2^*(say).$$

We claim that $f : (X, \tau_1^*, \tau_2^*) \to (X, \eta_1^*, \eta_2^*)$ defined by $f(x) = x, \forall x \in X$ is a fpfαc function.

We note that $1, x_{\frac{1}{4}}, A$ are η_1^*-fo q-neighbourhood of x_t where $t > \frac{3}{4}$. $\eta_2^* - cl(x_{\frac{1}{4}}) = B^c = \eta_2^* - cl(A), \eta_2^* - cl(1) = 1$. Clearly $1\ q\ (x_{\frac{1}{4}})^c$ *and* $B^c\ q\ (x_{\frac{1}{4}})^c$. So, $(\eta_1^*, \eta_2^*) - cl_\theta(x_{\frac{1}{4}})^c = 1$. Therefore $(x_{\frac{1}{4}})^c$ is not a $(\eta_1^*, \eta_2^*) - f\theta c$ set and so $x_{\frac{1}{4}}$ is not a $(\eta_1^*, \eta_2^*) - f\theta o$ set of X. But, A^c is a $(\eta_1^*, \eta_2^*) - f\theta c$ set. For, A is a η_1^*- fo q-neighbourhood of x_t, y_t where $t > \frac{1}{4}$. $\eta_2^* - cl(A) = A$ and $A\ \bar{q}\ A^c$. So $(\eta_1^*, \eta_2^*) - cl_\theta(A^c) = A^c$. Consequently A is a $(\eta_1^*, \eta_2^*) - f\theta o$ set of X. Similarly it can be proved that $x_{\frac{1}{8}}$ is not a $(\eta_2^*, \eta_1^*) - f\theta o$ set and B is a $(\eta_2^*, \eta_1^*) - f\theta o$ set. Also $f^{-1}(A) = A, f^{-1}(B) = B, \tau_1^* - int(\tau_2^* - cl(\tau_1^* - int\ A)) = A$ and $\tau_2^* - int(\tau_1^* - cl(\tau_2^* - int\ B)) = B$. So A is a $(\tau_1, \tau_2) - f\alpha o$ set and B is a $(\tau_2, \tau_1) - f\alpha o$ set. Therefore $f : (X, \tau_1^*, \tau_2^*) \to (X, \eta_1^*, \eta_2^*)$ is a fpfαc function.

Remark 3.5 *In the Example(3.2), we note that $f^{-1}(A)$ is not a (τ_1, τ_2)-fuzzy pre open set. For $A \not\leq \tau_1 - int(\tau_2 - cl\ A)$. So $f : (X, \tau_1, \tau_2) \to (X, \eta_1, \eta_2)$ is not a fpfpc function. Note that $f : (X, \tau_1^*, \tau_2^*) \to (X, \eta_1^*, \eta_2^*)$ is a fpfpc function.*

Remark 3.6 *If a function $f : (X, \tau_1, \tau_2) \to (Y, \eta_1, \eta_2)$ is fpfαc(resp. fpfpc) then $f : (X, \tau_1^*, \tau_2^*) \to (Y, \eta_1^*, \eta_2^*)$ need not be fpfαc(resp. fpfpc) where $\tau_i^*(resp.\ \eta_i^*)$ is weaker than $\tau_i(resp.\ \eta_i)$ where $i \in \{1, 2\}$.*

Theorem 3.1 *If a fbts (X, τ_1, τ_2) is FP-T_0 (resp. FP-T_1, FP-T_2), then it's biextension $(X, \tau_1(g_1), \tau_2(g_2))$ is also a FP-T_0 (resp. FP-T_1, FP-T_2) fbts.*

Proof: Trivial.

Remark 3.7 *The biextension of a FP-regular(resp. FP-connected)fbts need not be a FP-regular(resp. FP-connected) fbts. This can be seen from the following examples.*

Example 3.3

Let $X = \{x, y\}$. Define fuzzy topologies on X as follows.
$\eta_1 = \{0, 1, A_{x,y}^{\frac{3}{4},\frac{3}{4}}\}$, $\eta_2 = \{0, 1, B_{x,y}^{\frac{1}{4},\frac{1}{4}}\}$. We claim that (X, η_1, η_2) is FP-regular.

For A, which is a η_1-fo q-neighbourhood of x_t *and* y_t where $t > \frac{1}{4}$, $\eta_2 - cl(A) = A$. Also for B, which is a η_2-fo q-neighbourhood of x_t *and* y_t where $t > \frac{3}{4}$, $\eta_1 - cl(B) = B$. So, (X, η_1, η_2) is a FP-regular fbts.
Now we claim that $(X, \eta_1(x_{\frac{1}{4}}), \eta_2(x_{\frac{1}{8}}))$ is not a FP-regular fbts where $\eta_1(x_{\frac{1}{4}}) = \{0, 1, A, x_{\frac{1}{4}}\} = \eta_1^*(say)$, $\eta_2(x_{\frac{1}{8}}) = \{0, 1, B, x_{\frac{1}{8}}\} = \eta_2^*(say)$. For $x_{\frac{1}{4}}$, which is a η_1-fo q-neighbourhood of x_t where $t > \frac{3}{4}$, there does not exist any η_1-fo q-neighbourhood μ of x_t such that $\eta_2 - cl(\mu) \leq x_{\frac{1}{4}}$. So $(X, \eta_1(x_{\frac{1}{4}}), \eta_2(x_{\frac{1}{8}}))$ is not FP-regular.

Example 3.4

Let $X = \{x, y\}$. Let $\tau_1 = \{0, 1, A_{x,y}^{\frac{1}{2},\frac{1}{2}}\}, \tau_2 = \{0, 1, P_{x,y}^{\frac{1}{4},\frac{1}{4}}\}$. We claim that (X, τ_1, τ_2) is FP-connected. For any representation of the fuzzy set $1 = \mu \cup \nu$, where μ and ν are non-zero fuzzy subsets of X, either $\mu(x) = 1$ or $\nu(x) = 1$ where $x \in X$. Suppose $\mu(x) = 1(resp.\ \nu(x) = 1)$, then $\tau_2 - cl(\mu) = 1(resp.\ \tau_1 - cl(\nu) = 1)$ so that $\tau_2 - cl(\mu)\ q\ \nu,\ as\ \nu \neq 0(resp.\ \tau_1 - cl(\nu)\ q\ \mu,\ as\ \mu \neq 0)$. So 1 is not representable as $\mu \cup \nu$ where (μ, ν) is a fuzzy pairwise separation. Therefore (X, τ_1, τ_2) is FP-connected.
Consider the fuzzy sets $B_{x,y}^{0,1}$, $R_{x,y}^{1,0}$. Clearly the biextension fbts $(X, \tau_1(B), \tau_2(R))$ is not FP-connected. For, $1 = B \cup R$,

$$\tau_1(B)\text{-}cl\ R = R, \tau_2(R)\text{-}cl\ B = B,\ B\ \bar{q}\ R,\ B\ \bar{q}\ \tau_1(B)\text{-}cl\ R\ and\ R\ \bar{q}\ \tau_2(R)\text{-}cl\ B.$$

Theorem 3.2 *If the biextension $(X, \tau_1(g), \tau_2(h))$ of a fbts (X, τ_1, τ_2) is FP-connected then (X, τ_1, τ_2) is also FP-connected.*

Proof: Suppose (X, τ_1, τ_2) is FP-disconnected, then there will exist a pair (λ, μ) of non-null fuzzy sets of X such that $\lambda\ \bar{q}\ \mu$, $\lambda\ \bar{q}\ \tau_1 - cl(\mu)$,

$$\mu\ \bar{q}\ \tau_2 - cl(\lambda)\ and\ 1 = \lambda \cup \mu.$$

Now $\lambda\ \bar{q}\ \tau_1 - cl(\mu) \Rightarrow \lambda \leq (\tau_1 - cl(\mu))^c \Rightarrow \lambda \leq (\tau_1(g) - cl(\mu))^c$ (since, $\tau_1 - cl(\mu) \geq \tau_1(g) - cl(\mu)$). So, $\lambda\ \bar{q}\ \tau_1(g) - cl(\mu)$. Similarly it can be proved that $\mu\ \bar{q}\ \tau_2 - cl(\lambda) \Rightarrow \mu\ \bar{q}\ \tau_2(h) - cl(\lambda)$. Therefore $(X, \tau_1(g), \tau_2(h))$ is not FP-connected, which is a contradiction.

Theorem 3.3 *Let (X, τ_1, τ_2) be a FP-regular fbts. Let $g_1, g_2 \in I^X$ be such that $g_1 \notin \tau_1, g_1$ is τ_2 closed, $g_2 \notin \tau_2$ and g_2 is τ_1 closed. Then the biextension $(X, \tau_1(g_1), \tau_2(g_2))$ is also a FP-regular fbts.*

Proof: Let $\mu \in \tau_1(g_1)$ be a q-neighbourhood of a fuzzy point x_t of X. Let $\mu = f_1 \vee (f_2 \wedge g_1)$ where $f_1, f_2 \in \tau_1$.

$$\mu^c(x) < t \Rightarrow (f_1^c \wedge (f_2^c \vee g_1^c))(x) < t \Rightarrow f_1^c(x) < t \text{ or } (f_2^c \vee g_1^c)(x) < t.$$

Case 1: Suppose $f_1^c(x) < t$
Since f_1 is a τ_1 fuzzy open q-neighbourhood of x_t and (X, τ_1, τ_2) is FP-regular, there exists a τ_1 fuzzy open q-neighbourhood say f_1^* of x_t such that $\tau_2\text{-}cl(f_1^*) \leq f_1$. Since $\tau_2(g_2)\text{-}cl(f_1^*) \leq \tau_2\text{-}cl(f_1^*)$ and $f_1 \leq \mu$ we get $\tau_2(g_2)\text{-}cl(f_1^*) \leq \mu$.

Case 2: Suppose $(f_2^c \vee g_1^c)(x) < t$. So $f_2^c(x) < t$ and $g_1^c(x) < t$. Since f_2 is a τ_1 fuzzy open q-neighbourhood of x_t and (X, τ_1, τ_2) is FP-regular, there exists a τ_1 fuzzy open q-neighbourhood say
f_2^* of x_t such that $\tau_2\text{-}cl(f_2^*) \leq f_2$. Since g_1 is τ_2 fuzzy closed, $\tau_2\text{-}cl(g_1) = g_1$.

So, $\tau_2\text{-}cl(f_2^*) \wedge \tau_2\text{-}cl(g_1) \leq f_2 \wedge g_1$. Since $\tau_2\text{-}cl(f_2^* \wedge g_1) \leq \tau_2\text{-}cl(f_2^*) \wedge \tau_2\text{-}cl(g_1)$ and $f_2 \wedge g_1 \leq \mu$,we get $\tau_2\text{-}cl(f_2^* \wedge g_1) \leq \mu$ and so $\tau_2(g_2)\text{-}cl(f_2^* \wedge g_1) \leq \mu$ where $f_2^* \wedge g_1$ is clearly a $\tau_1(g_1)$ fuzzy open q-neighbourhood of x_t.

Similarly it can be proved that for any $\tau_2(g_2)$ fuzzy open q-neighbourhood μ^* of x_t, there exists a $\tau_2(g_2)$ fuzzy open q-neighbourhood ν^* of x_t such that $\tau_1(g_1)\text{-}cl(\nu^*) \leq \mu^*$. So, $(X, \tau_1(g_1), \tau_2(g_2))$ is FP-regular.

Theorem 3.4 *Let (X, τ_1, τ_2) be a fbts. Let $g, h \in I^X$ be such that $g \notin \tau_1$ and $h \notin \tau_2$. Suppose $\tau_2(h)\text{-}cl(f \wedge g) = \tau_2(h)\text{-}cl(f), \forall f \in \tau_1$. Then the biextension $(X, \tau_1(g), \tau_2(h))$ is not FP-regular.*

Proof: Suppose $(X, \tau_1(g), \tau_2(h))$ is a FP-regular fbts. Let x_t be a fuzzy point such that $x_t \notin g^c$ and $x_t \in \tau_1\text{-}cl(g^c)$. Since g is a $\tau_1(g)$ fuzzy open q-neighbourhood of x_t and $(X, \tau_1(g), \tau_2(h))$ is FP-regular, there exists a $\tau_1(g)$ fuzzy open q-neighbourhood say μ of x_t such that $\tau_2(h)\text{-}cl(\mu) \leq g$. Since $\mu \in \tau_1(g)$, there exists $f_1, f_2 \in \tau_1$ such that $\mu = f_1 \vee (f_2 \wedge g)$.
Now $\mu^c(x) < t \Rightarrow (f_1^c \wedge (f_2 \wedge g)^c)(x) < t \Rightarrow f_1^c(x) < t$ or $(f_2 \wedge g)^c(x) < t$.

Case 1: Let $f_1^c(x) < t$.
Since $f_1 \vee (f_2 \wedge g) \leq \tau_2(h)\text{-}cl(f_1 \vee (f_2 \wedge g)) \leq g$, we get $f_1 \leq g$ and so $f_1^c \geq g^c$. Since $x_t \notin f_1^c, x_t \notin \tau_1\ cl(g^c)$ which is a contradiction.

Case 2: Suppose $(f_2 \wedge g)^c(x) < t$. Then $(f_2^c \vee g^c)(x) < t \Rightarrow f_2^c(x) < t$.
Now $\tau_2(h)\text{-}cl(f_1 \vee (f_2 \wedge g)) \leq g \Rightarrow \tau_2(h)\text{-}cl(f_1) \vee \tau_2(h)\text{-}cl(f_2 \wedge g) \leq g \Rightarrow$
$\tau_2(h)\text{-}cl(f_2 \wedge g) \leq g \Rightarrow \tau_2(h)\text{-}cl(f_2) \leq g$ (by the given condition). So, $f_2 \leq g \Rightarrow f_2^c \geq g^c$. Since $x_t \notin f_2^c$, we get $x_t \notin \tau_1\text{-}cl(g^c)$ which is a contradiction. So $(X, \tau_1(g), \tau_2(h))$ is not a FP-regular fbts.

Remark 3.8 *Converse of the Theorem(3.4) need not be true. This can be seen in the following example.*

Example 3.5

In the Example(3.3), we note that $(X, \eta_1(x_{\frac{1}{4}}), \eta_2(x_{\frac{1}{8}}))$ is not FP-regular. Also the condition $\eta_2(x_{\frac{1}{8}})\text{-}cl(f \wedge x_{\frac{1}{4}}) = \eta_2(x_{\frac{1}{8}})\text{-}cl(f), \forall f \in \eta_1$ fails for $f = 1$. For, $\eta_2(x_{\frac{1}{8}})\text{-}cl(x_{\frac{1}{4}}) = (B^c)^{\frac{3}{4},\frac{3}{4}}_{x,y} \neq \eta_2(x_{\frac{1}{8}})\text{-}cl(1)$.

References

[1] K. K. Azad, On fuzzy semicontinuity,fuzzy almost continuity and fuzzy weakly continuity, *J.Math.Anal.Appl.*, **82**(1981) 14-32.

[2] C. L. Chang, Fuzzy topological spaces, *J.Math.Anal.Appl.*, **24**(1968), 182-190.

[3] A. Kandil,*Biproximities and fuzzy bitopological spaces, simon Stevin* **63**(1989) 45-66

[4] Y. M. Liu, Pao-ming and P. M. Pu, Fuzzy Topology I-Neighborhood structure of a fuzzy point and Moore-Smith convergence, *J.Math.Anal.Appl.* **76**(1980), 571-599.

[5] Manjari Srivastava and Rekha Srivastava, On Fuzzy Pairwise - T_0 and Fuzzy Pairwise - T_1 Bitopological Spaces, *Indian J. pure appl. Math.*, **32**(3), (2001), 387-396

[6] A. A. Ramadn,S. E. Abbas and Y. C. Kim, Separation axioms in terms of θ-closure and δ-closure operators, *Indian,J.pure.appl.Math.*, **34**(7),(2003) 1067-1083

[7] S. Sampath Kumar, On fuzzy pairwise α-continuity and fuzzy pairwise pre-continuity, *Fuzzy Sets Syst.* **62**(1994), 231-238

[8] S. Sampath Kumar, Semi-open sets, semi-continuity and semi-open mappings in fuzzy bitopological spaces, *Fuzzy Sets Syst.*, **64**(1994), 421-426

[9] S. Sampath Kumar, θ-Closure and δ-Closure operators on fuzzy bitopological spaces, *Fuzzy Sets Syst.*, **87**(1997), 235-242

[10] S. Sampath Kumar, Some near fuzzy pairwise continuous mappings between fuzzy bitopological spaces, *Fuzzy Sets Syst.*, **109**(2000) 251-262

[11] S. Sampath Kumar, θ conncetedness and δ connectedness in fuzzy bitopological spaces, *Fuzzy Sets Syst.*, **103**(1999) 535-540

[12] Sunil C. Mathew and T. P. Johnson, On c-fuzzy topological space and simple extensions, *J.Tri.Math.Soc.*, **3**(2001),55-58

[13] C. K. Wong, Fuzzy topology, product and quotient theorems, *J.Math.Anal.Appl*, **45**(1974), 512-521.

In: Mathematics, Game Theory and Algebra Compendium ISBN 978-1-60456-659-8
Editor: Jacob H. Mathias, pp. 437-443

Chapter 37

ON POLYNILPOTENT COVERING GROUPS OF A POLYNILPOTENT GROUP

Behrooz Mashayekhy**and Mahboobeh Alizadeh Sanati
Department of Mathematics, Ferdowsi University of Mashhad,
P.O.Box 1159-91775, Mashhad, Iran

Abstract

Let $\mathcal{N}_{c_1,\ldots,c_t}$ be the variety of polynilpotent groups of class row $(c_1,\ldots,c_t)$. In this paper, first, we show that a polynilpotent group G of class row $(c_1,\ldots,c_t)$ has no any $\mathcal{N}_{c_1,\ldots,c_t,c_{t+1}}$-covering group if its Baer-invariant with respect to the variety $\mathcal{N}_{c_1,\ldots,c_t,c_{t+1}}$ is nontrivial. As an immediate consequence, we can conclude that a solvable group G of length c with nontrivial solvable multiplier, $\mathcal{S}_nM(G)$, has no $\mathcal{S}_n$-covering group for all $n > c$, where $\mathcal{S}_n$ is the variety of solvable groups of length at most n. Second, we prove that if G is a polynilpotent group of class row $(c_1,\ldots,c_t,c_{t+1})$ such that $\mathcal{N}_{c'_1,\ldots,c'_t,c'_{t+1}}M(G) \neq 1$, where $c'_i \geq c_i$ for all $1 \leq i \leq t$ and $c'_{t+1} > c_{t+1}$, then G has no any $\mathcal{N}_{c'_1,\ldots,c'_t,c'_{t+1}}$-covering group. This is a vast generalization of the first author's result on nilpotent covering groups (Indian J. Pure Appl. Math. 29(7) 711-713, 1998).

PACS 20E10, 20F18, 20F16.

Keywords: $\mathcal{V}$-**covering group; polynilpotent groups; solvable-group**

1. Introduction and Motivation

Let $G \simeq F/R$ be a free presentation for G and $\mathcal{V}$ be a variety of groups. Then, after R. Baer [1], the Baer-invariant of G with respect to $\mathcal{V}$ is defined to be $\mathcal{V}M(G) = R \cap V(F)/[RV^*F]$, where $V(F)$ is the verbal subgroup of F with respect to $\mathcal{V}$ and
$[RV^*F] = \langle v(f_1,\cdots,f_{i-1},f_ir,f_{i+1},\cdots,f_n)v(f_1,\cdots,f_n)^{-1} \mid r \in R,\ f_i \in F,$

$$1 \leq i \leq n,\ v \in V,\ n \in \mathbf{N}\rangle\ .$$

*E-mail address: mashaf@math.um.ac.ir

In special case, if $\mathcal{V}$ is the variety of abelian groups, then the Baer-invariant of G will be the well-known notion the Schur-multiplier of G, denoted by $M(G) = R \cap F'/[R, F]$ (See [5,6] for further details).

It is easy to see that if $\mathcal{V} = \mathcal{N}_c$, the variety of nilpotent groups of class at most $c \geq 1$, then

$$\mathcal{N}_c M(G) = \frac{R \cap \gamma_{c+1}(F)}{[R,_c F]} ,$$

where $\gamma_{c+1}(F)$ is the $(c+1)$-st term of the lower central series of F and $[R,_1 F] = [R, F]$, $[R,_c F] = [[R,_{c-1} F], F]$, inductively. We shall also call $\mathcal{N}_c M(G)$ the c-nilpotent multiplier of G.

In a more general case, if $\mathcal{V} = \mathcal{N}_{c_1,\ldots,c_t}$, the variety of polynilpotent groups of class row $(c_1, \ldots, c_t)$, then

$$\mathcal{N}_{c_1,\ldots,c_t} M(G) = \frac{R \cap \gamma_{c_t+1} \circ \cdots \circ \gamma_{c_1+1}(F)}{[R,_{c_1} F,_{c_2} \gamma_{c_1+1}(F), \cdots ,_{c_t} \gamma_{c_{t-1}+1} \circ \cdots \circ \gamma_{c_1+1}(F)]} ,$$

where $\gamma_{c_t+1} \circ \cdots \circ \gamma_{c_1+1}(F) = \gamma_{c_t+1}(\gamma_{c_{t-1}+1}(\cdots(\gamma_{c_1+1}(F))\cdots))$ are the terms of iterated lower central series of F. See [4, corollary 6.14] for the following equality

$$[RN^*_{c_1,\ldots,c_t} F] = [R,_{c_1} F,_{c_2} \gamma_{c_1+1}(F), \ldots ,_{c_t} \gamma_{c_{t-1}+1} \circ \ldots \circ \gamma_{c_1+1}(F)].$$

We shall also call $\mathcal{N}_{c_1,\ldots,c_t} M(G)$, the $(c_1, \ldots, c_t)$-polynilpotent multiplier of G.

Let $\mathcal{V}$ be a variety of groups and G be an arbitrary group, then a $\mathcal{V}$-covering group of G (a generalized covering group of G with respect to the variety $\mathcal{V}$) is a group G^* with a normal subgroup A such that $G^*/A \simeq G$, $A \subseteq V(G^*) \cap V^*(G^*)$, and $A \simeq \mathcal{V}M(G)$, where $V^*(G^*)$ is the marginal subgroup of G^* with respect to $\mathcal{V}$ (see [6]).

Note that if $\mathcal{V}$ is the variety of abelian groups, then the $\mathcal{V}$-covering group of G will be ordinary covering group (sometimes it is called representing group) of G. Also if $\mathcal{V} = \mathcal{N}_{c_1,\ldots,c_t}$, then an $\mathcal{N}_{c_1,\ldots,c_t}$-covering group of G is a group G^* with a normal subgroup A such that

$G \simeq G^*/A$,
$A \simeq \mathcal{N}_{c_1,\ldots,c_t} M(G^*)$ and
$A \subseteq N^*_{c_1,\ldots,c_t}(G^*) \cap \gamma_{c_t+1}(\cdots(\gamma_{c_1+1}(G^*))\cdots)$.

We shall also call G^* a $(c_1, \ldots, c_t)$-polynilpotent covering group of G.

It is a well-known fact that every group has at least a covering group (see [5,13]). Also, the first author proved that every group has a $\mathcal{V}$-covering group if $\mathcal{V}$ is the variety of all groups, $\mathcal{G}$, or the variety of all abelian groups, $\mathcal{A}$, or the variety of all abelian groups of exponent m, $\mathcal{A}_m$, where m is square free (see [7,9]).

Moreover, C. R. Leedham-Green and S. Mckay [6] proved, by a homological method, that a sufficient condition for existence of a $\mathcal{V}$-covering group of G is that $G/V(G)$ should be a $\mathcal{V}$-splitting group.

Some people have tried to construct a covering group for some well-known structures of groups. For example, the generalized quaternion group $Q_{4n} = \langle a, b | a^{2n} = 1, b^2 = a^n, b^{-1}ab = a^{-1}\rangle$ is a covering group of the dihedral group $D_{2n} = \langle a, b | a^n = b^2 = 1, b^{-1}ab = a^{-1}\rangle$ (see [5]).

Also J. Wiegold [12] presented a covering group for a direct product of two finite groups. W. Haebich [2, 3] generalized the Wiegold's result and gave a covering group for a regular product of a family of groups and also for a verbal wreath product of two groups. Moreover, the first author [10] recently proved the existence and presented a structure of an $\mathcal{N}_c$-covering group for a nilpotent product of a family of cyclic groups.

It is interesting to mention that there are some groups which have no any $\mathcal{V}$-covering group, for some variety $\mathcal{V}$. The first author [8] gave an example of the group $G \simeq \mathbf{Z}_r \oplus \mathbf{Z}_s$, where $(r, s) \neq 1$, which has no $\mathcal{N}_c$-covering group for all $c \geq 2$. Moreover, the first author [7, 9] proved that a nilpotent group G of class n with nontrivial c-nilpotent multiplier $\mathcal{N}_cM(G)$, has no $\mathcal{N}_c$-covering group for all $c > n$.

Now, in this paper, we concentrate on nonexistence of polynilpotent covering groups of a polynilpotent group. More precisely, we show that if G is a polynilpotent group of class row $(c_1, \ldots, c_t)$ such that $\mathcal{N}_{c_1,\ldots,c_t,c_{t+1}}M(G) \neq 1$, then G has no $(c_1, \ldots, c_t, c_{t+1})$-polynilpotent covering group of G. Also, if $\mathcal{N}_{c'_1,\ldots,c'_t}M(G) \neq 1$ and $c'_i \geq c_i$ for all $1 \leq i \leq t-1$ and $c'_t > c_t$, then G has no $(c'_1, \ldots, c'_t)$-polynilpotent covering group of G.

2. The Main Result

Let G be a group and $\mathcal{V}$ be a variety of groups. It is clear, by definition, that if $\mathcal{V}M(G) = 1$, then G is the only $\mathcal{V}$-covering group of itself. So it is natural to put the condition $\mathcal{V}M(G) \neq 1$ for nonexistence of $\mathcal{V}$-covering group of G.

Theorem 2.1

Let G be a polynilpotent group of class row $(c_1, \ldots, c_t)$ and $\mathcal{N}_{c_1,\ldots,c_t,c_{t+1}}M(G)$ $\neq 1$, for some $c_{t+1} \geq 1$. Then G has no any $\mathcal{N}_{c_1,\ldots,c_t,c_{t+1}}$-covering group.

Proof.

Let G^* be a $(c_1, \ldots, c_t, c_{t+1})$-polynilpotent covering group of G with the normal subgroup A of G^* such that

$G \simeq G^*/A$,

$A \simeq \mathcal{N}_{c_1,\ldots,c_t,c_{t+1}}M(G^*)$ and

$A \subseteq N^*_{c_1,\ldots,c_t,c_{t+1}}(G^*) \cap \gamma_{c_{t+1}+1}(\gamma_{c_t+1}(\cdots(\gamma_{c_1+1}(G^*))\cdots))$.

We define ρ_t inductively, for any group M and $t \geq 0$, as follows:

$$\rho_0(M) = M \text{ and } \rho_i(M) = \gamma_{c_i+1}(\rho_{i-1}(M)), \text{ for } i > 1.$$

By hypothesis, $\rho_t(G) = 1$ and so $\rho_t(G^*/A) = 1$. Hence $\rho_t(G^*) \subseteq A$. Also $A \subseteq \rho_{t+1}(G^*)$, then $\rho_t(G^*) \subseteq \rho_{t+1}(G^*)$. Clearly $\rho_{t+1}(G^*) \subseteq \rho_t(G^*)$, so $\rho_{t+1}(G^*) = \rho_t(G^*)$. In particular,
$\rho_t(G^*) = \gamma_2(\rho_t(G^*)) = \cdots = \gamma_{c_{t+1}}(\rho_t(G^*)) = \gamma_{c_{t+1}+1}(\rho_t(G^*)) = \rho_{t+1}(G^*)$ (I).
Since $A \subseteq N^*_{c_1,\ldots,c_t,c_{t+1}}(G^*)$ and $\rho_t(G^*) \subseteq A$, so we have

$$[\cdots[[\rho_t(G^*),_{c_1} G^*],_{c_2} \gamma_{c_1+1}(G^*)], \cdots ,_{c_{t+1}} \gamma_{c_t+1}(\cdots(\gamma_{c_1+1}(G^*))\cdots)] = 1,$$

or by the above notation,

$$[\cdots[[\rho_t(G^*),_{c_1} G^*],_{c_2} \rho_1(G^*)], \cdots ,_{c_{t+1}} \rho_t(G^*)] = 1.$$

First, we show that $[M,_i N] \overset{(II)}{\supseteq} [\gamma_i(N), M]$ for each natural number i and normal subgroups M and N of any group. By Three Subgroups Lemma, we have

$$[M,_i N] = [M,_{i-2} N, N, N] \supseteq [N, N, [M,_{i-2} N]] = [[M,_{i-2} N, [N, N]]$$
$$= [[M,_{i-3} N], N, \gamma_2(N)] \supseteq [N, \gamma_2(N), [M,_{i-3} N]] = [[M,_{i-3} N], \gamma_3(N)]$$
$$= [[M,_{i-4} N], N, \gamma_3(N)] \supseteq \cdots \supseteq [M, \gamma_i(N)] = [\gamma_i(N), M].$$

Now, we claim
$$[\cdots[[\rho_t(G^*),_{c_1} G^*],_{c_2} \rho_1(G^*)], \cdots ,_{c_i} \rho_{i-1}(G^*)] \overset{(III)}{\supseteq} [\rho_i(G^*), \rho_t(G^*)],$$
for all $1 \le i \le t+1$.

Clearly the equality is valid for $i = 1$. Now for $i = 2$, we can write

$$[[\rho_t(G^*),_{c_1} G^*],_{c_2} \rho_1(G^*)] \supseteq [[\gamma_{c_1}(G^*), \rho_t(G^*)],_{c_2} \rho_1(G^*)] \qquad by(II)$$
$$\supseteq [[\rho_1(G^*), \rho_t(G^*)],_{c_2} \rho_1(G^*)]$$
$$\supseteq [\gamma_{c_2}(\rho_1(G^*)), [\rho_1(G^*), \rho_t(G^*)]] \qquad by(II)$$
$$= [[\rho_1(G^*), \rho_t(G^*)], \gamma_{c_2}(\rho_1(G^*))]$$
$$\supseteq [\gamma_{c_2}(\rho_1(G^*)), \rho_1(G^*), \rho_t(G^*)]$$
$$= [\gamma_{c_2+1}(\rho_1(G^*)), \rho_t(G^*)]$$
$$= [\rho_2(G^*), \rho_t(G^*)].$$

Suppose the inclusion (III) holds for $i = j$. Now, we prove it for $i = j+1$.

$$[[\cdots[[\rho_t(G^*),_{c_1} \rho_0(G^*)],_{c_2} \rho_1(G^*)], \cdots ,_{c_j} \rho_{j-1}(G^*)],_{c_{j+1}} \rho_j(G^*)]$$
$$\supseteq [\rho_j(G^*), \rho_t(G^*),_{c_{j+1}} \rho_j(G^*)]$$
$$\supseteq [\gamma_{c_{j+1}}(\rho_j(G^*)), [\rho_j(G^*), \rho_t(G^*)]] \qquad by(I)$$
$$= [\rho_j(G^*), \rho_t(G^*), \gamma_{c_{j+1}}(\rho_j(G^*))]$$
$$\supseteq [\gamma_{c_{j+1}}(\rho_j(G^*)), \rho_j(G^*), \rho_t(G^*)]$$
$$= [\gamma_{c_{j+1}+1}(\rho_j(G^*)), \rho_t(G^*)]$$
$$= [\rho_{j+1}(G^*), \rho_t(G^*)].$$

Now, we have

$$1 = [\cdots[[\rho_t(G^*),_{c_1} \rho_0(G^*)],_{c_2} \rho_1(G^*)], \cdots ,_{c_{t+1}} \rho_t(G^*)] \supseteq [\rho_{t+1}(G^*), \rho_t(G^*)].$$

Hence $[\rho_{t+1}(G^*), \rho_t(G^*)] = 1$. Since $\rho_{t+1}(G^*) = \rho_t(G^*)$, we can conclude $[\rho_t(G^*), \rho_t(G^*)] = 1$. i.e. $\gamma_2(\rho_t(G^*)) = 1$. Hence by (I), we have $\rho_{t+1}(G^*) = 1$. Therefore $A = 1$, which is a contradiction. □

Now we can state the following interesting corollary about nonexistence of solvable covering groups.

Corollary 2.2

Let G be a solvable group with derived length at most n. If the l-solvable multiplier of G, $\mathcal{S}_l M(G)$, is nontrivial, then G has no any $\mathcal{S}_l$-covering group, for all $l > n$.

Proof.

Note that, $\mathcal{S}_l$, the variety of solvable groups of derived length at most l is in fact the variety of polynilpotent groups of class row $\underbrace{(1, \ldots, 1)}_{l-times}$. Hence the result is a consequence of Theorem 2.1. □

In a different view, the following theorem is also about nonexistence of polynilpotent covering groups which is a vast generalization of a result of the first author (see [7, Theorem 3.1.6], [8,Theorem 2] and [9, Theorem 2.1]).

Theorem 2.3

Let G be a polynilpotent group of class row $(c_1, \ldots, c_t, c_{t+1})$ such that $\mathcal{N}_{c'_1,\ldots,c'_t,c'_{t+1}} M(G) \neq 1$ where $c'_i \geq c_i$ for all $1 \leq i \leq t$ and $c'_{t+1} > c_{t+1}$. Then G has no any $\mathcal{N}_{c'_1,\ldots,c'_t,c'_{t+1}}$-covering group.

Proof.

Let G^* be a $(c'_1, \ldots, c'_t, c'_{t+1})$-polynilpotent covering group of G with the normal subgroup A of G^* such that

$G \simeq G^*/A$,

$A \simeq \mathcal{N}_{c'_1,\ldots,c'_t,c'_{t+1}} M^*(G)$ and

$A \subseteq N^*_{c'_1,\ldots,c'_t,c'_{t+1}}(G^*) \cap \gamma_{c'_{t+1}+1}(\gamma_{c'_t+1}(\cdots(\gamma_{c'_1+1}(G^*)))\cdots))$.

We consider the following notations, inductively:

$\rho_0(G^*) = G^*$ and $\rho_i(G^*) = \gamma_{c_i+1}(\rho_{i-1}(G^*))$, for all $i \geq 1$,

$\rho'_0(G^*) = G^*$ and $\rho'_i(G^*) = \gamma_{c'_i+1}(\rho'_{i-1}(G^*))$, for all $i \geq 1$.

Since, $\rho_{t+1}(G) = 1$, so we have $\rho_{t+1}(G^*/A) = 1$, and hence $\rho_{t+1}(G^*) \subseteq A$. Also $A \subseteq \rho'_{t+1}(G^*)$, then $\rho_{t+1}(G^*) \subseteq \rho'_{t+1}(G^*)$. On the other hand, by $c'_i \geq c_i$ for all $1 \leq i \leq t$ and $c'_{t+1} > c_{t+1}$ we can imply that $\rho'_j(G^*) \subseteq \rho_j(G^*)$, for all $1 \leq j \leq t+1$. Therefore

$$\rho'_{t+1}(G^*) = \rho_{t+1}(G^*) \ (I).$$

Consider the following trivial inclusions:

$$\rho'_{t+1}(G^*) = \gamma_{c'_{t+1}+1}(\rho'_t(G^*)) \subseteq \gamma_{c'_{t+1}}(\rho'_t(G^*)) \subseteq \gamma_{c'_{t+1}-1}(\rho'_t(G^*)) \subseteq$$

$$\cdots \subseteq \gamma_{c_{t+1}+1}(\rho'_t(G^*)) \subseteq \gamma_{c_{t+1}+1}(\rho_t(G^*)) = \rho_{t+1}(G^*).$$

Thus by the equality (I), we can conclude that

$$\gamma_{c'_{t+1}+1}(\rho'_t(G^*)) = \gamma_{c_{t+1}+1}(\rho'_t(G^*)) \ (II).$$

Since $\rho_{t+1}(G^*) \subseteq A \subseteq N^*_{c'_1,\ldots,c'_t,c'_{t+1}}(G^*)$, we have

$$[\cdots[[\rho_{t+1}(G^*),_{c'_1} \rho'_0(G^*)],_{c'_2} \rho'_1(G^*)],\cdots,_{c'_{t+1}} \rho'_t(G^*)] = 1.$$

Clearly $\rho'_t(G^*) \subseteq \rho'_i(G^*)$ for all $0 \leq i \leq t$, so by (II), we can conclude that

$$[\cdots[[\gamma_{c_{t+1}+1}(\rho'_t(G^*)),_{c'_1} \rho'_t(G^*)],_{c'_2} \rho'_t(G^*)],\cdots,_{c'_{t+1}} \rho'_t(G^*)] = 1.$$

and then $\gamma_{c_{t+1}+1+c'_1+\cdots+c'_{t+1}}(\rho'_t(G^*)) = 1$. Put $c = c_{t+1} + 1 + c'_1 + \cdots + c'_{t+1}$ and $k = c'_{t+1} - c_{t+1}$. By division algorithm, there are $q, r \in \mathbf{Z}$ such that $c = kq + r$, where $r < k$. Put $j = \min\{i \in \mathbf{N} | ki + r \geq c'_{t+1} + 1\}$. Then $kj + r \geq c'_{t+1} + 1$ and $k(j-1) + r < c'_{t+1} + 1$. Now, using (II) we have

$$\begin{aligned}
1 &= \gamma_c(\rho'_t(G^*)) = [\gamma_{c'_{t+1}+1}(\rho'_t(G^*)),_{c-c'_{t+1}-1} \rho'_t(G^*)] \\
&= [\gamma_{c_{t+1}+1}(\rho'_t(G^*)),_{c-c'_{t+1}-1} \rho'_t(G^*)] \\
&= \gamma_{c-k}(\rho'_t(G^*)) \\
&\vdots \\
&= \gamma_{c-k(q-j)}(\rho'_t(G^*)) \\
&= \gamma_{kj+r}(\rho'_t(G^*)) \\
&= \gamma_{k(j-1)+r}(\rho'_t(G^*)) \\
&\supseteq \gamma_{c'_{t+1}+1}(\rho'_t(G^*)) \\
&= \rho'_{t+1}(G^*).
\end{aligned}$$

Hence $\rho'_{t+1}(G^*) = 1$ and so $A = 1$, which is a contradiction. □

Notes

(i) The condition $c'_{t+1} > c_{t+1}$ in the theorem 2.3 is essential, since the first author [10] showed that for any natural number n, there exists a nilpotent group G of class n such that $\mathcal{N}_c M(G) \neq 1$ and G has at least one $\mathcal{N}_c$-covering group for all $c \leq n$.

(ii) In a joint paper with the first author [11], it is shown that a finitely generated abelian group $G \cong \mathbf{Z}_{n_1} \oplus \mathbf{Z}_{n_2} \oplus \ldots \oplus \mathbf{Z}_{n_k}$, where $n_{i+1} | n_i$ for all $1 \leq i \leq k-1$, has a non-trivial polynilpotent multiplier, $\mathcal{N}_{c_1,\cdots,c_t} M(G)$, if $k \geq 3$. Hence we can find many groups satisfying in conditions of Theorems 2.1 and 2.3.

References

[1] R. Baer, Representations of Groups as Quotient Groups, I,II,III, *Trans. Amer. Math. Soc.*, **58**, (1945), 295-419.

[2] W. Haebich, The Multiplicator of a Regular Product of Groups, *Bull. Austral. Math. Soc.*, **7**, (1972), 279-296.

[3] W. Haebich, The multiplicator of a splitting extention, *Journal of Algebra,* **44**, (1977), 420-433.

[4] J. A. Hulse and J. C. Lennox, Marginal Series in Groups, *Proceedings of the Royal Society of Edinburgh,* **76**A, (1976), 139-154.

[5] G. Karpilovsky, The Schur Multiplier, *London Math. Soc. Monographs, New Series* no. **2**, (1987).

[6] C. R. Leedham-Green; S. McKay, Baer-invariant, Isologism, Varietal Laws and Homology, *Acta Math.,* **137**, (1976), 99-150.

[7] B. Mashayekhy, The Baer-invariant and Generalized Covering Groups, *Ph.D. Thesis, Ferdowsi University of Mashhad,* (1996).

[8] B. Mashayekhy, A remark on generalized covering groups, *Indian J. Pure Appl. Math.,* **29**, 7, (1998), 711-713.

[9] B. Mashayekhy, On the Existance of $\mathcal{V}$-covering groups, *Proceedings of 31-st Iranian Mathematics Conference,* (2000), pp. 227-232.

[10] B. Mashayekhy and A. Khaksar, On $\mathcal{N}_c$-Covering groups of a nilpotent product of Cyclic Groups, *International J. of Math.Game Theory and Algebra,* **13**, 2, (2003), 129-132.

[11] B. Mashayekhy and M. Parvizi, Polynilpotent Multipliers of Finitely Generated Abelian Groups, Submitted.

[12] J. Wiegold, The multiplicator of a direct product, *Quart. J. Math. Oxford,* **(2)**, 22 (1971), 103-105.

[13] J. Wiegold, The Schur multiplier: an elementary approach pp. 137-154 in: Groups-St Andrews (1981),*London Math. Soc. Lecture Note Ser.* vol. 71, Cambridge Univ. Press (1982).

In: Mathematics, Game Theory and Algebra Compendium
Editor: Jacob H. Mathias, pp. 445-462
ISBN 978-1-60456-659-8

Chapter 38

The Baer Invariant of Semidirect and Verbal Wreath Products of Groups*

Behrooz Mashayekhy[†]
Department of Mathematics, Ferdowsi University of Mashhad,
P.O.Box 1159-91775, Mashhad, Iran

Abstract

W. Haebich (1977, Journal of Algebra **44**, 420-433) presented some formulas for the Schur multiplier of a semidirect product and also a verbal wreath product of two groups. The author (1997, Indag. Math., (N.S.), **8(4)**, 529-535) generalized a theorem of W. Haebich to the Baer invariant of a semidirect product of two groups with respect to the variety of nilpotent groups of class at most $c \geq 1$, $\mathcal{N}_c$. In this paper, first, it is shown that $\mathcal{V}M(B)$ and $\mathcal{V}M(A)$ are direct factors of $\mathcal{V}M(G)$, where $G = B \rhd\!\!< A$ is the semidirect product of a normal subgroup A and a subgroup B and $\mathcal{V}$ is an arbitrary variety. Second, it is proved that $\mathcal{N}_c M(B \rhd\!\!< A)$ has some homomorphic images of Haebich's type. Also some formulas of Haebich's type is given for $\mathcal{N}_c M(B \rhd\!\!< A)$, when B and A are cyclic groups. Third, we will present a formula for the Baer invariant of a $\mathcal{V}$-verbal wreath product of two groups with respect to the variety of nilpotent groups of class at most $c \geq 1$, where $\mathcal{V}$ is an arbitrary variety. Moreover, it is tried to improve this formula, when $G = A\,Wr_{\mathcal{V}}\,B$ and B is cyclic. Finally, a structure for the Baer invariant of a free wreath product with respect to $\mathcal{N}_c$ will be presented, specially for the free wreath product $A\,Wr_{*}B$ where B is a cyclic group.

AMS Subject Classification 2000: 20E10, 20E22, 20F12.

Key Words : DBaer invariant; semidirect product; verbal wreath product; nilpotent variety

---

*This research was in part supported by a grant from the Research Council of Ferdowsi University of Mashhad (No. 241-1316)
[†]E-mail address: mashaf@math.um.ac.ir

1. Introduction and Motivation

As stated in my previous paper (2001, Journal of Algebra **235**, 15-26), there are some results on the *Schur multiplier* and the *Baer invariant* of the direct product of groups by I. Schur [16], J. Wiegold [18], M.R.R. Moghaddam [13], G.Ellis [4], and the author [12].

It is known that a semidirect product of two groups is a generalization of a direct product of groups. Also, we know that a verbal wreath product is a kind of semidirect product. Therefore, it is interesting to find some formulas for the Schur multiplier or the Baer invariant of a semidirect product and a verbal wreath product.

In 1972, K.I. Tahara [17, (Theorem 2.2.5)], using cohomological methods, gave a structure for a semidirect product of two groups. W. Haebich [6] in 1977, presented some formulas for the Schur multiplier of a semidirect product and a $\mathcal{V}$-verbal wreath product of two groups, where $\mathcal{V}$ is an arbitrary variety. His method was based on presentations of groups.

In 1972, N. Blackburn [2] gave an explicit formula for the Schur multiplier of a standard wreath product of two groups. Also E.W. Read [15] in 1976, found an explicit formula for the Schur multiplier of a wreath product (we know that a standard wreath product is a special kind of a wreath product). It should be mentioned that their formulas were more explicit than Haebich's one. But Haebich's results were more general than Blackburn and Read's formulas.

In 1997, the author [10] generalized a result of W. Haebich [6, (Theorem 1.7)] to the Baer invariant of a semidirect product with respect to the variety $\mathcal{N}_c$.

Now, in this paper, we concentrate on the Haebich's method in order to find some structures for the Baer invariant of a semidirect product and a verbal wreath product of two groups with respect to the variety of nilpotent groups of class at most c, $\mathcal{N}_c$.

More precisely, first, using the functorial property of the Baer invariant, we show that $\mathcal{V}M(B)$ and $\mathcal{V}M(A)$ are direct factors of $\mathcal{V}M(G)$, where $G = B \rhd\!< A$ is the semidirect product of A by B and $\mathcal{V}$ is an arbitrary variety of groups (Theorem 3.2). Also, in Section 3, we find some homomorphic images and in finite case subgroups with structures similar to Haebich's type [6,(Theorems 2.2 and 2.3)] for the Baer invariant of a semidirect product $B \rhd\!< A$ with respect to the variety $\mathcal{N}_c$ (Theorem 3.6). Moreover, if B and A are cyclic groups, then we show that the above structures are isomorphic to $\mathcal{N}_cM(B \rhd\!< A)$ (Theorem 3.7 and Corollary 3.11).

In Section 4, we concentrate on a verbal wreath product and first, we find a structure for $\mathcal{N}_cM(A\,Wr_{\mathcal{V}}\,B)$ similar to Haebich's type (Theorem 4.2). Second, we improve this structure for a homomorphic image and in finite case for a subgroup of $\mathcal{N}_cM(A\,Wr_{\mathcal{V}}\,B)$ (Theorem 4.5). Moreover, if B is cyclic, then we prove that the last structure is isomorphic to $\mathcal{N}_cM(A\,Wr_{\mathcal{V}}\,B)$ (Theorem 4.6).

Finally, in Section 5, using the previous results, we try to find a structure for the Baer invariant of a free wreath product with respect to the variety $\mathcal{N}_c$. Specially, we present a formula for $\mathcal{N}_cM(A\,Wr_*B)$, when B is cyclic and $\mathcal{N}_cM(A\,Wr_*B)$ is finite (Theorem 5.5), and also a formula when A and B are both cyclic groups (Theorem 5.6).

2. Notation and Preliminaries

We assume that the reader is familiar with the notions of variety of groups, verbal and marginal subgroups (see [14]). Let G be a group with a free presentation $G \cong F/R$. Then the Baer invariant of G with respect to the variety $\mathcal{V}$, denoted by $\mathcal{V}M(G)$ is defined to be

$$\mathcal{V}M(G) = \frac{R \cap V(F)}{[RV^*F]},$$

where $V(F)$ is the verbal subgroup of F and

$$[RV^*G] =< v(g_1, \ldots, g_{i-1}, g_i a, g_{i+1}, \ldots, g_n) v(g_1, \ldots, g_i, \ldots g_n)^{-1} \mid a \in R,$$

$$1 \leq i \leq n, v \in V, g_i \in G, n \in \mathbf{N} >,$$

which is independent of the choice of the presentation of G and it is always an abelian group (see [1,5,9] for further properties of the Baer invariant).

In special case, if $\mathcal{V}$ is the variety of abelian groups, $\mathcal{A}$, then the Baer invariant of a group G will be

$$\frac{R \cap F'}{[R, F]},$$

the *Schur multiplier* of G, where, in finite case, by Hopf's formula [7] is isomorphic to the *second cohomology group* of G.

Also, if $\mathcal{V}$ is the variety of nilpotent groups of class at most $c \geq 1$, $\mathcal{N}_c$, then the Baer invariant of a group G will be

$$\mathcal{N}_c M(G) = \frac{R \cap \gamma_{c+1}(F)}{[R, \ _cF]},$$

where $[R, \ _cF] = [R, \underbrace{F, F, \ldots, F}_{c-times}]$. The above notion is also called the *c-nilpotent multiplier* of G (see [3]).

In order to deal with the Baer invariant of a semidirect product we need a presentation for the semidirect product which is given as follows.

Lemma 2.1 [6] Suppose G is a semidirect product (or a splitting extension) of A by B under $\theta : B \longrightarrow Aut(A)$ and

$$1 \longrightarrow R_1 \longrightarrow F_1 \xrightarrow{\nu_1} A \longrightarrow 1 \quad , \quad 1 \longrightarrow R_2 \longrightarrow F_2 \xrightarrow{\nu_2} B \longrightarrow 1$$

are free presentations for A and B, respectively. Then

$$1 \longrightarrow R \longrightarrow F \longrightarrow G \longrightarrow 1$$

is a free presentation for G , where

(i) $F = F_1 * F_2$, the free product of F_1 and F_2 ;

(ii) $R = R_1^F R_2^F S$;

(iii) $S =< f_1^{-1}\overline{f_1}[f_2, f_1] | f_1, \overline{f_1} \in F_1; f_2 \in F_2; \nu_1(\overline{f_1}) = \theta(\nu_2(f_2))(\nu_1 f_1) >^F$.

W. Haebich [6] in 1977, using the above lemma prove the following theorem about the Schur multiplier of a semidirect product.

Theorem 2.2 [6] Let G be a semidirect product of A by B under θ and F_1/R_1 and F_2/R_2 be presentations for A and B, respectively. Then, by the above notation, the following isomorphism holds:

$$M(G) \cong M(B) \oplus \frac{S \cap F'}{[R_2, F_1][S, F]} .$$

A generalization of the above theorem was presented by the author in 1997 as follows.

Theorem 2.3 [10] Let G be a semidirect product of A by B under $\theta : B \longrightarrow Aut(A)$ and $\mathcal{N}_c$ be the variety of nilpotent groups of class at most c $(c \geq 1)$. Then

$$\mathcal{N}_c M(G) \cong \mathcal{N}_c M(B) \oplus \frac{S \cap \gamma_{c+1}(F)}{\prod [R_2, F_1, F_2]_c [S, {}_c F]} ,$$

where

$$\prod [R_2, F_1, F_2]_c = < [r_2, f_1 \ldots, f_c] \mid f_i \in F_1 \cup F_2, 1 \leq i \leq c, \exists k \; f_k \in F_1 >^F .$$

In particular, $\mathcal{N}_c M(B)$ can be regarded as a direct factor of $\mathcal{N}_c M(G)$.

We also assume that the reader is familiar with notions of a verbal product and the cartesian subgroup of a free product (see [11,14]). We need the following properties of the cartesian subgroup.

Lemma 2.4 [14] Let $\{A_i | i \in I\}$ be a family of groups and $\prod^*_{i \in I} A_i$ be the free product of A_i's, and let $[A_i]^*$ be the cartesian subgroup of the above free product. Then

(i) The cartesian subgroup avoids the constituents i.e $[A_i]^* \cap A_j = 1$, for all $j \in I$.

(ii) If $a \in \prod^*_{i \in I} A_i$, then $a = a_{i_1} a_{i_2} \ldots a_{i_m} c$, where $a_{i_j} \neq 1$ for all j, $i_1 < i_2 < \ldots < i_m$ and $c \in [A_i]^*$. The elements a_{i_j} and c are uniquely determined by a and chosen order of I.

Lemma 2.5 [14] Let $\mathcal{V}$ be a variety of groups defined by the set of laws V. Then

(i) If $V(A)$ is the verbal subgroup of $A = \prod^*_i A_i$, then $V(A) \cap A_i = V(A_i)$ for all $i \in I$.

(ii) If $A = \prod^*_i A_i$, then

$$V(A) = (\prod_i V(A_i))(V(A) \cap [A_i]^*) .$$

Now, in the following, you can find the definition of a *verbal wreath product.*

Definition 2.6 Given arbitrary groups A and B, let A_b be an isomorphic copy of A for each $b \in B$ and denote by a_b the element of A_b mapped to $a \in A$. Consider the $\mathcal{V}$-verbal product of A_b's, $G_V = \prod^V_{b \in B} A_b$ corresponding to the variety $\mathcal{V}$ with a set of words V, which is isomorphic to the quotient C/C_V, where $C = \prod^*_{b\in B} A_b$ and $C_V = V(C) \cap [A_b]^*$. The map $a_b \mapsto a_{bb'}$ for all $a \in A$, $b \in B$ and fixed $b' \in B$ induces automorphisms $\theta_* b'$ and $\theta_V b'$ of C and G_V, respectively, i.e.

$$\theta_* : B \longrightarrow Aut(C) \quad , \qquad \theta_V : B \longrightarrow Aut(G_V)$$

$$b' \mapsto \theta_* b' : C \to C \qquad\qquad b' \mapsto \theta_V b' : G_V \to G_V$$

$$a_b \mapsto a_{bb'} \qquad\qquad \overline{a}_b \mapsto \overline{a}_{bb'}$$

It can be proved that $\theta_* : B \to Aut(C)$ and $\theta_V : B \to Aut(G_V)$ are monomorphisms.

Now, the $\mathcal{V}$-*verbal wreath product* of A by B is the semidirect product of G_V by B under θ_V, denoted by $A\,Wr_V B$. The two special cases of the $\mathcal{V}$-verbal wreath product are important, *the free wreath product* and *the standard wreath product.*

(i) If $\mathcal{V}$ is the variety of all groups, then the $\mathcal{V}$-verbal wreath product is the free wreath product which is the semidirect product of C by B under θ_*, denoted by $A\,Wr_* B$.

(ii) If $\mathcal{V} = \mathcal{A}$ is the variety of abelian groups then the $\mathcal{V}$-verbal wreath product is the standard wreath product which is the semidirect product of the direct product $\prod^\times_{b\in B} A_b$ by B under

$$\theta : B \longrightarrow Aut(\prod_{b\in B}{}^{\times} A_b)$$

$$b' \mapsto \theta b' : \prod_b{}^\times A_b \longrightarrow \prod_b{}^\times A_b$$

$$\{a_b\}_b \mapsto \{a_{bb'}\}_b$$

denoted by $A\,Wr B$ or $A \wr B$.

Now, a presentation for a $\mathcal{V}$-verbal wreath product $A\,Wr_V B$ is given based on the presentations of A and B as follows.

Lemma 2.7 [6] Let A and B be two groups with the following presentations:

$$1 \longrightarrow R_1 \longrightarrow F_1 \xrightarrow{\nu_1} A \longrightarrow 1 \quad , \quad 1 \longrightarrow R_2 \longrightarrow F_2 \xrightarrow{\nu_2} B \longrightarrow 1 \, .$$

Then

$$1 \longrightarrow R \longrightarrow F \longrightarrow A\,Wr_V B \longrightarrow 1$$

is a free presentation for the $\mathcal{V}$-verba wreath product $A\,Wr_V B$, where

(i) $F = F_1 * F_2$, the free product of F_1 and F_2 ;
(ii) $R = R_2 S_V$;
(iii) $S_V = [R_2, F_1]^F R_1^F R_V$;
(iv) $R_V = V(H) \cap [(F_1^{f_2})^H]$ with $H = F_1^{F_2}$ (that is, F_1^F) and

$$[(F_1^{f_2})^H] = < [x,y] | x \in (F_1^u)^H, y \in (F_1^v)^H, \; u, v \in F_2 > \, .$$

Now, the following theorem gives a structure of the Schur multiplier of a $\mathcal{V}$-verbal wreath product.

Theorem 2.8 [6] In accordance with the above notation, the following isomorphism holds.

$$M(A\,Wr_V\,B) \cong M(A) \oplus M(B) \oplus \frac{C_V}{[C_V, A\,Wr_* B]} \ .$$

Corollary 2.9

$$M(A\,Wr_* B) \cong M(A) \oplus M(B) \ .$$

Proof. We know that $A\mathrm{wr}_* B = A\,Wr_V\,B$, where $V = \{1\}$. So $V(C) = 1$ and then $C_V = 1$. Now by Theorem 2.8 the result holds. □

Corollary 2.10

$$M(A \wr B) \cong M(A) \oplus M(B) \oplus \frac{[A_b]^*}{[[A_b]^*, A\,Wr_* B]} \ ,$$

where $[A_b]^*$ is the cartesian subgroup of the free product $\prod^*_{b \in B} A_b$.

Proof. As mentioned before $A \wr B = A\,Wr_{\mathcal{V}}\,B$, where $\mathcal{V}$ is the variety of abelian groups. Therefore $V(C) = C'$ and $C_V = V(C) \cap [A_b]^* = [A_b]^*$ since $[A_b]^* \subseteq C'$. Now, the result holds by Theorem 2.8 . □

We also need the following lemmas in our investigation.

Lemma 2.11 [16] Let A be a subgroup of a group G, and N be a normal subgroup of G and let $\{M_i | i \in I\}$ be a family of normal subgroups of G. Then

$$[A \prod_i M_i, N] = [A, N] \prod_i [M_i, N] \ .$$

Lemma 2.12

If H is a subgroup of a finite abelian group G, then G has a subgroup isomorphic to G/H. Consequently, if A is a homomorphic image of a finite abelian group G, then G has a subgroup isomorphic to A.

3. Some Results on the Baer Invariant of a Semidirect Product

First of all, by the functorial property of the Baer invariant, we are going to generalize somehow Theorems 2.2 and 2.3 to an arbitrary variety.

Theorem 3.1 Using the notion of the Baer invariant, we can consider the following functor from the category of all groups to the category of all abelian groups:

$$\mathcal{V}M(-) : \mathcal{G}roups \longrightarrow \mathcal{A}b \ ,$$

where $\mathcal{V}$ is an arbitrary variety of groups.

Proof. See [5,9].

THEOREM 3.2 Let $G = B \overset{\theta}{\ltimes} A$ be the semidirect product of A by B under θ. Then $\mathcal{V}M(B)$ and $\mathcal{V}M(A)$ are direct factors of $\mathcal{V}M(G)$, where $\mathcal{V}$ is an arbitrary variety.

Proof. By the property of a semidirect product, we have the following split exact sequence:

$$1 \longrightarrow A \overset{h}{\longrightarrow} G \overset{f}{\longrightarrow} B \longrightarrow 1 .$$

So there exists a homomorphism $g : B \to G$ such that $f \circ g = I_B$. Now, applying the functorial property of $\mathcal{V}M(-)$ (Theorem 3.1), the following exact sequence splits:

$$1 \longrightarrow Ker(\mathcal{V}M(f)) \overset{\subseteq}{\longrightarrow} \mathcal{V}M(G) \overset{\mathcal{V}M(f))}{\longrightarrow} \mathcal{V}M(B) \longrightarrow 1 ,$$

and $\mathcal{V}M(f) \circ \mathcal{V}M(g) = I_{\mathcal{V}M(B)}$. Hence, we have

$$\mathcal{V}M(G) \cong \mathcal{V}M(B) \oplus Ker(\mathcal{V}M(f)) .$$

i.e. $\mathcal{V}M(B)$ is a direct factor of $\mathcal{V}M(G)$.

To prove that $\mathcal{V}M(A)$ is a direct factor of $\mathcal{V}M(G)$, by a similar argument as above, we should consider the following split exact sequence:

$$1 \longrightarrow \mathcal{V}M(A)) \overset{\mathcal{V}M(h)}{\longrightarrow} \mathcal{V}M(G) \overset{nat}{\longrightarrow} \frac{\mathcal{V}M(G)}{Im(\mathcal{V}M(h))} \longrightarrow 1 . \square$$

Now, in the following, we state a theorem of W. Haebich which is improves somehow the structure of the complementary factor of $M(B)$ in $M(G)$ in Theorem 2.2 .

THEOREM 3.3 [6] By the assumptions and notations of Theorem 2.2, we have

$$M(G) \cong M(B) \oplus \frac{T \cap K'}{[T, K]} ,$$

where $K = F_1 * B$ and

$$T = < f_1^{-1}\overline{f_1}[b, f_1] \mid f_1, \overline{f_1} \in F_1; b \in B; \nu_1\overline{f_1} = (\theta\nu_2 b)(\nu_1(f_1)) >^K .$$

Now, we are going to find a structure similar to the above for the complementary factor, $\mathcal{N}_cM(B)$ of $\mathcal{N}_cM(G)$ in Theorem 2.3 .

First we need the following lemmas.

LEMMA 3.4 [10] considering the assumptions and notations of Lemma 2.1, the following statements hold:

(i) R_1 and $[R_2, F_1]$ are subgroups of S ;
(ii) $R = R_2S$;
(iii) $R \cap \gamma_{c+1}(F) = (R_2 \cap \gamma_{c+1}(F_2))(S \cap \gamma_{c+1}(F))$, for all $c \geq 1$;
(iv) $[R,\ _cF] = [R_2,\ _cF_2] \prod [R_2, F_1, F_2]_c [S,\ _cF]$, for all $c \geq 1$.

LEMMA 3.5 [6] Let C and $\overline{C}$ be two normal subgroups of A and let A be a subgroup of a group B. Let ϕ and $\overline{\phi}$ be two homomorphisms from B to any groups such that

$$(A \cap Ker\phi)C = (A \cap Ker\overline{\phi})\overline{C} .$$

Then the map $\psi : \phi(A)/\phi(C) \longrightarrow \overline{\phi}(A)/\overline{\phi}(\overline{C})$ given by

$$(\phi a)\phi(C) \longmapsto (\overline{\phi}a)\overline{\phi}(\overline{C})$$

is an isomorphism.

Now we are in a position to state and prove the main results of this section.

Theorem 3.6 Let G be a semidirect product of A by B under $\theta : B \to Aut(A)$, and $\mathcal{N}_c$ be the variety of nilpotent groups of class at most c. Then, using the notation of Theorem 3.3, there exists an epimorphism as follows:

$$\mathcal{N}_cM(G) \longrightarrow\!\!\!\!\!\!\succ \mathcal{N}_cM(B) \oplus \frac{T \cap \gamma_{c+1}(K)}{[T,\ _cK]} .$$

So, in finite case, $\mathcal{N}_cM(G)$ has a subgroup isomorphic to

$$\mathcal{N}_cM(B) \oplus \frac{T \cap \gamma_{c+1}(K)}{[T,\ _cK]} .$$

Proof. By the universal property of a free product, let δ be the natural homomorphism from $F = F_1 * F_2$ onto $K = F_1 * B$ induced by $\nu_2 : F_2 \to B$ and the identity on F_1. If $b = \nu_2(f_2)$ and $(\theta b)(\nu_1(f_1)) = \nu_1(\overline{f}_1)$, then

$$\delta(f_1^{-1}\overline{f}_1[f_2, f_1]) = \delta(f_1)^{-1}\delta(\overline{f}_1)[\delta(f_2), \delta(f_1)] = f_1^{-1}\overline{f}_1[b, f_1] .$$

Thus $\delta(S) = T$ and hence $\delta([S,\ _cF]) = [\delta(S),\ _c\delta(F)] = [T,\ _cK]$.
Also, we have

$$\delta(S \cap \gamma_{c+1}(F)) \subseteq \delta(S) \cap \delta(\gamma_{c+1}(F)) = T \cap \gamma_{c+1}(K) .$$

To show the reverse containment, let $u \in T \cap \gamma_{c+1}(K)$. Then there exists $x \in S$ and $y \in \gamma_{c+1}(F)$ such that $\delta(x) = u = \delta(y)$, and so $yx^{-1} \in Ker\delta$. It is easy to see that $Ker\delta = R_2^F$, so $yx^{-1} \in R_2^F$ and hence $y \in R_2^F S$.
By Lemma 3.4 parts (i) and (ii) we have $R = R_2^F S$. Therefore

$$y \in R \cap \gamma_{c+1}(F) .$$

Also, by Lemma 3.4 part (iii) $R \cap \gamma_{c+1}(F) = (R_2 \cap \gamma_{c+1}(F_2))(S \cap \gamma_{c+1}(F))$. Since $R_2 \cap \gamma_{c+1}(F_2) \subseteq Ker\delta$, $u = \delta(y) \in \delta(R \cap \gamma_{c+1}(F)) = \delta(S \cap \gamma_{c+1}(F))$, thus

$$\delta(S \cap \gamma_{c+1}(F)) = T \cap \gamma_{c+1}(K) .$$

Hence

$$\frac{T \cap \gamma_{c+1}(K)}{[T,\ _cK]} = \frac{\delta(S \cap \gamma_{c+1}(F)}{\delta([S,\ _cF])} .$$

Now, putting $\phi = \delta : F \to K, \overline{\phi} = identity : F \to F,\ A = S \cap \gamma_{c+1}(F)$ and

$$C = \prod [R_2, F_1, F_2]_c[S,\ _cF] \ ;$$

$$\overline{C} = (\gamma_{c+1}(F) \cap [R_2, F_1]^F)[S, {}_cF] \ ,$$

we have

$$(A \cap Ker\phi)C = (S \cap \gamma_{c+1}(F) \cap R_2^F) \prod [R_2, F_1, F_2]_c [S, {}_cF]$$

$$= (S \cap \gamma_{c+1}(F) \cap R_2^F)[S, {}_cF] \quad (since \prod [R_2, F_1, F_2]_c \subseteq R_2^F \cap \gamma_{c+1}(F) \ and \ S)$$

$$= (S \cap \gamma_{c+1}(F) \cap R_2^F[R_2, F_1]^F)[S, {}_cF]$$

$$= (S \cap \gamma_{c+1}(F) \cap [R_2, F_1]^F)[S, {}_cF] \ (since \ F = F_2 \rtimes F_1[F_1, F_2] \ and \ S \leq F_1[F_1, F_2])$$

$$= (\gamma_{c+1}(F) \cap [R_2, F_1]^F)[S, {}_cF] \qquad (By \ Lemma \ 3.4 \ (i))$$

$$(A \cap Ker\overline{\phi})\overline{C} \ .$$

So, by Lemma 3.5, we have the following isomorphism:

$$\frac{T \cap \gamma_{c+1}(K)}{[T, {}_cK]} = \frac{\delta(S \cap \gamma_{c+1}(F))}{\delta([S, {}_cF])} \cong \frac{S \cap \gamma_{c+1}(F)}{\gamma_{c+1}(F) \cap [R_2, F_1]^F)[S, {}_cF]}$$

(Note that $\prod [R_2, F_1, F_2]_c \subseteq Ker\delta$). Clearly

$$\prod [R_2, F_1, F_2]_c [S, {}_cF] \subseteq (\gamma_{c+1}(F) \cap [R_2, F_1]^F)[S, {}_cF] \ .$$

Therefore there exists a natural epimorphism

$$\frac{S \cap \gamma_{c+1}(F)}{\prod [R_2, F_1, F_2]_c [S, {}_cF]} \longrightarrow\!\!\!\!\!\rightarrow \frac{S \cap \gamma_{c+1}(F)}{(\gamma_{c+1}(F) \cap [R_2, F_1]^F)[S, {}_cF]} \ .$$

Now, by Theorem 2.3 there exists the following epimorphism

$$\mathcal{N}_c M(G) \longrightarrow\!\!\!\!\!\rightarrow \mathcal{N}_c M(B) \oplus \frac{T \cap \gamma_{c+1}(K)}{[T, {}_cK]} \ .$$

Moreover, if G is finite, then $\mathcal{N}_c M(G)$ is also finite and hence by Lemma 2.12

$$\mathcal{N}_c M(B) \oplus \frac{T \cap \gamma_{c+1}(K)}{[T, {}_cK]}$$

is isomorphic to a subgroup of $\mathcal{N}_c M(G)$. □

Now, using the above result, we can present an improved formula with respect to that of Theorem 2.3 for the Baer invariant of a semidirect product of an arbitrary group by a cyclic group with respect to the variety $\mathcal{N}_c$.

Theorem 3.7 Let G be a semidirect product of an arbitrary group A by a cyclic group B under θ and $\mathcal{N}_c$ be the variety of nilpotent groups of class at most c. Then, by the previous notation, there exists the following isomorphism

$$\mathcal{N}_c M(G) \cong \mathcal{N}_c M(B) \oplus \frac{T \cap \gamma_{c+1}(K)}{[T, {}_cK]} \cong \frac{T \cap \gamma_{c+1}(K)}{[T, {}_cK]} \ .$$

Proof. Applying the previous proof, it is enough to show that

$$\prod[R_2, F_1, F_2]_c = \gamma_{c+1}(F) \cap [R_2, F_1]^F \ .$$

Since B is cyclic, we can consider the following presentation for B

$$1 \longrightarrow R_2 =< y^m > \longrightarrow F_2 =< y > \xrightarrow{\nu_2} B \longrightarrow 1 \ .$$

So R_2 has no commutators and hence by commutator calculus, specially on basic commutators, we can conclude the required equality (see also [11] section 4).
Thus

$$\frac{T \cap \gamma_{c+1}(K)}{[T,\ _cK]} \cong \frac{S \cap \gamma_{c+1}(F)}{\prod[R_2, F_1, F_2]_c[S,\ _cF]} \ ,$$

and so, by Theorem 2.3, the result holds. □

Note 3.8 In order to find a relation between $\mathcal{N}_cM(A)$ and the quotient group $T \cap \gamma_{c+1}(K)/[T,\ _cK]$, it is enough to consider the following natural epimorphism:

$$\mathcal{N}_cM(A) = \frac{R_1 \cap \gamma_{c+1}(F_1)}{[r_1,\ _cF_1]} \twoheadrightarrow \frac{(R_1 \cap \gamma_{c+1}(F_1))[T,\ _cK]}{[T,\ _cK]} \leq \frac{T \cap \gamma_{c+1}(K)}{[T,\ _cK]}.$$

Note that $R_1 \leq T$ and $[R_1,\ _cF_1] \leq [T,\ _cK]$ and in finite case $T \cap \gamma_{c+1}(K)/[T,\ _cK]$ has a subgroup which is isomorphic to a subgroup of $\mathcal{N}_cM(A)$.

In the following theorem, we try to provide a considerable simplification of the quotient group $T \cap \gamma_{c+1}(K)/[T,\ _cK]$ and relate its structure more closely to that of A.

Theorem 3.9 With the previous notation and assumption the following isomorphism holds:

$$\frac{U \cap \gamma_{c+1}(D)}{[U,\ _cD]} \cong \frac{T \cap \gamma_{c+1}(K)/[T,\ _cK]}{(R_1 \cap \gamma_{c+1}(F_1))(\gamma_{c+1}(K) \cap [R_1, B]^K)[T,\ _cK]/[T,\ _cK]} \ ,$$

where $D = A * B$ and $U =< a^{-1}(\theta(b)(a))[b, a] | a \in A, b \in B >^D$.

Proof. let $\eta : K = F_1 * B \to D = A * B$ be the natural homomorphism induced by $\nu_1 : F_1 \to A$ and the identity on B. Clearly $Ker\eta = R_1^K$, $\eta(T) = U$ and hence $\eta([T,\ _cK]) = [U,\ _cD]$.
Also, similar to the proof of Theorem 3.6 we can show that

$$\eta(T \cap \gamma_{c+1}(K)) = U \cap \gamma_{c+1}(D) \ .$$

Now, using Lemma 3.5 and a similar method of the proof of Theorem 3.6, if we put $\phi = \eta : K \to D, \overline{\phi} = identity : K \to K,\ A = T \cap \gamma_{c+1}(K)$ and

$$C = [T,\ _cK] \ ;$$

$$\overline{C} = (R_1 \cap \gamma_{c+1}(F_1))(\gamma_{c+1}(K) \cap [R_1, B]^K)[T,\ _cK] \ ,$$

then we have

$$\frac{U\cap\gamma_{c+1}(D)}{[U,\ {}_cD]}=\frac{\eta(T\cap\gamma_{c+1}(K))}{\eta([T,\ {}_cK])}\cong\frac{T\cap\gamma_{c+1}(K)}{(R_1\cap\gamma_{c+1}(F_1))(\gamma_{c+1}(K)\cap[R_1,B]^K)[T,\ {}_cK]}$$

$$\cong\frac{T\cap\gamma_{c+1}(K)/[T,\ {}_cK]}{(R_1\cap\gamma_{c+1}(F_1))(\gamma_{c+1}(K)\cap[R_1,B]^K)[T,\ {}_cK]/[T,\ {}_cK]}.\ \Box$$

Corollary 3.10 If A is a cyclic group, then

$$\frac{U\cap\gamma_{c+1}(D)}{[U,\ {}_cD]}\cong\frac{T\cap\gamma_{c+1}(K)}{[T,\ {}_cK]}\ .$$

Proof. Since A is cyclic, we can consider $A=F_1/R_1$, where $F_1=<x>$ and $R_1=<x^n>$. So R_1 has no commutators and hence by commutator calculus we have

$$\gamma_{c+1}(K)\cap[R_1,B]^K\subseteq[T,\ {}_cK]\,.$$

Also, since A is cyclic, so $\mathcal{N}_cM(A)=1$ and hence

$$R_1\cap\gamma_{c+1}(F_1)=[R_1,\ {}_cF_1]\leq[T,\ {}_cK]\,.$$

Thus, by Theorem 3.9 the result holds. □

Corollary 3.11 If A and B are cyclic groups, then

$$\mathcal{N}_cM(B\overset{\theta}{\rtimes}A)\cong\frac{U\cap\gamma_{c+1}(D)}{[U,\ {}_cD]}\ .$$

Proof. By Theorem 3.7 and Corollary 3.10 the result holds. □

4. The Baer Invariant of a Verbal Wreath Product

In this section, first, using a presentation for a verbal wreath product given in [6], (Lemma 2.7), we are going to present a formula similar to that of Theorem 2.3 for the Baer invariant of a $\mathcal{V}$-verbal wreath product with respect to the variety of nilpotent groups, $\mathcal{N}_c$, where $\mathcal{V}$ is an arbitrary variety of groups.

The following theorem is vital in the proof of the results of this section.

THEOREM 4.1 By the notation of Lemma 2.7, the following equalities hold.
(i) $R \cap \gamma_{c+1}(F) = (R_2 \cap \gamma_{c+1}(F_2))(S_V \cap \gamma_{c+1}(F))$, for all $c \geq 1$;
(ii) $[R,\ _cF] = [R_2,\ _cF_2] \prod [R_2, F_1, F_2]_c [S_V,\ _cF]$, for all $c \geq 1$,
where $\prod [R_2, F_1, F_2]_c$ was defined in Theorem 2.3 .

Proof. First, by definition of S_V and R_V in Lemma 2.7, it is easy to see that

$$R_1 \leq S_V \ , \ [R_2, F_1] \leq S_V \ , \ R_V \leq V(H) \leq H = F_1^{F_2} = F_1[F_1, F_2] \ ,$$

$$S_V \leq F_1[F_1, F_2] \ , \ F = F_2 \rtimes F_1[F_1, F_2] \ \textit{and} \ S_V \trianglelefteq F \ .$$

(i) By Lemma 2.5

$$\gamma_{c+1}(F) = \gamma_{c+1}(F_1)\gamma_{c+1}(F_2) \prod [F_1, F_2]_{c+1} \ ,$$

where

$$\prod [F_1, F_2]_{c+1} = < [F_1, F_2, F_{i_1}, \ldots, F_{i_{c-1}}] \mid i_j \in \{1, 2\}, 1 \leq j \leq c - 1 >$$

and $\prod [F_1, F_2]_{c+1} \trianglelefteq F$ (see also [13]). Now, by Lemma 2.4, we have

$$R \cap \gamma_{c+1}(F) = R_2 S_V \cap \gamma_{c+1}(F_1)\gamma_{c+1}(F_2) \prod [F_1, F_2]_{c+1}$$

$$= (R_2 \cap \gamma_{c+1}(F_2))(S_V \cap \gamma_{c+1}(F_1) \prod [F_1, F_2]_{c+1})$$

$$= (R_2 \cap \gamma_{c+1}(F_2))(S_V \cap \gamma_{c+1}(F)) \ .$$

(ii) We use induction on c . If $c = 1$, then

$$[R, F] = [R_2 S_V, F]$$

$$= [R_2, F][S_V, F] \ \ , \ (\textit{since } S_V \trianglelefteq F)$$

$$\subseteq [R_2, F_2][R_2, F_1]^F [S_V, F] \ , \ (\textit{since } F =< F_1, F_2 >)$$

$$= [R_2, F_2][R_2, F_1][S_V, F] \ \ , \ (\textit{since } [R_2, F_1] \leq S_V) \ .$$

The reverse containment is clear. Hence $[R, F] = [R_2, F_2][R_2, F_1][S, F]$.
Now, suppose $[R,\ _kF] = [R_2,\ _kF_2] \prod [R_2, F_1, F_2]_k [S_V,\ _kF]$. Then we have

$$[R,\ _{k+1}F] = [[R,\ _kF], F]$$

$$= [[R_2,\ _kF_2] \prod [R_2, F_1, F_2]_k [S_V,\ _kF], F] \quad (\textit{by induction hypothesis})$$

$$= [[R_2,\ _kF_2], F][\prod [R_2, F_1, F_2]_k, F][[S_V,\ _kF], F]$$

$$(\textit{since } [S_V,\ _kF] \ , \ \prod [R_2, F_1, F_2]_k \trianglelefteq F)$$

$$\subseteq [[R_2,\ _kF_2], F_2] \prod [R_2, F_1, F_2]_{k+1} [S_V,\ _{k+1}F]$$

$$\subseteq [R,\ _{k+1}F] \ .$$

Therefore, by induction we have

$$[R,\ _cF] = [R_2,\ _cF_2] \prod [R_2, F_1, F_2]_c [S_V,\ _cF] \quad for\ all\ c \geq 1\ . \ \square$$

Now, we are in a position to prove one of the main result of this section.

THEOREM 4.2 Let G be the $\mathcal{V}$-verbal wreath product of A by B, where $\mathcal{V}$ is an arbitrary variety of groups. Then, using the previous notation, the following isomorphism exists.

$$\mathcal{N}_cM(G) = \mathcal{N}_cM(A\,Wr_V\,B) \cong \mathcal{N}_cM(B) \oplus \frac{S_V \cap \gamma_{c+1}(F)}{\prod [R_2, F_1, F_2]_c [S_V,\ _cF]}\ .$$

Proof. By the previous assumptions and notations the following natural homomorphisms exists

$$F \xrightarrow{\varphi} \frac{F}{[R_2,\ _cF_2]^F} \xrightarrow{\eta} \frac{F}{[R_2,\ _cF_2] \prod [R_2, F_1, F_2]_c [S_V,\ _cF]}\ .$$

By Theorem 4.1 the following isomorphisms hold:

$$\frac{R \cap \gamma_{c+1}(F)}{[R,\ _cF]} \cong (\eta\varphi)(R \cap \gamma_{c+1}(F))$$

$$\cong (\eta\varphi)(R_2 \cap \gamma_{c+1}(F_2))(\eta\varphi)(S_V \cap \gamma_{c+1}(F))\ . \quad (*)$$

Consider the following two natural homomorphisms:

$$\frac{F_1 * F_2}{[R_2,\ _cF_2]^F} \xrightarrow{h} F_1 * \frac{F_2}{[R_2,\ _cF_2]} \xrightarrow{g} \frac{F_1 * F_2}{[R_2,\ _cF_2]^F}\ ,$$

given by

$$\overline{f_1} \longmapsto f_1 \qquad f_1 \longmapsto \overline{f_1}$$

$$\overline{f_2} \longmapsto \overline{f_2} \qquad \overline{f_2} \longmapsto \overline{f_2}\ .$$

It is easy to see that $h \circ g = 1$ and $g \circ h = 1$ i.e h and g are isomorphisms. Thus, we have

$$\frac{F_1 * F_2}{[R_2,\ _cF_2]^F} = \varphi(F) \cong F_1 * \frac{F_2}{[R_2,\ _cF_2]}\ .$$

Also

$$\varphi(F_2) = \frac{F_2[R_2,\ _cF_2]^F}{[R_2,\ _cF_2]} = \frac{F_2[R_2,\ _cF_2][R_2,\ _cF_2, F_1]^F}{[R_2,\ _cF_2][R_2,\ _cF_2, F_1]^F}$$

$$\cong \frac{F_2}{F_2 \cap [R_2,\ _cF_2][R_2,\ _cF_2, F_1]^F} \quad (By\ the\ second\ isomorphism\ Theorem)$$

$$= \frac{F_2}{[R_2,\ _cF_2]} \quad (By\ Lemma\ 2.5)$$

and

$$\varphi(F_1[F_1, F_2]) \cong \varphi(F_1)[\varphi(F_1), \varphi(F_2)] \cong F_1[F_1, \frac{F_2}{[R_2,\ _cF_2]}]\ .$$

(Note that

$$\varphi(F_1) = \frac{F_1[R_2,\ _cF_2]^F}{[R_2,\ _cF_2]^F} \cong \frac{F_1}{F_1 \cap [R_2,\ _cF_2]^F} = \frac{F_1}{1} \cong F_1\ .)$$

Therefore

$$\varphi(F) \cong F_1 * \frac{F_2}{[R_2,\ _cF_2]} \cong \frac{F_2}{[R_2,\ _cF_2]} \ltimes F_1[F_1, \frac{F_2}{[R_2,\ _cF_2]}]$$
$$\cong \varphi(F_2) \ltimes \varphi(F_1)[\varphi(F_1), \varphi(F_2)]\ .$$

Clearly $Ker(\eta) = \varphi(\prod[R_2, F_1, F_2]_c[S_V,\ _cF])$

$$\leq \varphi(F_1)[\varphi(F_1), \varphi(F_2)]\ .\ (**)$$

So by [6, (Lemma 1.3)] we have

$$(\eta\varphi)(F) \cong \frac{\varphi(F)}{Ker(\eta)} = \frac{\varphi(F)}{\varphi(\prod[R_2, F_1, F_2]_c[S_V,\ _cF])}$$

$$\cong \varphi(F_2) \ltimes \frac{\varphi(F_1)[\varphi(F_1), \varphi(F_2)]}{Ker(\eta)}\ .$$

By $(**)$, it is easy to see that $(\eta\varphi)(F_2) \cong \varphi(F_2)$ and $(\eta\varphi)(F_1) \cong \varphi(F_1)/Ker(\eta)$. Thus we have

$$(\eta\varphi)(F) \cong \varphi(F_2) \ltimes \frac{\varphi(F_1)[\varphi(F_1), \varphi(F_2)]}{Ker(\eta)}$$
$$\cong (\eta\varphi)(F_2) \ltimes (\eta\varphi)(F_1)[(\eta\varphi)(F_1), (\eta\varphi)(F_2)]\ .$$

So, we can conclude that

$$(\eta\varphi)(R_2 \cap \gamma_{c+1}(F_2)) \cap (\eta\varphi)(S_V \cap \gamma_{c+1}(F))$$
$$\subseteq (\eta\varphi)(F_2) \cap (\eta\varphi)(F_1)[(\eta\varphi)(F_1), (\eta\varphi)(F_2)] = 1$$

Hence, by considering $(*)$, we have

$$\frac{R \cap \gamma_{c+1}(F)}{[R,\ _cF]} \cong (\eta\varphi)(R_2 \cap \gamma_{c+1}(F_2)) \oplus (\eta\varphi)(S_V \cap \gamma_{c+1}(F))\ .$$

On the other hand, we have the following isomorphisms.

$$(\eta\varphi)(R_2 \cap \gamma_{c+1}(F_2)) = \frac{(R_2 \cap \gamma_{c+1}(F_2))Ker(\eta\varphi)}{Ker(\eta\varphi)}$$

$$\cong \frac{R_2 \cap \gamma_{c+1}(F_2)}{(R_2 \cap \gamma_{c+1}(F_2)) \cap Ker(\eta\varphi)} \quad (by\ the\ second\ isomorphism\ Theorem)$$

$$\cong \frac{R_2 \cap \gamma_{c+1}(F_2)}{[R_2,\ _cF_2]} \quad (by\ Lemma\ 2.5)$$

$$\cong \mathcal{N}_cM(B)\ .$$

Also

$$(\eta\varphi)(S_V \cap \gamma_{c+1}(F)) = \frac{(S_V \cap \gamma_{c+1}(F))Ker(\eta\varphi)}{Ker(\eta\varphi)}$$

$$\cong \frac{S_V \cap \gamma_{c+1}(F)}{(S_V \cap \gamma_{c+1}(F)) \cap Ker(\eta\varphi)} \quad (by\ the\ second\ isomorphism\ Theorem)$$

$$= \frac{S_V \cap \gamma_{c+1}(F)}{\prod[R_2, F_1, F_2][S_V,\ {}_cF]} \quad (by\ Lemma\ 2.5)\ .$$

Therefore

$$\mathcal{N}_cM(B\,Wr_V A) \cong \frac{R \cap \gamma_{c+1}(F)}{[R,\ {}_cF]}$$

$$\cong \mathcal{N}_cM(B) \oplus \frac{S_V \cap \gamma_{c+1}(F)}{\prod[R_2, F_1, F_2]_c[S_V,\ {}_cF]}\ . \quad \square$$

Now, we can obtain the following corollary if we put $c = 1$.

Corollary 4.3 [6] Suppose $G = A\,Wr_V B$ is the $\mathcal{V}$-verbal wreath product of A by B, where $\mathcal{V}$ is a variety of groups. Then

$$M(A\,Wr_V B) \cong M(B) \oplus \frac{S_V \cap F'}{[R_2, F_1][S_V, F]}\ .$$

Now, in what follows, we are going to improve the structure of the complementary factor of $\mathcal{N}_cM(B)$ in $\mathcal{N}_cM(A\,Wr_V B)$.

Notation 4.4 By considering the previous notation, let $\delta : F_1 * F_2 \longrightarrow F_1 * B$ be the natural epimorphism, where $A \cong F_1/R_1$ and $B \cong F_2/R_2$ are presentations for A and B. Put $T_V = \delta(S_V)$ and $K = F_1 * B$. Using Lemmas 3.4 and 2.7, Theorem 4.1 and similar properties of S_V to that of S, we can rewrite the proof of Theorem 3.6 to get the following Theorem about the Baer invariant of a verbal wreath product.

Theorem 4.5 By the notation above, there exists the following epimorphism:

$$\mathcal{N}_cM(A\,Wr_V B) \longrightarrow\!\!\!\!\!\rightarrow \mathcal{N}_cM(B) \oplus \frac{T_V \cap \gamma_{c+1}(K)}{[T_V\ {}_c, K]}\ .$$

Also, in finite case, $\mathcal{N}_cM(A\,Wr_V B)$ has a subgroup isomorphic to

$$\mathcal{N}_cM(B) \oplus \frac{T_V \cap \gamma_{c+1}(K)}{[T_V\ {}_c, K]}\ .$$

Now, by putting a condition on B, and a similar proof to that of Theorem 3.7, we can present a formula for a verbal wreath product which is somehow better than that of Theorem 4.2 .

Theorem 4.6 By the above notation, let G be a $\mathcal{V}$-verbal wreath product of an arbitrary group A by a cyclic group B, where $\mathcal{V}$ is any variety of groups. Then

$$\mathcal{N}_cM(A\,Wr_V B) \cong \mathcal{N}_cM(B) \oplus \frac{T_V \cap \gamma_{c+1}(K)}{[T_V\ {}_c, K]} = \frac{T_V \cap \gamma_{c+1}(K)}{[T_V\ {}_c, K]}\ .$$

Note 4.7 Note that W. Haebich in [6] prove that

$$M(A\,Wr_V B) \cong M(B) \oplus \frac{T_V \cap K'}{[T_V, K]}\ ,$$

where A and B are two arbitrary groups.

5. The Baer Invariant of a Free Wreath Product

In this final section we try to find a structure for the Baer invariant of a free wreath product with respect to the variety of nilpotent groups. We recall that a free wreath product is in fact a $\mathcal{V}$-verbal wreath product where $\mathcal{V}$ is the variety of all groups, i.e. $V = \{1\}$. Then, by the previous notation, Lemma 2.7, we have $V(H) = 1$ and so $R_V = 1$ and $S_V = [R_2, F_1]^F R_1^F$. Now, by the definition of the natural epimorphism $\delta : F_1 * F_2 \longrightarrow F_1 * B$, $\delta(R_2) = 1$ and hence $T_V = \delta(S_V) = R_1^K = R_1[R_1, K]^K$.

Lemma 5.1 By the above notation, we have
(i) $[T_V,\ _cK] = [R_1,\ _cF_1]D_c$,
where $D_c = \prod_{\substack{G_i\in\{F_1,B\}\\ \exists j\ G_j=B}}[R_1, G_1, \ldots, G_c]^K$;
(ii) $T_V \cap \gamma_{c+1}(K) = (R_1 \cap \gamma_{c+1}(F_1))E_c$,
where $E_c = [R_1, K]^K \cap \gamma_{c+1}(K) \cap [F_1, B]$.

Proof.
(i) $[T_V,\ _cK] = [R_1,\ _cK]^K = [R_1,\ _K]^K = [R_1,\ _cF_1]D_c$ *(by commutator calculus and Lemma* 2.11).
(ii) $T_V \cap \gamma_{c+1}(K) = R_1^K \cap \gamma_{c+1}(K) = R_1[R_1, K]^K \cap \gamma_{c+1}(K)$
$\subseteq (R_1 \cap \gamma_{c+1}(F_1))([R_1, K]^K \cap \gamma_{c+1}(K))$ *(by Lemmas* 2.4 *and* 2.5)

$$= (R_1 \cap \gamma_{c+1}(F_1))E_c \ .$$

The reverse inclusion can be seen easily. □

Lemma 5.2 By the above notation, we have the following isomorphism:

$$\frac{T_V \cap \gamma_{c+1}(K)}{[T_V,\ _cK]} \cong \mathcal{N}_cM(A) \oplus \frac{E_c}{D_c} \ .$$

Proof. Put

$$\varphi : K \overset{nat}{\twoheadrightarrow} \frac{K}{[T_V,\ _cK]} \quad and$$

$$K = F_1 * B \overset{\xi=nat}{\twoheadrightarrow} \frac{F_1}{[R_1,\ _cF_1]} * B \overset{\rho}{\twoheadrightarrow} \frac{\xi(K)}{\xi(D_c)} \ .$$

Clearly $ker(\rho\xi) = [R_1,\ _cF_1]D_c = [T_V,\ _cK] = ker\varphi$, so $\varphi = \rho\xi$. Now similar to the proof of Theorem 3.9 in [6] we can show that

$$\varphi(T_V \cap \gamma_{c+1}(K)) = \varphi(R_1 \cap \gamma_{c+1}(F_1)) \oplus \varphi(E_c)$$

and

$$\varphi(R_1 \cap \gamma_{c+1}(F_1)) \cong \xi(R_1 \cap \gamma_{c+1}(F_1)) \cong \mathcal{N}_cM(A) \ .$$

Also

$$\varphi(E_c) = \frac{E_c([R_1,\ _cF_1]D_c)}{[R_1,\ _cF_1]D_c} \cong \frac{E_c}{[R_1,\ _cF_1]D_c \cap E_c} = \frac{E_c}{([R_1,\ _cF_1] \cap E_c)D_c} \ .$$

Since $[R_1, {}_cF_1] \cap E_c = 1$, so $\varphi(E_c) \cong E_c/D_c$. Hence the result holds. □

Now, we are in a position to find a structure for the c-nilpotent multiplier of a free wreath product.

Theorem 5.3 Let G be the free wreath product of A by B i.e. $G = AWr_*B$. Then there exists an epimorphism as follows:

$$\mathcal{N}_cM(AWr_*B) \longrightarrow\!\!\!\!\!\succ \mathcal{N}_cM(A) \oplus \mathcal{N}_cM(B) \oplus \frac{E_c}{D_c} \ .$$

Also, in finite case, the above structure is isomorphic to a subgroup of $\mathcal{N}_cM(AWr_*B)$.

Proof. By Theorem 4.5 and Lemma 5.2 the result holds. □

Theorem 5.4 Suppose A and B are two groups such that B is cyclic. Then

$$\mathcal{N}_cM(AWr_*B) \cong \mathcal{N}_cM(A) \oplus \mathcal{N}_cM(B) \oplus \frac{E_c}{D_c} \cong \mathcal{N}_cM(A) \oplus \frac{E_c}{D_c} \ .$$

Proof. By Theorem 4.6, Lemma 5.3 the result holds. □

References

[1] R. Baer, Representations of groups as quotient groups, I,II,III, *Trans. Amer. Math. Soc.* **58** (1945), 295-419.

[2] N. Blackburn, Some homology groups of wreath product, *Illinois J. Math.* **16** (1972), 116-129.

[3] J. Burns and G. Ellis, On the nilpotent multipliers of a group, *Math. Z.* **226** (1997), 405-428.

[4] G. Ellis, On groups with a finite nilpotent upper central quotient, *Arch. Math.* **70** (1998), 89-96.

[5] A. Fröhlich, Baer invariants of algebras, *Trans. Amer. Math. Soc.* **109** (1962), 221-244.

[6] W. Haebich, The multiplicator of a splitting extension, *Journal of Algebra* **44** (1977), 420-433.

[7] H. Hopf, Fundamental gruppe und zweiet bettische gruppe, *Comment. Math. Helvetici* **14** (1942), 257-309.

[8] G. Karpilovsky, "The Schur Multiplier," *London Math. Soc. Monographs* (N.S.) Vol. 2, Oxford Univ. Press, London, 1987.

[9] C.R. Leedham-Green and S. McKay, Baer-invariant, isologism, varietal laws and homology, *Acta Math.* **137** (1976), 99-150.

[10] B. Mashayekhy, The Baer-invariant of a semidirect product, *Indag. Math.* (N.S.) **8**(**4**) (1997), 529-535.

[11] B. Mashayekhy, Some notes on the Baer-invariant of a nilpotent product of groups, *Journal of Algebra* **235**, No. 1 (2001), 15-26.

[12] B. Mashayekhy and M.R.R. Moghaddam, Higher Schur multiplicator of a finite abelian group, *Algebra Colloq.* **4**, No. 3 (1997), 317-322.

[13] M.R.R. Moghaddam, The Baer-invariant of a direct product, *Arch. Math.* **33** (1979), 504-511.

[14] H. Neumann, "Varieties of Groups," Springer-Verlag, New York/Berlin, 1967.

[15] E.W. Read, On the Schur multiplier of a wreath product, *Illinois J. Math.* **20** (1976), 456-466.

[16] I. Schur, Untersuchungen über die darstellung der endlichen gruppen durch gebrochene lineare substitutionen, *J. Reine Angew. Math.* **132** (1907), 85-137.

[17] K.I. Tahara, On the second cohomology groups of semidirect products, *Math. Z.* **129** (1972), 365-379.

[18] J. Wiegold, The multiplicator of a direct product, *Quart. J. Math. Oxford* **2**, No. 22 (1971), 103-105.

In: Mathematics, Game Theory and Algebra Compendium ISBN 978-1-60456-659-8
Editor: Jacob H. Mathias, pp. 463-475

Chapter 39

A CHARACTERIZATION OF COMMUTATIVE CLEAN RINGS*

***Warren Wm. McGovern*†**
Department of Mathematics and Statistics,
Bowling Green State University,
Bowling Green, OH 43403, USA

Abstract

A commutative ring A is said to be *clean* if every element of A can be written as a sum of a unit and an idempotent. This definition dates back to 1977 where it was introduced by W. K. Nicholson [7]. In 2002, V. P. Camillo and D. D. Anderson [1] investigated commutative clean rings and obtained several important results. In [4] Han and Nicholson show that if A is a semiperfect ring, then $A[Z_2]$ is a clean ring. In this paper we generalize this argument (for commutative rings) and show that $A[Z_2]$ is clean if and only A is clean. We also show that if the group ring $A[G]$ is a commutative clean ring, then G must be a torsion group. Our investigations lead us to introduce the class of 2-clean rings.

Key Words : commutative clean ring, commutative group ring

1. On 2-Clean Rings

We start with the following definition:

The element $a \in A$ is said to be *clean* if there exists an idempotent $e \in A$ such that $a - e$ is invertible. If every element of A is clean, then A is said to be a *clean ring*.

---

*Communicated by Daneil Anderson

†E-mail address: warrenb@bgnet.bgsu.edu

The set of units and idempotents of A shall be denoted by $\mathcal{U}(A)$ and $Id(A)$, respectively. As usual, $Max(A)$ denotes the set of maximal ideals of A, and it is equipped with the hull-kernel topology. This means that the collection of sets of the form $V(a) = \{M \in Max(A) : a \in M\}$ for $a \in A$ form a base for the closed sets. The complement of $V(a)$ is denoted by $U(a)$.

It is well known that von Neumann regular rings, and more generally zero-dimensional rings, are clean. Also, local rings are clean, and, in fact, are precisely the indecomposable clean rings. Thus, if A is an integral domain, then A is clean if and only if A is local. Some fundamental facts regarding clean rings are that every homomorphic image of a clean ring is a clean ring, and that a ring A is clean if and only if $A/n(A)$ is clean where $n(A)$ denotes the nilradical of A. Also, a direct product of rings is clean if and only if each factor is clean. Most of these facts can be found in [4] or [1]. We urge the interested reader to check there for more information and selected proofs.

2. All Rings Are Commutative and with Identity

In [6] a list of several characterizations of commutative clean rings is given. Included in the list is one given by Johnstone [5] which we presently state.

Theorem 2.1 (Johnstone). *A is a clean ring if and only if $Max(A)$ is zero-dimensional and every prime ideal is contained in a unique maximal ideal.*

A topological space is said to be *zero-dimensional* if it has a base of clopen sets. It is known that if $e \in Id(A)$, then $U(e)$ is a clopen subset of $Max(A)$, but the reverse is not true in general. A clopen subset of $Max(A)$ of the form $U(e)$ for some $e \in Id(A)$ will be called an *idempotent clopen*. In [6], it is shown that $Max(A)$ is zero-dimensional and every clopen is an idempotent clopen precisely when A is a clean ring.

Theorem 2.2. *The following are equivalent for a ring A:*

(i) A is a clean ring.

(ii) For each $a \in A$ there exists an idempotent e such that $V(a) \subseteq U(e)$ and $V(a-1) \subseteq V(e)$.

(iii) The collection of idempotent clopen subsets of $Max(A)$ is base for the topology on $Max(A)$.

(iv) $Max(A)$ is zero-dimensional and every clopen subset is of the form $U(e)$ for some $e \in Id(A)$.

Here is another characterization of a clean ring.

Proposition 2.3. *A is clean if and only if for every $a \in A$, there is an idempotent $e \in A$ such that both $a+e$ and $a-e$ are invertible.*

Proof. The sufficiency is clear. As for the necessity, let $a \in A$. By Theorem 2.2, choose an idempotent $e \in A$ such that $V(a^2) \subseteq U(e)$ and $V(a^2-1) \subseteq V(e)$. Let M be any maximal ideal of A. If $a \in M$, then so is a^2 and hence $e \notin M$. It follows that $a+e, a-e \notin M$. If $a, e \notin M$, then neither $a+1$ nor $a-1$ belong to M. Thus, $a+e+M = a+1+M \neq M$ and $a-e+M = a-1+M \neq M$, whence $a+e, a-e$ are invertible. □

We call a commutative ring A 2-clean if for every pair of elements $a, b \in A$ there exists an idempotent $e \in A$ such that $a+b-e$ and $a-b-e$ are both invertible. Observe that letting $b=0$ we obtain that a 2-clean ring is clean. Also, if $\operatorname{char} A = 2$, then A is clean if and only if A is 2-clean.

The proof of our next proposition is straightforward.

Proposition 2.4. *1) A homomorphic image of a 2-clean ring is 2-clean.*

2) A direct product $A = \Pi_{i \in I} A_i$ is 2-clean if and only if each factor A_i is 2-clean.

Proposition 2.5. *The following are equivalent for a ring A.*

(i) A is 2-clean.

(ii) $A/n(A)$ is 2-clean.

(iii) A is clean and $A/\mathfrak{J}(A)$ is 2-clean.

(iv) $A/\mathfrak{J}(A)$ is 2-clean and idempotents lift modulo the Jacobson radical.

Proof. The proof is exactly the same as the one used to show that a ring A is clean if and only if $A/n(A)$ is clean. We leave the verification to the interested reader. □

Our next result gives an easier way of checking whether a clean ring is 2-clean.

Proposition 2.6. *A is 2-clean if and only if A is clean and $2 \in \mathfrak{J}(A)$.*

Proof. Necessity. If A is 2-clean, then it is clean. Let $a \in A$ be an arbitrary element. Using the 2-clean condition on the pair a, a there is an idempotent $e \in A$ such that $2a - e$ and $-e$ (and hence e) are invertible. The only idempotent that is invertible is $e = 1$. Therefore, $2a - 1$ is invertible, whence $2 \in \mathfrak{J}(A)$.

Sufficiency. Let $a, b \in A$. Choose an idempotent $e \in A$ such that $a+b-e$ is invertible. We claim that $a-b-e$ is also invertible. Let M be a maximal ideal of A. By hypothesis, $2 \in M$ so that $a-b-e+M = a+b-e+M \neq M$. Therefore, $a-b-e \notin M$ for any maximal ideal of A. It follows that both $a-b-e$ and $a+b-e$ are invertible. □

Theorem 2.7. *Let A be a zero-dimensional ring. Then the following are equivalent:*

(i) A is 2-clean.

(ii) $char\,A = 2^k$ for some natural k.

(iii) 2 is nilpotent.

(iv) For each maximal ideal M of A, $char\,A/M = 2$.

Proposition 2.8. *Suppose A is a local ring. Then A is 2-clean if and only if $char\,A/M = 2$.*

Proof. The necessity is clear, so suppose that $char\,A/M = 2$. This means that $a+b+M = a-b+M$ for all $a, b \in A$. Therefore, $a+b \in M$ if and only if $a-b \in M$. Thus, if both $a+b, a-b \in M$, then $a+b-1, a-b-1 \notin M$ and hence both are invertible. Otherwise, $a+b, a-b \notin M$ means that they are both invertible. □

3. Commutative Clean Group Rings

In this section we address the question of when a group ring is clean. Some work on this question was done in [4] where they showed that if A is a boolean ring and G is a torsion group, then $A[G]$ is a clean ring. We will only consider the case where the group in question is abelian. When a group ring is zero-dimensional or local has been characterized, and therefore we have some sufficient conditions for a group ring to be a clean ring. We list [3] as our main reference on commutative group rings, though we warn the casual reader that most of the information in [3] is done for semi-group rings. **All groups are abelian**.

Let A be a ring and $(G, +)$ be a group. The group ring is the set of formal sums of the form $\Sigma_{i=1}^{n} a_i X^{g_i}$ under pointwise addition and multiplication defined by $X^g X^h = X^{g+h}$ on monomials and extending this in the obvious way.

Theorem 3.1. *$A[G]$ is zero-dimensional if and only if A is zero-dimensional and G is a torsion group.*

For commutative rings and abelian groups we obtain the following corollary.

Corollary 3.2. *If A is a boolean ring and G is a torsion group, then $A[G]$ is a clean ring.*

Theorem 3.3. *$A[G]$ is local if and only if A is local, $char\,A/M = p$, and G is a p-group. In particular, if $A[G]$ is local, then G is torsion.*

Our aim is to show that for $A[G]$ to be a clean ring it is necessary that A be clean and G be torsion. The first part is clear as A is a homomorphic image of $A[G]$. To obtain the second part we need a series of lemmas. The first is well known.

Lemma 3.4. *$A[G]$ is an integral domain if and only if A is an integral domain and G is torsion free.*

Lemma 3.5. *For any ring A and group G, $A[G]$ is clean if and only if $A/n(A)[G]$ is clean.*

Proof. The necessity is obvious since $A/n(A)[G]$ is a homomorphic image of $A[G]$. As for the sufficiency it is known that $n(A)[G] \leq n(A[G])$, which means that $A[G]/n(A[G])$ is a homomorphic image of $A/n(A)[G]$ and therefore $A[G]$ is clean. □

Proposition 3.6. *If $A[G]$ is a clean ring, then A is clean and G is a torsion group.*

Proof. Clearly, A is a clean ring. Let $t(G)$ denote the torsion subgroup of G. Let P be a prime ideal of A. Then $(A/P)[G/t(G)]$ is a homomorphic image of $A[G]$ and hence is clean. Since $G/t(G)$ is a torsion-free group and A/P is an integral domain, it follows by Lemma 3.4 that $(A/P)[G/t(G)]$ is a clean integral domain and hence is local. By Theorem 3.3 it follows that $G/t(G)$ is torsion, whence $G = t(G)$. □

Remark 1. *The converse of the previous proposition is certainly not true. In [4] it is shown that if A is the localization of Z at the prime $7Z$ (A is clean because it is local) and $G = Z_3$ (clearly torsion), then $A[G]$ is not clean. In general, discovering when $A[G]$ is clean has eluded us except in certain cases. In the next section we consider the specific case when $G = Z_2$.*

Proposition 3.7. *Suppose A is zero-dimensional. Then the following are equivalent:*

(i) *$A[G]$ is clean.*

(ii) *$A[G]$ is zero-dimensional.*

(iii) *G is a torsion group.*

Proof. This is in fact a corollary to Proposition 3.6 and Theorem 3.1. □

Corollary 3.8. *Let F be a field. Then $F[G]$ is clean if and only if G is torsion.*

4. When $G = Z_2$

In this section we investigate the cleanliness of the group ring $A[Z_2]$.

Let e be an idempotent of A. The element $eX^0 \in A[G]$ is an idempotent of $A[G]$. Any idempotent of $A[G]$ of this form is said to be an idempotent in A. If every idempotent of $A[G]$ is in A, then we say that the idempotents of $A[G]$ are in A.

Lemma 4.1. *The element $rX^0+sX^1 \in A[Z_2]$ is invertible if and only if $r+s, r-s \in \mathcal{U}(A)$.*

Proof. Suppose $rX^0 + sX^1 \in \mathcal{U}(A[Z_2])$. Then there are $f, g \in A$ such that

$$(rX^0 + sX^1)(fX^0 + gX^1) = 1.$$

This means that $rf + sg = 1$ and $rg + sf = 0$. Adding these two equations gives us that $(r+s)(f+g) = 1$. Subtracting yields $(r-s)(f-g) = 1$. Therefore, $r+s, r-s \in \mathcal{U}(A)$.

Conversely, if $r+s, r-s \in \mathcal{U}(A)$, then so is $r^2 - s^2$. Let $f = \frac{r}{r^2-s^2}$ and $g = \frac{-s}{r^2-s^2}$. A quick check shows that $rf + sg = 1$ and $rg + sf = 0$. It follows that $rX^0 + sX^1$ is an invertible element of $A[Z_2]$. □

Lemma 4.2. *The element $cX^0 + dX^1 \in A[Z_2]$ is an idempotent if and only if $c^2 + d^2 = c$ and $2cd = d$. In this case, $c + d, c - d \in Id(A)$.*

Proof. The first statement follows from the equation $cX^0 + dX^1 = (cX^0 + dX^1)^2 = (c^2 + d^2)X^0 + 2cdX^1$. The second is obtained by adding (and subtracting) the equations. □

Proposition 4.3. *The following are equivalent for the ring A:*

(i) A is clean.

(ii) For every $a \in A$, the element $aX^0 \in A[Z_2]$ can be written as the sum of a unit and an idempotent in A.

(iii) For every $a \in A$, the element $aX^1 \in A[Z_2]$ can be written as the sum of a unit and an idempotent in A.

Proof. (i) ⇒ (ii) If A is clean, then for each $a \in A$ let $u \in \mathcal{U}(A)$ and $e \in Id(A)$ such that $a = u + e$. Then

$$aX^0 = uX^0 + eX^0$$

is a representation of aX^0 as a sum of a unit of and an idempotent in A.

(ii) ⇒ (i) Let $a \in A$. If

$$aX^0 = (rX^0 + sX^1) + eX^0$$

is a representation of aX^0 as a sum of a unit and an idempotent in A, then $s = 0$ and thus $r = r + s \in \mathcal{U}(A)$. Therefore, A is a clean ring.

(iii) ⇒ (i) Let $a \in A$. Suppose that aX^1 is the sum of a unit and an idempotent in A, say

$$aX^1 = (rX^0 + aX^1) + eX^0.$$

for some $e \in Id(A)$. By Lemma 4.1 $r + a, a - r \in \mathcal{U}(A)$. Since $r = -e$ it follows that $a - e$ and $a + e$ are both invertible and so by Proposition 2.3 A is clean.

(i) ⇒ (iii) Let $a \in A$. By Proposition 2.3 there exists an $e \in Id(A)$ such that $a - e, a + e$ are invertible. Set $r = -e$ so that

$$aX^1 = (rX^0 + aX^1) + eX^0.$$

Since $r + a = a - e$ and $r - a = -(a + e)$ are invertible it follows that aX^1 is the sum of a unit and an idempotent in A. □

Proposition 4.4. *For every $a \in A$, the element $aX^0 + aX^1 \in A[Z_2]$ can be written as the sum of a unit and an idempotent in A if and only if $2 \in \mathfrak{J}(A)$. In this case every idempotent of $A[Z_2]$ belongs to A.*

Proof. Suppose $aX^0 + aX^1 \in A[Z_2]$ can be written as the sum of a unit and an idempotent in A for every $a \in A$, say

$$ax^0 + aX^1 = (rX^0 + sX^1) + eX^0.$$

Since $s = a$ it follows that $r + a, r - a \in \mathcal{U}(A)$ by Lemma 4.1. Notice that $e = a - r$ which is invertible and hence $e = 1$ and $r = a - 1$. Thus $r + a = 2a - 1 \in \mathcal{U}(A)$ for every $a \in A$, whence $2 \in \mathfrak{J}(A)$.

Conversely, let $a \in A$ and $2 \in \mathfrak{J}(A)$. Since

$$aX^0 + aX^1 = ((a-1)X^0 + aX^1) + X^0$$

and because $(a-1) + a = 2a - 1$ and $(a-1) - a = -1$ are both invertible, the result follows from Lemma 4.1.

Finally, suppose $2 \in \mathfrak{J}(A)$ and let $aX^0 + bX^1$ be idempotent. This means that $a^2 + b^2 = a$ and $2ab = b$. The latter equality implies that $(2a-1)b = 0$. By hypothesis, $2 \in \mathfrak{J}(A)$ and therefore $2a - 1$ is invertible. It follows that $b = 0$ and so the idempotents of $A[Z_2]$ belong to A. □

Proposition 4.5. *The following are equivalent.*

(a) For each $a \in A$, the element $aX^0 + aX^1$ is clean.

(b) For each $a \in A$, there exists an $e \in Id(A)$ and $t, u \in A$ such that $(2a-1)t = (1-e)$ and $2au = e$.

(c) For each $a \in A$, there exists an $e \in Id(A)$ such that $V(2a) \subseteq V(e)$ and $V(2a-1) \subseteq U(e)$.

Proof. (b)⇒(a). Let $a \in A$ and choose $e \in Id(A)$ and $t, u \in A$ satisfying (b). Let $f = 1 - e$ so that $ef = 0$ and $e + f = 1$. Observe that since both e and f are idempotent we may, without loss of generality, assume that $ue = u$, and $tf = t$. Let $x = aue$ and observe that $2x = e$ and $ex = x$. Let $y = (a-1)f$, $z = (a-x)e$, and $r = y + z$. Next let $d = -x$ and $c = d + 1 = 1 - x$.

Now,

$$c^2 + d^2 = d^2 + 2d + 1 + d^2 = 2d^2 + 2d + 1 = 2x^2 - 2x + 1 = x + f.$$

Notice that $2x = e = 1 - f$ so that $c = 1 - x = x + f$ and therefore $c^2 + d^2 = c$. Next,

$$2cd = 2c(-x) = -ce = -(1-x)e = ex - e = x - e = x - 2x = -x = d$$

and so $cX^0 + dX^1$ is an idempotent. Now,

$$\begin{aligned} r + c &= y + z + c \\ &= (a-1)f + (a-x)e + 1 - x \\ &= af - f + ae - ex + 1 - x \\ &= a - f - ex + 1 - x \\ &= a + e - 2x \\ &= a \end{aligned}$$

so that

$$aX^0 + aX^1 = (rX^0 + (r+1)X^1) + (cX^0 + dX^1).$$

What is left to be shown is that $(rX^0 + (r+1)X^1)$ is invertible. To that end, first notice that

$$\begin{aligned} 2r+1 &= 2y + 2z + 1 \\ &= 2(a-1)f + 2(a-x)e + e + f \\ &= (2a-1)f + 2ae. \end{aligned}$$

Next,

$$\begin{aligned} (2r+1)(t+u) &= ((2a-1)f + 2ae)(t+u) \\ &= (2a-1)t + 2aue \\ &= f + e \\ &= 1. \end{aligned}$$

Therefore $rX^0 + (r+1)X^1$ is invertible by Lemma 4.1, whence $aX^0 + aX^1$ is clean.

(a)⇒(c). Suppose $aX^0 + aX^1$ is clean and write it as a sum of a unit and idempotent, say

$$aX^0 + aX^1 = (rX^0 + sX^1) + (cX^0 + dX^1).$$

Note that $a = r + c = s + d$. Since $(rX^0 + sX^1)$ is invertible, $r + s \in \mathcal{U}(A)$, and since $(cX^0 + dX^1)$ is an idempotent, it follows that $c + d \in Id(A)$. Now $2a = a + a = r + c + s + d = (r+s) + (c+d)$ and so $2a$ is a clean element. It follows that there exists an idempotent e such that $V(2a) \subseteq V(e)$ and $V(2a-1) \subseteq U(e)$.

The equivalence of (b) and (c) is straighforward. □

Theorem 4.6. *For a ring A, the following are equivalent:*

(i) A is 2-clean.

(ii) $A[Z_2]$ is clean and the idempotents of $A[Z_2]$ belong to A.

(iii) $A[Z_2]$ is clean and $2 \in \mathfrak{J}(A)$.

(iv) Every element of $A[Z_2]$ can be written as the sum of a unit and an idempotent in A.

(v) A is clean and for every $a, b \in A$, $V(a+b) \cap V(a-b-1) = \emptyset$.

Proof. (i)⇒ (ii) If A is 2-clean, then by Proposition 2.6 and Proposition 4.4 every idempotent of $A[Z_2]$ is in A. Let $aX^0 + bX^1 \in A[G]$. By the hypothesis there is an idempotent e such that $a + b - e$ and $a - b - e$ are both invertible. Set $r = a - e$ so that $r + b$ and $r - b$ are invertible. It follows that

$$aX^0 + bX^1 = (rX^0 + bX^1) + eX^0$$

is a clean representation,whence $A[Z_2]$ is a clean ring.

(ii) $\Rightarrow$ (iii) Suppose that $A[Z_2]$ is clean and that idempotent elements are in A. Let $a \in A$ and write

$$aX^0 + aX^1 = (rX^0 + aX^1) + eX^0$$

where $e^2 = e$, $r + e = a$, and $rX^0 + aX^0$ is invertible. As previously noted it follows that $r + a, r - a$ are both invertible. Since $r - a = -e$ is invertible it follows that $e = 1$. Therefore, the invertibility of $2a - 1 = r + a$ and arbitrariness of a imply that $2 \in \mathfrak{J}(A)$.

(iii) $\Rightarrow$ (ii) This is patent.

(ii) $\Rightarrow$ (i) Let $a, b \in A$. Then the element $aX^0 + bX^1$ is clean and thus we can write it as

$$aX^0 + bX^1 = (rX^0 + bX^1) + eX^0$$

where $e \in Id(A)$ and $(rX^0 + bX^1) \in \mathcal{U}(A[Z_2])$. Observe that $r + e = a$ and $r + b, r - b \in \mathcal{U}(A)$. Since

$$a + b - e = r + b \text{ and } a - b - e = r - b$$

we are done.

(ii) $\Rightarrow$ (iv). This is clear.

(iv) $\Rightarrow$ (i). Let $a, b \in A$. By hypothesis there is a unit $rX^0 + sX^1 \in A[Z_2]$ and an idempotent $e \in A$ such that

$$aX^0 + bX^1 = (rX^0 + sX^1) + eX^0.$$

It follows that $s = b$ and that $a - e = r$. It is straighforward from here to show that A is 2-clean.

(i) $\Rightarrow$ (v). If A is 2-clean, then $V(a - b - 1) = V(a + b - 1)$ which is always disjoint from $V(a + b)$.

(v) $\Rightarrow$ (1). For any $a \in A$ we have $\emptyset = V(a - a) \cap V(a + a - 1) = V(0) \cap V(2a - 1) = V(2a - 1)$ and therefore $2a - 1$ is invertible. It follows that $2 \in \mathfrak{J}(A)$. □

Corollary 4.7. *Let $char A = 2$. Then $A[Z_2]$ is clean if and only if A is clean.*

The next result was shown in [4]. We include a sketch of proof for completeness sake.

Proposition 4.8. *Let $2 \in \mathcal{U}(A)$. Then $A[Z_2] \cong A \times A$. In this case A is clean if and only if $A[G]$ is clean.*

Proof. Define $\psi : A[G] \to A \times A$ by $\psi(aX^0 + bX^1) = (a + b, a - b)$. Then ψ is a ring homomorphism. When 2 is a regular element this homomorphism is injective, and when 2 is invertible the map is a surjection. □

Remark 2. If $V(2)$ is a clopen subset of $Max(A)$, then there is a decomposition of $A = A_1 \oplus A_2$ where the rings A_1 and A_2 satisfy that $2 \in \mathcal{U}(A_1)$ and $2 \in \mathfrak{J}(A_2)$. It follows that $A[Z_2]$ is a clean ring, but not every element of $A[Z_2]$ is the sum of a unit and an idemptotent in A.

Lemma 4.9. *Suppose* $cX^0 + dX^1 \in Id(A[Z_2])$. *Then both* $V(c)$, $V(d) \setminus V(c)$ *are idempotent clopen subsets of* $Max(A)$. *Moreover,* $V(d)$ *is an idempotent clopen subset of* $Max(A)$.

Proof. The second statement clearly follows from the first. By hypothesis we know that $c^2 + d^2 = c$, $2cd = d$, and $c + d, c - d \in Id(A)$. The second equation implies that $V(2), V(c) \subseteq V(d)$. In particular, we obtain that $V(c) \subseteq V(c+d) \cap V(c-d)$. Now, if $M \in V(c+d) \cap V(c-d)$, then both $2c \in M$ and $c^2 - d^2 \in M$. If $2 \in M$, then the residue field has characteristic 2, and

$$c + M = c^2 + d^2 + M = c^2 - d^2 + M = M$$

so that $c \in M$. It follows that $V(c) = V(c+d) \cap V(c-d)$, and hence $V(c)$ is an idempotent clopen subset of $Max(A)$.

Next we show that $V(c+d) \cup V(c-d) = U(d) \cup V(c)$ from which it will follow that $V(d) \setminus V(c)$ is an idempotent clopen subset of $Max(A)$. Suppose that $d \notin M$. Then $2c + M = 1 + M$. From this and $c^2 + d^2 = c$ we conclude that $2d + M = 1 + M$ or $2d + M = -1 + M$. (Observe that in any field of characteristic different from 2 we have that $1/2 - (1/2)^2 = (1/2)^2$). Therefore, $c + M = d + M$ or $c + M = -d + M$ in which case $M \in V(c-d)$ or $M \in V(c+d)$, whence $U(d) \cup V(c) \subseteq V(c+d) \cap V(c-d)$. The reverse is the easy case. □

Theorem 4.10. *For a ring* A, $A[Z_2]$ *is clean if and only if for each* $a, b \in A$ *there exist idempotents* $e_1, e_2, e_3, e_4 \in Id(A)$ *such that* $U(e_i) \cap U(e_j) = \emptyset$, $i \neq j$,

$$\begin{aligned} 1 &= e_1 + e_2 + e_3 + e_4 && (1) \\ V(2) &\subseteq U(e_3) \cup U(e_4) && (2) \\ V(a+b) &\subseteq U(e_3) \cup U(e_1) && (3) \\ V(a+b-1) &\subseteq U(e_4) \cup U(e_2) && (4) \\ V(a-b) &\subseteq U(e_3) \cup U(e_2) && (5) \\ V(a-b-1) &\subseteq U(e_4) \cup U(e_1) && (6) \end{aligned}$$

Proof. Necessity. Suppose $A[Z_2]$ is clean and let $a, b \in A$. Then we can write

$$aX^0 + bX^1 = (rX^0 + sX^1) + (cX^0 + dX^1)$$

where $rX^0 + sX^1 \in U(A[Z_2])$ and $(cX^0 + dX^1) \in Id(A[Z_2])$. This means that $c^2 + d^2 = c$, $2cd = d$, $c + d, c - d \in Id(A)$ and both $r + s$ and $r - s$ are invertible. By Lemma 4.9 it follows that there are idempotents e_1, e_2, e_3, e_4 such that $U(e_4) = V(c), U(e_3) = V(d) \setminus V(c), U(e_1) = V(c-d) \setminus V(c)$ and $U(e_2) = V(c+d) \setminus V(c)$. Observe that $U(e_i) \cap U(e_j) = \emptyset$ for each $i \neq j$ and that $\bigcup_{1 \leq i \leq 4} U(e_i) = Max(A)$. Without loss of generality we can assume that $e_i e_j = 0$ and thus $e_1 + e_2 + e_3 + e_4$ is an invertible idempotent and hence (1) is satisfied. That (2) is satisfied follows from the fact that $V(2) \subseteq V(d) = U(e_3) \cup U(e_4)$.

Next, notice that $a+b-(c+d) = r+s$ and so (3) $V(a+b) \subseteq U(c+d) = U(e_1) \cup U(e_3)$. Similarly, $a - b - (c - d) = r - s$ and so (5) $V(a - b) \subseteq U(c - d) = U(e_2) \cup U(e_3)$. Let $M \in V(a+b-1)$. If $M \in U(e_1) = V(c-d) \setminus V(c)$, then $c+d \notin M$ (otherwise $2c \in M$ and $2 \notin M$ would imply $c \in M$.) Since $c + d$ is idempotent, $c + d + M = 1 + M$. Hence

$$M = a + b - 1 + M = a + b - (c + d) + M$$

contradicting that $r + s$ is invertible. Similarly, if $M \in U(e_3) = V(d) \setminus V(c)$ then $d + M = M$ and so $c + M = 1 + M$. Therefore

$$M = a + b - 1 + M = a + b - (c + d) + M.$$

Therefore, (4) is satisfied. A similar argument yields (6).

Sufficiency. Let $aX^0 + bX^1 \in A[Z_2]$ and suppose that e_1, e_2, e_3, e_4 are idempotents satisfying the hypothesis. Observe that $e_i e_j = 0$ for all $i \neq j$. We shall produce elements $r, s, c, d \in A$ such that $c^2 + d^2 = c$, $2cd = d$, and both $r + s = a + b - (c + d)$ and $r - s = a - b - (c - d)$ are invertible. Thus, it follows that

$$aX^0 + bX^1 = (rX^0 + sX^1) + (cX^0 + dX^1)$$

is a clean representation. To that end observe that (1) implies there is an $x \in (e_1 + e_2)A$ such that $2x = e_1 + e_2$. We further suppose that $x(e_1 + e_2) = x$. Let $y_1 = xe_1$ and $y_2 = -xe_2$. Set $c = x + e_3$ and $d = y_1 + y_2$. Observe that

$$V(c) = U(e_4) \text{ and } V(d) = U(e_4) \cup U(e_3).$$

Since

$$\begin{aligned} c^2 + d^2 &= (x + e_3)^2 + (y_1 + y_2)^2 \\ &= x^2 + e_3 + x^2 e_1 + x^2 e_2 \\ &= 2x^2 + e_3 \\ &= x + e_3 \\ &= c \end{aligned}$$

and

$$\begin{aligned} 2cd &= 2(x + e_3)(xe_1 - xe_2) \\ &= 2(x^2 e_1 - x^2 e_2) \\ &= 2x^2(e_1 - e_2) \\ &= x(e_1 + e_2)(e_1 - e_2) \\ &= x(e_1 - e_2) \\ &= d \end{aligned}$$

it follows that $cX^0 + dX^1 \in Id(A[Z_2])$. By construction $c + d = 2xe_1 + e_3$ so that

$$V(c + d) = U(e_2) \cup U(e_4).$$

Let $r = a - c$ and $s = b - d$. Then $r + s = a + b - (c + d)$ and $r - s = a + b - (c - d)$.

If $r + s$ is not invertible, then it lies in some maximal ideal M. If $M \in V(a + b)$, then $M \in V(c + d)$ contradicting (3). Therefore, $M \in U(a + b)$ and hence $M \in U(c + d)$. So $a + b + M = c + d + M = 1 + M$, i.e. $M \in V(a + b - 1)$. By (4) $M \in U(e_2)$ (since $M \notin V(c) = U(e_4)$). Our contradiction is

$$1 + M = c + d + M = xe_1 + xe_2 + e_3 + xe_1 - xe_2 + M = 2xe_1 + xe_3 + M = M$$

since $M \in V(e_1) \cap V(e_3)$. Therefore, $r + s$ is invertible. A similar argument shows that $r - s$ is invertible. □

Theorem 4.11. *$A[Z_2]$ is clean if and only if A is clean.*

Proof. As we have pointed so many times already, if $A[Z_2]$ is clean, then so is A. So suppose that A is clean and let $a, b \in A$. Choose idempotents $f_1, f_2 \in A$ such that $V(a + b) \subseteq U(f_1), V(a + b - 1) \subseteq V(f_1)$ and $V(a - b) \subseteq U(f_2), V(a - b - 1) \subseteq V(f_2)$. Let e_1, e_2, e_3, e_4 be idempotents such that

$$U(e_1) = U(f_1) \cap V(f_2),$$
$$U(e_2) = U(f_2) \cap V(f_1),$$
$$U(e_3) = U(f_1) \cap U(f_2),$$
$$U(e_4) = Max(A) \setminus (U(e_1) \cup U(e_2) \cup U(e_3)).$$

Observe that $U(e_i) \cap U(e_j) = \emptyset$ whenever $i \neq j$. Without loss of generality $e_ie_j = 0$ whenever $i \neq j$. Therefore, all the conditions of Theorem 4.10 are satisfied except for possibly (2).

We claim that $V(2) \cap U(e_1)$ and $V(a+b) \cup V(a-b-1)$ are disjoint closed sets. To see this let $M \in V(2) \cap U(e_1)$. Suppose that $M \in V(a+b)$. Since $M \in V(2)$, $M \in V(a-b)$ and hence $M \in U(f_2)$, contradicting that $M \in U(e_1)$. If $M \in V(a - b - 1)$, then $M \in V(a + b - 1)$ which implies that $M \in V(f_1)$, again contradicting that $M \in U(e_1)$. Similarly, $V(2) \cap U(e_2)$ and $V(a - b) \cap V(a + b - 1)$ are disjoint closed sets.

Since A is clean it follows that the collection of idempotent clopen sets form a base for the topology. It follows that there is an idempotent clopen subset $K_1 \subseteq U(e_1)$ such that $V(2) \cap U(e_1) \subseteq K_1$ and $(V(a + b) \cup V(a - b - 1)) \cap K_1 = \emptyset$. Similarly, there is an idempotent clopen subset $K_2 \subseteq U(e_2)$ such that $V(2) \cap U(e_2) \subseteq K_2$ and $(V(a - b) \cup V(a + b - 1)) \cap K_2 = \emptyset$. By combining $U(e_4)$ together with K_1 and K_2 and shrinking down $U(e_1)$ and $U(e_2)$ in the appropriate manner, we obtain idempotent clopen subsets which satisfy conditions (1)-(6) of Theorem 4.10. Therefore, $A[Z_2]$ is clean. □

Corollary 4.12. *Suppose G is an elementary 2-group. Then $A[G]$ is clean if and only if A is clean.*

Proof. Since G is an elementary 2-group, for each $f \in A[G]$ there exists a finite subgroup H of G such that $f \in A[H]$ and H is a finite direct product of copies of Z_2. It follows that $A[H]$ is a finite product of copies of $A[Z_2]$ which is clean. Therefore, $f \in A[H]$ can be written as a s um of a unit and an idempotent in $A[H]$, and hence can be written as a sum of a unit and an idempotent in $A[G]$. □

References

[1] D.D. Anderson, V.P. Camillo, Commutative rings whose elements are a sum of a unit and an idempotent, *Comm. Alg.* **30** no. 7, 3327-3336 (2002).

[2] I. Connell, On the group ring, *Can. J. Math.* **15**, 650-685 (1963).

[3] R. Gilmer, *Commutative Semi-Group Rings*, Univ. Chicago Press, Chicago (1984).

[4] J. Han, W.K. Nicholson, Extensions of clean rings, *Comm. Alg.* **29** no. 6, 2589-2595 (2001).

[5] P.T. Johnstone, Stone Spaces, *Cambridge Stud. Adv. Math.,* **3**, Cambridge Univ. Press, Cambridge (1982).

[6] W.Wm. McGovern, *Neat rings*, preprint.

[7] W.K. Nicholson, Lifting idempotents and exchange rings, *Trans. A.M.S.*, **229**, 269-278 (1977).

Received: September 2004
revised: November 2004

In: Mathematics, Game Theory and Algebra Compendium
Editor: Jacob H. Mathias, pp. 477-491
ISBN: 978-1-60456-659-8

Chapter 40

ON ABSOLUTE MATRIX SUMMABILITY OF FOURIER SERIES AND ITS ALLIED SERIES

Madan L. Mittal
Mathematics Department
Indian Institute of Technology Roorkee
Roorkee 247 667 , Uttranchal India.
Dedicated to Prof. O.P. Varshney

1. Let Σu_n be a given infinite series with the sequence of partial sum $\{s_n\}$. Let $T \equiv (a_{n,k})$ be an infinite triangular matrix with real constants. The sequence-to-sequence transformation

(1) defines the T- transform of the sequence $\{s_n\}$. The series Σu_n is said to be T-summable to s, if $\lim_{n\to\infty} t_n = s$. It is said to be absolutely summable-T, or summable |T|, if the infinite series $\Sigma |t_n - t_{n-1}|$ is convergent.

$$t_n = \sum_{k=0}^{n} a_{n,k}\ s_k\ , \qquad n = 0, 1, 2, \ldots,$$

The T-transform reduces to the Nörlund transform N_p, generated by the sequence of coefficients $\{p_n\}$, if

$$a_{n,k} = p_{n-k} / P_n\ , 0 \leq k \leq n\ ,$$

$$= 0\ k > n\ ,$$

where $P_n = \sum_{r=0}^{n} p_r \neq 0.$

2. Let f(t) be a periodic function, with period 2π, and be Lebesgue integrable over (-π,π). We may assume, with out loss of generality, that the constant term in the Fourier series of f(t) is zero, so that

$$f(t) \sim \sum_{n=1}^{\infty} (a_n \cos nt + b_n \sin nt) \equiv \sum_{n=1}^{\infty} A_n(t), \tag{2}$$

is the Fourier series of f(t). Then the series obtained by formally differentiating (2), at t = x, is

$$\sum_{n=1}^{\infty} n(b_n \cos nx - a_n \sin nx) \equiv \sum_{n=1}^{\infty} nB_n(x). \tag{3}$$

We write

$\phi(t) = \frac{1}{2}[f(x+t) + f(x-t)],$ $\psi(t) = \frac{1}{2}[f(x+t) - f(x-t)],$

$\Phi_0(t) = \phi(t),$ $g(t) = t^{-1}\psi(t)$,

$\Phi_\alpha(t) = \frac{1}{\Gamma(\alpha)}\int_0^t (t-u)^{\alpha-1}\phi(u)\,du, \alpha > 0,$ $\alpha_n = \sum_1^n k \cos kt,$

$\phi_\alpha(t) = \Gamma(\alpha+1)\, t^{-\alpha}\, \Phi_\alpha(t), \alpha \geq 0,$ $v_n = (n+1)\, p_n / P_n,$

$A_{n,k} = \sum_{r=k}^{n} a_{n,r} \neq 0,$ $V_{n,k} = (n-k+1)a_{n,k} \big/ A_{n,k},$

$B_{n,k} = A_{n-1,k} - (1 + a_{n,0})\, A_{n,k+1},$ $0 \leq k \leq n-1, \ n \geq 1,$

$D_n(t) = \sin(n + ½)\, t \,/\, 2\sin(t/2),$ $Q_{m,n} = \sum_{k=m}^{n} \alpha_k,$

$\tau = [1/t]$, the integral part of 1/t, $\Delta_k a_{n,k} = a_{n,k} - a_{n,k+1},$

and BV (a,b) denotes the bounded variation in the interval (a,b).

Furthermore, C will denote a absolute positive constant, not necessarily the same at each occurrence. Throughout this paper, we take $a_{n,k} \geq 0$ $(0 \leq k \leq n)$ and $A_{n,0} = 1, \forall n$.

3. Absolute Cesaro summability of a Fourier series has been studied by various investigators. In particular, the following theorems are well known :

Theorem A[2].

$$\phi(t) \in BV(0,\pi) \Rightarrow \Sigma A_n(x) \in |C,\alpha|, \quad \alpha > 0.$$

Theorem B [3].

$$\phi_\alpha(t) \in BV(0,\pi) \Rightarrow \Sigma A_n(x) \in |C,\beta|, \ \forall \beta > \alpha \geq 0.$$

Theorem C [6].

$$g(t) \in BV(0,\pi) \Rightarrow \Sigma n B_n(x) \in |C,\alpha|, \alpha > 1.$$

Varshney [16] has generalized Theorem A to Nörlund means while Bhatt [1] proved a theorem on summability-$|N_p|$ of Fourier series (2) assuming $\phi_1(t) \in BV(0,\pi)$ and monotonicity on $\{p_n\}$ and $\{p_{n+1} - p_n\}$ while other investigators such as Lal [10], Dikshit [5] and Varshney [15] have proved the theorem without monotonicity on both the sequences. Varshney proved :

Theorem D[15].

If $\phi_1(t)$ is of bounded variation in $(0,\pi)$ then the Fourier series (2), at t = x, is summable-$|N_p|$, provided the generating sequence $\{p_n\}$ satisfies conditions :

$$p_n \geq 0, \qquad P_n \to \infty \quad \text{as} \quad n \to \infty, \tag{4}$$

$$v_n \in BV, \tag{5}$$

$$\sum_{n=k}^{\infty} \frac{P_k}{(k+1)P_n} < C, \tag{6}$$

and

$$\sum_{n=k}^{\infty} (k+1)|\Delta^2 v_n| < C, \qquad k = 0,1,2,.... \tag{7}$$

Theorem E [15].

If $g(t)$ is of bounded variation in $(0, \pi)$, then the series (3) is summable-$|N_p|$, provided the generating sequence $\{p_n\}$ satisfies conditions (4), (5), (6) and (7).

Theorem D and Theorem E are generalisations respectively, of Theorems B (for $\alpha = 1$) and C to summability $|N_p|$. The problem of absolute matrix summabiility of series (2), where $\phi(t) \in BV(0, \pi)$, was studied by Kishore and Hotta [8] , Varshney [14] ,Kuttner and Sahney [9], Jurkat et al. [7], and Mittal [11] .

Recently the author and Kumar [12] have extended a Theorem of Dikshit [5].We have proved :

Theorem F [12].

Let $T \equiv (a_{n,k})$ be an infinite triangular matrix satisfying :

$$a_{n-1,k} \geq (1+a_{n,0})a_{n,k+1}, \quad \text{for} \quad 0 \leq k < n, n \geq 1, \tag{8}$$

$$\sum_{n=k}^{\infty} A_{n,n-k}/(n+1) \leq C, \quad \text{for every } k, \tag{9}$$

$$\sum_{k=1}^{n-1} |\Delta_k a_{n,k}| = O(n^{-1}), \tag{10}$$

$$\sum_{k=0}^{n-2} (k+1)|\Delta_k^2 a_{n,k}| = O(n^{-1}), \tag{11}$$

$$V_{n,k} \leq C, \quad \text{uniformly for } 0 \leq k \leq n, \tag{12}$$

$$\sum_{n=k}^{\infty} \frac{1}{n+1} \sum_{k=0}^{n-1} |\Delta_k V_{n,k}| A_{n,k+1} < \infty. \tag{13}$$

If $\phi_1(t) \in BV(0,\pi)$, then the series $\Sigma e_n A_n(t)$, at $t = x$, is summable - $|T|$, provided $\{e_n\}$ satisfies

$$\sum_{k=1}^{\infty} |\Delta e_k| < \infty, \tag{14}$$

and

$$\sum_{k=1}^{\infty} k|\Delta^2 e_k| < \infty. \tag{15}$$

Theorem F has also generalised partially (Cf. [12, p-532] for $e_n = 1, \forall n$) the earlier work (which itself includes the theorem of Bhatt (Cf. [1])) of the author, Prasad and Kumar (Cf. [13]).

The object of this paper is to generalise Theorems D and E. We prove:

Theorem 1.

Let $T \equiv (a_{n,k})$ be infinite triangular matrix. If $\phi_1(t) \in BV(0,\pi)$, then the Fourier series (2) of f(t), at t = x, is summable -|T| , provided the generating sequence $\{a_{n,k}\}$ satisfies (8), (12), (13) and

$$\sum_{n=k}^{\infty} A_{n,n-k} \le C\,(k+1), \qquad \forall k, \tag{16}$$

$$\sum_{n=k+1}^{\infty} |\, a_{n,[n/2]} - a_{n,[n/2]+1} \,| \le C/(k+1), \;\; \forall k, \tag{17}$$

$$\sum_{n=k+3}^{\infty} \sum_{r=k+1}^{n-2} |\, \Delta_r^2\, a_{n,n-r} \,| \; < C/(k+2), \; k = 0,1,2,.... \tag{18}$$

Theorem 2

Let $T \equiv (a_{n,k})$ be infinite triangular matrix. If $g(t) \in BV(0,\pi)$ then the differentiated series (3) of f(t), at t = x, is summable-|T|, provided the generating sequence $\{a_{n,k}\}$ satisfies (8), (12), (13) and (16)-(18).

Remarks : In case of the N_p-transform, we will discuss some consequences of the conditions we have imposed on the generating sequence $\{a_{n,k}\}$:

1) Condition (17) reduces to a known (Cf. [15,p-3]) condition

$$\sum_{n=k}^{\infty} (k+1)\,|\,\Delta p_k\,| \;/\; P_{k+1} < C, \qquad k = 0,1,2,...,$$

which itself is obtained (Cf. [16,p-24]) from conditions (4) and (5). Besides it is known (Cf. [15, p-3]) that these conditions, together with (7), also imply

$$\sum_{n=k}^{\infty} (k+1)\,|\,\Delta^2 p_n\,|\,(n+1)/P_{n+1} \; < C, \qquad k = 0,1,2,.... \tag{19}$$

2) Conditions (6) and (19) imply condition (18)

$$\sum_{n=k+3}^{\infty} \sum_{r=k+1}^{n-2} |\,\Delta_r^2 a_{n,n-r}\,| = \sum_{n=k+3}^{\infty} \frac{1}{P_n} \sum_{r=k+1}^{n-2} |\,\Delta_r^2 p_r\,| \; = \sum_{r=k+1}^{\infty} |\,\Delta_r^2\, p_r\,| \sum_{n=r+2}^{\infty} \frac{1}{P_n}$$

$$< C \sum_{r=k+1}^{\infty} |\Delta^2 p_r| \frac{(r+3)}{P_{r+2}} < \frac{C}{k+1}.$$

3) Condition (12), for $k = 0$, implies

$$a_{n,0} \leq C/(n+1), \ \forall n \geq 0.$$

Thus Theorems 1 and 2 generalise Theorems D and E respectively (which in turn generalize Theorems B and C respectively) to absolute matrix summability. Also Theorem 1 generalises Theorem F and includes the result of Bhatt [1] and Mittal et al. [13].

Further note that Dikshit and Rhoades [4] have proved a general inclusion

Theorem on $|N_p|$-summability from which they deduced Theorem D besides many other known results.

4. For proving our Theorems, we need the following Lemmas:

Lemma 1[14]. Let $T \equiv (a_{n,k})$ be an infinite triangular matrix satisfying (9) (12) and (13). Then

$$\sum_{n=r+1}^{\infty} \frac{1}{n+1} \sum_{k=0}^{n-r-1} \left|\Delta_k a_{n,k}\right| \leq \frac{C}{r+1}, \quad \text{uniformly for } r \geq 0.$$

Lemma 2. Let $T \equiv (a_{n,k})$ be an infinite triangular matrix satisfying conditions (8), (12) (13) and (16). Then the series $\Sigma (-1)^k$ is summable - $|T|$.

Proof. Using (1), we have

$$t_n - t_{n-1} = \sum_{k=0}^{n} a_{n,k} s_k - \sum_{k=0}^{n-1} a_{n-1,k} s_k = \sum_{k=1}^{n} (A_{n,k} - A_{n-1,k}) u_k, A_{n-1,n} = 0$$

$$= \sum_{k=1}^{n} (a_{n,k} - a_{n,0} A_{n,k+1}) u_k - \sum_{k=1}^{n-1} \left[A_{n-1,k} - (1 + a_{n,0}) A_{n,k+1}\right] u_k$$

$$= \frac{1}{n+1} \sum_{k=1}^{n} (V_{n,k} - V_{n,0}) A_{n,k} u_k + \frac{1}{n+1} \sum_{k=1}^{n} k a_{n,k} u_k +$$

$$+ a_{n,0} \sum_{k=1}^{n} a_{n,k} u_k - \sum_{k=1}^{n-1} B_{n,k} u_k$$

$$= M_1 + M_2 + M_3 - M_4 .$$

Hence

$$\sum_n |t_n - t_{n-1}| = \sum_n |M_1 + M_2 + M_3 - M_4|$$

$$\leq \sum_n |M_1| + \sum_n |M_2| + \sum_n |M_3| + \sum_n |M_4|. \tag{20}$$

In order to prove the lemma, we shall prove that $\Sigma_n |M_i|$, $1 \leq i \leq 3$, are convergent, since $\Sigma_n |M_4|$ is convergent in view of bounded partial sums of the series $\Sigma_{k=0} (-1)^k$ and [14, p-1019]. Let us take them one by one.

Using Abel's transformation, we have

$$\sum_n |M_1| = \sum_{n=1}^{\infty} \frac{1}{n+1} \left| \sum_{k=1}^{n} (V_{n,k} - V_{n,0}) A_{n,k} u_k \right| \tag{21}$$

$$= \sum_{1}^{\infty} \frac{1}{n+1} \left| \sum_{k=0}^{n-1} (\Delta_k V_{n,n-k}) \left(\sum_{r=0}^{k} A_{n,n-r} (-1)^r \right) \right|$$

$$\leq \sum_{1}^{\infty} \frac{1}{n+1} \sum_{k=0}^{n-1} |\Delta_k V_{n,n-k}| A_{n,n-k} \leq C,$$

in view of (13) and $A_{n,k} \geq A_{n,k+1}$ $\forall\, 0 \leq k \leq n$. Again, since

$$q_k = \sum_{j=1}^{k} \sum_{i=1}^{j} (-1)^i\, i = -\frac{(k+1)}{4} \left[1 + (-1)^{k+1}\right], \quad q_0 = 0 = q_{-1},$$

we obtain by employing Abel's transformation twice

$$M_2 = \frac{1}{n+1} \sum_{k=1}^{n} k a_{n,k} u_k = \frac{1}{n+1} \sum_{k=1}^{n} (-1)^k k a_{n,k}$$

$$= \frac{1}{n+1} \left[\sum_{k=1}^{n-1} (\Delta_k a_{n,k}) \left(\sum_{r=1}^{k} (-1)^r r\right) + a_{n,n} \left(\sum_{r=1}^{n} (-1)^r r\right) + a_{n,1} \right]$$

$$= \frac{1}{n+1} \left[\sum_{k=1}^{n-2} \Delta_k^2 a_{n,k} \left(\sum_{s=1}^{k} \sum_{r=1}^{s} (-1)^r r\right) + (a_{n,n-1} - a_{n,n}) \left(\sum_{s=1}^{n-1} \sum_{r=1}^{s} (-1)^r r \right) + \right.$$

$$+a_{n,n}(\sum_{r=1}^{n}(-1)^r r)+a_{n,1}\Bigg]$$

$$=\frac{1}{n+1}\left[\sum_{k=1}^{n-2}(\Delta_k^2 a_{n,k})q_k+(a_{n,n-1}-a_{n,n})q_{n-1}+a_{n,n}(\sum_{r=1}^{n}(-1)^r r)+a_{n,1}\right]$$

Hence

$$\sum_n |M_2|=\sum_n\left|\frac{1}{n+1}\sum_{k=1}^{n} k\, a_{n,k}u_k\right| \tag{22}$$

$$\leq\sum_{n>2}\frac{1}{n+1}\sum_{k=1}^{n-2}|q_k(\Delta_k^2 a_{n,k})|+\sum_{n=1}^{\infty}\frac{1}{n+1}\left|(a_{n,n-1}-a_{n,n})q_{n-1}\right|+$$

$$+\sum_{1}^{\infty}\frac{a_{n,n}}{n+1}\left|\sum_{r=1}^{n}(-1)^r r\right|+\sum_{n>2}\frac{a_{n,1}}{n+1}$$

$$<\sum_{n>2}\frac{1}{n+1}\sum_{k=1}^{n-2}k\,|\Delta_k^2 a_{n,k}|\;+C\sum_{n=1}^{\infty}|a_{n,n-1}-a_{n,n}|$$

$$+C\sum_{n=1}^{\infty}\frac{a_{n,n}}{n+1}+C\sum_{n=1}^{\infty}\frac{1}{(n+1)^2}$$

$$<C\sum_{n>2}\sum_{k=1}^{n-2}|\Delta_k^2 a_{n,k}|+C\sum_{n=1}^{\infty}A_{n,n-1}+C\sum_{n=1}^{\infty}\frac{A_{n,n}}{n+1}<C,$$

in view of remark 3), condition (16) and (18). Again using Abel's transformation, we get

$$\sum_n |M_3|=\sum_{n=1}^{\infty}a_{n,0}\,|\sum_{k=1}^{n}a_{n,k}u_k| \tag{23}$$

$$\leq\sum_{n=1}^{\infty}\frac{1}{n+1}\left|\sum_{k=1}^{n}(-1)^k a_{n,k}\right|$$

$$=\sum_{n=1}^{\infty}\frac{1}{n+1}\left|\sum_{k=1}^{n-1}(\Delta_k a_{n,k})\left(\sum_{r=1}^{k}(-1)^r\right)+a_{n,n}\sum_{r=1}^{n}(-1)^r+a_{n,1}\right|$$

$$\leq C\sum_{n=1}^{\infty}\frac{1}{n+1}\sum_{k=1}^{n-1}|\Delta_k a_{n,k}| + C\sum_{n=1}^{\infty}\frac{a_{n,n}}{n+1} + C\sum_{n=1}^{\infty}\frac{1}{n(n+1)}$$

$$\leq C + C\sum_{n=1}^{\infty}\frac{A_{n,n}}{n+1} \leq C,$$

in view of Lemma 1 (for r = 0), remark 3) and condition (16). Finally, combining (21) to (23), the proof of Lemma 2 is complete.

5. Proof of Theorem 1. We have

$$u_n(x) = A_n(x) = \frac{2}{\pi}\int_0^{\pi}\phi(t)\cos nt\,dt$$

$$= \frac{2}{\pi}\int_0^{\pi}\phi_1(t)\cos nt\,dt + \frac{2}{\pi}\int_0^{\pi} t\cos nt\,d\phi_1(t)$$
$$= e_n + E_n, \text{say.}$$

Since $\phi_1(t) \in BV(0,\pi)$ and condition (16) implies (9), the series $\Sigma\, e_n$ is summable -$|T|$ in view of Theorem 1 of Varshney [14]. Thus, in order to complete the proof of Theorem 1, it is sufficient to prove that the series $\Sigma\, E_n$ is also summable-$|T|$. In view of (20) for $0 < t < \pi$, we have to prove that

$$\sum_n |M_1'| = \sum_{n=1}^{\infty}\frac{t}{n+1}\left|\sum_{k=1}^{n}(V_{n,k} - V_{n,o})A_{n,k}\cos kt\right| < C, \tag{24}$$

$$\sum_n |M_2'| = \sum_{n=1}^{\infty}\frac{t}{n+1}\left|\sum_{k=1}^{n} k a_{n,k}\cos kt\right| < C, \tag{25}$$

$$\sum_n |M_3'| = \sum_{n=1}^{\infty} t\, a_{n,0}\left|\sum_{k=1}^{n} a_{n,k}\cos kt\right| < C, \tag{26}$$

and

$$\sum_n |M_4'| = \sum_{n=1}^{\infty} t\left|\sum_{k=1}^{n-1} B_{n,k}\cos kt\right| < C. \tag{27}$$

Since $A_{n,k} \geq A_{n,k+1}$ and for $0 < t < \pi$,

$$\left| \sum_{r=0}^{k} A_{n,n-r} \cos (n-r)t \right| < C t^{-1} A_{n,n-k}, \tag{28}$$

we have, by Abel's transformation

$$\sum_n |M_1'| = \sum_n \frac{t}{n+1} \left| \sum_{k=1}^{n} (V_{n,k} - V_{n,0}) A_{n,k} \cos kt \right|$$

$$= \sum_n \frac{t}{n+1} \left| \sum_{k=0}^{n-1} (V_{n,n-k} - V_{n,0}) A_{n,n-k} \cos (n-k)t \right|$$

$$= \sum_n \frac{t}{n+1} \left| \sum_{k=0}^{n-1} (\Delta_k V_{n,n-k}) \sum_{r=0}^{k} A_{n,n-r} \cos (n-r)t \right|$$

$$\leq C \sum_n \frac{1}{n+1} \sum_{k=0}^{n-1} |\Delta_k V_{n,n-k}| A_{n,n-k} \leq C,$$

in view of condition (13).

For proving (25), we need an estimate of $\Sigma\, k\, a_{n,k} \cos kt$. Thus, we have

$$\left| \sum_{k=1}^{n} k a_{n,k} \cos kt \right| = \left| \sum_{k=0}^{n-1} (n-k) a_{n,n-k} \cos (n-k)t \right|$$

$$\leq \left| \sum_{k=0}^{\tau} (n-k) a_{n,n-k} \cos (n-k)t \right| + \left| \sum_{k=0}^{n-\tau-1} k a_{n,k} \cos kt \right|$$

$$= E_1 + E_2, \text{ say}, \tag{29}$$

where

$$E_1 = \left| \sum_{0}^{\tau} (n-k) a_{n,n-k} \sin (n-k)t \right| \leq n A_{n,n-\tau} \tag{30}$$

and, using Abel's transformation

$$E_2 = \left| \sum_{k=1}^{n-\tau-1} k\, a_{n,k} \cos kt \right|$$

$$= \left| \sum_{k=1}^{n-\tau-2} (\Delta_k a_{n,k}) (\sum_{n=1}^{k} r \cos rt) + a_{n,n-\tau-1} (\sum_{r=1}^{n-\tau-1} r \cos rt) \right|$$

$$= \left| \sum_{k=1}^{n-\tau-2} \alpha_k \Delta_k a_{n,k} + \alpha_{n-\tau-1}\, a_{n,n-\tau-1} \right|$$

$$= \left| \alpha_{n-\tau-1}\, a_{n,n-\tau-1} + \sum_{k=1}^{m} \alpha_k \Delta_k a_{n,k} + \sum_{k=m+1}^{n-\tau-2} \alpha_k\, \Delta_k\, a_{n,k} \right|$$

$$= \left| \alpha_{n-\tau-1}\, a_{n,n-\tau-1} + \sum_{k=1}^{m-1} (\Delta_k^2\, a_{n,k})(\sum_{r=1}^{k} \alpha_r) + (a_{n,m} - a_{n,m+1}) \left(\sum_{r=1}^{m} \alpha_r \right) \right.$$

$$\left. + \sum_{k=\tau+1}^{n-m-1} (a_{n,n-k-1} - a_{n,n-k}) \alpha_{n-k-1} - (a_{n,m} - a_{n,m+1}) \alpha_m \right|$$

$$= \left| \alpha_{n-\tau-1}\, a_{n,n-\tau-1} + \sum_{k=1}^{m-1} Q_{1,k}\, (\Delta_k^2\, a_{n,k}) + (a_{n,m} - a_{n,m+1}) Q_{1,m} \right.$$

$$\left. - \sum_{k=\tau+1}^{n-m-2} (\Delta_k^2\, a_{n,n-k})(\sum_{r=n-k-1}^{n-\tau-2} \alpha_r) + (a_{n,m} - a_{n,m+1})(\sum_{r=m+1}^{n-\tau-2} \alpha_r) - (a_{n,m} - a_{n,m+1}) \alpha_m \right|$$

$$= \left| \alpha_{n-\tau-1}\, a_{n,n-\tau-1} + \sum_{k=1}^{m-1} Q_{1,k} (\Delta_k^2 a_{n,k}) + (a_{n,m} - a_{n,m+1}) Q_{1,m} \right.$$

$$\left. - \sum_{k=\tau+1}^{n-m-2} Q_{n-k-1,n-\tau-2} (\Delta_k^2 a_{n,n-k}) + (a_{n,m} - a_{n,m+1}) Q_{m+1,n-\tau-2} - (a_{n,m} - a_{n,m+1}) \alpha_m \right|$$

$$\leq C\,(nt^{-1})\, a_{n,n-\tau-1} + C(nt^{-2})\, |\, a_{n,m} - a_{n,m+1} \,|$$

$$+ C(nt^{-1})\, |\, a_{n,m} - a_{n,m+1} \,| + C(nt^{-2}) \left[\sum_{k=\tau+1}^{n-m-2} + \sum_{k=n-m+1}^{n-1} \right] |\, \Delta_k^2 a_{n,n-k} \,|$$

$$= C n\, t^{-2} \left[t a_{n,n-\tau-1} + t\, |\, a_{n,m} - a_{n,m+1} \,| + |\, a_{n,m} - a_{n,m+1} \,| + \sum_{k=\tau+1}^{n-3} |\, \Delta_k^2\, a_{n,n-k} \,| \right], \tag{30}$$

in view of $|\alpha_n| < C\, nt^{-1}$, $m = [n/2]$ and $|Q_{1,n}| < C\, nt^{-2}$.

Now

$$\Sigma|M_2'| \le \sum_{n\le 3\tau} + \sum_{n\ge 3\tau} = \Sigma_1 + \Sigma_2,\ \text{say},$$

where

$$\Sigma_1 = \sum_{n\le 3\tau} \frac{t}{n+1} \left|\sum_{k=0}^{n-1} (n-k)a_{n,n-k} \cos(n-k)t\right|$$

$$\le \sum_{n\le 3\tau} \frac{t}{n+1} \sum_{k=0}^{n-1} (n-k)a_{n,n-k}$$

$$\le \sum_{n\le 3\tau} \frac{nt}{n+1} \sum_{k=0}^{n-1} a_{n,n-k} < \sum_{n<3\tau} t \le C.$$

Thus, using (27) (28) and (29) we have

$$\Sigma_2 = \sum_{n>3\tau} \frac{t}{n+1} \left|\sum_{k=0}^{n-1} (n-k)a_{n,n-k} \cos(n-k)t\right|$$

$$\le \sum_{n>3\tau} \frac{t}{n+1}(E_1 + E_2)$$

$$\le C\sum_{n>3\tau} t A_{n,n-\tau-1} + C t^{-1} \sum_{n>3\tau} |a_{n,[n/2]} - a_{n,[n/2]+1}|$$

$$+ C\sum_{n>3\tau} |a_{n,[n/2]} - a_{n,[n/2]+1}| + C t^{-1} \sum_{n>3\tau} \sum_{k=\tau+1}^{n-3} |\Delta_k^2 a_{n,n-k}|$$

$$< C t \sum_{n>3\tau} A_{n,n-\tau-1} + C t^{-1}\left(\frac{1}{3\tau+1}\right) + C\left(\frac{1}{3\tau+1}\right) + C t^{-1} \sum_{n>3\tau} \sum_{k=\tau+1}^{n-3} |\Delta_k^2 a_{n,n-k}|$$

$$< C t (3\tau+1) + C t^{-1}\left(\frac{1}{3\tau+1}\right) + C\left(\frac{1}{3\tau+1}\right) + C$$

$$< C + \frac{C(\tau+1)}{3\tau+1} < C,$$

in view of conditions (16), (17) and (18) and $0 < t < \pi$. This completes the proof of (25). Again, using Abel's transformation, we get

$$\sum_n |M_3'| \le \sum_{n=1}^{\infty} \frac{t}{n+1} \left| \sum_{k=1}^{n} a_{n,k} \cos kt \right|$$

$$\le \sum_{n=1}^{\infty} \frac{t}{n+1} \left| \sum_{k=1}^{n-1} (\Delta_k a_{n,k}) (\sum_{r=1}^{k} \cos rt) + a_{n,n} (\sum_{r=1}^{n} \cos rt) \right|$$

$$\le \sum_{n=1}^{\infty} \frac{1}{n+1} \sum_{k=1}^{n-1} |\Delta_k a_{n,k}| + \sum_{n=1}^{\infty} \frac{a_{n,n}}{n+1}$$

$$\le C + \sum_{n=1}^{\infty} \frac{A_{n,n}}{n+1} < C,$$

in view of Lemma 1 , condition (16) and remark 3). Hence this completes the proof of (26).

Finally, by Abel's Lemma, we have

$$\sum_n |M_4'| = \sum_n t \left| \sum_{k=1}^{n-1} B_{n,k} \cos kt \right|$$

$$\le \sum_n t B_{n,0} \max_{1 \le r \le n-1} \left| \sum_{k=1}^{r} \cos rt \right| < C \sum_n \frac{1}{(n+1)^2} < C,$$

in view of condition (12) [14, p – 1018]. This completes the proof of (27) and also of Theorem 1.

Proof of Theorem 2. Integrating by parts

$$nB_n(x) = \frac{2}{\pi} \int_0^{\pi} g(t)\, nt \sin nt\, dt$$

$$= \frac{2}{\pi} \int_0^{\pi} g(t) \cos nt\, dt - \frac{2}{\pi} \int_0^{\pi} g(t)\, d(t \cos nt)$$

$$= \frac{2}{\pi} \int_0^{\pi} g(t) \cos nt\, dt - \left[\frac{2}{\pi} t g(t) \cos nt \right]_0^{\pi} + \frac{2}{\pi} \int_0^{\pi} t \cos nt\, dg(t)$$

$$= \frac{2}{\pi} \int_0^{\pi} g(t) \cos nt\, dt - 2g(\pi)(-1)^n + \frac{2}{\pi} \int_0^{\pi} \cos nt\, dg(t)$$

$$= \lambda_n + \mu_n + \delta_n, \text{ say.}$$

The series $\Sigma\mu_n$ is summable |T| in view of our Lemma 2. The series $\Sigma\lambda_n$ is summable |T| by a Theorem of Varshney [14], since $g(t) \in BV(0,\pi)$. The discussion of $\Sigma\,\delta_n$ is similar to that of $\Sigma\,E_n$ in the proof of Theorem 1 with g(t) in place of $\phi_1(t)$ and thus the proof of Theorem 2 is complete.

Acknowledgement

The author is grateful to Prof. B.E. Rhoades, Bloomington University, Indiana, U.S.A. for the valuable suggestions which has improved the presentation of the paper.

References

[1] Bhatt, S.N., On the absolute Nörlund summability of Fourier series, *Indian J. Math.* **10** (1968),157 - 166.

[2] Bosanquet, L.S., Note on the absolute summability (C) of a Fourier series, *J. London Math. Soc.* **11** (1936), 11 - 15.

[3] _____, The absolute Cesaro summability of Fourier series, *Proc. London Math. Soc.* **41** (1936), 517 - 528.

[4] Dikshit, G.D., and Rhoades, B.E., An inclusion relation between Cesaro and Nörlund matrices for absolute summability, *J. Math. Anal. Appl.* **170** (1992), 171 -195.

[5] Dikshit, H.P., On the absolute summability $|N,p_n|$ of a Fourier series at a point, *Proc. Camb. Phil. Soc.* **65** (1969), 495-505.

[6] Hyslop, J.M., On the absolute summability of the successively derived series of a Fourier series and its allied series, *Proc. London Math. Soc.* **46** (1940), 55 - 80.

[7] Jurkat, W.B., Kumar, A. and Peyerimhoff, A., On the absolute Fourier effectiveness of summability method, *J. London Math. Soc.* **31** (1985), 101 - 114.

[8] Kishore, N. and Hotta, G.C., On absolute matrix summability of Fourier series Indian *J. Math.* **13** (1971), 99 - 110.

[9] Kuttner, B. and Sahney, B.N., On the absolute matrix summability of Fourier series, *Pacific J. Math.* **43** (1972), 407 - 419.

[10] Lal, S.N., On the absolute Nörlund summability of Fourier series, *Indian J. Math.* **13**, (1971), 51 - 67.

[11] Mittal, M.L., On absolute matrix summability factors of Fourier series, J. Math., *Game Theory and Algebra U.S.A.* **13**,**5**(2003),387-398.

[12] Mittal, M.L., and Kumar, R., On Multipliers for the absolute matrix summability, *J. Math. Anal. Appl.* **191** (1995), 528 - 539.

[13] Mittal, M.L., Prasad, G. and Kumar, N., On multipliers for the absolute matrix summability, *Indian J. Pure Appl. Math.* **19** (1989), 183 - 193.

[14] Varshney, O.P., On the absolute matrix summability of Fourier series, *Indian J. Pure Appl. Math.* **21** (1990), 1015 - 1023.

[15] Varshney, O.P., On the absolute Nörlund summability of Fourier series, *Boll. U.M.I.* **4** (1968), 500 - 507.

[16] Varshney, O.P., On the absolute Nörlund summability of Fourier series, *Math. Z.* **83** (1964), 18-24.

In: Mathematics, Game Theory and Algebra Compendium ISBN 978-1-60456-659-8
Editor: Jacob H. Mathias, pp. 493-504

Chapter 41

SEMIREGULAR ASSOCIATIVE PAIRS

Inmaculada de las Peñas Cabrera*
Departamento de Matemática Aplicada.
Universidad de Málaga. E.U. Politécnica.
Campus Universitario de El Ejido
29071 Málaga. Spain

Abstract

A new class of associative pairs, which can be considered as an extension of semiregular rings, is defined and studied in terms of its elements. A new characterization for a ring to be semiregular is provided and it allows studying the semiregularity of the standard embedding of a pair. It is proved that an associative pair A with Jacobson radical J is semiregular if and only if A/J is regular and idempotents can be lifted modulo J.

AMS Subject Classification: 16U99

Keywords: Associative pairs, Semiregular rings, (von Neumann) regular element.

Introduction

The class of F-semiperfect rings are introduced by Oberst and Schneider [13] as a generalization of the left perfect and semiperfect rings of Bass [2]. A ring R with Jacobson radical $Rad\ A$ is called *semiregular* if $R/Rad\ A$ is (von Neumann) regular an idempotents can be lifted modulo $Rad\ A$. Several authors [3, 12, 9, 1, 14] look further at these rings and have the aim of generalizing this concept to modules from various points of view. This contribution attempts to study this theory in the frame of associative pairs as it is close to other topics like semiperfectness, perfectness and lifting idempotents, which were studied in previous works [6, 7, 8]. Our benchmark is a work of Nicholson [12]. In this paper, semiregular modules are defined and the rings which are semiregular when they are considered as modules over themselves, are shown to coincide with F-semiperfect rings. The

*E-mail address: ipcabrera@uma.es

starting point is defining the concept of semiregular element of a pair and then a semiregular associative pair will be an associative pair in which each of its elements is semiregular. Some properties of this class of pairs are deduced in the first section. For a ring R and an idempotent $e \in R$, if eRe and $(1-e)R(1-e)$ are semiregular, then R does not have to be semiregular. However, if the pair $(eR(1-e), (1-e)Re)$ is also semiregular, then R is so. As a consequence, it is obtained the relationship between the semiregularity of a pair and its standard embedding. In the last section, it is established a theorem of characterization for semiregular associative pairs, which combines conditions in terms of the elements of the pair with others which are concerned with module theory.

1. Definition and Basic Concepts

Let $A = (A^+, A^-)$ be a pair of modules over a commutative unital ring K and

$$\begin{array}{rcl} \langle,,\rangle : A^\sigma \times A^{-\sigma} \times A^\sigma & \longrightarrow & A^\sigma \\ (x^\sigma, y^{-\sigma}, z^\sigma) & \mapsto & \langle x^\sigma y^{-\sigma} z^\sigma \rangle \end{array}$$

for $\sigma \in \{+,-\}$, two K-trilinear mappings called *triple products*. A is said to be an *associative K-pair* if the identities

$$\langle\langle x^\sigma y^{-\sigma} z^\sigma\rangle u^{-\sigma} v^\sigma\rangle = \langle x^\sigma \langle y^{-\sigma} z^\sigma u^{-\sigma}\rangle v^\sigma\rangle = \langle x^\sigma y^{-\sigma} \langle z^\sigma u^{-\sigma} v^\sigma\rangle\rangle$$

are satisfied for $x^\sigma, z^\sigma, v^\sigma \in A^\sigma, y^{-\sigma}, u^{-\sigma} \in A^{-\sigma}$ and $\sigma \in \{+,-\}$.

Associative pairs are introduced in [11] and since then different aspects of them have been studied [4, 5, 6, 7, 8]. The previous definitions and basic facts on associative pairs and modules that we need in this paper can be found there.

Nevertheless, in order to make this exposition as self-contained as possible, we recall some of them.

For any associative pair (A^+, A^-), it can be considered R_{11} as the subalgebra of $End_K(A^+) \times End_K(A^-)^{op}$ generated by (Id, Id) and the pairs of operators $\{x^+x^- := (L(x^+, x^-), R(x^+, x^-)) : x^\sigma \in A^\sigma\}$ and, analogously, R_{22} is generated by (Id, Id) and the elements $\{x^-x^+ := (L(x^-, x^+), R(x^-, x^+)) : x^\sigma \in A^\sigma\}$ in $End_k(A^-) \times End_k(A^+)^{op}$.

Then A^+ is a R_{11}-R_{22}-bimodule for the actions $(x^+x^-)a^+ := \langle x^+x^-a^+\rangle$ and $a^+(x^-x^+) := \langle a^+x^-x^+\rangle$. Similarly A^- is a R_{22}-R_{11}-bimodule for the actions $(x^-x^+)a^- := \langle x^-x^+a^-\rangle$ and $a^-(x^+x^-) := \langle a^-x^+x^-\rangle$.

For an associative pair (A^+, A^-), the Jacobson radical can be defined in the following way. An element $(x^+, x^-) \in (A^+, A^-)$ is said to be *quasi-invertible* if there exists $z^+ \in A^+$ such that $x^+ + z^+ = \langle x^+x^-z^+\rangle = \langle z^+x^-x^+\rangle$. An element x^σ is called *properly quasi-invertible* if for each $x^{-\sigma} \in A^{-\sigma}$ the element (x^+, x^-) is quasi-invertible. The *Jacobson radical* of (A^+, A^-) is the pair (J^+, J^-), where J^σ is the set of all properly quasi-invertible elements of A^σ, for all $\sigma \in \{+,-\}$.

In this section, we define the semiregular associative pairs and study some of their properties.

Proposition 1.1. *Let $A = (A^+, A^-)$ be an associative pair, $J = (J^+, J^-)$ its Jacobson radical and $a^+ \in A^+$. Then the following conditions are equivalent:*

i) There exist $u^+ \in A^+$ and $x^- \in A^-$ such that $u^+ = \langle u^+x^-u^+\rangle \in a^+R_{22}$ and $(Id - u^+x^-)a^+ \in J^+$.

ii) There exists an idempotent (e^+, e^-) of A such that $e^+ \in a^+R_{22}$ and $(Id - e^+e^-)a^+ \in J^+$.

iii) There exists an idempotent (f^+, f^-) of A such that $f^+ \in R_{11}a^+$ and $a^+(Id - f^-f^+) \in J^+$.

iv) There exist $u^+ \in A^+$ and $x^- \in A^-$ such that $u^+ = \langle u^+x^-u^+\rangle \in R_{11}a^+$ and $a^+(Id - x^-u^+) \in J^+$.

v) There exists $b^+ \in A^+$ regular such that $a^+ - b^+ \in J^+$.

vi) There exists an idempotent (g^+, g^-) of A satisfying $g^+ = \langle a^+g^-a^+\rangle$ and $g^- = \langle g^-a^+g^-\rangle$ with $a^+ - g^+ \in J^+$.

Proof. i)⇒ ii). Let $u^+ \in A^+$ and $x^- \in A^-$ be such that $u^+ = \langle u^+x^-u^+\rangle \in a^+R_{22}$ and $(Id - u^+x^-)a^+ \in J^+$. The element $(e^+, e^-) = (u^+, \langle x^-u^+x^-\rangle)$ is an idempotent and $(Id - e^+e^-)a^+ = (Id - u^+x^-)a^+ \in J^+$.
ii)⇒ iii). Given an idempotent (e^+, e^-) of A such that $e^+ = a^+\beta$ and $(Id - e^+e^-)a^+ \in J^+$, we consider $(f^+, f^-) = (\langle e^+e^-a^+\rangle, \beta e^-)$, which is an idempotent, $f^+ \in R_{11}a^+$ and $\langle a^+f^-f^+\rangle = \langle e^+e^-a^+\rangle$ whence $a^+(Id - f^-f^+) \in J^+$. iii)⇒ iv). It is trivial.
iv)⇒ v). Let u^+ and x^- be the elements given in iv). As $u^+ = \alpha a^+ \in R_{11}a^+$ for $b^+ = \langle a^+x^-u^+\rangle \in R_{11}a^+$, we have $\langle b^+x^-\alpha\, b^+\rangle = b^+$. Hence, b^+ is regular and $a^+ - b^+ = a^+(Id - x^-u^+) \in J^+$.
v)⇒ vi). First, we show that if b^+ is a regular element such that $a^+ - b^+ \in J^+$, then we can assume that $b^+ \in R_{11}a^+$: indeed, if $b^+ = \langle b^+x^-b^+\rangle$, then

$$a^+ - \langle b^+x^-a^+\rangle = a^+ - b^+ + \langle b^+x^-(b^+ - a^+)\rangle \in J^+.$$

As the element $b^+x^- - a^+x^- = (b^+ - a^+)x^- \in J^+A^- \subset Rad\, R_{11}$ is quasi-invertible, there exists $\alpha_0 \in R_{11}$ such that $(Id - b^+x^- + a^+x^-)\alpha_0 = Id$. Therefore $b^+x^-a^+x^-\alpha_0 = b^+x^-$, which implies that the element $\langle b^+x^-a^+\rangle \in R_{11}a^+$ is regular. Let $b^+ = \alpha a^+$ be a regular element with $a^+ - b^+ \in J^+$. Taking $b^- \in A^-$ such that (b^+, b^-) is an idempotent, the element $g^- = b^-\alpha$ satisfies $\langle g^-a^+g^-\rangle = g^-$ and if $g^+ = \langle a^+g^-a^+\rangle = \langle a^+b^-b^+\rangle$, we obtain that (g^+, g^-) is an idempotent such that

$$a^+ - g^+ = (a^+ - b^+) + \langle (b^+ - a^+)b^-b^+\rangle \in J^+.$$

vi)⇒ i). Let (g^+, g^-) be the idempotent given in vi). Then $\langle g^+g^-a^+\rangle = \langle g^+g^-a^+g^-a^+\rangle = g^+$, whence $a^+ - \langle g^+g^-a^+\rangle \in J^+$. □

Given $A = (A^+, A^-)$, an associative pair, an element $a^+ \in A^+$ is *semiregular* if the conditions in Proposition 1.1 are satisfied. An element $a^- \in A^-$ is *semiregular* if a^- is semiregular in the opposite pair (A^-, A^+). An associative pair is called *semiregular* if all the elements of A^+ and A^- are semiregular. As an immediate consequence of this definition we obtain that a regular associative pair is semirregular. Moreover, a semiprimitive pair is semiregular if and only if it is regular.

The next example shows that there exist associative pairs (A^+, A^-) for which A^+ is semiregular in spite of the fact that A^- is not: define the ring

$$S = \{(a_1, a_2, \ldots, a_k, a, a, \ldots) : a_i, a \in \mathbb{R}\},$$

where the operations are componentwise, and let R denote the ideal of S given by $R = \{(a_1, a_2, \ldots, a_k, 0, 0, \ldots) : a_i \in \mathbb{R}\}$. Observe that the associative pair $A = (R, S)$ is semiprimitive, $A^+ = R$ is (von Neumann) regular while $A^- = S$ is not, since the elements $(a_1, a_2, \ldots, a_k, a, a, \ldots)$ where $a \neq 0$ are not regular.

A reasonable task to regard is that the definition of semiregular associative pair is coherent with ring theory.

Proposition 1.2. *A ring with unit R es semiregular if and only if the associative pair (R, R) is semiregular.*

Proof. An easy verification shows that an element $a \in R$ is regular if and only if it is regular in the associative pair (R, R). Besides, $Rad\,(R, R) = (Rad\,R, Rad\,R)$. □

Lemma 1.3. *Homomorphic images of semiregular associative pairs are semiregular.*

Proof. If $(\phi^+, \phi^-) : (A^+, A^-) \to (B^+, B^-)$ is an associative pair homomorphism, then $\phi^\sigma(Rad^\sigma\, A) \subset Rad^\sigma\, B$ for all $\sigma \in \{+, -\}$ and if a^σ is regular, then $\phi^\sigma(a^\sigma)$ is also regular. The proof immediately follows from these assertions. □

Let A be an associative pair and e be an idempotent of A. We have the *Peirce decomposition* given by $A = A_{00} \oplus A_{01} \oplus A_{10} \oplus A_{11}$, where $A_{ij} = (A^+_{ij}, A^-_{ji})$ and $A^\sigma_{ij} = \{x^\sigma \in A^\sigma : \langle e^\sigma e^{-\sigma} x^\sigma\rangle = ix^\sigma \text{ and } \langle x^\sigma e^{-\sigma} e^\sigma\rangle = jx^\sigma\}$ for $i, j \in \{0, 1\}$.

Lemma 1.4. *If e is an idempotent of a semiregular associative pair A, then $A_{ii}(e)$ is semiregular for $i \in \{0, 1\}$.*

Proof. For $a^+ \in A^+_{ii}(e)$, we take (f^+, f^-) to be an idempotent of A verifiying $f^+ = \langle a^+ f^- a^+\rangle$ and $f^- = \langle f^- a^+ f^-\rangle$ with $a^+ - f^+ \in Rad^+\, A$. The element $f^+ \in A^+_{ii}(e)$ satisfies $\langle f^+ f^-_{ii} f^+\rangle = f^+$ where f^-_{ii} is the ii-Peirce component of f^- and $a^+ - f^+ \in Rad^+\, A \cap A^+_{ii}(e) = Rad^+\, A_{ii}(e)$. □

We point out that the results given below are also true in A^-, by taking the opposite pair.

Lemma 1.5. *For an associative pair A, if $a^+ - b^+ \in J^+$ and b^+ is semiregular, then a^+ is semiregular.*

Proof. It is clear from Proposition 1.1. □

The following lemma might be considered as a version of Mc Coy' s Lemma in associative pairs for semiregular elements.

Lemma 1.6. *Let A be an associative pair and $a^+ \in A^+$. If there exists an idempotent $(e^+, e^-) \in A$ such that $e^+ \in a^+R_{22}$ and $(Id - e^+e^-)a^+$ is semiregular, then a^+ is semiregular.*

Proof. Let (f^+, f^-) be an idempotent of A such that $f^+ \in (Id - e^+e^-)a^+R_{22}$ and $(Id - f^+f^-)(Id - e^+e^-)a^+ \in J^+$. Consider the operator $\alpha = (Id - f^+f^-)e^+e^- + f^+f^- \in R_{11}$. As $\langle e^+e^-f^+\rangle = 0$, then $e^+e^-(Id - f^+f^-) = e^+e^-$. Both conditions together with that (e^+, e^-) and (f^+, f^-) are two idempotents, imply that α is an idempotent. Taking into account that $e^+ \in a^+R_{22}$ and $f^+ \in (Id - e^+e^-)a^+R_{22} \subseteq a^+R_{22}$, we obtain that there exists $b^- \in A^-$ such that $\alpha = a^+b^-$. So, $\langle a^+b^-a^+\rangle \in a^+R_{22}$ is regular and $(Id - a^+b^-)a^+ = (Id - f^+f^-)(Id - e^+e^-)a^+ \in J^+$. □

Though the definitions of primitive and local idempotents in associative pairs are the usual ones, we refer the reader to [7, 3.].

Proposition 1.7. *In a semiregular associative pair, every primitive idempotent is local.*

Proof. First of all, we want to recall a standard notation in Jordan pairs. For an arbitrary element a^σ, we use $Q(a^\sigma)$ to denote the operator $Q(a^\sigma)x^{-\sigma} = \langle a^\sigma x^{-\sigma} a^\sigma\rangle$. Now, let (e^+, e^-) be a primitive idempotent of A. We need to prove that $Q(\overline{a^\sigma})$ is surjective, for all $\overline{a^\sigma} \in A^\sigma_{11}/Rad^\sigma A_{11}$. To this end, let a^+ be any element in $\langle e^+A^-e^+\rangle$ such that $a^+ \notin Rad^+ A$. Choose an idempotent $f = (f^+, f^-)$ of A such that $f^+ \in a^+R_{22}$ and $(Id - f^+f^-)a^+ \in Rad^+ A$. Then $f^+ \neq 0$, since $a^+ \notin Rad^+ A$, and $f^+ = \langle e^+e^-f^+\rangle$. The element $\alpha = f^+f^-e^+e^- \in R_{11}$ is an idempotent satisfying $f^+f^-\alpha = \alpha$ and $\alpha e^+e^- = \alpha$. This implies that $(\alpha e^+, e^-\alpha)$ and $(e^+ - \alpha e^+, e^- - e^-\alpha)$ are orthogonal idempotents of $A_{11}(e)$ whose sum is e, hence $(\alpha e^+, e^-\alpha) = (0,0)$ or $(\alpha e^+, e^-\alpha) = (e^+, e^-)$. If we suppose $\alpha e^+ = 0$, then $\alpha = \alpha e^+e^- = 0$ so $f^+ = \alpha f^+ = 0$, which is a contradiction. Therefore, $\alpha e^+ = e^+$ and $e^-\alpha = e^-$. Besides, we can consider an idempotent $g = (g^+, g^-)$ of A such that $g^+ \in R_{11}a^+$ and $a^+(Id - g^-g^+) \in Rad^+ A$. It follows, as before that $e^+ = e^+\beta$ for $\beta = e^-e^+g^-g^+ \in R_{22}$. As a consequence, $e^+ = \alpha e^+\beta \in \langle a^+A^-a^+\rangle$, whence $Q(e^+) = Q(a^+)Q(z^-)Q(a^+)$ for some $z^- \in A^-$. This implies that $Q(a^+)$ is surjective and hence so it is $Q(\overline{a^+})$. Similar arguments applied to an element $a^- \in \langle e^-A^+e^-\rangle$ such that $a^- \notin Rad^- A$, show that e is local. □

2. Semiregularity of the Standard Embedding

The next result provides a characterization of semiregular rings from semiregular associative pairs.

Proposition 2.1. *Let R be a ring with unit and $e \in R$ be an idempotent. Then R is semiregular if and only if $eRe, (1-e)R(1-e)$ are semiregular rings and the associative pair $A = (eR(1-e), (1-e)Re)$ is semiregular.*

Proof. If R is semiregular, given $a \in eR(1-e)$, by [12, Proposition 2.2] there exists an idempotent $f = ar$ such that $(1-f)a \in Rad\, R$. By taking $f^+ = fa$ and $f^- = (1-e)rfe$, we have that (f^+, f^-) is an idempotent of the pair $A = (eR(1-e), (1-e)Re)$ verifying

$$a - \langle f^+f^-a\rangle = a - fa \in Rad\, R \cap eR(1-e) = Rad^+ A.$$

Analogously it is proved that each element of $(1-e)Re$ is semiregular. The rings eRe and $(1-e)R(1-e)$ are semiregular, by [12, Corollary 2.3]. Conversely, let $x \in R$ and $a^- = (1-e)xe$. Choose an idempotent (u^+, u^-) of A such that $u^- = a^-\gamma$ and $(Id - u^-u^+)a^- = a^- - u^-u^+a^- \in Rad^- A$. Then $f = x\gamma u^+ \in xR$, is an idempotent of R (take into account that $\gamma u^+ \in eR(1-e)$) with

$$(1-e)(1-f)xe = a^- - u^-u^+a^- \in Rad\, R.$$

Set $z = e(1-f)xe + (1-f)x(1-e)$ which verifies $(1-e)ze = 0$. Therefore, $(1-f)x - z \in Rad\, R$. By [12, Corollary 2.4] and [12, Lemma 2.6], we may assume that $x \in R$ satisfies $xe = exe \in eRe$. As eRe is a semiregular ring, we can take an idempotent $g \in eRe$ such that $g = exere \in xR$ and $(e-g)exe \in Rad\, R$. If $z = (1-g)x(1-e)$, we have that $(1-g)x - z = (e-g)xe \in Rad\, R$ hence, we may assume $x \in R$ with $xe = 0$. Given $ex \in eR(1-e)$, there exists (v^+, v^-) an idempotent of A such that $v^+ = ex\beta$ and $(Id - v^+v^-)ex \in Rad^+ A$. Then, $w = x\beta v^-$ is an idempotent of R such that $e(1-w)x \in Rad\, R$ so $(1-w)x - z \in Rad\, R$, for $z = (1-e)(1-w)x$ a semiregular element (since $(1-e)R(1-e)$ is a semiregular ring). Applying [12, Corollary 2.4 and Lemma 2.6], we can conclude that x is semiregular. □

Observe that R can fail to be semiregular when eRe and $(1-e)R(1-e)$ are semiregular. Consider S and R the rings defined in section 2. Consider $T = \begin{pmatrix} S & R \\ S & S \end{pmatrix}$, with usual sum and product, and the idempotent $e = \begin{pmatrix} 1 & 0 \\ 0 & 0 \end{pmatrix}$. As both eRe and $(1-e)R(1-e)$ are isomorphic to S (which is regular), they are semiregular. However, T is not so, since the associative pair (R, S) is not semiregular.

Any associative pair can be identified with a pair of the type $(e_1\mathcal{U}_A e_2, e_2\mathcal{U}_A e_1)$, where

$$\mathcal{U}_A = \begin{pmatrix} R_{11} & A^+ \\ A^- & R_{22} \end{pmatrix}$$

is the standard embedding of A and

$$e_1 = \begin{pmatrix} 1_{R_{11}} & 0 \\ 0 & 0 \end{pmatrix} \qquad e_2 = \begin{pmatrix} 0 & 0 \\ 0 & 1_{R_{22}} \end{pmatrix}$$

are two orthogonal idempotents such that $1_{\mathcal{U}_A} = e_1 + e_2$ ([11]). Applying the last result to the algebra $\mathcal{U}_A$, it is deduced the following obvious corollary.

Corollary 2.2. *Let A be an associative pair and $\mathcal{U}_A$ the standard embedding. Then, $\mathcal{U}_A$ is semiregular if and only if R_{11}, R_{22} are semiregular and A is semiregular.*

The conditions that R_{11} and R_{22} are semiregular cannot be omitted in general. However, they are not needed in the case of the pair A is unital, that is, the idempotents e_1 and e_2 are full.

Before dealing with this issue, we show the next lemma, which will be used later on.

Lemma 2.3. *Let A be a semiregular associative pair and $(u^+, u^-) \in A$ be an idempotent. Then, for every element $\alpha \in R_{11}$, we have that $\alpha u^+ u^-$ $(u^+ u^- \alpha)$ is a semiregular element of R_{11}.*

Proof. Given $\alpha u^+ \in A^+$, consider an idempotent (f^+, f^-) such that $f^+ \in \alpha u^+ R_{22}$ and $(Id - f^+ f^-)\alpha u^+ \in J^+$. Then, $f^+ f^-$ is an idempotent of $\alpha u^+ u^- R_{11}$ and $(Id - f^+ f^-)\alpha u^+ u^- \in J^+ A^- \subseteq Rad\ R_{11}$ which implies that $\alpha u^+ u^-$ is semiregular. □

Proposition 2.4. *Let A be a unital semiregular associative pair. Then, the rings R_{11} and R_{22} are semiregular.*

Proof. As A is unital, for all $\alpha \in R_{11}$ we have $\alpha = \sum_{i=1}^{n} a_i^+ a_i^-$, for $a_i^\sigma \in A^\sigma$. We proceed by induction on the number of nonzero summands in this expression: if there exists just one nonzero summand, i.e, $\alpha = a^+ a^-$ we consider an idempotent (u^+, u^-) such that $u^- \in R_{22} a^-$ and $a^-(Id - u^+ u^-) \in J^-$. Thus, $\alpha - \alpha u^+ u^- \in A^+ J^- \subseteq Rad\ R_{11}$. By lemma 2.3, the element $\alpha u^+ u^- \in R_{11}$ is semiregular, then α is semiregular [12, Corollary 2.4].

Assume now that the result holds whenever the number of nonzero summands is less than or equal to n and let $\alpha = \sum_{i=1}^{n+1} a_i^+ a_i^-$ where all the summands are nonzero. Given $a_1^+ \in A^+$ take an idempotent (u^+, u^-) such that $u^+ \in a_1^+ R_{22}$ and $(Id - u^+ u^-) a_1^+ \in J^+$. The operator $u^+ u^- \in R_{11}$ is an idempotent and $(Id - u^+ u^-)\alpha = (Id - u^+ u^-) a_1^+ a_1^- + \sum_{i=2}^{n+1} (Id - u^+ u^-) a_i^+ a_i^-$. Therefore,

$$(Id - u^+ u^-)\alpha - \sum_{i=2}^{n+1} (Id - u^+ u^-) a_i^+ a_i^- \in J^+ A^- \subseteq Rad\ R_{11}.$$

By induction hypothesis, $\sum_{i=2}^{n+1} (Id - u^+ u^-) a_i^+ a_i^-$ is semiregular, so by [12, Corollary 2.4], we obtain that $(Id - u^+ u^-)\alpha$ is semiregular. Let $\gamma \in R_{11}(Id - u^+ u^-)\alpha$ be an idempotent such that $(Id - u^+ u^-)\alpha(Id - \gamma) \in Rad\ R_{11}$. Observe that $\gamma \in R_{11}\alpha$ and satisfies

$$\alpha(Id - \gamma) - u^+ u^- \alpha(Id - \gamma) = (Id - u^+ u^-)\alpha(Id - \gamma) \in Rad\ R_{11}$$

As the element $u^+ u^- \alpha(Id - \gamma)$ is semiregular (lemma 2.3), therefore, by [12, Corollary 2.4], $\alpha(Id - \gamma)$ is also semiregular, which implies that α is semiregular, by [12, Lemma 2.6]. One can proceed analogously to prove that R_{22} is semiregular. □

Proposition 2.5. *Let A be a unital semiregular associative pair. Then $\mathcal{U}_A$ is a semiregular ring.*

Proof. Since A is unital, the rings R_{11}, R_{22} and $\mathcal{U}_A$ are (Morita) equivalent and semiregularity is a Morita invariant [12, Corollary 2.8]. □

The last result is not true if the hypothesis that A is unital is deleted. The $\mathbb{Z}$-associative pair $(\mathbb{Z}, \mathbb{Z})$ with trivial product is semiregular in spite of the fact that its standard embedding is not, since $\mathbb{Z}$ is not semiregular.

Corollary 2.6. *Let R be a ring with unit. Then R is a semiregular ring if and only if $(\mathcal{M}_{p\times q}(R), \mathcal{M}_{q\times p}(R))$ is a semiregular pair.*

Proof. $A = (\mathcal{M}_{p\times q}(R), \mathcal{M}_{q\times p}(R))$ is semiregular if and only if $\mathcal{U}_A = \mathcal{M}_{p+q}(R)$ is a semiregular ring if and only if R is a semiregular ring [12, Proposition 2.7]. □

3. Preliminary Facts on Modules

We recall that Nicholson established the identification between semiregular and F-semiperfect rings. Since the definition of semiregular associative pair is provided by adapting the one given by Nicholson for rings, it is expected that the same holds for associative pairs. This is a first issue to show. After setting this fact, we also try to generalize other characterizations given by Oberst and Schneider in the paper [13]. This alternative way of characterizing semiregularity stems from module theory. We first have to face a noticeable difference with ring theory. The concept of module in the frame of associative pairs does not allow considering a pair as a module over itself. We try to deal with this obstacle using the standard embedding.

The previous definitions and basic facts on modules over associative pairs we need in this paper can be found in [7]. In this section, we develop the new concepts and results we will use in this note.

Let A be an associative pair. A submodule (L^+, L^-) of a given right A-module (M^+, M^-) is called *small* (*superfluous*) and denoted $(L^+, L^-) \overset{o}{\hookrightarrow} (M^+, M^-)$ if for any submodule (N^+, N^-) of (M^+, M^-) such that $L^\sigma + N^\sigma = M^\sigma$ for all $\sigma \in \{+, -\}$,we have that $N^\sigma = M^\sigma$ for all $\sigma \in \{+, -\}$. Following [10, Theorem 9.1.1], it is not difficult to obtain the following lemma, whose proof is omitted because it runs parallel to the one for rings.

Lemma 3.1. *Let $M = (M^+, M^-)$ be a right A-module.*

i) If M contains maximal submodules,

$$\bigcap_{\substack{B \hookrightarrow M \\ B\ maximal}} B = \sum_{N \overset{o}{\hookrightarrow} M} N.$$

ii) If M contains no maximal submodules,

$$M = \sum_{N \overset{o}{\hookrightarrow} M} N.$$

In this section, we develop results that are satisfied for both modules (A^+, R_{11}) and (R_{22}, A^-) in a similar way. In order to avoid repetition, we will establish them only for the first one.

Lemma 3.2. *If M is a right $\mathcal{U}_A$-module whose radical, $Rad\ M$, is a proper submodule, we have*

$$((Rad\ M)e_2, (Rad\ M)e_1) = \bigcap_{\substack{B \hookrightarrow (Me_2, Me_1) \\ B\ maximal}} B.$$

Proof. It is based on the fact that a submodule (B^+, B^-) of (Me_2, Me_1) is maximal if and only if $(B^+, B^-) = (Le_2, Le_1)$ where L is a maximal submodule of M. □

Let A be an associative pair with Jacobson radical $J = (J^+, J^-)$. If we apply the last lemma to the $\mathcal{U}_A$-module $e_1\mathcal{U}_A = R_{11} \oplus A^+$, as $Rad\ (e_1\mathcal{U}_A) = e_1 Rad\ \mathcal{U}_A = Rad\ R_{11} \oplus J^+$, we obtain

$$(J^+, Rad\ R_{11}) = \bigcap_{\substack{B \hookrightarrow (A^+, R_{11}) \\ B \text{ maximal}}} B.$$

By Lemma 3.1, we have

$$(J^+, Rad\ R_{11}) = \sum_{N \overset{\circ}{\hookrightarrow} (A^+, R_{11})} N.$$

If (I^+, U) is a submodule of (A^+, R_{11}), it can be easily check that I^+ is a right ideal of A^+ and U is a right ideal of R_{11}. Hence we can obtain results on the A-module (A^+, R_{11}) from properties of the unital ring R_{11}.

Lemma 3.3. *For an associative pair A with Jacobson radical $J = (J^+, J^-)$ we have*

i) $(J^+, Rad\ R_{11})$ is a small submodule of (A^+, R_{11}),

ii) (I^+, U) is a direct summand of (A^+, R_{11}) if and only if there exists an idempotent $\alpha \in R_{11}$ such that $(I^+, U) = (\alpha A^+, \alpha R_{11})$.

Proof. i): Suppose $(L^+, U) + (J^+, Rad\ R_{11}) = (A^+, R_{11})$. Then $U + Rad\ R_{11} = R_{11}$ and since $Rad\ R_{11}$ is a small submodule of R_{11}, we have that $U = R_{11}$. This leads to $A^+ = R_{11}A^+ = UA^+ \subset L^+$, whence $(L^+, U) = (A^+, R_{11})$.
ii): For every idempotent $\alpha \in R_{11}$, we have

$$(A^+, R_{11}) = (\alpha A^+, \alpha R_{11}) \oplus ((Id - \alpha)A^+, (Id - \alpha)R_{11}).$$

Conversely, if $(A^+, R_{11}) = (I^+, U) \oplus (L^+, V)$ then, $R_{11} = U \oplus V$. Therefore there exists an idempotent $\alpha \in R_{11}$ such that $U = \alpha R_{11}$ and $V = (Id - \alpha)R_{11}$. It follows that $I^+ = \alpha A^+$. □

Let $M = (M^+, M^-)$ be a right A-module. We denote the *A-submodule of M generated by an element $m^+ \in M^+$* by $\langle m^+ \rangle = (Km^+ + (m^+A^-)A^+, m^+A^-)$ and the *A-submodule generated by $m^- \in M^-$* by $\langle m^- \rangle = (m^-A^+, Km^- + (m^-A^+)A^-)$. For $a^+ \in A^+$, we can consider the submodule $\langle a^+ \rangle$ of (A^+, R_{11}), which can be also expressed by (a^+R_{11}, a^+A^-).

Let $N \subseteq M$ be modules. A submodule L is said to be a *complement* of N in M if $L + N = M$ and $L \cap N$ is a small submodule of L. Observe that under the hypothesis $L + N = M$, we have for a submodule L_0 of L that $L \cap N + L_0 = L$ is equivalent to $L_0 + N = M$. As a consequence, L is a complement of N if and only if $L + N = M$ and L is minimal with respect this property, that is, for a submodule L_0 of L, if $L_0 + N = M$ then $L_0 = L$.

A submodule N of a module M is said to *lie over a summand* P of M if there exists a direct decomposition $M = P \oplus Q$ with $P \subseteq N$ and $N \cap Q$ small in M.

4. Characterization of Semiregular Associative Pairs

Theorem 4.1. *For an associative pair A with Jacobson radical J, the following conditions are equivalent:*

i) A is semiregular.

ii) $\overline{A} = A/J$ is (von Neumann) regular and idempotents modulo J can be lifted.

iii) For every $(a^+, a^-) \in A$, the A-submodules $\langle a^+\rangle$ and $\langle a^-\rangle$ lie over summands in (A^+, R_{11}) and (R_{22}, A^-) respectively.

iv) For every $(a^+, a^-) \in A$, the A-submodules $\langle a^+\rangle$ and $\langle a^-\rangle$ have complements in (A^+, R_{11}) and (R_{22}, A^-), respectively, that are direct summands.

Proof. i)$\Rightarrow$ ii). If A is semiregular, it is clear that $\overline{A}$ is regular. Let (a^+, a^-) be an idempotent modulo J and (f^+, f^-) be an idempotent of A such that $f^+ = \langle a^+a^-a^+\rangle\beta \in \langle a^+a^-a^+\rangle R_{22}$ and $a^+ \equiv \langle a^+a^-a^+\rangle \equiv f^+ \,(\mathrm{mod}\, J^+)$. Then the element $(\langle a^+\beta f^-a^+\rangle, \langle a^-a^+\beta f^-a^+a^-\rangle)$ is an idempotent of A such that

$$\langle a^+\beta f^-a^+\rangle \equiv \langle f^+f^-f^+\rangle \equiv a^+ \,(\mathrm{mod}\, J^+) \quad \text{and}$$
$$\langle a^-a^+\beta f^-a^+a^-\rangle \equiv \langle a^-f^+f^-f^+a^-\rangle \equiv \langle a^-a^+a^-\rangle \equiv a^- \,(\mathrm{mod}\, J^-).$$

ii)$\Rightarrow$ i). Given $a^+ \in A^+$, as $\overline{a^+} \in \overline{A^+}$ is regular, there exists $\overline{a^-} \in \overline{A^-}$ such that $(\overline{a^+}, \overline{a^-})$ is idempotent. Then, there exists an idempotent (e^+, e^-) of A with $e^\sigma \equiv a^\sigma \,(\mathrm{mod}\, J^\sigma)$. The element $e^+ \in A^+$ is regular and $a^+ - e^+ \in J^+$.
i)$\Rightarrow$ iii). Let $a^+ \in A^+$ and take an idempotent (e^+, e^-) of A such that $e^+ \in a^+R_{22}$ with $(Id - e^+e^-)a^+ = y^+ \in J^+$. We have

$$(A^+, R_{11}) = (\langle e^+e^-A^+\rangle, e^+e^-R_{11}) \oplus ((Id - e^+e^-)A^+, (Id - e^+e^-)R_{11})$$

with $(\langle e^+e^-A^+\rangle, e^+e^-R_{11}) \subset \langle a^+\rangle$ and thus $\langle a^+\rangle \cap \langle Id - e^+e^-\rangle \subset (J^+, Rad\, R_{11})$, which is a small submodule. Indeed, if $(Id - e^+e^-)x^+ = \langle e^+e^-a^+\beta\rangle + y^+\beta \in \langle a^+\rangle$, then $\langle e^+e^-a^+\beta\rangle = -\langle e^+e^-y^+\beta\rangle \in J^+$. One can prove that $a^+A^- \cap (Id - e^+e^-)R_{11} \subset Rad\, R_{11}$ in a similar way.
iii)$\Rightarrow$ iv). If $\langle a^+\rangle$ lies over the direct summand (I^+, U), we have $(A^+, R_{11}) = (I^+, U) \oplus (L^+, V)$, for $(I^+, U) \subset \langle a^+\rangle$, and $(L^+, V) \cap \langle a^+\rangle$ being a small submodule of (A^+, R_{11}). If $(L^+, V) = (L_0^+, V_0) + ((L^+, V) \cap \langle a^+\rangle)$ we have

$$(A^+, R_{11}) = (I^+, U) \oplus ((L_0^+, V_0) + ((L^+, V) \cap \langle a^+\rangle)).$$

Then $(A^+, R_{11}) = (I^+, U) \oplus (L_0^+, V_0)$. Hence, $(L_0^+, V_0) = (L^+, V)$. This implies that $(L^+, V) \cap \langle a^+\rangle$ is a small submodule of (L^+, V). Therefore, (L^+, V) is a complement of $\langle a^+\rangle$.
iv)$\Rightarrow$ i). Given $a^+ \in A^+$, let (I^+, U) be the direct summand of (A^+, R_{11}) that is complement of $\langle a^+\rangle$. As $R_{11} = a^+A^- + U$ and U is a direct summand of the ring R_{11}, by [12, Lemma 1.16] there exists a right ideal $V \subset a^+A^-$ such that $R_{11} = V \oplus U$. Then $Id = \alpha + \gamma$, for $\alpha = a^+b^- \in V$ being an idempotent and $\gamma \in U$. Hence, αa^+ is a regular element satisfaying $a^+ - \alpha a^+ = \gamma a^+ \in a^+R_{22} \cap I^+$. As $\langle a^+\rangle \cap (I^+, U)$ is a small submodule, it is contained in $(J^+, Rad\, R_{11})$, whence $a^+ - \alpha a^+ \in J^+$. □

Lemma 4.2. *Let $A = (A^+, A^-)$ be an associative pair with Jacobson radical (J^+, J^-) and $a^+ \in A^+$. If there exist two idempotents (e^+, e^-) and (f^+, f^-) of A such that $e^+ \in a^+R_{22}$with $(Id - e^+e^-)a^+ \in J^+$ and $f^+ \in a^+R_{22}$ with $(Id - f^+f^-)a^+ \in J^+$, then the A-modules eA and fA are isomorphic.*

Proof. Given two idempotents e and f satisfaying the required condition, it is clear that $\langle e^+e^-A^+\rangle + J^+ = a^+R_{22} + J^+ = \langle f^+f^-A^+\rangle + J^+$. Write $\overline{x^\sigma} = x^\sigma + J^\sigma$ and $\overline{A} = A/J$. Then $\langle\overline{e^+}\,\overline{e^-}\,\overline{A^+}\rangle = \langle\overline{f^+}\,\overline{f^-}\,\overline{A^+}\rangle$. We can consider the $\overline{A}$-modules homomorphism given by

$$\begin{array}{ccc} (\langle\overline{e^+e^-}\,\overline{A^+}\rangle, \ \langle\overline{e^-}\,\overline{e^+}\,\overline{A^-}\rangle) & \longrightarrow & (\langle\overline{f^+}\,\overline{f^-}\,\overline{A^+}\rangle, \ \langle\overline{f^-}\,\overline{f^+}\,\overline{A^-}\rangle) \\ (\overline{x^+}, \ \overline{x^-}) & \longmapsto & (\overline{x^+}, \ \langle\overline{f^-}\,\overline{e^+}\,\overline{x^-}) \end{array}$$

which is biyective. Applying [7, Proposition 9], we obtain $eA \cong fA$. □

Corollary 4.3. *If $A = (A^+, A^-)$ is a semiregular associative pair, for every $a^\sigma \in A^\sigma$, the A-submodule $\langle a^\sigma\rangle$ lies over a summand that is unique up to isomorphism, for all $\sigma \in \{+, -\}$.*

Proof. Given $a^+ \in A^+$, an idempotent (e^+, e^-) of A verifies $e^+ \in a^+R_{22}$ and $(Id - e^+e^-)a^+ \in J^+$ if and only if $\langle a^+\rangle$ lies over the summand $\langle e^+\rangle$ of (A^+, R_{11}). Now, it suffices to apply Lemma 4.2. □

Acknowledgments

The author would like to thank Prof. Dr. Esperanza Sánchez Campos, for her interesting suggestions which have enriched this paper. The author wish to thank Prof. Dr. Pedro A. García Sánchez too, for her generous help with translating this text into English.

References

[1] G. Azumaya, *F-Semiperfect modules*, Journal of Algebra **136** (1991), 73–85.

[2] H. Bass, Finitistic dimension and a homological generalization of semiprimary rings, *Trans. Amer. Soc.* **95** (1960), 466–488.

[3] M. Contessa and L. Lesieur, D-Anneux et anneux F-semi-parfaits, *Journal of Algebra* **18** (1971), 103–111.

[4] J. A. Cuenca, A. García and C. Martín, Jacobson density for associative pairs and its applications, *Communications in Algebra* **17 (10)** (1989), 2595–2610.

[5] M. A. Fortes, I. de las Peñas and E. Sánchez, *Elementary proofs of the Wedderburn-Artin theorems for associative pairs, International Journal of Mathematics, Game Theory and Algebra* **8**, Number 2/3 (1998), 107–114.

[6] ______, *Lifting idempotents in associative pairs, Journal of Algebra* **222** (1999), 511–523.

[7] ______, On semiperfect associative pairs, *Communications in Algebra* **27 (12)** (1999), 6015–6038.

[8] ______, *On perfect associative pairs. Application to non unital perfect rings*, Communications in Algebra **29(2)** (2001), 625–638.

[9] W. G. Jansen, FSP rings and modules and local modules, Comunications in Algebra **6(6)** (1978), 617–637.

[10] F. Kasch, Modules and rings, *London Mathematical Society, Monograph* No. **17**, Amer. Math. Soc, 2'nd ed., 1982.

[11] O. Loos, Jordan pairs, Springer- Verlag, Berlin, 1975.

[12] W. K. Nicholson, Semiregular modules and rings, *Can. J. Math* **28** No 5 (1976), 1105–1120.

[13] U. Oberst and H. J. Schneider, Die struktur von projektiven moduln, *Inventiones Math.* **13** (1971), 295–304.

[14] W. Xue, Semiregular modules and F-semiperfect modules, *Comunications in Algebra* **23(3)** (1995), 1035–1046.

In: Mathematics, Game Theory and Algebra Compendium ISBN 978-1-60456-659-8
Editor: Jacob H. Mathias, pp. 505-516

Chapter 42

MAGNETOHYDRODYNAMIC FLOW AND RADIATION OF A RAREFIED GAS

C. Perdikis and E. Rapti
University of Ioannina
Department of Mathematics
Ioannina 451 10, Greece

Abstract

Analytical solutions for the velocity, temperature and magnetic fields of an electrically conducting rarefied gas by the presence of radiation are given, when the magnetic Reynolds number of the flow is not small and the effect of the radiation on the temperature is discuss.

1. Introduction

The fluctuating flow of an incompressible rarefied gas past an infinite porous plate has been investigated by Reddy [1]. Soundalgekar and Aranake [2] have studied the free convection effects on the oscillatory flow of an electrically conducting rarefied gas past an infinite vertical plate with constant suction, when the magnetic Reynolds number of the flow is small. Seddeek [3] have investigated the radiation and magnetohydrodynamic flow by the presence of free convection, without taking into account that the fluid is a rarefied gas.

The purpose of this paper is to get analytical solutions of the velocity, temperature and magnetic fields for an electrically conducting rarefied gas past an infinite vertical plate with constant suction, when the magnetic Reynolds number of the flow is not small. The fluid is a gray, absorbing-emmiting radiation but non-scattering medium and the Rosseland approximation is used to describe the radiative heat flux in the energy equation.

2. Mathematical Analysis

We consider the two-dimensional unsteady flow with the origin of the coordinate system at an arbitrary point on an infinite vertical porous plate. The x'-axis is along the plate in the upward direction and the y'-axis normal to it. It is assumed that the plate is electrically non-conducting and the applied magnetic field is uniform and perpendicular to the plate, so that in the region of the plate it is, $\vec{\mathbf{H}}$=$(H_{x'}, H_0^*, 0)$, where H_0^* is the externally constant applied transverse magnetic field. Let $u^{'}$,$v^{'}$ be the components of the velocity parallel and perpendicular to the plate. The rarefied gas is a gray gas, emmiting and absorbing radiation but non-scattering medium and it is also a viscous, incompressible and electrically conducting one. All the gas properties are considered constant except that the influence of density variation with temperature has been considered only in the body-force term. Under the above assumptions, the hydromagnetic flow relevant to the problem is governed by the following equations:

Continuity equation:

$$\frac{\partial v'}{\partial y'} = 0 \tag{(1)}$$

Equation of motion:

$$\frac{\partial u'}{\partial t'} + v'\frac{\partial u'}{\partial y'} = \frac{dU'}{dt'} + g\beta_1(T^{'} - T^{'}_{\infty}) + \nu\frac{\partial^2 u'}{\partial y'^2} + \frac{\mu_0}{\rho}H_0^*\frac{\partial H_{x'}}{\partial y'} \tag{(2)}$$

Equation of induced magnetic field:

$$\frac{\partial H_{x'}}{\partial t'} + v'\frac{\partial H_{x'}}{\partial y'} = H_0^*\frac{\partial u'}{\partial y'} + \frac{1}{\sigma\mu_0}\frac{\partial^2 H_{x'}}{\partial y'^2} \tag{(3)}$$

Energy equation:

$$\frac{\partial T'}{\partial t'} + v'\frac{\partial T'}{\partial y'} = \frac{k}{\rho c_p}\frac{\partial^2 T^{'}}{\partial y'^2} + \frac{\nu}{c_p}(\frac{\partial u'}{\partial y'})^2 + \frac{1}{\rho\sigma^* c_p}(\frac{\partial H_{x'}}{\partial y'})^2 - \frac{1}{\rho c_p}\frac{\partial q_r}{\partial y'}, \tag{(4)}$$

where U' is the free-stream velocity of the form $U'(t') = U_0(1 + \varepsilon e^{i\omega^{'} t^{'}})$, U_0 is the mean free-stream velocity, ε a positive constant ($\ll 1$), ω' the angular velocity, $t^{'}$ the time, g the acceleration due to gravity, β_1 the coefficient of volume expansion, $T^{'}$ the fluid temperature near the porous plate, $T^{'}_{\infty}$ the fluid temperature at infinity, ν the kinematic viscosity, μ_0 the magnetic permeability, ρ the fluid density, σ^* the electrical conductivity, c_p the specific heat at constant pressure and q_r the radiative heat flux.

The boundary conditions of the problem are :

$$u' = \frac{2 - f_1}{f_1}l\frac{\partial u^{'}}{\partial y^{'}} = L_1\frac{\partial u^{'}}{\partial y^{'}}$$

$$T^{'} - T^{'}_w = \frac{2 - f_2}{f_2}\frac{2\gamma}{\gamma + 1}\frac{l}{P}\frac{\partial T^{'}}{\partial y^{'}} = L_2\frac{\partial T^{'}}{\partial y^{'}}$$

$$H_{x'} = 0, \qquad H_{y'} = H_0^* \qquad at \qquad y' = 0$$

and ((5))

$$u' \to U'(t')$$

$$T' \to T'_\infty$$

$$H_{y'} \to H_0^* \qquad as \qquad y' \to \infty,$$

where f_1 is the Maxwell's reflection coefficient, l the mean free path, T'_w the plate temperature, f_2 the thermal accommondation coefficient, γ the specific heat ratio, P the Prandtl number, $L_1 = \frac{(2-f_1)l}{f_1}$, and $L_2 = \frac{2-f_2}{f_2}\frac{2\gamma}{\gamma+1}\frac{l}{P}$.

For constant suction we get from (1)

$$v' = -v_0 \tag{(6)}$$

where the negative sign indicates that the suction velocity is towards the plate.

By using Rosseland approximation for radiation [3] we take

$$q_r = -\frac{4\sigma}{3k^*}\frac{\partial T'^4}{\partial y'} \tag{(7)}$$

where σ is the Stefan-Bolzmann constant and k^* the mean absorption coefficient. We assume that the temperature differences within the flow are sush that T'^4 may be expressed as a linear function of temperature. This is accomplished by expanding T'^4 in a Taylor series about T'_∞ and neglecting higher-order terms, thus

$$T'^4 \simeq 4T'^3_\infty T' - 3T'^4_\infty \tag{(8)}$$

By using the transformations

$$y = \frac{v_0 y'}{\nu}$$

$$t = \frac{t' v_0^2}{4\nu}$$

$$\theta = \frac{T' - T'_\infty}{T'_w - T'_\infty}$$

$$u = \frac{u'}{U_0}$$

$$U = \frac{U'}{U_0}$$

$$H = (\frac{\mu_0}{\rho})^{1/2}\frac{H_{x'}}{U_0}$$

$$M = (\frac{\mu_0}{\rho})^{1/2}\frac{H_0^*}{\upsilon_0} \qquad ((9))$$

$$P_m = \nu\sigma^*\mu_0$$

$$E = \frac{U_0^2}{c_p(T'_w - T'_\infty)}$$

$$G = \frac{\nu g\beta_1(T'_w - T'_\infty)}{U_0\upsilon_0^2}$$

$$P = \frac{\rho\nu c_p}{k}$$

$$N = \frac{k^*k}{4\sigma T'^3_\infty}$$

$$\omega = \frac{4\nu\omega'}{\upsilon_0^2}$$

equations (2), (3) and (4) through (6), (7) (8) and (9) become

$$\frac{\partial u}{\partial t} - 4\frac{\partial u}{\partial y} = \frac{dU}{dt} + 4G\theta + 4\frac{\partial^2 u}{\partial y^2} + 4M\frac{\partial H}{\partial y} \qquad ((10))$$

$$\frac{\partial H}{\partial t} - 4\frac{\partial H}{\partial y} = 4M\frac{\partial u}{\partial y} + \frac{4}{P_m}\frac{\partial^2 H}{\partial y^2} \qquad ((11))$$

$$3NP\frac{\partial \theta}{\partial t} - 12NP\frac{\partial \theta}{\partial y} = 4(3N+4)\frac{\partial^2 \theta}{\partial y^2} + 12NPE(\frac{\partial u}{\partial y})^2 + \frac{12NPE}{P_m}(\frac{\partial H}{\partial y})^2 \qquad ((12))$$

where M is the ratio of the Alven wave velocity to the suction velocity, P_m the magnetic Prandtl number, P the Prandtl number, E the Eckert number, G the Grashof number and N the radiation parameter.

The corresponding boundary conditions (5) in non-dimensional form are

$$u = h_1\frac{\partial u}{\partial y}, \quad \theta = 1 + h_2\frac{\partial \theta}{\partial y}, \quad H = 0 \quad at \quad y = 0$$

$$u \rightarrow U(t), \quad \theta \rightarrow 0, \quad H \rightarrow 0 \quad as \quad y \rightarrow \infty \qquad ((13))$$

where $h_1 = \dfrac{L_1 v_0}{\nu}$, $h_2 = \dfrac{L_2 v_0}{\nu}$.

In order to solve the non-linear system of equations (10)-(12), under the boundary conditions (13), we assume that

$$u(y,t) = u_0(y) + \varepsilon u_1(y) e^{i\omega t} \qquad ((14))$$

$$\theta(y,t) = \theta_0(y) + \varepsilon \theta_1(y) e^{i\omega t} \qquad ((15))$$

$$H(y,t) = H_0(y) + \varepsilon H_1(y) e^{i\omega t} \qquad ((16))$$

and the non-dimensional free-stream velocity is

$$U(t) = 1 + \varepsilon e^{i\omega t} \qquad ((17))$$

where u_0, θ_0, H_0 are the mean velocity, mean temperature and mean magnetic fields respectively and u_1, θ_1, H_1 the corresponding components in the case of oscillatory motion.

Inserting equations (14)-(16) into equations (10)-(12) we obtain the following differential equations:

$$\begin{aligned} u_0'' + u_0' &= -G\theta_0 - MH_0' \\ H_0'' + P_m H_0' &= -MP_m u_0' \\ (3N+4)\theta_0'' + 3NP\theta_0' &= -3NPE(u_0'^2 + \frac{1}{P_m} H_0'^2) \end{aligned} \qquad ((18))$$

for the mean steady flow, and

$$\begin{aligned} u_1'' + u_1' - \frac{i\omega}{4} u_1 &= -\frac{i\omega}{4} - G\theta_1 - MH_1' \\ H_1'' + P_m H_1' - \frac{i\omega}{4} P_m H_1 &= -MP_m u_1' \\ (3N+4)\theta_1'' + 3NP\theta_1' - \frac{3i\omega}{4} NP\theta_1 &= -6NPE(u_0' u_1' + \frac{1}{P_m} H_0' H_1') \end{aligned} \qquad ((19))$$

for the oscillatory one. The corresponding boundary conditions are

$$u_0 = h_1 u_0', \quad \theta_0 = 1 + h_2 \theta_0', \quad H_0 = 0,$$

$$u_1 = h_1 u_1^{'}, \quad \theta_1 = h_2 \theta_1^{'}, \quad H_1 = 0 \quad at \quad y = 0$$

and ((20))

$$u_0 \to 1, \quad \theta_0 \to 0, \quad H_0 \to 0,$$

$$u_1 \to 1, \quad \theta_1 \to 0, \quad H_1 \to 0 \quad as \quad y \to \infty$$

where primes denote differentiation with respect to y.

The systems of equations (18) and (19) are still coupled and non-liner and in order to solve these we expand u_i , θ_i and H_i (i=0,1) as

$$\begin{aligned} u_i(y,t) &= u_{i1}(y) + E u_{i2}(y) + O(E^2) \\ \theta_i(y,t) &= \theta_{i1}(y) + E \theta_{i2}(y) + O(E^2) \\ H_i(y,t) &= H_{i1}(y) + E H_{i2}(y) + O(E^2) \end{aligned} \quad ((21))$$

where u_i, θ_i and H_i are the mean (i=0) and the fluctuating (i=1) parts of the velocity, temperature and induced magnetic fields respectively and E is the Eckert numbert which is very small for incompressible fluids ($E \ll 1$).

Introducing equations (21) into (18) and (19) we obtain the following systems (22) and (23), when $P_m = 1$:

$$\begin{aligned} u_{01}^{''} + u_{01}^{'} &= -G\theta_{01} - MH_{01}^{'} \\ u_{02}^{''} + u_{02}^{'} &= -G\theta_{02} - MH_{02}^{'} \\ H_{01}^{''} + H_{01}^{'} &= -Mu_{01}^{'} \\ H_{02}^{''} + H_{02}^{'} &= -Mu_{02}^{'} \\ (3N+4)\theta_{01}^{''} + 3NP\theta_{01}^{'} &= 0 \\ (3N+4)\theta_{02}^{''} + 3NP\theta_{02}^{'} &= -3NP(u_{01}^{'2} + H_{01}^{'2}) \end{aligned} \quad ((22))$$

and

$$u''_{11} + u'_{11} - \frac{i\omega}{4}u_{11} = -\frac{i\omega}{4} - G\theta_{11} - MH'_{11}$$

$$u''_{12} + u'_{12} - \frac{i\omega}{4}u_{12} = -G\theta_{12} - MH'_{12}$$

$$H''_{11} + H'_{11} - \frac{i\omega}{4}H_{11} = -Mu'_{11} \qquad ((23))$$

$$H''_{12} + H'_{12} - \frac{i\omega}{4}H_{12} = -Mu'_{12}$$

$$(3N+4)\theta''_{11} + 3NP\theta'_{11} - \frac{3i\omega}{4}NP\theta_{11} = 0$$

$$(3N+4)\theta''_{12} + 3NP\theta'_{12} - \frac{3i\omega}{4}NP\theta_{12} = -6NP(u'_{01}u'_{11} + H'_{01}H'_{11}),$$

with the corresponding boundary conditions:

$$u_{01} = h_1 u'_{01}, \quad \theta_{01} = 1 + h_2\theta'_{01}, \quad H_{01} = 0,$$

$$u_{02} = h_1 u'_{02}, \quad \theta_{02} = h_2\theta'_{02}, \quad H_{02} = 0,$$

$$u_{11} = h_1 u'_{11}, \quad \theta_{11} = h_2\theta'_{11}, \quad H_{11} = 0,$$

$$u_{12} = h_1 u'_{12}, \quad \theta_{12} = h_2\theta'_{12}, \quad H_{12} = 0, \quad at \quad y = 0$$

and ((24))

$$u_{01} \to 1, \quad \theta_{01} \to 0, \quad H_{01} \to 0,$$

$$u_{02} \to 0, \quad \theta_{02} \to 0, \quad H_{02} \to 0$$

$$u_{11} \to 1, \quad \theta_{11} \to 0, \quad H_{11} \to 0,$$

$$u_{12} \to 0, \quad \theta_{12} \to 0, \quad H_{12} \to 0, \quad as \quad y \to \infty\,.$$

The solutions for the mean velocity, mean temperature and mean induced magnetic fields are:

$$\begin{aligned}u_0(y) &= u_{01} + Eu_{02} = 1 - \frac{1}{2}[\frac{2(A-1)G}{A(1+h_2A)(A-M-1)(A+M-1)}e^{-Ay} - \\ &L_3e^{-(1-M)y} - L_4e^{-(1+M)y}] + \frac{E}{2}[(L_{12}+L_{19})e^{-Ay} + (L_{13}+L_{20})e^{-2(M+1)y} + \\ &(L_{14}+L_{21})e^{-2(1-M)y} + (L_{15}+L_{16}+L_{22}+L_{23})e^{-2Ay} + \\ &(L_{17}+L_{24})e^{-(M+1+A)y} + (L_{18}+L_{25})e^{-(1-M+A)y} + L_{29}e^{-(M+1)y} + \\ &L_{30}e^{-(1-M)y}] \qquad ((25))\end{aligned}$$

$$\begin{aligned}\theta_0(y) &= \theta_{01} + E\theta_{02} = \frac{1}{(1+h_2A)}e^{-Ay} + E[L_5e^{-2(M+1)y} + L_6e^{-2(1-M)y} + \\ &(L_7+L_8)e^{-2Ay} + L_9e^{-(M+1+A)y} + L_{10}e^{-(1-M+A)y} + L_{11}e^{-Ay}] \qquad ((26))\end{aligned}$$

$$\begin{aligned}H_0(y) &= H_{01} + EH_{02} = -\frac{1}{2}(L_3e^{-(1-M)y} - L_4e^{-(M+1)y} + \\ &\frac{2MG}{A(1+h_2A)(A-M-1)(A+M-1)}e^{-Ay}) + \frac{E}{2}[(L_{12}-L_{19})e^{-Ay} + \\ &(L_{13}-L_{20})e^{-2(M+1)y} + (L_{14}-L_{21})e^{-2(1-M)y} + (L_{15}+L_{16} - \\ &L_{22}-L_{23})e^{-2Ay} + (L_{17}-L_{24})e^{-(M+1+A)y} + (L_{18}-L_{25})e^{-(1-M+A)y} + \\ &L_{29}e^{-(1+M)y} - L_{30}e^{-(1-M)y}] \qquad ((27))\end{aligned}$$

where

$$A = \frac{3NP}{3N+4}$$

$$L_1 = \frac{G[2A-2+Ah_1(A+M-1)]}{A(1+h_2A)(A-M-1)(A+M-1)}$$

$$L_2 = \frac{2GM}{A(1+h_2A)(A-M-1)(A+M-1)}$$

$$L_3 = \frac{L_1 - [h_1(M+1)+1]L_2}{2(h_1+1)}$$

$$L_4 = L_2 + L_3$$

$$L_5 = \frac{AL_4^2(M+1)}{4(A-2M-2)}$$

$$L_6 = \frac{AL_3^2(1-M)}{4(A+2M-2)}$$

$$L_7 = -\frac{G^2}{4A(1+h_2A)^2(A-M-1)^2}$$

$$L_8 = -\frac{G^2}{4A(1+h_2A)^2(A+M-1)^2}$$

$$L_9 = \frac{GAL_4}{(1+h_2A)(A-M-1)(A+M+1)}$$

$$L_{10} = \frac{GAL_3}{(1+h_2A)(A+M-1)(A-M+1)}$$

$$L_{11} = \frac{-(L_5+L_6+L_7+L_8+L_9+L_{10}) - h_2[2(M+1)L_5 + 2(1-M)L_6]}{1+h_2A} - \frac{h_2[2AL_7 + 2AL_8 + (M+1+A)L_9 + (1-M+A)L_{10}]}{1+h_2A}$$

$$L_{12} = -\frac{GL_{11}}{A(A-M-1)}$$

$$L_{13} = -\frac{GL_5}{2(M+1)^2}$$

$$L_{14} = \frac{GL_6}{2(1-M)(3M-1)}$$

$$L_{15} = \frac{GL_7}{2A(M+1-2A)}$$

$$L_{16} = \frac{GL_8}{2A(M+1-2A)}$$

$$L_{17} = -\frac{GL_9}{A(M+1+A)}$$

$$L_{18} = \frac{GL_{10}}{(1-M+A)(2M-A)}$$

$$L_{19} = -\frac{GL_{11}}{A(A+M-1)}$$

$$L_{20} = -\frac{GL_5}{2(M+1)(3M+1)}$$

$$L_{21} = -\frac{GL_6}{2(1-M)^2}$$

$$L_{22} = -\frac{GL_7}{2A(2A+M-1)}$$

$$L_{23} = -\frac{GL_8}{2A(2A+M-1)}$$

$$L_{24} = -\frac{GL_9}{(M+A+1)(2M+A)}$$

$$L_{25} = -\frac{GL_{10}}{A(1-M+A)}$$

$$L_{26} = -\sum_{i=12}^{25} L_i$$

$$L_{27} = -h_1[A(L_{12}+L_{19}) + 2(M+1)(L_{13}+L_{20}) + 2(1-M)(L_{14}+L_{21}) + 2A(L_{15}+L_{16}+L_{22}+L_{23}) + (M+1+A)(L_{17}+L_{24}) + (1-M+A)(L_{18}+L_{25})]$$

$$L_{28} = \sum_{i=12}^{18} L_i - \sum_{j=19}^{25} L_j$$

$$L_{29} = \frac{L_{26}+L_{27}-[1+h_1(1-M)]L_{28}}{2(1+h_1)}$$

$$L_{30} = L_{28} + L_{29}$$

3. Discussion

The variations of the mean temperature θ_0 when E=0.002, P=0.71 , G=5 , M=5 $h_1 = 0.05$ and $h_2 = 0.02$. We conclude that the mean temperature decreases when the radiation parameter increases.

References

[1] REDDY, K.C. : Fluctuating flow past a porous infinite flat plate in slip flow regime. *Quart. J. Mech. Math.*, **17** (1964) 381-393.

[2] SOUNDALGEKAR,V.M. and ARANAKE, R.N. : Free convection effects on the oscillatory flow of an electrically conducting rarefied gas past an infinite vertical plate with constant suction. *Appl. Sci. Res.*, **34**, (1978) 49-77.

[3] SEDDEEK, M. A. .: Effects of radiation and variable viscosity on a MHD free convection flow past a semi-infinite flat with an aligned magnetic field in the case of unsteady. *International Journal of Heat and Mass Transfer*, **45** (2002) 931-935.

In: Mathematics, Game Theory and Algebra Compendium ISBN: 978-1-60456-659-8
Editor: Jacob H. Mathias, pp. 517-523

Chapter 43

FIRST STRONGLY GRADED MODULES[#]

Mashhoor Refai*
Faculty of Information, Technology and Computer Sciences
Yarmouk University, Irbid – Jordan
Fida A. M. Moh'd**
Department of Mathematics, Yarmouk University
Irbid – Jordan

Abstract

In this paper, we study first strongly graded modules which are another generalization of the srongly graded modules side by side with flexible graded modules.

2000 Mathematics Subject Classification: 13 A 02

Key words and phrases: Graded rings, Graded modules, Strongly graded modules

0. Introduction

Let G be a group with identity e. A ring R is said to be G-graded ring if there exist additive subgroups R_g of R such that $R = \bigoplus_{g\in G} R_g$ and $R_gR_h \subseteq R_{gh}$ for all g, h $\in$ G. The G-graded ring R is denoted by (R,G). We denote by supp(R,G) the support of G which is defined to be $\{g \in G: R_g \neq 0\}$. The elements of R_g are called homogeneous of degree g. For x $\in$ R, x can be written uniquely as $\sum_{g\in G} x_g$ where x_g is the component of x in R_g. Also, we write h(R) $= \bigcup_{g\in G} R_g$.

[#] This research was supported in part by Yarmouk University.
[*] E-mail address: mrefai@yu.edu.jo
[**] E-mail address: feda4444@yahoo.com

We have already known strongly graded module and recognized their properties as well after we had deled with strongly graded rings. Also, the relationship between those notions had been completely indicated. In [5], Refai established the notion of first strongly graded rings and study many of their properties. Then we had a questionable point " Can we construct a like-wise concept of first strongly graded modules which are a generalization of strongly graded modules and have a strong relationship to first strongly rings as the same of that between strongly graded modules and strongly graded rings". This paper comes to answer this question and to release many ideas concerning it. Moreover, we indicate a slight relationship of this new concept and flexible graded modules which are an another generalization of strongly graded modules.

1. Preliminaries

In this section we give some preliminaries about graded rings and graded modules which are necessary in our work. For more details one can look in [4, 5].

Lemma 1.1. Let R be a G-graded ring and x, y $\in$ R, g $\in$ G. Then

$(x+y)_g = x_g + y_g$.

$(xy)_g = \sum_{h\in G} x_h y_{h^{-1}g}$.

Proposition 1.2. Let R be a G-graded ring. Then

R_e is a subring of R and 1 $\in$ R_e.

R_g and R are left (resp. right) R_e-modules, for all g $\in$ G.

Definition 1.3. A G-graded ring (R,G) is said to be strongly graded if $R_g R_h = R_{gh}$ for all g, h $\in$ G.

Proposition 1.4. Let R be a G-graded ring. Then (R,G) is strong iff $R_g R_{g^{-1}} = R_e$ for all g $\in$ G.

Corollary 1.5. (R,G) is strong iff $1 \in R_g R_{g^{-1}}$ for all g $\in$ G.

Definition 1.6. Let R be a G-graded ring. Then (R,G) is first strong if $1 \in R_g R_{g^{-1}}$ for all g$\in$ supp(R,G) or equivalently if $R_g R_{g^{-1}} = R_e$ for all g $\in$ supp(R,G).

Remark 1.7. Every strongly graded ring is first strong but the converse is not true in general (see [5]). Indeed, (R,G) is strong iff it is first strong and supp(R,G) is a subgroup of G.

Proposition 1.8. If (R,G) is first strong then supp(R,G) is a subgroup of G.

Definition 1.9. Let R be a G-graded ring. Then (R,G) is said to be second strong if supp(R,G) is a monoid in G and $R_g R_h = R_{gh}$ for all g, h $\in$ supp(R,G).

Remark 1.10. Every first strongly graded ring is second strong but the converse is not true in general (see [5]).

Definition 1.11. Let M be a left R-module and $N \subseteq M$ be an R-submodule with $N \neq 0$. Then N is said to be essential R-submodule in M if $N \cap L \neq 0$ for each R-submodule L of M such that $L \neq 0$.

Clearly if $M \neq 0$ is an R-module then M is essential in M.

Lemma 1.12. Let M be a left R-module and L, N are submodules of M. If $X \subseteq R$ is a subset, then $X(L + N) = XL + XN$.

Proof: Obvious.

Definition 1.13. Let R be a G-graded ring and M be a left R-module. We say that M is G-graded if there exist additive subgroups M_g of M indexed by $g \in G$ such that $M = \bigoplus_{g \in G} M_g$ and $R_g M_h \subseteq M_{gh}$ for all g, h $\in$ G. If $x \in M_g$ we say that x is a homogeneous of degree g. If $x \in M$ then $x = \sum_{g \in G} x_g$ where $x_g \in R_g$ is the gth – component of x and this presentation is unique. We let $h(M) = \bigcup_{g \in G} M_g$ the set of all homogeneous elements of M and $supp(M,G) = \{g \in G: M_g \neq 0\}$.

Remark 1.14. Suppose M is a G-graded left R-module. Then M_g is left R_e-submodule of M.

Definition 1.15. Let M be a G-graded left R-module and $N \subseteq M$ be an R-submodule. Then N is called G-graded R-submodule of M if $N = \bigoplus_{g \in G} (N \cap M_g)$, or equivalently, if for any $x \in N$, the homogeneous components of x are again in N.

Definition 1.16. Let M be a G-graded left R-module. We say that M is a gr-Noetherian if it satisfies the ascending chain condition on the graded R-submodules of M, or equivalently, if each nonempty family of graded R-submodules of M has a maximal element.

We let $L_{gr}(M)$ to be the family of all graded R-submodules of M. We define the relation "$\leq$" by $A \leq B$ iff $A \subseteq B$ for A, $B \in L_{gr}(M)$. The next proposition shows that this relation makes $L_{gr}(M)$ into a lattice.

Proposition 1.17. Let M be a G-graded left R-module and N, $L \in L_{gr}(M)$. Then $L \cap N$ and $L + N$ are G-graded R-submodules of M.

Proof: Suppose $N = \bigoplus_{g \in G} N_g$ and $L = \bigoplus_{g \in G} L_g$. It is easy to see that N + L is G-graded by $(N + L)_g = (N_g + L_g)$ for all $g \in G$ and $N \cap L$ is G-graded by $(N \cap L)_g = N_g \cap L_g$ for all $g \in G$.

Definition 1.18. Let M be a G-graded left R-module. Then M is called strongly graded if $R_g M_h = M_{gh}$ for all g, h $\in$ G.

Lemma 1.19. Suppose R is first strongly graded ring and M is left R-module. Then $R_\sigma N = N$ for all $\sigma \in$ supp(R,G) and R-submodule N of M.
Proof: Obvious.

Lemma 1.20. Let R be a first strongly graded ring and M be a left R-module. If N and L are two R_e-submodules of M then for any g $\in$ G we have $R_g(N \cap L) = R_g N \cap R_g L$.
Proof: The case where g $\notin$ supp(R,G) is obvious. Suppose g $\in$ supp(R,G) then $N \cap L = R_{g^{-1}} R_g (N \cap L) \subseteq R_{g^{-1}} (R_g N \cap R_g L) \subseteq N \cap L$. Thus $R_g(N \cap L) = R_g N \cap R_g L$.

Definition 1.21. Let R be a G-graded ring and $M = \bigoplus_{g\in G} M_g$ be a G-graded left R-module. Then M is called flexible R-module if $M_g = R_g M_e$ for all g $\in$ G.

Proposition 1.22. Let R be a first strongly graded ring and M $\neq 0$ be a graded left R-module. Then M is flexible iff supp(R,G) = supp(M,G).
Proof: Suppose M is flexible. Clearly, supp(M,G) $\subseteq$ supp(R,G). Let g $\in$ supp(R,G). Hence $R_g{}^{-1} M_g \neq 0$. Since $M_g \neq 0$, g $\in$ supp(M,G). Therefore, supp(R,G) = supp(M,G).

Conversely, assume supp(R,G) = supp(M,G). If g $\in$ supp(R,G). Then $M_g = R_e M_g = R_g R_g{}^{-1} M_g \subseteq R_g M_e \subseteq M_g$. Hence $M_g = R_g M_e$, the case where g $\notin$ supp(R,G) is obvious. Therefore, M is flexible R-module.

2. First Strongly Graded Modules

In this section we introduce the concept of first strongly graded modules and give some of its properties. Also, results depending on this concept are presented.

Definition 2.1. Let R be a G-graded ring and M be a G-graded left R-module. Then M is said to be first strong if the following is satisfied

1. supp(R,G) is a subgroup of G, and
2. $R_g M_h = M_{gh}$ for all (g,h) $\in$ supp (R,G) $\times$ G.

Proposition 2.2. Let R be a G-graded ring. Then (R,G) is first strong iff every G-graded left R-module is first strong.
Proof: Suppose (R,G) is first strong and let M be a G-graded R-module. Let g $\in$ supp(R,G) and h $\in$ G then $M_{gh} = R_e M_{gh} = R_g R_g{}^{-1} M_{gh} \subseteq R_g M_h \subseteq M_{gh}$. By Proposition 1.8, M is first strong. The Converse follows if we set M = R.

Corollary 2.3. It follows from Proposition 2.2 if R is G-graded ring then (R,G) is first strong iff $R_g R_h = R_{gh}$ for all (g,h) $\in$ supp(R,G) $\times$ G.

Remark 2.4. **(1)** Every strongly graded left R-module is first strong. However, the converse is not always true where one can take G = $\mathbf{Z}_4$, R = **C**, the field of complex numbers, and M = **C**[x]. Then R is G-graded by $R_0 = \mathbf{R}$, $R_2 = i\mathbf{R}$ and $R_1 = R_3 = 0$. Also, M = **C**[x] is G-graded R-module by $M_0 = \mathbf{R}[x]$, $M_2 = i\mathbf{R}[x]$ and $M_1 = M_3 = 0$. It is easy to see that (R,G) is first strong and M is first strong but M is not strong for $R_1M_1 = 0 \neq M_2$.

(2) (R,G) is strong iff every G-graded R-module is strong, see [4]. Indeed, this characterization was the idea of first strongly graded modules.

Lemma 2.5. Let R be a G-graded ring and M be a first strongly G-graded left R-module. If supp(R,G) $\bigcap$ supp (M,G) $\neq \Phi$ then e $\in$ supp (M,G).

Proof: Obvious.

Proposition 2.6. Let R be a G-graded ring and M $\neq 0$ be a G-graded left R-module. Then M is strong iff M is first strong and supp (R,G) = G.

Proof: Suppose M is strong. Clearly, M is first strong and supp(M,G) $\subseteq$ supp(R,G). Suppose for the contrary G – supp(R,G) $\neq \Phi$. Then there exists g $\in$ G with $R_g = 0$ which implies e $\notin$ supp(M,G) and this contradicts Lemma 2.5. Thus G = supp(R,G). The converse is obvious.

Proposition 2.7. Let R be a first strongly G-graded ring and M be a G-graded left R-module with supp(R,G) $\bigcap$ supp(M,G) $\neq \Phi$. Then supp(R,G) $\subseteq$ supp(M,G).

Proof: Assume (R,G) is first strong and M is G-graded R-module such that supp(R,G) $\bigcap$ supp(M,G) $\neq \Phi$. By Proposition 2.2, $M_g = R_gM_e$ for all g $\in$ supp(R,G). Let g $\in$ supp(R,G) and assume for the contrary g $\notin$ supp(M,G). Then $R_gM_e = 0$ and e $\notin$ supp (M,G) which is a contradiction to Lemma 2.5. Therefore, supp(R,G) $\subseteq$ supp(M,G).

Definition 2.8. Let R be a G-graded ring and M be a first strongly graded left R-module. Let N $\subseteq$ M be a G-graded R-submodule of M. We say N is first strong R-submodule if $R_\sigma N_g = N_{\sigma g}$ for all $(\sigma,g) \in$ supp (R,G) $\times$ G where $N = \bigoplus_{g\in G} N_g$.

Notation 2.9. We let $L_{F.S}(M)$ denotes the family of all first strongly G-graded R-submodules of M and we define the relation "$\leq$" by N $\leq$ L iff N $\subseteq$ L for all N, L $\in L_{F.S}(M)$. Then we can see that $(L_{F.S}(M), \leq)$ is a partial ordered set.

The next proposition is a characterization of first strongly graded modules in the case "supp(R,G) is a subgroup of G".

Proposition 2.10. Let R $\neq 0$ be a G-graded ring such that supp(R,G) is a subgroup of G. Then the following are equivalent:

(R,G) is first strong.

2. $L_{gr}(M) = L_{F.S}(M)$ for any G-graded left R-module M $\neq$ R.

Proof: (1) gives (2): Suppose (R,G) is first strong and M is any G-graded left R-module. By Proposition 2.2, M is first strong. Let X $\in L_{gr}(M)$, g $\in$ supp(R,G) and s $\in$ G. Then X_{gs} =

$R_eX_{gs} = R_g R_g^{-1} X_{gs} \subseteq R_g X_s \subseteq X_{gs}$ and hence $X_{gs} = R_gX_s$ for all $(g,s) \in supp(R,G) \times G$, i.e., $X \in L_{F.S}(M)$. So, $L_{gr}(M) = L_{F.S}(M)$.

(2) gives (1): Suppose $L_{gr}(M) = L_{F.S}(M)$ for any graded R-module $M \neq R$. Suppose for the contrary (R,G) is not first strong. Then by Proposition 2.2, there exists a G-graded R-module $M \neq R$ such that M is not first strong. So, there are $g \in supp(R,G)$ and $h \in G$ with $M_{gh} \not\subset R_gM_h$. Let $x \in M_{gh} - R_g M_h$ then Rx is G-graded R-submodule of M where $(Rx)_g = R_{gh^{-1}g^{-1}} x$; $g \in G$. But $Rx \notin L_{F.S}(M)$. Thus we get a contradiction to the fact that $L_{gr}(M) = L_{F.S}(M)$. Therefore (R,G) is first strong.

Corollary 2.11. Let R be a G-graded ring. Then (R,G) is first strong iff $L_{gr}(M) = L_{F.S}(M)$ for all graded left R-module M.

Corollary 2.12. If (R,G) is first strong then $L_{F.S}(M)$ is a lattice for any left R-module M.

Proposition 2.13. Let R be a G-graded ring and M be a first strongly graded left R-module. If $N, L \in L_{F.S}(M)$ then $N + L \in L_{F.S}(M)$.

Proof: Follows from Proposition 1.17 and Definition 2.8.

Now we give the complete answer to the question" when does $L_{F.S}(M)$ become a lattice".

Proposition 2.14. Let R be a G-graded ring and M be a first strongly G-graded left R-module. Then $L_{F.S}(M)$ is a lattice [i.e., if $N, L \in L_{F.S}(M)$ then $N \cap L, N + L \in L_{F.S}(M)$] iff $R_\sigma (N \cap L) = R_\sigma N \cap R_\sigma L$ for all $\sigma \in supp(R,G)$ (or $\sigma \in G$) and $N, L \in L_{F.S}(M)$.

Proof: Assume $L_{F.S}(M)$ is a lattice and let $N, L \in L_{F.S}(M)$. Then $N \cap L$ is first strong. Thus for each $(\sigma, h) \in supp(R,G) \times G$ we have $R_\sigma (N \cap L)_h = (N \cap L)_{\sigma h} = N_{\sigma h} \cap L_{\sigma h} = R_\sigma N_h \cap R_\sigma L_h$.

Now, $R_\sigma (N \cap L) = R_\sigma \bigoplus_{g\in G} (N \cap L)_g = \bigoplus_{g\in G} R_\sigma (N \cap L)_g = \bigoplus_{g\in G} (R_\sigma N_g \cap R_\sigma L_g) = (\bigoplus_{g\in G} R_\sigma N_g) \cap (\bigoplus_{g\in G} R_\sigma L_g) = (R_\sigma \bigoplus_{g\in G} N_g) \cap (R_\sigma \bigoplus_{g\in G} L_g) = R_\sigma N \cap R_\sigma L$.

Conversely, assume $R_\sigma (N \cap L) = R_\sigma N \cap R_\sigma L$ for all $N, L \in L_{F.S}(M)$ and $\sigma \in supp(R,G)$. Let $\gamma \in G$. Then $[R_\sigma (N \cap L)]_{\sigma\gamma} = [R_\sigma N \cap R_\sigma L]_{\sigma\gamma}$ implies $R_\sigma (N \cap L)_\gamma = R_\sigma N_\gamma \cap R_\sigma L_\gamma = N_{\sigma\gamma} \cap L_{\sigma\gamma} = (N \cap L)_{\sigma\gamma}$. Thus $N \cap L$ is first strong. Recalling Proposition 2.13, we get $L_{F.S}(M)$ is a lattice.

Now, we give some applications of first strongly graded modules. Some of this applications are similar to that of strongly graded modules.

Proposition 2.15. Let R be a first strongly G-graded ring and M be a G-graded left R-module. Then M is gr-Noetherian iff M_e is Noetherian R_e-module.

Proof: Suppose M is gr-Noetherian and Let $X^{(1)} \subseteq X^{(2)} \subseteq \ldots$ be an ascending chain of R_e-submodules of M_e. Then $RX^{(1)} \subseteq RX^{(2)} \subseteq \ldots$ is an ascending chain of graded R-submodules of M. Thus there exists $n \in \mathbf{N}$ such that $RX^{(n)} = RX^{(n+1)} = \ldots$. So $(RX^{(n)})_e = (RX^{(n+1)})_e = \ldots$ or $X^{(n)} = X^{(n+1)} = \ldots$. Hence M_e is Noetherian.

Conversely, assume that M_e is Noetherian. By Proposition 2.3, M is first strong. Let $X^{(1)} \subseteq X^{(2)} \subseteq \ldots$ be an ascending chain of graded R-submodules of M. Then $X^{(1)} \bigcap M_e \subseteq X^{(2)} \bigcap M_e \subseteq \ldots$ is an ascending chain of R_e-submodules of M_e and hence there exists $n \in N$ such that $X^{(n)} \bigcap M_e = X^{(n+1)} \bigcap M_e = \ldots$ or $X_e^{(n)} = X_e^{(n+1)} = \ldots$. By Corollary 2.11, $R_gX_e^{(n)} = R_gX_e^{(n+1)} = \ldots$ and so $X_g^{(n)} = X_g^{(n+1)} = \ldots$ for all $g \in supp(R,G)$. Hence $X^{(n)} = X^{(n+1)} = \ldots$. Therefore, M is gr-Noetherian.

Remark 2.16. It follows from Proposition 1.22, if (R,G) is first strong and M is G-graded left R-module, then M is flexible iff M is first strong and supp(R,G) = supp(M,G).

References

[1] M. Cohen and L. Rowen, Group graded rings, *Comm. In Algebra*, (**11**)(1983), 1253-1270.

[2] E. C. Dade, Group-graded rings and modules, *Math Z.*, **174** (1980), 241-262.

[3] C. Nastasescu, Strongly graded rings of finite groups, *Comm. In Algebra*, **11**(10) (1983), 1033-1071.

[4] C. Nastasescu and F. Van Oystaeyen, Graded Ring Theory, *Mathematical Library* **28**, North Holand, Amsterdam, (1982).

[5] M. Refai, Various types of strongly graded rings, *Abhath Al-Yarmouk Journal (Pure Sciences and Engineering Series)*, **4** (2) (1995), 9-19.

[6] M. Refai and M. Obeidat, On a strongly-support graded rings, *Mathematica Japonica Journal*, **39**(3) (1994), 519-522.

In: Mathematics, Game Theory and Algebra Compendium
Editor: Jacob H. Mathias, pp. 525-540
ISBN 978-1-60456-659-8

Chapter 44

SOLVING SOME HIGHER-ORDER DISCRETE DYNAMIC SYSTEMS AND APPLICATIONS

R. Ben Taher[1,*], W. Iraoui[1†] and M. Rachidi[2‡]
[1] Département de Mathématiques et Informatique,
Faculté des Sciences,Université Moulay Ismail
B.P. 4010, Beni M'hamed, Meknès, Morocco
[2] Département de Mathématiques et Informatique,
Faculté des Sciences, Université Mohammed V,
B.P. 1014, Rabat-Morocco

Abstract

We are interested here in the solutions of some higher-order discrete dynamic systems in the algebra of square matrices. To this aim, we consider the combinatorial method for the computation of the powers A^n ($n \geq 0$), where A is a square matrix of order ≥ 2. Therefore, various classes of discrete higher-order (matrix) dynamic systems are studied. Examples and applications are discussed.

Key Words : Nonhomogeneous linear recurrence relations, discrete dynamic system, combinatorial powers, higher-order discrete dynamic systems.

2000 Mathematical Subject Classifications : Primary 15A99, 40A05 Secondary 15A18, 40A25, 45M05.

1. Introduction

The scalar nonhomogeneous linear recurrence equations of order $r \geq 2$ are defined as follows,

$$v_{n+1} = a_0 v_n + \cdots + a_{r-1} v_{n-r+1} + c_n, \quad \text{for } n \geq r-1, \tag{1.1}$$

*E-mail address: bentaher@fsmek.ac.ma
†E-mail address: iraoui@fsmek.ac.ma
‡E-mail address: rachidi@fsr.ac.ma

where $a_0, \cdots, a_{r-1}$, $v_0, \cdots, v_{r-1}$ are specified as the coefficients and the initial values, and $\{c_n\}_{n\geq 0}$ is a given sequence of real or complex numbers. Recently, many studies and applications of Equation (1.1) have been improved. Indeed, their solutions are studied by various methods and technics (see [1, 2, 3, 6, 10, 12, 24] for example). In particular, the recent matrix methods of [10] are founded on the following matrix formulation $U_{n+1} = AU_n + C_n$ $(n \geq r-1)$ of Equation (1.1), where A is a companion matrix and $U_n = ^t (v_n, \cdots, v_{n-r+1})$, $C_n = ^t (c_n, 0, \cdots, 0)$ are two columns vectors. Therefore, Equations (1.1) can be considered as a particular case of a discrete dynamic system of the first order,

$$U_{n+1} = AU_n + F_n, \quad \text{for } n \geq 0, \tag{1.2}$$

where $A \in GL(r;\mathbb{C})$ $(r \geq 2)$ the algebra of square matrices $r \times r$, and $\{F_n\}_{n\geq 0}$ is a (fixed) sequence of $\mathbb{R}^r$ (or $\mathbb{C}^r$). Moreover, the system (1.2) is subjected to some specific initial condition $U_0 \in \mathbb{R}^r$ (or $\mathbb{C}^r$)). Equation (1.2) has been extensively studied in the literature, since it is related to many important fields of applied science and engineering (see [12, 17, 18, 24] for example).

Solutions of (1.2) depend on the computation of the explicit expression of the powers A^n $(n \geq 0)$ of $A \in GL(r;\mathbb{C})$. For solving (1.1), using its matrix formulation, the powers of the associated companion matrix are computed in terms of the combinatorial expression of a class of r-generalized Fibonacci sequences (r-GFS for short) (see [7, 10]). For the computation of the powers A^n of an arbitrary $A \in GL(r;\mathbb{C})$, many methods and algorithms have been developped (see [4, 5, 9, 11, 15, 19, 24] for example). The combinatorial method improved by Ben Taher-Rachidi (see [4, 5, 21]) is among those considered recently for studying some expressions of A^n $(n \geq r)$, with the aid of the combinatorial aspect of r-GFS sequences (see [16, 20, 22] for example).

The main purpose of this paper is to apply the combinatorial method (see [4, 5, 21]), for studying some properties of the general higher-order discrete dynamic systems defined by,

$$U_{n+1} = A_0U_n + \cdots + A_{s-1}U_{n-s+1} + F_n, \text{ for } n \geq s-1, \tag{1.3}$$

where $s \geq 2$, $A_0, \cdots, A_{s-1}$ is a finite sequence of $GL(r;\mathbb{C})$ and $\{F_n\}_{n\geq 0}$ is a sequence of $\mathbb{R}^r$ (or $\mathbb{C}^r$). The solution of this general system (1.3) is subjected to the prescribed initial conditions $U_0, \cdots, U_{s-1} \in \mathbb{R}^r$ (or $\mathbb{C}^r$). For $s = 1$ Equation (1.3) is reduced to (1.2). We note also that Expression (1.3) represents a generalization of the scalar nonhomogeneous linear recurrence equation (1.1), to the case of the dynamic systems in the algebra of square matrices $GL(r,\mathbb{C})$.

To study properties of the system (1.3), we proceed in a progressive way as follows. In Section 2. we are interested in the formulation of the solutions of (1.2), in the Putzer's algorithm (see [23, 14]) and in a basis $\{q_{i,j}\}$ of Taylor type (see [24]). Results of this Section are used in the next Sections. In Section 3. we find explicit solutions of a simple second-order discrete system, corresponding to $s = 2$ with $A_0 = 0$. In Section 4. we extend properties of Section 3. to the simple case of (1.3) associated to $A_{s-1} \neq 0$ and $A_0 = \cdots = A_{s-2} = 0$. Section 5. is devoted to the combinatorial solutions of the general discrete systems (1.3). Finally, some concluding remarks and problems are given in the last Section 6.

2. Some Expressions of Solutions of the System (1.2)

Consider the discrete system (1.2), where $A \in GL(r;\mathbb{C})$ and $U(n) =^t (u_1(n), \cdots, u_r(n))$, $F_n =^t (f_1(n), \cdots, f_r(n))(n \geq 0)$ are two columns vectors of$\mathbb{C}^r$. By a simple induction, we can show that the solution of (1.2) is given by,

$$U(n) = A^n U_0 + \sum_{s=0}^{n-1} A^{n-s-1} F(s), \tag{2.4}$$

where $U_0 =^t (u_0, \cdots, u_{r-1})$ is the column vector associated to the prescribed initial conditions (see [10, 17] for example). Expression (2.4) shows that the knowledge of the powers of the matrix A, is important for deriving an explicit solution of Equation (1.2). Let $P_A(X) = X^r - a_0 X^{r-1} - \cdots - a_{r-1}$ $(a_{r-1} \neq 0)$ be the characteristic polynomial of A. It was established in [4] that the combinatorial powers of A are given as follows,

$$A^n = \sum_{k=0}^{r-1} \left(\sum_{j=0}^{k} a_{r-k+j-1} \rho(n-j, r) \right) A^k, \; n \geq r, \tag{2.5}$$

where $\rho(r,r) = 1$, $\rho(n,r) = 0$ for $n \leq r-1$ and

$$\rho(n,r) = \sum_{k_0+2k_1+...+rk_{r-1}=n-r} \frac{(k_0 + ... + k_{r-1})!}{k_0!...k_{r-1}!} a_0^{k_0} ... a_{r-1}^{k_{r-1}}, \tag{2.6}$$

for every $n \geq r$. Therefore, the substitution of (2.5) in (2.4), allows us to derive the following proposition.

Proposition 2.1 *The solution of the discrete system (1.2), with the prescribed initial condition vector $U_0 =^t (v_0, \cdots, v_{r-1})$, is given as follows*

$$U(n) = \sum_{k=0}^{r-1} A^k \left[\rho_k(n) U_0 + \sum_{s=0}^{n-r-1} \rho_k(n-s-1) F(s) + F(n-k-1) \right], \tag{2.7}$$

where $\rho_k(n) = \sum_{j=0}^{k} a_{r-k+j-1} \rho(n-j, r)$.

Example 2.2 We consider a model of a population structured by the following age groups $i = 0, \cdots, r-1$, with $P(t)_i$ representing the i age groups at time t. If we consider that S_i is the probability of survival until the $i+1$ age, we have $P(t)_{i+1} = S_i P(t-1)_i$, for $i = 0, \cdots, r-2$. Let f_i denote the fertility rate of the i age group, then we have $f_i P(t-1)_i$ birth rate born from parents of i age, which implies that $(P(t))_0 = \sum_{i=0}^{r-1} f_i (P(t-1))_i$. If we set

$$A = \begin{pmatrix} f_0 & f_1 & f_2 & \cdots & f_{r-1} \\ S_0 & 0 & 0 & \cdots & 0 \\ 0 & S_1 & 0 & \cdots & 0 \\ \vdots & \ddots & \ddots & & \vdots \\ 0 & \cdots & 0 & S_{r-2} & 0 \end{pmatrix},$$

the population of a given species age at time t satisfies the homogeneous discrete system $P(t) = AP(t-1)$. The matrix A is known in the literature as the Leslie matrix (see[8, 13]). For example, if $r = 3$ the indices $i = 0, 1$ and 2 may represent: the birth, maturity and old age (respectively). In this late case, suppose that $f_0 = f_2 = 0$, then $A = \begin{pmatrix} 0 & f_1 & 0 \\ S_0 & 0 & 0 \\ 0 & S_1 & 0 \end{pmatrix}$. *Let $F(t)$ be another specific phenomenon related to the evolution of the preceding population of a given species (for example, $F(t)$ is the rural exodus rate or other). Therefore, its evolution can be described by the nonhomogeneous discrete system $P(t) = AP(t-1) + F(t-1)$ ($t \geq 1$), with a prescribed initial condition vector P_0. Solutions of this later system, obtained by a straightforward computation, using Proposition 2.1, are given by,*

$$P(2t) = A^2W_1(t) + AW_2(t) + F(2t-1),$$

where $W_1(t) = (S_0f_1)^{t-1}P_0 + \sum_{s=0}^{t-1}(S_0f_1)^{t-s-2}F(2s+1) + F(2t-3)$ *and* $W_2(t) = \sum_{s=0}^{t-1}(S_0f_1)^{t-s-1}F(2s) + F(2t-2)$,
and

$$P(2t+1) = A^2W_3(t) + AW_4(t) + F(2t),$$

for every $t \geq r$, where $W_3(t) = \sum_{s=0}^{t-1}(S_0f_1)^{t-s-1}F(2s) + F(2t-2)$ *and* $W_4(t) = (S_0f_1)^t P_0 + \sum_{s=0}^{t-1}(S_0f_1)^{t-s-1}F(2s+1) + F(2t-1)$.

Solutions (2.4) of the system (1.2) are given in terms of the powers A^k $(0 \leq k \leq r-1)$. Now we are interested in the formulation of these solutions in the Putzer's algorithm (see [17, 14, 23] for example). Suppose that $\lambda_1, \lambda_2, ..., \lambda_r$ are the characteristic roots of $P_A(X)$. Following the method of Putzer the exponential e^{tA} of A, can be written as follows,

$$e^{tA} = \sum_{j=0}^{r-1} V_{j+1}(t)P_j(A), \tag{2.8}$$

where $P_0(A) = I_r$, $P_j(A) = \prod_{k=1}^{j}(A - \lambda_k I_r)$ $(1 \leq j \leq r-1)$ and $V_j(t)$ $(0 \leq j \leq r-1)$ are some functions satisfying $\frac{dV_1}{dt}(t) = a_{r-1}V_r(t)$ and $\frac{dV_j}{dt}(t) = V_{j-1}(t) + a_jV_r(t)$ for $j = 2, \cdots, r-1$ (see [14, 23]). The preceding result of Putzer-Kolodner has been improved recently by the authors. Indeed, the class of functions $V_j(t)$ $(0 \leq j \leq r-1)$ of the Putzer decomposition (2.8) is described as follows. For $j = 1$ we have

$$V_1(t) = B_{00}\Omega_0(t) + B_{01}\Omega_1(t) + ... + B_{0r-1}\Omega_{r-1}(t),$$

where $B_{00} = 1, B_{01} = \lambda_1, B_{0i} = -\lambda_1^i$ $(2 \leq i \leq r-1)$. For $j = 2$ we have

$$V_2(t) = B_{11}\Omega_1(t) + B_{12}\Omega_2(t) + ... + B_{1r-1}\Omega_{r-1}(t),$$

where $B_{11} = 1, B_{12} = \lambda_1 + \lambda_2, B_{1i} = \lambda_2 B_{1i-1} + B_{0i}$ $(3 \leq i \leq r-1)$. And for $2 \leq j \leq r-1$, we have

$$V_{j+1}(t) = B_{jj}\Omega_j(t) + \cdots + B_{jr-1}\Omega_{r-1}(t),$$

such that $B_{jj} = 1$, $B_{j,j+1} = \lambda_1 + ... + \lambda_j$ and $B_{j,i} = \lambda_{j+1} B_{ji-1} + B_{j-1,i}$ $(j+1 \leq i \leq r-1)$ and $\Omega_k(t) = \frac{t^k}{k!} + \sum_{p=o}^{k} a_{r-k+p-1} w_p(t)$, where $w_p(t) = \sum_{n=r}^{\infty} \rho(n-p,r) \frac{t^n}{n!}$ $(0 \leq p \leq r-1)$. Therefore, Expression (2.8) allows us to derive the following proposition.

Proposition 2.3 *Under the preceding hypothesis, the powers of A are given by,*

$$A^n = \sum_{i=0}^{r-1} \sum_{j=i}^{r-1} B_{ij} \rho_j(n) P_i(A), \text{ for every } n \geq r,$$

where $\rho_j(n) = \sum_{k=o}^{j} a_{r-j+k-1} \rho(n-k,r)$ *and*

$$A^n = \sum_{i=0}^{n} B_{in} P_i(A) \quad , \text{ for every } n \leq r-1,$$

where $P_0(A) = I_r$ *and* $P_j(A) = \prod_{k=1}^{j} (A - \lambda_k I_r)$ $(0 \leq j \leq r-1)$.

Combining result of Proposition 2.3 and the solution (2.4) of the discrete systems (1.2), we obtain the following result.

Proposition 2.4 *Under the preceding hypothesis the solution of the discrete system (1.2) is given by,*

$$U(n) = \sum_{i=0}^{r-1} P_i(A) K_i(n) + \sum_{k=0}^{r-1} \Pi_k(A) F(n-k-1), \tag{2.9}$$

such that $K_i(n) = C_{i+1}(n) U_0 + \sum_{k=r}^{n-1} C_{i+1}(k) F(n-k-1)$ *with* $C_{i+1}(n) = \sum_{i=j}^{r-1} B_{ij} \rho_j(n)$ *and* $\Pi_k(A) = \sum_{i=0}^{k} B_{ik} P_i(A)$, *where* $P_0(A) = I_r$ *and* $P_j(A) = \prod_{k=1}^{j} (A - \lambda_k I_r)$ $(0 \leq j \leq r-1)$.

The sequence $\{C_{i+1}(n)\}_{0 \leq i \leq r-1}$ is important in the study of some properties of stability of the discrete systems (see [17] for example).

Proposition 2.3 shows that solutions (2.9) of the discrete system (1.2) are expressed in the basis $\{P_j\}_{0 \leq j \leq r-1}$, defined by $P_0(X) = 1$ and $P_j(X) = \prod_{k=1}^{j} (X - \lambda_k)$, of the space $\mathbb{C}_r[X]$ of polynomials of degree $\leq r-1$. Meanwhile, we can also express these solutions in the basis $\{q_{i,j}\}$ of $\mathbb{C}_r[X]$ defined as follows. Suppose that $P_A(z) = \prod_{k=1}^{s} (z - \lambda_k)^{m_k}$, where $\lambda_k \neq \lambda_j (k \neq j)$ and consider the sequence of polynomials $\{q_{i,j}\}_{(i,j) \in J}$, where $J = \{(i,j), 1 \leq i \leq s, \ 0 \leq j \leq m_i - 1\}$, defined by

$$q_{i,j}(z) = w_i(z)(z - \lambda_i)^j, \text{ for } (i,j) \in J, \tag{2.10}$$

with $w_i(z) = P_A(z)/(z-\lambda_i)^{m_i}$. It is easy to verify that the class of polynomials $\{q_{i,j}\}_{(i,j)\in J}$, defined by (2.10), is a basis of the vector space $\mathbb{C}_r[X]$. Following results of [5], the powers of the matrix A can be written in the basis $\{q_{i,j}\}_{(i,j)\in J}$ as follows,

$$A^n = \sum_{i=1}^{s} \sum_{j=0}^{min(n,m_i-1)} \sum_{k=0}^{m_i-j-1} \frac{q_{i,j+k}(A)}{\prod_{d=1,d\neq i}^{s}(\lambda_i-\lambda_d)^{md}} \binom{n}{j} \lambda_i^{n-j} \alpha_{k,i}, \tag{2.11}$$

where

$$\alpha_{i,j} = -\frac{1}{a_{0j}}\left(a_{1j}\alpha_{i-1,j} + a_{2j}\alpha_{i-2,j} + ... + a_{i-1j}\alpha_{1,j} + a_{i,j}\alpha_{0,j}\right), \text{ with } \alpha_{0,j}=1, \tag{2.12}$$

and

$$a_{ij} = \sum_{h_1+h_2+...+h_{j-1}+h_{j+1}+h_s=i, h_d\leq m_d} \prod_{d=1_{d\neq j}}^{s} \binom{h_d}{m_d} (\lambda_j-\lambda_d)^{m_d-h_d}. \tag{2.13}$$

Expressions (2.11), (2.12) and (2.13) allow us to derive the following result on the solutions of the discrete system (1.2).

Theorem 2.5 *Under the preceding hypothesis, the expression of the unique solution of the discrete system (1.2), subjected to the prescribed initial condition U_0, in the basis $\{q_{i,j}\}_{(i,j)\in J}$ of $\mathbb{C}_r[X]$, is given by*

$$U(n) = \sum_{i=1}^{s}\sum_{j=0}^{m_i-1}\sum_{k=0}^{m_i-j-1} \Gamma_{i,j,n} H_i^{(1)}(A)U_0 + \sum_{l=0}^{n-1}\sum_{i=1}^{s}\sum_{k=0}^{m_i-j-1} \Gamma_{i,j,l} H_i^{(2)}(A)F(n-l-1),$$

for every $n \geq r$, *where* $H_i^{(1)}(A) = \frac{q_{i,j+k}(A)}{\prod_{d=1_{d\neq j}}^{s}(\lambda_i-\lambda_d)^{m_d}}$, $H_i^{(2)}(A) = \sum_{j=0}^{min(m_i-1,l)} \frac{q_{i,j+k}(A)}{\prod_{d=1_{d\neq j}}^{s}(\lambda_i-\lambda_d)^{m_d}}$ *and* $\Gamma_{i,j,n} = \binom{n}{j}\lambda_i^{n-j}\alpha_{k,i}$.

Example 2.6 *Consider the discrete system (1.2) defined by a matrix $A \in GL(r,\mathbb{C})$, with initial value $U(0)$. Suppose that λ_1 and λ_2 are two distinct eigenvalues of A of multiplicity m_1 and m_2 respectively. A straightforward computation, using Theorem 2.5, implies that* $U(n) = S_{1,j,n}(A)U_0 + \sum_{l=0}^{n-1} S_{1,j,l}(A)F(n-l-1) + S_{2,j,n}(A)U_0 + \sum_{l=0}^{n-1} S_{2,j,l}(A)F(n-l-1)$,

where $S_{i,j,n} = \frac{1}{(\lambda_i-\lambda_d)^{m_d}} \sum_{j=0}^{min(n,m_i-1)} \sum_{k=0}^{m_i-k-1} q_{i,j+k}\lambda_i^{n-j}\binom{n}{j}\alpha_{k,i}$, *with* $d=1$ *for* $i=2$ *and* $d=2$ *for* $i=1$.

Fundamental Remark. In this Section we had considered the characteristic polynomial $P_A(X)$. This choice is due to the practical computation of $P_A(X) = det(XI_r - A)$, by the usual technics of the determinant properties. Meanwhile, results of this Section are still valid if we consider any polynomial $R(X) = X^s - \sum_{i=0}^{s-1} b_j X^{s-j-1}$ satisfying $R(A) = \Theta_r$, for example the minimal polynomial of A. Moreover, in the next Sections we continue to use $P_A(X)$, but all the results are also still valid if we replace this polynomial by any $R(X)$ such that $R(A) = \Theta_r$.

3. The Second-Order Discrete Systems: Simple Case

This Section concerns the particular case of the discrete system (1.3) associated to $s = 2$ such that $A_0 = 0$. For reason of convenience we write this system as follows,

$$U(n+2) = AU(n) + F(n), \tag{3.14}$$

where $U(n) =^t (u_1(n), \cdots, u_r(n))$, $F(n) =^t (f_1(n), \cdots, f_r(n))$. The prescribed initial values of the system (3.14) are the two column vectors $U(0) =^t (u_1(0), \cdots, u_r(0))$ and $U(1) =^t (u_1(1), \cdots, u_r(1))$.

As usual, for solving (3.14) we proceed by studying separately its homogeneous and nonhomogeneous parts (respectively).

3.1. Solutions of the Homogeneous Part

In this subsection we suppose that $F(n) =^t (0, \cdots, 0)$. Thus, we show easily that Equation (3.14) is equivalent to the following homogeneous system of the first order,

$$W(n+1) = BW(n), \quad \text{for every} \quad n \geq 1, \tag{3.15}$$

where $W(n) =^t (U(n+1), U(n))$ and $B = \begin{pmatrix} 0 & A \\ I_r & 0 \end{pmatrix}$ (I_r is the identity matrix). A straightforward computation shows that

$$B^{2k} = \begin{pmatrix} A^k & 0 \\ 0 & A^k \end{pmatrix} \text{ and } B^{2k+1} = \begin{pmatrix} 0 & A^{k+1} \\ A^k & 0 \end{pmatrix}, \quad \text{for every } k \geq 0. \tag{3.16}$$

Proposition 3.1 *The unique solution of the homogeneous part of the discrete system (3.15), with the prescribed initial conditions $U(0)$ and $U(1)$, is given by*

$$U(2n) = A^n U(0) \quad \textit{and} \quad U(2n+1) = A^n U(1) \tag{3.17}$$

Hence, similarly to Proposition 2.1, the substitution of (2.5)-(2.6) in Expression (3.17) permits us to derive the following combinatorial solution of (3.15),

Proposition 3.2 *The unique solution of the homogeneous part of the discrete system (3.15), with the prescribed initial conditions $U(0)$ and $U(1)$, is given by*

$$U(2n) = \sum_{p=0}^{r-1} \rho_p(n) A^p U(0) \quad \textit{and} \quad U(2n+1) = \sum_{p=0}^{r-1} \rho_p(n) A^p U(1),$$

such that $\rho_p(n) = \sum_{j=0}^{p} a_{r-p+j-1}\rho(n-j, r)$, where a_j $(0 \leq j \leq r-1)$ are the coefficients of the characteristic polynomial $P_A(X) = X^r - a_0 X^{r-1} - \cdots - a_{r-1}$ of A.

Similarly to Theorem 2.5, solutions of the second-order discrete system (3.15), can be also expressed as follows.

Proposition 3.3 *Suppose that the characteristic polynomial of A is given by $P_A(X) = \prod_{j=1}^{s}(X-\lambda_j)^{m_j}$. Then, the unique solution of the homogeneous part of the discrete system (3.15), with the prescribed initial conditions $U(0)$ and $U(1)$, is*

$$U(2n) = \sum_{j=1}^{s} \sum_{k=0}^{min(n,m_j-1)} \Gamma_{j,k,n}(A)H_j(A)U(0),$$

and

$$U(2n+1) = \sum_{j=1}^{s} \sum_{k=0}^{min(n,m_j-1)} \Gamma_{j,k,n}(A)H_j(A)U(1),$$

with $H_j(A) = \prod_{d=1,d\neq j}(A - \lambda_d I_r)^{m_d}/(\lambda_j - \lambda_d)^{m_d}$ and $\Gamma_{j,k,n}(A) = \sum_{i=0}^{m_j-k-1}\binom{n}{k}\lambda_j^{n-k}\alpha_{i,j}(A-\lambda_j I_r)^{k+i}$, where the $\alpha_{i,j}$ given by (2.12)-(2.13).

Since the eigenvalues of the matrix B are $\pm\sqrt{\lambda_1},\cdots,\pm\sqrt{\lambda_s}$, where $\lambda_1,\cdots,\lambda_s$ are the eigenvalues of the matrix A (see [14]), Propositions 3.2 and 3.3 can be formulated as follows.

Proposition 3.4 *Suppose that the characteristic polynomial of A is given by $P_A(X) = \prod_{j=1}^{s}(X-\lambda_j)^{m_j}$. Then, for the homogeneous part of (3.15), we have*

$$W(n) = \sum_{p=0}^{2r-1}(\sum_{j=0}^{p} b_{2r-p+j-1}\rho(n-j,2r))B^p W(0),$$

such that $W(n) =^t (U(n+1),U(n))$ and $W(0) =^t (U(1),U(0))$, $b_{2i}=0$, $b_{2i+1}=a_i$ $(0\leq i\leq r-1)$ and the $\rho(n-j,r)$ are given by (2.6), where the a_i and r are replaced by b_i and $2r$ (respectively). And using the eigenvalues of B, we have

$$W(n) = \sum_{j=1}^{s}\sum_{k=0}^{min(n,m_j-1)}\sum_{i=0}^{m_j-k-1}\binom{n}{k}\lambda_j^{(n-k-m_j)/2}Q_{i,j,k}(B)\frac{H_j(B)}{2^{m_j}}W(0),$$

where $H_j(B) = \prod_{d=1,d\neq j}(B^2-\lambda_d I_{2r})^{m_d}/(\lambda_j-\lambda_d)^{m_d}$ and $Q_{i,j,k}(B) = (B - \sqrt{\lambda_j}I_{2r})^{k+i}(B+\sqrt{\lambda_j}I_{2r})^{m_j}\alpha_{ij}^{(1)} + (-1)^{n-k-m_j}(B+\sqrt{\lambda_j}I_{2r})^{k+i}(B-\sqrt{\lambda_j}I_{2r})^{m_j}\alpha_{ij}^{(2)}$ such that the $\alpha_{i,j}^{(k)}$ $(k=1,\ 2)$ are defined by $\alpha_{i,j}^{(k)} = -\frac{1}{b_{0,j}^{(k)}}[b_{1,j}^{(k)}\alpha_{i-1,j}^{(k)}+\cdots+b_{i,j}^{(k)}\alpha_{0,j}^{(k)}]$ $(k=1,\ 2)$ with $b_{i,j}^{(k)}=0$ (k=1, 2) for $i > m_j + 2\sum_{1\leq i\neq j\leq d} m_j$, if not

$$b_{i,j}^{(k)} = \sum\prod_{d=1,d\neq j}\binom{h_{d,h}}{m_d}(\lambda_{j,k}^{1/2}-z_{d,h})^{m_d-h_{d,h}}2^{m_j-h_j}\lambda_{j,k}^{(m_j-h_j)/2},$$

where the summation $\sum$ is taken over the set $\{h_{1,1}+h_{1,2}+\cdots+h_{j-1,1}+h_{j-1,2}+h_j+h_{j+1,1}+h_{j+1,2}+\cdots+h_{s,1}+h_{s,2}=i; h_{d,1},h_{d,2}\leq m_d\}$, $h=1,2$ with $z_{d,1}=\lambda_d^{1/2}$, $z_{d,2}=-\lambda_d^{1/2}$ and $\lambda_{j,1}=\lambda_j^{1/2}$, $\lambda_{j,2}=-\lambda_j^{1/2}$.

Example 3.5 *Consider the homogeneous part of the discrete system (3.15) defined by the matrix* $A = \begin{pmatrix} 2 & 1 \\ 0 & -2 \end{pmatrix}$ *and initial values* $U(0) =^t (0,1)$, $U(1) =^t (1,1)$. *A direct verification, using Proposition 3.2, gives,*

$$U(4n) =^t (0, 2^{2n}), U(4n+1) =^t (2^{2n}, 2^{2n}), U(4n+2) =^t (2^{2n}, -2^{2n}2),$$

and

$$U(4n+3) =^t (2^{2n}3, -2^{2n+1}).$$

Example 3.6 *Consider the homogeneous part of the discrete system (3.15) defined by a matrix* $A \in GL(2,\mathbb{C})$, *with initial values* $U(0)$, $U(1)$. *Suppose that* λ *and* μ *are two distinct eigenvalues of A. A straightforward computation, using Proposition 3.3, implies*

$$U(2n) = \frac{1}{\lambda - \mu} \left[\lambda^n (A - \mu I_2) - \mu^n (A - \lambda I_2)\right] U(0),$$

and

$$U(2n+1) = \frac{1}{\lambda - \mu} \left[\lambda^n (A - \mu I_2) - \mu^n (A - \lambda I_2)\right] U(1).$$

3.2. Solutions of the System (3.14)

Suppose now that $F(n) =^t (f_1(n), \cdots, f_r(n))$ is not trivial. The system (3.14) takes the analogous expression of (1.2) : $W(n) = BW(n-1) +^t (F(n-1), 0)$, where B is the matrix considered in the preceding Subsection 3.1. Since the powers of B^n are described in (3.16), Expression (2.4) implies the following proposition.

Proposition 3.7 *The unique solution of (3.14), subjected to the given initial conditions* $U(0)$ *and* $U(1)$, *is*

$$U(2n) = A^n U(0) + \psi_1(A, F(n)) \text{ and } U(2n+1) = A^n U(1) + \psi_2(A, F(n)),$$

where $\psi_1(A, F(n)) = \sum_{k=0}^{n-1} A^{n-k-1} F(2k)$ *and* $\psi_2(A, F(n)) = \sum_{k=0}^{n-1} A^{n-k-1} F(2k+1)$.

With the aid of Proposition 3.2 and results of [5], we obtain the analogous result of Proposition 2.1.

Proposition 3.8 *The unique solution of (3.14), subjected to the given initial conditions* $U(0)$ *and* $U(1)$, *is*

$$U(2n) = \sum_{p=0}^{r-1} A^p S_{p,n}^{(1)}[U(0)] \text{ and } U(2n+1) = \sum_{p=0}^{r-1} A^p S_{p,n}^{(2)}[U(1)], \tag{3.18}$$

such that

$$S_{p,n}^{(1)}[U(0)] = \sum_{j=0}^{r-1} F(2n - 2p - 2) + \sum_{j=0}^{p} a_{r-p+j-1} K_j(n,p),$$

where $K_j(n,p) = \rho(n-j,r)U(0) + \sum_{k=r}^{n-1} \rho(k-j,r)F(2n-2k-2)$ *and*

$$S^{(2)}_{p,n}[U(1)] = \sum_{j=0}^{r-1} F(2n-2p-1) + \sum_{j=0}^{p} a_{r-p+j-1} L_j(n,p),$$

where $L_j(n,p) = \rho(n-j,r)U(1) + \sum_{k=r}^{n-1} \rho(k-j,r)F(2n-2k-1)$.

Moreover, Proposition 3.4 shows that the solutions of (3.14) can be expressed in terms of the eigenvalues of the matrix B.

4. Study of the Discrete System (1.3): Simple Case

In this Section we are interested in the extension of the results of Section 3., to the discrete system (1.3) associated to the case $A_0 = \cdots = A_{s-2} = 0$. Let write this kind of systems as follows,

$$U(n+p) = AU(n) + F(n). \tag{4.19}$$

For $p = 2$ we recover the system (3.14). Similarly, the system (4.19) is equivalent to the following discrete system of the first order,

$$W(n) = BW(n-1) + R(n-1) \tag{4.20}$$

where $W(n) =^t (U(n+p-1), U(n+p-2), \cdots, U(n))$, $R(n) =^t (F(n), 0, \cdots, 0)$ and

$$B = \begin{pmatrix} 0 & 0 & 0 & \cdots & 0 & A \\ 1 & 0 & 0 & \cdots & 0 & 0 \\ 0 & 1 & 0 & \cdots & 0 & 0 \\ \vdots & \ddots & \ddots & \ddots & \ddots & \vdots \\ 0 & 0 & 0 & \cdots & 1 & 0 \end{pmatrix}.$$

Therefore, solutions of (4.19) can be given by (2.4), where appears the powers of B, instead of those of A. By induction we verify easily that such powers of B, are given by

$$B^{pn} = \begin{pmatrix} A^n & 0 & \cdots & 0 \\ 0 & \ddots & \cdots & 0 \\ \vdots & \ddots & \ddots & \vdots \\ 0 & 0 & \cdots & A^n \end{pmatrix}, B^{pn+i} = \begin{pmatrix} 0 & \cdots & A^{n+1} & \cdots & 0 \\ \vdots & \ddots & \ddots & \ddots & \vdots \\ 0 & 0 & \cdots & 0 & A^{n+1} \\ A^n & 0 & \ddots & \ddots & \vdots \\ \vdots & \ddots & \ddots & \ddots & \vdots \\ 0 & \cdots & A^n & \cdots & 0 \end{pmatrix}, \tag{4.21}$$

for every $n \geq 0$ and $1 \leq i \leq p-1$, where the power A^{n+1} in the first line is located at the $(p-i+1)^{th}$ column and the power A^n, of the first column is located at the $(i+1)^{th}$ line. For $p = 2$ we recover expression (3.16) of Section 3. Thus, using Expressions (4.21) and results of [4], we have the following extension of Propositions 3.1, 3.2.

Proposition 4.1 *The solution of the homogeneous part of the discrete system (4.19), subjected to the prescribed initial conditions* $U(0), \cdots, U(p-1)$, *is given by,*

$$U(pn+k) = A^n U(k), \quad \text{for every } n \geq 1 \text{ and } 0 \leq k \leq p-1, \tag{4.22}$$

and

$$U(pn+k) = \sum_{j=0}^{r-1} \rho_j(n) A^j U(k), \quad \text{for every } n \geq 1 \text{ and } 0 \leq k \leq p-1,$$

where $\rho_j(n) = \sum_{h=0}^{j} a_{r-j+h-1} \rho(n-h, r)$.

Moreover, using results of [5] we derive the following extension of Proposition 3.3.

Proposition 4.2 *Suppose that the characteristic polynomial of* A *is* $P_A(X) = \prod_{j=1}^{s}(X - \lambda_j)^{m_j}$. *Then, the unique solution of the homogeneous part of the discrete system (4.19), subjected to the prescribed initial conditions* $U(0), \cdots, U(p-1)$, *is*

$$U(np+k) = \sum_{j=1}^{s} \sum_{k=0}^{min(n,m_j-1)} \Gamma_{j,h,n}(A) H_j(A) U(k),$$

where $H_j(A) = \prod_{d=1, d\neq j}(A - \lambda_d I_r)^{m_d}/(\lambda_j - \lambda_d)^{m_d}$ *and* $\Gamma_{j,h,n}(A) = \sum_{i=0}^{m_j-h-1} \binom{n}{h} \lambda_j^{n-h} \alpha_{i,j} (A - \lambda_j I_r)^{h+i}$, *with the* $\alpha_{i,j}$ *given by (2.12)-(2.13).*

Let $\lambda_1, \cdots, \lambda_s$ be the distinct eigenvalues of the matrix $A \in GL(r, \mathbb{C})$ of multiplicities $m_1, \cdots, m_s$ (respectively). Set $\lambda_j = r_j e^{i\theta_j}$ (here $i^2 = -1$), since the eigenvalues of the matrix B defined by (4.20)are $\lambda_j^{1/p} = r_j^{1/p} e^{i\theta_j + 2k\pi/p}$ of multiplicities $m_1, \cdots, m_s$ (respectively) where $0 \leq k \leq p-1$ and $0 \leq j \leq s$. Then the extension of the preceding Proposition 3.4, in terms of the eigenvalues of the matrix B, is given as follows.

Proposition 4.3 *Under the preceding hypothesis, the solution of the homogeneous part of (4.20), subjected to the initial condition* $W(0) =^t (U(p-1), \cdots, U(0))$, *is*

$$W(n) = \sum_{j=1}^{s} \sum_{k=0}^{min(n,m_j-1)} \sum_{d=0}^{m_j-k-1} \binom{n}{k} r_j^{(n-k-(p-1)m_j)/p} Q_{d,j,k}(B) \frac{H_j(B)}{p^{m_j}} W(0),$$

for every $n \geq 0$, *such that* $H_j(B) = \prod_{d=1, d\neq j}(B^p - \lambda_d I_{pr})^{m_d}/(\lambda_j - \lambda_d)^{m_d}$ *and* $Q_{d,j,k}(B) = \sum_{h=0}^{p-1} e^{\frac{ih\theta_j}{p}[(n-k)-(p-1)m_j]} \left(B^{p-1} + \lambda_{j,h} B^{p-2} + \cdots + \lambda_{j,h}^{p-1} B^0\right)^{m_j} (B - \lambda_{j,h} I_{pr})^{k+d} \alpha_{d,j}^{(h)}$, *where* $\lambda_{j,h} = r_j^{1/p} e^{i(\theta_j + 2h\pi)/p}$ *and the* $\alpha_{d,j}^{(h)}$ $(0 \leq h \leq p-1)$ *are given by similar formula as of those of Proposition 3.4 :* $\alpha_{d,j}^{(h)} = -\frac{1}{b_{0,j}^{(h)}}[b_{1,j}^{(h)} \alpha_{i-1,j}^{(h)} + \cdots + b_{i,j}^{(h)} \alpha_{0,j}^{(h)}]$ $(h = 1, \cdots, p-1)$ *such that* $b_{i,j}^{(h)} = 0$ *for* $d > pm_1 + \cdots + pm_{j-1} + (p-1)m_j + pm_{j+1} + \cdots + pm_d$, *if not*

$$b_{i,j}^{(h)} = \sum \prod_{d=1, d\neq j} \binom{h_{d,t}}{m_d} (\lambda_{j,h} - \lambda_{d,t})^{m_d - h_{d,t}} \prod_{t=1, t\neq h} \binom{h_{j,t}}{m_j} (\lambda_{j,h} - \lambda_{j,t})^{m_j - h_{j,t}},$$

where the summation $\sum$ *is taken over the set* $\{h_{1,1} + \cdots + h_{1,p} + \cdots + h_{j-1,1} + \cdots + h_{j-1,p} + h_{j,1} + \cdots + h_{j,h-1} + h_{j,h+1} + \cdots + h_{j,p} + h_{j+1,1} + \cdots + h_{j+1,p} + \cdots + h_{s,1} + \cdots + h_{s,p} = i; h_d \leq m_d\}$.

Moreover, we have the following extension of Propositions 3.7 and 3.8, for the nonhomogeneous case of (4.19).

Theorem 4.4 *The unique solution of the discrete system (4.19), subjected to the prescribed initial conditions* $U(0), \cdots, U(p-1)$, *is*

$$U(pn+j) = A^n U(j) + \sum_{d=0}^{n-1} A^d F((n-d-1)p+j), \tag{4.23}$$

for every $n \geq 1$ *and* $0 \leq j \leq p-1$,
and

$$U(pn+k) = \sum_{h=0}^{r-1} \rho_j(n) A^h \left[F((n-h)p-k-1) + \sum_{j=0}^{h} a_{r-h+j-1} \Omega_{(j,k,n)}[U(k)] \right], \tag{4.24}$$

for every $n \geq 1$ *and* $0 \leq k \leq p-1$,
where $\Omega_{(j,k,n)}[U(k)] = \rho(n-j,r)U(k) + \sum_{d=0}^{n-1} \rho(d-j,r)F((n-d)p-k-1)$.

Proof. In view of Expression (4.20) – (4.21), we have $w(n) = B^n w(0) + \sum_{s=0}^{n-1} B^{n-s-1} R(s)$. In the sides, by direct computation, we obtain that $B^{pn+i}R(s) =^t (0, ..., A^n F(s), ..., 0)$, where $A^n F(s)$ is located at the $(i+1)^{th}$ column. So according from (4.22), we have the expression (4.23). And Proposition 2.1 gives (4.24). ■

5. Combinatorial Solutions of (1.3)

Consider the general discrete system (1.3), subjected to some prescribed initial conditions $U(s-1), \cdots, U(0)$. Suppose that $A_0, \cdots, A_{s-1} \in GL(r,\mathbb{C})$ satisfy $A_{s-1} \neq 0$ and $A_j A_k = A_k A_j$ $(0 \leq j, k \leq s-1)$. A straightforward computation shows that (1.3) is equivalent to the first-order discrete system,

$$W(n) = BW(n-1) +^t (F(n+s-2, 0, \cdots, 0), \quad \text{for every} \quad n \geq 1, \tag{5.25}$$

where $W(n) =^t (U(n+s-1), \cdots, U(n))$ and $B \in GL(rs;\mathbb{C})$ is the companion matrix

$$B = \begin{pmatrix} A_0 & A_1 & & \cdots & & A_{s-1} \\ I_r & 0 & 0 & \cdots & 0 & 0 \\ 0 & I_r & 0 & \cdots & 0 & 0 \\ \vdots & \ddots & \ddots & \ddots & \ddots & \vdots \\ 0 & 0 & & \cdots & I_r & 0 \end{pmatrix}. \tag{5.26}$$

The initial value of the system (5.25) is the vector column $W(0) =^t (U(s-1), \cdots, U(0))$. The unique solution of the system (1.3) can be obtained from that of (5.25), if we succeed to compute the powers of the matrix (5.26) in terms of $A_0, \cdots, A_{r-1}$. To this aim, we consider the recent result for computing the powers of the usual companion matrix (see [7, 10]) applied to the matrix (5.26). Indeed, consider the class $\{Y_n^{(d)}\}_{n\geq 0}$ $(0 \leq d \leq$

$s-1$) of sequences defined by the linear homogeneous part of (1.3) in $GL(r,\mathbb{C})$ as follows : $Y_n^{(d)} = S_n^{(d)} = \delta_{n,d} I_r$ for $0 \le n \le s-1$ and

$$Y_{n+1}^{(d)} = A_0 Y_n^{(d)} + A_1 Y_{n-1}^{(d)} + \cdots + A_{s-1} Y_{n-s+1}^{(d)}, \text{ for } n \ge s-1. \tag{5.27}$$

For $k \le -1$, we set $Y_k^{(d)} = 0_r$ (the zero matrix of $GL(r,\mathbb{C})$) . As for the matrices associated to (3.2) and (4.1), a direct computation allows us to have,

$$B^n = \begin{pmatrix} Y_{n+s-1}^{(s-1)} & Y_{n+s-1}^{(s-2)} & \cdots & Y_{n+s-1}^{(0)} \\ Y_{n+s-2}^{(s-1)} & Y_{n+s-2}^{(s-2)} & \cdots & Y_{n+s-2}^{(0)} \\ \vdots & \vdots & \ddots & \vdots \\ Y_n^{(s-1)} & Y_n^{(s-2)} & \cdots & Y_n^{(0)} \end{pmatrix}, \tag{5.28}$$

for every $n \ge 0$. Thus, using (5.27)-(5.28), solutions of the homogeneous part of (1.3) are given by

$$U(n) = Y_n^{(s-1)} U(s-1) + Y_n^{(s-2)} U(s-2) + \cdots + Y_n^{(0)} U(0), \quad \text{for every } \; n \ge s.$$

Hence, the combinatorial aspect of sequences (5.27) given in [21] yields the explicit solutions of the homogeneous part of (5.25) as follows.

Proposition 5.1 *Under the preceding hypothesis, the unique solution of the homogeneous part of (5.25), subjected to initial conditions $(U(s-1), \cdots, U(0))$, is given by*

$$U(n) = \sum_{k=0}^{s-1} A_k \left(\sum_{j=0}^{k} \rho(n-j,s) U(s-k-1+j) \right), \quad \textit{for every } \; n \ge s,$$

where $\rho(j,s) = 0_r$ for $0 \le j \le s-1$, and

$$\rho(n,s) = \sum_{k_0+2k_1+...+sk_{s-1}=n-s} \frac{(k_0+...+k_{s-1})!}{k_0!...k_{s-1}!} A_0^{k_0} ... A_{s-1}^{k_{s-1}}, \tag{5.29}$$

for every $n \ge s$.

For $s = 2$ we have the following corollary,

Corollary 5.2 *Consider the homogeneous part of the second-order discrete system (1.3) such that $A_0 A_1 = A_1 A_0$. Then, its solution is given by*

$$U(n) = A_0 \rho(n,2) U(1) + A_1 (\rho(n-1,2) U(1) + \rho(n,2) U(0)),$$

for every $n \ge 2$, where $\rho(n,2) = \sum_{k=0}^{[(n-2)/2]} \frac{(n-k-2)!}{k!(n-2k-2)!} A_0^{n-2k-2} A_1^k$.

Application of Expression (2.4) shows that the unique solution of Equation (5.25) is given as follows $W(n) = B^n W(0) + \sum_{j=0}^{n-1} B^{n-j-1} \mathcal{F}(j)$, where $\mathcal{F}(j) = {}^t(F(j+s-1), 0, \cdots, 0)$. Since $Y_n^{(s)} = A_0 \rho(n,s) + A_1 \rho(n-1,s) + \cdots + A_{s-1} \rho(n-s+1,s)$, where $\rho(n,s)$ is given by (5.29), the solution of the system (1.3) is as follows.

Proposition 5.3 *Under the preceding hypothesis, the unique solution of the discrete system (1.3), subjected to initial conditions $U(s-1),\cdots,U(0))$, is given as follows,*

$$U(n)=F(n-1)+\sum_{k=0}^{s-1}A_k\left(\sum_{j=0}^{k}\rho(n-j,s)U(s-k-1+j)+\sum_{j=s-1}^{n-k-2}\rho(n-k-j-2+s,s)F(j)\right),$$

for every $n\geq s$.

Example 5.4 Consider the nonhomogeneous second-order discrete system (1.3) such that $A_0=\begin{pmatrix}1&1\\-1&-1\end{pmatrix}$ and $A_1=I$, subjected to the initial conditions $U(1)$, $U(0)$. A straightforward computation, using Proposition 5.3 implies

$$U(2n+1)=U(1)+nA_0U(0)+\sum_{j=1}^{n}F(2j)+\sum_{j=0}^{n-1}(n-j)A_0F(2j+1)$$

and

$$U(2n)=U(0)+nA_0U(1)+\sum_{j=1}^{n-1}F(2j+1)+\sum_{j=1}^{n-1}(n-j)A_0F(2j)$$

6. Some Concluding Remarks

For the preceding discrete systems (1.2) and (1.3) we can consider the coefficients, the initial conditions and the sequence $\{F_n\}_{n\geq 0}$ in $GL(r;\mathbb{C})$ (or $GL(r;\mathbb{R})$) instead of$\mathbb{C}^r$ (or $\mathbb{R}^r$). More generally, instead of the algebra of square matrices $GL(r;\mathbb{C})$, the preceding results on (1.2)-(1.3) may be expressed for algebraic of unitary topological algebra (over$\mathbb{C}$).

On the other hand, for the discrete system (1.3) viewed as a nonhomogeneous linear recurrence in the algebra of square matrices $GL(r;\mathbb{C})$ (or $GL(r;\mathbb{R})$), results of Sections 2 and 3 can be formulated and interpreted in the spirit of [1, 2, 3, 6, 10, 12]. Indeed, it seems for us that analogous results on the solutions of scalar linear nonhomogeneous Equation (1.1), obtained in [6, 10] can be extended to the matrix Equation (1.3).

In Section 5. solutions of Equations (1.3) are expressed under the combinatorial form, in terms of the powers of $A_0,\cdots,A_{s-1}$. As it was pointed before, if $A_0,\cdots,A_{s-1}$ are algebraic such that $P_k(A)=0$, with $P_k(X)=X^{s_k}-a_0^{(k)}X^{s_k-1}-\cdots-a_{s_k-1}$. Then, these solutions can be given in terms of the powers $A_0^{j_1},\cdots,A_{s-1}^{j_{s-1}}$ $(0\leq j_k\leq s_k)$. It seems for us important to study solutions of (1.3), using the spectral aspects of the matrices $A_0,\cdots,A_{s-1}$, as it was done for the two particular cases of Section 3..

Another interesting problem is to study the discrete system (1.2) (or more generally (1.3)), when the matrix $A=A(n)$ is a function of n, taking into account properties of the matrices $A(n)$ and $F(n)$ $(n\geq 0)$. For example, the periodic property $A(n+p)=A(n)$ will lead to a closed result obtained for the discrete system (1.3). In particular, the discrete Floquet's theorem and some of its applications, may be written in adequate and precise terms, taking into account the algebraic expressions of the matrices $A(0),\cdots,A(p-1)$.

References

[1] A. Andrade and S.P. Pethe, On the rth-order Nonhomogeneous Recurrence Relation and Some Generalized Fibonacci Sequences, *Fibonacci Quart.* **30.3** (1992), pp. 256-62.

[2] P.R.J. Asveld, A Family of Fibonacci-Like Sequences, *Fibonacci Quart.* **25.1** (1987), pp. 81-83.

[3] P.R.J. Asveld, Another Family of Fibonacci-Like Sequences, *Fibonacci Quart.* **25.4** (1987), pp. 361-64.

[4] R. Ben Taher and M. Rachidi, Linear recurrence relations in the algebra of matrices and applications, *Linear Algebra And Its Applications*, Vol. **330** (1-3) (2001), pp. 15-24.

[5] R. Ben Taher and M. Rachidi, Some explicit formulas for the polynomial decomposition of the matrix exponential and applications, *Linear Algebra And Its Applications*, **350**, Issue 1-3 (2002), pp. 171-184.

[6] R. Ben Taher, M. Mouline and M. Rachidi, Solving some general nonhomogeneous recurrence relations of order r by linearization method and application to polynomial and factorial polynomial case, *Fibonacci Quarterly* **40-1** (2002), pp. 79-84.

[7] R. Ben Taher and M. Rachidi, On the matrix powers and exponential by the r-generalized Fibonacci sequences methods. The companion matrix case, *Linear Algebra and Applications*, **370** (2003), pp. 341-353.

[8] H. Caswell, *Matrix population models*, Sinauer Associates, Inc. Sunderland, Massachussetts, USA (1989).

[9] H.W. Cheng and S.S.-T. Yau, On more explicit formulas for matrix exponential, *Linear Algebra And Its Applications*. **262** (1997), pp. 131-163.

[10] B. El Wahbi, M. Mouline and M. Rachidi, Solving nonhomogeneous recurrence relations by matrix methods, *The Fibonacci Quarterly* (2002), in press.

[11] F.R. Gantmacher, *Theory of matrices*, Chelsea Publishing Company, New York (1959).

[12] A.F. Horadam and A.G. Shannon, Asveld's Polynomials $p_j(n)$, *Applications of Fibonacci*. Numbers 1, Eds G.E. Bergum, A.N. Philippou and A.F. Horadam, Dordrecht : Kluwer (1988), pp. 163-176.

[13] N. Keyfitz, *Introduction to the Mathematics of population, with Revisions*, Addison-Wesley, Reading, Massachussetts (1977).

[14] I.I. Kolodner, On $exp(tA)$ with A satisfying a polynomial, *J. Math.Anal. and Appl.* **52** (1975), pp. 514-524.

[15] I. E. Leonardo, The matrix exponential, *SIAM Review* Vol. 38, No. 3 (1996), pp. 507-512.

[16] C. Levesque, On the $m-th$ order linear recurrences, *Fibonacci Quarterly* **23** (4), (1985), pp. 290-295.

[17] W. G. Kelly and A. C. Peterson, *Difference equations : an introduction with application*. Academic Press, Second Edition (2001).

[18] K. S. Miller, *Linear difference equations*, W.A. Benjamin Inc., New York-Amsterdam (1968).

[19] C. Moler and C. Van Loan, Nineteen dubious ways to compute the exponential of a matrix, *SIAM Review* Vol. 20 (4), (1978), pp. 801-836.

[20] M. Mouline and M. Rachidi, Application of Markov Chains properties to r-Generalized Fibonacci Sequences, *Fibonacci Quarterly* **37** (1999), pp. 34-38.

[21] M. Mouline and M. Rachidi, *Suites de Fibonacci généralisées, Théorème de Cayley-Hamilton et chaines de Markov*, Rendiconti del Seminario Matematico di Messina Serie II, Tomo XIX, No. 4 (1996/97), pp. 107-115.

[22] G.N. Philippou, On the k-th order linear recurrence and some probability applications, Applications of Fibonacci Numbers 1, Eds G.E. Bergum, A.N. Philippou and A.F. Horadam, Dordrecht: Kluwer Academic Publishers (1988).

[23] E.J. Putzer, Avoiding the Jordan canonical form in the discussion of linear systems with constant coefficients, *Amer. Math. Monthly* **73** (1966), pp. 2-7.

[24] L. Verde-Star, Operator identities and the solution of linear matrix difference and differential equations, *Studies in Applied Mathematics* **91** (1994), pp. 153-177

In: Mathematics, Game Theory and Algebra Compendium ISBN: 978-1-60456-659-8
Editor: Jacob H. Mathias, pp. 541-546

Chapter 45

A PROPERTY OF THE CYCLOTOMIC POLYNOMIAL

Tang Tao
SP XuanWei Power Generation Co., Ltd.
Xuan Wei, Yun Nan
The People's Republic of China

Abstract

The polynomial $\Phi(x) = \sum_{j=0}^{p-1} x^j$ is called the cyclotomic polynomial, where p is an odd prime. We prove an interested result:

$$4\,\Phi(x) = [C(x)]^2 - (-1)^n p[B(x)]^2,$$

$$C(x) = \sum_{j=0}^{n} c_j x^j \text{ , } B(x) = \sum_{j=0}^{n} b_j x^j \text{ , } n = \frac{p-1}{2}$$

where c_j and b_j are rational integers (j = 0, 1, 2, …, n).

Keywords: prime; quadratic residue; cyclotomic polynomial

In this paper, we use the following notations:

p is an odd prime, $n = \dfrac{p-1}{2}$;

$k_1, k_2, \ldots, k_n$ is a complete system of quadratic residues of p;
$l_1, l_2, \ldots, l_n$ is a complete system of quadratic non-residues of p;
$e_k = 1$ (when k is a quadratic residues of p), $e_k = 0$ (when k is a quadratic non-residues of p);

Legendre's symbol $v_k = \left(\dfrac{k}{p}\right)$;

ω is the primitive pth root of 1, $Q(\omega)$ is the cyclotomic field, $\zeta = \sum_{j=1}^{n} \omega^{k_j}$, $\lambda = \sqrt{(-1)^n p}$;

polynomials: $\Phi(x) = \sum_{j=0}^{p-1} x^j = f(x)g(x)$,

$$f(x) = \prod_{j=1}^{n}(x-\omega^{k_j}) = \sum_{j=0}^{n} \beta_{n-j} x^j \text{ , } g(x) = \prod_{j=1}^{n}(x-\omega^{l_j}).$$

Lemma 1. $\lambda = 1 + 2\zeta$;

$$\zeta^2 + \zeta + \frac{(-1)^n p - 1}{4} = 0.$$

Proof. Since $\lambda = 1 + 2\sum_{j=1}^{n} \omega^{j^2}$ (see Ref.[1], pp.85~87 and Ref.[2], p.195),

so $\lambda = 1 + 2\zeta$,

$$\zeta^2 + \zeta + \frac{(-1)^n p - 1}{4} = 0.$$

Lemma 2. $\beta_k = a_{n-k} + b_{n-k}\zeta$, where a_{n-k} and b_{n-k} are rational integers ($k = 1, 2, \ldots, n$), $\beta_0 = 1$.

Proof. Let $S_k = \sum_{j=1}^{n} (\omega^{k_j})^k$ ($k = 1, 2, \ldots, n$),

then $S_k = \zeta$ (when k is a quadratic residues of p),

$S_k = -1 - \zeta$ (when k is a quadratic non-residues of p),

namely $S_k = d_k + f_k \zeta$,

where d_k and f_k are rational integers ($k = 1, 2, \ldots, n$).

From $f(x) = \prod_{j=1}^{n}(x-\omega^{k_j}) = \sum_{j=0}^{n} \beta_{n-j} x^j$

and *Newtonian* formula $\sum_{j=0}^{k-1} S_{k-j}\beta_j + k\beta_k = 0$.

We have:

since $S_1 + \beta_1 = 0$, $S_1 = d_1 + f_1\zeta$, d_1 and f_1 are rational integers

we obtain $\beta_1 = a_{n-1} + b_{n-1}\zeta$, $a_{n-1} = -d_1$ and $b_{n-1} = -f_1$ are rational numbers;

since $\sum_{j=0}^{1} S_{2-j}\beta_j + 2\beta_2 = 0$, $\beta_1 = a_{n-1} + b_{n-1}\zeta$ (a_{n-1} and b_{n-1} are rational numbers),

$S_j = d_j + f_j\zeta$, d_j and f_j are rational integers ($j = 1, 2$),

we obtain $\beta_2 = a_{n-2} + b_{n-2}\zeta$, $a_{n-2} = -\frac{1}{2}[d_2 + d_1 a_{n-1} + \frac{f_1 b_{n-1}((-1)^n p - 1)}{4}]$,

$b_{n-2} = -\frac{1}{2}[f_2 + d_1 b_{n-1} + f_1(a_{n-1} - b_{n-1})]$,

a_{n-2} and b_{n-2} also are rational numbers;

generally, we use the inductive method, since $\sum_{j=0}^{k-1} S_{k-j}\beta_j + k\beta_k = 0$,

$\beta_j = a_{n-j} + b_{n-j}\zeta$, a_{n-j} and b_{n-j} are rational numbers ($j = 1, 2, \ldots, k-1$),

$S_j = d_j + f_j\zeta$, d_j and f_j are rational integers ($j = 1, 2, \ldots, k$),

we obtain $\beta_k = a_{n-k} + b_{n-k}\zeta$,

$$a_{n-k} = -\frac{1}{k}\{d_k + \sum_{j=1}^{k-1}[d_{k-j}a_{n-j} + \frac{f_{k-j}b_{n-j}((-1)^n p - 1)}{4}]\},$$

$$b_{n-k} = -\frac{1}{k}\{f_k + \sum_{j=1}^{k-1}[d_{k-j}b_{n-j} + f_{k-j}(a_{n-j} - b_{n-j})]\},$$

a_{n-k} and b_{n-k} also are rational numbers;

hence $\beta_i = a_{n-i} + b_{n-i}\zeta$, a_{n-i} and b_{n-i} are rational numbers ($i = 1, 2, \ldots, n$).

By *Viete* theorem, we know

$$\beta_i = (-1)^i \sum_{j1<j2<\ldots<ji} \prod_{v=1}^{i} \omega^{k_{jv}}, \ (i = 1, 2, \ldots, n),$$

because $\omega^{k_1}, \omega^{k_2}, \ldots, \omega^{k_n}$ are algebraic integers over $Q(\omega)$,

so $\beta_i = \sum_{j=0}^{p-2} e_j^{(i)}\zeta^j$, where $e_0^{(i)}, e_1^{(i)}, \ldots, e_{p-2}^{(i)}$ are rational integers.

Since $\beta_i = a_{n-i} + b_{n-i}\zeta = a_{n-i} + b_{n-i}\sum_{j=0}^{p-2}(e_j - e_{p-1})\omega^j$, so $e_0^{(i)} = a_{n-i} + b_{n-i}(e_0 - e_{p-1})$, $e_j^{(i)} = b_{n-i}(e_j - e_{p-1})$ ($j = 1, 2, \ldots, p-2$).

When p-1 is a quadratic residues of p, $e_{p-1}= 1$, $e_{l_1} = 0$, $e_{l_1}^{(i)} = -b_{n-i}$, b_{n-i} is a rational integer, $a_{n-i} = e_0^{(i)} - b_{n-i}(e_0 - e_{p-1})$, a_{n-i} also is a rational integer;

When p-1 is a quadratic non-residues of p, $e_{p-1}= 0$, $e_{k_1} = 1$, $e_{k_1}^{(i)} = b_{n-i}$, b_{n-i} is a rational integer, $a_{n-i} = e_0^{(i)} - b_{n-i}(e_0 - e_{p-1})$, a_{n-i} also is a rational integer.

Theorem 1.

$$f(x) = A(x) + B(x)\zeta \tag{1}$$

$$2f(x) = C(x) + B(x)\lambda \tag{2}$$

$$4\Phi(x) = [C(x)]^2 - (-1)^n p[B(x)]^2 \tag{3}$$

$$A(x) = \sum_{j=0}^{n} a_j x^j,\ B(x) = \sum_{j=0}^{n} b_j x^j,\ C(x) = \sum_{j=0}^{n} c_j x^j,$$

$$a_n = 1,\ b_n = 0,\ c_n = 2,$$

where a_j, b_j and c_j are rational integers ($j = 0, 1, 2, \ldots, n$).

Proof. Indeed Eq.(1) and Eq. (2), $a_n=1$, $b_n= 0$, $c_n=2$.

Since polynomials $[C(x)]^2 - (-1)^n p[B(x)]^2$ and $\Phi(x)$ have common roots $\omega^{k_1}, \omega^{k_2}, \ldots, \omega^{kn}$; degrees of polynomials $[C(x)]^2 - (-1)^n p[B(x)]^2$ and $\Phi(x)$ all are p-1; $\Phi(x)$ is an irreducible polynomial over rational number field;

so $[C(x)]^2 - (-1)^n p[B(x)]^2 = e\,\Phi(x)$, e is a rational number.

From $b_n= 0$, $c_n=2$, we obtain $e = 4$.

For coefficients c_k and b_k of $C(x)$ and $B(x)$ ($k = 0, 1, 2, \ldots, n$), we have:

Theorem 2. Let matrices:

$C = (c_0, c_1, c_2, \ldots, c_n)$, $B = (b_0, b_1, b_2, \ldots, b_n)$,

$$U = \begin{pmatrix} -1, & m_1, & 0, & \ldots & \ldots & & \ldots & & \ldots, & 0, & 0 \\ -1, & -1, & m_2, & 0, & \ldots & & \ldots & & \ldots, & 0, & 0 \\ \ldots & & \ldots & & \ldots & & \ldots & & \ldots & & \ldots \\ -1, & -1, & -1, & \ldots & \ldots, & -1, & m_k, & 0, & \ldots, & 0, & 0 \\ \ldots & & \ldots & & \ldots & & \ldots & & \ldots & & \ldots \\ -1, & -1, & -1, & \ldots & \ldots & & \ldots & & \ldots, & -1, & m_n \\ -1, & -1, & -1, & \ldots & \ldots & & \ldots & & \ldots, & -1, & -1 \end{pmatrix}$$

$$V=\begin{pmatrix} v_1, & 0, & \cdots & \cdots & \cdots & & \cdots & & \cdots, & 0 \\ v_2, & v_1, & 0, & \cdots & \cdots & & \cdots & & \cdots, & 0 \\ \cdots & & \cdots & & \cdots & & \cdots & & \cdots & \\ v_k, & v_{k-1}, & \cdots & \cdots, & v_2, & v_1, & 0, & \cdots & \cdots, & 0 \\ \cdots & & \cdots & & \cdots & & \cdots & & \cdots & \\ v_{n+1}, & v_n, & \cdots & & \cdots & & \cdots & \cdots, & v_2, & v_1 \end{pmatrix}$$

V^{-1} be the inverse matrix of V, where $m_k = 2n\text{-}2k\text{+}2$ ($k = 1, 2, \ldots, n$), then

$$C = -BUV^{-1} \tag{4}$$

$$(-1)^n pB = -CUV^{-1} \tag{5}$$

Proof. Let $S_k = \sum_{j=1}^{n} (\omega^{k_j})^k$ ($k = 1, 2, \ldots, n$).

Since $S_k = \zeta$ (when k is a quadratic residues of p),

$S_k = -1 - \zeta$ (when k is a quadratic non-residues of p),

$$2\zeta = \lambda - 1$$

so we obtain $2S_k = -1 + v_k \lambda$ ($k = 1, 2, \ldots, n$).

Because $2f(x) = \sum_{j=0}^{n} 2\beta_{n-j} x^j = C(x) + B(x)\lambda$,

$$2\beta_i = c_{n-i} + b_{n-i}\lambda \ (i = 1, 2, \ldots, n),\ \beta_0 = 1,\ b_n = 0,\ c_n = 2,$$

so from *Newtonian* formula $\sum_{j=0}^{k-1} S_{k-j}\beta_j + k\beta_k = 0$ ($k = 1, 2, \ldots, n$),

$$\sum_{j=0}^{k-1} (2S_{k-j})(2\beta_j) + (2k)(2\beta_k) = 0,\ (k = 1, 2, \ldots, n),$$

$$\sum_{j=0}^{k-1} (-1 + v_{k-j}\lambda)(c_{n-j} + b_{n-j}\lambda) + 2k(c_{n-k} + b_{n-k}\lambda) = 0\ (k = 1, 2, \ldots, n),$$

$$\sum_{j=0}^{k-1} (-c_{n-j} + (-1)^n p v_{k-j} b_{n-j}) + 2kc_{n-k} = 0\ (k = 1, 2, \ldots, n),$$

$$\sum_{j=0}^{k-1}(-b_{n-j}+v_{k-j}c_{n-j})+2kb_{n-k}=0\ (k=1,2,\ldots,n),$$

we obtain Eq.(4) and Eq.(5).

References

[1] Serge Lang, *Algebraic Number Theory*, Springer-Verlag, New York, 1986
[2] T. M. Apostol, *Introduction to Analytic Number Theory*, Springer-Verlag, New York, 1976

In: Mathematics, Game Theory and Algebra Compendium ISBN: 978-1-60456-659-8
Editor: Jacob H. Mathias, pp. 547-556

Chapter 46

ON STRONGER FORMS OF (1,2)* QUOTIENT MAPPINGS IN BITOPOLOGICAL SPACES

M. Lellis Thivagar[1,*] and O. Ravi[2,**]

[1] Department of Mathematics, Arul Anandar College, Karumathur, Madurai Dt., PIN-625 514, Tamil Nadu, India

[2] Department of Mathematics, P.M.Thevar College, Usilampatti, Madurai Dt., PIN-625 532, Tamil Nadu, India

Abstract

This paper is devoted to introduce and study a new type of sets called (1,2)* α–open set, (1,2)* semi-open set and (1,2)* preopen set and a new class of mappings called (1,2)* α–continuous mapping, (1,2)* α–open mapping and (1,2)* α–irresolute mapping in bitopological spaces. We obtain several characterizations of this class and study its properties and investigate the relationships with other mappings like (1,2)* α–quotient mapping, strongly (1,2)* α–quotient mapping and (1,2)* α^*-quotient mapping.

2000 AMS Subject Classification : 54E55

Key words and Phrases :

(1,2)* α–open set, (1,2)* semi-open set,
(1,2)* preopen set, (1,2)* α–continuous mapping, (1,2)* α–open mapping,
(1,2)* α–irresolute mapping, (1,2)* α–quotient mapping, strongly
(1,2)* α–quotient mapping and (1,2)* α^*-quotient mapping.

* E-mail address: mlthivagar@yahoo.co.in
** E-mail address: siingam@yahoo.com

1. Introduction

Njastad [7] introduced the concept of an α - set in (X, τ). The notions of semi-open set and preopen set were introduced by Levine [3] and Mashhour et al. [5] respectively. Reilly [10] and Noiri [8] have obtained some important results on α–continuity, α–open mappings and α–irresolute mappings. Lellis Thivagar [2] introduced the concept of α–quotient mappings and α^*-quotient mappings in (X, τ). In this paper, we introduce (1,2)* α–continuous mappings and (1,2)* α^*-quotient mappings in bitopological spaces based on the concept of α^*-quotient mappings in (X, τ). The generalization, in most of the cases, are substantiated by suitable examples.

2. Preliminaries

Throughout the present paper, (X, τ_1, τ_2), (Y, σ_1, σ_2) and (Z, U_1, U_2) (or simply X, Y, Z) denote bitopological spaces.

Definition 2.1. A subset S of X is called

$\tau_{1,2}$ open if $S = A \cup B$ where $A \in \tau_1$ and $B \in \tau_2$.
The complement of $\tau_{1,2}$ open set is $\tau_{1,2}$ closed.

Example 2.2. Let $X = \{a, b, c\}$, $\tau_1 = \{\varphi, X, \{a\}\}$ and $\tau_2 = \{\varphi, X, \{b\}\}$. Then the sets in $\{\varphi, X, \{a\}, \{b\}, \{a, b\}\}$ are called $\tau_{1,2}$ open and the sets in $\{\varphi, X, \{b, c\}, \{a, c\}, \{c\}\}$ are called $\tau_{1,2}$ closed.

Definition 2.3. Let S be a subset of X. Then

i. The $\tau_1\tau_2$ closure of S, denoted by $\tau_1\tau_2$ cl S, is defined by $\cap \{F/S \subset F$ and F is $\tau_{1,2}$ closed$\}$ and
ii. The $\tau_1\tau_2$ interior of S, denoted by $\tau_1\tau_2$ int S, is defined by $\cup \{F/F \subset S$ and F is $\tau_{1,2}$ open$\}$.

Note 2.4. $\tau_{1,2}$ open sets need not necessarily form a topology.

Here we introduce a new type of sets as follows.

Definition 2.5. A subset S of X is said to be

i. (1,2)* α-open set if $S \subseteq \tau_1\tau_2$ int $(\tau_1\tau_2$ cl $(\tau_1\tau_2$ int S$))$
ii. (1,2)* semi-open set if $S \subseteq \tau_1\tau_2$ cl $(\tau_1\tau_2$ int S$)$ and
iii. (1,2)* preopen set if $S \subseteq \tau_1\tau_2$ int $(\tau_1\tau_2$ cl S$)$.

The family of all (1,2)* α–open sets [resp. (1,2)* semi-open sets, (1,2)* preopen sets] of X is denoted by (1,2)* αO (X) [resp. (1,2)* SO (X), (1,2)* PO (X)].

Example 2.6. Let X = {a, b, c}, τ_1 = {X, ϕ, {b}} and τ_2 = {X, ϕ, {a}}. So the sets in {X, ϕ, {a}, {b}, {a, b}} are $\tau_{1,2}$ open and the sets in {ϕ, X, {a, c}, {b, c}, {c}} are $\tau_{1,2}$ closed. The following table shows the easy way of finding (1,2)* α–open sets, (1,2)* semi-open sets and (1,2)* preopen sets of X.

Subset of X	(1,2)* α– open set	(1,2)* semi-open set	(1,2)* preopen set
ϕ	ϕ	ϕ	ϕ
X	X	X	X
{a}	{a}	{a}	{a}
{b}	{b}	{b}	{b}
{c}	---	---	---
{a,b}	{a,b}	{a,b}	{a,b}
{a,c}	---	{a,c}	---
{b,c}	---	{b,c}	---

Here we introduce new concept like (1,2)* α–continuous and study the conditions under various properties.

Definition 2.7. A mapping f : X $\rightarrow$ Y is called

i. (1,2)* α–continuous [resp. (1,2)* semi-continuous, (1,2)* pre continuous] if the inverse image of each $\sigma_{1,2}$ open set in Y is an (1,2)* α–open set [resp. (1,2)* semi-open set, (1,2)* preopen set] in X.
ii. (1,2)* α–open [resp. (1,2)* semi-open, (1,2)* preopen] mapping if the image of each $\tau_{1,2}$ open set in X is an (1,2)* α–open [resp. (1,2)* semi-open, (1,2)* preopen] set in Y.

Example 2.8.

Let X = {a, b, c}, τ_1 = {ϕ, X, {a}} and τ_2 = {ϕ, X, {b}}. So the sets in {ϕ, X, {a}, {b}, {a, b}} are called $\tau_{1,2}$ open and the sets in {ϕ, X, {b, c}, {a, c}, {c}} are called $\tau_{1,2}$ closed. We have (1,2)* αO(X) = {ϕ, X, {a}, {b}, {a,b}} and (1,2)*SO(X) = {ϕ, X, {a}, {b}, {a,b}, {a,c}, {b,c}}. Let Y = {p, q, r}, σ_1 = {ϕ, Y, {p}, {p,q}} and σ_2 = {ϕ, Y, {q}}. So the sets in {ϕ, Y, {p}, {q}, {p,q}} are called $\sigma_{1,2}$ open and the sets in {ϕ, Y, {r}, {q,r}, {p,r}} are called $\sigma_{1,2}$ closed. We have

(1,2)* αO(Y) = {ϕ, Y, {p}, {q}, {p,q}} and

(1,2)* SO(Y) = {ϕ, Y, {p}, {q}, {p,q}, {p,r}, {q,r}}

Define f : X $\rightarrow$ Y by f(a) = p; f(b) = q and f(c) = r.

Since the inverse image of each $\sigma_{1,2}$ open in Y is (1,2)*α-open set, clearly f is (1,2)*α-continuous.

Remark 2.9. Every $\tau_{1,2}$ open set is (1,2)*α-open but not conversely.

We quote the following theorems, without proof, as they are very useful in studying the definition 2.7.

Theorem 2.10. A subset S of X is an (1,2)* α–open iff S is (1,2)* semi-open and (1,2)* preopen.

Theorem 2.11. The mapping $f : X \to Y$ is (1,2)* α–continuous iff it is (1,2)* semi-continuous and (1,2)* pre continuous.

Theorem 2.12. $f : X \to Y$ is an (1,2)* α–open mapping iff it is (1,2)* semi-open mapping and (1,2)* preopen mapping.

Definition 2.13. A mapping $f : X \to Y$ is called (1,2)* α–irresolute [resp. (1,2)* semi-irresolute, (1,2)* pre irresolute] if the inverse image of every (1,2)* α–open [resp. (1,2)* semi-open, (1,2)* preopen] set in Y is an (1,2)* α–open [resp. (1,2)* semi-open, (1,2)* preopen] set in X.

3. (1,2)* α–Quotient Mappings

Definition 3.1. A surjective mapping $f : X \to Y$ is called

i. an (1,2)* α–quotient mapping if f is (1,2)* α–continuous and $f^{1}(V)$ is $\tau_{1,2}$ open in X implies V is an (1,2)* α–open set in Y.
ii. a (1,2)* semi-quotient mapping if f is (1,2)* semi-continuous and $f^{1}(V)$ is $\tau_{1,2}$ open in X implies V is a (1,2)* semi-open set in Y.
iii. a (1,2)* pre quotient mapping if f is (1,2)* pre continuous and $f^{1}(V)$ is $\tau_{1,2}$ open in X implies V is a (1,2)* preopen set in Y.

Example 3.2. As per Example: 2.8 clearly f is an (1,2)* α–quotient mapping.

Theorem 3.3. If $f : X \to Y$ is surjective, (1,2)* α–continuous and (1,2)* α–open then f is an (1,2)* α–quotient mapping.

Proof. Suppose $f^{1}(V)$ is any $\tau_{1,2}$ open set in X. Then $f(f^{1}(V))$ is an (1,2)* α–open set in Y as f is (1,2)* α–open. Since f is surjective, $f(f^{1}(V))=V$. Thus V is an (1,2)* α–open set in Y. Hence f is (1,2)* α–quotient mapping.

Theorem 3.4. If $f : X \rightarrow Y$ is (1,2)* open surjective and (1,2)* α–irresolute, and $g : Y \rightarrow Z$ is an (1,2)* α– quotient mapping then gof is an (1,2)* α–quotient mapping.

Proof. Let V be any $u_{1,2}$ open set in Z. Since g is (1,2)* α–continuous, $g^{-1}(V)$ is an (1,2)* α–open set in Y. Then $f^{1}(g^{-1}(V)) = (gof)^{-1}(V)$ is an (1,2)* α–open set in X as f is (1,2)* α–irresolute. Thus gof is (1,2)* α–continuous. Also suppose $f^{1}(g^{-1}(V))$ is $\tau_{1,2}$ open in X. $f(f^{1}(g^{-1}(V)))$ is $\sigma_{1,2}$ open in Y as f is (1,2)* open. But $f(f^{1}(g^{-1}(V))) = g^{-1}(V)$ as f is surjective. Since g is (1,2)* α–quotient mapping, V is an (1,2)* α–open set in Z. Hence gof is an (1,2)* α–quotient mapping.

Corollary 3.5. If $f : X \rightarrow Y$ is (1,2)* open surjective and (1,2)* semi-[(1,2)* pre] irresolute and $g : Y \rightarrow Z$ is (1,2)* semi- [(1,2)* pre] quotient mapping then gof is (1,2)* semi- [(1,2)* pre] quotient mapping.

Theorem 3.6. A mapping $f : X \rightarrow Y$ is an (1,2)* α–quotient iff it is (1,2)* semi-quotient mapping and (1,2)* prequotient mapping.

Proof. Let f be an (1,2)* α–quotient mapping. Let V be any $\sigma_{1,2}$ open in Y. Since f is (1,2)* α–continuous, $f^{-1}(V) \in$ (1,2)* αO (X) = (1,2)* SO (X) ∩ (1,2)* PO (X). As $f^{1}(V) \in$ (1,2)* SO (X), f is (1,2)* semi-continuous. Also suppose $f^{-1}(V)$ is any $\tau_{1,2}$ open set in X. Since f is an (1,2)* α–quotient mapping, V is an (1,2)* α–open set in Y. Then V ∈ (1,2)* αO (Y) = (1,2)* SO (Y) ∩ (1,2)* PO (Y). As V ∈ (1,2)* SO (Y), V is (1,2)* semi-open set in Y. Thus f is (1,2)* semi-quotient mapping. Similarly f is (1,2)* pre quotient mapping.

Conversely let f be (1,2)* semi-quotient mapping and (1,2)* pre quotient mapping. Since f is (1,2)* semi-continuous and (1,2)* pre continuous, by Theorem 2.11, f is (1,2)* α–continuous. Suppose $f^{1}(V)$ is $\tau_{1,2}$ open in X. By definition 3.1, V ∈ (1,2)* SO(Y) ∩ (1,2)* PO(Y) = (1,2)* αO(Y). Thus f is (1,2)* α–quotient mapping.

4. Strongly (1,2)* α-Quotient Mappings

Definition 4.1.Let $f : X \rightarrow Y$ be surjective mapping. Then f is called strongly (1,2)* α–quotient [resp. strongly (1,2)* semi-quotient, strongly (1,2)* prequotient] mapping provided a set S of Y is $\sigma_{1,2}$ open in Y iff

$f^{1}(S)$ is an (1,2)* α–open [resp. (1,2)* semi-open, (1,2)* preopen] set in X.

Theorem 4.2. Every strongly (1,2)* α–quotient mapping is an (1,2)* α–quotient mapping.

Proof. Let f be strongly (1,2)* α–quotient mapping. By definition 4.1, clearly f is (1,2)* α–continuous. Suppose $f^{1}(V)$ is $\tau_{1,2}$ open in X. $f^{1}(V)$ is an (1,2)* α–open set in X. By assumption V is $\sigma_{1,2}$ open in Y. V is (1,2)* α–open set in Y. Thus f is (1,2)* α–quotient mapping.

Example 4.3. The converse of Theorem 4.2 is false.

Let X = {a,b,c}, τ_1 = {ϕ, X, {b}, {b,c}} and τ_2 ={ϕ, X}. So the sets in {ϕ, X, {b}, {b,c}} are called $\tau_{1,2}$ open and the sets in {ϕ, X, {a}, {a,c}} are called $\tau_{1,2}$ closed. Let Y = {p, q, r}, σ_1 = {ϕ, Y, {q}} and σ_2 = {ϕ, Y, {q,r}}. So the sets in {ϕ, Y, {q}, {q,r}} are called $\sigma_{1,2}$ open and the sets in {ϕ, Y, {p}, {p,r}} are called $\sigma_{1,2}$ closed. We have (1,2)*αO(X) = {ϕ, X, {b}, {a,b}, {b,c}} and (1,2)*αO(Y) = {ϕ, Y, {q}, {p,q}, {q,r}}.

Define f : X → Y by f(a) = p, f(b) = q & f(c) = r. Clearly f is (1,2)* α–quotient mapping. Since f^{1}({p,q}) = {a,b} is (1,2)* α–open set and {p,q} is not $\sigma_{1,2}$ open, f is not strongly (1,2)* α–quotient mapping.

Theorem 4.4. If f : X → Y is strongly (1,2)* semi-quotient and strongly (1,2)* pre quotient then f is strongly (1,2)* α–quotient mapping.

Proof. Let V be any $\sigma_{1,2}$ open set in Y. Since f is strongly (1,2)* semi-quotient and strongly (1,2)* pre quotient, f^{1}(V) ∈ (1,2)* SO (X) ∩ (1,2)* PO (X) = (1,2)* αO (X). Thus f^{-1}(V) is an (1,2)* α–open set in X. Conversely, Let f^{-1}(V) be an (1,2)* α–open set in X.

Then f^{-1}(V) ∈ (1,2)* αO (X) = (1,2)* SO (X) ∩ (1,2)* PO (X).

Since f^{-1}(V) ∈ (1,2)* SO (X) and f^{1}(V) ∈ (1,2)* PO (X) and f is strongly (1,2)* semi-quotient and strongly (1,2)* pre quotient, V is $\sigma_{1,2}$ open set in Y. Hence f is strongly (1,2)* α–quotient mapping.

Example 4.5. The converse of Theorem 4.4 is false.

Refer Example 2.8. Clearly f is strongly (1,2)*α–quotient mapping. Since f^{1}({p ,r}) = {a, c} ∈ (1,2)*SO(X) and {p,r} is not $\sigma_{1,2}$ open, f is not strongly (1,2)* semi-quotient.

5. (1,2)* α^* – Quotient Mappings

Definition 5.1. Let f : X → Y be surjective mapping. Then f is called

i. (1,2)* α^*–quotient mapping if f is (1,2)* α–irresolute and f^{1}(S) is (1,2)* α–open in X implies S is $\sigma_{1,2}$ open set in Y.
ii. (1,2)* semi-*quotient mapping if f is (1,2)* semi-irresolute and f^{1}(S) is (1,2)* semi-open in X implies S is $\sigma_{1,2}$ open set in Y.
iii. (1,2)* pre*quotient mapping if f is (1,2)* pre irresolute and f^{1}(S) is (1,2)* preopen in X implies S is $\sigma_{1,2}$ open set in Y.

Example 5.2.

Refer Example 2.8. Clearly f is (1,2)*α–irresolute and (1,2)* α^*–quotient mapping.

Example 5.3. An (1,2)* α–irresolute mapping need not be an (1,2)*α^*–quotient mapping.

As per Example 4.3, clearly f is an (1,2)* α–irresolute mapping. Since f^{1}({p,q}) = {a,b} ∈ (1,2)* αO (X) and {p,q} is not $\sigma_{1,2}$ open, f is not (1,2)* α^*-quotient mapping.

Definition 5.4. A mapping $f : X \to Y$ is called strongly (1,2)* α–open if the image of every (1,2)* α–open set in X is an (1,2)* α–open set in Y.

Example 5.5.
Refer Example 2.8. Clearly f is strongly (1,2)* α–open mapping.

Theorem 5.6. Let $f : X \to Y$ be surjective strongly (1,2)* α–open and (1,2)* α–irresolute and $g : Y \to Z$ be an (1,2)* α^*–quotient mapping. Then gof is an (1,2)* α^*–quotient mapping.

Proof. Let V be any (1,2)* α–open set in Z. Then $g^{-1}(V)$ is an (1,2)* α–open set in Y as g is an (1,2)* α^*–quotient mapping. Then $f^1(g^{-1}(V)) = (gof)^{-1}(V)$ is an (1,2)* α–open set in X as f is (1,2)* α–irresolute. This shows that gof is (1,2)* α–irresolute. Also suppose $(gof)^{-1}(V) = f^{1}(g^{-1}(V))$ is an (1,2)* α–open set in X. Since f is strongly (1,2)* α–open, $f(f^1(g^{-1}(V)))$ is an (1,2)* α–open set in Y. Since f is surjective, $f(f^1(g^{-1}(V))) = g^{-1}(V)$ is an (1,2)* α–open set in Y. Since g is an (1,2)* α^*–quotient mapping, V is $\mathcal{U}_{1,2}$ open in Z. Hence the theorem.

Theorem 5.7.
If $f : X \to Y$ is (1,2)* semi- * quotient mapping and (1,2)* pre * quotient mapping then f is (1,2)* α^*–quotient mapping.

Proof. Let V be any (1,2)* α–open set in Y. Then V $\in$ (1,2)* αO (Y) = (1,2)* SO (Y) $\cap$ (1,2)* PO (Y). Since f is (1,2)* semi-*quotient mapping and (1,2)* pre * quotient mapping, f is (1,2)* semi-irresolute and (1,2)* pre irresolute implies $f^1(V) \in$ (1,2)* SO (X) $\cap$ (1,2)* PO (X) = (1,2)* αO (X). This shows that f is (1,2)* α–irresolute. Suppose $f^1(V)$ is an (1,2)* α–open set in X. Then $f^1(V)$ is (1,2)* semi-open and (1,2)* preopen in X implies V is $\sigma_{1,2}$ open set in Y as f is (1,2)* semi-*quotient mapping and (1,2)* pre * quotient mapping. Thus f is (1,2)* α*-quotient mapping.

Example 5.8. The converse of Theorem 5.7 is false.
Refer Example 2.8 and 5.2. Clearly f is (1,2)*α*-quotient mapping. Also clearly f is (1,2)*semi-irresolute and since $f^1(\{q,r\}) = \{b,c\} \in$ (1,2)*SO(X) and {q,r} is not $\sigma_{1,2}$ open, f is not (1,2)*semi-*quotient mapping.

6. Comparisons

Theorem 6.1.
Every (1,2)* quotient mapping is an (1,2)* α–quotient mapping.

Proof. Let V be any $\sigma_{1,2}$ open set in Y. Then V is an (1,2)* α–open set in Y. Since f is (1,2)* quotient mapping, $f^1(V)$ is $\tau_{1,2}$ open in X and $f^1(V) \in$ (1,2)* αO (X). This shows that f is (1,2)* α–continuous. Suppose $f^1(V)$ is any $\tau_{1,2}$ open in X. Then $f^{-1}(V) \in$

(1,2)*αO(X). Since f is (1,2)* quotient mapping, V $\in$ (1,2)* αO (Y). Thus f is (1,2)* α–quotient mapping.

Example 6.2. The converse of Theorem 6.1 is false.

Let X = {a, b, c}, τ_1 = {ϕ, X, {a}} and τ_2 = {ϕ, X}. So the sets in {ϕ, X, {a}} are $\tau_{1,2}$ open and the sets in {ϕ, X, {b, c}} are $\tau_{1,2}$ closed. Let Y = {p, q, r}, σ_1 = {ϕ, Y, {p}, {p, q}, {p, r}} and σ_2 = {ϕ, Y}. So the sets in {ϕ, Y, {p}, {p, q}, {p, r}} are $\sigma_{1,2}$ open and the sets in {ϕ, Y, {q}, {r}, {q, r}} are $\sigma_{1,2}$ closed. We have (1,2)* αO(X) = {ϕ, X, {a}, {a, b}, {a, c}} and (1,2)*αO(Y) = {ϕ, Y, {p}, {p, q}, {p, r}}. Let f : X $\rightarrow$ Y be defined as f(a) = p, f(b) = q and f(c) = r. Clearly f is (1,2)* α-quotient mapping. Since f^{-1}({p, q}) = {a, b} is not $\tau_{1,2}$ open and {p, q} is $\sigma_{1,2}$ open, f is not (1,2)* quotient mapping.

Definition 6.3. The mapping f : X $\rightarrow$ Y is called quasi (1,2)* α–open if the image of every (1,2)* α–open set in X is $\sigma_{1,2}$ open in Y.

Theorem 6.4. Every quasi (1,2)* α–open mapping is a strongly (1,2)* α–open mapping.

Proof. Let V be any (1,2)* α–open set in X. Since f is quasi (1,2)* α–open, f(V) is $\sigma_{1,2}$ open in Y. f(V) is (1,2)* α–open set in Y. Thus f is a strongly (1,2)* α–open mapping.

Theorem 6.5. Every (1,2)* α*-quotient mapping is strongly (1,2)* α–quotient mapping.

Proof. Let V be any $\sigma_{1,2}$ open set in Y. Then V $\in$ (1,2)* αO (Y). Since f:X $\rightarrow$ Y is (1,2)* α–irresolute, f^{-1}(V) is an (1,2)* α–open set in X. Conversely Let f^{-1}(V) be an (1,2)* α–open set in X. Since f is (1,2)*α*-quotient mapping, V is $\sigma_{1,2}$ open set in Y. Hence f is strongly (1,2)*α–quotient mapping.

Theorem 6.6.

Every strongly (1,2)* α-quotient mapping is (1,2)* α*-quotient mapping.

Proof.

Let V be any $\sigma_{1,2}$ open set in Y. Therefore V is an (1,2)*α-open set in Y. Since f is strongly (1,2)*α-quotient mapping, f^{-1}(V) is an (1,2)* α-open set in X. Thus f is (1,2)*α-irresolute. Also since f is strongly (1,2)*α-quotient mapping, f^{-1}(V) is (1,2)*α-open set in X implies V is $\sigma_{1,2}$ open set in Y. Thus f is (1,2)*α*-quotient mapping.

Example 6.7.

Refer Example 6.2., Clearly f is (1,2)*α-irresolute, (1,2)*α*-quotient mapping and strongly (1,2)*α-quotient mapping since f^{-1}({p,q}) = {a,b} is not $\tau_{1,2}$ open where {p, q} is $\sigma_{1,2}$ open, f is not (1,2)*quotient mapping.

Example 6.8.

Let X = {a,b,c}, τ_1 = {ϕ, X, {a}, {a,b}} and τ_2 = {ϕ, X}. Then the sets in {ϕ, X, {a}, {a,b}} are called $\tau_{1,2}$ open and the sets in {ϕ, X, {c}, {b,c}} are called $\tau_{1,2}$ closed. Let Y =

{p,q,r}, σ_1 = {ϕ, Y, {p}, {p,q}} and σ_2 = {ϕ, Y}. Then the sets in {ϕ, Y, {p}, {p,q}} are called $\sigma_{1,2}$ open and the sets in {ϕ,Y,{r},{q,r}} are called $\sigma_{1,2}$ closed. We have (1,2)*αO(X) = {ϕ, X, {a}, {a,b}, {a,c}} and (1,2)*αO(Y) = {ϕ, Y, {p}, {p,q}, {p,r}}.

Define f : X→Y by f(a) = p; f(b) = q and f(c) = r. Clearly f is (1,2)*α-quotient mapping, (1,2)*α-irresolute and (1,2)* quotient mapping but f is neither strongly (1,2)*α-quotient mapping nor (1,2)*α^*-quotient mapping since f^{1}({p,r}) = {a,c} ∈ (1,2)*αO(X) and {p,r} is not $\sigma_{1,2}$ open.

Theorem 6.9. Let f : X → Y be surjective mapping. Consider the following statements.

(i) f is (1,2)* α*-quotient mapping.
(ii) f is strongly (1,2)* α–quotient mapping.
(iii) f is (1,2)* α–quotient mapping.

Proof. (i) ⇒ (ii) Theorem 6.5, (ii) ⇒ (iii) Theorem 4.2.

By the previous theorems and examples, we obtain the following diagram:

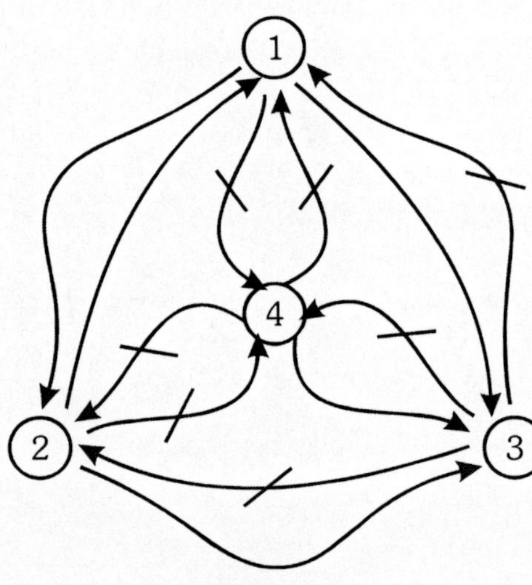

Where A ↛ B means that A does not necessarily imply B and

(1) = (1,2)* α*-quotient mapping.
(2) = Strongly (1,2)* α–quotient mapping.
(3) = (1,2)* α–quotient mapping.
(4) = (1,2)* quotient mapping.

Theorem 6.10. Let f : X→Y be a strongly (1,2)* α–quotient mapping and g : Y → Z be an (1,2)*α*-quotient mapping then gof is an (1,2)*α*-quotient mapping.

Proof. Let V be any (1,2)* α–open set in Z. Since g is (1,2)* α–irresolute, $g^{-1}(V)$ is an (1,2)* α–open set in Y. Since f is strongly (1,2)* α–quotient mapping, $f^{1}(g^{-1}(V))$ is an

(1,2)*α–open set in X. This shows that gof is (1,2)* α–irresolute. Suppose $(gof)^{-1}(V) = f^{-1}(g^{-1}(V))$ is an (1,2)* α–open set in X. Since f is strongly (1,2)* α–quotient mapping, $g^{-1}(V)$ is an $\sigma_{1,2}$ open in Y. Since g is an (1,2)*α*-quotient mapping and $g^{-1}(V)$ is an (1,2)* α-open set in Y, V is an $\mathcal{U}_{1,2}$ open in Z. Hence gof is (1,2)*α*-quotient mapping.

References

[1] Gene Crossley, S; Hildebrand, S.K. Semi-topological properties. *Fund. Math.* **74** (1972), 233-254.

[2] Lellis Thivagar, M. A note on Quotient mappings. *Bull. Malays. Math. Sci. Soc.* (Second Series). **14** (1991), 21-30.

[3] Levine, N. Semi-open sets and semi-continuity in topological spaces. *Amer. Math. Monthly.* **70** (1963), 36-41.

[4] Maheswari, S.N; Thakur, S.S. On α–irresolute mappings. *Tamkang J.Math.* **11** (1980), 209-214.

[5] Mashhour, A.S; Abd El-Monsef, M.E; El–Deeb, S.N. On pre continuous and weak pre continuous mappings. *Proc. Math. Phys. Soc. Egypt.* **53** (1982), 47-53.

[6] Mashhour, A.S; Hasanein, I.A; El–Deeb, S.N. α–continuous and α–open mappings. *Acta Math. Hung.* **41** (1983), 213-218.

[7] Njastad, O. On some classes of nearly open sets. *Pacific J. math.* **15** (1965), 961-970.

[8] Noiri, T. On α–continuous mappings. *Cas. Pest. Math.* **109** (1984), 118-126.

[9] Ravi, O; Lellis Thivagar, M. A Bitopological (1,2)* semi-generalized continuous maps. *Bull. Malayas. Math. Sci. Soc.* (To appear)

[10]Reilly, I.L; Vamanamurthy, M.K. On α–continuity in topological spaces. *Acta Math. Hung.* **45** (1985), no.1-2, 27-32.

In: Mathematics, Game Theory and Algebra Compendium
Editor: Jacob H. Mathias, pp. 557-565
ISBN 978-1-60456-659-8

Chapter 47

A Further Instability Result for a Certain Vector Differential Equation of Fourth Order

Cemil Tunç*
Faculty of Arts and Sciences, Department of Mathematics,
Yüzüncü Yıl University, 65080 ,Van, TURKEY

Abstract

This paper determines sufficient conditions under which the trivial solution of the fourth order nonlinear vector differential equation

$$X^{(4)} + \Psi(\dot{X}, \ddot{X})\dddot{X} + \Phi(X, \dot{X})\ddot{X} + H(\dot{X}) + F(X) = 0$$

is unstable.

AMS Subject Classification: 34D05, 34D20

Keywords: Vector differential equation, Fourth order, Instability, Trivial solution, Lyapunov Function

1. Introduction and Statement of the Result

It is well-known that the true creator of stability theory is A. M. Liapunov at the close of the 19th century. The technique discovered by himself is called Liapunov's second method or the direct method. This technique can be applied directly to the differential equation under investigation, without any knowledge of the solutions, provided the person using the method is clever enough to construct the right axuilary functions. So far, by using this well-known method, a good deal of work has been done and many interesting results have been obtained concerning the instability of solutions of certain scalar nonlinear differential equations of third-, fourth-, fifth-, sixth-, seventh and eighth order. One may refer to the

*E-mail address: cemtunc@yahoo.com

books of Bellman [2], Michel et al [10], Reissig et al [11] and papers of Bereketo uglu [1], Ezeilo ([3], [4], [5], [6], [7]), Li and Yu [8], Li and Duan [9], Sadek [13], Skrapek ([14], [15]), Tiryaki ([16], [17], [18]), C.Tunc and E. Tunc[19], Tunc [20] and the references cited therein for some publications on these topics. But, according to our observations in the relevant literature, except that the results of Sadek [11] and Tunc [21], there is no result on analysis of the instability of solutions of certain nonlinear vector differential equations of fourth order . Namely, recently, Sadek [12] and Tunc [21] investigated the instability of the trivial solution $X = 0$ of the fourth vector differential equations as follows:

$$X^{(4)} + A\dddot{X} + H(X, \dot{X}, \ddot{X}, \dddot{X})\ddot{X} + G(X)\dot{X} + F(X) = 0$$

and

$$X^{(4)} + \Psi(\ddot{X})\dddot{X} + \Phi(\dot{X})\ddot{X} + H(\dot{X}) + F(X) = 0,$$

respectively.

In this paper, we consider the fourth order non-linear vector differential equation of the form

$$X^{(4)} + \Psi(\dot{X}, \ddot{X})\dddot{X} + \Phi(X, \dot{X})\ddot{X} + H(\dot{X}) + F(X) = 0 \tag{1.1}$$

in the real Euclidean space R^n (with the usual norm denoted in what follows by $\|.\|$) , where $X \in R^n$, Ψ and Φ are $n \times n$ - symmetric matrices depending, in each case, on the arguments shown; $H : R^n \to R^n$, $F : R^n \to R^n$ and $H(0) = F(0) = 0$. Let the matrices Ψ and Φ and the functions H and F be continuous and so constructed such that the uniqueness theorem is valid. Equation (1.1) represents systems of real fourth order nonlinear differential equations of the form:

$$x_i^{(4)} + \sum_{k=1}^{n} \psi_{ik}(\dot{x}_1, \dot{x}_2, ..., \dot{x}_n; \ddot{x}_1, \ddot{x}_2, ..., \ddot{x}_n)\dddot{x}_k + \sum_{k=1}^{n} \phi_{ik}(x_1, x_2, ..., x_n; \dot{x}_1, \dot{x}_2, ..., \dot{x}_n)\ddot{x}_k$$

$$+h_i(\dot{x}_1, \dot{x}_2, ..., \dot{x}_n) + f_i(x_1, x_2, ..., x_n) = 0, (i = 1, 2, ..., n).$$

We shall assume, as basic throughout in what follows, that the derivatives $\frac{\partial \psi_{ik}}{\partial \dot{x}_j}$, $\frac{\partial \psi_{ik}}{\partial \ddot{x}_j}$, $\frac{\partial \phi_{ik}}{\partial x_j}$, $\frac{\partial \phi_{ik}}{\partial \dot{x}_j}$, $\frac{\partial h_i}{\partial \dot{x}_j}$ and $\frac{\partial f_i}{\partial x_j}$ $(j, k = 1, 2, ..., n)$ exist and are continuous.

The motivation for the present study has come from the papers of Sadek [12], Tunc [21] and the papers mentioned above. Our aim is to acquire a similar result for a certain nonlinear vector differential equation (1.1).

It will be convenient to consider, instead of the equation (1.1), the equivalent system

$$\begin{aligned} &\dot{X} = Y, \dot{Y} = Z, \dot{Z} = W, \\ &\dot{W} = -\Psi(Y, Z)W - \Phi(X, Y)Z - H(Y) - F(X) \end{aligned} \tag{1.2}$$

obtained as usual by setting $\dot{X} = Y, \ddot{X} = Z, \dddot{X} = W$ in (1.1).

It will be assumed that the Jacobian matrices $J_F(X)$, $J_H(Y)$, $J(\Phi(X,Y)Z \mid X)$, $J(\Phi(X,Y)Z \mid Y)$, $J(\Psi(Y,Z)Z \mid Y)$ and $J(\Psi(Y,Z)Z \mid Z)$ corresponding to $F(X)$, $H(Y)$ and $\Phi(X,Y)$ and $\Psi(Y,Z)$, respectively, are symmetric and given by:

$$J_F(X) = \left(\frac{\partial f_i}{\partial x_j}\right), J_H(Y) = \left(\frac{\partial h_i}{\partial y_j}\right),$$

$$J(\Phi(X,Y)Z \mid X) = \left(\frac{\partial}{\partial x_j}\sum_{k=1}^{n}\phi_{ik}z_k\right) = \left(\sum_{k=1}^{n}\frac{\partial \phi_{ik}}{\partial x_j}z_k\right),$$

$$J(\Phi(X,Y)Z \mid Y) = \left(\frac{\partial}{\partial y_j}\sum_{k=1}^{n}\phi_{ik}z_k\right) = \left(\sum_{k=1}^{n}\frac{\partial \phi_{ik}}{\partial y_j}z_k\right),$$

$$J(\Psi(Y,Z)Z \mid Y) = \left(\frac{\partial}{\partial y_j}\sum_{k=1}^{n}\psi_{ik}z_k\right) = \left(\sum_{k=1}^{n}\frac{\partial \psi_{ik}}{\partial y_j}z_k\right),$$

$$J(\Psi(Y,Z)Z \mid Z) = \left(\frac{\partial}{\partial z_j}\sum_{k=1}^{n}\psi_{ik}z_k\right) = \Psi(Y,Z) + \left(\sum_{k=1}^{n}\frac{\partial \psi_{ik}}{\partial z_j}z_k\right).$$

Apart from these, it is also assumed that the Jacobian matrices $J_F(X)$, $J_H(Y)$, $J(\Phi(X,Y)Z \mid X)$, $J(\Phi(X,Y)Z \mid Y)$, $J(\Psi(Y,Z)Z \mid Y)$, and $J(\Psi(Y,Z)Z \mid Z)$ exist and are continuous. The symbol $\langle X,Y\rangle$ corresponding to any pair X,Y in R^n stands for the usual scalar product $\sum_{i=1}^{n} x_i y_i$; thus $\langle X,X\rangle = \|X\|^2$. The matrix A is said to be negative-definite, when $\langle AX,X\rangle < 0$ for all nonzero X in R^n and $\lambda_i(A)$ $(i = 1,2,...,n)$ are the eigenvalues of the $n \times n$ -matrix A.

Now, we consider, in the case $n = 1$, the linear constant coefficients differential equation of the form:

$$x^{(4)} + a_1\dddot{x} + a_2\ddot{x} + a_3\dot{x} + a_4 x = 0, \tag{1.3}$$

It is well-known from the qualitative behaviour of solutions of linear differential equations that if $a_4 > \frac{1}{4}a_2^2$, then the trivial solution $x = 0$ of equation (1.3) is unstable.

The following theorem is the main result.

Theorem: Assume that the functions Ψ, Φ, H and F that appeared in *(1.1)* are continuously differentiable and there are positive constants a_1, a_2, a_3 and $a_4 (\neq 0)$ with $a_4 > \frac{1}{4}a_2^2$ such that

(i) $\Psi(Y,Z)$ and $J(\Psi(Y,Z)Z \mid Y)$ are symmetric $\lambda_i(\Psi(Y,Z) \geq a_1$ and $(J(\Psi(Y,Z)Z \mid Y)$ is positive-definite for all $Y,Z \in R^n$.

(ii) $\Phi(X,Y)$ and $J(\Phi(X,Y)Z \mid X)$ are symmetric , $\lambda_i(\Phi(X,Y) \geq a_2$ and $(J(\Phi(X,Y)Z \mid X))$ is positive-definite for all $X,Y,Z \in R^n$

(iii) $J_H(Y)$ is symmetric $\lambda_i(J_H(Y) \geq a_3$ for all $Y \in R^n$.

(iv) $J_F(X)$ is symmetric $\lambda_i(J_F(X) \geq a_4$ for all $X(\neq 0) \in R^n$ $(i = 1,2,...,n)$.

Then the zero solution $X = 0$ of the system *(1.2)* is unstable.

In order to reach our main result we'll need the following lemma which will be required in the proof.

Lemma: Let A be a real symmetric $n \times n$ -matrix and

$$a' \geq \lambda_i(A) \geq a > 0 \;\; (i = 1, 2, ..., n), \text{ where } a', a \text{ are constants.}$$

Then

$$a' \langle X, X \rangle \geq \langle AX, X \rangle \geq a \langle X, X \rangle$$

and

$$a'^2 \langle X, X \rangle \geq \langle AX, AX \rangle \geq a^2 \langle X, X \rangle .$$

Proof: See [2].

2. Proof of the Theorem

The proof of the theorem is based on the use of Ceatev's instability criterion in [10]. Namely, it depends on certain properties of a continuously differentiable Liapunov function $V = V(X, Y, Z, W)$. This function and its total time derivative satisfy some fundamental inequalities. We define the function V as follows:

$$V = \langle W, Z \rangle + \langle Y, F(X) \rangle + \int_0^1 \langle \sigma \Psi(Y, \sigma Z) Z, Z \rangle \, d\sigma + \int_0^1 \langle \Phi(X, \sigma Y) Z, Y \rangle \, d\sigma + \int_0^1 \langle H(\sigma Y), Y \rangle \, d\sigma. \tag{2.1}$$

We claim that V is a positive function. It is obvious that $V(0, 0, 0, 0) = 0$. Because of $\frac{\partial}{\partial \sigma} \langle H(\sigma Y), Y \rangle = \langle J_H(\sigma Y) Y, Y \rangle$, $H(0) = 0$ and $\lambda_i(J_H(Y)) \geq a_3 > 0$ we have

$$\langle H(Y), Y \rangle = \int_0^1 \langle J_H(\sigma Y) Y, Y \rangle \, d\sigma \geq \int_0^1 \langle a_3 Y, Y \rangle \, d\sigma = a_3 \langle Y, Y \rangle .$$

Therefore, we obtain

$$\int_0^1 \langle H(\sigma Y), Y \rangle \, d\sigma \geq a_3 \int_0^1 \langle \sigma Y, Y \rangle \, d\sigma = \frac{1}{2} a_3 \left\| Y \right\|^2 . \tag{2.2}$$

Next, in view of (2.1), (2.2) and the assumptions of the theorem, it follows that

$$V(X,Y,Z,W) \geq \frac{1}{2}a_1\|Z\|^2 + \frac{1}{2}a_3\|Y\|^2 + \langle Y, F(X)\rangle + \langle W, Z\rangle + \int_0^1 \langle \Phi(X,\sigma Y)Z, Y\rangle\, d\sigma.$$

and hence

$$\begin{aligned} V(0,\varepsilon,\varepsilon,0) &\geq \tfrac{1}{2}a_1\|\varepsilon\|^2 + \tfrac{1}{2}a_3\|\varepsilon\|^2 + \int_0^1 \langle \Phi(0,\sigma\varepsilon)\varepsilon, \varepsilon\rangle\, d\sigma \\ &\geq \tfrac{1}{2}(a_1 + 2a_2 + a_3)\|\varepsilon\|^2 > 0 \end{aligned}$$

for all arbitrary $\varepsilon \in R^n$. It is also clear that in every neighbourhood of $(0,0,0,0)$ there exists a point (ξ,η,ζ,μ) such that $V(\xi,\eta,\zeta,\mu) > 0$ for all ξ,η,ζ and μ in R^n. Let $(X,Y,Z,W) = (X(t),Y(t),Z(t),W(t))$ be any arbitrary solution of (1.2). A straightforward calculation from (2.1) and (1.2) will verify that

$$\begin{aligned} \dot{V} = &\ \tfrac{d}{dt}V(X,Y,Z,W) \\ &= \langle W,W\rangle - \langle \Psi(Y,Z)W,Z\rangle - \langle \Phi(X,Y)Z,Z\rangle - \langle H(Y),Z\rangle + \langle Y, J_F(X)Y\rangle \\ &+ \tfrac{d}{dt}\int_0^1 \langle \sigma\Psi(Y,\sigma Z)Z,Z\rangle\, d\sigma + \tfrac{d}{dt}\int_0^1 \langle \Phi(X,\sigma Y)Z,Y\rangle\, d\sigma + \tfrac{d}{dt}\int_0^1 \langle H(\sigma Y),Y\rangle\, d\sigma. \end{aligned} \tag{2.3}$$

But

$$\begin{aligned} \tfrac{d}{dt}\int_0^1 \langle H(\sigma Y),Y\rangle\, d\sigma &= \int_0^1 \sigma \langle J_H(\sigma Y)Z,Y\rangle\, d\sigma + \int_0^1 \langle H(\sigma Y),Z\rangle\, d\sigma \\ &= \int_0^1 \sigma\tfrac{\partial}{\partial\sigma}\langle H(\sigma Y),Z\rangle\, d\sigma + \int_0^1 \langle H(\sigma Y),Z\rangle\, d\sigma \\ &= \sigma\langle H(\sigma Y),Z\rangle \Big|_0^1 = \langle H(Y),Z\rangle, \end{aligned} \tag{2.4}$$

$$
\begin{aligned}
\frac{d}{dt}\int_0^1 \langle \Phi(X,\sigma Y)Z, Y\rangle\, d\sigma = {} & \int_0^1 \langle \Phi(X,\sigma Y)Z, Z\rangle\, d\sigma + \int_0^1 \langle \Phi(X,\sigma Y)W, Y\rangle\, d\sigma \\
& + \int_0^1 \langle J(\Phi(X,\sigma Y)Z \mid X)Y, Y\rangle\, d\sigma \\
& + \int_0^1 \langle \sigma J(\Phi(X,\sigma Y)Z \mid Y)Z, Y\rangle\, d\sigma \\
& = \int_0^1 \langle \Phi(X,\sigma Y)Z, Z\rangle\, d\sigma + \int_0^1 \langle \Phi(X,\sigma Y)Y, W\rangle\, d\sigma \\
& + \int_0^1 \sigma \tfrac{\partial}{\partial\sigma} \langle \Phi(X,\sigma Y)Z, Z\rangle\, d\sigma \\
& + \int_0^1 \langle J(\Phi(X,\sigma Y)Z \mid X)Y, Y\rangle\, d\sigma \\
& = \langle \Phi(X,Y)Z, Z\rangle + \int_0^1 \langle \Phi(X,\sigma Y)Y, W\rangle\, d\sigma \\
& + \int_0^1 \langle J(\Phi(X,\sigma Y)Z \mid X)Y, Y\rangle\, d\sigma
\end{aligned} \tag{2.5}
$$

and

$$
\begin{aligned}
\frac{d}{dt}\int_0^1 \langle \sigma\Psi(Y,\sigma Z)Z, Z\rangle\, d\sigma & = \int_0^1 \langle \sigma\Psi(Y,\sigma Z)Z, W\rangle\, d\sigma + \int_0^1 \langle \sigma\Psi(Y,\sigma Z)W, Z\rangle\, d\sigma \\
& + \int_0^1 \sigma^2 \langle J(\Psi(Y,\sigma Z) \mid Z)ZW, Z\rangle\, d\sigma \\
& + \int_0^1 \sigma \langle J(\Psi(Y,\sigma Z)Z \mid Y)Z, Z\rangle\, d\sigma \\
& = \int_0^1 \langle \sigma\Psi(Y,\sigma Z)W, Z\rangle\, d\sigma + \int_0^1 \sigma \tfrac{\partial}{\partial\sigma} \langle \sigma\Psi(Y,\sigma Z)W, Z\rangle\, d\sigma \\
& + \int_0^1 \sigma \langle J(\Psi(Y,\sigma Z)Z \mid Y)Z, Z\rangle\, d\sigma \\
& = \sigma^2 \langle \Psi(Y,\sigma Z)W, Z\rangle \Big|_0^1 \\
& + \int_0^1 \sigma \langle J(\Psi(Y,\sigma Z)Z \mid Y)Z, Z\rangle\, d\sigma \\
& = \langle \Psi(Y,Z)W, Z\rangle + \int_0^1 \sigma \langle J(\Psi(Y,\sigma Z)Z \mid Y)Z, Z\rangle\, d\sigma.
\end{aligned} \tag{2.6}
$$

On substituting the estimates (2.4)-(2.6) in (2.3) we obtain

$$\dot{V} = \langle W, W\rangle + \int_0^1 \langle \Phi(X,\sigma Y)Y, W\rangle\, d\sigma + \langle Y, J_F(X)Y\rangle$$
$$+ \int_0^1 \langle J(\Phi(X,\sigma Y)Z \mid X)Y, Y\rangle\, d\sigma + \int_0^1 \sigma \langle J(\Psi(Y,\sigma Z)Z \mid Y)Z, Z\rangle\, d\sigma$$

Let

$$\Phi_1(X,Y) = \int_0^1 \Phi(X,\sigma Y)Y\, d\sigma.$$

Then

$$\int_0^1 \langle \Phi(X,\sigma Y)Y, W\rangle\, d\sigma = \Phi_1(X,Y)W.$$

Hence, in view of the assumptions of the theorem and the lemma, we conclude that

$$\dot{V} \geq \left\| W + \tfrac{1}{2}\Phi_1(X,Y)\right\|^2 + \langle Y, J_F(X)Y\rangle - \tfrac{1}{4}\langle \Phi_1(X,Y), \Phi_1(X,Y)\rangle$$
$$\geq \langle Y, J_F(X)Y\rangle - \tfrac{1}{4}\langle \Phi_1(X,Y), \Phi_1(X,Y)\rangle \geq \left(a_4 - \tfrac{1}{4}a_2^2\right)\|Y\|^2 > 0.$$

Thus, the assumption $a_4 > \frac{1}{4}a_2^2$ indicates that $\dot{V}_0$ is positive semi - definite. Furthermore $\dot{V}_0 = 0$ for all $t \geq 0$ necessarily implies that $Y(t) = 0$ for all $t \geq 0$, and therefore also that $X(t) = \xi$ (a constant vector), $Z = \dot{Y} = 0$, $W = \ddot{Y} = 0, \dddot{Y} = \dot{W} = 0$ for all $t \geq 0$. The substitutions:

$$X = \xi, Y = Z = W = 0$$

in (1.2) give that $F(\xi) = 0$ which, in view of the fact that $F(0) = 0$, necessarily implies that $\xi = 0$. Thus $\dot{V}(X,Y,Z,W) = 0$ for all $t \geq 0$ necessarily implies that

$$X = Y = Z = W = 0 \text{ for all } t \geq 0.$$

So the basic properties of $V(X,Y,Z,W)$, which we have proved just above, justify that the function V has all the requisite Ceatev criterion proved in [10] subject to the conditions of the theorem, which now follows. This proves that the zero solution of system (1.2) is unstable. (See Theorem 1.15 in Reissig et al [11] and Miller and Michel [10].) The system of equations (1.2) is equivalent to the differential equation (1.1). Consequently, it follows the original statement of the theorem.

References

[1] Bereketoğlu, H. On the instability of trivial solutions of a class of eighth-order differential equations. *Indian J. Pure. Appl. Math.* **22** (1991), no.3, 199-202.

[2] Bellman, R. Introduction to matrix analysis. Reprint of the second (1970) edition. With a foreword by Gene Golub. Classics *in Applied Mathematics*, **19**. Society for Industrial and Applied Mathematics (SIAM), Philadelphia, PA, 1997.

[3] Ezeilo, J.O.C. An instability theorem for a certain fourth order differential equation. *Bull. London Math. Soc.* **10** (1978), no.2, 184-185

[4] Ezeilo, J.O.C. Instability theorems for certain fifth-order differential equations. *Math. Proc. Cambridge. Philos. Soc.* **84** (1978), no.2, 343-350.

[5] Ezeilo, J.O.C. A further instability theorem for a certain fifth-order differential equation. *Math. Proc. Cambridge Philos. Soc.* **86** (1979), no.3, 491-493.

[6] Ezeilo, J.O.C. Extension of certain instability theorems for some fourth and fifth order differential equations. *Atti. Accad. Naz. Lincei. Rend. Cl. Sci. Fis. Mat. Natur.* (8) **66** (1979), no.4, 239-242.

[7] Ezeilo, J.O.C. An instability theorem for a certain sixth order differential equation. *J. Austral. Math. Soc. Ser. A* **32** (1982), no.1, 129-133.

[8] Li, W. J.; Yu, Y.H. Instability theorems for some fourth-order and fifth-order differential equations. (Chinese) *J. Xinjiang Univ. Natur. Sci.* **7** (1990), no.2, 7-10.

[9] Li, W. J.; Duan, K. C. Instability theorems for some nonlinear differential systems of fifth order. *J. Xinjiang Univ. Natur. Sci.* **17** (2000), no.3, 1-5.

[10] Miller, Richard K.; Michel, A. N. *Ordinary differential equations*. Academic Press, Inc. [Harcourt Brace Jovanovich, Publishers], New York-London, 1982.

[11] Reissig, R.; Sansone, G.; Conti, R. *Non-linear differential equations of higher order*. Translated from the German. Noordhoff International Publishing, Leyden, (1974)

[12] Sadek, A.I. Instability results for certain systems of fourth and fifth order differential equations. *Appl. Math. Comput.* **145** (2003), no. 2-3, 541-549.

[13] Sadek, A. I. An instability theorem for a certain seventh-order differential equation. *Ann. Differential Equations* **19** (2003), no. 1, 1-5.

[14] Skrapek, W.A. Instability results for fourth-order differential equations. *Proc. Roy. Soc. Edinburgh Sect. A* **85** (1980), no. 3-4, 247-250.

[15] Skrapek, W.A. Some instability theorems for third order ordinary differential equations. *Math. Nachr.* **96** (1980), 113-117.

[16] Tiryaki, A. Extension of an instability theorem for a certain fourth order differential equation. *Bull. Inst. Math. Acad. Sinica.* **16** (1988), no.2, 163-165.

[17] Tiryaki, A. An instability theorem for a certain sixth order differential equation. *Indian J. Pure. Appl. Math.* **21** (1990), no.4, 330-333.

[18] Tiryaki, A. Extension of an instability theorem for a certain fifth order differential equation. National Mathematics Symposium (Trabzon, 1987). *J. Karadeniz Tech. Univ. Fac. Arts Sci. Ser. Math. Phys.* **11** (1988), 225-227 (1989).

[19] Tunç, C. Tunç, E. An instability theorem for a certain eighth order differential equation. *Differential Equations* (*Differ. Uravn.*), (2004) (in press).

[20] Tunç, C. An instability result for certain system of sixth order differential equations.*Appl. Math. Comput.*, (2004) (in press).

[21] Tunç, C. An instability theorem for a certain vector differential equation of the fourth order. *Journal of Inequalities in Pure and Applied Mathematics*, **5** (2004), no.1, 1-5.

In: Mathematics, Game Theory and Algebra Compendium ISBN 978-1-60456-659-8
Editor: Jacob H. Mathias, pp. 567-583

Chapter 48

ON THE REGULARITY OF MAGNETIC FLUX FUNCTION IN NUCLEAR FUSION

Takayuki Yamauchi*
Department of Mathematics and Computer Science
Lincoln University, 1570 Baltimore Pike,
Lincoln University, PA19352, U.S.A.

Abstract

The Grad-Shafranov equation is a well known equation that is satisfied in a state of equilibrium of a plasma flow in a nuclear fusion reactor. Its rigorous derivation based on the Kruskal-Kulsrud Principle, however, has been unknown since the optimal regularity of a minimizer of the magnetic flux energy over a class of continuous rearrangements has not been shown to be smooth for a general domain. It will be shown that a rigorous derivation of the Grad-Shafranov equation based on the Kruskal-Kulsrud Principle can be obtained when the constraint functional satisfies a certain sufficient condition.

AMS subject class: 76W05,35Q72,35J35,35J60,49K30,49K20

Keywords: Magnetohydrodynamics, Grad-Shafranov equation, Lagrange Multiplier Rule, Rearrangement of functions, Partial differential equations of elliptic type, Optimization.

1. Introduction

The **Grad-Shafranov equation** is a well known equation in axi-symmetric magneto-hydrodynamics (MHD) that has been derived **assuming that all the functions associated with a state of equilibrium are smooth**, based on the Kruskal-Kulsrud Principle which was obtained in 1958 by Kruskal and Kulsrud [25]. The Kruskal-Kulsrud Principle states that an equilibrium state in a torus-shaped nuclear fusion reactor is achieved as an energy minimizer (possibly a stationary point) of the energy functional subject to (i) the conservation of the total toroidal flux, (ii)the conservation of the toroidal flux inside each magnetic surface, and

*E-mail address: tyamauchi@lincoln.edu

(iii) the conservation of the mass inside each magnetic surface. the constraints (ii)(iii) are expressed as two sets of continuously many constraints, which will be explicitly described below.

However, so far, the above-stated assumption has not been established in general except for a special case in which the meridian cross section Ω of the nuclear fusion reactor is a circle because the above-stated two sets of continuously many constraints are in $H_0^1(\Omega)$. For this reason, no rigorous derivation of the Grad-Shafranov equation based on the Kruskal-Kulsrud Principle has been obtained.

The actual minimization problem based on the Kruskal-Kulsrud Principle is given by

$$E(\psi, f, g) \equiv \frac{1}{2}\int_{\Omega}\left[|r^{-1}\nabla\psi|^2 + r^{-1}\cdot f^2 + r\cdot g^2\right] drdz \to \min, \tag{1.1}$$

subject to

$$\int_{\Omega} r^{-1}\cdot f drdz = F_0^* \tag{1.2}$$

$$\int_{\{\Psi>\sigma\}} r^{-1}\cdot f\cdot drdz = F^*(\sigma), \forall\sigma\in[\sigma_0, b] \tag{1.3}$$

$$\int_{\{\Psi>\sigma\}} r\cdot g^{\frac{2}{\gamma}}\cdot drdz = G^*(\sigma), \forall\sigma\in[\sigma_0, b] \tag{1.4}$$

where $\sigma_0 > 0, b > \sigma_0, F_0^*$ are prescribed constants, $F^*(\sigma), G^*(\sigma)$ are prescribed smooth functions, r is the horizontal axis, z is the vertical axis, Ω is an open bounded set on the rz-plane, $\Psi = \psi + \bar{\psi}$ is the total flux function. $\gamma = 5/3$. $b > 0$ is the maximum of the flux function Ψ. $\bar{\psi}$ is a prescribed smooth flux function, and $\psi \in H_0^1(\Omega)$ is the flux function induced by the current density supported in the plasma region, and is an unknown. $f \in L^2(\Omega)$ is the toroidal flux, $g = (2/(\gamma-1))^{1/2}p^{1/2}$, and p is the pressure, which are also unknown (cf: see [20] for more detail). (1.2) , (1.3), (1.4) represent the conservation of toroidal flux and mass inside each magnetic surface $\{\Psi > \sigma\}, \forall\sigma\in[\sigma_0, b]$, and the conservation of the total toroidal flux, respectively.

In the special case of Ω = a circle, P. Laurence and E. Stredulinsky showed in Section 3 of [27] that a minimizer is real analytic C^ω. They also conjectured that (on page 151 of [27]) an analytic minimizer (as a minimizer of the modified variational problem that can be obtained as a limit of free boundary problems) should exist when the cross section Ω of the nuclear fusion reactor is convex with C^2 boundary. The authors so far have shown that **a Lipschitz continuous minimizer exists when the cross section Ω of the nuclear fusion reactor is convex with C^2 boundary** in [31][26].

The standard optimization methods with finitely many equality constraints (Lagrange Multiplier Rule) cannot be used for (1.1)–(1.4) since the constraints (1.3)(1.4) mean that the number of constraints is continuously infinite. Currently, no practical method for handling an optimization problem with continuously infinitely many constraints has been established. The problem of Lagrange Multiplier Rule with continuously infinitely many constraints has been solved in [42] as follows.

Since the constraints (1.3)(1.4) are only right continuous functions of σ, they cannot be used to create a smooth variation structure that is needed in using the fundamental idea of calculus of variation. These constraints (1.3)(1.4) are first transformed into smooth functions of σ, and the set of constraints (1.2)(1.3)(1.4) is regarded as one functional constraint in a closed subspace of the product space $\mathbf{R} \times C([\sigma_0, b]) \times C([\sigma_0, b])$.

The study of an equilibrium solution of axi-symmetric MHD problems and other fluid dynamics problems has been closely tied with the development of the theory of rearrangement of functions [1] [2] [10] [12] [13] [14] [15] [16] [11] [17] [28] [29] [27] [31] [26] [30] [32] [33] [34] [35] [37] [36] [41]. Statistical equilibrium models of an equilibrium solution to axi-symmetric MHD problems have been studied in [9] [23] [24] without any reference to the regularity of the energy minimizer.

In 1991, Turkington, Eydeland, and Spruck [19] and Turkington, Eydeland, Lifshitz, and Spruck [20] discretized the minimization problem (1.1)-(1.4) by introducing finitely many constraints which replace the 2 continuously infinitely many constraints (1.3)(1.4), thereby relaxing the Kruskal-Kulsrud Principle. They also showed that the slow evolution problem for an axi-symmetric plasma-vacuum system in MHD can be reduced to a quasi-equilibrium problem. These finitely many constraints are defined in [20] by

$$\int_{\Omega} r^{-1} \cdot f \cdot [(\Psi - \sigma_{i-1})_+^2 - (\Psi - \sigma_i)_+^2] drdz = F^*(\sigma), 1 \leq i \leq n \tag{1.5}$$

$$\int_{\Omega} r \cdot g^{2/\gamma} \cdot [(\Psi - \sigma_{i-1})_+^2 - (\Psi - \sigma_i)_+^2] drdz = G^*(\sigma), 1 \leq i \leq n. \tag{1.6}$$

Turkington, Eydeland, and Spruck [19] and Turkington, Eydeland, Lifshitz, and Spruck [20] successfully showed that a minimizer of the discretized problem is at least $C^{2,\alpha}(\Omega), 0 < \alpha \leq 1$. They first derived a set of Euler-Lagrange equations satisfied by the minimizer using the Lagrange Multiplier Rule with Finitely Many Constraints. Using these Euler-Lagrange equations, they showed that a minimizer of the above-stated discretized problem satisfies the Grad-Shafranov equation. Since the question of the optimal regularity of a minimizer remained unsolved, the question of rigorous derivation of the Grad-Shafranov equation also remained open in [19][20].

The study of the optimal regularity of an equilibrium solution in fluid dynamics arises in the study of vortices [12] [13] [14] [15] [16] [11]. In particular, Burton and McLeod [11], and Alvino, Trombetti, and Lions [2] studied the problem:

$$E(u) = \int_{\Omega} |\nabla u|^2 dx \rightarrow max, \text{ subject to } -\Delta u \sim f$$

for a prescribed $f \in L^2(\Omega)$. Burton and McLeod [11] and Alvino, Trombetti, and Lions [2] independently established the existence and uniqueness of a maximizer. But **the regularity properties of the maximizer are still unknown** .

Laurence and Stredulinsky [31][30] also discretized the problem by introducing a sequence of free boundary problems associated with n-layers

$$A_i^{(n)} = \{x \in \Omega : t_{i-1}^{(n)} \leq \psi_n < t_i^{(n)}\}, 1 \leq i \leq n-1, A_n^{(n)} = \{x \in \Omega : \psi_n = t_{n-1}^{(n)}\}.$$

They obtained a gradient bound for certain minimizers, not for all minimizers, of the energy functional when Ω is convex. They then obtained a uniform gradient bound on minimizers of the approximating problems

$$|\nabla\psi|(x) \leq \inf J, a.e. x \in \Omega. \tag{1.7}$$

Laurence and Stredulinsky showed in their early works [27][26] that an energy minimizer is in $C^\infty(\Omega)$ if it is in $C^2(\Omega)$ using a bootstrap argument. So, the gap between the $C^{0,1}(\Omega)$ regularity and the $C^2(\Omega)$ regularity still remains open even with the assumption that Ω is convex. So, the rigorous derivation of the Grad-Shafranov equation within the framework of the Kruskal-Kulsrud Principle is still an open problem.In Section 2, the physics of an axi-symmetric equilibrium state for the plasma in a nuclear fusion reactor is described to derive the minimization problem defined by (1.1)–(1.4).In Section 3, smooth variation structures which replace the class of equimeasurable functions are constructed. As a result, the 2 sets of continuously infinitely many constraints (1.3)(1.4) become differentiable functions of σ. The original problem (1.1)-(1.4) will then be reformulated as a minimization problem with a functional constraint. The classical Lagrange Multiplier Rule with a Functional Constraint (cf: [3] [4] [5] [6] [7] [18] [38] [39]) will then be modified into a practical form.

In Section 4, a sufficient condition for the existence of non-degenerate Euler-Lagrange equations satisfied by the state of equilibrium will be stated. These Euler-Lagrange equations are then used to obtain a rigorous derivation of the Grad-Shafranov equation based on the Kruskal-Kulsrud Principle.

2. Mathematical Model of an Axi-symmetric Equilibrium State for the Plasma in a Tokamak

The slow evolution problem for an axi-symmetric plasma-vacuum system that arises in MHD will be investigated in this section. The problem will then be reduced to a variational problem with 2 smooth functional constraints. The construction of the mathematical model in this section is based on [8] and [20].

Under the assumption of ideal magnetohydrodynamics (MHD), the plasma region and the vacuum region are governed by the following equations.

Plasma Region

$$\frac{D\rho}{Dt} + \rho\nabla\cdot\mathbf{V} = Q_M \tag{2.1}$$

$$\frac{D\zeta}{Dt} + \zeta\nabla\cdot\mathbf{V} = Q_S \tag{2.2}$$

$$\rho\frac{D\mathbf{V}}{Dt} + \nabla p - \mathbf{J}\times\mathbf{B} = \mathbf{F} \tag{2.3}$$

$$\frac{\partial\mathbf{B}}{\partial t} - \nabla\times(\mathbf{V}\times\mathbf{B}) = 0 \tag{2.4}$$

$$\nabla\times\mathbf{B} = \mathbf{J} \text{ (Ampere's Theorem)} \tag{2.5}$$

$$\nabla\cdot\mathbf{B} = 0 \text{ (Conservation of Magnetic Induction)} \tag{2.6}$$

Vacuum Region

$$\nabla \times \mathbf{B} = \bar{\mathbf{J}} \tag{2.7}$$

$$\nabla \cdot \mathbf{B} = 0 \tag{2.8}$$

where

$$\rho = \text{Mass} \tag{2.9}$$

$$\zeta = \text{ Entropy Density Per Unit Volume} \tag{2.10}$$

$$p = \text{Pressure} \tag{2.11}$$

$$\mathbf{V} = \text{ Velocity of Plasma Flow} \tag{2.12}$$

$$\mathbf{B} = \text{ Magnetic Induction} \tag{2.13}$$

$$\mathbf{J} = \text{ Current Density} \tag{2.14}$$

$$\frac{D}{Dt} = \frac{\partial}{\partial t} + \mathbf{V} \cdot \nabla = \text{ Convective Derivative} \tag{2.15}$$

$$Q_M = \text{ Distributed Sources of Mass} \tag{2.16}$$

$$Q_S = \text{ Entropy} \tag{2.17}$$

$$\mathbf{F} = \text{ Momentum Production} \tag{2.18}$$

$$\bar{\mathbf{J}} = \text{ Given Current Density of External Field Coils} \tag{2.19}$$

where all the equations (2.1) -(2.19) are expressed in non-dimensional variables.

The equation of state for an ideal gas and the internal energy density are given by the following.

Equation of State for Ideal Gas

$$\bar{p} = e^{\zeta/\rho}\rho^{\gamma} \tag{2.20}$$

Internal Energy Density

$$U(\rho, \zeta) = e^{\zeta/\rho}\rho^{\gamma}/(\gamma - 1) \tag{2.21}$$

where $\gamma = 5/3$ is the adiabatic index.

$$D_{tor} = \text{Toroidal Region} \tag{2.22}$$

where (2.1)-(2.21) hold in D_{tor}, and the fixed boundary ∂D_{tor} is assumed to be a perfectly conducting shell on which the normal component of **B** vanishes.

$$S_{free} = \text{Plasma-Vacuum Interface} \tag{2.23}$$

On S_{free}, the plasma-vacuum interface conditions hold

$$p = 0, \mathbf{n} \cdot \mathbf{B} = 0 \text{ on } S_{free} \tag{2.24}$$

$$\mathbf{n} \times \mathbf{B} = 0 \text{ is continuous across } S_{free}. \tag{2.25}$$

2.1. Equilibrium of Plasma in Magnetic Field

The equilibrium of a plasma in the presence of an external magnetic field is governed by Maxwell's equations (2.5)(2.6)

and

$$\nabla p = \mathbf{J} \times \mathbf{B}. \tag{2.26}$$

(2.26) shows a state of equilibrium between the force ∇p due to the kinetic pressure and the force $\mathbf{J} \times \mathbf{B}$ of the magnetic pressure.

(2.26) then yields

$$\mathbf{B} \cdot \nabla p = 0 \tag{2.27}$$

and

$$\mathbf{J} \cdot \nabla p = 0. \tag{2.28}$$

(2.27) and (2.28) show that the field lines generated by **B** and the current lines generated by **J** lie on isobaric surfaces ($p =$ constant), called **magnetic surfaces**. The magnetic surfaces form a family of nested tori, and each magnetic surface is traversed ergodically by the flux lines. The inner most magnetic surface degenerates to a simply closed curve, called the **magnetic axis**.

Under the hypothesis of axial symmetry, the magnetic induction **B** is independent of the toroidal angle ω. Let z be the vertical axis, and r be the horizontal axis. Then, the rz coordinate system represents the meridian (vertical) cross section Ω of the toroidal region D_{tor}. Then, $\nabla \cdot \mathbf{B} = 0$ with $\mathbf{B} = (B_r, B_z)$ becomes

$$\frac{1}{r}\frac{\partial}{\partial r}(rB_r) + \frac{\partial B_z}{\partial z} = 0. \tag{2.29}$$

Let C be a circle of radius r centered at the axis of the torus, and $D_0 = B_0(r)$ be the disk enclosed by C. Define the variable Ψ by

$$\Psi(r,z) = \int_0^r tB_z dt = \frac{1}{2\pi}\int_{D_0} B_z dS = \frac{1}{2\pi}\int_{D_0} \mathbf{B} \cdot d\mathbf{S} \tag{2.30}$$

showing that $2\pi\Psi$ is the poloidal flux of **B** across D_0.

In (2.30), $\Psi(r,z) = \int_0^r tB_z dt$ implies that Ψ is differentiable with respect to r at $a.e.r$ so that

$$\frac{\partial\Psi}{\partial r} = rB_z.$$

Hence,

$$B_z = \frac{1}{r}\frac{\partial\Psi}{\partial r}, a.e.r. \tag{2.31}$$

(2.29) implies that

$$B_r = -\frac{1}{r}\frac{\partial\Psi}{\partial z}, a.e.z. \tag{2.32}$$

Let $\mathbf{B}_T$ be the toroidal component of **B**, and define f by

$$\mathbf{B}_T = \frac{f}{r}\mathbf{e}_T \tag{2.33}$$

where $\mathbf{e}_T$ is the unit vector perpendicular to Ω. Then, the magnetic induction **B** can be written as

$$\mathbf{B} = \mathbf{B}_T + \mathbf{B}_P \tag{2.34}$$

where $B_P = [\nabla\Psi \times \mathbf{e}_T]/r$ is the poloidal component of **B**.

Since $\nabla \times \mathbf{B} = \mathbf{J}$, one obtains

$$\mathbf{J} = \mathbf{J}_T + \mathbf{J}_P \tag{2.35}$$

where

$$\mathbf{J}_P = [\nabla f \times \mathbf{e}_T]/r = \text{ poloidal component} \tag{2.36}$$

$$\mathbf{J}_T = (L\psi)\mathbf{e}_T = \text{ toroidal component, } L = -\frac{\partial}{\partial r}\left(\frac{1}{r}\frac{\partial}{\partial r}\right) - \left(\frac{1}{r}\frac{\partial}{\partial z}\right) \tag{2.37}$$

Now writing $\mathbf{J}_T = J_T\mathbf{e}_T$, (2.37) yields

$$L\psi = \mathbf{J}_T. \tag{2.38}$$

2.2. Slow Evolution of a Quasi-Equilibrium Plasma-Vacuum System

Next, approximate equations that describe the slow evolution of a quasi-equilibrium plasma-vacuum system will be derived. It will be assumed that

(1) the external sources are small and vary slowly, and

(2) the external currents vary slowly.

These assumptions are expressed in the form

$$Q_M = \epsilon\hat{Q}_M(x,\epsilon t), Q_S = \epsilon\hat{Q}_S(x,\epsilon t), \mathbf{F} = \epsilon^2\hat{F}(x,\epsilon t), \mathbf{J} = \hat{J}(x,\epsilon t) \tag{2.39}$$

where $\epsilon > 0$ is a dimensionless parameter that is sufficiently small such that the quantities associated with ϵ^2 become physically negligible. Using these expressions, the unknowns can be scaled as follows.

$$\rho = \hat{\rho}(\hat{x},\bar{t}), \zeta = \hat{\zeta}(\hat{x},\bar{t}), p = \hat{p}(\hat{x},\bar{t}), \mathbf{V} = \epsilon\hat{\mathbf{V}}(\hat{x},\bar{t}), \mathbf{B} = \hat{\mathbf{B}}(\hat{x},\bar{t}) \tag{2.40}$$

where $\hat{x} = x, \bar{t} = \epsilon t$. Using these expressions in the governing equations with associated fixed and free boundary conditions, after dropping the hats, only the force balance equation changes to

$$\epsilon^2 \rho \frac{D\mathbf{V}}{Dt} + \nabla p - \mathbf{J} \times \mathbf{B} = \epsilon^2 \mathbf{F} \tag{2.41}$$

By the assumption of ϵ, the $O(\epsilon^2)$ terms can be neglected in (2.41). Thus, the adiabatically slow evolution of the plasma-vacuum system with external sources and external currents is governed by a system of reduced equations. By combining the conservation laws for ρ and ζ, this system can be reduced to

$$\frac{Dp^{1/\gamma}}{Dt} + p^{1/\gamma} \nabla \cdot \mathbf{V} = Q$$

where Q denoted a heating source density. Thus, the plasma region and vacuum region are governed by the following equations.

Plasma Region

$$\frac{Dp^{1/\gamma}}{Dt} + p^{1/\gamma} \nabla \cdot \mathbf{V} = Q \tag{2.42}$$

$$-r^{-2} f \nabla f + r^{-2} (L\Psi) \nabla \Psi = \nabla p \tag{2.43}$$

$$\nabla f \times \nabla \Psi = 0 \tag{2.44}$$

$$\frac{D\Psi}{Dt} = 0 \tag{2.45}$$

Vacuum Region

$$\nabla f = 0 \tag{2.46}$$

Since the **unknown profile functions** $f(\Psi, t), p(\Psi, t)$ are constant on the magnetic surfaces $\{\Psi = \sigma\}$, they are called "**surface quantities**".

Note: (2.45) means that each magnetic surface $\{\Psi = \sigma\}$ moves with the plasma flow $\mathbf{V}$. Hence, **the poloidal flux is conserved**.
Thus, the fixed boundary condition and free boundary conditions (at the plasma-vacuum interface) can be restated as follows.

Fixed Boundary Condition

$$\Psi = 0 \text{ on } \partial\Omega \tag{2.47}$$

Free Boundary Conditions

$$p = 0, \Psi = \sigma_0 \text{ on } S_{free} \tag{2.48}$$

$$\nabla\Psi, f \text{ are continuous across } S_{free} \tag{2.49}$$

where $\sigma_0 > 0$ is a flux constant that determines the free boundary surface S_{free}.

Plasma Region

Since the total plasma current is positive, the plasma region can be expressed by the set

$$\{\Psi > \sigma_0\} = \text{Plasma Region}. \tag{2.50}$$

External Current

The prescribed external current $\bar{\mathbf{J}}$ is assumed to be purely poloidal, i.e.,

$$\bar{\mathbf{J}} = \bar{J}\mathbf{e}_P \tag{2.51}$$

where $\mathbf{e}_P$ is the unit vector in the poloidal direction.

Assume that the meridian cross section Ω of the reactor is in a rectangle, i.e.,

$$\Omega \subset [R_1, R_2] \times [Z_1, Z_2] \tag{2.52}$$

where $R_1 > 0, R_2, Z_1, Z_2$ are given.

Poloidal Flux Function

Let

$$\Psi = \psi + \bar{\psi} \tag{2.53}$$

be the poloidal function, such that

$$\sigma_0 \leq \Psi(x) \leq b, a.e. x \in \Omega \text{ for some given constants } \sigma_0 > 0, b > 0 \tag{2.54}$$

where $\bar{\psi}$ is the poloidal flux function generated by the external current, which is a prescribed smooth function, and
$\psi \in H_0^1(\Omega)$ is an unknown flux function generated by the plasma flow.

Toroidal Flux Function

The toroidal flux function is denoted by $f \in L_+^2(\Omega) \bigcap L_+^\infty(\Omega)$ such that $||f||_{L^2(\Omega)} \leq C_1$ for some constant $C_1 > 0$.

Pressure Function

The pressure function is denoted by $p \in L_+^1(\Omega) \bigcap L_+^\infty(\Omega)$ such that $||f||_{L^1(\Omega)} \leq C_2$ for some constant $C_2 > 0$. The pressure function will be converted to a non-negative bounded function in $L^2(\Omega)$ by the following set-up.

$$g = \left(\frac{2p}{\gamma - 1}\right)^{1/2}, \gamma = 5/3 \tag{2.55}$$

Constraints of Motion

The general constraints of motion governed by axisymmetric ideal MHD are the conservation of toriodal flux and mass within each magnetic surface, and the conservation of total toriodal flux (plasma and vacuum).

$$C_F(\sigma) = \int_{\Psi>\sigma} fr^{-1} drdz, \forall \sigma_0 \leq \sigma \leq b, \text{ (conservation of toriodal flux)} \tag{2.56}$$

$$C_M(\sigma) = \int_{\Psi>\sigma} p^{1/\gamma} rdrdz, \forall \sigma_0 \leq \sigma \leq b, \text{ (conservation of mass)} \tag{2.57}$$

$$F_0^* = \int_\Omega fr^{-1} drdz, \text{ (conservation of total toroidal flux)} \tag{2.58}$$

3. Smooth Variational Structures and Modified Lagrange Multiplier Rule with a Functional Constraint

Notations
(1) $\Omega \subset \mathbf{R}^2$ is an open bounded set.
(2) dx denotes the Lebesgue measure on Ω.
(3) For any Lebesgue measurable set $A \subset \Omega, |A| \equiv \int_A dx$.
(4) $\langle \cdot, \cdot \rangle$ denotes the inner product on $L^2(\Omega)$.
(5) $H_0^1(\Omega) = W_0^{1,2}(\Omega)$
(6) $H^{-1}(\Omega)$ is the dual of $H_0^1(\Omega)$ with respect to the pairing $\langle \cdot, \cdot \rangle$.
(7) For $s \in \mathbf{R}$, $s_+ \equiv \max\{0, s\}$.
(8) χ_A is the characteristic function of set $A \subset \Omega$.
(9) L is the Laplacian

$$L \equiv -r\frac{\partial}{\partial r}\left(\frac{1}{r}\frac{\partial}{\partial r}\right) - \frac{\partial^2}{\partial z^2}.$$

Definition 3.1. *Let $f : \Omega \to \mathbf{R}$ and $g : \Omega \to \mathbf{R}$ be measurable functions. f is said to be* **equimeasurable** *with g or f is said to be a* **rearrangement** *with g, if $|\{x \in \Omega : f(x) > \sigma\}| = |\{x \in \Omega : g(x) > \sigma\}|, \forall \sigma \in \mathbf{R}$. A set of equimeasurable functions forms an equivalence relation, which is denoted by $\sim$. That is,*

$$f \sim g \Longleftrightarrow |\{x \in \Omega : f(x) > \sigma\}| = |\{x \in \Omega : g(x) > \sigma\}|, \forall \sigma \in \mathbf{R}. \tag{3.1}$$

Hereafter,

$$d_f(\sigma) \equiv |f > \sigma| \equiv |\{x \in \Omega : f(x) > \sigma\}| \tag{3.2}$$

Using this definition, $M_\infty(\bar{u})$ is defined as

$$M_\infty(\bar{u}) \equiv \{u \in H_0^1(\Omega) : u \sim \bar{u}\} \tag{3.3}$$

where $u, \bar{u}$ are assumed to be bounded and non-negative functions satisfying $0 \le u(x) \le b, \forall x \in \Omega, 0 \le \bar{u}(x) \le b, \forall x \in \Omega$.

(Note 6)(3.1) does not have a smooth variational structure in general. A smooth variational structure can be constructed in (3.1) using the following lemma. Its proof is given in [42].

Proposition 3.1. *Assume that $u, \bar{u}$ are bounded and non-negative functions satisfying $0 \le u(x) \le b, \forall x \in \Omega$, $0 \le \bar{u}(x) \le b, \forall x \in \Omega$. Then,*

$$u \in M_\infty(\bar{u}) \Longleftrightarrow \int_\Omega (u - \sigma)_+^k dx = \int_\Omega (\bar{u} - \sigma)_+^k dx, \forall \sigma \in [0, b], \forall k \in \mathbf{N}. \tag{3.4}$$

$$d_u(\sigma) \equiv |u > \sigma| = \frac{(-1)^k}{k!}\frac{d^k}{d\sigma^k}\int_\Omega (u - \sigma)_+^k dx. \tag{3.5}$$

Let $A_k(\sigma) = \int_\Omega (u - \sigma)_+^k dx$ and $B_k(\sigma) = \int_\Omega (\bar{u} - \sigma)_+^k dx$. Then, $A_k(\sigma), B_k(\sigma) \in C^{k-1}([0, b])$ such that $A_k^{(k)}(\sigma), B_k^{(k)}(\sigma)$ are right continuous on $[0, b]$.

Proof. See [42]. □

A level set having a smooth variational structure is now created as follows. Let

$$K_k(u;\sigma) \equiv \frac{1}{k}[A_k(\sigma) - B_k(\sigma)] = \frac{1}{k}\int [(u-\sigma)_+^k - (\bar{u}-\sigma)_+^k]dx \tag{3.6}$$

and

$$S_k \equiv K_k^{-1}(\{0\}). \tag{3.7}$$

Then, $K_k(u;\sigma)$ is Frechét C^{k-1} for $k \geq 2$. Hence, S_k is the level set of K_k through 0 having a Frechét C^{k-1} structure.

Definition 3.2. *(cf: [40]) Let $BV([a,b])$ be the set of all functions of bounded variation on $[a,b]$. Let $\widehat{BV}([a,b])$ be the set of left continuous functions on $(a,b]$ such that $v(b) = 0$, i.e.,*

$$\widehat{BV}([a,b]) \equiv \{v \in BV([a,b]) | v(b) = 0, v(t-) = v(t), \forall t \in (a,b]\}. \tag{3.8}$$

$\widehat{BV}([a,b])$ is a closed proper subspace of $BV([a,b])$.

Proposition 3.2. *(cf: [40])Let $(C([a,b]))^*$ be the dual of $C([a,b])$. Then $(C([a,b]))^*$ is congruent with $\widehat{BV}([a,b])$ in the sense that $\forall\lambda \in (C([a,b]))^*, \exists! v \in \widehat{BV}([a,b])$ such that $\langle\lambda, h\rangle = \int_a^b h(t)dv(t), \forall h \in (C([a,b]))$, where $dv(t)$ is a signed measure (Riemann-Stieltjes integrator).*

Definition 3.3.

$$C^k([a,b])_R = \{v \in C^k([a,b]) | v(b) = 0\}, k \geq 0 \tag{3.9}$$

Next, the classical Lagrange multiplier rule with a functional constraint (cf: [21][22]) is modified suitably so as to derive a set of Euler-Lagrange equations satisfied by a state of equilibrium.

Now the classical version of Lagrange multiplier rule with a functional constraint due to Flett [21] will be modified so as to achieve the main result.

Proposition 3.3. *(***Modified Lagrange Multiplier Rule with a Functional Constraint** *Let X, Y be Banach Spaces. Let U be an open subset of X, and $f : U \to \mathbf{R}, F : U \to Y$ be Frechét C^1 at x_0.*

Assume that: (1) f has a local extremum at x_0 subject to $F(x) = 0$, and $Im(F'(x_0))$ is closed in Y.

(2) F is **non-degenerate at** *x_0 in the sense that*

$$(F'(x_0))^* y_0^* = 0 \Rightarrow y_0^* = 0 \in Y^* \tag{3.10}$$

and ker $F'(x_0) \setminus \{0\}$ *is non-empty.*

Then, $\exists! y_0^* \in Y^* \setminus \{0\}$ such that

$$f'(x_0) = (F'(x_0))^* y_0^*. \tag{3.11}$$

Proof. See [42].

4. A State of Equilibrium of Plasma Flow in Nuclear Fusion

In this section, the variational problem defined by (1.2) , (1.3), (1.4) will be reformulated in the framework of the Lagrange Multiplier Rule with a Functional Constraint.

Let $\Omega \subset \mathbf{R}^2$ be an open bounded set and $x = (r, z) \in \Omega$, such that

$$R_1 \le r \le R_2, Z_1 \le z \le Z_2 \tag{4.1}$$

$$X = H_0^1(\Omega) \times L^2(\Omega) \times L^2(\Omega) \tag{4.2}$$

$$Y = \mathbf{R} \times C([\sigma_0, b])_R \times C([\sigma_0, b])_R \tag{4.3}$$

where $C([\sigma_0, b])_R \equiv \{v \in C([\sigma_0, b]) | v(b) = 0\}$.

Let

$$u = (\psi, f, g) \in X, ||u||_X = ||\psi||_{H_0^1(\Omega)} + ||f||_{L^2(\Omega)} + ||g||_{L^2(\Omega)} \tag{4.4}$$

$$K_k(\psi, f, g) \equiv (K_{k,1}(f), K_{k,2}(\psi, f; \sigma), K_{k,3}(\psi, g; \sigma)) \in Y \tag{4.5}$$

where

$$K_{k,1}(f) = \int_\Omega r^{-1} \cdot f drdz - F_0^* \in \mathbf{R} \tag{4.6}$$

$$K_{k,2}(\psi, f; \sigma) = \frac{1}{k+1} \int_\Omega r^{-1} \cdot f \cdot (\Psi - \sigma)_+^{k+1} drdz - F^*(\sigma) \in C([\sigma_0, b]) \tag{4.7}$$

$$K_{k,3}(\psi, g; \sigma) = \frac{1}{k+1} \int_\Omega r \cdot g^{2/\gamma} \cdot (\Psi - \sigma)_+^{k+1} drdz - G^*(\sigma) \in C([\sigma_0, b]) \tag{4.8}$$

Note: (4.7)(4.8) are direct consequences of Proposition 3.1.

$$B_X = \{u \in X : ||u||_X \le (2E_0)^{1/2}, f, g \in L^\infty(\Omega), \sigma_0 \le \psi(x) \le b, a.e. x \in \Omega\} \tag{4.9}$$

$$S_k \equiv K_k^{-1}(\{0\}) \bigcap B_X \tag{4.10}$$

With (4.2) –(4.10), the original variational problem (1.1) – (1.4) can be concisely stated as follows.

$$E(u) \to \min, u \in S_k \tag{4.11}$$

Proposition 4.1. *(Existence of a Minimizer) The variational problem (4.11) has a solution* $u_{op} = (\psi_{op}, f_{op}, g_{op})$.

Proof. See [42]. □

Now the main result will be stated.

Proposition 4.2. *(Sufficient Condition for the Existence of a Smooth Minimizer)*
(i) Suppose that $F_0^*, F^*(\sigma), G^*(\sigma)$ be given as described in (4.2) –(4.10).
Suppose that the following condition (A2) is satisfied:
(A2) *Suppose that, corresponding to these prescribed data $F_0^*, F^*(\sigma), G^*(\sigma)$, and $\forall u \in B_X$, there exist a closed subspace V of X and a closed subspace W of Y such that $K_k'(u)V$ is closed in W.*

Let

$$S_{k,V} \equiv S_k \bigcap V \tag{4.12}$$

Let $u_{op,k} \equiv (\psi_{op,k}, f_{op,k}, g_{op,k})$ be a solution of the problem

$$E(u) \to \min, u \in S_{k,V} \tag{4.13}$$

Then, $\psi_{op,k}, f_{op,k}, g_{op,k}$ satisfy the following system of Euler-Lagrange equations.

$$L\psi_{op,k} = f_{op,k} \int_{\sigma_o}^{b} (\psi_{op,k} + \bar{\psi} - \sigma)_+^k d\lambda_k(\sigma) + r^2 (g_{op,k})^{2/\gamma} \int_{\sigma_o}^{b} (\psi_{op,k} + \bar{\psi} - \sigma)_+^k d\mu_k(\sigma) \tag{4.14}$$

for some signed measures $\lambda_k, \mu_k \in \widehat{BV}([\sigma_0, b])$,

$$f_{op,k} = c_k + \frac{1}{k+1} \int_{\sigma_o}^{b} (\psi_{op,k} + \bar{\psi} - \sigma)_+^{k+1} d\lambda_k(\sigma), \exists c_k \in \mathbf{R} \tag{4.15}$$

$$g_{op,k} = \left[\frac{1}{\gamma(k+1)} \int_{\sigma_0}^{b} (\psi_{op} + \bar{\psi} - \sigma)_+^{k+1} d\mu_k(\sigma) \right]^{\frac{\gamma}{2\gamma-2}}, g_{op}(\sigma_0) = 0. \tag{4.16}$$

(ii) Moreover, $\psi_{op,k}, f_{op,k}, g_{op,k} \in C^{k,1}(\Omega)$.

Outline of the proof

The complete proof is given in [42]. The subscript k will be dropped in the proof. E, K are Frechét C^1 at any $u \in X$ with

$$E'(u) = (r^{-1} L\psi, r^{-1} f, rg) \in H^{-1}(\Omega) \times L^2(\Omega) \times L^2(\Omega) \tag{4.17}$$

First, K_k is shown to be non-degenerate at u_{op}, i.e.,
Let $h = (h_1, h_2, h_3) \in X$.

$$y^* K'(u_{op})h = 0, y^* \in Y^*, \forall h \in X \Rightarrow y^* = 0 \in Y^*. \tag{4.18}$$

Let

$$y^* = (c, \lambda, \mu) \in \mathbf{R} \times (C([\sigma_0, b])^* \times (C([\sigma_0, b])^*. \tag{4.19}$$

Then, by Fubini's Theorem,

$$\langle y^*, K'(u_{op})h \rangle = \langle (K'(u_{op}))^* y^*, h \rangle. \tag{4.20}$$

The rest is very similar to the proof of Theorem 4.1 of [42]. □

Theorem 4.1. *A minimizer* $\psi_{op} = \psi_{op,k}$ *satisfies the Grad-Shafranov equation*

$$L\psi_{op} = f(\psi_{op})f'(\psi_{op}) + r^2 \cdot p'(\psi_{op}) \tag{4.21}$$

where

$$p \equiv (\gamma - 1)(g(\psi))^2/2. \tag{4.22}$$

Proof.
By (4.15),

$$f_{op}^k(\psi_{op}) = \int_{\sigma_o}^{b} (\psi_{op} + \bar{\psi} - \sigma)_+^k d\lambda(\sigma). \tag{4.23}$$

(4.16)(4.22) then yield

$$p'(\psi_{op}) = (\gamma - 1)g(\psi_{op})g'(\psi_{op}) = (g_{op})^{2/\gamma} \int_{\sigma_o}^{b} (\psi_{op} + \bar{\psi} - \sigma)_+^k d\mu(\sigma) \tag{4.24}$$

Substituting (4.23)(4.24) into (4.14), one obtains (4.21). □

References

[1] A. Alvino, V. Ferone, G. Trombetti,and P. L. Lions Convex symmetrization and applications, *Ann. Inst. Henri Pincaré,* Vol. 14, No. 2, 1997, p. 275-293.

[2] A. Alvino, G. Trombetti, and P. L. Lions, On Optimization Problems with Prescribed Rearrangements, *Nonlinear Analysis*, Vol. 13, No. 2, 1989, pp185–220.

[3] A. V. Arutyunov, On Necessary Optimization Conditions in a Problem with Phase Constraints, *Soviet Math. Dokl.* Vol. 31, No. 1, pp174-177, 1985.

[4] A. V. Arutyunov, On the Theory of Quadratic Mappings in Banach Spaces, *Soviet Math. Dokl.* Vol. 42, No. 2, pp. 621-623, 1991.

[5] A. V. Arutyunov, Higher Order Conditions in Anormal Extremal Problems with Constraints of Equality Type, *Soviet Math. Dokl.* Vol. 42, No. 3, pp. 799-804, 1991.

[6] E. R. Avakov, Extremum Conditions for Smooth Anormal Problems with Equality and Inequality Type Constraints, *Math. Notes* **47** (1990), pp. 431-437.

[7] M.S. Bazaraa and J. J. Goode, Necessary Optimality Criteria in Mathematical Programming in Normed Linear Spaces, *Journal of Optimization Theory and Applications*, Vol. 111, No. 3, pp. 235-244, 1973.

[8] J. Blum, *Numerical Simulation and Optimal Control in Plasma Physics with Applicaton to Tokamaks*, John Wiley & Sons, 1989.

[9] C. Boucher, R. Ellis, and B. Turkington

Derivation of Maximum Entropy Principles in Two-Dimensional Turbulence via Large Deviation, *Journal of Statistical Physics*, Vol. 98, Nos.5/6, 2000.

[10] Friedemann Brock, Continuous rearrangement and symmetry of solutions of elliptic problems, *Proc. Indian Acad. Sci. (Math. Sci.)*, Vol. 110, No.2, May 2000, pp. 157-204.

[11] G. R. Burton and J. B. McLeod Maximization and Minimization on Classes of Rearrangements, *Proceedings of the Royal Society of Edinburgh*, Vol. 119A, pp. 287-300, 1991.

[12] G. R. Burton, Vortex Rings in a Cylinder and Rearrangements, *Journal of Differential Equations*, Vol. 70, pp. 333-348, 1987.

[13] G. R. Burton, Rearrangements of Functions, Maximization of Convex Functionals, and Vortex Rings, *Mathematische Annalen*, Vol. 276, pp. 225-253, 1987.

[14] G. R. Burton, Steady Symmetric Vortex Pairs and Rearrangements, *Proceedings of the Royal Society of Edinburgh*, Vol. 108A, pp. 269-290, 1988.

[15] G. R. Burton, Rearrangements and Steady Vortices, *A.I.H.P. Analyse Nonlineaire*, Vol. 6, pp. 295-319, 1989.

[16] G. R. Burton, Rearrangements of Functions, Saddle Points, and Uncountable Families of Steady Configurations for a Vortex, *Acta Math.* Vol. 163, pp. 291-309, 1989.

[17] A. Cianchi and N. Fusco, *Functions of bounded variation and rearrangements*, Preprint.

[18] B. D. Craven, A Generalization of Lagrange Multipliers, *Bull. Australia. Math. Soc.*, Vol. 3, pp. 353-362, 1970.

[19] A. Eydeland, J. Spruck, and B. Turkington, Multiconstrained Variational Problems of Nonlinear Eigenvalue Type: New Formulations and Algorithms, *Mathematics of Computation*, Vol. 55, No. 192, 1990, pp. 509-535.

[20] A. Eydeland, J. Spruck, B. Turkington, and A. Lifshitz, Multiconstrained Variational Problems in Magnetohydrodynamics, *Journal of Computational Physics*, Vol. 106, No. 2, 1993, pp. 269-285.

[21] T. M. Flett, On Differentiation in Normed Vector Spaces, *Journal.London. Math. Soc.* Vol. 42, 1967, pp. 523-533.

[22] A. D. Ioffe and V. M. Tihomirov, *Theory of Exremal Problems*, Elsevier North Holland, New York, 1979.

[23] R. Jordan A statistical equilibrium model of coherent structures in magnetohydrodynamics, *Nonlinearity*, Vol. 8, 1995, pp. 585-613

[24] R. Jordan and B. Turkington, Ideal magnetofluid turbulence in two dimensions, *Journal of Statistical Physics*, Vol. 87, pp. 661-695 (1997)

[25] M. Kruskal and R. Kulsrud, Equilibrium of a Magnetically Confined Plasma in a Toroid , *Physics of Fluids*, Vol. 1, No. 1, 1958, pp. 265-274.

[26] Peter Laurence and E. Stredulinsky A New Approach to Queer Differential Equations, *Comm. Pure Appl. Math*, Vol. 38, 1985, pp. 333-355.

[27] P. Laurence and E. Stredulinsky, A survey of recent regularity results for second order queer differential equations, *Proceedings of the Conference on the Calculus of Variations and Nonlinear Partial Differential Equations*, in Honor of Hans Lewy, Trento, Italy, June 1986, Springer Verlag.

[28] P. Laurence and E. Stredulinsky, Existence of Regular Solutions with Convex Level Sets for Semilinear Elliptic Equations with Monotone L^1 nonlinearities, Part I: An Approximating Free Boundary Problems, Indiana University *Math. Journal*, Vol. 39, No. 4, 1990, pp. 1081-1114.

[29] P. Laurence and E. Stredulinsky, Existence of Regular Solutions with Convex Level Sets for Semilinear Elliptic Equations with Monotone L^1 nonlinearities, Part II: Passage to the Limit, *Indiana University Math. Journal*, Vol. 3, No. 2, 1990, pp. 485-489.

[30] Peter Laurence and E. Stredulinsky A GRADIENT BOUND FOR THE GRAD-KRUSKAL-KULSRUD FUNCTIONAL, *Mathematical Research Letters* Vol. 1, pp. 377-387 (1994).

[31] Peter Laurence and E. Stredulinsky A Gradient Bound for the Grad Kruskal-Kulsrud Variational Problem, *Communications on Pure and Applied Math*, vol. 49, n3, pp. 237-286, 1996.

[32] Mossino, J, A priori estimates for a model Mercier Type in Plasma Physics, *Appl.Anal.* Vol. 13 (1982), No. 3, pp. 185-207.

[33] J. M. Rakotoson, A Differentiability Result for the Relative Rearrangement, *Differential and Integral Equations*, Vol. 2, No. 3, pp. 366-377, 1989.

[34] J. M. Rakotoson, Some Properties of the Relative Rearrangement, *Journal of Mathematical Analysis and Applications*, Vol. 135, pp. 488-500.

[35] J. M. Rakotoson, Relative Rearrangement and interpolation inequalities, *Rev. R. Acad. Cien. Series A. Mat.* Vol. 97 (1), 2003, pp. 133-145.

[36] J.-M. Rakotoson and M.L. Seoane, Numerical approximations of the relative rearrangement: The piecewise linear case. Application to some Nonlocal Problems, p. 477, *Mathematical Modelling and Numerical Analysis*, M2AN, Volume 34, Num 2, March/April 2000.

[37] J.-M. Rakotoson and D. Serre, *Sur un probléme d'optimisation lie aux équations de Navier-Stokes*, Preprint.

[38] D. Schechter, Derivatives of Mappings with Applications to Nonlinear Differential Equations, *Transactions of the AMS*, Vol. 293, pp. 53-69, 1986.

[39] M. Schechter, Non-uniqueness in the Equilibrium Shape of a Confined Plasma, *Communications in Partial Differential Equations*, Vol. 2, No. 6, pp. 587-600.

[40] A. E. Taylor, *General Theory of Functions and Integration*, Blaisdell Publishing Company, 1965.

[41] R. Temam, Monotone Rearrangements of a Function and the Grad-Mercier Equation of Plasma Physics, *Proc. International Meeting on Recent Methods in Nonlinear Analysis*, Rome, May 1978, pp. 83-98.

[42] T. Yamauchi, On Modified Lagrange Multiplier Rule and its Application to the Regularity of Magnetic Flux Function in Nuclear Fusion, submitted to *Communications on Pure and Applied Mathematics*.

In: Mathematics, Game Theory and Algebra Compendium
Editor: Jacob H. Mathias, pp. 585-593
ISBN: 978-1-60456-659-8

Chapter 49

COLLISION ORBITS FOR N-BODY PROBLEMS WITH FIXED ENERGY

Shiqing Zhang
Department of Mathematics, Chongqing University,
Chongqing 400044, P.R.China

Abstract

For N-body problems with weak-force type homogeneous potentials in R^k ($k+1 \geq N$) we prove that the constrained variational minimizing critical points of the fixed energy or fixed mean potential from a given initial position to the final simultaneous collision position must be such orbits that the configurations of the N-bodies have similar equilateral homographic shapes.

Key words: N-body problems, simultaneous collision solutions, Lagrangian action, constrained minimizing methods.

Mathematics Subject Classification: 34C25, 34C15, 70F10

1. Introduction and Main Results

N-body problems are related with the motion of N point masses $m_1, \ldots, m_N$ in R^k ($k \geq 1$) under the action of a homogeneous potential $-U(q)$ given by

$$U(q) = \frac{G}{2} \sum_{1 \leq i \neq j \leq N} \frac{m_i m_j}{\left|q_i - q_j\right|^{\alpha}} \tag{1.1}$$

where $G>0$, $\alpha>0$, $q=(q_1, \ldots, q_N)$, $q_i \in R^k$.

The equations of the motion for the N-body problems with a potential $-U(q)$ are given by

$$`m_i\ddot{q}_i = \frac{\partial U(q)}{\partial q_i} \quad i=1, \ldots, N \tag{1.2}$$

Note that $-U(q)$ and (1.2) are the classical Newtonian potential and Newtonian equations respectively when $\alpha=1$.

Coti Zelati [4] used variational methods to study the existence of a non-simultaneous collision periodic solution for the classical Newtonian N-body problems under some restrictions on the masses m_i ($i=1, \ldots, N$).

Serra-Terracini ([12]) used variational methods to study the existence of a noncollision periodic solution for 3-body problems with the classical Newtonian potential and a radial perturbation potential in R^3.

Long and Zhang ([10]) and Chenciner and Desolneux ([3]) proved independently that the shapes of the solutions from minimizing the Lagrangian action integral on the T/2 –antiperiodic or zero mean loop space of class $W^{1,2}$ (R/TZ, R^k) for 3-body problems with a homogeneous potential must be the planar equilateral triangle circulars.

Long and Zhang [11] studied the shapes for constrained variational minimizing solutions for 2 and 3 body problems with a fixed energy or mean potential. Zhang [16] studied the shapes for the simultaneous solutions of minimizing the Lagrangian action integral in $R^k(k+1\geq N)$ with a fixed period.

In this paper, we study the variational properties for simultaneous collision solutions of N-body problems with a fixed energy. We have the following theorems.

Theorem 1.1 For N-body problems (1.1)-(1.2) in R^k ($k+1\geq N$) with $0<\alpha<2$, and $h<0$, we consider

$$f(q) = \frac{1}{2}\int_0^1 \sum_{i=1}^N m_i |\dot{q}_i|^2 \, dt \int_0^1 \{h + U(q(t))\} dt, \quad \forall q \in M \tag{1.3}$$

where $M = \{q \in W^{1,2}([0,1],(R^k)^N),\ q(0)=0,\ q(1)=\text{constant vector} \neq 0,$

$$\int_0^1 \left[-U(q) - \frac{1}{2} U'(q) \cdot q \right] dt = h \} \tag{1.4}$$

Then the orbits which minimize $f(q)$ must satisfy that the configurations of the N-bodies are similar equilateral geometric graphs, and the momentum of inertial is

$$I(t) = \sum_{i=1}^N m_i |q_i(t)|^2 = At^{\frac{4}{\alpha+2}},\ A > 0 \tag{1.5}$$

If we define a constant $T>0$ by

$$T = \left[\frac{\frac{1}{2}\int_0^1 \sum_{i=1}^{N} m_i \, |\dot{q}_i(t)|^2 \, dt}{-\int_0^1 U'(q)q dt} \right]^{\frac{1}{2}} \tag{1.6}$$

Then
$$x(t) = q(t/T) \tag{1.7}$$

is a simultaneous collisioni solution of (1.1)–(1.2) with the energy h:

$$\frac{1}{2}\sum_{i=1}^{N} m_i |\dot{x}_i(t)|^2 - U(x(t)) = h, \ \ 0 < t \le T \tag{1.8}$$

Theorem 1.2 For N-body problems (1.1)-(1.2) in R^k ($k+1 \geq N$) with $0<\alpha<2$, and $P<0$, we define

$$\psi_R(q) = \int_0^1 U(Rq(t))dt, \ \ \forall q \in S, R > 0 \tag{1.9}$$

where $S = \{q \in W^{1,2}([0,1], (R^k)^N),$ $q(0)=0$, $q(1)$=constant vector $\neq 0$,

$$\left. \int_0^1 \sum_{i=1}^{N} m_i \, |\dot{q}_i|^2 \, dt = 1 \right\} \tag{1.10}$$

Then the orbits which minimize $\psi_R(q)$ have the properties that the configurations of the N-bodies are similar equilateral geometric graphes, and the momentum of inertial has the form of (1.5).

If we define a constant $\omega = \omega(R) > 0$ depending on R by

$$\omega^2 = -\frac{1}{R^2}\int_0^1 U'(Rq) \ Rq \, dt \tag{1.11}$$

Then $p(t)=Rq(\omega t)$ is a simultaneous collision solution of (1.1) and (1.2) with mean potential energy

$$P = -\psi_R(q) = -\frac{1}{\omega}\int_0^{\frac{1}{\omega}} U(q)dt \tag{1.12}$$

2. The Proof of Theorem 1.1 and Theorem 1.2

In order to prove Theorem 1.1 and Theorem 1.2, we need the following Lemmas.

Lemma 2.1 1° ([5], [10]) For any $q_i \in W^{1,2}([a,b],R^k)$, $-\infty \le a \le b \le +\infty$, i=1, …, N.

There holds,

$$\frac{1}{2}\int_a^b \left|\frac{d}{dt}[(\sum_{i=1}^N m_i |q_i(t)|^2)^{1/2}]\right|^2 dt \le \frac{1}{2}\int_a^b \sum_{i=1}^N m_i |\dot{q}_i(t)|^2 \, dt \tag{2.1}$$

2° Furthermore, (2.1) becomes equality when

$$q_i(t) = a_i g(t),\ a_i \in R^k,\ g(t) \ge 0,\ t \in [a,b] \tag{2.2}$$

Lemma 2.2 1° (Lemma 2.1 of Coti Zelati [4]) For any positive integer k, $\alpha>0$, N≥2, $m_i>0$, $q_i \in R^k$, we have

$$\frac{1}{2}\sum_{1\le i\ne j\le N} \frac{m_i m_j}{|q_i - q_j|^\alpha} \ge D_\alpha(N,m)(\sum_{i=1}^N m_i |q_i|^2)^{-\alpha/2} \tag{2.3}$$

where $$D_\alpha(N,m) = 2^{-(2+\alpha)/2} M^{-\alpha/2}(\sum_{1\le i\ne j\le N} m_i m_j)^{(2+\alpha)/2} \quad \text{and } M = \sum_{i=1}^N m_i \tag{2.4}$$

2° (Long-Zhang [10]) when N≥3 the equality holds in (2.3) if and only if there hold

$$N \le k+1 \tag{2.5}$$

$$|q_i - q_j| = |q_l - q_h|,\ \forall i \ne j, l \ne h \tag{2.6}$$

$$\sum_{i=1}^N m_i q_i = 0 \tag{2.7}$$

3° (Long-Zhang [10]) when N=2, the equality holds in (2.3) if and only if there hold

$$k \ge 1 \text{ and } m_1 q_1 + m_2 q_2 = 0 \tag{2.8}$$

Lemma 2.3 1° (Hardy-Littlewood Theorem 254 of [8]) If $\mu>4$, $\xi<+\infty$, $y(0)=0$, $y(\xi)=c$, $\dot{y} \in L^2[0,\xi]$ then

$$J(y) = \int_0^{\xi} (\mu \dot{y}^2 - \frac{y^2}{x^2})dx \geq \frac{2}{1-2a}\frac{c^2}{\xi} \tag{2.9}$$

where a satisfies

$$a = (\frac{1}{4} - \frac{1}{\mu})^{1/2} < \frac{1}{2} \tag{2.10}$$

2° The inequality (2.8) becomes the equality if and only if there holds

$$y = Ax^{\frac{1}{2}+a}, \quad A = \xi^{-(\frac{1}{2}+a)} \cdot y(\xi), \xi < +\infty, x \in [0, \xi] \tag{2.11}$$

Lemma 2.4 (Jensen inequality, Theorem 204 of [8]) 1° Assume ϕ is a convex function on [r, R], $-\infty \leq r \leq R \leq +\infty$, f, p are integrable functions on $[c,. d]$, $-\infty \leq c \leq d \leq +\infty$ $r \leq f(x) \leq R$, $p(x) \geq 0$, $\forall x \in [c,d]$ and $\int_c^d p(x)dx > 0$ then

$$\phi(\frac{\int_c^d p(x)f(x)dx}{\int_c^d p(x)dx}) \leq \frac{\int_c^d p(x)\phi(f(x))dx}{\int_c^d p(x)dx} \tag{2.12}$$

2° Inequality (2.13) becomes equality if and only if there holds

$$f(x) \equiv const \tag{2.13}$$

Now we prove Theorem 1.1.

Assume

$$\sum_{i=1}^{N} m_i \,|\, q_i(1) \,|^2 = c \tag{2.14}$$

For $q \in M$, by the homogeneous property for U we have

$$\int_0^1 (h + U(q))dt = -\frac{1}{2}\int_0^1 U'(q)q dt = \frac{\alpha}{2}\int_0^1 U(q)dt \tag{2.15}$$

$$\int_0^1 U(q)dt = \frac{-2h}{2-\alpha} \tag{2.16}$$

By Coti Zelati's inequality and Jensen's inequality we have

$$\frac{-2h}{2-\alpha}=\int_0^1\frac{G}{2}\sum_{1\le i\ne j\le N}\frac{m_i m_j}{|q_i-q_j|^{\alpha}}dt \tag{2.17}$$

$$\ge G\cdot D_\alpha(N,m)\int_0^1(\sum_{i=1}^N m_i\,|q_i|^2)^{-\alpha/2}dt \tag{2.18}$$

$$\ge G\cdot D_\alpha(N,m)\int_0^1 t^{-\frac{2\alpha}{\alpha+2}}(t^{\frac{-4}{\alpha+2}}\sum_{i=1}^N m_i\,|q_i|^2)^{-\alpha/2}dt \tag{2.19}$$

$$\ge G\cdot D_\alpha(N,m)\left[\int_0^1 t^{\frac{-2\alpha}{\alpha+2}}dt\right]^{1+\frac{\alpha}{2}}\left[\int_0^1 t^{-2}\sum_{i=1}^N m_i\,|q_i|^2\,dt\right]^{\frac{-\alpha}{2}} \tag{2.20}$$

$$=G\cdot D_\alpha(N,m)\cdot(\frac{2+\alpha}{2-\alpha})^{(2+\alpha)/2}\left[\int_0^1(\sum_{i=1}^N m_i\,|q_i|^2)t^{-2}dt\right]^{-\alpha/2} \tag{2.21}$$

Now we estimate the lower bound for $f(q)$:

$$f(q)=\frac{1}{2}\int_0^1\left(\sum_{i=1}^N m_i\,|\dot{q}_i|^2\right)dt\int_0^1(h+U(q))dt$$

$$\ge\frac{1}{2}\int_0^1\left|\frac{d}{dt}\left(\sum_{i=1}^N m_i|q_i(t)|^2\right)^{1/2}\right|^2 dt\int_0^1(h+U(q))dt \tag{2.22}$$

$$=\frac{-h\alpha}{2(2-\alpha)}\left\{\frac{1}{4}-\left[\frac{2-\alpha}{2(\alpha+2)}\right]^2\right\}\int_0^1\left(\sum_{i=1}^N m_i\,|q_i|^2\right)t^{-2}dt$$

$$+\frac{1}{2}\left\{\frac{1}{4}-\left[\frac{2-\alpha}{2(2+\alpha)}\right]^2\right\}\left(1+\frac{2}{\alpha}\right)\frac{c^2}{T} \tag{2.23}$$

Notice that we used inequality (2.1) in (2.22) and Hardy-Littlewood's inequality (2.9) with $\mu=\left\{\frac{1}{4}-\left[\frac{2-\alpha}{2(\alpha+2)}\right]^2\right\}^{-1}$ and $a=\frac{2-\alpha}{2(\alpha+2)}$ and equality (2.16) in (2.23)

By (2.21) and (2.23) we have

$$f(q)\ge\frac{-h\alpha}{2(2-\alpha)}\left\{\frac{1}{4}-\left[\frac{2-\alpha}{2(\alpha+2)}\right]^2\right\}\left[\frac{(2-\alpha)G\cdot D_\alpha(N,m)}{-2h}\left(\frac{2+\alpha}{2-\alpha}\right)^{(2+\alpha)/2}\right]^{2/\alpha}$$

$$+\frac{1}{2}\left\{\frac{1}{4}-\left[\frac{2-\alpha}{2(\alpha+2)}\right]^2\right\}\left(1+\frac{2}{\alpha}\right)\frac{c^2}{T} \tag{2.24}$$

Note that we used inequalities (2.3) in (2.8 and used Jensen's inequality (2.12) with the weight function $p(t) = t^{-2\alpha/(\alpha+2)}$ in (2.20).

Notice that $f(q)$ attains its global minimum if and only if inequalities (2.18)–(2.24) hold equalities simultaneously, hence by the neccessary and sufficient conditions for making these equalities hold, we have that

(1) $$N \le k+1 \tag{2.25}$$

(2) $$|q_i(t)-q_j(t)| = |q_l(t)-q_h(t)|,\ \forall i \ne j, l \ne h \tag{2.26}$$

(3) $$\sum_{i=1}^{N} m_i q_i(t) \equiv 0,\ \forall t \in [0,T] \tag{2.27}$$

(4) $$\sum_{i=1}^{N} m_i |q_i(t)|^2 = A \cdot t^{\frac{4}{\alpha+2}},\ A>0 \tag{2.28}$$

Note that (2) can be conceled when N=2.

Using Lagrangian multiplier methods and the similar proof of [1], we know that $x(t)=q(t/T)$ has the properties as stated in Theorem 1.1.

Hence Theorem 1.1 is true.

Now we prove Theorem 1.2:

Proof Assume $\sum_{i=1}^{N} m_i |q_i(1)|^2 = c^2 > 0$. We estimate the lower bound for $\psi_R(q)$:

$$\psi_R(q) = \frac{G}{2}\int_0^1 \sum_{1\le i\ne j\le N} \frac{m_i m_j}{(R|q_i-q_j|)^\alpha} dt \tag{2.29}$$

$$\ge \frac{G}{R^\alpha} D_\alpha(N,m) \int_0^1 (\sum_{i=1}^{N} m_i |q_i|^2)^{-\alpha/2} dt \tag{2.30}$$

$$= \frac{G}{R^\alpha} D_\alpha(N,m) \int_0^1 t^{\frac{-2\alpha}{\alpha+2}} (t^{\frac{-4}{\alpha+2}} \sum_{i=1}^{N} m_i |q_i|^2)^{-\alpha/2} dt \tag{2.31}$$

$$\ge \frac{G}{R^\alpha} D_\alpha(N,m) \left[\int_0^1 t^{\frac{-2\alpha}{\alpha+2}} dt\right]^{1+\frac{\alpha}{2}} \left[\int_0^1 t^{-2} \sum_{i=1}^{N} m_i |q_i|^2\, dt\right]^{-\alpha/2} \tag{2.32}$$

$$= \frac{C_1}{R^\alpha}\left[\int_0^1\left(\sum_{i=1}^N m_i\,|\,q_i\,|^2\right)t^{-2}dt\right]^{-\alpha/2} \tag{2.33}$$

where
$$C_1 = GD_\alpha(N,m)\left(\frac{2+\alpha}{2-\alpha}\right)^{(2+\alpha)/2} \tag{2.34}$$

Note that we used Coti Zelati's inequality (2.3) in (2.30) and used Jensen's inequality (2.12) with the weight function $p(t) = t^{-2\alpha/(\alpha+2)}$ in (2.32).

For any $q \in S$, we have

$$1 = \int_0^1 \sum_{i=1}^N m_i\,|\,\dot{q}_i\,|^2 dt$$

$$\geq \int_0^1 \left|\frac{d}{dt}\left(\sum_{i=1}^N m_i\,|\,q_i\,|^2\right)^{1/2}\right|^2 dt \tag{2.35}$$

$$\geq C_2 \int_0^1\left(\sum_{i=1}^N m_i\,|\,q_i\,|^2\right)t^{-2}dt + C_3 \tag{2.36}$$

where
$$C_2 = \frac{1}{4} - \left[\frac{2-\alpha}{2(\alpha+2)}\right]^2 \tag{2.37}$$

$$C_3 = \frac{1}{2}\left\{\frac{1}{4} - \left[\frac{2-\alpha}{2(\alpha+2)}\right]^2\right\}\left(1+\frac{2}{\alpha}\right)\frac{c^2}{T} \tag{2.38}$$

Note that we used (2.1) and Hardy-Littlewood' inequality (2.9) in (2.35) and (2.36)..

Hence by (2.33) and (2.36) we have

$$\psi_R(q) \geq \frac{C_1}{R^\alpha}\left(\frac{1-C_3}{C_2}\right)^{-\alpha/2}$$

We note that $\psi_R(q)$ attains its global minimum if and only if (2.30) –(2.36) hold equalities, Hence we still have (2.25) – (2.28)

We take

$$R = \left(\frac{C_1}{-P}\right)^{\frac{1}{\alpha}} \cdot \left(\frac{1-C_3}{C_2}\right)^{-1/2}$$

and define $\omega = \omega(R)$ by (1.11).

Then $y(t) = Rq(\omega t)$ is a simultaneous collisioni solution with given mean potential energy P<0 by the similar proof of [14].

References

[1] A. Ambrosetti and V. Coti Zelati, *Periodic solutions of singular Lagrangian systems*, Birkhäuser, Basel, 1993.

[2] Arnold V., Kozlov, V. and Neishtadt A., Dynamical Systems, III, *Mathematical aspects of classical and celestial mechanics*, Russian ed. 1985, Springer, Berlin, English ed. 1988.

[3] Chenciner A. and Desolneux N., Minima de l'intégrale d'action et équilibres relatifs de n corps, C. R. *Acad. Sci. Paris, t.***326**, serie I (1998), 1209-1212.

[4] Coti Zelati V., The periodic solutions of N-body type problems, *Ann. IHP. Anal. nonlinéaire* **7** (1990), 477-492.

[5] Degiovanni M. and Giannoni F., Dynamical systems with Newtonian type potentials, *Ann. Scuola Norm. Super. Pisa* **15**(1989), 467-494.

[6] Euler, L., De motu rectilineo trium corpörum se mutuo attrahentium, *Novi. Comm. Acad Sci. Imp. Petropll* (1767),145-151.

[7] Gordon W., A minimizing property of keplerian orbits, *Amer. J. Math.* **99**(1977), 961-971

[8] Hardy G., Littlewood J. and Polya G., *Inequalities*, 2nd ed. University Press, Cambridge,1952.

[9] Lagrange, J., Essai sur le probléme des trois corps. (1772). *Ouvres*, Vol. 3 (1783), 229-331.

[10] Long Y. and Zhang S. Q., Geometric characterization for variational minimization solutions of the 3-body problm, *Nankai Inst. of Math. Preprint*, No. 1998-M-005, June (1998). *Acta Math. Sinica*, to appear

[11] LongY. And Zhang S. Q., Geometric characterization for variational minimization solutions of the 3-body problem with fixed energy, *J. Diff. Equ.*, to appear.

[12] Serra E. and Terracini S., Collisionless periodic solutions to some three-body problems, *Arch. Rational Mech. Anal.* **120**(1992), 305-325.

[13] Siegel C. and Moser J., *Lectures on celestial mechanics*, Springer, Berlin, 1971

[14] Vitillaro E., Closed orbits of prescribed mean potential energy for a class of N-body type problems, *Dynamical Systems and Applications* **4**(1995), 225-235

[15] Wintner A., *Analytical foundations of celestial mechanics*, Princeton University Press, Princeton, 1941.

[16] Zhang S. Q., *The simultaneous collision orbits for N-body problems*, preprint.

In: Mathematics, Game Theory and Algebra Compendium
Editor: Jacob H. Mathias, pp. 595-605
ISBN 978-1-60456-659-8

Chapter 50

GRAPH SEARCHING PROBLEMS WITH THE COUNTERACTION

V.Y. Andrianov and Nikolay N. Petrov
Faculty of Mathematics and Mechanics
St. Petersburg State University
St. Petersburg, Russia

1. Introduction

Graph searching problems have attracted the attention of many researchers because of numerous (sometimes unexpected) applications in various fields of Discrete Mathematics. These problems arise in some computer-virus protection models ([1]) and in the VLSI theory where many important parameters of graph layout such as the cutwidth ([2]), the bandwidth ([3]), the vertex separation number ([4]) have the search-theoretic interpretation. There are important connections of graph searching problems with "pebble games" ([5]) that model sequential calculations and the theory of graph minors developed by Robertson and Seymour ([1]).

More detailed information on search problems can be found in surveys ([1,6,7]).

The main problem in the graph searching theory is to find (in the framework of some formalization) the search number of the graph, i.e. the minimum number of pursuers (searchers) needed "to catch" the invisible evader (the fugitive). In the present paper we survey some theorems on search problems with "the counteraction". The counteraction means the following: under appropriate conditions the evader can "destroy" a pursuer. At this moment the evader becomes "visible" for pursuers and those may use this information in the future (the precise statement of the problem is given below).

2. Statement of the Problem

The board of the game is a finite connected undirected graph G without loops and multiple edges containing at least one edge. On this graph a multistage pursuit-evasion game with m pursuers (searchers) $P_1, P_2 ... P_m$ and the invisible evader E is considered. We consider that $m < n = |VG|$. At each step of the game all the players (pursuers and the evader) occupy nodes of the graph G, each node being occupied by at most one player.

A move of a player is the transition to an adjacent node (we denote this transition by the arrow $\longrightarrow$). *A move of the pursuers team*, by definition, is a move of exactly one pursuer. The pursuers team and the evader move in turns (pursuers begin) Each step in the game consists of a move of the pursuers team and an answering move of the evader. If a pursuer passes to a node occupied by the evader the game ends (in this case we say the evader is *caught* and pursuers win). On the other hand, if the evader passes to a node occupied by a pursuer the latter is removed from the graph and the game goes on. The aim of the pursuers team is to catch the evader, the aim of the evader is to prevent it. If the latter achieves its aim we say the *evasion* takes place. At each step the evader receives the complete information on the arrangement of pursuers (i.e., on the set of all the nodes occupied by them). Moreover, the evader learns the program of actions chosen by the pursuers team before the start of the game. So, in this formalization we deal with a problem of guaranteed search. As for pursuers, during the game they receive the following information. At the first step they do not know the location of the evader that may be in any node which is not occupied by pursuers.

At the step with number $k \geq 2$ they receive the information of the following three types.

1. At the $(k-1)$-th step the evader was caught, in this case the game ends with the victory of pursuers.

2. At the $(k-1)$-th step the evader removed a pursuer, occupied its node and became "visible" for pursuers (the pursuers team receives the information on the location of the evader only after a move "with removal").

3. At the $(k-1)$-th step the evader moved "without removal".

A position at the k-th step in the game considered is the information received by pursuers at this step on the arrangement of players together with the "prehistory" which contains the complete description of the course of the game in "the past", i.e., at all the steps with numbers $i < k$.

A strategy of the pursuers team is a triplet $S = (p, D_0, U)$, where p is the number of steps, D_0 is the initial arrangement of pursuers, and U is a map which determines the move for any position.

A strategy of the pursuers team is called *winning* if it leads to the capture of the evader for any actions of the latter.

The minimum cardinality of the pursuers team having a winning strategy is called the *α-search number* of the graph G and denoted by $\alpha(G)$.

3. Theorems on α-search Numbers

Here we state some assertions concerning α-search numbers without proofs. In what follows

P_m is the path with m nodes,

C_m is the cycle with m nodes,

K_m is the complete graph with m nodes,
$K_{m,l}$ is the complete bipartite graph.

Theorem 3.1. ([8]) *For any graph G ($n \geq 2$)*
$\alpha(G) \leq n-1$, *and*
$\alpha(G) = n-1$ *iff either* $n = 2$ *or* $G = C_3$.

Theorem 3.2. ([8]) *For any graph G ($n \geq 4$)*
$\alpha(G) \leq n-2$, *and*
$\alpha(G) = n-2$ *iff G satisfies one of the following conditions:*

1. $n = 4$, $G \neq K_{1,3}$,
2. $G = C_5$,
3. $G = C_6$,
4. $G = K_n$, $n \geq 4$.

The description of all the graphs G with $\alpha(G) = n-3$ is an open problem. The complete solution for $n \geq 7$ is known if $|\{v \in VG | deg\ v \leq n-3\}| \leq 2$.

Theorem 3.3. ([8]) *Let* $V = \{v \in VG | deg\ v \leq n-3\}$, $|V| \leq 2$ *and* $n \geq 7$. *Then* $\alpha(G) = n-3$ *iff V contains an edge.*

Now we consider graphs with "small" α-search numbers.

Theorem 3.4. ([9]) $\alpha(G) = 1$ *iff* $G = K_{1,n-1}$.

The case of trees is rather complicated. Here is a statement concerning trees with "small" α-search numbers.

Theorem 3.5. ([9]) *Let G be a tree ($n \geq 3$). Then the following assertions are equivalent:*

1. $\alpha(G) \geq 3$;
2. *G contains one of the following trees:*

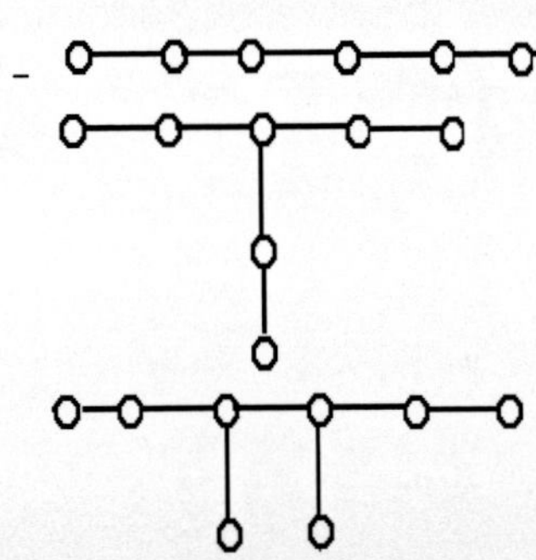

3. *for any triplet T of nodes containing an edge the graph $G\backslash T$ is empty (i.e., contains no edges).*

The complete solution of the main problem for some classes of graphs is given in the following theorems.

Theorem 3.6. $\alpha(P_2) = \alpha(P_3) = 1$, $\alpha(P_4) = \alpha(P_5) = \alpha(P_6) = 2$, $\alpha(P_n) = 3$ *for* $n \geq 7$.

It is an easy consequence of Theorem 5.

Theorem 3.7. ([10]) $\alpha(C_3) = \alpha(C_4) = 2$, $\alpha(C_5) = 3$, $\alpha(C_6) = \alpha(C_7) = \alpha(C_8) = 4$, *for* $n \geq 9$

$$\alpha(C_n) = \begin{cases} 4, \; if \; n \; is \; odd, \\ 5, \; if \; n \; is \; even. \end{cases}$$

The solution reveals an unexpected phenomenon: for a cycle with the lesser number of nodes the α-search number may be greater.

Theorem 3.8. ([11]) $\alpha(K_2) = 1$, $\alpha(K_3) = 2$, $\alpha(K_n) = n - 2$, *for* $n \geq 4$.

Let G be the complete n-partite graph with "parts" $M_1, M_2 ... M_n$. We consider that $\mid M_1 \mid \leq \mid M_2 \mid \leq ... \leq \mid M_n \mid$ and let $M = \bigcup_{i=1}^{n-1} M_i$. Then the following statement is valid.

Theorem 3.9. ([11]) *For any complete n-partite graph G*

1. $\alpha(G) = |M_1|$ *if* $n = 2$;

2. $\alpha(G) = 2$, *if* $n = 3$ *and* $|M_2| = 1$;

3. $\alpha(G) = |M| - 1$, *if* $n = 3$ *and* $|M_2| \geq 2$;

4. $\alpha(G) = |M| - 1$, *if* $n \geq 4$.

4. Related Problems

Consider the main problem under the following additional condition: the evader *must* remove a pursuer at the first opportunity and denote the corresponding search number by $\alpha_0(G)$. It is obvious, that $\alpha_0(G) \leq \alpha(G)$ for any graph G. For example, for $n \geq 4$ $\alpha(K_n) = n - 2$, but $\alpha_0(K_n) = 2$. However, as the following theorem shows, for cycles $\alpha_0(G)$ is slightly different from $\alpha(G)$.

Theorem 4.1. ([10]) $\alpha_0(C_3) = \alpha_0(C_4) = 2$, $\alpha_0(C_5) = \alpha_0(C_6) = 3$. $\alpha_0(C_i) = 4$ *for* $i = 7, 8, 9, ...12$ *and for* $n \geq 13$

$$\alpha(C_n) = \begin{cases} 4, \; if \; n \; is \; odd, \\ 5, \; if \; n \; is \; even. \end{cases}$$

Now consider the main problem "with prescribed duration": the number of steps in the game is bounded by an integer k and let $\alpha_k(G)$ be the corresponding search number.

In the following theorem the solution is given for $\alpha_2(G)$. We say that a set $M \subset VG$ satisfies property A if the subgraph F generated by M has no isolated nodes and at least one of the following assertions takes place:

1. EF is not a matching in G;
2. EF contains "pendant" (in G) edges.

Let $W = \{M \subset VG : M$ satisfies property A, $|M| \geq 3$, $G \backslash M = \oslash\}$.

Theorem 4.2. ([11]) *For any graph G, $G \neq K_{1,n-1}$,*

$$\alpha_2(G) = \min_{M \in W} |M| - 1.$$

5. Cycles

In this section we present a sketch of the proof of the most striking result concerning cycles: for $n > 13$

$$\alpha(C_n) = \alpha_0(C_n) = \begin{cases} 5, & if\ n\ is\ even, \\ 4, & if\ n\ is\ odd. \end{cases}$$

Proposition 5.1. *If n is even, $n > 13$, then $\alpha_0(C_n) \geq 5$.*

Suppose there exists a winning strategy S of the pursuers team consisting of P_1, P_2, P_3, P_4 numbered clockwise according to their location on the cycle. One can consider S is a winning strategy with the minimum number of steps.

Show how the evader can avoid the capture. We need some definitions concerning strategy S.

A moment (in S) is the number of a step together with indicating the side to move.

We call a player *isolated* at a moment t iff both the nodes adjacent to the node occupied by this player at the moment t are free of pursuers.

Let H be a quadruple of players. An *H-arrangement* at a moment t is the set of all the nodes occupied by players of H at this moment.

Two players of H are said to be *neighbors* at a moment t iff nodes occupied by them at this moment may be joined with a path free of other players of H.

The *parity* of a player of H at a moment t means the parity of the sum of distances between this player and its neighbors at the moment t.

We call an H-arrangement at a moment t *exact* iff all players of H have the same parity. Otherwise, we call it *unexact*.

Let H be $\{P_1, P_2, P_3, P_4\}$. It is not hard to see that after any move of P the exact H-arrangement becomes unexact and vice versa.

First show that the evader can remove an isolated pursuer in an unexact H-arrangement.

Let D_0 be the initial H-arrangement in S and D_k be the H-arrangement realized after the k-th move of P, $k \geq 1$. If D_1 is unexact there is an isolated pursuer that may be removed by the evader at the first step. Indeed, if the isolated pursuer is located in a node $v \in D_1$, then at the initial moment the evader occupies a node $v_1 \notin D_1$ adjacent to v and moves : $v_1 \longrightarrow v$.

Now let D_1 be exact. Denote by d the maximum distance between neighbor pursuers and let l_1 be the path of length d joining nodes occupied by these pursuers in D_1. By assumption, $d \geq 4$. Denote nodes of l_1 by $v_0, v_1, ..., v_d$ numbered clockwise and assume the nodes v_0 and v_d are occupied in D_1 by pursuers P_1 and P_2 respectively. Nodes of the cycle following v_d are denoted by $v_{d+1}, v_{d+2}, ...$ etc. Let Q_0 be the set of all the nodes v of l_1 such that $dist(v, v_0) \geq 2$ and $dist(v, v_d) \geq 2$. If is obvious that $|Q_0| \geq 1$.

We show how the evader can ensure the evasion starting from a node in Q_0. Denote by l_k the path joining nodes occupied by P_1 and P_2 in D_k free of the other pursuers and let Q_k be the set of all the nodes of l_k different from its ends. Paths l_k and sets Q_k are considered for $k \geq 1$ while sets D_k are defined. It is easy to prove the following assertion: if $|Q_0| > 1$ then for any node $u \in Q_k$ $(k \geq 1)$ there exists a "trajectory" of the evader

$$u_0 \longrightarrow u_1 \longrightarrow u_2 \longrightarrow ... \longrightarrow u_{k-1} \longrightarrow u$$

such that $u_i \in Q_i$ for $i = 0, 1, ..., k-1$. Any such trajectory is called *admissible*. So, at the k-th step the evader can occupy any node in Q_k.

First consider the case when P_1 and P_2 are isolated in D_1. Then one of them will be isolated in D_2 and the evader can remove it at the second step. So, in this case the evader removes P_1 or P_2 in the unexact H-arrangement D_2.

Now assume that in D_1 at least one of pursuers P_1, P_2 is not isolated (say, P_1 is a neighbor of P_4). In this case $d \geq 5$ and hence $|Q_0| > 1$.

If at the second step P_1 moves $v_0 \longrightarrow v_1$, then E removes P_1 after $v_3 \longrightarrow v_2 \longrightarrow v_1$.

If at the second step P_4 moves, then P_1 becomes isolated and may be removed by E after

$$v_2 \longrightarrow v_1 \longrightarrow v_0.$$

If at the second step P_3 moves, then P_2 becomes isolated (D_2 is unexact) and may be removed by E after

$$v_{d-2} \longrightarrow v_{d-1} \longrightarrow v_d.$$

If at the second step P_2 moves $v_d \longrightarrow v_{d-1}$, then it becomes isolated and may be removed by E after

$$v_{d-3} \longrightarrow v_{d-2} \longrightarrow v_{d-1}.$$

It is easy to see all the trajectories indicated above are admissible. So, in all the cases considered the evader removes one of the pursuers in the unexact H-arrangement D_2.

Now we have to consider the remaining case: at the second step P_2 moves: $v_d \longrightarrow v_{d+1}$ and becomes isolated in D_2 because D_2 is the unexact H-arrangement. At this step E can't remove a pursuer and has to "look through" the 3-d and the 4-th moves of P in strategy S. If at the 3-d step P_1 or P_4 moves, then in the exact H-arrangement D_3 two pursuers (P_1 and P_4) will be isolated. In this case E achieves its aim as well as after the first move of P. Namely, E removes the isolated pursuer P_2:

$$v_{d-3} \longrightarrow v_{d-2} \longrightarrow v_{d-1} \longrightarrow v_d \longrightarrow v_{d+1}$$

if at the 4-th step either P_1 or P_4 moved. If at the 4-th step either P_2 or P_3 moves, E removes the isolated pursuer P_1:

$$v_2 \longrightarrow v_3 \longrightarrow v_2 \longrightarrow v_1 \longrightarrow v_0$$

(if at the 3-d step P_4 moved) or

$$v_3 \longrightarrow v_2 \longrightarrow v_3 \longrightarrow v_2 \longrightarrow v_1$$

(if at the 3-d step P_1 moved).

Let at the 3-d step P_3 move. If at the 4-th step P_3 also moves, then E removes the isolated (in D_4) pursuer P_2 after

$$v_{d-3} \longrightarrow v_{d-2} \longrightarrow v_{d-1} \longrightarrow v_d \longrightarrow v_{d+1}.$$

If at the 4-th step either P_1 or P_4 moved, then E removes the isolated (in D_4) pursuer P_1 after

$$v_2 \longrightarrow v_3 \longrightarrow v_2 \longrightarrow v_1 \longrightarrow v_0$$

(if at the 4-th step P_4 moved) or after

$$v_3 \longrightarrow v_2 \longrightarrow v_3 \longrightarrow v_2 \longrightarrow v_1$$

(if at the 4-th step P_1 moved).

Finally, if at the 4-th step P_2 moves: $v_{d+1} \longrightarrow v_d$, it becomes isolated in D_4 and may be removed by E after

$$v_{d-2} \longrightarrow v_{d-3} \longrightarrow v_{d-2} \longrightarrow v_{d-1} \longrightarrow v_d.$$

However, if at the 4-th step P_2 moves: $v_{d+1} \longrightarrow v_{d+2}$ E can't remove any pursuer without being captured.

Now let at the 3-d step P_2 move. If at the 4-th step either P_1 or P_4 moves, then P_1 becomes isolated in D_4 and may be removed by E after

$$v_2 \longrightarrow v_3 \longrightarrow v_2 \longrightarrow v_1 \longrightarrow v_0.$$

(if at the 4-th step P_4 moved) or after

$$v_3 \longrightarrow v_2 \longrightarrow v_3 \longrightarrow v_2 \longrightarrow v_1$$

(if at the 4-th step P_1 moved). If at the 4-th step P_3 moves, then P_2 becomes isolated in D_4 and may be removed by E after

$$v_{d-2} \longrightarrow v_{d-3} \longrightarrow v_{d-2} \longrightarrow v_{d-1} \longrightarrow v_d$$

(if at the 3-d step P_2 moved: $v_{d+1} \longrightarrow v_d$) or after

$$v_{d-2} \longrightarrow v_{d-1} \longrightarrow v_d \longrightarrow v_{d+1} \longrightarrow v_{d+2}$$

(if at the 3-d step P_2 moved: $v_{d+1} \longrightarrow v_{d+2}$).

If P_2 moves at the 3-d and 4-th steps, we consider the following four cases:

1. P_2 moves: $v_{d+1} \longrightarrow v_d \longrightarrow v_{d-1}$; in this case P_2 may be removed by E after

$$v_{d-3} \longrightarrow v_{d-2} \longrightarrow v_{d-3} \longrightarrow v_{d-2} \longrightarrow v_{d-1};$$

2. P_2 moves: $v_{d+1} \longrightarrow v_{d+2} \longrightarrow v_{d+1}$; in this case P_2 may be removed by E after

$$v_{d-3} \longrightarrow v_{d-2} \longrightarrow v_{d-1} \longrightarrow v_d \longrightarrow v_{d+1};$$

3. P_2 moves: $v_{d+1} \longrightarrow v_d \longrightarrow v_{d+1}$; such moves are impossible in the winning strategy with the minimum number of steps.

4. P_2 moves: $v_{d+1} \longrightarrow v_{d+2} \longrightarrow v_{d+3}$; in this case E can't remove any pursuer without being captured.

It is easy to see that all the trajectories of E described above are admissible. So, the pursuers team can avoid the removal in an unexact H-arrangement only in case 4 and if at the 3-d step P_3 moves and at the 4-th step P_2 moves $v_{d+1} \longrightarrow v_{d+2}$.

However, the situation arising after the 4-th move of P in these cases is similar to one arising after their second move, the distance between P_1 and P_2 just increased.

We have $|Q_4| = |Q_2| + 2$ or $|Q_4| = |Q_2| + 1$. Therefore, looking through the 5-th and 6-th moves of P the evader may choose its trajectory in a similar way except for the case when $|Q_6| = |Q_4| + 2$ or $|Q_6| = |Q_4| + 1$.

In this case the evader looks through the 7-th and 8-th moves of P in strategy S etc.

Since the distance between P_1 and P_2 in any H-arrangement is bounded one of the pursuer (say, P_2) will be removed by E in an unexact H-arrangement D_{2q}. After this removal we have a H_1-arrangement where $H_1 = \{P_1, E, P_3, P_4\}$. It is easy to see that after any move of E the H_1-arrangement is unexact and after any move of P it becomes exact.

Show how the evader has to act to avoid the capture. If at any step E can become isolated (possibly, removing a pursuer), then the evasion is obvious. Suppose that at the m-th step E can't become isolated. Then before the m-th move of the evader the distance between E and P_1 (P_3) is equal to 2. This H_1-arrangement is exact and the evader is the even player. Hence, all the pursuers are even players as well and therefore those are isolated. Since after its m-th move the evader becomes invisible again the pursuers before their $(m + 1)$-th move do not know what node was occupied by E at the previous step. Their $(m + 1)$-th move according to strategy S is known to the evader and the latter may use this information choosing its m-th move. If at the $(m + 1)$-th step $P_1(P_3)$ moves, then E removes $P_3(P_1)$ at the same step. If at the $(m+1)$-th step P_4 moves, then after this move in the H_1-arrangement P_1 or P_3 is isolated. This isolated pursuer may be removed by E at the $(m + 1)$-th step. It is obvious that two pursuers can't catch the evader. The latter can avoid the capture acting as described above. This assertion is valid for any cycle C_n, $n \geq 5$.

If from the very beginning of the game the number of pursuers is less then 4, the proof of the evasion is essentially simplified. However, in general, it is unknown whether the evasion from k pursuers implies the evasion from s pursuers for $s < k$.

Proposition 5.2. *If n is even, $n > 13$, then $\alpha(C_n) = \alpha_0(C_n) = 5$.*

Describe a winning strategy for pursuers $P_1, P_2, ..., P_5$. Let $v_1, ..., v_5$ be nodes such that (v_1, v_2), (v_2, v_3), (v_3, v_4), (v_4, v_5) are edges and let pursuers $P_1, ..., P_5$ occupy nodes $v_1, ..., v_5$ respectively at the initial moment. Now let P_1 move to P_5. One can consider the evader doesn't remove a pursuer that is not isolated. Then P_1 catches E if the distance between v_1 and the node occupied by E at the initial moment is odd. In the opposite case, E removes P_1 in a node u. If the distance between u and P_2 is odd, P_2 can catch E. Otherwise, the distance between u and P_5 is odd and in this case P_5 catches E. So, $\alpha(C_n) \leq 5$.

By Proposition 1, $\alpha(C_n) \geq \alpha_0(C_n) \geq 5$ and, hence, $\alpha(C_n) = a_0(C_n) = 5$.

Proposition 5.3. *If n is odd, $n > 13$, then $\alpha(C_n) = \alpha_0(C_n) = 4$.*

Show $\alpha_0(C_n) \geq 4$. Suppose there exists a winning strategy S for three pursuers. If after the first move of P according to strategy S one of pursuers becomes isolated, it may be removed by the evader at the first step. Otherwise, it removes a pursuer at the second step.

The inequality $\alpha(C_n) \leq 4$ follows from the arguments in Proposition 2. So,

$$4 \leq \alpha_0(C_n) \leq \alpha(C_n) \leq 4.$$

6. A Chess Problem

In this section we consider a chess problem with "the invisible" Black's King composed by the second author ("special prize" at the competition "Chess composition" in 1998-1999, Russia). In fact, it is a graph searching problem under some additional conditions defined by chess rules. Here is a new version of this problem.

(3+0)

Checkmate in 5 moves (the lonely Black's King is invisible). This problem has three twins:

a) Kg7,

b) Kg6,

c) Kf7.

In all the cases the first move is
1. Qe8.
If the reply of Black is a move "without removal", the sequence of white moves
2.Qh3-d7. 3. Qe8-c8. 4. Qd7-b7. 5. Qc8-a8
leads to the aim.
Now let the reply of Black's King be
1. ...K:e8
(in case c) this move is impossible).
Then in the position on the diagram: 2. Qc8 checkmate. In case a)
2. Qd3. 3. Qd5. 4. Kg6. 5. Qf7(d8, a8) checkmate.
In case b)
2. Qh7. 3.Qb7. 4. Qc7. 5. Qd8 checkmate.
In case c) there are many solutions with the first moves:
1. Qe8. 1. Qd7. 1. Qh4. 1.Qh8.

Those were found by making use of a computer program, designed by the first author for solving many graph searching problems with the counteraction. It is essentially different from known chess programs. The solution in case c) required considering about 500 000 000 variants and 8 hours of calculations for the computer (AMD K-7, 850, 128 Mb RAM) that the authors used in 2000.

Acknowledgment

This research was supported by the Russian program "Leading scientific schools" (project NSH-2271.2001.1).

References

[1] D. Bienstock, Graph searching, path-width, tree-width and related problems (a survey), *DIMACS Series in Discrete Mathematics and Theoretical Computer Science* **5** (1991), 239–245.

[2] F.S. Makedon, I.H. Sudborough, On minimizing width in linear layouts, *Discrete Applied Mathematics* **23** (1989), 243–265.

[3] F.V. Fomin, Helicopter search problems, bandwidth and pathwidth, *Discrete Applied Mathematics* **85** (1998), 70–79.

[4] J.A. Ellis, I.H. Sudborough and J. Turner, The vertex separation and search number of a graph, *Information and Computation* **113** (1994), 50–79.

[5] L.M. Kirousis, C.H. Papadimitriou, Searching and pebbling, *Theoretical Computer Science* **47** (1986), 205–218.

[6] F.V. Fomin, N.N. Petrov, Persuit-evasion and search problems on graphs, *Congressus Numerantium* **22** (1996), 47–58.

[7] R.H. Möhring, Graph problems related to gate matrix layout and PLA folding, in *Computational Graph Theory, Computing Suppl.*7, Mayr, E. Noltemeier, H., and Syslo, M., eds., Berlin: Spriner, 1990, 17–51.

[8] N.N. Petrov, M.A. Teteryatnikova, On some search problems on graphs with the counteraction, *Vestnik of St.- Petersburg University* **1(2)** (2004), (to appear).

[9] V.O. Kapilevich, N.N. Petrov, Search problems on graphs with the counteraction, *Vestnik of St.- Petersburg University* **1(4)** (2003), (to appear).

[10] A.B. Zelenevskaya, N.N. Petrov, On search problems on graphs with the counteraction, *Vestnik of St.- Petersburg University* **1(2)** (2004), (to appear).

[11] N.N. Petrov, A.V. Chumanova, On some search problems on graphs with the counteraction, *Vestnik of St.- Petersburg University* **1(1)** (2004), (to appear).

In: Mathematics, Game Theory and Algebra Compendium ISBN 978-1-60456-659-8
Editor: Jacob H. Mathias, pp. 607-619

Chapter 51

Competition for Staff between Two Departments

Vic J. Baston
School of Mathematics
University of Southampton
Southampton United Kingdom
Andrey Y. Garnaev*
Department of Computer Modelling and Multiprocessor Systems
Faculty of Applied Mathematics and Control Processes
St.Petersburg State University
St.Petersburg, Russia

1. Introduction

The Secretary Problem is a well-known problem in the decision sciences. It can be stated as follows. An organization wishes to appoint a secretary from n applicants who appear in random order with all $n!$ permutations equally likely. At any stage the applicants interviewed up to and including that stage can be ranked in a preference order and, at each stage, the organization has to decide whether or not to make the candidate an offer. If an applicant is made an offer, it will be accepted but any applicant who is rejected will not be available for consideration at a later stage. The organization wants to maximize the probability of choosing the best candidate. There has been a lot of interest in this problem and variants of it which attempt to make the problem more realistic; see, for instance Tamaki (1991), who has considered the variation in which there is a fixed probability (independent of rank) of the applicant accepting an offer. Other papers dealing with sequential selection problems include Chun (1992) and Kwan and Yuan (1988). Less attention has been paid to the game theory aspects of hiring an employee although Sakaguchi (1985, 1989) is a notable exception. A selection panel usually involves at least two members so there is the potential for its members to have different interests. In particular Mazalov, Sakaguchi and Zabelin (2002) have modelled a situation in which a labor union and management jointly employ

*Research of this author was supported in part by Grant SS-2174.2003.1

a secretary. We will investigate a variant of their problem (see also Sakaguchi, Mazalov (2004)).

The main purpose of this paper is to analyze the following scenario. Two departments in a large organization are each seeking to make an appointment within the same area of expertise, for instance, a computer science specialist. To avoid duplication it has been decided that the heads of the two departments should together interview the applicants in turn and make their decisions on one applicant before interviewing any others. The organization has a very good reputation for treating its staff well so it can be assumed that an applicant who is offered an appointment will accept it. If a candidate is rejected by both departmental heads, the candidate cannot be considered for either post at a later date. If both heads decide to make an offer, the candidate will be allowed to choose which department he or she joins; from past experience it is known that applicants do not have a preference for one department over the other so each department is equally likely to be accepted. The departmental heads know that there are precisely n applicants and that each applicant has an expertise which is random over a known range; the expertise of an applicant can only be accurately assessed during the interviewing process and it is assumed that the heads will agree on this value. If no appointment is made to a department from these n applicants, then the department will suffer from a shortfall of expertise.

Although the problem is a simply stated and natural one, we will see that it is not at all straightforward to analyze it using the methods of game theory. As stated earlier, the competitive hiring of secretaries has already been analyzed in the literature but we shall see that our analysis differs markedly from that in previous papers. Because these papers have concentrated on the situation in which one secretary has to be hired, the players are essentially playing the same game at each stage; from one stage to the next the only essential difference is that the number of applicants decreases by one. However, in our scenario, this is not true; as soon as the decision to appoint is made, the problem becomes a decision problem rather than a game. Thus our problem is a bridge between the classical sequential selection problems and the game theoretic ones. A similarity between our paper and previous ones is that, if there are a comparatively large number of applicants, the interviewers can hold out for an applicant of comparatively high quality when the interviewing process is in its early stages; this is because there is a high probability that a good, or two good, applicants can be found from the remaining ones. A further similarity is that arguments of a backward induction nature are used. However a major difference is that our problem gives rise to many Nash equilibria and this raises the question of equilibrium selection. We will argue that there are comparatively few natural ones in the sense that they could be explained and justified to non-mathematical personnel. We will also argue that it is reasonable to have several Nash equilibrium solutions as different dynamics within the firm can result in different outcomes. Thus, if one departmental head is aggressive and one passive, we might expect a different outcome to one in which both are of a similar temperament. In the former case we would not necessarily expect a symmetric outcome even though the scenario does not give one player an advantage over the other. Thus, although it may be natural to expect a solution to be symmetric, we will also investigate non-symmetric solutions These non-symmetric equilibria have the advantage that the players have pure actions whereas, in our symmetric solution, the players are called upon to employ actions with complicated probabilities.

Before analyzing the above scenario, we will consider the modification in which one department has better prospects than the other; these prospects could take several forms such as being able to offer a higher salary, better training or faster promotion than the other. In this case it can be assumed that, if a candidate is accepted by both departmental heads, then the applicant will choose the one with the better prospects. Generally speaking the analysis of this problem follows the pattern of previous papers and provides a good lead in to the main problem.

2. Preliminary Notions

By choosing the units appropriately we can suppose that each applicant has an expertise drawn at random from the interval $[0, 1]$ and that the shortfall in expertise of a department not employing an applicant is c. We can think of the situation as a multi-stage game with the r-th stage being one in which rejection of the current applicant would leave $r - 1$ applicants available for interview. However we will adopt a naive approach and not use the general theory of such games. Consider first the decision problem in which a player knows that

there are r applicants remaining and the other player has filled the post in his department. We will denote the expectation of the player in this case by u_r. Clearly, if $r = 1$, the player will accept the applicant and $u_r = 1/2$. Further, when there are $r - 1$ applicants available if the current one is rejected, he will accept the applicant he is interviewing if the applicant has expertise x where $x \geq u_{r-1}$ and reject otherwise. Thus

$$u_r = \int_0^{u_{r-1}} u_{r-1}\, dx + \int_{u_{r-1}}^{1} x\, dx = u_{r-1}^2 + (1 - u_{r-1}^2)/2 = (1 + u_{r-1}^2)/2. \quad (2.1)$$

The recurrence relation $u_r = (1 + u_{r-1}^2)/2$ is known in the literature as Moser's relation (see, for example, Petrosjan and Zenkevich (1996)).

3. The Dominant Firm Problem

Suppose one department is dominant so that an applicant will always choose it in preference to the other. We will assume the dominant department is represented by player 1.

The one applicant game is trivial because player 1 will always accept and receive an expectation of 1/2 so that player 2 will always get $-c$.

Now consider the $n \geq 2$ applicant game and the stage where an applicant is being interviewed and there will be $r - 1$ applicants available if the current one is rejected by both players. If one player accepts and the other player rejects, or both players accept, the current applicant, the players clearly know what their expectations are. However, if both reject, it is not so clear that the players know what their expectations are because these expectations depend on actions in subsequent stages. Indeed, in the other scenario we consider, this is in fact the case and assumptions have to be made for an analysis of it to be undertaken. Thus let us assume that the players do know what their expectations are if they both reject the current applicant, say v_{r-1} and w_{r-1}, for player 1 and player 2 respectively. The situation can then be represented by a bimatrix given by

$$M_r(x) = \begin{array}{c} \\ \text{Accept} \\ \text{Reject} \end{array} \begin{array}{c} \quad\text{Accept} \qquad\qquad \text{Reject} \\ \begin{pmatrix} (x, u_{r-1}) & (x, u_{r-1}) \\ (u_{r-1}, x) & (v_{r-1}, w_{r-1}) \end{pmatrix} \end{array}.$$

We now show by induction that $v_r = u_r$ for all r. The result holds for $r = 1$ so suppose it holds for $r = k$. Thus, at the $k+1$-stage player 1 gets x if he accepts the current applicant and u_k if he doesn't which leads to expression for u_r in (2.1) and the result is established. Not surprisingly the dominant player can behave as though he has no rival. Player 2 will accept a candidate with expertise at least w_{r-1} so, when $w_k \geq 0$, his expectation w_{k+1} is given by

$$w_{k+1} = \int_{u_k}^{1} u_k \, dx + \int_{w_k}^{u_k} x \, dx + \int_{0}^{w_k} w_k \, dx = u_k - u_k^2/2 + w_k^2/2$$

and, when $w_k < 0$, by

$$w_{k+1} = u_k - u_k^2/2.$$

It is easy to see that the latter case arises only when $k = 1$. This reflects the fact that player 2 can ensure a non-negative expectation by always accepting the final two applicants. The following table shows that, even when there are 14 applicants, the dominant player does considerably better.

Stage	*Dominant*	*NotDominant*
1	0.5	0
2	0.625	0.375
3	0.6953	0.5
4	0.7417	0.5786
5	0.7750	0.6340
6	0.8004	0.6757
7	0.8203	0.7084
8	0.8364	0.7347
9	0.8498	0.7565
10	0.8611	0.7749
11	0.8707	0.7906
12	0.8791	0.8042
13	0.8864	0.8160
14	0.8929	0.8265

4. The One and Two Applicant Games with No Candidate Preference

We now investigate the scenario described in Section 1 in which an applicant is indifferent between the two departments. The case when there is only one applicant presents no difficulties because both players will want to appoint as they will not want a shortfall of expertise in their department. Thus there is a unique Nash equilibrium and the expected payoff for each player is $(1/2 - c)/2 = (1 - 2c)/4$.

The position is very different when there are two applicants. The game is now a two-stage one rather than single shot and what is meant by a Nash equilibrium has to be considered. We will take a naive view and assume the players will play a Nash equilibrium when interviewing an individual applicant. In essence we are restricting attention to subgame perfect Nash equilibria. When the first applicant is interviewed, each player knows that he can expect 1/2 if he rejects and the other player accepts and $(1-2c)/4$ if both reject. Hence both players will accept an applicant with expertise $x \geq 1/2$ and reject an applicant with expertise $x < (1 - 2c)/4$. For $x \in [(1 - 2c)/4, 1/2]$, each player would prefer the other player to accept the applicant so that he could reject and obtain an expectation of 1/2 at the second stage. We can represent the situation at the first stage by a game bimatrix $M_2(x)$ given by

$$M_2(x) = \begin{array}{c} \\ \text{Accept} \\ \text{Reject} \end{array} \begin{array}{c} \begin{array}{cc} \text{Accept} & \text{Reject} \end{array} \\ \begin{pmatrix} ((x + 1/2)/2, (x + 1/2)/2) & (x, 1/2) \\ (1/2, x) & ((1 - 2c)/4, (1 - 2c)/4) \end{pmatrix} \end{array}$$

Note that, for $x \geq 1/2$ Accept dominates Reject for both players whereas for $x \leq (1 - 2c)/4$, Reject dominates Accept for both players. For $x \in [\max\{0, (1 - 2c)/4\}, 1/2]$ there are two pure Nash equilibria, namely (Accept, Reject) and (Reject, Accept) and one mixed in which each player accepts an applicant with expertise x with probability $(4x - 1 + 2c)/(2x + 2c)$. The payoff vector for the mixed strategy pair is given by (v, v) where

$$v = \frac{4cx + 2c + 6x - 1}{8(x + c)} = \frac{2c + 3}{4} - \frac{(2c + 1)^2}{8(x + c)}. \tag{3.1}$$

Thus v is an increasing function of x in $[(1 - 2c)/4, 1/2]$ for fixed $c \in [0, 1/2]$.

Of course neither of the pure Nash equilibria are symmetric and this could be seen as a drawback because the scenario does not present any differences between the players. However we should not dismiss these equilibria out of hand as they could arise from different dynamics in the organization. For instance, if one departmental head is somewhat aggressive and the other more easy-going, it is quite likely that the more aggressive head would state what he was going to do; in essence the aggressive head assumes the role of a Stackleberg leader. The mixed Nash equilibrium has the advantage that each player adopts the same strategy so that it can be argued that it is fairer than either of the pure equilibria but it has the drawback that the players use complicated probabilities to decide their actions.

So far we have looked at the situation where the players are actually confronted with the applicant's expertise but, from a game point of view, the strategies should be announced before the interview takes place; in other words we should be providing a function $s : [0,1] \to [0,1]$ for each player where $s(x)$ represents the probability of accepting an applicant with expertise x. However this means that, unless restrictions are imposed, the set of Nash equilibria becomes unmanageable. Given any subset S of $I = [(1-2c)/4, 1/2]$ suppose player 1 accepts $x \in S \cap I$ and rejects $x \in I \setminus S$ while player 2 accepts $x \in I \setminus S$ and rejects $x \in S \cap I$; neither player benefits from unilaterally deviating so this strategy pair is a Nash equilibrium. Clearly most such Nash equilibria are unsatisfactory. From the purely mathematical standpoint, one would not be able to calculate an expectation for our problem for non-measurable S. However most measurable S would have undesirable properties from the modelling viewpoint. If a player accepts an applicant with expertise$x$, then it would appear unreasonable to expect the player to reject an applicant of expertise greater than x; indeed, in many countries, in doing so, an organization might find itself in conflict with the law. If the condition is imposed that acceptance of an applicant with expertise x implies acceptance of every applicant with expertise greater than$x$,, there are just two pure Nash equilibria for the players; one player accepts the first applicant if and only if the applicant has expertise at least $1/2$ and the other player accepts if and only if the applicant has expertise at least $(1-2c)/4$. Note that, with these Nash equilibria, both applicants will be accepted if $c \geq 1/2$ but that, for $c < 1/2$, there is a probability that only one applicant will be accepted.

We now calculate the expectations for the pure Nash equilibrium in which player 1 accepts an applicant if and only if his expertise x is at least $1/2$ and player 2 accepts if and only if $x \geq (1-2c)/4$. Assuming $c \leq 1/2$, for player 1 we have

$$\begin{aligned} v_2 &= \int_0^{(1-2c)/4} (1-2c)/4\, dx + \int_{(1-2c)/4}^{1/2} 1/2\, dx + \int_{1/2}^{1} (x+1/2)/2\, dx \\ &= \tfrac{1}{2} + \tfrac{c^2}{4}. \end{aligned}$$

For player 2 we have

$$\begin{aligned} w_2 &= \int_0^{(1-2c)/4} (1-2c)/4\, dx + \int_{(1-2c)/4}^{1/2} x\, dx + \int_{1/2}^{1} (x+1/2)/2\, dx \\ &= \frac{(1-2c)^2}{32} + \frac{7}{16} = \tfrac{15}{32} - \tfrac{c}{8} + \tfrac{c^2}{8}. \end{aligned}$$

Clearly, for $c \geq 1/2$, the integrals and therefore the expectations are the same as for$c = 1/2$. Thus they are independent of$c$ and $v_2 = 9/16$ and $w_2 = 7/16$.

It is interesting to note that, as a function of c, v_2 is increasing whereas w_2 is decreasing so the Stackleberg leader gains more the larger the value of c. As the leader expects more than the follower when $c = 0$, the most unfair value of c is 1/2. Furthermore as c goes from 0 to 1/2, the leader gains more than the follower loses because the leader's expectation rises from 1/2 to 9/16 while the follower's expectation declines from 15/32 to 7/16. In a sense, the situation can be regarded as a game of Chicken. For $x \leq 1/2$, the leader rejects the applicant and dares the follower to also reject the applicant. The larger the value of c, the more pressure there is on the follower to accept. We shall see in the next section that a similar position arises in the general game but that the effects become insignificant when there are still a large number of applicants remaining.

So far we have looked at the situation from the heads of department viewpoint but their interests do not necessarily coincide with that of the organization as a whole. If we think of the organization's interest as the totality of expertise employed, this interest is effectively the sum of the expectations of the two players. Thus it would clearly be better for the organization if both applicants are accepted. We have seen that both applicants are accepted if $c \geq 1/2$ so there is no conflict of interest in this case. Whether the interests of the organization are represented by an increasing function of c as c ranges over the interval [0,1/2] is not so intuitively obvious. We have seen that, as c goes from 0 to 1/2, the Stackleberg leader gains more than the follower loses over the interval so, as a whole, the function increases. However, the smaller the value of c, the more likely it is that the first applicant will be rejected and the penalty c brought into operation. The interest of the organization E is given by

$$E = v_2 + w_2 = 1 + \frac{(1-2c)(3-6c-4)}{32} = 1 - \frac{(1-2c)(1+6c)}{32}$$

which is a convex quadratic with a minimum at $c = 1/6$ giving $E = 23/24$ at the minimum. Thus the worst value of c for the organization is 1/6. Further the most unfair value of c, namely 1/2, is actually the best value for the organization.

There is also a mixed Nash equilibrium for the players. Let $p(x) = (1-2x)/(2x+2c)$ then $p(x)$ is the probability that a player will reject $x \in [(1-2c)/4, 1/2]$ so, using (3.1), the payoff Q for each player in the symmetric Nash equilibrium is given by

$$\begin{aligned} Q &= \int_{1/2}^{1} \frac{(x+1/2)}{2}\,dx + \int_{(1-2c)/4}^{1/2} \left(\frac{2c+3}{4} - \frac{(2c+1)^2}{8(x+c)}\right) dx \\ &+ \int_{0}^{(1-2c)/4} \frac{1-2c}{4}\,dx \\ &= \frac{5}{16} + \frac{(2c+3)(2c+1)}{16} - \frac{(2c+1)^2 \ln 2}{8} + \frac{(1-2c)^2}{16} \\ &= \frac{9+4c+8c^2}{16} - \frac{(2c+1)^2 \ln 2}{8} \end{aligned}$$

Note that this is also a convex quadratic with a minimum in (0, 1/2).

Intuition tells us that a player in the role of Stackleberg leader in a pure Nash equilibrium does better than a player in the symmetric Nash equilibrium because his position is essentially unchanged if the first applicant has expertise outside of $[(1-2c)/4, 1/2]$ and he does better if the applicant has expertise within that range because he is guaranteed 1/2 as the Stackleberg leader. To check this, consider, for $c \leq 1/2$,

$$\begin{aligned} D_1 = v_2 - Q &= \frac{1}{2} + \frac{c^2}{4} - \frac{9+4c+8c^2}{16} + \frac{(2c+1)^2 \ln 2}{8} \\ &= \frac{2\ln 2 - 1}{16}(1+2c)^2 > 0 \end{aligned}$$

However it is not clear intuitively whether a player in the role of Stackleberg follower does better or worse than a player in the symmetric Nash equilibrium. To see whether he does so, for $c \leq 1/2$ we consider

$$\begin{aligned} D_2 = w_2 - Q &= \frac{15}{32} - \frac{c}{8} - \frac{9+4c+8c^2}{16} + \frac{(2c+1)^2 \ln 2}{8} \\ &= \frac{4\ln(2) - 3}{32}(1+2c^2) < 0. \end{aligned}$$

Note that

$$\frac{v_2 + w_2}{2} - Q = \frac{8\ln(2) - 5}{64}(1+2c)^2 > 0. \tag{4.1}$$

When $c \geq 1/2$, we have

$$Q = \frac{11+4c}{16} + \frac{(1+2c)^2}{8}\ln\frac{2c}{1+2c}$$

and

$$D_1 = 9/16 - Q > 0 \qquad D_2 = 7/16 - Q < 0$$

so we have the same properties as the case $c = 1/2$. Thus a player in the symmetric game does better than a Stackleberg follower in a pure Nash equilibrium whatever the value of c. From (4.1) the sum of the expectations for a pure Nash equilibrium is greater than twice that for the symmetric one. This accords with intuition in that $x \in ((1-2c)/4, 1/2)$ is always accepted by the pure strategy pair but only with a probability less than 1 by the symmetric strategy pair so there is a positive probability that such values will lead to a shortfall of expertise in one of the departments when the symmetric strategy pair is used. The players' actions for x not in the range are the same for both strategy pairs. Thus the players would have better expectations by choosing one of the pure Nash equilibria at random than using the symmetric Nash equilibrium. From a practical standpoint randomizing the role of Stackleberg leader seems a more attractive option than using the symmetric Nash equilibrium. It is a lot easier to understand and to put into practice; a simple toss of a coin rather than a complicated randomization if $x \in (1-2c)/4, 1/2]$.

5. The General Game with No Candidate Preference

We saw in the last section that the two applicant game presented difficulties so we must expect similar ones to occur in the general game. However, when there are three or more applicants, further complications arise. In the two applicant game, the players know their expectations if both players reject the first applicant but, when there are three applicants and both reject the first applicant, their expectations will depend on which Nash equilibrium is adopted at the second stage. We will assume that there is some kind of consistency in the actions of the players at the various stages and investigate a number of scenarios to see how the payoffs differ and how they are affected by the value of c.. In the two applicant case, we have suggested that, it might benefit the players to agree to randomize who takes the role of Stackleberg leader so we will consider the situation in which, at each stage, the players agree to randomize who takes that role. By doing this the players create a symmetric situation where, at each stage, the players have the same expectation if both players reject the current applicant. It is therefore very different from the situation in which the players agree to take it in turns to assume the rȯle of Stackleberg leader; in this case, at any given stage, the players will in general have different expectations. If there is no agreement to randomize concerning who takes the role of Stacklberg leader, it seems natural to assume that a player who adopts the role of a Stackleberg leader for one stage would take that role at every stage; we investigate how unfair this is on the Stackleberg follower. Similarly, if a symmetric (mixed) strategy is used at one stage then one might expect it to be used at every stage. Intuitively we would expect that, when the number of applicants is large, there would not be very much difference in the expectations of the players from any of these scenarios.

Assuming players 1 and 2 believe their expectations are respectively v_{r-1} and w_{r-1} if they both reject the applicant interviewed at stage r, the r-stage matrix is

$$M_r(x) = \begin{array}{c} \\ \text{Accept} \\ \text{Reject} \end{array} \begin{array}{c} \begin{array}{cc} \text{Accept} & \text{Reject} \end{array} \\ \begin{pmatrix} ((x+u_{r-1})/2, (x+u_{r-1})/2) & (x, u_{r-1}) \\ (u_{r-1}, x) & (v_{r-1}, w_{r-1}) \end{pmatrix} \end{array}$$

The result of the following lemma enables us to carry out a similar analysis to that of the two applicant case.

Lemma. *For all* $r,\ u_r \geq \max\{v_r, w_r\}$.

Proof. For any strategy pair in a two-person non-zero sum game, the expectations in a Nash equilibrium are non-decreasing functions of the elements in the matrix. Because $u_1 = 1/2 \geq (1-2c)/4$, the result follows by induction.

If player 1 assumes the role of Stackleberg leader and Player 2 that of Stackleberg follower the respective expectations are

$$\begin{aligned} v_r &= \int_{u_{r-1}}^{1} (x+u_{r-1})/2\,dx + \int_{w_{r-1}}^{u_{r-1}} u_{r-1}\,dx + \int_0^{w_{r-1}} v_{r-1}\,dx \\ &= \frac{(1+u_{r-1})^2}{4} - w_{r-1}(u_{r-1} - v_{r-1}) \end{aligned}$$

and

$$\begin{aligned} w_r &= \int_{u_{r-1}}^{1} (x+u_{r-1})/2\,dx + \int_{w_{r-1}}^{u_{r-1}} x\,dx + \int_{0}^{w_{r-1}} w_{r-1}\,dx \\ &= \frac{1-u_{r-1}^2+2u_{r-1}+2w_{r-1}^2}{4}. \end{aligned}$$

We now consider the case in which the players use the symmetric Nash equilibrium at each stage. At stage r the players will have a common expectation which will be denoted by q_r. The probability of a player accepting $x \in [q_{r-1}, u_{r-1}]$ at stage r is given by $2(x - v_{r-1})/(x + u_{r-1} - 2v_{r-1})$ and the corresponding payoff is

$$2u_{r-1} - q_{r-1} - \frac{2(u_{r-1}-q_{r-1})^2}{x+u_{r-1}-2q_{r-1}}$$

Hence the expectation of the symmetric Nash equilibrium is

$$\begin{aligned} q_r &= \int_{u_{r-1}}^{1} \frac{x+u_{r-1}}{2}\,dx \\ &+ \int_{q_{r-1}}^{u_{r-1}} (2u_{r-1} - q_{r-1} - \frac{2(u_{r-1}-q_{r-1})^2}{x+u_{r-1}-2q_{r-1}})\,dx \\ &+ \int_{0}^{q_{r-1}} q_{r-1}\,dx = \frac{(1+u_{r-1})^2}{4} \\ &+ (u_{r-1}-q_{r-1})(u_{r-1}-2q_{r-1}) + 2(u_{r-1}-q_{r-1})^2 \ln 2. \end{aligned}$$

When $c = 0$, the values are

Stage	*U value*	*Leader*	*Follower*	*Sum/2*	*Symmetric*
1	0.5	0.25	0.25	0.25	0.25
2	0.625	0.5	0.4688	0.4844	0.4759
3	0.6953	0.6016	0.5747	0.5881	0.5806
4	0.7417	0.6646	0.6419	0.6533	0.6468
5	0.7750	0.7089	0.6894	0.6991	0.6935
6	0.8004	0.7421	0.7250	0.7335	0.7286
7	0.8203	0.7681	0.7528	0.7605	0.7561
8	0.8364	0.7891	0.7753	0.7822	0.7782
9	0.8498	0.8064	0.7939	0.8001	0.7965
10	0.8611	0.8210	0.8095	0.8152	0.8119
11	0.8707	0.8335	0.8228	0.8281	0.8250
12	0.8791	0.8442	0.8343	0.8393	0.8364
13	0.8864	0.8537	0.8444	0.8490	0.8463
14	0.8929	0.8620	0.8533	0.8576	0.8551

When $c = 1/2$, the values are

Stage	*U value*	*Leader*	*Follower*	*Sum*/2	*Symmetric*
1	0.5	0	0	0	0
2	0.625	0.5625	0.4375	0.5	0.4659
3	0.6953	0.6328	0.5605	0.5967	0.5763
4	0.7417	0.6835	0.6339	0.6587	0.6444
5	0.7750	0.7215	0.6842	0.7028	0.6921
6	0.8004	0.7511	0.7214	0.7362	0.7276
7	0.8203	0.7748	0.7503	0.7625	0.7553
8	0.8364	0.7942	0.7734	0.7838	0.7777
9	0.8498	0.8104	0.7924	0.8014	0.7961
10	0.8611	0.8243	0.8083	0.8163	0.8116
11	0.8707	0.8362	0.8218	0.8290	0.8248
12	0.8791	0.8465	0.8335	0.8400	0.8362
13	0.8864	0.8558	0.8437	0.8497	0.8462
14	0.8929	0.8636	0.8527	0.8582	0.8550

These tables demonstrate that the general case follows a similar pattern to that of the two applicant case. The Stackleberg leader benefits from a higher value of$c$ whereas the follower benefits from a lower value of c. However, by the time there are fourteen applicants, the value of c makes very little difference to either of the expectations. The tables also show that, for fourteen applicants, the expectations of the players are very close to each other. Thus aggression manifested by taking the role of Stackleberg leader is of little benefit when there are a comparatively large number of applicants.

When $c = 1/6$, the values are

Stage	*U value*	*Leader*	*Follower*	*Sum*/2	*Symmetric*
1	0.5	0.1667	0.1667	0.1667	0.1667
2	0.625	0.5069	0.4514	0.4792	0.4640
3	0.6953	0.6069	0.5667	0.5868	0.5754
4	0.7417	0.6684	0.6374	0.6529	0.6440
5	0.7750	0.7117	0.6864	0.6991	0.6918
6	0.8004	0.7442	0.7230	0.7376	0.7274
7	0.8203	0.7697	0.7514	0.7605	0.7552
8	0.8364	0.7904	0.7742	0.7823	0.7776
9	0.8498	0.8075	0.7930	0.8002	0.7960
10	0.8611	0.8219	0.8088	0.8153	0.8115
11	0.8707	0.8342	0.8223	0.8282	0.8247
12	0.8791	0.8449	0.8339	0.8394	0.8362
13	0.8864	0.8542	0.8440	0.8491	0.8461
14	0.8929	0.86250	0.8530	0.8577	0.8549

This table confirms the analysis of Section 4 that, when there are just two applicants the organization can expect less expertise for $c = 1/6$ than for $c = 0$. We will have further comments on this table in the next section.

Now consider the case in which, at each stage, the players randomize for who should play the role of leader, then $v_{r-1} = w_{r-1}$ in the above matrix and the expectation v_r^* is

$$v_r^* = \frac{1 + 2u_{r-1} + v_{r-1}^*(3v_{r-1}^* - 2u_{r-1})}{4}.$$

Stage	*U value* $c = 0$	*value* $c = 1/2$
1	0.25	0
2	0.4844	0.5
3	0.5938	0.6240
4	0.6599	0.6780
5	0.7057	0.7177
6	0.7397	0.7483
7	0.7662	0.7727
8	0.7876	0.7927
9	0.8052	0.8091
10	0.8200	0.8232
11	0.8326	0.8352
12	0.8435	0.8457
13	0.8530	0.8549
14	0.8614	0.8630

6. Discussion of the Results

The tables in the previous section suggest some interesting conclusions. Perhaps the most illuminating is that it appears that, whatever the value of c in [0,1/2], the players have a better expectation by randomizing who should adopt the role of Stackleberg leader at each stage other than the first than by using any of the other Nash equilibria considered. In particular, for the two "fair" solutions in the sense that they give the same expectation to each player at each stage, the simpler method which can be implemented by a simple toss of a fair coin at each stage appears preferable to the one in which a complicated randomization is necessary at each stage. Furthermore the simple fair method is best from the organizations standpoint in that a higher level of expertise is appointed than in any of the others, certainly in the cases when $c = 0$ and $c = 1/2$ and there are more than two applicants. There is a striking difference between the symmetric Nash equilibrium and the others. In the former the expectation for a given stage is higher when $c = 0$ than it is when $c = 1/2$ whereas, for the others, the expectation is higher when $c = 1/2$ than when $c = 0$.

In Section 4 we saw that, when there are just two applicants, the average expectation of the players has a minimum w.r.t c in $(0, 1/2)$ for both the symmetric and the pure Nash equilibrium. The tables for $c = 0$ and $c = 1/6$ illustrate a difference between these Nash equilibria in the general case. The symmetric Nash equilibrium has a lower expectation for

$c = 1/6$ than for $c = 0$ for all but the one applicant case whereas the pure Nash equilibrium has this property for only a small number of applicants.

References

[1] Y. H. Chun, On the Sequential Selection Problem, *Decision Sciences* **23** (1992), 1251–1262.

[2] C.C.Y. Kwan, Y. Yuan, A Sequential Selection Problem, *Decision Sciences* **19** (1988), 762–770.

[3] V.V. Mazalov, M. Sakaguchi and A.A. Zabelin, Multistage Arbitration Game with Random Offers, *International Journal of Mathematics, Game Theory and Algebra* **12** (2002), 409–417

[4] L.A. Petrosjan, N.A. Zenkevich, *Game Theory*, World Scientific, London, 1996.

[5] M. Sakaguchi, The Bilateral Secretary Problem with a Random Population Size, *Mathematica Japonica* **30** (1985), 457–469.

[6] M. Sakaguchi, Multiperson Secretary Problems, *Mathematica Japonica* **34** (1989), 459–473.

[7] M. Sakaguchi, V.V. Mazalov, A non–zero–sum no–information best–choice game, *Math. Meth. of Oper. Res.* **60** (2004).

[8] M. Tamaki, A Secretary Problem with Uncertain Employment and Best Choice of Available Candidates, *Operations Research* **39** (1991), 274–284.

In: Mathematics, Game Theory and Algebra Compendium ISBN 978-1-60456-659-8
Editor: Jacob H. Mathias, pp. 621-629

Chapter 52

A Competitive Prediction Number Game

D.V. Belkovskii and A.Y. Garnaev
Department of Computer Modelling and Multiprocessor
Systems Faculty of Applied Mathematics and
Control Processes St. Petersburg State University
St. Petersburg Russia

1. Introduction

The paper deals with a generalization of a Prediction Number Game suggested by Sakaguchi and Szajowski [4] in the case of unsymmetrical conditions. Sakaguchi and Szajowski investigated the following game. Two players want to guess a realization of a random variable whose distribution $H(t)$ is apriori known to both players. Each player suggests his predicted value. The winner is a player whose prediction is a better approximation of the realization t of the random variable than the rival's guess but does not overestimate t. This game has something in common with the card game Twenty Ones where it is essential not only to exceed the score of the rival but not to break the threshold value. Sakaguchi and Szajowski considered the zero-sum as well as the non-zero-sum cases of the game. Note that, as it will be shown, this problem is closely related to the silent duel (Garnaev [1], Karlin [2]).

Let us formulate the zero-sum Sakaguchi and Szajowski in detail as well as their received results. There are two players (1 and 2) who want to predict a realization t of a random variable having uniform distribution in $[0, 1]$. Let x and y be the players's prediction. Then the payoffs of the players are given by

(a) if $x < y < t$ or $y < t < x$, then player 2 wins 1,

(b) if $y < x < t$ or $x < t < y$, then player 1 wins 1,

(c) in all the other cases it is a draw and the payoffs are 0.

Denote by $K(x, y)$ the expectation payoff of player 1, if they predicted values x and y. Then

$$K(x,y) = \begin{cases} -x+2y-1 & \text{for } x<y, \\ 0 & \text{for } x=y, \\ -2x+y+1 & \text{for } x>y. \end{cases}$$

A pure strategy for a player is a predicted number x and a mixed strategy is a cumulative probability distribution function F in $[0,1]$.
This game is a symmetric one. It is known (see, for example [3]) that the value of any symmetric game is zero. The optimal strategy of a player is also optimal for his rival. In the zero-sum Sakaguchi and Szajowski there is no pure equilibrium but there is a unique equilibrium (F,F) in mixed strategies

$$F(x) = \begin{cases} (1-x)^{-1/2}-1 & \text{for } x \in [0,3/4], \\ 1 & \text{for } x>3/4. \end{cases}$$

Sakaguchi and Szajowski [4] also considered a non-zero-sum variant of Prediction Number Game. In this game let $K_1(x,y)$ and $K_2(x,y)$ be the expected payoffs of player 1 and 2, if their predicted values are x and y. Then

$$K_1(x,y) = \begin{cases} y-x & \text{for } x<y, \\ 0 & \text{for } x=y, \\ 1-x & \text{for } x>y, \end{cases}$$

$$K_2(x,y) = \begin{cases} x-y & \text{for } y<x, \\ 0 & \text{for } x=y, \\ 1-y & \text{for } y>x. \end{cases}$$

In this non-zero sum game there is no Nash equilibrium in pure strategies but there is a unique equilibrium in mixed strategies. Besides, the optimal strategy of a player is also optimal for his rival. Thus, (F,F) is Nash equilibrium where

$$F(x) = \begin{cases} -\ln(1-x) & \text{for } x \in [0,1-1/e], \\ 1 & \text{for } x>1-1/e. \end{cases}$$

The payoffs vector corresponding to the Nash equilibrium (F,F) is $(1/e,1/e)$.

In this paper we consider a generalization of the Sakaguchi and Szajowski game in the case where the player's payoff depends on whether he can make his rival overestimate the realization of the random variable with distribution $H(t)$ which is strictly increasing differentiable in $[0,1]$ and it is known to both players. Namely, if the player suggested a better approximation of the random variable's realization and both of them do not overestimate this value then the player who suggested a better approximation wins a. If his rival overestimated the random variable's realization then the player who does not overestimate wins b. We also consider the case where players have different payoffs under the same circumstances. This game plot can be motivated by the fact that players can be differently well-off and thus the value of the prize can have different meaning for them. Say, for a rich person the prize 1000 dollars means nothing but for for a student it could mean buying a new computer. For these variants of Prediction Number Game we consider zero-sum and non-zero-sum cases.

2. Zero-Sum Game

In the zero-sum game the expected payoff $K(x, y)$ of player 1, if the players' predicted values are x and y, is given by

$$K(x,y) = \begin{cases} b(H(y) - H(x)) - a(1 - H(y)) & \text{for } x < y, \\ 0 & \text{for } x = y, \\ a(1 - H(x)) - b(H(x) - H(y)) & \text{for } x > y, \end{cases}$$

For a given mixed strategy F of player 2 and a pure strategy x of player 1, the payoff to player 1 is given by

$$K(x, F) = \int_{[0,x)} a(1 - H(x)) - b(H(x) - H(y))\, dF(y) + \int_{(x,1]} b(H(y) - H(x)) - a(1 - H(y))\, dF(y).$$

Theorem 1. *The zero-sum Prediction Number Game is a symmetric one. So, its value is zero. The optimal strategy of a player is also optimal for his rival. In this game there is no pure equilibrium but there is a unique equilibrium (F, F) in mixed strategies where*

$$F(x) = \begin{cases} k(\frac{1}{\sqrt{1 - H(x)}} - 1) & \text{for } x \in [0, x_*], \\ 1 & \text{for } x > x_*, \end{cases}$$

where $x_* = H^{-1}\left(1 - \left(\frac{k}{k+1}\right)^2\right)$ *and* $k = b/a$.

Proof. Let (F, F) be an equilibrium. Then

(a) If F is constant in $(c, d) \subset [0, 1]$ then $M(x, F)$ is strictly decreasing in (c, d), since

$$K(x, F) = \int_{[0,c)} a(1 - H(x)) - b(H(x) - H(y))\, dF(y) + \int_{(d,1]} b(H(y) - H(x)) - a(1 - H(y))\, dF(y) \text{ for } x \in (c, d).$$

(b) F is continuous in $[0,1]$. Suppose there is $c \in [0,1)$ such that $q(c) = F(c) - F(c-0) > 0$. So, $c \in \operatorname{supp}(F)$. We remind the reader that the support $\operatorname{supp}(F)$ of the distribution function is given by

$$\begin{aligned}\operatorname{supp}(F) = \{x \in (0,1) : F(x+\varepsilon) - F(x-\varepsilon) > 0 \\ \text{for any sufficient small positive } \varepsilon\} \\ \cup \{1, \text{ if } F(1-\varepsilon) < 1 \text{ for any sufficient small positive } \varepsilon\} \\ \cup \{0, \text{ for } F(0) > 0\}\end{aligned}$$

Thus, $K(c, F)$ (see, for example [1]) is equal to the value of the game (in our case it is zero). But

$$K(c+0, F) - K(c, F) = a(1 - H(c))q(c) > 0.$$

This contradiction proves that $q(c) = 0$ and so F is continuous in $[0,1]$.

(c) (a) and (b) imply that there is x_* such that

$$F(x) \text{ is } \begin{cases} \text{continuous and strictly increasing} & \text{for } x \in [0, x_*], \\ 1 & \text{for } x > x_*. \end{cases}$$

(d) Since zero is the value of the game then (c) implies that $K(x, F) = 0$ for $x \in [0, x_*]$ and so

$$\int_{[0,x)} a(1 - H(x)) - b(H(x) - H(y))\, dF(y) + \int_{(x,1]} b(H(y) - H(x)) - a(1 - H(y))\, dF(y) = 0.$$

Thus

$$2(1 - H(x))F'(x) = (k + F(x))H'(x) \quad \text{for} \quad x \in [0, x_*]$$

and $F(0) = 0$ where $k = b/a$. So,

$$F(x) = k(\frac{1}{\sqrt{1 - H(x)}} - 1) \quad \text{for} \quad x \in [0, x_*].$$

(c) The turning point x_* can be found from the equation $F(x_*) = 1$.

This completes the proof of Theorem 1.

Corollary 1. *Let $H(t) = t$, i.e. the random variable has a uniform distribution then*

$$F(x) = \begin{cases} k((1-x)^{-1/2} - 1) & \text{for } x \in [0, x_*^1], \\ 1 & \text{for } x > x_*^1, \end{cases}$$

where $x_*^1 = 1 - \left(\frac{k}{k+1}\right)^2$.

Remark 1. *The optimal strategy F is a linear function on k within the interval $[0, x_*]$. The turning point $x_*^1 = 1 - (k/(1+k))^2$ is a decreasing function on k and*

$$x_*^1 = \begin{cases} 1 & \text{for } k = 0, \\ 3/4 & \text{for } k = 1, \\ \to 0 & \text{for } k \to \infty. \end{cases}$$

Now let us consider the case where the random variable has a distribution function $H(t) = 1 - \exp(-\lambda t)$ and the game is played over the interval $[0, \infty]$. Then analogously to Theorem 1 we can prove the following theorem.

Corollary 2. *The value of the zero-sum Prediction Number Game played over the interval $[0, \infty)$ with the distribution function $H(t) = 1 - \exp(-\lambda t)$ is zero. The optimal strategy of a player is also optimal for his rival. In this game there is no pure equilibrium but there is a unique equilibrium (F, F) in mixed strategies where*

$$F(x) = \begin{cases} k(e^{\lambda x/2} - 1) & \text{for } x \in [0, x_*^2], \\ 1 & \text{for } x > x_*^2, \end{cases}$$

where $x_*^2 = \frac{2}{\lambda} \ln(\frac{k+1}{k})$.

Remark 2. *Let us find k for which the turning points of the zero-sum Prediction Number with the distribution functions $H(t) = t$ and $H(t) = 1 - \exp(-\lambda t)$ coincide with each other, i.e. $x_*^1 = x_*^2$. For that we have to solve the following equation:*

$$m^2 - 1 = \frac{2}{\lambda} \ln(m)$$

where $m = \frac{k}{k+1}$. In the next table the root of this equation for λ changing from 0.5 to 4.5 with step 0.5 are given

λ	0.5	0.1	1.5,	2	2.5	3	3.5	4	4.5
m	0.534	1	1.548	2.218	3.052	4.099	5.422	7.102	9.241

3. Non-Zero-Sum Game

For non-zero-sum game the expectation payoffs $K_1(x, y)$ and $K_2(x, y)$ of player 1 and 2, if their predicted values are x and y, are given by

$$K_1(x, y) = \begin{cases} b(H(y) - H(x)) & \text{for } x < y, \\ 0 & \text{for } x = y, \\ a(1 - H(x)) & \text{for } x > y, \end{cases}$$

$$K_2(x, y) = \begin{cases} b(H(x) - H(y)) & \text{for } y < x, \\ 0 & \text{for } x = y, \\ a(1 - H(y)) & \text{for } y > x, \end{cases}$$

Theorem 3. *In the non-zero-sum game there is a unique Nash equilibrium and both players have the same optimal strategy F where*

$$F(x) = \begin{cases} \frac{k}{1-k}((1-H(x))^{k-1}-1) & \text{for } x \in [0, x_*], \\ 1 & \text{for } x > x_*. \end{cases}$$

where $x_* = H^{-1}(1-k^{1/(1-k)})$.
The payoffs corresponding to this strategies are given by

$$v_1 = v_2 = v = a(1-H(x_*)).$$

Proof. The theorem can be proved analogously to Theorem 1.

Corollary 3. *Let $H(t) = t$, i.e. the random variable has a uniform distribution then*

$$F(x) = \begin{cases} \frac{k}{1-k}((1-x)^{k-1}-1) & \text{for } x \in [0, x_*^3], \\ 1 & \text{for } x > x_*^3. \end{cases}$$

$$v = ak^{1/(1-k)}.$$

where $x_*^3 = 1-k^{1/(1-k)}$.

Remark 3. *(a)* $\lim_{k\to 1} v = a(1-1/e)$.
(b) The turning point x_*^3 is strictly increasing function on k and

$$\lim x_*^3 = \begin{cases} 0 & \text{for } k \to 0, \\ 1/e & \text{for } k \to 1, \\ 1 & \text{for } k \to \infty. \end{cases}$$

(c) $\lim_{k\to 1} F(x) = -\ln(1-x)$ *for* $x \in [0, x_*^3]$.

Now consider the case where the random variable has a distribution function $H(t) = 1-\exp(-\lambda t)$ and the game is played over the interval $[0, \infty]$. Then analogously to Theorem 1 we can prove the following theorem.

Corollary 4. *In the non-zero-sum Prediction Number Game with the distribution function $H(t) = 1-\exp(-\lambda t)$ there is a unique Nash equilibrium and both players have the same optimal strategy F where*

$$F(x) = \begin{cases} \frac{k}{k-1}(1-e^{\lambda(1-k)x}) & \text{for } x \in [0, x_*^4], \\ 1 & \text{for } x > x_*^4, \end{cases}$$

and the payoff vector corresponding this Nash equilibrium is $(a(1-x_*^4), a(1-x_*^4))$ where $x_*^4 = \ln(k)/(\lambda(k-1))$.

Remark 4. *(a)* $\lim_{k\to 1} F(x) = \lambda x$ *for each* $x \in [0, x_*^4]$.

(b) The turning point x_*^4 is strictly decreasing function on k from ∞ for $k = 0+$ to 0 for $x \to \infty$. Besides, $x_*^4 = 1/\lambda$ for $k = 1$.

Remark 5. *Let us consider the relation between the turning points of the zero-sum and non-zero-sum games with the distribution function $H(t) = t$, i.e. x_*^1/x_*^3. Then*

$$\frac{x_*^1}{x_*^3} = L(k) = \frac{1 - (k/(1+k))^2}{1 - k^{1/(1-k)}}.$$

There is $k_*(= 0.728)$ such that L reaches its maximum at $k = k_*$ on $[0, \infty)$. Also, $L(k_*) = 1.194$, $\lim_{k \to 0+} L(k) = 1$ and $\lim_{k \to \infty} L(k) = 0$.

Remark 6. *It is interesting that the relation between the turning points of the zero-sum and non-zero-sum games with the distribution function $H(t) = 1 - \exp(-\lambda t)$ (i.e. x_*^2/x_*^4) depends only on k but it is independent of λ and it is given by*

$$\frac{x_*^2}{x_*^4} = R(k) = \frac{2(k-1)\ln(1 + 1/k)}{\ln k}$$

R is a strictly decreasing function on $[0, \infty)$ such that $R(0) = 2$, $R(1) = 2\ln(2)$ and $\lim_{k \to \infty} R(k) = 0$.

4. Non-Zero-Sum Game

In this section we consider the case where players have different payoffs under the same circumstances. This game plot can be motivated by the fact that players can be differently well-off and thus, the value of the prize can have different meaning for them. Say, for a rich person the prize 1000 dollars means nothing but for a student it could mean buying a new computer. Let us move on to describing the game plot.

(a) If player i ($i = 1, 2$) suggested a better approximation of the random variable realization and both of them do not overestimate this value then the payoff to player i is a_i.

(b) If player i did not overestimate the random variable realization but his rival overestimated then the payoff to player i is b_i.

(c) In all the rest of the cases the payoff to player i is 0.

Thus, the expected payoffs of player 1 and 2 (if their predicted values are x and y) are given by

$$K_1(x,y) = \begin{cases} b_1(H(y) - H(x)) & \text{for } x < y, \\ 0 & \text{for } x = y, \\ a_1(1 - H(x)) & \text{for } x > y, \end{cases}$$

$$K_2(x,y) = \begin{cases} b_2(H(x) - H(y)) & \text{for } y < x, \\ 0 & \text{for } x = y, \\ a_2(1 - H(y)) & \text{for } y > x. \end{cases}$$

For a given mixed strategy F_2 for player 2 and a pure strategy x for player 1 the payoff to player 1 is given by

$$K_1(x, F_2) = \int_{[0,x)} a_1(1 - H(x))\, dF_2(y) + \int_{(x,1]} b_1(H(y) - H(x))\, dF_2(y).$$

A similar expression holds for $K_2(F_1, y)$.

Theorem 4. *In the non-zero-sum Prediction Number Game there is a Nash equilibrium (F_1, F_2) if and only if*

$$k_i = b_i/a_i \text{ for } i = 1, 2. \tag{1}$$

If (1) *holds then*

$$F_i(x) = \begin{cases} \frac{k_i}{1-k_i}((1 - H(x))^{k_i - 1} - 1) & \text{for } x \in [0, x_*], \\ 1 & \text{for } x > x_* \end{cases} \tag{2}$$

with the payoff vector $(a_1(1 - H(x_*)), a_2(1 - H(x_*)))$ and $x_* = H^{-1}(1 - k_1^{1/(1-k_1)})$.

Proof. Following the proof of Theorem 1 we can show that if (F_1, F_2) be a Nash equilibrium then for $j = 3 - i$ and $i = 1, 2$

(a) $q_1(x)q_2(x) = 0$ for $x \in (0, 1]$ where $q_i(x) = F_i(x) - F_i(x - 0)$,

(b) $q_i(0) = 0$,

(c) if $(c, d) \subseteq (0, 1)$ and F_j is constant in (c, d), then F_i also is constant in(a, b),

(d) if $x \in (0, 1)$ such that $q_j(x) = 0$ then $q_i(x) = 0$,

(e) if there is $c > 0$ such that $F_2(x) = 0$ for $x \in [0, c]$ then $K_1(x, F_2)$ is strictly decreasing in $[0, c]$,

(f) if there is $[c, d) \subseteq (0, 1)$ such that F_i is strictly increasing continuous function $[c, d)$ then

$$a_i(1 - H(x))F_i'(x) = H'(x)(b_i + (a_i - b_i)F(x)) \quad \text{for} \quad x \in (a, b).$$

From (a)-(f) we have that if (F_1, F_2) be a Nash equilibrium with the payoff vector (v_1, v_2) then there is x_* such that

$$F_i(x) \text{ is } \begin{cases} \text{continuous and strictly increasing} & \text{for } x \in [0, x_*), \\ 1 & \text{for } x \in (x_*, 1), \end{cases}$$

$$v_i = a_i(1 - H(x_*)),$$

where $i = 1, 2$. Also,

$$K_1(x, F_2) = \begin{cases} v_1 & \text{for } x \in [0, x_*], \\ v_1 + \int_{[0,x_*)} a_1(H(x) - H(x_*))\, dF_2(y) & \\ +a_1(1 - H(x))q_2(x_*) & \text{for } x > x_*. \end{cases}$$

A similar expression holds for $K_2(F_1, x)$. Thus, (F_1, F_2) is a Nash equilibria if $q_1(x_*) = q_2(x_*) = 0$. Then according to (f) it follows that (F_1, F_2) given by (2) is the unique Nash equilibrium. This completes the proof of Theorem 3.

Corollary 5. *Let $H(t) = t$, i.e. the random variable has a uniform distribution and and (1) holds then for $i = 1, 2$*

$$F_i(x) = \begin{cases} \frac{k_i}{1 - k_i}((1 - x)^{k_i - 1} - 1) & \text{for } x \in [0, x_*], \\ 1 & \text{for } x > x_*. \end{cases}$$

where $x_* = 1 - k_1^{1/(1-k_1)}$.

Corollary 6. *Let $H(t) = 1 - \exp(-\lambda t)$ and (1) holds then for $i = 1, 2$*

$$F_i(x) = \begin{cases} \frac{k_i}{k_i - 1}(1 - e^{\lambda(1-k_i)x}) & \text{for } x \in [0, x_*], \\ 1 & \text{for } x > x_*, \end{cases}$$

and the payoff vector corresponding this Nash equilibrium is $(a_1(1-x_*), a_2(1-x_*))$ where $x_* = \ln(k_1)/(\lambda(k_1 - 1))$.

References

[1] S.C. Karlin, *Mathamatical Methods and Theory in Games, Programming and Economics*, **II**, Pergamon Press, London, 1959.

[2] A.Y. Garnaev, *Search Games and Other Applications of Game Theory*, Springer, New York, 2000.

[3] L.A. Petrosjan, N.A. Zenkevich, *Game Theory*, World Scientific, London, 1996.

[4] M. Sakaguchi, K. Szajowski, Competitive Predoction A Random Variable, *Mathematica Japonica* **43(3)** (1996), 461–472.

In: Mathematics, Game Theory and Algebra Compendium ISBN 978-1-60456-659-8
Editor: Jacob H. Mathias, pp. 631-639

Chapter 53

EFFICIENCY OF BERTRAND AND COURNOT UNDER PRECOMMITMENT

Michèle Breton
CREF and GERAD, Canada
Abdalla Turki
IEA and GERAD, Canada
Georges Zaccour
GERAD & Marketing Department
HEC Montreal Canada

1. Introduction

This note compares the relative efficiency of Cournot and Bertrand equilibria in a differentiated duopoly. In a Cournot game, players compete by choosing their outputs while in a Bertrand game, the strategies are the prices of the products. In a seminal paper, Singh and Vives (1984) show that (i) Bertrand competition is always more efficient than Cournot competition, (ii) Bertrand prices (quantities) are lower (higher) than Cournot prices (quantities) if the goods are substitutes (complements), and (iii) it is a dominant strategy for a firm to choose quantity (price) as its strategic variable provided that the goods are substitutes (complements).

These findings attracted the economists' attention and gave rise to two main streams in the literature. The first one extends the above model in different ways. However, it treats the firms' costs of production as constants and assumes that firms face the same demand and cost structure in both types of competition (Vives (1985), Okuguchi (1987), Dastidar (1997), Hackner (2000), Amir and Jin (2001), and Tanaka (2001)).

The second stream of research aims at examining the robustness of the findings in Singh and Vives (1984) in a duopoly where firms' strategies include investments in research and development (R&D), in addition to prices or quantities (see for instance, Delbono and Denicolo (1990), Motta (1993), Qiu (1997) and Symeonidis (2003)).

This note belongs to this second stream. Our objective is to derive and compare Bertrand and Cournot equilibria when firms are investing in R&D to reduce their unit production cost (i.e., process R&D). The literature has adopted typically a two-stage game formalism where the firms choose their levels of R&D investments in the first stage and compete with either price or quantity in the second stage. A general result is that the findings in Singh and Vives (1984) may not always hold. We shall consider here a one-stage game where the players select at the same time their investments in R&D and either prices or quantities. This is equivalent to assume that the players play a "two-stage" game with precommitment, or open-loop information structure. We show in this setting that the results in Singh and Vives (1984) are robust to considering R&D as an additional strategic variable.

The rest of this note is organized as follows. In Section 2, we outline the model. In Section 3 and 4, we respectively characterize Cournot and Bertrand equilibria. In Section 5, we compare the results and we briefly conclude in Section 6.

2. The Model

Consider an industry formed of two firms producing differentiated but substitutable goods. Each firm independently undertakes cost-reducing R&D investments, and chooses the price p_i or the quantity q_i of its product, so as to maximize its profits. If the firms choose quantity (price) levels, then it is said that they engage in a Cournot (Bertrand) game.

Following Singh and Vives (1984), the representative consumer's preferences are described by the following utility function:

$$U(q_i, q_j) = A(q_i + q_j) - \frac{1}{2}(q_i^2 + q_j^2) - \eta q_i q_j, \quad i, j = 1, 2, i \neq j, \tag{1}$$

where A is positive constant and $0 \leq \eta < 1$. The parameter η represents the degree of product differentiation; products become closer to substitutes as this parameter approaches 1. The resulting market inverse demands are linear and given by

$$p_i = A - q_i - \eta q_j, \quad i, j = 1, 2, i \neq j. \tag{2}$$

Denote by x_i firm i's R&D investment. The unit production cost is assumed to depend on both firms' investments in R&D as well as on the quantity produced and has the following form

$$C_i(q_i, x_i, x_j) = (c + \frac{1}{2}q_i - x_i - \eta\beta x_j), i, j = 1, 2, i \neq j, \tag{3}$$

where $0 < c < A$ and $0 \leq \beta \leq 1$. The parameter β is the industry degree of R&D spillover and $\eta\beta$ represents the effective R&D spillover. We assume that the unit cost is strictly positive.

The cost specification in (3) differs from the one proposed by Qiu (1997), which is the paper most related to this one, in two aspects. First, the benefit in cost reduction from rival's R&D depend on the effective spillover rate $\eta\beta$ and not only on β. An explanation of our assumption lies in the fact that the products must be related, i.e., $\eta \neq 0$, for spillover to occur. Second, firm i's production cost function is assumed to be quadratic in its level of output rather than linear, which allows to model decreasing returns to scale.

The investment cost incurred by player i is assumed quadratic in R&D effort, i.e.,

$$F_i(x_i) = \frac{[1 + d\eta\beta]}{2} x_i^2, i = 1, 2, \tag{4}$$

where d is a dichotomous variable. For $d = 1$, the cost depends on the level of investment in R&D and on the effective spillover level $\eta\beta$. Thus higher R&D effective spillover leads to higher R&D costs for each firm. For $d = 0$, the cost is independent of the spillover. We shall assess the impact of the two cost specifications on the results. When $d = 1$, the cost function is steeper in R&D and is similar to the one proposed by Amir (2000)*.

The total profit of firm i, to be maximized, is given by

$$\pi_i = (p_i - C_i(q_i, x_i, x_j))\, q_i - F_i(x_i). \tag{5}$$

We shall in the sequel compare consumer surplus (CS) and total welfare (TW) under Bertrand and Cournot modes of play. Recall that consumer surplus is defined as consumer's utility minus total expenditures evaluated at equilibrium, i.e.,

$$CS = U(q_i^*, q_j^*) - p_i^* q_i^* - p_j^* q_j^*, \quad i, j = 1, 2, i \neq j. \tag{6}$$

Total welfare is defined as the sum of consumer surplus plus industry's profit, i.e.,

$$TW = CS + \pi_i^* + \pi_j^*, \quad i, j = 1, 2, i \neq j, \tag{7}$$

where the superscript $*$ refers to (Bertrand or Cournot) equilibrium values.

Remark 1. *Our model is symmetric, i.e., all parameters involved in the model, namely, A, β, η and c are the same for both players. This assumption allows us to compare Bertrand and Cournot equilibria in a setting where any difference would be due to the choice of the strategic variables and nothing else. We shall confine our interest to symmetric equilibria in both Cournot and Bertrand games.*

3. Cournot Equilibrium

In the non-cooperative Cournot game firms select their output and R&D investment levels, independently. The following proposition characterizes Cournot equilibrium.

[0]* Notice that following d'Aspremont and Jacquemin (1988, 1990), the models used in Qui (1997) postulate that the possible R&D spillovers take place in the final R&D outcomes. However, Kamien et al. (1992), among others, presume that such spillovers take place in the R&D dollars (spending). Amir (2000) makes an extensive comparison between these two models (i.e. d'Aspremont and Jacquemin (1988, 1990), and Kamien et al. (1992)), and assesses their validity. He concludes that the latter's predictions and results are more valid and robust. Furthermore, in order to make the above two models equivalent, Amir (2000) suggests a certain specification of the R&D cost function of the first above-mentioned models. As he shows, such specification does make the R&D cost functions steeper, and hence do correct for the biases introduced by postulating additive spillovers on R&D outputs rather than inputs.

Proposition 1. *In the unique symmetric Cournot equilibrium, output and R&D investment strategies are given by*

$$q^C = \frac{(A-c)(1+d\eta\beta)}{(1+d\eta\beta)(3+\eta)-(1+\eta\beta)}, \quad (8)$$

$$x^C = \frac{(A-c)}{(1+d\eta\beta)(3+\eta)-(1+\eta\beta)}. \quad (9)$$

Proof. Player i's optimization problem is given by

$$\max_{q>0,x>0} \pi_i = \left((A-q_i-\eta q_j)-\left(c+\frac{1}{2}q_i-x_i-\eta\beta x_j\right)\right)q_i-\frac{[1+d\eta\beta]}{2}x_i^2.$$

Assuming interior solution, necessary equilibrium conditions are given by:

$$\frac{\partial \pi_i}{\partial q_i} = A-3q_i-\eta q_j-c+x_i+\eta\beta x_j=0,$$

$$\frac{\partial \pi_i}{\partial x_i} = q_i-(1+d\eta\beta)x_i=0.$$

To obtain a symmetric equilibrium, it suffices to take $q_i = q_j$ and $x_i = x_j$ in the above equations and the result follows. ■

The equilibrium values q^C and x^C are strictly positive for all admissible parameter values. Note that player i's optimization problem is concave in (q_i, x_i), $i = 1, 2$. Thus second order sufficient conditions are satisfied and the unique Cournot equilibrium is well defined.

The equilibrium Cournot price is given by

$$p^C = \frac{A(1+2d\eta\beta-\eta\beta)+c(1+\eta)(1+d\eta\beta)}{(1+d\eta\beta)(3+\eta)-(1+\eta\beta)}.$$

Inserting Cournot equilibrium values in (6) and (7) provides the following consumer surplus and total welfare

$$CS^C = \frac{(1+\eta)(A-c)^2(1+d\eta\beta)^2}{((1+d\eta\beta)(3+\eta)-(1+\eta\beta))^2}, \quad (10)$$

$$TW^C = \frac{(A-c)^2(1+d\eta\beta)\left(3+4d\eta\beta+\eta+d\eta^2\beta\right)}{((1+d\eta\beta)(3+\eta)-(1+\eta\beta))^2}. \quad (11)$$

The following proposition compares the results under the two specifications of the investment cost function in R&D, i.e., for $d = 1$ and $d = 0$.

Proposition 2. *Cournot's output, investment in R&D, consumer surplus and total welfare are lower when the cost function is steeper in R&D investments* $(i.e., d = 1)$.

Proof. Let the superscripts $C0$ and $C1$ refer respectively to Cournot equilibrium with $d = 0$ and $d = 1$. Straightforward computations lead to

$$q^{C0} - q^{C1} = \frac{(A-c)\eta\beta}{(2+\eta-\eta\beta)(2+\eta)} > 0,$$

$$x^{C0} - x^{C1} = \frac{(A-c)\eta\beta(3+\eta)}{(2+\eta-\eta\beta)+(1+\eta\beta)(2+\eta)} > 0,$$

$$CS^{C0} - CS^{C1} = \frac{4\eta\beta(1+\eta)(A-c)^2}{(2+\eta)^2(2+\eta-\eta\beta)^2} > 0,$$

$$TW^{C0} - TW^{C1} = \frac{(A-c)^2\eta\beta\left(8+6\eta+\eta^2+\eta\beta\left(13+11\eta-4\eta\beta+\eta^2\right)\right)}{(\eta+2-\eta\beta)^2(1+\eta\beta)(2+\eta)^2} > 0.$$

■

A consequence of the first item of the Proposition is that the equilibrium price when $d = 0$ is lower than when $d = 1$. Indeed,

$$p^{C0} - p^{C1} = -\frac{(A-c)\eta\beta(1+\eta)}{(2+\eta-\eta\beta)(2+\eta)} < 0.$$

4. Bertrand Equilibrium

In the non-cooperative Bertrand game, firms select independently their price and R&D investment levels. From (2), we define the demand functions

$$q_i = \frac{(1-\eta)A - p_i + \eta p_j)}{(1-\eta^2)}, i, j = 1, 2, i \neq j. \tag{12}$$

The following proposition characterizes Bertrand equilibrium.

Proposition 3. *In the unique symmetric Bertrand equilibrium, price and R&D investment strategies are given by*

$$p^B = \frac{\left(-\eta^2 - d\eta^3\beta - \eta\beta + 2d\eta\beta + 1\right)A + \left(d\eta\beta + d\eta^2\beta + 1 + \eta\right)c}{\Gamma}. \tag{13}$$

$$x^B = \frac{A-c}{\Gamma}. \tag{14}$$

where

$$\Gamma = \left(-d\eta^3\beta + d\eta^2\beta + 3d\eta\beta - \eta\beta - \eta^2 + \eta + 2\right) > 0.$$

Proof. Player i's optimization problem is given by

$$\max_{p_i>0, x_i>0} \pi_i = \left((A - q_i - \eta q_j) - \left(c + \frac{1}{2}q_i - x_i - \eta\beta x_j\right)\right) q_i - \frac{[1+d\eta\beta]}{2}x_i^2,$$

where q_i is defined in (12).

Assuming interior solution, necessary equilibrium conditions are given by:

$$\begin{aligned}\frac{\partial \pi_i}{\partial p_i} &= \left(\frac{(-1+\eta)(-2+\eta^2)A}{(-1+\eta^2)^2}\right) - \frac{(c - x_i - \eta\beta x_j)}{(-1+\eta^2)} \\ &+ \frac{(-3+\eta^2)p_i - \eta(-2+\eta^2)p_j}{(-1+\eta^2)^2} = 0, \\ \frac{\partial \pi_i}{\partial x_i} &= \frac{(-1+\eta)}{(-1+\eta^2)}A + \frac{(p_i - \eta P_j)}{(-1+\eta^2)} - (1+d\eta\beta)x_i = 0.\end{aligned}$$

For a symmetric equilibrium we have $p_i = p_j = p$ and $x_i = x_j = x$. Straightforward computations lead to the result. ■

The equilibrium values p^B and x^B are strictly positive for all admissible parameters' values. Note that each player i's optimization problem is concave in (p_i, x_i) , $.i = 1, 2$. Thus second order conditions are satisfied and the unique Bertrand equilibrium is well defined.

The equilibrium Bertrand output is given by

$$q^B = \frac{(A-c)(d\eta\beta+1)}{\Gamma}. \tag{15}$$

Evaluating (6) and (7) at Bertrand equilibrium values leads to the following consumer surplus and total welfare:

$$CS^B = \frac{(A-c)^2(1+\eta)(1+d\eta\beta)^2}{\Gamma^2}, \tag{16}$$

$$TW^B = \frac{-(1+d\eta\beta)(A-c)^2(4d\eta\beta - 2d\eta^3\beta - 2\eta^2 + 3 + d\eta^2\beta + \eta)}{\Gamma^2}. \tag{17}$$

The following proposition compares the Bertrand equilibrium for $d = 0$ to its counterpart with $d = 1$.

Proposition 4. *Bertrand's output, investment in R&D, consumer surplus and total welfare are lower when the cost function is steeper in R&D investments* (*i.e.*, $d = 1$) .

Proof. Let the superscripts $B0$ and $B1$ refer respectively to Bertrand equilibrium with $d = 0$ and $d = 1$. Straightforward computations lead to

$$q^{B0} - q^{B1} = \frac{(A-c)\eta\beta}{(-2+\eta\beta+\eta^2-\eta)(\eta+1)(\eta-2)} > 0,$$

$$x^{B0} - x^{B1} = \frac{(A-c)(3-\eta^2+\eta)\eta\beta}{(-2+\eta\beta+\eta^2-\eta)(\eta+1)(\eta-2)(1+\eta\beta)} > 0,$$

$$CS^{B0} - CS^{B1} = \frac{(-\eta\beta - 2\eta^2 + 2\eta + 4)\eta\beta(A-c)^2}{(-2+\eta\beta+\eta^2-\eta)^2(\eta+1)(\eta-2)^2} > 0,$$

$$\begin{aligned}TW^{B0} - TW^{B1} &= \frac{\eta\beta(A-c)^2}{(-2+\eta\beta+\eta^2-\eta)^2(\eta+1)^2(\eta-2)^2(1+\eta\beta)}(2\eta^4\beta^2 - \eta^3\beta^2 \\ &-4\eta^2\beta^2 + 4\eta^5\beta - 6\eta^4\beta - 12\eta^3\beta + 11\eta^2\beta + 13\eta\beta + 8 - 4\eta^3 - 9\eta^2 + 3\eta^4 + 6\eta) > 0.\end{aligned}$$

■

As in the Cournot case, a consequence of the first item of the Proposition is that the price is lower when $d = 0$. Indeed,

$$p^{B0} - p^{B1} = \frac{(A-c)\eta\beta}{(2-\eta\beta-\eta^2+\eta)(\eta-2)} < 0.$$

5. Comparison of Equilibria

In this section we compare Cournot outputs, prices, profits, R&D levels, consumer surplus and total welfare with their Bertrand counterparts.

Proposition 5. *Bertrand and Cournot outputs, R&D investments and profits compare as follows:*
$(i) : q^B - q^C > 0.$
$(ii) : x^B - x^C > 0.$
$(iii) : \pi^C - \pi^B > 0.$

Proof. (i): Use (8) and (15) to get

$$q^B - q^C = \frac{(1+d\eta\beta)^2(A-c)\eta^2}{\Lambda\Psi} > 0$$

where

$$\Lambda = \left(-d\beta\eta^3 + d\beta\eta^2 + 3d\beta\eta - \eta\beta - \eta^2 + \eta + 2\right) > 0,$$
$$\Psi = \left(d\eta^2\beta + 3d\eta\beta - \eta\beta + \eta + 2\right) > 0.$$

(ii) From (9) and (14) we have

$$x^B - x^C = \frac{(A-c)\eta^2(1+d\eta\beta)}{\Lambda\Psi} > 0.$$

(iii) Substituting for equilibrium values in profit functions (5) and computing the difference between Cournot and Bertrand profits lead to:

$$\pi^C - \pi^B = \frac{(1+d\eta\beta)^2(A-c)^2\eta^3}{2\Lambda^2\Psi^2}\left(d^2\eta^2\beta^2(5\eta+6) + 2d\eta\beta(5-2\eta\beta-3\beta) + 2\eta\beta^2 + 9d\eta^2\beta - 4\eta\beta - 4\beta + 4 + 4\eta\right) > 0.$$

■

The first result in the above proposition replicates the one in Singh and Vives (1984) who assessed the comparative efficiency of Bertrand and Cournot equilibria in a game with fixed cost and no investment in R&D. An implication is that Cournot's price is higher than Bertrand's price. Item (ii) shows that firms will invest more in R&D under Bertrand competition which leads to lower unit production cost. As in Singh and Vives, the last result shows that profits are higher under Cournot competition. Given the choice between (both) playing Bertrand or Cournot, the answer is clear cut.

Proposition 6. *Bertrand and Cournot consumer surplus and total welfare compare as follows:*

$(i) : CS^B - CS^C > 0.$

$(ii) : TW^B - TW^C > 0.$

Proof. (i) Substituting in turn for Bertrand and Cournot equilibrium values in (6) and computing the difference lead to

$$\begin{aligned} CS^B - CS^C &= \frac{(1+d\eta\beta)^3 (A-c)^2 (1+\eta)\eta^2}{\Lambda^2 \Psi^2} \left(-d\eta^3\beta + 6d\eta\beta \right. \\ &\left. +2d\eta^2\beta - \eta^2 + 2\eta + 4 - 2\eta\beta\right), \end{aligned}$$

which is clearly positive.

(ii) Substituting in turn for Bertrand and Cournot equilibrium values in (7) and computing the difference lead to

$$\begin{aligned} TW^B - TW^C &= \frac{(1+d\eta\beta)^2 (A-c)^2 \eta^2}{\Lambda^2 \Psi^2} \left(d\eta\beta \left(10 + \eta \left(2\eta\beta + 4\beta - 2\eta^2 - 7\eta + 4\right)\right)\right. \\ &\left. +d^2\eta^2\beta^2 \left(-\eta^3 - 4\eta^2 + 2\eta + 6\right) - 2\eta\beta\left(\eta\beta - \eta + 1\right) - 3\eta^2 + 4 - \eta^3 + 2\eta\right) \end{aligned}$$

which is clearly positive. ∎

Again these results are similar to those in Singh and Vives (1984).

Remark 2. *It can be easily verified that if* $\eta = 0$*, i.e., the products are not substitutes, then Bertrand and Cournot equilibria coincide. This means that each firm becomes a monopoly in its market and it does not matter if the players choose prices or quantities as strategic variables.*

6. Concluding Remarks

The general conclusion is that including process R&D as an additional strategic variable in a duopoly game will not change the standard result, namely that Bertrand competition is more efficient than Cournot competition. This conclusion is reached under the assumption that the game is played as a one stage game. Note that the result holds true for any positive levels of product differentiation (η) and industry spillover (β) and independently of the form of the investment cost function.

Extensions include the analysis of the relative efficiency of Bertrand and Cournot, (i) in a dynamic game setting, and (ii) in a case where the degree of R&D spillover is endogenous and depends on the firms' absorptive capacities.

References

[1] R. Amir, Modelling imperfectly appropriable R&D via spillovers,*International Journal of Industrial Organization* **18** (2000), 1013–1032.

[2] R. Amir, J.Y. Jin, Cournot and Bertrand equilibria compared: substitutability, complementarity, and concavity, *International Journal of Industrial Organization* **19** (2001), 303–317.

[3] C. d'Aspremont, A. Jacquemin, Cooperative and non cooperative R&D in duopoly with spillovers, *American Economic Review* **78** (1988), 1133–1137.

[4] C. d'Aspremont, A. Jacquemin, Cooperative and Noncooperative R&D in Duopoly with Spillovers: Erratum, *American Economic Review* **80** (1990), 641–642.

[5] K.G. Dastidar, Comparing Cournot and Bertrand in a homogeneous product market, *Journal of Economic Theory* **75** (1997), 205–212.

[6] F. Delbono, V. Denicolo, R&D in a symmetric and homogeneous oligopoly,*International Journal of Industrial Organization* **8** (1990), 297–313.

[7] J. Hackner, A note on price and quantity competition in differentiated oligopolies, *Journal of Economic Theory* **93** (2000), 233–239.

[8] M. Kamien, E. Muller and I. Zang, Research Joint Ventures and R & D Cartels, *American Economic Review* **82** (1992), 1293–1306.

[9] M. Motta, Endogenous quality choice: price versus quality competition,*Journal of Industrial Economics* **41** (1993), 113–131.

[10] K. Okuguchi, Equilibrium prices in Bertrand and Cournot oligopolies, *Journal of Economic Theory* **42** (1987), 128–139.

[11] L. Qiu, On the dynamic efficiency of Bertrand and Cournot equilibria, *Journal of Economic Theory* **75** (1997), 213–229.

[12] N. Singh, X. Vives, Price and quantity competition in a differentiated duopoly,*Rand Journal of Economics* **15** (1984), 546–554.

[13] G. Symeonidis, Comparing Cournot and Bertrand equilibria in a differentiated duopoly with product R&D, *International Journal of Industrial Organization* **21** (2003), 39–55.

[14] Y. Tanaka, Profitability of price and quantity strategies in an oligopoly, *Journal of Mathematical Economics* **35** (2001), 409–418.

[15] X. Vives, On the efficiency of Bertrand and Cournot equilibria with product differentiation, *Journal Economic Theory* **36** (1985), 166–175.

In: Mathematics, Game Theory and Algebra Compendium ISBN 978-1-60456-659-8
Editor: Jacob H. Mathias, pp. 641-649

Chapter 54

ONE APPROACH TO SOLUTION OF COMPLEX GAME PROBLEMS

G.Ts. Chikrii
Department of Economic Cybernetics
Institute of Cybernetics NAS
Kyiv, Ukraine

1. Introduction

In recent years certain interest was shown in study of the game problems of control for systems with Volterra evolution. These systems encompass a wide range of processes including, in their number, the systems of integral, integral-differential, and differential-difference equations.

One of the pioneer papers in this direction is the study (A. Chikrii and Eidelman, 2000). The method, used there for solving the problem of pursuit, is the method of resolving functions. This method as well as the Pontryagin's First Direct Method are based on so-called Pontryagin's condition (Pontryagin, 1988), reflecting some advantage of the pursuer over the evader in control resources. But there are classes of games for which this condition fails. The problem of soft meeting of two moving objects is an example. The in-depth analysis of this condition in the case the differential dynamics is given in (Nikolskii, 1972).

To begin with the problem of pursuit for system with Volterra evolution is treated here under assumption for variable time-delay of information on the evader's behavior.

It is shown that just as in the case of differential dynamics (G. Chikrii, 1995) this problem is equivalent to a certain perfect-information game with somewhat changed both the dynamics and the terminal set.

On the one hand, this equivalence makes it feasible in the case of delayed information to apply the methods, developed for the games with perfect information. In this paper we apply the First Direct Method.

On the other hand, on the basis of this equivalence an approach to solution of perfect-information games of pursuit, for which fundamental Pontryagin's condition fails, is developed. This approach consists in a passage from original game with perfect information to an auxiliary one with special kind information delay and subsequent analysis of the game, equivalent to the latter. Using this approach, sufficient conditions for the game termination are derived. In so doing, the idea of Zonnevend, consisting in construction of current control of the pursuer on the basis of control of the evader in the past, is used (D. Zonnevend, 1973). Earlier this scheme was developed in the case of differential dynamics (G. Chikrii, 1999).

In conclusion, one model game of soft meeting of two moving objects with second-order differential dynamics is analyzed in detail under the initial condition for finding the 'track' of the evader by the pursuer. In such setting this problem first appeared in [Zonnevend, 1972]. Also an attempt to obtain its full solution was made there. Here an explicit formula for the pursuer's control, ensuring soft meeting with the evader in a certain time and constructed on the basis of control of his adversary in the past, is obtained.

2. Using Effect of Information Delay for Solution of the Linear Evolutionary Game with Perfect Information

Let the evolution of a conflict-controlled process be described by the equation

$$z(t) = g(t) + \int_0^t \Omega(t,\theta)(u(\theta) - v(\theta))\, d\theta, t \geq 0. \tag{1}$$

Here $z(t) \in R^n$, where R^n is the n-dimensional euclidean space; $g(t)$ – vector-function, taking its values in R^n, which is Lebegue measurable and bounded at $t > 0$; $\Omega(t,\theta)$, $t \geq \theta \geq 0$, is a matrix-function, measurable in θ at each $t \geq 0$.

At each current instant of time the pursuer and the evader choose their controls u and v from the control domains U and V in such a way that their realizations in time are measurable functions. The sets U and V, respectively, are assumed to be convex compacts from R^n.

Also the terminal set M, $M \subset R^n$, having cylindrical form, is given:

$$M = M_0 + M_1, \tag{2}$$

where M_0 is a linear subspace from R^n and the set M_1 belongs to the subspace L being an orthogonal complement to M_0 in R^n. Denote by π the operator of orthogonal projection from R^n onto L.

The goal of the pursuer is to steer a trajectory of the system to the terminal set in the shortest time. The goal of the evader is the opposite one: to deviate the system trajectory from the terminal set as long as possible.

We will treat the game standing on the pursuer's side. The problem, facing the pursuer, is complicated by the fact that current information about the evader's behavior is available to the former with some delay in time. Let the game evolves on the interval of time $[0,t]$. During some initial half-interval $[0,\tau_0)$ of time the pursuer is unaware of the evader's control and therefore applies an arbitrary admissible control. Denote its realization on $[0,\tau_0)$ by $u^{\tau_0}(\cdot)$. The pair $(g(t), u^{\tau_0}(\cdot))$ will be referred to as the initial position of the game.

Then, at each current instant of time $\theta \geq \tau_0$ the pursuer chooses his control on the basis of knowledge of $g(t)$ and fresh information on the evader's control. It is assumed that this information is available to the pursuer with time delay $\tau(t), t \geq \tau_0$, where $\tau(t)$ is a positive, continuosly differentiable function, such that $\tau(\tau_0) = \tau_0$. In addition, the condition for its growth rate holds: $\dot{\tau}(t) < 1$. This condition provides for continuous updating of the information on system state in time.

Thus, at the initial instant τ_0 the pursuer comes to know $v(0)$ and then at each current instant of time $\theta \in [\tau_0, t]$ he knows $v(\theta - \tau(\theta))$. From the pursuer's point of view, the attainability set of the process (1), under the initial position $(g(t), u^{\tau_0}(\cdot))$ and chosen controls $u(\cdot)$ and $v(\cdot)$ on $[\tau_0, t)$ and $[0, t-\tau(t)]$ of the pursuer and the evader respectively, can be presented in the following form

$$Z(t) = \left\{ z : z = g(t) + \int_0^{\tau_0} \Omega(t,\theta)\, u^{\tau_0}(\theta) d\theta + \int_{\tau_0}^{t} \Omega(t,\theta)\, u(\theta) d\theta \right.$$
$$\left. - \int_0^{t-\tau(t)} \Omega(t,\theta)\, v(\theta) d\theta - \int_{t-\tau(t)}^{t} \Omega(t,\theta)\, V d\theta \right\} \quad (3)$$

The last term in the above expression presents itself the integral of a set, that is a union of all integrals of the form $\int_{t-\tau(t)}^{t} \Omega(t,\theta)\, v(\theta) d\theta$ for all measurable functions $v(\cdot)$ with values from the compact V.

We shall say that the game of pursuit (1),(2) with delay of information $\tau(t)$ can be terminated at instant t if the pursuer, choosing approriately his control, can achieve the inclusion

$$Z(t) \subset M \text{ or } \pi Z(t) \subset M_1.$$

Denote

$$\tilde{z}(t) = g(t) + \int_0^{\tau_0} \Omega(t,\theta)\, u^{\tau_0}(\theta) d\theta$$

$$+ \int_{\tau_0}^{t} [\Omega(t,\theta)\, u(\theta) - (1-\dot{\tau}(\theta))\, \Omega(t, \theta - \tau(\theta))\, v(\theta - \tau(\theta))] d\theta. \quad (4)$$

$$M_1(\tau(t)) = M_1 \overset{*}{-} \int_0^{\tau(t)} -\Omega(t, t-\theta) V d\theta, \tilde{M} = M_0 + M_1(\tau(t)). \quad (5)$$

Condition 1. *Set $M_1(\tau(t))$ is nonempty for all $t \geq \tau_0$.*

Theorem 1. *Let Condition 1 hold. If the game (1), (2) with variable delay of information $r(t)$ can be terminated at time T then the game (4), (5) with perfect information can be terminated at the same time and vice versa.*

In the case of the linear differential game, when $g(t) = e^{tA}z_0$, $\Omega(t,\theta) = e^{(t-\theta)A}$, this fact was fruitfully used for solution of the games with various kind information delay, see f.e. (G. Chikrii, 1995).

Here we shall apply the classic scheme of the Pontryagin's First Direct Method, originally created for solution of the linear differential games of pursuit, to the linear evolutionary system of general kind (1). It should be noted that for the dynamic game (1) with perfect information the analog of Pontryagin's condition, which is the basis of the First Direct Method, has the following form.

Condition 2. *Set $W(t,\theta) = \Omega(t,\theta)U \overset{*}{-} \Omega(t,\theta)V$ is nonempty for all $0 \leq \theta \leq t$.*

Denote

$$W_*(t,\theta) = \pi\Omega(t,\theta)U \overset{*}{-} (1-\dot{\tau}(\theta))\,\pi\Omega(t,\theta-\tau(\theta))\,V$$

Condition 3. *Set $W_*(t,\theta)$ is nonempty for all θ, $\tau_0 \leq \theta \leq t < +\infty$.*

Let us define the Pontryagin's time for the game under study.

$$t^* = t^*\left(g(\cdot), u^{\tau_0}(\cdot)\right) = \min\left\{ t : \pi\left(g(\tau_0+t) + \int_0^{\tau_0} \Omega(\tau_0+t,\theta)\,u^{\tau_0}(\theta)d\theta \right)\right.$$

$$\left. \in M_1\left(\tau(\tau_0+t)\right) - \int_{\tau_0}^{\tau_0+t} W_*(\tau_0+t,\theta)\,d\theta \right\}. \tag{6}$$

Theorem 2. *Let Conditions 1 and 3 hold and there exist finite t^*, defined by (6). Then from the initial position $(g(\cdot), u^{\tau_0}(\cdot))$ the game (1), (2) with delayed information may be terminated at the time t^*.*

As in the case of differential game, the impact of information delay to the dynamics and the terminal set of the equivalent game (4), (5) can be used for solution of some evolutional games (1),(2) with perfect information, for which fundamental Condition 2 fails. The main idea consists in the artificial worsening of information conditions for the pursuer as if information on the evader's behavior is available to the pursuer with a time delay, decreasing in time and vanishing at the moment when the system trajectory hits the terminal set. Furthur analysis is conducted on the basis of equivalence of obtained game with delayed information to certain game with perfect information (Theorem 2).

Condition 4. *There exists a positive function $I(\theta)$, $\theta \in [0, \infty)$, $I(\theta) \in C^1[0, +\infty)$, $I(\theta) \geq \theta$, $I(0) = 0$, such that for all $0 \leq \theta \leq t$ the set*

$$W_1(t, \theta) = \pi\Omega(I(t), I(t) - \theta) U \overset{*}{-} \dot{I}(\theta)\pi\Omega(I(t), I(t) - I(\theta)) V$$

is nonempty.

Denote

$$t_1(g(\cdot)) = \min\left\{ t : \pi\left(g(I(t)) + \int_0^{I(t)-t} \Omega(I(t), \theta) U d\theta \right) \cap \left(-M_1 + \int_0^t W_1(t, \theta) d\theta \right) \neq \emptyset \right\}. \quad (7)$$

Theorem 3. *Let for the game (4), (5) with perfect information Condition 3 hold and there exist finite $t_1(g(\cdot))$. Then the game can be terminated at the time $I(t_1)$.*

Proof. From the inclusion in (7) it follows that a measurable selection $\omega_1(t, \theta) \in W_1(t, \theta)$, a point $m_1 \in M_1$, and a control $u^{\tau_0}(\cdot)$ on $[0, \tau_0)$ exist, where $\tau_0 = I(t_1) - t_1$, such that

$$-\pi\left(g(I(t_1)) + \int_0^{\tau_0} \Omega(I(t_1), \theta) u^0(\theta) d\theta \right) = -m_1 + \int_0^{t_1} \omega_1(t_1, \theta) d\theta.$$

We assign the pursuer to use on $[0, \tau_0)$ the control $u^{\tau_0}(\cdot)$ and beginning with the instant τ_0 to construct control $u(\tau_0 + \theta)$, $\theta \in [0, t_1]$, defined by the formula

$$\pi\Omega(I(t_1), \tau_0 + \theta) u(\tau_0 + \theta) =$$
$$= \pi\left(\dot{I}(t_1 - \theta)\Omega(I(t_1), I(t_1) - I(t_1 - \theta)) v(I(t_1) - I(t_1 - \theta)) \right) + \omega_1(t_1, t_1 - \theta).$$

Thus, at current instant $\tau_0 + \theta$, $\theta \in [0, t_1]$, the pursuer constructs his control on the basis of the evader's control on time $\tau(\tau_0 + \theta)$ earlier, where function $\tau(\cdot)$ has the form

$$\tau(\tau_0 + \theta) = I(t_1 - \theta) - (t_1 - \theta).$$

Note that $\tau(\tau_0 + t_1) = 0$ that is discrepancy in time dissappears at the instant of trajectory hitting the terminal set, therefore $\tilde{M} = M$ and Condition 1 is readily fulfilled.

3. Case of differential dynamics. Model game of soft meeting of two moving objects

In the case of differential dynamics of a conflict-controlled process:

$$\dot{z} = Az + u - v, \quad (8)$$

where $z \in R^n$ and A is a quadratic matrix of order n, Condition 4 reduces to the following form.

Condition 5 (Zonnevend). *There exists a positive, monotonically increasing function* $I(t)$, $t \in [0, +\infty)$, *such that* $I(0) = 0$, $I(t) \geq t$, *and*

$$W(t) = \pi e^{tA} U \overset{*}{-} \dot{I}(t) \pi e^{I(t)A} V \neq \emptyset \quad \forall t \geq 0. \tag{9}$$

From (7) one can deduce formula for the time of game termination

$$t_1 = t_1(z_0) = \min \left\{ t \geq 0 : \pi \left(e^{I(t)A} z^0 + \int_0^{I(t)-t} e^{(I(t)-\theta)A} U \right) \bigcap \left(-M_1 + \int_0^t W(\theta) d\theta \right) \neq \emptyset \right\}. \tag{10}$$

Now we can formulate sufficient conditions for the game termination in a finite time.

Theorem 4. *Let in the game (8), (2) Condition 5 hold and let for the given initial position the time* t_1 *is finite. Then, from this position the game can be terminated at the instant of time* $I(t_1)$.

Consider one specific case when the intersection of sets in (10) contains the zero vector $\{0\}$;

$$\{0\} \in \pi \left(e^{I(t_1)A} z^0 + \int_0^{I(t_1)} e^{(I(t_1)-\theta)A} U d\theta \right) \bigcap \left(-M_1 + \int_0^{t_1} W(\theta) d\theta \right), \tag{11}$$

as applied to the model example, analyzed in (Nikolskii, 1972).

Let the dynamics of objects be described by the equations

$$\ddot{x} + \alpha \dot{x} = \rho u, \quad \ddot{y} + \beta \dot{y} = \sigma v, \quad x, y \in R^n, \quad \|u\| \leq 1, \quad \|v\| \leq 1,$$

where x and y are geometric coordinates, α and β are coefficients of friction and ρ , σ coefficients of control resources of the pursuer and the evader respectively.

The goal of the pursuer is to achieve simultaneous coincidence of geometric coordinates and velocities of the players at some finite instant of time.

With the help of substitution $x_1 = x$, $y_1 = y$, $x_2 = \dot{x}$, $y_2 = \dot{y}$ let us pass to the equations

$$\begin{aligned} \dot{x}_1 &= x_2 \\ \dot{x}_2 &= -\alpha x_2 + \rho u \end{aligned} \tag{12}$$

$$\begin{aligned} \dot{y}_1 &= y_2 \\ \dot{y}_2 &= -\beta y_2 + \sigma v. \end{aligned} \tag{13}$$

Thus, the terminal set M appears as a linear subspace in R^{4n}:

$$M = \{(x_1 \, x_2 \, y_1 \, y_2) \, , \; x_i \in R^n , \; y_i \in R^n : x_1 = y_1 , \; x_2 = y_2\} \, ..$$

From Condition 5 (formula (9)) one can deduce an explicit formula for function $I(t)$

$$I(t) = \frac{1}{\beta} \ln \left\{ \frac{\beta}{\alpha} \left(e^{\alpha t} - 1 \right) + 1 \right\} \tag{14}$$

under the following conditions on the game parameters, providing existence of this function:

$$\alpha \geq \beta, \quad \frac{\rho}{\sigma} \geq \frac{\alpha^2}{\beta^2}. \tag{15}$$

Formula (11) implies two inclusions:

$$\{0\} \in \pi e^{I(t_1)A} z^0 + \int_0^{I(t_1)} e^{(I(t_1)-\theta)A} U d\theta \tag{16}$$

and

$$\{0\} \in \int_0^{t_1} \left(\rho \pi e^{tA} U \stackrel{*}{-} \sigma \dot{I}(t) \pi e^{I(t)A} V \right) d\theta. \tag{17}$$

In the turn, condition (16) for the game under study means that there exists a measurable control $u^0(\cdot), \quad u^0(t) \in U, \quad t \in [0, I(t_1) - t_1]$, such that the following equations are true

$$- \left(x_1^0 - y_1^0\right) - \left(\frac{1 - e^{-\alpha I(t_1)}}{\alpha} x_2^0 - \frac{1 - e^{-\beta I(t_1)}}{\beta} y_2^0 \right) = \int_0^{I(t_1)-t_1} \frac{1 - e^{-\alpha(I(t_1)-\theta)}}{\alpha} u^0 \left(\theta\right) d\theta \tag{18}$$

$$- \left(e^{-\alpha I(t_1)} x_2^0 - e^{-\beta I(t_1)} y_2^0 \right) = \int_0^{I(t_1)-t_1} e^{-\alpha(I(t_1)-\theta)} u^0(\theta) d\theta. \tag{19}$$

It can be shown that they fully correspond to the condition for finding the 'track' of the evader (G. Chikrii, 2000) when at some initial instant of time $\tau_0 > 0 \quad x_1(\tau_0) = y_1(0)$ and $x_2(\tau_0) = (1+\varepsilon) y_2^0$, where $\varepsilon > 0$. Here we can set $I(t_1) - t_1 = \tau_0$, $\dot{I}(t_1) = 1 + \varepsilon$, since for $t > 0 \quad I(t) > t$ and $1 < \dot{I}(t) < \frac{\alpha}{\beta}$, while at $t = 0 \quad I(0) = 0$ and $\dot{I}(t) = 1$. In so doing, the following formulas for the function $I(t)$ (14) are used

$$\frac{1 - e^{-\beta I(t)}}{\beta} = \frac{1 - e^{-\alpha t}}{\alpha} \dot{I}(t), \quad e^{-\beta I(t)} = e^{-\alpha t} \dot{I}(t). \tag{20}$$

The condition (17) for the game under study has the form

$$\{0\} \in \int_0^{t_1} \left\{ \rho \begin{pmatrix} \frac{1-e^{-\alpha\theta}}{\alpha} U \\ e^{-\alpha\theta} U \end{pmatrix} \stackrel{*}{-} \sigma \dot{I}(\theta) \begin{pmatrix} \frac{1-e^{-\beta I(\theta)}}{\beta} V \\ e^{-\beta I(\theta)} V \end{pmatrix} \right\} d\theta$$

and, in view of formulas (20), is equivalent to the inequalities (15) since $\max\limits_{t\in[0,\infty)} \dot{I}(t) = \frac{\alpha}{\beta}$. What is more, from here there follows formula for the control of the pursuer on the half-interval $[\tau_0, \tau_0 + t_1)$

$$u(\tau_0 + t) = \frac{\sigma}{\rho} \left(\dot{I}(t_1 - t)\right)^2 v(\tau_0 + t_1 - I(t_1 - t)), \quad t \in [0, t_1]. \tag{21}$$

It can be shown (G. Chikrii, 2000) that using this control the pursuer moves with decreasing velocity along the geometric trajectory of the evader up to the instant of soft meeting. To prove this it suffice to pass in system (13) to new variables $\tilde{y}_1(t+\tau_0) = y_1(\tau_0 + t_1 - I(t_1 - t))$, $\tilde{y}_2(t+\tau_0) = \dot{\tilde{y}}_1(t+\tau_0)$. We then come to the system for $\tilde{y}_1, \tilde{y}_2$:

$$\begin{aligned} \dot{\tilde{y}}_1(t+\tau_0) &= \tilde{y}_2(t+\tau_0) \\ \dot{\tilde{y}}_2(t+\tau_0) &= -\left(\beta \dot{I}(t_1 - t) + \frac{\ddot{I}(t_1 - t)}{\dot{I}(t_1 - t)}\right) \tilde{y}_2(t+\tau_0) + \\ &\quad + \sigma \left(\dot{I}(t_1 - t)\right)^2 v(\tau_0 + t_1 - I(t_1 - t)) \end{aligned} \tag{22}$$

$$\tilde{y}_1(\tau_0) = y_1(0), \; \tilde{y}_2(\tau_0) = (1+\varepsilon) y_2(0).$$

Then the initial condition for finding the evader's 'track' reduces to the form: $x_1(\tau_0) = \tilde{y}_1(0), x_2(\tau_0) = \tilde{y}_2(0)$. Comparing solutions of (12) $x_1(t+\tau_0), x_2(t+\tau_0)$ with the solutions of (22) $\tilde{y}_1(t+\tau_0), \tilde{y}_2(t+\tau_0)$ on the interval $[0, t_1]$ and taking into account the explicit formulas for function $I(t)$ (14) and control $u(t+\tau_0)$ (21) we obtain the desired result. Note that the time t_1 (during which the pursuer achieves the soft meeting with the evader) can be easily evaluated in specific case under study, using formula (14) and the above-mentioned conditions (Chikrii, 2000).

References

[1] A.A. Chikrii, S.D. Eidelman, Game problems for systems with Volterra evolution. Fractal games, *J. Game Theory and Appl.* **6** (2000).

[2] L.S. Pontryagin, *Selected scientific papers*, **II**, Nauka, Moscow, 1988, (in Russian).

[3] M.S. Nikolskii, Application of the Pontryagin's first direct method, *Izv. Akad. Nauk SSSR* **6** (1972), (in Russian).

[4] G.Ts. Chikrii, An approach to the solution of linear differential games with variable information delay, *J. Autom. Inform. Sci.* **27**, No. **3&4** (1995).

[5] D. Zonnevend, (1973) On one kind of advantage of a player, *Doklady SSSR* **208**, No. 3 (1973).

[6] G.Ts. Chikrii, Using impact of information delay for solution of game problems of pursuit, *Dopovidi NANU* **12** (1999).

[7] D. Zonnevend, One method of Pursuit, *Doklady SSSR* **204**, No. 6 (1972).

[8] G. Ts. Chikrii, On one method of pursuit in tracks, *Dopovidi NANU* **6** (2000).

In: Mathematics, Game Theory and Algebra Compendium ISBN 978-1-60456-659-8
Editor: Jacob H. Mathias, pp. 651-663

Chapter 55

A DIFFERENTIAL GAME WITH INVESTMENT IN TRANSPORT AND COMMUNICATION R&D

Luca Colombo, Luca Lambertini and Andrea Mantovani
Department of Economics
University of Bologna
Bologna, Italy

1. Introduction

Transport and communication technologies (TC) have always represented a relevant feature in the literature on the *New Trade Theory* (see Helpman and Krugman, 1985, *inter alia*). However, most of the contributions in this field tend to consider the transportation plan and its related technology as exogenously given characteristics.

Even if firms devote substantial amounts to research and development (R&D), only process innovation and product innovation have been widely explored.* The activity of product innovation consists in the development of technologies either for producing new products or for increasing the quality of the existing ones. On the other hand, process innovation aims at decreasing the costs of producing existing products. The literature in this field has considered the different degree of efficiency of process innovating R&D between the Cournot and the Bertrand setting. An established result states that there is an excess of process-innovating R&D efforts under Cournot competition as compared to the social optimum, while the opposite holds under Bertrand competition (Brander and Spencer, 1983; Dixon, 1985).

A more comprehensive analysis requires then the study of R&D activities that allow firms to reach markets in a more efficient way and be more competitive in serving their customers. Launhardt (1885), whose contribution has been recently acknowledged,** proposed a simple spatial duopoly model with both horizontal and vertical product differentiation. Furthermore, he paid attention to the influence of differences in transportation costs.

0* See Reinganum (1989) and Martin (2001) for exhaustive surveys.

0** See Dos Santos Ferreira and Thisse (1996) and Dos Santos Ferreira (1998), *inter alia*

He thus recognized the possibility of different form of heterogeneity among firms, associated either to location or to transportation technology. Recently, Dos Santos Ferreira and Thisse (1996) have expanded the Launhardt model by allowing firms to choose their transportation cost technologies.

So far, however, the topic of strategic investment to reduce the burden of TC costs has been rather neglected. In this paper, we will consider investments in R&D that concern mainly transport and communication needed to let the product reach the final buyer. The related R&D may be figured out as an expenditure that is going to improve the technology of the last stage of the production process (investment in the Internet, in advanced logistics, or in faster transportation technologies). We define this sort of activity transport and communication R&D (TCRD). Transport and communication costs are assumed to be of the 'iceberg' form introduced by Samuelson (1954) and widely used in trade theory (Helpman and Krugman, 1985; Krugman, 1990). When a quantity q_i of product i is produced, only a portion q_i/s_i, $s_i > 1$ of the product reaches the consumer. By investing in communication and transport-specific R&D, a firm may increase such a portion, thus enlarging its effective market share.

Lambertini, Mantovani and Rossini (2003) analyse R&D activity in transport and communication technology (TCRD), in a static Cournot duopoly. They find that a variety of equilibria may appear as a result of the different levels of TCRD efficiency, when the TCRD investment takes the form of a discrete effort. If the analysis is extended to allow for a continuous choice space, equilibria exist only if production costs are low *vis à vis* market size and transport costs. Finally, Lambertini and Rossini (2001) investigate the role of TCRD in a Cournot duopoly with trade.

We propose a differential game where firms invest to increase the percentage of the product that arrives at destination. Although several applications of differential games can be found in different fields of industrial organization (see Dockner *et al.,* 2000; Cellini and Lambertini, 2003), to the best of our knowledge the problem of TCRD investments has never been considered in such a framework.

We present a dynamic model of duopoly with differentiated products, where firms compete either in prices or quantities in a unique market and invest in TCRD. Alternatively, as usual in the international trade literature (see Spencer and Brander, 1983, *inter alia*), one can think of two firms located in different countries that export to a third market. We consider both open-loop and closed loop Nash equilibria, with a closer attention to closed loop ones. Finally, we assess the solutions emerging from the two market settings against the social optimum and proceed with a welfare appraisal. This allows to draw some conclusions on the efficiency of Bertrand and Cournot oligopolies. In particular, in line with the existing literature, we prove that Bertrand competition yields the socially optimal amount of TCRD effort, while Cournot competition involves excess investment.

The paper is organized as follows. The basic model is laid out in Section 1. Section 2 considers Bertrand competition while Section 3 deals with the Cournot setting. Section 4 deals with the social optimum and the welfare appraisal. Section 5 gives the conclusion.

2. The Model

We employ a quadratic utility function for a representative consumer as in Bowley (1924), Dixit (1979) and Singh and Vives (1984):

$$U(t) = A\widehat{q}_1(t) + A\widehat{q}_2(t) - \frac{1}{2}\left[\widehat{q}_1(t)^2 + 2\gamma\widehat{q}_1(t)\,\widehat{q}_2(t) + \widehat{q}_2(t)^2\right] \tag{1}$$

whose maximization under the budget constraint $Y(t) \geq \sum_i p_i(t)\,\widehat{q}_i(t)$, $i = \{1, 2\}$ where $Y(t)$ is nominal income, yields demand functions:

$$p_i(t) = A - \widehat{q}_i(t) - \gamma\widehat{q}_j(t) \quad \forall i \neq j,\ i,\ j = \{1, 2\} \tag{2}$$

where $\gamma \in [0, 1]$ stands for the symmetric degree of substitutability between the two goods and A is the market-size, both supposed to be constant over time; $\widehat{q}_i(t) \equiv \dfrac{q_i(t)}{s_i(t)}$ represents the share of firm i's good that is available for consumption at price $p_i(t)$, i.e., $1 - 1/s_i(t)$, with $s_i(t) > 1\ \forall t \in [0, \infty)$, indicates the fraction of firm i's good that is lost during the transportation phase (see Samuelson, 1954).

The direct demand functions can be obtained by solving (2) for $q_i(t)$:

$$q_i(t) = \frac{s_i(t)\left[A(\gamma - 1) + p_i(t) - \gamma p_j(t)\right]}{\gamma^2 - 1} \tag{3}$$

On the supply side, we make the assumption that production entails constant marginal costs, normalized to zero for the sake of simplicity. Instantaneous profits are then given by:

$$\pi_i(t) = \frac{q_i(t)}{s_i(t)} p_i(t) - \beta\,[k_i(t)]^2 \tag{4}$$

where $k_i(t)$ represents the amount of effort made by firm i at time t in order to reduce $s_i(t)$ and parameter β is an inverse measure of TCRD productivity. By reducing $s_i(t)$ through capital accumulation over time, firm i increases the fraction of $q_i(t)$ that reaches the final market. We assume that $s_i(t)$ evolves over time according to the following kinematic equation:

$$\frac{\partial s_i(t)}{\partial t} = [\alpha k_i(t) - \delta s_i(t)]\,[1 - s_i(t)] \tag{5}$$

where $\delta \in [0, 1]$ denotes the depreciation rate, which is common to both firms and constant over time; α is a time-invariant parameter positively affecting the accumulation process. It is worth noting that equation (5) accounts for the fact that $s_i(t)$ has to be always greater than one for our model to be meaningful: when $s_i(t) = 1$ what is produced corresponds to what is offered, and capital accumulation stops, otherwise an increase in the capital stock yields a decrease in $s_i(t)$ as long as $k_i(t) > \dfrac{\delta}{\alpha} s_i(t)$.

We assume that the two firms behave alternatively as either price or quantity setters. Each firm i aims at maximizing the discounted profit flow:

$$\Pi_i(t) = \int_0^\infty \pi_i(t)\,e^{-\rho t} dt \tag{6}$$

w.r.t. controls $k_i(t)$ and the market variable, either $p_i(t)$ or $q_i(t)$, under the constraint given by the state dynamics (5). The discount rate $\rho > 0$ is assumed to be constant and common to both firms. The corresponding current value Hamiltonian function of firm i is:

$$\mathcal{H}_i(t) = e^{-\rho t}\left[\pi_i(t) + \lambda_{ii}(t)\dot{s}_i(t) + \lambda_{ij}(t)\dot{s}_j(t)\right] \tag{7}$$

where $\lambda_{ii}(t) = \mu_{ii}(t)e^{\rho t}$ and $\lambda_{ij}(t) = \mu_{ij}(t)e^{\rho t}$, $\mu_{ii}(t)$ being the co-state variable associated to $s_i(t)$.

For future reference, we also define consumer surplus $CS(t) \equiv U(t) - \sum_i p_i(t)\,\widehat{q}_i(t)$. Under the symmetry assumption $\widehat{q}_i(t) = \widehat{q}_j(t) = \widehat{q}(t) = q(t)/s$, consumer surplus writes as:

$$CS(t) = \widehat{q}(t)^2(1+\gamma) = \left[\frac{q(t)}{s(t)}\right]^2(1+\gamma). \tag{8}$$

Disregarding the issue of surplus distribution among agents, we can define social welfare as $SW(t) \equiv 2\pi(t) + CS(t)$, with $\pi(t) = q(t)\{(As(t) - q(t)[1+\gamma]\}/[s(t)]^2$:

$$SW(t) = 2A\widehat{q}(t) - \widehat{q}(t)^2(1+\gamma) = 2A\frac{q(t)}{s(t)} - \left[\frac{q(t)}{s(t)}\right]^2(1+\gamma). \tag{9}$$

Note that (9) is decreasing in s as long as $q < As/(1+\gamma)$.

3. Bertrand Competition

Let us move from the case of firms competing in prices. By substituting (3) in (6) we get the relevant objective function for firm i:

$$\Pi_i(t) = \int_0^\infty e^{-\rho t}\left\{\left[\frac{A(1-\gamma) - p_i(t) + \gamma p_j(t)}{1-\gamma^2}\right]p_i(t) - \beta\left[k_i(t)\right]^2\right\}dt \tag{10}$$

Using the Hamiltonian function (7), first order conditions on controls are (we omit the indication of time for brevity):

$$\frac{\partial\mathcal{H}_i}{\partial p_i} = 0 \Rightarrow p_i = \frac{A(1-\gamma) + \gamma p_j}{2} \tag{11}$$

$$\frac{\partial\mathcal{H}_i}{\partial k_i} = 0 \Rightarrow \lambda_{ii} = \frac{2\beta k_i}{\alpha(1-s_i)} \tag{12}$$

Optimal prices can be easily derived from (11) by imposing symmetry ($p_i = p_j$):

$$p^B = \frac{A(1-\gamma)}{2-\gamma} \tag{13}$$

which can be plugged into (3) to get optimal quantities:

$$q^B = \frac{A}{(\gamma+1)(2-\gamma)} \tag{14}$$

Note, first, that (12) does not contain λ_{ij} since the present game features separated dynamics. Therefore, we set $\lambda_{ij} = 0$ for all $t \in [0, \infty)$. Secondly, (11-12) do not contain s_j, i.e., $\partial^2\mathcal{H}_i/\partial u_i \partial s_j = 0$, $u_i = p_i, k_i$. Therefore, the open-loop solution and the closed-loop memoryless solution coincide. Second order conditions are always met throughout the paper. They are omitted for brevity.

3.1. Degenerate Markov Perfect Nash Equilibrium

According to the closed-loop memoryless solution concept, we specify the firm i's co-state equation as follows:

$$-\frac{\partial \mathcal{H}_i}{\partial s_i} = \lambda_{ii} \left(\alpha k_i + \delta - 2\delta s_i \right) = \dot{\lambda}_{ii} - \rho \lambda_{ii} \tag{15}$$

along with the transversality and initial conditions:

$$\lim_{t \to \infty} \mu_i s_i = 0, \quad s_i(0) > 1. \tag{16}$$

Now, by using (12), (11) and the co-state equation (15), we write:

$$\dot{k}_i = k_i \left(\rho + \delta - \delta s_i \right) \tag{17}$$

The steady state equilibrium requires $\left\{ \dot{k}_i = 0, \dot{s}_i = 0 \right\}$, yielding:

$$k_i^{BSS} = \frac{\rho + \delta}{\alpha} > 0; \quad s_i^{BSS} = \frac{\rho + \delta}{\delta} > 1 \tag{18}$$

We are interested in the dynamics of the system formed by (17) and (5) in the $\{k_i, s_i\}$ space, which can be represented in Figure 1:

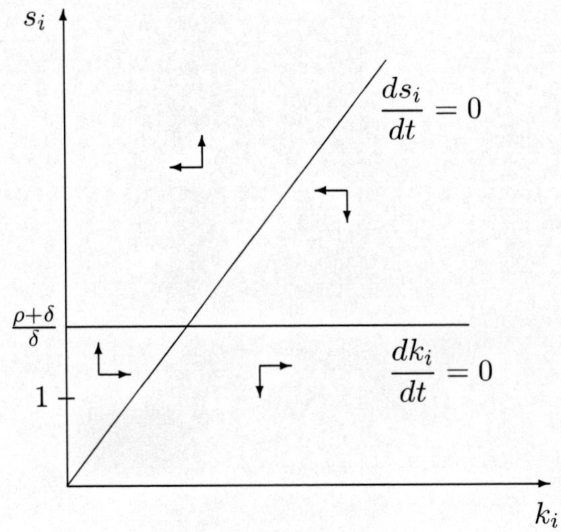

Figure 1. Phase Diagram.

The above phase diagram suggests that the equilibrium is a saddle, and that it can be approached along the north-est arm of the path. It is worth noting that if $s_i(0) >> \frac{\rho + \delta}{\delta}$, then the system never converges to the equilibrium, the reason being that the effort required to increase the share of the good arriving at destination is too costly. We are now in a position to write:

Proposition 0.1. *The steady state defined by* $\left\{k_i^{BSS}, s_i^{BSS}\right\}$ *is a saddle point.*

Proof. See Appendix 1.

As to the comparative statics of the steady state w.r.t. all involved parameters, we have the following properties: $\frac{\partial s_i^{BSS}}{\partial \rho} > 0$ and $\frac{\partial s_i^{BSS}}{\partial \delta} < 0$. First, when the rate of time preference ρ increases, firms invest more at the steady state level but a lower fraction arrives at destination. Firms are impatient to deliver a higher fraction and spend substantial amounts on TCRD without waiting for the beneficial effect coming from the accumulation dynamics. As to δ, at the steady state a higher depreciation rate increases the level of capital and consequently the fraction of output that reaches the market. To verify the intuition behind this, observe that:

$$\frac{\partial s_i}{\partial \delta} = \frac{\partial s_i}{\partial k_i} \cdot \frac{\partial k_i}{\partial \delta}$$

where:

$$\frac{\partial s_i}{\partial k_i} < 0 \text{ and } \frac{\partial k_i}{\partial \delta} > 0.$$

4. Cournot Competition

The relevant objective function for firm i in case of quantity competition is:

$$\Pi_i = \int_0^\infty e^{-\rho t} \left\{ \frac{q_i}{s_i} \left[A - \frac{q_i}{s_i} - \gamma \frac{q_j}{s_j} \right] - \beta \left[k_i\right]^2 \right\} dt \tag{19}$$

Using the Hamiltonian function (7), first order conditions on controls are:

$$\frac{\partial \mathcal{H}_i}{\partial q_i} = 0 \Rightarrow q_i = \frac{s_i \left(A s_j - \gamma q_j\right)}{2 s_j} \tag{20}$$

$$\frac{\partial \mathcal{H}_i}{\partial k_i} = 0 \Rightarrow \lambda_{ii} = \frac{2\beta k_i}{\alpha \left(1 - s_i\right)} \tag{21}$$

One can find optimal quantities by imposing symmetry ($q_i = q_j$; $s_i = s_j$) and by rearranging (20):

$$q^C = \frac{As}{2+\gamma} \tag{22}$$

Optimal prices are then given by:

$$p^C = \frac{A}{2+\gamma} \tag{23}$$

Note that, first, (21) is equivalent to (12), so it does not contain λ_{ij} (therefore, we set $\lambda_{ij} = 0$ for all $t \in [0, \infty)$). However, (20) contains s_j, i.e., the state variable of the rival, meaning that the open-loop solution and the closed-loop memoryless solution does not coincide anymore. As a consequence, we deal with both solution concepts.

4.1. Open-Loop Nash Equilibrium

Under the open-loop solution concept, we can specify the firm i's co-state equation as follows:

$$-\frac{\partial \mathcal{H}_i}{\partial s_i} = \frac{2q_i^2}{s_i^3} + \frac{q_i (\gamma q_j - A s_j)}{s_i^2 s_j} - \lambda_{ii} (\alpha k_i + \delta - 2\delta s_i) = \dot{\lambda}_{ii} - \rho \lambda_{ii} \tag{24}$$

along with the transversality and initial conditions:

$$\lim_{t\to\infty} \mu_i s_i = 0, \;\; s_i(0) > 1. \tag{25}$$

Now, by using (21), (20) and the co-state equation (24), we write:

$$\dot{k}_i = k_i (\rho + \delta - \delta s_i) \tag{26}$$

The steady state equilibrium requires $\left\{\dot{k}_i = 0, \dot{s}_i = 0\right\}$, yielding:

$$k_i^{OL} = \frac{\rho + \delta}{\alpha}; \;\; s_i^{OL} = \frac{\rho + \delta}{\delta} \tag{27}$$

It is evident that (27) are the same as (18), albeit it may be quickly checked that equilibrium profits are different. Nonetheless, such a comparison has a limited interest in that it involves the open loop solution, which is only weakly time consistent in the Cournot case.

4.2. Closed-Loop Nash Equilibrium

In order to perform a meaningful comparison between market regimes, we need to solve the game in closed-loop form, taking into account the feedback between player i's strategy and player j's state variable. This will lead to an equilibrium characterized by subgame perfection.

We specify the firm i's co-state equation:

$$-\frac{\partial \mathcal{H}_i}{\partial s_i} - \frac{\partial \mathcal{H}_i}{\partial q_j}\frac{\partial q_j^*}{\partial s_i} \equiv \Phi = \dot{\lambda}_{ii} - \rho \lambda_{ii} \tag{28}$$

with

$$\Phi = \frac{2q_i^2}{s_i^3} + \frac{q_i (\gamma q_j - A s_j)}{s_i^2 s_j} - \lambda_{ii} (\alpha k_i + \delta - 2\delta s_i) - \left(-\frac{\gamma q_i}{s_i s_j}\right)\left(\frac{\gamma q_i s_j}{2 s_i^2}\right)$$

along with the transversality and initial conditions:

$$\lim_{t\to\infty} \mu_i s_i = 0, \;\; s_i(0) > 1. \tag{29}$$

Now, by using (21), (20) and the co-state equation (28), we write:

$$\dot{k}_i = \frac{-A^2\gamma^2 (s_i - 1)\alpha + 4\beta (2+\gamma)^2 k_i s_i (\rho - \delta s_i + \delta)}{4\beta (2+\gamma)^2 s_i} \tag{30}$$

The steady state conditions $\left\{\dot{k}_i = 0,\ \dot{s}_i = 0\right\}$ yield the following real solutions:

$$k_i^{CL} = \frac{1}{6\alpha}\left[2(\delta+\rho) + \frac{3A^2\gamma^2\alpha^2 - 4\beta(2+\gamma)^2(\delta+\rho)^2}{\Psi} - \frac{\Psi}{\beta(2+\gamma)^2}\right] \tag{31}$$

$$s_i^{CL} = \frac{\alpha}{\delta}k_i^{CL} \tag{32}$$

where the expression corresponding to Ψ is provided in Appendix 2.

Proposition 0.2. *The steady state defined by* $\{k_i^{CL}, s_i^{CL}\}$ *is a saddle point.*

Proof. See Appendix 3.

However, expression (31) is hardly manageable in order to assess the relative performance of the two market regimes. Therefore, we proceed as follows. We impose $\dot{k}_i = 0$ to determine an equilibrium relation between k_i^{CL} and s_i^{CL}:

$$k_i^{CL}(s_i^{CL}) = \frac{A^2\gamma^2\left(s_i^{CL}-1\right)\alpha}{4\beta(2+\gamma)^2 s_i^{CL}\left(\rho - \delta s_i^{CL} + \delta\right)} \tag{33}$$

We are now in a position to compare the optimal open-loop and closed-loop levels of R&D investments under Cournot competition. This is equivalent to compare the closed-loop solution arising in Cournot *vis à vis* Bertrand's, given that the open-loop solution with quantities as a control variable corresponds to the closed-loop solution with prices as a control variable.

From a direct comparison between (27) and (33), we have:

Proposition 0.3. *The optimal effort in TCRD is higher under Cournot competition than under Bertrand competition.*

Proof. $k_i^{CL}(s_i) = \dfrac{A^2\gamma^2(s_i-1)\alpha}{4b(2+\gamma)^2 s_i(\rho-\delta s_i+\delta)} > 0$ in order to be economically meaningful. Since the numerator is always positive by definition, it has to be true that $\rho - \delta s_i + \delta > 0 \Rightarrow s_i < \dfrac{\rho+\delta}{\delta} \equiv s_i^{OL}$. This amounts to saying that $k_i^{CL} > k_i^{OL}$.

This result is in line with the kind of R&D activity at stake, which aims at increasing the percentage of the produced good that reaches the market. Moreover, we confirm the conventional wisdom that firms invest more when using closed-loop decision rules than open-loop ones.

5. Social Optimum and Welfare Appraisal

The aim of this section is threefold: (i) to characterize the first best solution; (ii) to compare the private optima with the social optimum; (iii) to compare the welfare generated by

Bertrand competition with that generated by Cournot competition. The objective function of an hypothetical benevolent planner is:

$$SW(t) = \int_0^{\infty} e^{-\rho t} \left\{ 2A\frac{q\,(t)}{s\,(t)} - \left[\frac{q\,(t)}{s\,(t)}\right]^2 (1+\gamma) - 2\beta\,[k\,(t)]^2 \right\} dt \tag{34}$$

to be maximized w.r.t. $q\,(t)$ and $k\,(t)$ under the dynamic constraint:

$$\frac{\partial s(t)}{\partial t} = [\alpha k(t) - \delta s(t)]\,[1 - s(t)] \tag{35}$$

The current value Hamiltonian function writes:

$$\mathcal{H}^{SP}(t) = e^{-\rho t}\left[SW(t) + \lambda(t)\dot{s}(t)\right] \tag{36}$$

First-order conditions on controls are (we omit the indication of time for brevity):

$$\frac{\partial \mathcal{H}^{SP}}{\partial q} = 0 \Rightarrow q = \frac{As}{1+\gamma} \tag{37}$$

$$\frac{\partial \mathcal{H}^{SP}}{\partial k} = 0 \Rightarrow \lambda = \frac{4\beta k}{\alpha\,(1-s)} \tag{38}$$

$$-\frac{\partial \mathcal{H}^{SP}}{\partial s} = \frac{q\,[2As - 2q\,(1+\gamma)] + \lambda s^3\,[\delta\,(1-2s) + \alpha k]}{s^3} = -\rho\lambda + \dot{\lambda} \tag{39}$$

By differentiating (38) w.r.t. time we obtain:

$$\dot{\lambda} = \frac{4\beta\left(\dot{k}\,(1-s) + k\dot{s}\right)}{\alpha\,(1-s)^2} \tag{40}$$

By plugging (40), (38) and (37) into (39) and by using (35) we have:

$$\dot{k} = k_i\,(\rho + \delta - \delta s_i) \tag{41}$$

The steady state equilibrium requires $\left\{\dot{k} = 0, \dot{s} = 0\right\}$, yielding:

$$k^{SP} = \frac{\rho+\delta}{\alpha}; \quad s^{SP} = \frac{\rho+\delta}{\delta} \tag{42}$$

On the basis of the steady state solutions previously obtained, without further proof, we can write:

Proposition 0.4. *Consider closed-loop memoryless equilibria. Under price competition, the amount of effort in TCRD is socially optimal; under quantity competition, the amount of effort in TCRD is socially excessive.*

Now, we proceed with a comparison between market regimes in terms of equilibrium welfare. Steady state welfare levels, gross of investment efforts, are:

$$SW^C = A^2 \frac{3+\gamma}{(2+\gamma)^2} \tag{43}$$

$$SW^B = A^2 \frac{3-2\gamma}{(1+\gamma)(2-\gamma)^2} \tag{44}$$

where superscript C and B stand for Cournot and Bertrand, respectively. Note that the state variable does not enter the above welfare expressions, in that it cancels out once equilibrium quantities are plugged into (9). From a direct comparison between (43) and (44) it is straightforward to conclude that $SW^B > SW^C$ always in the relevant parameter range, since the quantity firms decide to produce under Bertrand competition is always higher than the one they decide to produce under Cournot's, no matter the share of the good that reaches the market. *A fortiori*, taking into account that, as we know from Proposition 3, $k^{CL} > k^{OL} = k^{BSS}$, the welfare performance of the Bertrand game is superior to that of the Cournot game, net of steady state investments.

One could ask himself whether the fact that a planner would prefer firms to be price setters depends upon the fact that in this regime market coverage is larger. Reasoning for given quantities produced, meaning that production plans are independent of market regimes, one can investigate upon the effect of TCRD investments on welfare under both kinds of competition. As we have introduced before, (9) is decreasing in the state variable when $q < As/(1+\gamma)$. It can be easily proved that such a condition is always satisfied at equilibrium quantities (22) and (13).

Provided that firms invest more in TCRD under quantity competition, it is now true that $SW^C > SW^B$. Net of investment efforts, this inequality may indeed take either sign. A planner willing to ask firms to produce a given amount of substitute goods and ship it to the final market for consumption, is not indifferent over the choice of variable, rather, he might have a strict preference towards quantity competition. The reason lies in the incentives for firms to invest in TCRD, which are higher under Cournot than under Bertrand competition.

6. Concluding Remarks

An established result states that there is an excess of process-innovating R&D under Cournot competition, while the opposite holds under Bertrand competition (Brander and Spencer,1983; Dixon, 1985). In this paper we have taken a differential game approach to investments in transport and communication technology, confirming the acquired wisdom. Comparing the closed loop private optima with the social optimum, we have shown that the unique distortion that arises under price competition involves equilibrium prices, above marginal cost due to the presence of market power, while under Cournot competition, together with the usual downward market distortion, the amount of effort in TCRD turns out to be upward distorted. Finally, dealing with a welfare appraisal, we have argued that a planner willing to ask firms to produce a given amount of substitute goods and ship it to the final market for consumption, might have a strict preference towards quantity competition. Once accounted for different production levels, no matter the share of the good that reaches the market, the same planner would always opt for price competition.

Appendices

Appendix 1: Proof of Proposition 1

We consider the system composed by (5) in combination with the appropriate kinematics of the control variable k_i, that is, (17). We define the Jacobian matrix of the system and evaluate it at the steady state:

$$J^B = \begin{bmatrix} \frac{\partial \dot{s_i}}{\partial s_i} & \frac{\partial \dot{s_i}}{\partial k_i} \\ \frac{\partial \dot{k_i}}{\partial s_i} & \frac{\partial \dot{k_i}}{\partial k_i} \end{bmatrix}_{(s_i^{SS}, k_i^{SS})} = \begin{bmatrix} \rho & -\frac{\alpha\rho}{\delta} \\ -\frac{\delta(\rho+\delta)}{\alpha} & 0 \end{bmatrix}$$

Since the determinant of the above 2×2 matrix is $\Delta\left(J^B\right) = -\rho(\rho+\delta) < 0$, the equilibrium we have obtained is a saddle. From the phase diagram, it is clear that this saddlepoint equilibrium can be approached only along the north-est arm of the saddle path. ■

Appendix 2: Expression Ψ

$$\Psi = \left\{(3A\alpha\beta\gamma)^2(2+\gamma)^4(\rho-2\delta) - \left[2\beta(2+\gamma)^2(\rho+\delta)\right]^3 + 3\sqrt{3}\sqrt{\Sigma}\right\}^{\frac{1}{3}}$$

$$\begin{aligned} \Sigma &= \beta\left[A\alpha\beta\gamma(2+\gamma)^3\right]^2 \cdot \\ &\cdot \left[(A\alpha\gamma)^4 + \left(4\beta(2+\gamma)^2\right)^2\delta(\rho+\delta)^3 - \beta(A\alpha\gamma(2+\gamma))^2\left(\rho^2+20\rho\delta-8\delta^2\right)\right] \end{aligned}$$

Appendix 3: Proof of Proposition 2

On the basis of the two differential equations (5) and (30), we may write the Jacobian matrix:

$$J^C = \begin{bmatrix} \frac{\partial \dot{s_i}}{\partial s_i} & \frac{\partial \dot{s_i}}{\partial k_i} \\ \frac{\partial \dot{k_i}}{\partial s_i} & \frac{\partial \dot{k_i}}{\partial k_i} \end{bmatrix} = \begin{bmatrix} -\delta + 2\delta s_i - \alpha k_i & -\alpha(-1+s_i) \\ -k_i\delta - \frac{A^2\gamma^2\alpha}{4\beta(2+\gamma)^2 s_i^2} & (\rho+\delta-\delta s_i) \end{bmatrix}$$

The saddle path stability of the system requires the negativity of the determinant of the above 2×2 matrix evaluated at the steady state point. However, using expressions (31) and (32) turns out to be cumbersome. Hence, we proceed as follows. First, we compute the determinant of the Jacobian matrix:

$$\Delta\left(J^C\right) = (-\delta + 2\delta s_i - \alpha k_i)(\rho+\delta-\delta s_i) - \alpha(s_i - 1)\left[k_i\delta + \frac{A^2\gamma^2\alpha}{4\beta(2+\gamma)^2 s_i^2}\right]$$

Then, we define as $\widehat{\beta}$ the value of β such that $\Delta\left(J^C\right) = 0$:

$$\widehat{\beta} = \frac{A^2\gamma^2\alpha^2}{(\rho+\delta)(2+\gamma)^2 - 2s_i^{SS}\delta(4(1+\gamma)+\gamma^2)}$$

Saddle path stability requires:

$$\begin{aligned}\beta &> \widehat{\beta} \text{ iff } s_i^{CL} > \frac{\rho+\delta}{2\delta};\\ \beta &< \widehat{\beta} \text{ iff } s_i^{CL} < \frac{\rho+\delta}{2\delta}.\end{aligned}$$

However, the only admissible inequality is $s_i^{CL} > \frac{\rho+\delta}{2\delta}$, which, rewritten in terms of k_i^{CL}, turns out to be $k_i^{CL} > \frac{\rho+\delta}{2\alpha}$. From Proposition 3, in fact, we know that $k^{CL} > k^{OL} = \frac{\rho+\delta}{\alpha}$. This implies that $s_i^{CL} < \frac{\rho+\delta}{2\delta}$ is not admissible. Finally, straightforward algebra suffices to prove that, when $s_i^{SS} > \frac{\rho+\delta}{2\delta}$ then $\widehat{\beta} < 0$, meaning that $\beta > \widehat{\beta}$ always holds. ■

References

[1] A.L. Bowley, *The mathematical groundwork of economics*, Oxford University Press, Oxford, 1924.

[2] J. Brander, B. Spencer, Strategic commitment with R&D: the symmetric case, *Bell Journal of Economics* **14** (1983), 225–235.

[3] R. Cellini, L. Lambertini, Differential oligopoly games, In P. Bianchi and L. Lambertini (eds.) *Technology, Information and Market Dynamics: Topics in Advanced Industrial Organization*, Edward Elgar, Cheltenham, 2003.

[4] A. Dixit, A model of duopoly suggesting a theory of entry barriers, *Bell Journal of Economics* **10** (1979), 20–32.

[5] H.D. Dixon, Strategic investment in a competitive industry, *Journal of Industrial Economics* **33** (1985), 205–212.

[6] E.J. Dockner, S. Jorgensen, N.V. Long and G. Sorger, *Differential Games in Economics and Management Science*, Cambridge University Press, Cambridge, 2000.

[7] R. Dos Santos Ferreira, Launhardt and Hotelling, *Recherches Economiques de Louvain* **64** (1998), 97–109.

[8] R. Dos Santos Ferreira, J.F. Thisse, Horizontal and vertical differentiation: The Launhardt model, *International Journal of Industrial Organization*, **14** (1996), 485–506.

[9] E. Helpman, P. Krugman, *Market Structure and Foreign Trade*, MIT Press, Cambridge MA, 1985.

[10] P. Krugman, *Rethinking International Trade*, MIT Press, Cambridge, MA, 1990.

[11] L. Lambertini, A. Mantovani and G. Rossini, R&D in Transport and Communication in a Cournot Duopoly, *Rivista Internazionale di Scienze Economiche e Commerciali* **50** (2003), 185–98.

[12] L. Lambertini, G. Rossini, *Investment in transport and communication technology in a Cournot duopoly with trade*, Working Paper n. 407, Dipartimento di Scienze Economiche, Università degli Studi di Bologna, 2001.

[13] W. Launhardt, *Mathematische Begründung der Volkswirtschaftslehere*, B.G. Teubner, Leipzig, 1885.

[14] S. Martin, *Advanced Industrial Economics. Second Edition*, Blackwell, Oxford, 2001.

[15] J. Reinganum, The timing of innovation: Research, development and diffusion, In R. Schmalensee and R. Willig (eds.) *Handbook of Industrial Organization* **1**, North-Holland, Amsterdam, 1989.

[16] P. Samuelson, The transfer problem and transport costs, II: Analysis of effects of trade impediments, *Economic Journal* **64** (1954), 264–289.

[17] N. Singh, X. Vives, Price and quantity competition in a differentiated duopoly, *Rand Journal of Economics*, **15** (1984), 546–554.

[18] B.J. Spencer, J.A. Brander, International R&D rivalry and industrial strategy, *Review of Economic Studies*, **50** (1983), 707–722.

In: Mathematics, Game Theory and Algebra Compendium ISBN 978-1-60456-659-8
Editor: Jacob H. Mathias, pp. 665-679

Chapter 56

TIME–CONSISTENCY AND THE PROBLEM OF MINIMAL REDUCTION

Maria Dementieva, Pekka Neittaanm
Department of Mathematical Information Technology
University of Jyväskylä, Jyväskylä Finland
and Victor Zakharov
Faculty of Applied Mathematics and Control Processes
St. Petersburg State University
St. Petersburg, Russia

1. Introduction

In this paper we deal with cooperative utility games (TU-games). We formulate the problem of minimal reduction to apply this to the regularization of dynamic TU-games. One important problem in dynamic cooperative games is time-consistency of a solution. Time-consistency provides the optimality of the imputation at any moment of the game with respect to relevant initial states of the process. Otherwise, the absence of time-consistency involves the possibility that the previous "optimal" decision is left at some current moment of time. For more details see Chistiakov (1992), Danilov (1986), Kuzutin (1996), Petrosjan (1977), Simaan and Cruz (1973), Tolwinski (1983), *etc*.

Up to now there were two approaches to the regularization of games to provide time-consistency of a solution: the first one is based on the regularization of the optimality principle, and the second is based on a delay of total payoff to the players (Petrosjan and Zenkevich (1996), Villiger and Petrosjan (2001), Zakharov and Dementieva (2002, 2003)). In this work we use **a** reduced game due to Davis and Maschler (Davis and Maschler (1965)) and a modified Davis-Maschler reduced game to get time-consistent imputations from the core.

The work is organized as follows. Part 2 includes the preliminary information and the main definitions. In Part 3 we formulate the minimal reduction problem. In Part 4 this problem is considered under the restriction of a feasible removing coalition. Conditionally minimal coalitions are defined with respect to the core and DM-reduction (Davis and Maschler (1965)) and with respect to the subcore and MDM-reduction (Zakharov (1998)).

Part 5 is devoted to minimal reduction in a general case (when the removing coalition is chosen from the whole of the player set). Example 4 in Part 6 illustrates the application of minimal reduction in a multistage cooperative game.

2. Preliminaries

Let (N, v) be a classical cooperative TU-game; here N is the finite set of players, $v : 2^N \to \mathbb{R}$ is a characteristic function with $v(\emptyset) = 0$. Let us denote by n the number of players in the set N.

A vector $\xi = (\xi_1, \ldots, \xi_n)$ is called imputation in a cooperative game (N, v) if its components satisfy the following conditions

$$1)\ \xi_i \geq v(\{i\}),\ \forall\, i \in N,$$

$$2)\ \sum_{i \in N} \xi_i = v(N).$$

The set of all imputations is noted by $I(N, v)$.

In [6] it was proved that the following definition of the core is equivalent to the classical one ([4]).

Definition 1. *A subset $C(N, v)$ of imputation set is called core of cooperative game (N, v) if all its elements satisfy the conditions*

$$1)\ \sum_{i \in S} \xi_i \geq v(S),\ \forall\, S \subset N,$$

$$2)\ \sum_{i \in N} \xi_i = v(N).$$

To guarantee that the game is balanced , that is the nonemptyness of the core, we can use the condition based on the solution set $X^0(N, v)$ of the following linear programming problem

$$\min \sum_{i \in N} \xi_i, \tag{1}$$

$$\text{subject to} \sum_{i \in S} \xi_i \geq v(S),\ S \subset N,\ S \neq N. \tag{2}$$

The following solutions based on the set $X^0(N, v)$ are the subsets of the core.

Definition 2. *We call a set*

$$SC(v, \xi^0) = \{\xi = (\xi_1, \ldots, \xi_n) \mid \xi_i \geq \xi_i^0,\ \sum_{i \in N} \xi_i = v(N)\} =$$

$$\{\xi = (\xi_1, \ldots, \xi_n) \mid \xi = \xi^0 + \alpha\,(v(N) - \sum_{i \in N} \xi_i^0),$$

$$\text{where } \alpha = (\alpha_1, \ldots, \alpha_n):\ \ \alpha_i \geq 0, \sum_{i \in N} \alpha_i = 1\ \}$$

subcore of TU-game (N, v) *with respect to vector* $\xi^0 \in X^0(N, v)$*, if* $\sum\limits_{i \in N} \xi_i^0 \leq v(N)$*. The set*

$$GSC(N, v) = \bigcup_{\xi^0 \in X^0(N,v)} SC(v, \xi^0)$$

is called grand subcore of (N, v).

By Definitions 2. it is easy to see that the grand subcore is determined if and only if the inequality $\sum\limits_{i \in N} \xi_i^0 \leq v(N)$ holds for $\xi^0 \in X^0(N, v)$. Obviously, this condition is equivalent to $C(N, v) \neq \emptyset$ (see Definition 1.).

Given a cooperative game (N, v), a coalition $K \subset N$, any optimality principles ϕ, which is a subset of imputations, and any imputation $\xi \in \phi(N, v)$, there are various ways to define a reduced game $(N \setminus K, v_\xi^K)$ with respect to ξ. The characteristic function $v_\xi^K(S)$ of coalition $S \subset N \setminus K$ represents the total benefit that the players of S may achieve by mutual cooperation. The removed coalition K is paid according to the vector ξ.

Let us consider two special types of reduction game. The original DM-reduced game (Davis and Maschler (1965)) for a given coalition $K \subset N$, and a payoff vector ξ is the following

$$v_\xi^K(S) = \begin{cases} 0, & S = \emptyset, \\ v(N) - \sum_{j \in K} \xi_j, & S = N \setminus K, \\ \max_{R \subseteq K}\{v(S \cup R) - \sum_{j \in R} \xi_j\}, & \text{otherwise.} \end{cases}$$

The set $R = \emptyset$ is admissible in the operation of taking the maximum and in that case $\sum_{j \in R} \xi_j = 0$. An analogous remark refers to the notion of MDM-reduction.

Let us denote by $\xi_S \in \mathbb{R}^S$ the restriction of the vector $\xi \in \mathbb{R}^N$ to a set $S \subset N$.

Definition 3. *We call an optimality principle* ϕ *to be consistent with respect to the reduced game* $(N \backslash K, v_\xi^K)$*, if for all* $\xi \in \phi(N, v)$ *the following condition is fulfilled* $\xi_{N \backslash K} \in \phi(N \setminus K, v_\xi^K)$.

It is well-known that the core is consistent with respect to the DM-reduced game (see [7] and [8]). It can be shown that the subcore appears to be inconsistent with respect to the DM-reduced game and consistent with respect to a modification of the Davis-Maschler reduced game (MDM-reduced game) ([16]). For a given removing coalition $K \subset N$, $\xi^0 \in X^0(N, v)$ and a payoff vector ξ the characteristic function of the MDM-reduced game is the following

$$v_{\xi^0}^K(\cdot) = v_{\xi^0}^K(S, \xi_K) = \begin{cases} 0, & \text{if } S = \emptyset, \\ v(N) - \sum_{j \in K} \xi_j, & \text{if } S = N \setminus K, \\ \max_{R \subseteq K}\{v(S \cup R) - \sum_{j \in R} \xi_j^0\}, & \text{otherwise.} \end{cases}$$

3. Formulation of Minimal Reduction Problem

Let us consider the following problem of minimal reduction of a balanced TU-game. Let (N, v) be a cooperative game with non-empty core, vector $\xi = (\xi_1, \ldots, \xi_n)$ be an imputation, which does not belong to the core. What is the minimal coalition $K \subset N$ such that $\xi_{N \setminus K} \in C(N \setminus K, v_\xi^K)$? Obviously, this problem has a solution. Indeed, a set $N \setminus \{i\}$, for all $i \in N$, could be taken as removing coalition K. Then the characteristic function of DM-reduced game is

$$v_\xi^{N \setminus \{i\}}(\{i\}) = v(N) - \sum_{j \in N \setminus \{i\}} \xi_j = \xi_i.$$

That is ξ_i belongs to the core $C(\{i\}, v_\xi^{N \setminus \{i\}})$. Since ξ is an imputation and the set of coalitions is finite, there exists at least one minimal size coalition. This problem is called problem of minimal reduction with respect to the imputation $\xi \notin C(N, v)$ and DM-reduced game.

Let us denote by $\{T_r\}_{r=\overline{1,m}}$ the set of coalitions T_r, $r = \overline{1,m}$, for which the following inequalities hold

$$\sum_{i \in T_r} \xi_i < v(T_r), \ \ r = \overline{1,m} \tag{3}$$

$$\sum_{i \in S} \xi_i \geq v(S), \ \forall S \subset N, \ S \neq T_r, \ \ r = \overline{1,m}. \tag{4}$$

In this paper we discuss two approaches to minimal reduction. On the one hand, it is natural to find a feasible (for reduction) coalition K as a subset of $U = \bigcup_{r=\overline{1,m}} T_r$. The notion of a conditional minimal coalition is introduced (see Part 4). On the other hand, in a general case $(K \subset N)$ the number of players in a minimal coalition could be less, than in a conditionally minimal coalition (Part 5).

Analogically, we can formulate and consider the problem of minimal reduction with respect to an imputation $\xi \notin GSC(N, v)$ and MDM-reduced game.

4. Conditionally Minimal Coalition

Definition 4. *Coalition K is called conditionally minimal coalition with respect to an imputation $\xi \notin \phi(N, v)$ (ϕ is an optimality principle), coalition U and reduction rule $(N \setminus K, v_\xi^K)$ if*

1. $K \subseteq U \subset N$;
2. $\xi_{N \setminus K} \in \phi(N \setminus K, v_\xi^K)$;
3. *there is no such $K' \subset K$, that $\xi_{N \setminus K'} \in \phi(N \setminus K', v_\xi^{K'})$.*

Here U is a collection of players, which could depend on the optimality principle. In this paper we consider the following coalitions $U(C, \xi) = \bigcup_{r=1}^{m} T_r$ for the core and

$U(GSC, \xi, \xi^0) = \{i \in N : \xi_i < \xi_i^0\}$ for the grand subcore[a] .

The following example illustrates how to choose a conditionally minimal coalition with respect to $\xi \notin C(N,v)$, $U(C,\xi)$ and DM-reduced game in a cooperative game (N,v).

Example 1. *Let us consider a three person TU-game with* $v(\{i\}) = 0$ *for all* $i \in N$, *and* $0 \leq v(\{i,j\}) \leq v(N)$, $i,j \in N$. *Assume that for* $\xi = (\xi_1, \xi_2, \xi_3)$ *the following conditions hold*

$$\xi_1 + \xi_2 \geq v(\{1,2\});$$
$$\xi_1 + \xi_3 \geq v(\{1,3\});$$
$$\xi_2 + \xi_3 < v(\{2,3\}).$$

That is the vector ξ *does not belong to the core* $C(N,v)$ *because of the coalition* $\{2,3\}$ *(*$U(C,\xi) = \{2,3\}$*).*

Let us construct DM-reduced game $v_\xi^K(\cdot)$, $K = \{2,3\}$. *It is clear that* $v_\xi^K(\{1\}) = \xi_1$. *Hence, the removing of* $\{2,3\}$ *leads to imputation from the set* $C(\{1\}, v_\xi^{\{2,3\}}(\cdot))$.

If we set $K' := \{2\}$, *then the characteristic function of the reduced game is as follows*

$$v_\xi^{\{2\}}(\{1\}) = \max\{0, v\{1,2\} - \xi_2\};$$
$$v_\xi^{\{2\}}(\{3\}) = \max\{0, v\{2,3\} - \xi_2\} = v\{2,3\} - \xi_2;$$
$$v_\xi^{\{2\}}(\{1,3\}) = v(N) - \xi_2.$$

Vector ξ *is an imputation, therefore* $\xi_1 \geq 0$. *Moreover, from the inequality* $\xi_1 + \xi_2 \geq v(\{1,2\})$ *we have* $\xi_1 \geq v(\{1,2\}) - \xi_2$. *Then* $\xi_1 \geq v_\xi^{\{2\}}(\{1\})$. *However, the condition* $\xi_3 \geq v_\xi^{\{2\}}(\{3\})$ *does not apply in view of* $\xi_2 + \xi_3 < v(\{2,3\})$. *The situation is similar, if* $K' := \{3\}$.

Consequently, we obtained that the coalition $\{2,3\}$ *is conditionally minimal with respect to* $\xi \notin C(N,v)$, $U(C,\xi)$ *and DM-reduced game.*

Assume that there is such a coalition T in the set $\{T_r\}_{r=\overline{1,m}}$, that $T_r \subseteq T$ for all $r = \overline{1,m}$. In that case T is the conditionally minimal with respect to $\xi \notin C(N,v)$, $U(C,\xi)$ and DM-reduced game.

Theorem 1. *Coalition* $K(\xi) = T$ *is conditionally minimal with respect to* $\xi \notin C(N,v)$, $U(C,\xi)$ *and DM-reduced game.*

Proof. We should show that

1.$K \subseteq U(C,\xi)$,

2.$\xi_{N \setminus K(\xi)} \in C(N \setminus K(\xi), v_\xi^{K(\xi)})$,

3. there is no such $K' \subset K(\xi)$ that $\xi_{N\setminus K'} \in C(N \setminus K', v_\xi^{K'})$.

[0a] In case of an investigation of conditionally minimal coalition with respect to the grand subcore a set $U(GSC, \xi, \xi^0)$ depends on the vector ξ^0, then by "good enough" choice of $\xi^0 \in X^0(N,v)$ we can decrease a conditionally minimal coalition.

Since $U(C,\xi) = \bigcup_{r=1}^{m} T_r$ and $T_r \subseteq T$ for $r = \overline{1,m}$ then $T = U(C,\xi)$ (and $K(\xi) = U(C,\xi)$). By definition of the reduced game due to Davis and Maschler we have

$$v_\xi^{K(\xi)}(N \setminus K(\xi)) = v(N) - \sum_{i \in K(\xi)} \xi_i = \sum_{i \in N \setminus K(\xi)} \xi_i.$$

Moreover, for any $S \subsetneqq N \setminus K(\xi)$, $S \neq \emptyset$, and $R \subseteq K(\xi)$ the coalition $S \cup R$ does not include to T, hence

$$\sum_{i \in S \cup R} \xi_i \geq v(S \cup R),$$

consequently

$$\sum_{i \in S} \xi_i \geq v_\xi^{K(\xi)}.$$

That is $\xi_{N \setminus K(\xi)} \in C(N \setminus K(\xi), v_\xi^{K(\xi)})$.

Let us suppose that there exists a coalition $K' \subset K(\xi)$ and $\xi_{N \setminus K'} \in C(N \setminus K', v_\xi^{K'})$. That is for any coalition $S \subset N \setminus K'$, $S \neq N \setminus K'$, $S \neq \emptyset$ the inequality

$$\sum_{i \in S} \xi_i \geq v_\xi^{K'}(S) = \max_{R \subseteq K'} \left\{ v(S \cup R) - \sum_{i \in R} \xi_i \right\}$$

applies. Consequently, for $S = K(\xi) \setminus K'$ and $R = K'$ we can write

$$\sum_{i \in K(\xi) \setminus K'} \xi_i \geq v\left((K(\xi) \setminus K') \cup K'\right) - \sum_{i \in K'} \xi_i.$$

It is equivalent to

$$\sum_{i \in K(\xi)} \xi_i \geq v(K(\xi)).$$

This condition contradicts the original assumption.

The condition $\xi \notin C(N,v)$ implies that $\xi \notin GSC(N,v)$. The following example shows that a conditional minimal coalition with respect to $\xi \notin C(N,v)$, $U(GSC,\xi,\xi^0)$ and MDM-reduced game can be less than a conditional minimal coalition with respect to $\xi \notin C(N,v)$, $U(C,\xi)$ and DM-reduced game.

Example 2. *Let us fix a vector $\xi^0 = (\xi_1^0, \xi_2^0, \xi_3^0)$ from the solution set $X^0(N,v)$ of linear programming problem (1), (2) for the game in Example 1.. We assume that $\xi_1 \geq \xi_1^0$, $\xi_2 < \xi_2^0$, $\xi_3 \geq \xi_3^0$, then $U(GSC,\xi,\xi^0) = \{2\}$. It is evident that MDM-reduced game $(\{1\}, v_{\xi^0}^{\{2,3\}})$ with respect to ξ and ξ^0 coincides with DM-reduced game $(\{1\}, v_\xi^{\{2,3\}})$. If we take $K := \{2\}$ then*

$$v_{\xi^0}^{\{2\}}(\{1\}) = \max\{0, v\{1,2\} - \xi_2^0\};$$
$$v_{\xi^0}^{\{2\}}(\{3\}) = \max\{0, v\{2,3\} - \xi_2^0\}.$$

Since $\xi_1^0 + \xi_2^0 \geq v\{1,2\}$ and $\xi_1 \geq \xi_1^0 \geq 0$, then $\xi_1 \geq v_{\xi^0}^{\{2\}}(\{1\})$. Analogically, from $\xi_2^0 + \xi_3^0 \geq v\{2,3\}$ and $\xi_2 \geq \xi_2^0 \geq 0$ we have $\xi_3 \geq v_{\xi^0}^{\{2\}}(\{3\})$. That is the vector (ξ_1, ξ_3) belongs to the core of MDM-reduced game $(\{1,3\}, v_{\xi^0}^{\{2\}})$.

The following lemma ([16]) will be used in the proof of Theorem 2., which states a conditional minimal coalition for the grand subcore and MDM-reduction.

Lemma 1. *If* (N, v) *is a balanced TU-game and*

$$\sum_{i \in N \setminus K} \eta_i^0 \geq v(N \setminus K), \ \ \eta^0 \in X^0(N \setminus K, v_{\xi^0}^K(\cdot)), \tag{5}$$

then $\xi_{N \setminus K}^0 \in X^0(N \setminus K, v_{\xi^0}^K(\cdot))$, $\xi^0 \in X^0(N, v)$.

For some point $\xi^0 \in X^0(N, v)$ and an imputation $\xi \notin GSC(N, v)$ let us denote by $K(\xi^0, \xi)$ such a proper coalition of N, that

$$\xi_i^0 > \xi_i, \ \ i \in K(\xi^0, \xi), \tag{6}$$

$$\xi_i^0 \leq \xi_i, \ \ i \in N \setminus K(\xi^0, \xi). \tag{7}$$

Theorem 2. *Assume that* (N, v) *is balanced,* $\xi^0 \in X^0(N, v)$*,* $\xi \notin GSC(N, v)$*, and conditions (5), (6) and (7) hold. Vector* $\xi_{N \setminus K}$ *belongs to the set* $SC(v_{\xi^0}^K(\cdot), \xi_{N \setminus K}^0)$ *if* $K = K(\xi^0, \xi)$.

Proof. Directly from Lemma 1. we have $\xi_{N \setminus K}^0 \in X^0(N \setminus K, v_{\xi^0}^K(\cdot))$. Hence to show that the reduced game $(N \setminus K, v_{\xi^0}^K(\cdot))$ is balanced we should prove that the inequality

$$\sum_{i \in N \setminus K} \xi_i^0 \leq v_{\xi^0}^K(N \setminus K, \xi_K)$$

takes place. As the original game is balanced we obtain that

$$\sum_{i \in N} \xi_i^0 \leq v(N).$$

It is equivalent to

$$\sum_{i \in N \setminus K} \xi_i^0 \leq v(N) - \sum_{i \in K} \xi_i^0.$$

Taking into account the limitation $\xi_i^0 > \xi_i$ for all $i \in K$, we can write

$$\sum_{i \in N \setminus K} \xi_i^0 \leq v(N) - \sum_{i \in K} \xi_i,$$

that is

$$\sum_{i \in N \setminus K} \xi_i^0 \leq v_{\xi^0}^K(N \setminus K, \xi_K).$$

Consequently, the set $C(N \setminus K, v_{\xi^0}^K(\cdot))$ is not empty.

Since $\xi_{N \setminus K}^0 \in X^0(N \setminus K, v_{\xi^0}^K(\cdot))$ and for all $i \in N \setminus K$ the following is true $\xi_i^0 \leq \xi_i$ the inclusion $\xi_{N \setminus K} \in SC(v_{\xi^0}^K(\cdot), \xi_{N \setminus K}^0)$ holds by Definition 2..

It is easy to see that in this case coalition K is conditionally minimal with respect to the grand subcore and MDM-reduction of the original game.

5. Acceptable Coalitions

The goal of this part is to determine the collection of all such coalitions K, that $\xi_{N\setminus K} \in C(N \setminus K, v_\xi^K)$ for an imputation $\xi \notin C(N, v)$. The following example demonstrates that in a general case when we choose the removed coalition from the set of all players it can be less than a conditionally minimal coalition.

Example 3. *Let us come again to the TU-game which was described in Example 1. and consider player* 1 *as a removing coalition (*$K := \{1\}$*). DM-reduced game is as follows*

$$v_\xi^{\{1\}}(\{2,3\}) = v(N) - \xi_1,$$
$$v_\xi^{\{1\}}(\{2\}) = \max\{0, v(\{1,2\}) - \xi_1\},$$
$$v_\xi^{\{1\}}(\{3\}) = \max\{0, v(\{1,3\}) - \xi_1\}.$$

Vector (ξ_2, ξ_3) *belongs to the core of the reduced game* $(\{2,3\}, v_\xi^{\{1\}})$ *if and only if the inequalities*

$$\xi_2 \geq v_\xi^{\{1\}}(\{2\}),$$
$$\xi_3 \geq v_\xi^{\{1\}}(\{3\})$$

hold. Since ξ *is an imputation, then* $\xi_2 \geq 0$ *and* $\xi_3 \geq 0$*. That is, it remains to check the conditions*

$$\xi_2 \geq v(\{1,2\}) - \xi_1,$$
$$\xi_3 \geq v(\{1,3\}) - \xi_1.$$

It is equivalent to

$$\xi_1 + \xi_2 \geq v(\{1,2\}),$$
$$\xi_1 + \xi_3 \geq v(\{1,3\}).$$

Hence, $(\xi_2, \xi_3) \in C(\{2,3\}, v_\xi^{\{1\}})$.

Let us again fix a balanced TU-game (N, v) and an imputation $\xi \notin C(N, v)$.

Definition 5. *Coalition* K *is called acceptable for reduction with respect to* $\xi \notin C(N, v)$ *and DM-reduced game* $(N \setminus K, v_\xi^K)$*, if* $\xi_{N\setminus K} \in C(N \setminus K, v_\xi^K)$.

Theorem 3. *Coalition* $K \subset N$ *(*$|K| < |N| - 1$*) is acceptable for reduction with respect to* $\xi \notin C(N, v)$ *and DM-reduced game if and only if for all* T_r*,* $r = \overline{1,m}$ *there does not exist a couple of subcoalitions* $R \subseteq K$ *and* $S \subset N \setminus K$ *such that* $S \cup R = T_r$*,* $S \neq \emptyset$*,* $R \neq \emptyset$.

Proof. Assume that K is a coalition of N, $|K| < |N| - 1$. Davis-Maschler reduced game $(N \setminus K, v_\xi^K)$ is the following

$$v_\xi^K(N \setminus K) = v(N) - \sum_{i \in K} \xi_i, \tag{8}$$

$$v_\xi^K(S) = \max_{R \subseteq K} \left\{ v(S \cup R) - \sum_{i \in R} \xi_i \right\}, \ \forall S \subset N \setminus K,\ S \neq N \setminus K. \tag{9}$$

By Definition 5. coalition K is acceptable for reduction if and only if $\xi_{N\setminus K} \in C(N \setminus K, v_\xi^K)$, i.e. $\xi_{N\setminus K}$ is an imputation of corresponding DM-reduced game and $\sum_{i\in S} \xi_i \geq v_\xi^K(S)$ for all $S \subset N \setminus K$, $S \neq N \setminus K$. It is equal to the requirement

$$\sum_{i \in S} \xi_i \geq \max_{R \subseteq K} \left\{ v(S \cup R) - \sum_{i \in R} \xi_i \right\}, \ \forall S \subset N \setminus K,\ S \neq N \setminus K \iff$$
$$\sum_{i \in S} \xi_i \geq v(S \cup R) - \sum_{i \in R} \xi_i, \ \forall S \subset N \setminus K,\ S \neq N \setminus K,\ \forall R \subseteq K \iff$$
$$\sum_{i \in S \cup R} \xi_i \geq v(S \cup R), \ \forall S \subset N \setminus K,\ S \neq N \setminus K,\ \forall R \subseteq K. \tag{10}$$

To satisfy the condition (10) it is necessary and sufficient that $S \cup R \neq T_r$ for all $r = \overline{1, m}$, and for any proper subcoalition $S \subset N \setminus K$, and for any $R \subseteq K$ ($S \neq \emptyset$, $R \neq \emptyset$). Differently, the inequality $\sum_{i \in S \cup R} \xi_i \geq v(S \cup R)$ is tantamount to $\sum_{i \in T_r} \xi_i \geq v(T_r)$, which contradicts (3).

Corollary 1. *If $m = 1$ then there are only two coalitions $K_1 = T_1$ and $K_2 = N \setminus T_1$ which are acceptable for reduction.*

Proof. Let us assume that coalition $K \subset N$ is acceptable for reduction. We denote the intersection $K \cap T_1$ by R and $K \cap (N \setminus T_1)$ by S. Then $S \cup R \neq T_1$ if and only if either $S = T_1$ or $R = T_1$. Otherwise K is not acceptable for reduction by Theorem 3..

Corollary 2. *If $m = 2$, $T_1 \cap T_2 = \emptyset$ and $T_1 \cup T_2 = N$ then there are only two coalitions $K_1 = T_1$ and $K_2 = T_2$ which are acceptable for reduction.*

The **proof** is similar to the previous one.

Definition 6. *Coalition $K^* \in \{K\}$ is called minimal with respect to $\xi \notin C(N, v)$ and DM-reduced game $(N \setminus K, v_\xi^K)$, if $|K^*| = \min_{\{K\}} |K|$. Here $\{K\}$ is the set of all acceptable coalitions.*

Let us denote by $\min_{|\cdot|} A$ the minimal size coalition of a collection A. Corollaries 1. and 2. imply that under its conditions the minimal coalition with respect to the core and DM-reduced game is $K_{min} = \min_{|\cdot|}\{K_1, K_2\}$.

Corollary 3. *If $\bigcap_{r=1}^m T_r \neq \emptyset$ then $K = N \setminus \bigcap_{r=1}^m T_r$ is acceptable for reduction.*

Proof. Suppose that $K = N \setminus \bigcap_{r=1}^{m} T_r$. There is no proper coalition $S \subset N \setminus K$ that $S \cup R = T_r$ (for any $R \subseteq K$). Hence K is an acceptable coalition by Theorem 3..

Corollary 4. *If* $\bigcup_{r=1}^{m} T_r \neq N$ *then* $K = \bigcup_{r=1}^{m} T_r$ *is acceptable for reduction.*

Proof. Suppose that $K = \bigcup_{r=1}^{m} T_r$. Since $N \setminus K$ does not contain any player $i \in T_r$, for all $r = \overline{1, m}$, there is no such $S \subset N \setminus K$ that $S \cup R = T_r$ for any $R \subseteq K$ and $r = \overline{1, m}$. Consequently, K is acceptable for reduction.

Corollary 5. *If* $T_1 \subset T_2 \subset \cdots \subset T_m$ *then the minimal coalition is* $K_{min} = \min_{|\cdot|}\{T_m, N \setminus T_1\}$.

Proof. It immediately follows from Corollary 3. and 4. that $K_1 = N \setminus T_1$ and $K_2 = T_m$ are acceptable coalitions. Another acceptable coalitions could be presented in the forms $N \setminus T$, $T \subset T_1$, or $\overline{T}$, $T_m \subset \overline{T}$. However this coalition consists of more players than K_1 and K_2 correspondingly. This implies that the statement of the corollary is true.

The evidence of the following corollary is shown in Part 3.

Corollary 6. *The coalition* $K_i = N \setminus \{i\}$ *is acceptable for reduction for all* $i \in N$.

6. Dynamic Example

Let us denote a division of time period $t_0 < t_1 < \cdots < t_l$ by $\mathbb{T} = \{t_r\}_{r=\overline{0,l}}$, $l \in \mathcal{N}$. We call a pair $(N, v(\cdot))$ a multistage cooperative game. Here N is a finite set of players and $v : 2^N \times \mathbb{T} \mapsto \mathbb{R}$ is a characteristic function of the game, $v(\emptyset, t) = 0$ for all $t \in \mathbb{T}$, $v(S, t_l) = 0$ for all $S \subset N$. The sign $(N, v(t^*))$ means the subgame at a moment $t^* \in \mathbb{T}$. We assume that $v(N, t)$ is the decreasing function with respect to t.

Definition 7. *A vector* $\xi = (\xi_1, \ldots, \xi_n)$ *is called an imputation in a cooperative game* $(N, v(t))$ *if its components satisfy the following conditions*

$$1)\ \xi_i \geq v(\{i\}, t),\ \ \forall i \in N,$$

$$2)\ \sum_{i \in N} \xi_i = v(N, t).$$

Definition 8. *A subset* $C(N, v(t))$ *of the imputation set is called core of a cooperative game* $(N, v(t))$ *if all its elements satisfy inequalities*

$$\sum_{i \in S} \xi_i \geq v(S, t),\ \forall S \subset N.$$

Let us denote by $X^0(v(\cdot), t)$ the solution set of the linear programming problem

$$\min \sum_{i \in N} \xi_i, \tag{11}$$

$$\text{subject to} \sum_{i \in S} \xi_i \geq v(S, t),\ S \subset N,\ S \neq N \tag{12}$$

for any fixed $t \in \mathbb{T}$. Then we can introduce the notion of subcore in dynamic TU-games.

Definition 9. *We call a set*

$$SC(v(t),\xi^0(t)) = \{\xi(t) = (\xi_1(t),\ldots,\xi_n(t)) \mid \xi_i(t) \geq \xi_i^0(t),\ \sum_{i\in N}\xi_i(t) = v(N,t)\} =$$
$$\{\xi(t) = (\xi_1(t),\ldots,\xi_n(t)) \mid \xi(t) = \xi^0(t) + \alpha\,(v(N,t) - \sum_{i\in N}\xi_i^0(t)),$$
$$\text{where } \alpha = (\alpha_1,\ldots,\alpha_n):\ \alpha_i \geq 0, \sum_{i\in N}\alpha_i = 1\}$$

a subcore of multistage cooperative game $(N, v(t))$ *with respect to* $\xi^0(t)$, $\xi^0(t) \in X^0(v(\cdot),t)$, $t \in \mathbb{T}$, $\sum_{i\in N}\xi_i^0(t) \leq v(N,t)$.

In Petrosjan and Zenkevich (1996) the following definition was introduced for an optimality principle ϕ, which is given at every moment $t \in \mathbb{T}$.

Definition 10. *Suppose that* $\xi = (\xi_1,\ldots,\xi_n) \in\phi(t_0)$. *Any matrix* $\alpha = \{\alpha_{ik}\}$, $i = \overline{1,n}$, $k = \overline{0,l}$, *such that*

$$\xi_i = \sum_{k=0}^{l}\alpha_{ik},\quad \alpha_{ik} \geq 0,$$

is called the imputation distribution procedure (IDP).

We denote $\alpha_k = (\alpha_{1k},\ldots,\alpha_{nk})$ and $\alpha(t_k) = \sum_{r=0}^{k}\alpha_r$.

Definition 11. *The optimality principle* $\phi(t_0)$ *is called time-consistent if for every* $\xi \in \phi(t_0)$ *there exists IDP* α, *such that*

$$\xi - \alpha(t_k) \in \phi(t_k),\quad k = \overline{0,l}.$$

The following example demonstrates the use of a conditionally minimal coalition and a minimal coalition to provide time-consistency of given imputation.

Example 4. Let us consider the balanced multistage TU-game $(N, v(\cdot))$, $N = \{1,2,3\}$ (the characteristic function of the game see in Table 1. We assume $v(\{i\},t) = 0$ for all $t \in \mathbb{T}$.

Table 1. Multistage TU-game $(N, v(\cdot))$ for Example 4.

t	$v(\{1,2,3\},t)$	$v(\{1,2\},t)$	$v(\{2,3\},t)$	$v(\{1,3\},t)$
t_0	80	34	68	58
t_1	60	30	50	40
t_2	40	35	18	19
t_3	0	0	0	0

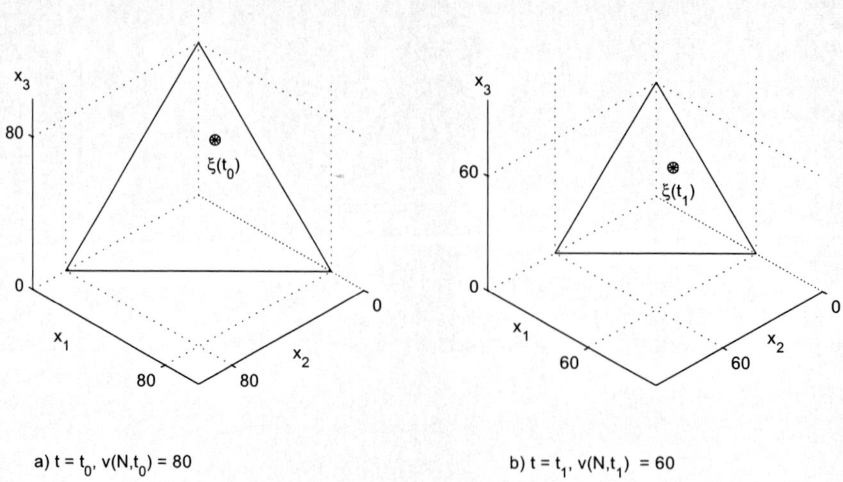

Figure 1. Imputation simplex for subgames at $t = t_0$ and at $t = t_1$.

At the moment $t = t_0$ the core, the set $X^0(\cdot)$ and the grand subcore coincide (see Figure 1a)

$$C(N, v(t_0)) = GSC(N, v(t_0)) = X^0(v(\cdot), t_0) = \{(12, 22, 46)\}.$$

Let us also compute the core, $X^0(\cdot)$ and the grand subcore at $t = t_1$ and $t = t_2$ (see Figure 1b and 2).

$$C(N, v(t_1)) = GSC(N, v(t_1)) = X^0(v(\cdot), t_1) = \{(10, 20, 30)\}$$

The core at $t = t_2$ is the convex capsule of vectors $(14, 21, 5)$, $(22, 13, 5)$, $(22, 18, 0)$ and $(19, 21, 0)$ (see Figure 2, bold line)

$$C(N, v(t_2)) = Co\{(14, 21, 5); (22, 13, 5); (22, 18, 0); (19, 21, 0)\}.$$

The set $X^0(v(\cdot), t_2)$ is the unique solution ξ^0

$$X^0(v(\cdot), t_2) = \{\xi^0\} = \{(18, 17, 1)\}.$$

Grand subcore coincides with subcore $SC(v(t_2), \xi^0)$ (see Figure 2, "dash-dot" line)

$$GSC(N, v(t_2)) = Co\{(22, 17, 1); (18, 21, 1); (18, 17, 5)\}.$$

The imputation in $C(N, v(t_0))$ is not time-consistent (i.e. there is no such IDP that can transfer the players from the point $(12, 22, 46)$ into the set $C(N, v(t_2))$) because time-consistency is broken at t_2. However, we can fix the following IDP (note, that it only provides non-negative payoffs during every step of the game):

$$\alpha = \begin{pmatrix} 0 & 2 & 0 & 10 \\ 0 & 2 & 2 & 18 \\ 0 & 16 & 18 & 12 \end{pmatrix}.$$

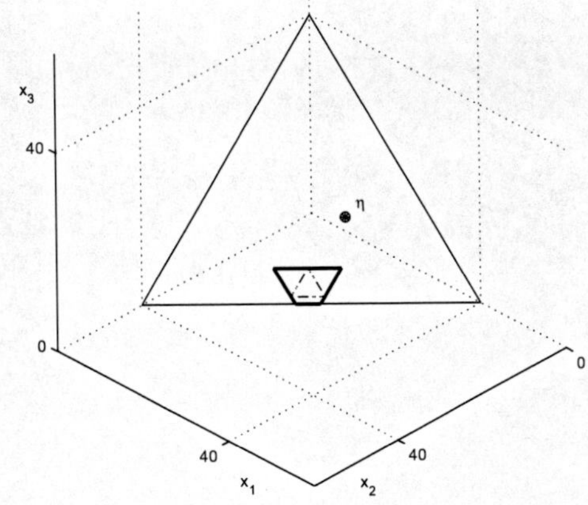

Figure 2. Imputation simplex for subgames at $t = t_2$, $v(N, t_2) = 40$.

The corresponding trajectory of the payoofs is

$$(12, 22, 46) \to (10, 20, 30) \to (10, 18, 12) \to (0, 0, 0).$$

We denote an intermediate vector $(10, 18, 12)^b$ by η. Then $U(C(N, v(t_2)), \eta) = \{1, 2\}$ and $U(GSC(t_2), \eta) = \{1\}$.

Conditional minimal coalition $K(\xi^0, \eta)$ with respect to the subcore and MDM-reduction is $\{1\}$. MDM-reduced subgame $(\{2, 3\}, v_{\xi^0}^{\{1\}}(\cdot))$ at $t = t_2$ is the following

$$v_{\xi^0}^{\{1\}}(\{2, 3\}, t_2) = 30,$$
$$v_{\xi^0}^{\{1\}}(\{2\}, t_2) = 17,$$
$$v_{\xi^0}^{\{1\}}(\{3\}, t_2) = 1.$$

The solution set of linear programming problem (1), (2) for this subgame is

$$X^0(v_{\xi^0}^{\{1\}}(\cdot), t_2) = \{(17, 1)\}.$$

A conditional minimal coalition with respect to the core and DM-reduced game is $K(\eta) = \{1, 2\}$. The characteristic function of DM-reduced game $(\{3\}, v_\eta^{\{1,2\}}(\cdot))$ is

$$v_\eta^{\{1,2\}}(\{3\}, t_2) = 12.$$

[0b] Let us underline that here we could take any vector η satisfying the assumptions $\eta \le (10, 20, 30)$ and $\eta_1 + \eta_2 + \eta_3 = 40$ and consider an appropriate IDP.

Minimal coalition K with respect to the core and DM-reduction is $\{3\}$. Corresponding DM-reduced subgame $(\{1,2\}, v_\eta^{\{3\}}(\cdot))$ is

$$v_\eta^{\{3\}}(\{1,2\},t_2) = 28,$$
$$v_\eta^{\{3\}}(\{1\},t_2) = 7,$$
$$v_\eta^{\{3\}}(\{2\},t_2) = 6.$$

Then the solution set of linear programming problem (1), (2) for the game $(\{1,2\}, v_\eta^{\{3\}}(\cdot))$ is

$$X^0(v_\eta^{\{3\}}(\cdot),t_2) = \{(7,6)\},$$

and

$$\eta_{\{1,2\}} = (10,18) \in GSC(\{1,2\}, v_\eta^{\{3\}}(t_2)) \subset C(\{1,2\}, v_\eta^{\{3\}}(t_2)).$$

This example shows the possibility of the minimal reduction use to guarantee time-consistency of a chosen solution. The number of the removed players depends on the fixed optimality principle and reduced game rule as well as on the set $U(\cdot)$.

7. Conclusion

The minimal reduction problem was considered in two ways. On the one hand, to regularize the original game it is natural to remove a coalition from the collection of the disturber players, and seeing this the notion of conditionally minimal coalition was introduced and investigated in Part 4. On the other hand, the minimal coalition in a general case could be less (in terms of number of players) than the conditionally minimal coalition, that is why the acceptable coalitions were discussed. Example 4 illustrated the method of the practical use of a minimal reduction to provide time-consistency of the imputation from the core of a multistage cooperative TU-game.

References

[1] S. Chistiakov, About construction strong time consistent solutions of cooperative differential games, *Vestnik St. Petersburg State Univ.* **1**, N 1 (1992), (in Russian)

[2] N.N. Danilov, A connection between a dynamic programming and time consistency in cooperative games, *Multistage, hierarchical, differential and cooperative games: Sborn. Nauchn. Trudov*, Kalinin, 1986, (in Russian)

[3] M. Davis, M. Maschler, The kernel of a cooperative game, *Naval Res. Logist. Quart.* **12** (1965).

[4] D.B. Gillis, *Some theorems on n-person games*, Ph. D. thesis, Princeton Univ. Press, Princeton, NJ, 1953.

[5] D. Kuzutin, One approach to the construction of time consistent optimality principles in n-person differential games, *Game Theory and Applications* (1996), NY: Nova Science Publishers.

[6] G. Owen, *Game theory*, W.B.Saunders Company, 1968.

[7] B. Peleg, On the reduced game property and its converse, *Internat. J. Game Theory* **15**, no. 3 (1986), 187–200.

[8] B. Peleg, An axiomatization of the core of market games, *Mathematics of Operations Research* **14**, no. 3 (1989), 448–456.

[9] L. Petrosjan, Stability of solutions in n-person differential games, *Bulletin of Leningrad University* **19** (1977), 46–52, (in Russian)

[10] L.A. Petrosjan, N.A. Zenkevich, *Game Theory*, World Scientific Publishing, 1996.

[11] N. Simaan, J.B.Cruz, On the Stackelberg strategy in non-zero sum games, *Journal of Optimization Theory and Applications* **11** (1973), 533–535.

[12] B. Tolwinski, A Stackelberg solution of dynamic games, *IEEE Trans. Automat. Contr.* **28** (1983), 85–93.

[13] R. Villiger, L. Petrosjan, Construction of time-consistent imputations in differential games, *Proceedings of the 2nd International Conference Logic, Game Theory and Social Choice*, (2001), St. Petersburg, Russia, 235–240.

[14] V.V. Zakharov, Consistent optimality principles in cooperative games, Optimal Control and Appendices. *International Conference Dedicated to the 90th Anniversary of L.S. Pontryagin*, 1998.

[15] V. Zakharov, M. Dementieva, Time-Consistent Imputations in Subcores of Dynamic Cooperative Games, *Proceedings of the 10th ISDG Symposium on Dynamic Games and Applications*, (2002), St.Petersburg, Russia.

[16] V. Zakharov, M. Dementieva, Multistage Cooperative Games and Problem of Time Consistency, *International Game Theory Review* (2003), (accepted).

In: Mathematics, Game Theory and Algebra Compendium ISBN 978-1-60456-659-8
Editor: Jacob H. Mathias, pp. 681-697

Chapter 57

A CONCEPT OF SOLUTION FOR A STRATEGIC COOPERATIVE GAME INVOLVING UNKNOWN PARAMETERS

Moussa Larbani
Department of Business Administration
Faculty of Economics and Management Sciences
IIUM University Kuala
Lumpu, Malaysia
and Youcef Askoura
Department of Mathematics Faculty of Sciences
University of Tizi-Ouzou
Tizi-Ouzou, Algeria

1. Introduction

Real decision making problems usually involve unknown parameters which are due to lack of information about environmental data. According to the available information on the unknown parameters, we distinguish two cases Luce and Raiffa (1957): the case of partial ignorance of the unknown parameters, which we label (PI), and the case of complete ignorance of these parameters labeled (CI). In the case of PI some characteristics of the unknown parameters are given. For example, the experts establish that they have stochastic, fuzzy or fuzzy-stochastic behavior. In the case of CI, the experts only succeed in giving a domain where the unknown parameters can vary. In both cases, it is necessary for the decision maker to use a criterion of decision making under uncertainty.

The problem of decision making involving one decision maker in presence of uncertainty in the case of CI has been studied by Hurwicz (1951); the case of PI has been investigated by Savage (1954), Bayes (1763), Arrow (1951)and Milnor (1951). The problem of multicriteria decision making under uncertainty in the case of CI has been investigated in Zhukovskii and Molostvov (1990); the case of PI has been studied in Slowinsky and Teguem (1990).

The problem of decision making involving unknown parameters and more than one decision maker in conflict is, generally, considered in the context of the theory of games with incomplete information Fudenberg and Tirole (1995). The case of PI has been investigated using the Bayesian Games by Harsany (1967). The case of CI has not received enough attention in the literature. The first important result in this field has been obtained by Zhukovskii (1994). He has introduced the concept of NS- equilibrium for a non cooperative n-person game under uncertainty. This concept is based on the concept of Nash equilibrium and the notion of weak Pareto optimal solution of a multiple criteria optimization problem Steuer (1986). Note that Zhukovskii used the weak Pareto optimality to adapt the Wald criterion (or maximin criterion) to the case of several decision makers in order to handle the aspect of decision making under uncertainty involved in the problem. Practical sufficient conditions of its existence are given in Larbani (1997). In Zhukovskii (1999), some concepts of solution for strategic cooperative games under uncertainty, in the case of CI, are introduced using the notion of imputation; the bargaining problem under uncertainty is also studied. In Larbani and Lebbah (1999), following the Zhukovskii's approach, we have introduced the concept of ZS-equilibrium for a cooperative strategic game under uncertainty in the case of CI. This concept is based on the idea that if any player deviates from his strategy in this equilibrium the remaining players can punish him; in addition to this, it is Pareto optimal and takes into account the interests of players against the possible realization of the unknown parameters.

In this paper, following the Zhukovskii's approach, we introduce the concept of αw-equilibrium for an n-person cooperative strategic game under uncertainty in the case of CI. This concept is a generalization of the concept of ZS-equilibrium. Indeed, in addition to the properties of the ZS-equilibrium, it is stable against the deviation of any coalition of players i.e. if any coalition of players deviates from its strategy in the αw-equilibrium, the remaining players can punish it. The remainder of the paper is organized as follows: the section 2 is devoted to the statement of the problem; in section 3 we introduce the concept of α w-equilibrium and describe its properties; in section 4 we study the problem of its existence; in section 5 we address the problem of its determination and finally, we end the work with a conclusion.

2. The Statement of the Problem

Let us consider the following n-person strategic game under uncertainty :

$$< I, X, Y, f(x,y) > \tag{1.1}$$

where $I = \{1, 2, ..., n\}$ is the set of players, $X = \prod_{i=1}^{n} X_i$ is the set of issues of the game, $X_i \subset \Re^{n_i}$ $n_i \geq 1$, is the set of strategies of the ith player, $i = \overline{1,n}$; Y is the set where the unknown parameters can vary $Y \subset \Re^p, p \geq 1$; $f = (f_1, f_2, ..., f_n)$,

$f_i : X \times Y \rightarrow \Re$, is the objective function of the ith player; $x = (x_1, x_2, ..., x_n) \in X$ is an issue of the game, $x_i \in X_i$ is the strategy of the ith player; y is the vector of unknown parameters, $y \in Y$. In this game we assume that the following conditions are met :

1) Players are rational and each of them knows the objective function and the set of strategies of each of the remaining players ;

2) Enforceable agreements are permitted between players of any coalition i.e. the game (2.1) is cooperative ;

3) The strategic possibilities of any coalition $K \in T$ are limited to the Cartesian product $\prod_{i \in K} X_i$, side payments are not permitted ;

4) All players only know the domain Y where the vector of the unknown parameters can take its values. Hence, we are in the case of complete ignorance.

Taking into account conditions 1)-3) and the complete ignorance of the behavior of unknown parameters , the aim of each player in this game, by choosing his strategy, is to maximize his payoff function. The game takes place as follows : when all players have chosen their respective strategies, we obtain a profile of strategies $x = (x_1, x_2, ..., x_n)$. If the unknown parameters take the value y then each player $i \in I$ receives the payoff $f_i(x, y)$.

As an example of a class of real problems which may have the game (2.1) with conditions 1)-4) as a mathematical model, let us consider the following. Let a_1,a_2,....,a_n be n firms (or countries) in competition for a given domain in economy. In this competition, firms (or countries) a_1,a_2,..,a_n have a collective agreement including conditions 2)-3) . This is on the one hand. On the other hand, this competition involves an unknown parameter which influences the payoff of each firm (or country) while all firms (or countries) know the set where it can vary but completely ignore its behavior. Note that this unknown parameter may be the behavior of a competitor not being in the community $\{a_1,a_2,..,a_n\}$, the payoff function of whom is unknown.

From this example, it is clear that a concept of solution for the game (2.1) should take into account both the aspect of cooperation between players and the aspect of decision making under uncertainty in the case of CI related to the presence of the unknown parameters. The solution we propose in the next section reflects these two aspects.

3. The Concept of the αw-Equilibrium

In this section we introduce the concept of solution we propose for the game (2.1):the αw-equilibrium. This concept is a combination of the concept of an element of the α-core Aumann (1961) of a cooperative strategic game and the notion of a weak Pareto optimal solution of a multiple criteria optimization problem Steuer (1986). The α-core is used to handle the aspect of game while the weak Pareto optimality is used to handle the aspect of decision making under uncertainty involved in the game. Before giving a formal definition of our solution, let us specify some notations.

Notation. Let $T = \{K/K \subset I, K \neq \emptyset, K \neq I\}$ be the set of coalitions of players not containing all players. For any coalition K, we will denote by $X_K = \prod_{i \in K} X_i$ the set of its strategies, by x_K an element of X_K and by $I - K$ the coalition of the remaining players. Thus, $\forall K \in T, \forall x_K \in X_K, \forall x_{I-K} \in X_{I-K}, x = (x_K, x_{I-K}) \in X$.

Definition 1.1. *A pair (x^0, y^0) is called αw-equilibrium for the game (2.1) if it satisfies the following conditions*

1) $\forall K \in T, \forall x_K \in X_K, \exists x_{I-K} \in X_{I-K}$ such that the system of inequalities

$$f_i(x^0, y^0) \leq f_i((x_K, x_{I-K}), y^0), \;\; i \in K,$$

where at least one is stringent, is impossible ;

2) $\forall x \in X$, *the system of inequalities*

$$f_i(x^0, y^0) \leq f_i(x, y^0), \quad i \in I$$

where at least one is stringent, is impossible ;

3) $\forall y \in Y$, *the system of inequalities*

$$f_i(x^0, y) < f_i(x^0, y^0), \quad i \in I$$

is impossible .

The notion of αw-equilibrium is fairly wide. Indeed ,

1) if $Y = \emptyset$ or $Y = \{y\}$ in the game (2.1), we obtain an n-person cooperative game, the α-core Aumann (1961) of which is the set of issues satisfying conditions 1) and 2) of definition 3.1;

2) if $X = \emptyset$ or $X = \{x\}$ in the game (2.1), we obtain a multiple criteria optimization problem, weak Pareto solutions of which are given by issues satisfying the condition 3) of definition 3.1.

3) The condition 1) of definition 3.1 means that for any deviation x_K of any coalition K from its αw-equilibrium bundle of strategies x_K^0 , the coalition $I - K$ of the remaining players can punish it by choosing a bundle of strategies so that in the resulting issue, the payoff of at least one player of the coalition K cannot be better than in the αw-equilibrium issue i.e. no coalition K find it " beneficial " to pursue a bundle of strategies other than its one in the αw-equilibrium x^0. Thus, the condition 1) guarantees the stability of this equilibrium .

4) The condition 2) of definition 3.1 means that x^0 is Pareto optimal for the players in the game obtained from the game (2.1) with $y = y^0$. In terms of theory of games, the condition 2) expresses the " cooperative " behavior of the players (considering the interests of all players at once).

5) The conditions 1)-2) of definition 3.1 mean that x^0 is an element of the α-core of the n-person game obtained from the game (2.1) with $y = y^0$.

6) The condition 3) of the definition 3.1 means that y^0 is a weak Pareto optimal solution for the multiple criteria optimization problem obtained from the game (2.1) with $x = x^0$. This condition expresses the collective interests of players against all possible realizations of the unknown parameters. Here it is to be noted that we adopted the "maximin" criterion or Wald criterion Luce an Raiffa (1951) of decision making under uncertainty in the case of CI adapted to the case of multiple criteria decision

making by Zhukovskii and Molostvov (1990). Our approach is based on the following consideration: since the game is cooperative, when dealing with the unknown parameters, the community of players should behave as one decision maker facing these parameters; as a result we obtain a problem of multiple criteria decision making under uncertainty in the case of CI, the criteria of which are the payoff functions of the players.

4. The Existence of the αw-Equilibrium

In this section we give sufficient conditions for the existence of the αw-equilibrium of the game (2.1). We construct a multi-valued function the fixed points of which are αw-equilibrium of the game (2.1). Then from the the Kakutani's fixed point theorem Kakutani (1941) we derive sufficent conditions for the existence the αw-equilibrium of the game (2.1).

In this section we suppose that :

A) The sets $X_i, i = \overline{1,n}$ and Y are nonempty, convex and compact.

B) The functions $f_i : X \times Y \to \Re$ are continuous on $X \times Y$.

Let us introduce the following functions :

$$\forall K \in T, \qquad \varphi_K = \sum\nolimits_{i\in K} f_i \,, \qquad \varphi = \sum\nolimits_{i\in I} f_i \tag{1.2}$$

$$\forall K \in T, \ \ \forall y \in Y, \qquad \alpha_K(y) = \max\nolimits_{x_K \in X_K} \min\nolimits_{x_{I-K} \in X_{I-K}} \varphi_K((x_K, x_{I-K}), y) \tag{1.3}$$

From assumptions A) and B), we deduce that the functions (4.2) are continuous and the functions (4.3) are well defined and continuous. For a fixed $K \in T$ and $y \in Y$ the number $\alpha_K(y)$ represents the payoff that the coalition K can guarantee against all possible strategies of the coalition of the remaining players $I - K$.

Let us introduce the following sets :

for each $y \in Y$

$$B(y) = \{\overline{x} \,/\, \overline{x} \in X, \ \alpha_K(y) \le \varphi_K(\overline{x}, y), \ \forall K \in T\} \tag{1.4}$$

$$P(y) = \left\{ \begin{array}{c} \overline{x} \,/\, \overline{x} \in X, \ \forall x \in X, \text{the system of inequalities} \\ f_i(\overline{x}, y) \le f_i(x, y), i = \overline{1,n} \text{ with } f_{i_0}(\overline{x}, y) \ne f_{i_0}(x, y) \text{ is impossible} \end{array} \right\} \tag{1.5}$$

$$C(y) = \{\overline{x} \,/\, \overline{x} \in B(y), \ \max\nolimits_{x \in B(y)} \varphi(x, y) = \varphi(\overline{x}, y)\} \tag{1.6}$$

for each $x \in X$

$$Y(x) = \left\{ \begin{array}{c} \overline{y} \mathrel{/} \overline{y} \in Y, \ \forall y \in Y, \text{the system of inequalities} \\ f_i(x,y) < f_i(x,\overline{y}), i = \overline{1,n} \text{ is impossible} \end{array} \right\} \quad (1.7)$$

for each $S \subseteq I$, $S \neq \emptyset$ and $x \in X$

$$Y_s(x) = \left\{ \overline{y} \mathrel{/} \overline{y} \in Y, \min_{y \in Y} \sum_{i \in S} f_i(x,y) = \sum_{i \in S} f_i(x,\overline{y}) \right\} \quad (1.8)$$

1- For a fixed y, an issue $\overline{x} \in B(y)$ guarantees the value $\alpha_K(y)$ for each coalition $K \in T$.
2- For a fixed $y \in Y$, $P(y)$ is the set of Pareto optimal issues for the community of players. The elements of $P(y^0)$ verify the condition 2) of definition 3.1.
3- For a fixed issue $x \in X$, $Y(x)$ is the set of weak Pareto optimal solutions of the multiple criteria optimization problem $< Y, f(x,y) >$. The elements of $Y(x^0)$ verify the condition 3) of definition 3.1.
4- Obviously we have :

$$\forall x \in X, \ \forall S \subset I, \ S \neq \emptyset, \ \ Y_S(x) \subset Y(x) \ \text{ and } \ \forall y \in Y, \ \ C(y) \subset B(y). \quad (1.9)$$

Proposition 1.2. *We have*
1) $\forall \overline{y} \in Y, \forall \overline{x} \in B(\overline{y})$, *the issue* $\overline{x}$ *verifies the condition 1) of definition 3.1 with* $y^0 = \overline{y}$;
2) if $(x^0, y^0) \in (P(y^0) \cap B(y^0)) \times Y(x^0)$ *then* (x^0, y^0) *is an* αw*-equilibrium of the game (2.1).*

Proof. 1) We have

$$\overline{x} \in B(\overline{y}) \Longleftrightarrow \forall K \subset T, \ \alpha_K(\overline{y}) \leq \varphi_K(\overline{x}, \overline{y}) \quad (1.10)$$

Suppose that the condition 1) of definition 3.1 with $y^0 = \overline{y}$ is not satisfied. Then $\exists S \in T, \exists \widetilde{x}_S \in X_S$, such that $\forall x_{I-S} \in X_{I-S} \ \ f_i(\overline{x}, \overline{y}) \leq f_i((\widetilde{x}_S, x_{I-S}), \overline{y}), \forall i \in S$ and $\exists i_0 \in S$ such that $f_{i_0}(\overline{x}, \overline{y}) < f_{i_0}((\widetilde{x}_S, x_{I-S}), \overline{y})$.It follows that $\varphi_S(\overline{x}, \overline{y}) \leq \varphi_S((\widetilde{x}_S, x_{I-S}), \overline{y})$, $\forall x_{I-S} \in X_{I-S}$.
Using the continuity of φ_S and the compacity of the sets X_{I-S}, we obtain $\varphi_S(\overline{x}, \overline{y}) < \min_{x_{T-S}} \varphi_S((\widetilde{x}_S, x_{I-S}), \overline{y})$ and $\varphi_S(\overline{x}, \overline{y}) < \max_{x_S} \min_{x_{I-S}} \varphi_S((x_S, x_{I-S}), \overline{y})$ i.e. $\varphi_S(\overline{x}, \overline{y}) < \alpha_S(\overline{y})$.
This inequality is in contradiction with (4.10). Consequently, the part 1) of the proposition is proved. The part 2) is a direct consequence of the part 1), remark 4.2 , (4.5) and (4.7).

Example 1.1. Suppose that in the game (2.1): $n = 3$, $X_i = [0,1]$, $i = \overline{1,3}$, $Y = [0,1]^3$, $x = (x_1, x_2, x_3)$, $y = (y_1, y_2, y_3)$.
$f_1(x,y) = x_1 y_1 + x_1 - x_2^2 - x_3^2$.
$f_2(x,y) = -x_2 y_2 + 2x_1 + 2x_3 - x_3^2$.
$f_3(x,y) = x_3 y_3 - x_1 - 2x_2 + x_3^2$.
We have $\alpha_1(y) = y_1 - 1$, reached at $x = (1,1,1)$, $\alpha_2(y) = 0$, reached at $x = (0,0,0)$,

$\alpha_3(y) = y_3 - 2$, reached at $x = (1,1,1)$,
$\varphi_{\{1,2\}}(x,y) = y_1x_1 - y_2x_2 + 3x_1 - x_2^2 + 2x_3 - 2x_3^2$,
$\varphi_{\{1,3\}}(x,y) = y_1x_1 + y_3x_3 - x_2^2 - 2x_2$,
$\varphi_{\{2,3\}}(x,y) = -y_2x_2 + y_3x_3 + x_1 - 2x_2 + 2x_3$,
$\alpha_{\{1,2\}}(y) = y_1 + 3$, reached at $x = (1,0,0)$.
$\alpha_{\{1,3\}}(y) = y_1 + y_3 - 3$, reached at $x = (1,1,1)$.
$\alpha_{\{2,3\}}(y) = y_3 + 2$, reached at $x = (0,0,1)$.
$\varphi(x,y) = y_1x_1 - y_2x_2 + y_3x_3 + 2x_1 - 2x_2 - x_2^2 + 2x_3 - x_3^2$.
Let us prove that $(\overline{x},\overline{y}) = ((1,0,1),(0,1,0))$ is an αw-equilibrium of the game (2.1).
i) $\overline{x} \in P(\overline{y})$.
We have $\varphi(x,\overline{y}) = -3x_2 - x_2^2 + 2x_1 + 2x_3 - x_3^2$.
We can easily verify that $\max_{x\in X}\varphi(x,\overline{y}) = \varphi(\overline{x},\overline{y})$. Thus , $\overline{x} \in P(\overline{y})$ Steuer (1986).
ii)$\overline{x} \in B(\overline{y})$.
We have $\alpha_1(\overline{y}) = -1 < \varphi_1(\overline{x},\overline{y}) = 0$, $\alpha_2(\overline{y}) = 0 < \varphi_2(\overline{x},\overline{y}) = 3$, $\alpha_3(\overline{y}) = -2 < \varphi_3(\overline{x},\overline{y}) = 0$,
$\alpha_{\{1,2\}}(\overline{y}) = 3 \leq \varphi_{\{1,2\}}(\overline{x},\overline{y}) = 3$, $\alpha_{\{1,3\}}(\overline{y}) = -3 < \varphi_{\{1,3\}}(\overline{x},\overline{y}) = 0$, and $\alpha_{\{2,3\}}(\overline{y}) = 2 < \varphi_{\{2,3\}}(\overline{x},\overline{y}) = 3$. Hence $\overline{x} \in B(\overline{y})$.
iii)$\overline{y} \in Y(\overline{x})$.
We have $\varphi(\overline{x},y) = y_1 + y_3 + 3$, then $\min_{y\in Y}\varphi(\overline{x},y) = \varphi(\overline{x},\overline{y})$. Hence $\overline{y} \in Y(\overline{x})$.
Using i)-iii) and the proposition 4.1, we deduce that $(\overline{x},\overline{y}) = ((1,0,1),(0,1,0))$ is an αw-equilibrium for the game (2.1).

Now before giving the main result of this paper, we need to prove some preliminary results.

Lemma 1.1. *Suppose that*

$$\forall y \in Y, \exists \overline{x} \in X \textit{such that} \,\forall K \in T,\ \ \alpha_K(y) < \varphi_K(\overline{x},y) \tag{1.11}$$

and $\forall y \in Y$, *the functions* $x \mapsto f_i(x,y)$ *are concave on* X. *Then*

1) $\forall y \in Y$, $B(y)$ *is nonempty, convex and compact.*

2) *The multivalued function* $y \mapsto B(y)$ *is continuous (lower and upper semi continuous).*

Proof. Let us prove 1).
– The fact that $\forall y \in Y$, $B(y)$ is nonempty is a consequence of the condition (4.11).
–$B(y)$ is compact, $\forall y \in Y$.
Let $y \in Y$, we will prove that $B(y)$ is closed. Let $(x_l)_{l\geq 1} \subset B(y)$ be a sequence such that $\lim_{l\to+\infty}x_l = x_0$. We have to prove that $x_0 \in B(y)$. We have the equivalence $\forall l \geq 1,\ x_l \in B(y) \Leftrightarrow \forall l \geq 1,\ \forall K \in T,\ \alpha_K(y) \leq \varphi_K(x_l,y)$.
Hence $\lim_{l\to+\infty}\alpha_K(y) \leq \lim_{l\to+\infty}\varphi_K(x_l,y)$, $\forall K \in T$.
Using the continuity of $\varphi_K(.)$, $\forall K \in T$, we obtain $\forall K \in T,\ \alpha_K(y) \leq \varphi_K(x_0,y)$. Hence $x_0 \in B(y)$.
Since $B(y) \subset X$ and X is compact, the last result means that $B(y)$ is compact.

$-B(y)$ is convex $\forall y \in Y$.
Let $y \in Y$, $\overline{x}$ and $\widetilde{x} \in B(y)$ and $\lambda \in [0,1]$.
The second condition of the lemma implies that the function $x \mapsto \varphi_K(x,y)$ is concave on $X, \forall y \in Y, \forall K \in T$. Hence,

$\varphi_K(\lambda\overline{x} + (1-\lambda)\widetilde{x}, y) \geq \lambda\varphi_K(\overline{x}, y) + (1-\lambda)\varphi_K(\widetilde{x}, y) \geq \lambda\alpha_K(y) + (1-\lambda)\alpha_K(y) = \alpha_K($
$\forall K \in T$, i.e. $\varphi_K(\lambda\overline{x} + (1-\lambda)\widetilde{x}, y) \geq \alpha_K(y)$, $\forall K \in T$.

It follows that $\lambda\overline{x} + (1-\lambda)\widetilde{x} \in B(y)$. Consequently, $B(y)$ is convex.

2) The multivalued function $y \mapsto B(y)$ is continuous.

– The multivalued function $y \mapsto B(y)$ is upper semi continuous.
Since Y and $B(y)$ are compact, it suffices to prove that if $y_0 \in Y$ and $(x_l)_{l\geq 1} \subset X$, $(y_l)_{l\geq 1} \subset Y$ are two sequences such that $\lim_{l\to+\infty} y_l = y_0$, $\lim_{l\to+\infty} x_l = x_0$ and $x_l \in B(y_l), \forall l \geq 1$, then $x_0 \in B(y_0)$.
Indeed, $x_l \in B(y_l), \forall l \geq 1 \Leftrightarrow \forall l \geq 1, \alpha_K(y_l) \leq \varphi_K(x_l, y_l), \forall K \in T$.
Hence, $\lim_{l\to+\infty} \alpha_K(y_l) \leq \lim_{l\to+\infty} \varphi_K(x_l, y_l), \forall K \in T$.
Using the continuity of functions $\varphi_K(.)$ and $\alpha_K(.)$, $\forall K \in T$ (remark 4.1 and assumption B), we deduce that $\alpha_K(y_0) \leq \varphi_K(x_0, y_0), \forall K \in T$, consequently $x_0 \in B(y_0)$.
–The multivalued function $y \mapsto B(y)$ is lower semi continuous on Y.
Let us introduce the set $B^\circ(y) = \{\widetilde{x} / \ \widetilde{x} \in X \text{ and } \forall K \in T, \ \alpha_K(y) < \varphi_K(\widetilde{x}, y)\}$ for any $y \in Y$.

We will prove that

$$\overline{B^\circ(y)} = B(y) \tag{1.12}$$

(where $\overline{A}$ is the closure of the set A, $A \subseteq R^s$, $s \geq 1$). Let $y \in Y$, since $B(y)$ is compact and $B^\circ(y) \subset B(y)$, we have $\overline{B^\circ(y)} \subset B(y)$.

It remains to prove that $B(y) \subset \overline{B^\circ(y)}$.

Let $\overline{x}$ be the corresponding element to y in (4.11), we have $\overline{x} \in B(y) \cap B^\circ(y)$.

Hence, it remains to prove that $B(y) - \{\overline{x}\} \subset \overline{B^\circ(y)}$, i.e. $\forall x \in B(y) - \{\overline{x}\}$, $\forall \varepsilon > 0$, $B(x,\varepsilon) \cap B^\circ(y) \neq \emptyset$, where $B(x,\varepsilon)$ is the open ball of radius ε and center x. Let $x \in B(y) - \{\overline{x}\}$ and $\varepsilon > 0$. From the convexity of $B(x,\varepsilon)$ and $B(y)$, we deduce that for fairly small λ

$$\widetilde{x} = x + \lambda(\overline{x} - x) \in B(y) \cap B(x,\varepsilon) \tag{1.13}$$

Using the concavity of functions $\varphi_K(.)$ and (4.11), we obtain $\varphi_K(\widetilde{x}, y) = \varphi_K(\lambda\overline{x} + (1-\lambda)x, y) \geq \lambda\varphi_K(\overline{x}, y) + (1-\lambda)\varphi_K(x,y) > \lambda\alpha_K(y) + (1-\lambda)\alpha_K(y) = \alpha_K(y), \forall K \in T$.
Hence, $\forall K \in T$, $\varphi_K(\widetilde{x}, y) > \alpha_K(y)$. From the definition of $B^\circ(y)$, we get $\widetilde{x} \in B^\circ(y)$.

Taking into account (4.13), we obtain : $\widetilde{x} \in B^\circ(y) \cap B(x,\varepsilon) \neq \emptyset$.

Let $y^\circ \in Y$, we will prove that the multivalued function $y \mapsto B(y)$ is lower semi continuous on y°

Since X, Y are compact and $B(y)$ is compact $\forall y \in Y$ one has only to prove that :

$$\forall \varepsilon > 0, \exists \delta > 0, \forall y \in Y, \ \|y - y_0\| < \delta, \quad B(y_0) \subset V_\varepsilon(B(y)) \tag{1.14}$$

Where $V_\varepsilon(A)$ is the $\varepsilon-$neighborhood of any bounded set A, $\forall A \subset R^s$, $s \geq 1$.
Suppose that (4.14) is not true. Then

$$\exists \varepsilon_0 > 0, \forall \delta > 0, \exists y_\delta \in Y, \ \|y_\delta - y_0\| < \delta, \quad B(y_0) \not\subset V_{\varepsilon_0}(B(y_\delta)) \tag{1.15}$$

Let $(\delta_l)_{l\geq 1}$ be a sequence of real numbers such that : $\delta_l > 0, \forall l \geq 1, \lim_{l\to+\infty}\delta_l = 0$
Using (4.15) to this sequence we can associate a sequence $(y_l)_{l\geq 1} \subset Y$, such that

$$\|y_l - y_0\| < \delta_l, \quad B(y_0) \not\subset V_{\varepsilon_0}(B(y_l)), \quad \forall l \geq 1 \tag{1.16}$$

From (4.16) we deduce that there exists a sequence $(x_l)_{l\geq 1} \subset B(y_0)$, such that :

$$x_l \notin V_{\varepsilon_0}(B(y_l)), \quad \forall l \geq 1. \tag{1.17}$$

Since $B(y_0)$ is compact, we deduce that the sequence $(x_l)_{l\geq 1}$ has a convergent subsequence which we even denote by $(x_l)_{l\geq 1}$. Then let

$$\lim_{l\to+\infty} x_l = x_0. \tag{1.18}$$

From (4.16), we obtain :

$$\lim_{l\to+\infty} y_l = y_0. \tag{1.19}$$

We have
$x_l \in B(y_0), \forall l \geq 1 \Longleftrightarrow \alpha_K(y_0) \leq \varphi_K(x_l, y_0), \forall l \geq 1, \forall K \in T.$
Using the continuity of functions $\varphi_K(.)$, $K \in T$ and (4.18), we obtain :
$\alpha_K(y_0) \leq \varphi_K(x_0, y_0), \forall K \in T$, i.e. $x_0 \in B(y_0)$.
By (4.12), we have : $x_0 \in \overline{B^\circ(y_0)}$. Hence,

$$\exists \widetilde{x} \in V_{\frac{\varepsilon_0}{2}}(x_0) \cap B^\circ(y_0) \tag{1.20}$$

Consequently, $\alpha_K(y_0) < \varphi_K(\widetilde{x}, y_0), \forall K \in T$.
Using the continuity of functions $\varphi_K(.), \alpha_K(.), K \in T$, we deduce that : $\alpha_K(y_l) < \varphi_K(\widetilde{x}, y_l), \forall K \in T$, for fairly large l. Hence, $\widetilde{x} \in B(y_l)$ for fairly large l. From (4.20) and the last inequalities, we obtain $x_0 \in V_{\frac{\varepsilon_0}{2}}(\widetilde{x})$ and $x_l \in V_{\varepsilon_0}(\widetilde{x}) \subset V_{\varepsilon_0}(B(y_0))$, for fairly large l. Hence, $x_l \in V_{\varepsilon_0}(B(y_0))$, for fairly large l. But by (4.17), we have : $x_l \notin V_{\varepsilon_0}(B(y_0)), \forall l \geq 1$. This contradiction proves that the multivalued function $y \mapsto B(y)$ is lower semi continuous on y_0. Since y_0 is arbitrarily chosen, this multivalued function is lower continuous on Y.
Thus, from the forth going we deduce that the multivalued function $y \mapsto B(y)$ is continuous on Y.

Lemma 1.2. *Suppose that conditions of lemma 4.1 are met. Then*

a) $\forall y \in Y, C(y)$ *is nonempty, convex and compact.*

b) $C(y) \subset P(y), \forall y \in Y$.

c) The multivalued function $y \mapsto C(y)$ *is upper semi continuous on* Y.

Proof. a) is a consequence of the assumptions A), B) and lemma 4.1.
b) $C(y) \subset P(y), \forall y \in Y$.
Let $\overline{x} \in C(y)$ and suppose that $\overline{x} \notin P(y)$. Then, $\exists \widetilde{x} \in X$ such that

$$f_i(\overline{x}, y) \leq f_i(\widetilde{x}, y), \forall i \in I \text{ and } \exists i_0 \in I \text{ such that } f_{i_0}(\overline{x}, y) < f_{i_0}(\widetilde{x}, y) \tag{1.21}$$

From (4.21) and the inclusion $C(y) \subset B(y)$ (remark 4.2), we deduce that if $i_0 \in K \in T$ then

$$\alpha_K(y) \leq \varphi_K(\overline{x}, y) < \varphi_K(\widetilde{x}, y). \quad (1.22)$$

Using (4.21), (4.22), (4.4), and (4.2), we obtain :

$$\widetilde{x} \in B(y) \text{ and } \varphi(\overline{x}, y) < \varphi(\widetilde{x}, y).$$

This fact is in contradiction with the fact that $\overline{x} \in C(y)$. Hence $\overline{x} \in P(y)$ and consequently $C(y) \subset P(y)$.
c) $y \mapsto C(y)$ is upper semi continuous on Y.
Since by lemma 4.1. the multivalued function $y \mapsto B(y)$ is continuous and the function $\varphi(.)$ is continuous, we deduce that the multivalued function $y \mapsto C(y)$ is upper semi continuous on Y Aubin and Ekland (1984).

Lemma 1.3. *Let S be a nonempty subset of I such that the function $y \mapsto \sum_{i\in S} f_i(x, y)$ is convex on $Y, \forall x \in X$. Then*

1) $\forall x \in X$, $Y_S(x)$ is nonempty convex and compact,

2) The multivalued function $x \mapsto Y_S(x)$ is upper semi continuous on X.

The proof of this lemma is similar to the one of lemma 4.1.
Let us now introduce the following multivalued functions

$$\begin{array}{rcl} H(.,.) & : & X \times Y \to X \times Y. \\ (x, y) & \mapsto & H(x, y) = C(y) \times Y(x). \end{array} \quad (1.23)$$

$\forall S \subset I, S \neq \emptyset,$

$$\begin{array}{rcl} H_S(.,.) & : & X \times Y \to X \times Y. \\ (x, y) & \mapsto & H_S(x, y) = C(y) \times Y_S(x). \end{array} \quad (1.24)$$

The following lemma gives the relation between the αw- equilibria of the game (2.1) and the fixed points of multivalued functions (4.23) and (4.24).

Lemma 1.4. *1) If $(x^\circ, y^\circ) \in X \times Y$ is a fixed point of a multivalued function $H_S(.,.)$, $S \subset I, S \neq \emptyset$, then (x°, y°) is an αw- equilibrium of the game (2.1).*

2) If $(x^\circ, y^\circ) \in X \times Y$ is a fixed point of a multivalued function $H(.,.)$, then (x°, y°) is an αw-equilibrium of the game (2.1).

Proof. We will prove 1), the proof of 2) is similar.
Suppose that $(x^\circ, y^\circ) \in X \times Y$ is a fixed point of the function $H_S(.,.)$, $S \subset I$, $S \neq \emptyset$. i.e. $(x^\circ, y^\circ) \in H_S(x^\circ, y^\circ)$. By (4.24), we obtain $(x^\circ, y^\circ) \in C(y^\circ) \times Y_S(x^\circ)$.

Then, $x^\circ \in C(y^\circ)$ and $y^\circ \in Y_S(x^\circ)$.
Using remark 4.2, lemma 4.2 and (4.6), we obtain:
$x^\circ \in C(y^\circ) \subset P(y^\circ) \cap B(y^\circ)$
and $y^\circ \in Y_S(x^\circ) \subset Y(x^\circ)$.
i.e. $(x^\circ, y^\circ) \in (P(y^\circ) \cap B(y^\circ)) \times Y(x^\circ)$.
By proposition 4.0, (x°, y°) is an αw-equilibrium of the game (2.1)

Let us now give the main result of this paper.

Theorem 1.1
Suppose that the following conditions are met :

1) $\forall y \in Y, \exists \overline{x} \in X$ *such that* $\alpha_K(y) < \varphi_K(\overline{x}, y), \forall K \in T$.

2) $\forall y \in Y$, *the functions* $x \mapsto f_i(x, y)$ *are concave on* X.

3) $\exists S \subset I$ *such that the function* $y \mapsto \sum_{i \in S} f_i(x, y)$ *is convex,* $\forall x \in X$.

Then, the game (2.1) has at least one α*w-equilibrium.*

Proof. From lemmas 4.2 and 4.3, we deduce that the multivalued functions $y \mapsto C(y)$ and $x \mapsto Y_S(x)$ are upper semi continuous and $\forall (x, y) \in X \times Y$, $C(x)$ and $Y_S(x)$ are nonempty convex and compact. Thus the multivalued function $(x, y) \mapsto H_S(x, y)$ is upper semi continuous on $X \times Y$ and $\forall (x, y) \in X \times Y$, $H_S(x, y)$ is nonempty, convex and compact, i.e. it verifies the conditions of the Kakutani's fixed point theorem Kakutani (1941). Consequently, it has a fixed point (x°, y°), which, by lemma 4.4, is an αw- equilibrium of the game (2.1).

If in the theorem 4.1, we substitute the condition 3) for the condition: $Y_S(x)$ is convex $\forall x \in X$, we can likewise prove the same result.

5. The Determination of αw-Equilibrium

Finding fixed points of a multi-valued function is a difficult task, hence, in the general case, the approach used in the theorem 4.1 is not adequate for the practical determination of an αw- equilibrium of the game (2.1). However, in this section we show that this approach is an effective tool for the determination of an αw- equilibrium for two classes of games: the fist one, denoted by **Class 1** is when the variables x and y are separated in the payoff functions i.e. when
$f_i(x, y) = g_i(x) + h_i(y), i = \overline{1, n}$;
the second, denoted by **Class 2**, is when the unknown parameter y has no effect on the payoff of a coalition of players i.e. $\exists S \subset I, \exists \lambda_S \in \Re^{\text{Card(S)}}, \quad \lambda_i \geq 0,\ i \in S$ and $\sum_{i \in S} \lambda_i = 1$ such that $\sum_{i \in S} \lambda_i f_i(x, y) = g_S(x), \forall (x, y) \in X \times Y$, the last relation

means that the payoff function $\sum_{i\in S}\lambda_i f_i(x,y)$ of the coalition S depends only on the variable x, in this sum the coefficient λ_i represents the power or the weight of the player i in this coalition.

Now let us study these two classes.

5.1. Class 1

We have the following corollary.

Corollary 1.1. *Suppose that in the game (2.1), the payoff functions are of the form*

$$f_i(x,y) = g_i(x) + h_i(y), i = \overline{1,n}.$$

and the following conditions are satisfied

1) *Sets* $X_i, i = \overline{1,n}$, Y *are nonempty and compact.*

2) *The functions* $x \mapsto g_i(x), y \mapsto h_i(y)$, $i = \overline{1,n}$ *are continuous.*

3) $\exists \overline{x} \in X$ *such that* $\overline{\alpha_K} = \max_{x_K\in X_K} \min_{x_{I-K}\in X_{I-K}} \overline{\varphi_K}(x_K, x_{I-K}) \le \overline{\varphi_K}(\overline{x})$, $\forall K \in T$, *where* $\overline{\varphi_K} = \sum_{i\in K} g_i$, $\forall K \in T$.
Then the game (2.1) has at least one αw*-equilibrium.*

Proof. It is similar to the one of the theorem 4.1. We have
$B(y) = B = \{\overline{x}/\ \overline{x} \in X$ and $\overline{\alpha_K} = \max_{x_K\in X_K} \min_{x_{I-K}\in X_{I-K}} \overline{\varphi_K}(x_K, x_{I-K}) \le \overline{\varphi_K}(\overline{x}), \forall K \in T\}$
$C(y) = C = \{\overline{x}/\ \overline{x} \in X$ and $\max_{x\in B} \sum_{i\in I} g_i(x) = \sum_{i\in I} g_i(\overline{x})\ \}$
$Y_S(x) = Y_S = \{\overline{y}/\ \overline{y} \in Y$ and $\min_{y\in Y} \sum_{i\in S} h_i(y) = \sum_{i\in S} h_i(\overline{y})\}$, for any $S \subset I$, $S \ne \emptyset$.
Hence,

$$S \subset I,\ S \ne \emptyset, \forall (x,y) \in X \times Y,\quad H_S(x,y) = C \times Y_S. \tag{1.25}$$

i.e. the multivalued function $(x,y) \mapsto H_S(x,y)$ is constant,$\forall S \subset I,\ S \ne \emptyset$.
Using the conditions 1)-3)of the corollary we can easily prove that $\forall S \subset I,\ S \ne \emptyset, C \times Y_S \ne \emptyset$.
Consequently, from lemma 4.4 and (5.25), we deduce that the game (2.1) has at least one αw- equilibrium.

Note that in this case, the convexity of sets $X_i, i = \overline{1,n}, Y$, the concavity of the functions $x \mapsto g_i(x)$ and the convexity of at least one function $y \mapsto h_i(y)$ are not needed. The condition 3) corollary 5.1 is less difficult than the similar one in the theorem 4.1.

The corollary 5.1 provides the following procedure for the determination of αw- equilibrium of the game (2.1).

Step 1. Calculate the numbers $\overline{\alpha_K} = \max_{x_K \in X_K} \min_{x_{I-K} \in X_{I-K}} \overline{\varphi_K}(x_K, x_{I-K}),\ K \in T$.

Step 2. Consider the multiple criteria optimization problem

$$< X, g(x) > \tag{1.26}$$

where $g = (g_1, g_2, .., g_n)$.
Determine a Pareto optimal solution $\overline{x}$ for the problem (5.26) (Miettinen, 1999) such that

$$\overline{\alpha_K} \leq \sum\nolimits_{i \in K} g_i(\overline{x}), \quad \forall K \in T$$

Note that the existence of such element $\overline{x}$ is guaranteed by conditions of corollary 5.1.

Step 3. Choose $S \subset I$, $S \neq \emptyset$, and determine $\overline{y} \in Y$ such that

$$\min_{y \in Y} \sum\nolimits_{i \in S} h_i(y) = \sum\nolimits_{i \in S} h_i(\overline{y}).$$

Then $(\overline{x}, \overline{y})$ is an αw-equilibrium for the game (2.1).

5.1. Example Suppose that in the game (2.1) : $n = 3$, $X_i = [0, 1]$, $Y = [-1, 1]^3$,
$f_1(x, y) = x_1 - x_2 - x_3^2 + y_1$,
$f_2(x, y) = -2x_1 + x_2^2 - x_3 + y_2$,
$f_3(x, y) = -3x_1 - 2x_2 + x_3^2 + y_3$,
$x = (x_1, x_2, x_3)$ and $y = (y_1, y_2, y_3)$. The conditions 1) and 2) of the corollary are obviously verified. Let us prove that the condition 3) is verified by $\overline{x} = (0, 1/2, 1/2) \in X$.
We have $\overline{\varphi_K}(x) = \sum_{i \in K} g_i(x),\ K \in T$.
$\overline{\alpha}_1 = -1,\ \overline{\alpha}_2 = -2,\ \overline{\alpha}_3 = -2,\ \overline{\alpha}_{\{1,2\}} = -2,\ \overline{\alpha}_{\{1,3\}} = -3,\ \overline{\alpha}_{\{2,3\}} = -5.$
$\overline{\varphi}_1(0, 1/2, 1/2) = -3/4 > \overline{\alpha}_1 = -1,$
$\overline{\varphi}_2(0, 1/2, 1/2) = -1/4 > \overline{\alpha}_2 = -2,$
$\overline{\varphi}_3(0, 1/2, 1/2) = -3/4 > \overline{\alpha}_3 = -4,$
$\overline{\varphi}_{\{1,2\}}(0, 1/2, 1/2) = -1 > \overline{\alpha}_{\{1,2\}} = -2,$
$\overline{\varphi}_{\{1,3\}}(0, 1/2, 1/2) = -3/2 > \overline{\alpha}_{\{1,3\}} = -3,$
$\overline{\varphi}_{\{2,3\}}(0, 1/2, 1/2) = -1 > \overline{\alpha}_{\{2,3\}} = -5.$
Hence $\overline{x} = (0, 1/2, 1/2) \in X$ véverifies the condition 3) of corollary 5.1.

Let us use the described above procedure.

Step 1. $\overline{\alpha_K},\ K \in T$ have been calculated above.

Step 2. $\overline{\varphi}(x) = \sum_{i \in I} g_i(x) = -(4x_1 + 3x_2 - x_2^2 + x_3)$.
It is well known that the elements of the set

$$\overline{P} = \left\{ \overline{x} / \ \overline{x} \in X \text{ and } \max_{x \in X} \overline{\varphi}(x) = \overline{\varphi(\overline{x})} \right\}$$

are Pareto optimal for the problem (5.26) Miettinen (1999). We have : $\overline{\varphi}(x) \leq 0, \forall x \in X$ and $\overline{\varphi}(0,0,0) = 0$. Hence $\overline{P} = \{(0,0,0)\}$. Furthermore, we have $\forall K \in T, \overline{\varphi_K}(0,0,0) > \overline{\alpha_K}$. Then $\overline{x} = (0,0,0)$ is an element satisfying the conditions of the step 2.

Step 3. Let $S = I$, we have $\min_{y\in Y}\sum_{i\in S} h_i(y) = \min_{y\in Y}\sum_{i\in S} y_i = \sum_{i\in S} h_i(-1,-1,-1) = -3$.
Then $(\overline{x}, \overline{y}) = ((0,0,0),(-1,-1,-1))$ is an αw-equilibrium of the considered game.

5.2. Class 2

We have the following corollary

Corollary 1.2. *Suppose that the game (2.1) is of Class 2 and the following conditions are satisfied*

1) *Sets $X_i, i = \overline{1,n}$, Y are nonempty and compact and the payoff functions are continuous.*
2) *There exists $y^0 \in Y$ and $\overline{x} \in X$ such that $\alpha_K(y^0) \leq \varphi_K(\overline{x}, y^0), \forall K \in T$.*

Then the game (2.1) has at least one αw-equilibrium.

Proof. From the condition 2) of the corollary we deduce that $B(y^0) \neq \emptyset$. Since the game is of Class 2, we have $\exists S \subset I, \exists \lambda_S \in \Re^{Card(S)}, \lambda_i \geq 0, i \in S$ such that $\sum_{i\in S}\lambda_i = 1$ and $\sum_{i\in S}\lambda_i f_i(x,y) = g_S(x), \forall (x,y) \in X \times Y$. Then using the continuity of the payoff functions and the compacity of the sets $X_i, i = \overline{1,n}$ and Y(conditions 1)), we deduce that $B(y^0)$ is compact and $C(y^0) \neq \emptyset$. Let x^0 be any issue in $C(y^0)$. From the equality $\sum_{i\in S}\lambda_i f_i(x,y) = g_S(x), \forall (x,y) \in X \times Y$ we deduce that the function $\sum_{i\in S}\lambda_i f_i(x,y)$ doesn't depend on y, hence we have $\{= y/ = y \in Y, \min_{y\in Y}\sum_{i\in S}\lambda_i f_i(x,y) = \sum_{i\in S}\lambda_i f_i(x, = y)\} = Y$. Since we have the inclusion $\{= y/ = y \in Y, \min_{y\in Y}\sum_{i\in S}\lambda_i f_i(x,y) = \sum_{i\in S}\lambda_i f_i(x, = y)\} \subset Y(x)$ Miettinen (1999), then $Y(x) = Y, \forall x \in X$. Consequently, $(x^0, y^0) \in H(x^0, y^0) = C(y^0) \times Y(x^0)$ i.e. (x^0, y^0) is a fixed point of the multivalued function $H(x,y)$ (defined in (2.23)) then, according to lemma 4.4, (x^0, y^0) is an α w-equilibrium of the game (2.1).

Remark 1.1. Note that in corollary 5.2 , the convexity of sets of strategies $X_i, i = \overline{1,n}, Y$, and the concavity or convexity of the payoff the functions are not required. The condition 2) of corollary 5.2 is much more easier to verify than the condition 1) in theorem 4.1 for we have to verify it only in one value y^0 of the unknown parameter.

Suppose that sets $X_i, i = \overline{1,n}$, Y are nonempty and compact and the payoff functions are continuous. The corollary 5.2 provides the following effective procedure for the determination of an α w-equilibrium of the game (2.1).

Step 1. Find out $S \subset I$ and $\lambda_S \in \Re^{Card(S)}, \lambda_i \geq 0, i \in S$ and $\sum_{i \in S} \lambda_i = 1$ such that $\sum_{i \in S} \lambda_i f_i(x, y) = g_i(x), \forall (x, y) \in X \times Y$;

Step 2. Find out $y^0 \in Y$ such that $B(y^0) \neq \emptyset$.

Step 3. Find out a solution of the following optimization problem

$$\max_{x \in B(y^0)} \varphi(x, y^0).$$

Let x^0 be a solution of this problem then (x^0, y^0) is an α w-equilibrium of the game (2.1).

Now let us illustrate this procedure by an example.

Example 5.2. Suppose that in the game (2.1): $n = 3$, $X_i = [0, 1]$, $i = \overline{1, 3}$, $Y = [0, 1]^3$, $x = (x_1, x_2, x_3)$, $y = (y_1, y_2, y_3)$.
$f_1(x, y) = x_1 y_1 + x_1 - x_2^2 - x_3^2 + 1/2 x_2 y_2$.
$f_2(x, y) = -x_2 y_2 + 2x_1 + 2x_3 - x_3^2 - 2x_1 y_1$.
$f_3(x, y) = x_3 y_3 - x_1 - 2x_2 + x_3^2$.
It is obvious that sets $X_i, i = \overline{1, n}$, Y are non empty and compact and the payoff functions are continuous. Next, we have $2/3 f_1(x, y) + 1/3 f_2(x, y) = 2/3(x_1 - x_2^2 - x_3^2) + 1/3(2x_1 + 2x_3 - x_3^2), \forall (x, y) \in X \times Y$, hence this game is of the Class II with $S = \{1, 2\}$ and $\lambda_1 = 2/3$, $\lambda_2 = 1/3$. Let us verify the condition of the step 2 with $y^0 = 0$. we have $\alpha_1(0) = -1$, reached at $x = (1, 1, 1)$, $\alpha_2(0) = 0$, reached at $x = (0, 0, 0)$, $\alpha_3(0) = -2$, reached at $x = (1, 1, 1)$,.
$\varphi_{\{1,2\}}(x, 0) = 3x_1 - x_2^2 + 2x_3 - 2x_3^2$,
$\varphi_{\{1,3\}}(x, 0) = -x_2^2 - 2x_2$,
$\varphi_{\{2,3\}}(x, 0) = x_1 - 2x_2 + 2x_3$,
$\alpha_{\{1,2\}}(0) = 3$, reached at $x = (1, 0, 0)$.
$\alpha_{\{1,3\}}(0) = -3$, reached at $x = (1, 1, 1)$.
$\alpha_{\{2,3\}}(0) = 2$, reached at $x = (0, 0, 1)$.
$\varphi(x, 0) = 2x_1 - 2x_2 - x_2^2 + 2x_3 - x_3^2$.
We can easily verify that $\overline{x} = (1, 0, 1)$ verifies $\alpha_K(0) \leq \varphi_K(\overline{x}, 0), \forall K \in T$. This means $B(0) \neq \emptyset$. Next, it is easy to verify that $\max_{x \in X} \varphi(x, 0) = \varphi((1, 0, 1), 0)$, since $(1, 0, 1)$ is also in $B(0)$, then it is also a solution of the optimization problem $\max_{x \in B(0)} \varphi(x, 0)$. Thus,$(x, y) = ((1, 0, 1), (0, 0, 0))$ is an α w-equilibrium of this game.

6. Conclusion

In this paper we have introduced the concept of αw-equilibrium for the cooperative strategic game under uncertainty (2.1). Then we described some of its properties and provided sufficient conditions for its existence through theorem 4.1. From this theorem, we deduced an adequate procedure for the practical determination of this equilibrium for two special classes of games. We think that the problem of the existence of αw-equilibrium when the payoff functions are of the general form is a potential direction of research.

References

[1] K.J. Arrow, Alternative Approaches to the Theory of Choice in Risk-Taking situations, *Econometrica* **19** (1951), 404–437.

[2] Aubin, Ekland, *Non linear Analysis and Application*, John Willey and Sons, New York, 1984.

[3] R.J. Aumann,The Core of a Cooperative Game without Side Payments, *Trans. Amer. Math. Soc.* **98** (1961), 539–552.

[4] T. Bayes, *An essay Towards Solving a Problem on the Doctrine of Chances*, Philosophical Translations, 1763.

[5] D. Fudenberg, J. Tirole, *Game Theory*, The MIT Press, Cambridge, Massachusetts, London, England, 1995.

[6] J.C. Harsanyi, Games with Incomplete Information Played by Bayesian Players, I, The Basic Model, *Management Science* **14** (1967), 317–334.

[7] L. Hurwicz, Optimality Criteria for Decision Making under Ignorance, Cowle Commission, Discussion Paper, *Statistics* **370** (1951).

[8] S. Kakutani,A generalization of Browers fixed point theorem, *Duke Math. Journal* **8** (1941), 457–458.

[9] M. Larbani, About the Existence of the NS-Equilibrium for a Non-cooperative Game under Uncertainty, in *Advances in Multiobjective Programming and Goal Programming*, R. Caballero and R. Steuer (eds), Springer Verlag, Berlin, 1997 .

[10] M. Larbani, Lebbah, A concept of Equilibrium for a Game under Uncertainty, *European Journal of Operational Research* **117**, no. 1 (1999), 145–156.

[11] R.D. Luce, H. Raiffa, *Games and Decisions*, John Wiley and sons INC New York, 1951.

[12] K. Miettinen, *Multiobjective Nonlinear Optimization* , Kluwer Academic Publisher, North Holland, 1999.

[13] J.H. Milnor, *Games Against Nature*, Research Memorandum RM-679, The Rand Corporation, Santa Monica, 1951.

[14] H. Moulin, *Game Theory for the Social Sciences*, New York ,University Press, 1982.

[15] J. Nash, Non-cooperative games, *Ann. Math.* **54**, no.5 (1951), 286–295.

[16] L.J. Savage, *The Foundations of Statistics*, John Wiley & Sons New York And Chapman & Hall London, 1954.

[17] R. Slowinski, J. Teghem, *Stochastic Versus Fuzzy Approaches to Multiobjective Mathematical Programming under Uncertainty*, Kluwer Academic Publishers, 1990.

[18] R. Steuer, *Multiple Criteria Optimization : Theory , Computation and Application*, Wiley, New York, 1986.

[19] V.I. Zhukovskii, *Cooperative Games under Uncertainty and Their Applications*, Editorial, Moscow, 1999.

[20] V.I. Zhukovskii, V.S. Molostvov, *Multiple Criteria Optimization under Incomplete Information*, International Research Institute for Management Sciences, Moscow, 1990.

[21] V.I. Zhukovskii, A.A. Tchikry, *Linear–Quadratic Differential Games*, Naukova Dumka, Kiev, 1994.

In: Mathematics, Game Theory and Algebra Compendium ISBN 978-1-60456-659-8
Editor: Jacob H. Mathias, pp. 699-703

Chapter 58

ON A DISCRETE ARBITRATION PROCEDURE IN THREE POINTS

Vladimir V. Mazalov
Institute of Applied Mathematical Research
Karelian Research Centre, Petrozavodsk, Russia
Alexander E. Mentcher and Julia S. Tokareva
Zabaikalie State Pedagogical University, Chita, Russia

1. Introduction

We consider a non-cooperative zero-sum game related with a model of final-offer arbitration procedure where two players L and M, interpreted here as Labour and Management, respectively, have a dispute on an improvement in the wage rate. Player L makes an offer x, and player M - an offer y. We shall assume that x and y are arbitrary real numbers from the interval $[A, B]$.

To solve the conflict we use the so-called final-offer arbitration scheme [1-3] developed by Farber (1980). If $x \leq y$, there is no conflict, and the players agree on a payoff equal to $(x+y)/2$. If otherwise, $x > y$, the parties call in the arbitrator (A). Assume that the arbitrator has a settlement he would like to impose, denoted by α. Then, after observing the offers, x and y, the arbitrator simply chooses the offer that is closer to α. We suppose that α is a random variable. Assume, that the Manager wants to minimize the expected wage settlement imposed by the Arbitrator and the Labour wants to maximize it.

If $\alpha = a$ almost sure it is evident that the equilibrium is the pair of strategies (a, a). If α is a random variable with continuous distribution the equilibrium often consists of pure strategies [2-3]. If the distribution support of α is concentrated in two points the solutions were derived in [4-5]. In this paper we analyse a case where the arbitrator's settlement is concentrated in three points a_1, a_2 and a_3.

2. Problem Statement

For simplicity suppose that α is a random variable that assumes the values $a_1 = -1, a_2 = 0$ and $a_3 = 1$ with equal probabilities $p = 1/3$. A non-cooperative game where the strategies of players L and M are arbitrary numbers $x, y \in [-a, a]$ is considered. The payoff in the game has form $H(x, y) = EH_\alpha(x, y)$, where

$$H_\alpha(x,y) = \begin{cases} \frac{x+y}{2}, & if \quad x \le y \\ x, & if \quad x > y, |x-\alpha| < |y-\alpha| \\ y, & if \quad x > y, |x-\alpha| > |y-\alpha| \\ a, & if \quad x > y, |x-\alpha| = |y-\alpha| \end{cases} \tag{1}$$

We find that the equilibrium in the game lies among the mixed strategies. Denote $f(x)$ and $g(y)$ as player L's and M's mixed strategies respectively. Suppose that the distribution $g(y)(f(x))$ support lies on the negative (positive) semiaxis. That is,

$$f(x) \ge 0, x \in [0,a], \int_0^a f(x)dx = 1, \quad g(y) \ge 0, y \in [-a,0], \int_{-a}^0 g(y)dy = 1.$$

By symmetry, it follows that the value of the game is equal to zero, and the optimum strategies must be symmetric in respect to the axis of ordinate, i.e. $g(y) = f(-y)$. It therefore suffices to build an optimum strategy for one of the players, e.g. L.

3. Optimal Strategies

Theorem. *For* $a \in (0, 8/3]$ *the optimal strategy is*

$$f(x) = \begin{cases} 0, & 0 \le x < \frac{a}{4} \\ \frac{\sqrt{a}}{2\sqrt{x^3}}, & \frac{a}{4} \le x \le a, \end{cases} \tag{2}$$

and for $a \in (8/3, \infty)$

$$f(x) = \begin{cases} 0, & 0 \le x < \frac{2}{3} \\ \sqrt{\frac{2}{3}} \frac{1}{\sqrt{x^3}}, & \frac{2}{3} \le x \le \frac{8}{3}, \\ 0, & \frac{8}{3} < x \le a \end{cases} \tag{3}$$

Proof. First, consider the case $a \in (0, 2]$. According to (1) the payoff of M for $y \in [-a, 0]$ is equal to

$$H(f,y) = \frac{1}{3}\int_0^a yf(x)dx + \frac{1}{3}\left(\int_0^{-y} xf(x)dx + \int_{-y}^{a} yf(x)dx\right) + \frac{1}{3}\int_0^a xf(x)dx.$$

We shall be looking the strategy f in the following form

$$f(x) = \begin{cases} 0, & 0 \le x < \alpha \\ \varphi(x), & \alpha \le x \le \beta, \\ 0, & \beta < x \le a, \end{cases} \tag{4}$$

where $\varphi(x) > 0, \quad x \in [\alpha, \beta]$ and φ has a continuous derivative in (α, β).

The strategy (4) will be optimal if $H(f,y) = 0$ for $y \in [-\beta, -\alpha]$ and $H(f,y) \geq 0$ for $y \in [-a, -\beta) \cup (-\alpha, 0]$. Notice that $H(f,0) = \frac{1}{3}\int_0^a xf(x)dx > 0$.

By $H(f,-\alpha) = H(f,-\beta) = 0$ it yields $\beta = 4\alpha$ and $\int_\alpha^\beta x\varphi(x)dx = 2\alpha$. Evidently, $0 < \alpha \leq \frac{a}{4}$. But, $H(f,-a) = \frac{1}{3}[-a + 4\alpha]$. Consequently, $H(f,a) \geq 0$ iff $a \leq 4\alpha$. Hence, $\alpha = \frac{a}{4}$ and $\beta = a$.

Let us find the function $\varphi(x)$. The condition $H(f,y) = 0, \quad y \in [\beta, -\alpha]$ yields $H'(f,y) = H''(f,y) = 0$. So,

$$H'(f,y) = 1 + 2yf(-y) + \int_{-y}^{2} f(x)dx = 0, H''(y) = 3f(-y) - 2yf'(-y) = 0.$$

Letting $y = -x$ we obtain the differential equation

$$3f(x) + 2xf'(x) = 0. \tag{5}$$

The solution is

$$f(x) = \frac{c}{\sqrt{x^3}}. \tag{6}$$

Because,

$$1 = \int_0^a f(x)dx = \int_{a/4}^a \frac{c}{\sqrt{x^3}} = \frac{2c}{\sqrt{a}},$$

we find c:

$$c = \frac{\sqrt{a}}{2}.$$

Thus,

$$f(x) = \begin{cases} 0, & 0 \leq x < a/4 \\ \frac{\sqrt{a}}{2\sqrt{x^3}}, & a/4 \leq x \leq a. \end{cases}$$

Let us check the optimality conditions. For $y \in [-a, -a/4]$ we have

$$3H(f,y) = y + \int_{a/4}^{-y} \frac{\sqrt{a}}{2\sqrt{x}}dx + y\int_{-y}^{a} \frac{\sqrt{a}}{2\sqrt{x^3}}dx + \int_{a/4}^{a} \frac{\sqrt{a}}{2\sqrt{x}}dx$$

$$= y + \sqrt{a}\sqrt{-y} - \frac{a}{2} - y - \sqrt{a}\sqrt{-y} + \frac{a}{2} = 0.$$

For $y \in (-a/4, 0]$

$$H(f,y) = y + y\int_{a/4}^{a} \frac{\sqrt{a}}{2\sqrt{x^3}}dx + \frac{a}{2} = 2(y + \frac{a}{4}) > 0.$$

It proves the optimality of (2).

Let $a \in (2, \frac{8}{3}]$. Consider $H(f,y)$ for $y \in [-a, -a/4]$ with f satisfying (2). Because the support of f is $[a/4, a]$ and $a \leq \frac{8}{3}$ then $-1 + (-1 - y) \leq a/4$ and $-y \geq a/4$ for all $y \in [-a, -a/4]$.

Hence, for $y \in [-a, -a/4]$

$$3H(f,y) = \int_{\frac{a}{4}}^{a} yf(x)dx + \left(\int_{\frac{a}{4}}^{-y} xf(x)dx + \int_{-y}^{a} yf(x)dx \right) + \int_{\frac{a}{4}}^{a} xf(x)dx.$$

Differentiating it we obtain again the differential equation (5). It's solution is $f(x)$ of the form (2). Then, arguing like above we show that $H(f,y) > 0$ for $y \in (-a/4, 0]$. So, (2) is optimal also for $a \in (2, \frac{8}{3}]$.

Suppose now that $a \in (\frac{8}{3}, \infty)$. In this case the form of $H(f,y)$ is more complicated. As an example let say about the case $a = 4$. For $y \in [-4,-2]$

$$3H(f,y) = \left[\int_0^{-2-y} xf(x)dx + \int_{-2-y}^{4} yf(x)dx\right] +$$

$$\left[\int_0^{-y} xf(x)dx + \int_{-y}^{4} yf(x)dx\right] + \int_0^4 xf(x)dx, \tag{7}$$

for $y \in [-2,0]$

$$3H(f,y) = \int_0^4 yf(x)dx + \left[\int_0^{-y} xf(x)dx + \int_{-y}^{4} yf(x)dx\right] +$$

$$\left[\int_0^{2-y} xf(x)dx + \int_{2-y}^{4} yf(x)dx\right], \tag{8}$$

We find f in the form (4) with $\beta = \alpha + 2$. The conditions $H(f,\beta) = H(f,\alpha) = 0$ gives us

$$\int_\alpha^\beta xf(x)dx = 2\alpha = \frac{\beta}{2}.$$

Consequently, $\beta = 4\alpha$ and with $\beta = \alpha + 2$ it yields $\alpha = 2/3, \quad \beta = 8/3$.

By (7) the condition $H''(f,y) = 0$ on $[-\beta,-2]$ is equivalent to

$$[3f(-y) - 2yf'(-y)] + [3f(-2-y) - (2+2y)f'(-2-y)] = 0. \tag{9}$$

If $y \in [-\beta,-2]$ then $x = -y \in [2,\beta]$ and $-2-y \in [0,\beta-2]$ or $-2-y \in [0,\alpha]$. But for $x \in [0,\alpha]$ $f(x) = 0, \quad f'(x) = 0$. Hence, the second term in sqare brackets in (9) is equal to zero, and we obtain the equation

$$[3f(-y) - 2yf'(-y)] = 0$$

to determine $f(-y)$. It coinsides with (5) for $x = -y$.

Analogously, we can rewrite $H''(f,y)$ for $y \in [-2,-\alpha]$ as

$$[3f(-y) - 2yf'(-y)] + [3f(2-y) + (2-2y)f'(2-y)] = 0.$$

Here $-y \in [\alpha, 2]$ and $2-y \in [\alpha+2, 4]$. So, $f(2-y) = f'(2-y) = 0$ and we obtain the same equation (5) for $f(x)$.

The solution of (5) on the interval $(2/3, 8/3)$ gives

$$f(x) = \begin{cases} 0, & 0 \le x < 2/3 \\ \frac{\sqrt{\frac{2}{3}}}{1\sqrt{x^3}}, & 2/3 \le x \le 8/3 \\ 0, & 8/3 < x \le 4 \end{cases} \tag{10}$$

The optimality of (10) we can check analogously to the case $a \in (0, \frac{8}{3}]$.

The full proof of the theorem for $a \in (\frac{8}{3}, \infty)$ is derived by analysis of the intervals $a \in (8/3, 4]$, $a \in [4, 14/3]$ and $a \in (14/3, \infty)$. Function $H(f, y)$ has the same form as (7) and (8).

References

[1] H. Farber, An analysis of final-offer arbitration, *Journal of conflict resolution* **35** (1980), 683–705.

[2] K. Chatterjee, Comparison of arbitration procedures: Models with complete and incomplete information, *IEEE Transactions on Systems, Man, and Cybernetics* **smc-11**, no. 2 (1981), 101–109.

[3] R. Gibbons, *A Primer in Game Theory*, Prentice Hall, 1992.

[4] D.M. Kilgour, Game-theoretic properties of final-offer arbitration, *Group Decision and Negot.* **3** (1994), 285–301.

[5] V.V. Mazalov, A.A. Zabelin, Equilibrium in an arbitration procedure, *Annals of Dynamic Games* **7** (2004), Birkhauser, 151–162.

In: Mathematics, Game Theory and Algebra Compendium
Editor: Jacob H. Mathias, pp. 705-715
ISBN 978-1-60456-659-8

Chapter 59

REPEATED GAME WITH CONSTRAINT ON THE TIME OF OBSERVATION*

Elena Z. Mokhonko
Computing Center named by A. A. Dorodnitsyn
of Russian Academy of Sciences
Moscow, Russia

1. Introduction

There are many papers devoted to the dynamic games with incomplete information, with memory, with delay of information, with hindrances, with inexact information. Let make at least mention the monographs of Krasovskii N.N., Subbotin A.I. [1], Kurzhanskii A.B. [2], Petrosjan L.A. [3], § 3 of Chapter 1 of dissertation of Kleimenov A.F. [4], the work of Portaller O., Durand G., Tomasin I.B., and Maynard de Lavalette G. [5]. Chernous'ko F.L. and Melikjan A.A. wrote [6] about the possibility to receive discrete information in antagonistic dynamic games. But nonantagonistic dynamic games of such types almost did not considered. During some period of time there was the only one paper of Kononenko A.F. [7]. Dissertation of Mokhonko E.Z. [8] filed the gap. Mathematical apparatus was developed for the solution of new problems which arose under investigation of acceptable and optimum information flows of different quality in dynamic nonantagonistic games. The results which are connected with sufficiency of reception of discrete, inexact information instead of exact and continuously receiving information are able to state with the help of the terms which are usual for the researchers working in the region of aggregation of information. The paper is devoted to the clarification how limitations on the possibility of receiving information influence the result of a dynamic process. A repeated continuous time game is considered on the segment $[0, 1]$. Players use the class of strategies of the fixed type. The strategies stipulate the continuous receipt of information during the game. Some conditions are fulfilled that provide for the existence of the equilibrium situation in the strategies. This

*The work was supported by the Russian Fund of Fundamental researches under the grant 0001–00688.

equilibrium may disappear if the time $T \in [0,1)$ of possible observation of the game going become less than the duration of the game. In the paper for every time $T \in [0,1)$ of possible observation the whole set of the equilibrium situations is found which is preserved in despite of the lack of continuous observation for the duration of the game.

The problem is considered under the following condition. The players are able to receive information in both discrete and continuous way. A fixed time is spent on the separate observation. Separate observation means that the player chooses the moment of time $\hat{t} \in [0,1]$, begins his observation at the moment of time $\hat{t}$-d and at the moment of time $\hat{t}$ receives information about the choices of his partner for all $t \in [0,\hat{t})$. After this procedure the time of possible observation become less on the value d.

This problem can be solved in the following way.

Some number q, $0 \le q \le 1$, is put in correspondence with every situation of the equilibrium. For each q the minimum time $T(q)$ is found which the player has to spend on the observation of the partner for the conservation of the situation of equilibrium. The type of function $T(q)$, which determines such time, depends on "the cost" of one observation only. The properties of this function are investigated. For given T it permits to determine the whole set of the equilibrium situations that is preserved despite of the lack of continuous observation for the duaration of the game.

2. Preservation of Equilibrium under Absence of Continuous Observation

Following [1] let us consider a two person's game of with continuous time defined on the segment $[0,1]$.

The sets of choices X_i^t, $i = 1,2$ of the players are described by the functions of the type $x_i(t) = x_i^1$, $t \in [0,1]$,

$$x_i(t) = \begin{cases} x_i^1, & t \in [0,\theta), \\ x_i^2, & t \in [\theta,1]. \end{cases}$$

Here $x_i^1 \in X_i$, $x_i^2 \in X_i$, X_i is a closed bounded set. The goal functions of players are

$$F_i = \int_0^1 M_i(x(t))\,dt.$$

Here $x(t) = (x_1,x_2) \in X = X_1 \times X_2$, the functions M_i, $i = 1,2$, are continuous.

Let us consider the class of strategies $\phi_i' = \varphi_i(x^i(.,t),t)$, $t > 0$, where $x^i = \begin{cases} x_1, & i = 2 \\ x_2, & i = 1, \end{cases}$, $x^i(.,t) = \{x^i(\tau),\ 0 \le \tau < t\}$, $\phi_i = x_i$ when $t = 0$ by definition. This strategy means that in every moment of time t each player knows the behaviour of the partner on the segment $[0,t]$. Let us denote

$$F_i(\phi) = F_i(\phi_1,\phi_2) = \int_0^1 M_i(\phi_1(x_2(.,t),t),\phi_2(x_1(.,t),t))\,dt.$$

The strategies $\phi^0 = (\phi_1^0, \phi_2^0)$ form the situation of equilibrium if $F_1(\phi^0) \geq F_1(\phi_1, \phi_2^0)$, $F_2(\phi^0) \geq F_2(\phi_1^0, \phi_2)\phi_i, i = 1, 2$.

If the set $D = \{x \in X, M_i(x_1, x_2) > \min\limits_{x^i} \max\limits_{x_i} M_i(x_1, x_2) = L_i, i = 1, 2\}$ is not empty, then the strategies of the type

$$\phi_1^0(x^i(.,t),t) = \begin{cases} x_i^0, & x^i(.,t) \equiv x^{0i}, \\ x_i^n \in \arg\min\limits_{x_i}\max\limits_{x^i} M^i(x_1,x_2), x^i(.,t) \not\equiv x^{0i}, \end{cases} \tag{2.1}$$

form the equilibrium situation, where

$$M^i(x_1, x_2) = \begin{cases} M_1(x_1, x_2), & i = 2, \\ M_2(x_1, x_2), & i = 1, \end{cases} \qquad x^0 = (x_1^0, x_2^0) \in D.$$

Really, let t_1 is the time, when the player 2 begin to deviate from the choice x_2^0, that is $x_2 = x_2^0$ when $t \in [0, t_1]$. Then the choice of the second player is estimated by the inequality

$$M_2(x^0)\,dt + \int_{t_1}^{1} \max_{x_2} M_2(x_1^n, x_2)\,dt = M_2(x^0)t_1 + L_2(1 - t_1) \leq$$

$$\leq M_2(x^0)t_1 + M_2(x^0)(1 - t_1) = M_2(x^0) = F_2(\phi^0).$$

The breach of the agreement about the choice x_2^0 at the moment t_1 leads to the loss of the gain.

Using the strategy $x_1' = \varphi_i(x^i(.,t),t)$ the player receives the information continuously. It is found that this continuous reception of information is unnecessary. It is possible to perceive the information at some separate moments of time. If the situation of equilibrium exists on the class of strategies $x_i' = \varphi_i(x^i(.,t),t)$ and the trajectory $x(t) \equiv x^0 \in D$ is realised then the situation of equilibrium exists on the class of strategies

$$x_i' = \varphi_i(x^i(.,t_k^i),t), 1 \geq \ldots t_{k+1}^i > t_k^i > \ldots > t_1^i > 0, \quad k = 1, 2, \ldots, \tag{2.2}$$

where $x^i(.,t_k^i) = \{x(\tau), 0 \leq \tau < t_k^i\}$,

$$t_k^i = \begin{cases} t_k^1, & i = 1, \\ t_k^2, & i = 2, \end{cases}$$

and the same trajectory $x(t) \equiv x^0 \in D$ is realised. In particular the next strategies are equilibrium strategies:

$$\varphi_i^0(x^i(.,t_k^i),t) = \begin{cases} x_i^0, & x^i(.,t_k^i) \equiv x^{0i}, \\ x_i^n \in \operatorname{Arg}\min\limits_{x_i}\max\limits_{x^i} M^i(x_1,x_2), x^i(.,t_k^i) \not\equiv x^{0i}. \end{cases} \tag{2.3}$$

The moments of time t_k^i are defined in the next way:

$$t_k^i = 1 - (q^i)^k, \quad q^i = q^i(x^0) = \begin{cases} q_1(x^0), & i = 2, \\ q_2(x^0), & i = 1, \end{cases}$$

$$q_i(x^0) = \frac{M_i^* - M_i^0}{M_i^* - L_i}, \qquad i = 1, 2,$$

$$M_i^0 = M_i(x^0), \quad M_1^*(x^0) = \max_{x_1} M_i(x_1, x_2^0), \quad M_2^* = \max_{x_2} M_2(x_1^0, x_2).$$

Let us show that $\varphi_i^0(x^i(., t_k^i), t)$, $i = 1, 2$, are indeed the equilibrium strategies. Let $j = 2$, if $i = 1$, and $j = 1$, if $i = 2$. Let us choose the moment of time of the first observation t_1^i according to the equation $M_j^* t_1^i + L_j(1 - t_1^i) = M_j^0$, that is $t_1^i = (M_j^0 - L_j)/(M_j^* - L_j) = 1 - q_j = 1 - q^i$. If we choose t_1^* in this way then the deviated on $[0, t_1)$ player will be punished by the partner after the revealing the player's deviation. So that the player will have no more, then he would have if he not declined. Similarly for the next moment we have

$$M_j^*(t_2^i - t_1^i) + L_j(1 - t_2^i) = M_j^0(1 - t_1^i),$$

$$t_2^i = (M_j^0 - L_j)/(M_j^* - L_j) + ((M_j^* - M_j^0)/(M_j^* - L_j))((M_j^0 - L_j)/(M_j^* - L_j)).$$

Taking into account that $1 - q_j = (M_j^0 - L_j)/(M_j^* - L_j)$, we obtain $t_2^i = 1 - (q_j)^2 = 1 - (q^i)^2$.

Choosing the moments of time t_k^i according such algorithm we come to the formula

$$t_k^i = 1 - (q^i)^k.$$

In what follows, the game is considered under condition that player are able to perform the observations in continuous way as well as discrete way. Every discrete observation and the beginning of the continuous observation costs $d_i > 0$ time units for player i. So, it is need to begin the observation at the moment $(t_k^i - d_i)$ in order to know the behaviour of the partner on the interval $[0, t_k^i)$. If t_n^i such moment of time that the distance between t_n^i and the moment of observation t_{n+1}^i is less than d_i, then beginning from this moment there is no sense to check the player by the discrete way. It is need to begin the continuous observation. So, when $0 < q^i < 1 - d_i$, the minimum expense of time is calculated in the next way

$$T_{i0}(q^i) = n_i d_i + (q^i)^{n_i}.$$

Here n_i is the number of observation, which depends on $q^i(x^0)$ and satisfies the system

$$\begin{cases} (q^i)^{n_i} \quad (1 - q^i) \le d_i, \\ (q^i)^{n_i - 1} \quad (1 - q^i) > d_i. \end{cases}$$

Recall that $q^i = q^i(x^0) = \begin{cases} q_1(x^0), & i = 2, \\ q_2(x^0), & i = 1. \end{cases}$

If $1 > q^i \ge 1 - d_i$ then the minimum time, which need to spend is equal to one for every strategy (2.1).

When $q^i = 0$ then $T_{i0}(q^i) = 0$.

3. Dependence of Set of Situations of Equilibrium on Limitation on Time of Observation

In this point such question is investigated. Let limitation exist on the sum of time, expendable on the observation by player i.

Player i is able to do any n discrete observations and any m continuous observations with duaration Θ_k^i units $(k = 1, \ldots, m)$ such that

$$nd_i + md_i + \sum_{k=1}^{m} \Theta_k^i = (n+m)d_i + \sum_{k=1}^{m} \Theta_k^i \le T_i \quad (0 < T_i \le 1).$$

We will call T_i as the reserve of time of observation, which player i has. Let us define the region D_{T_1,T_2}, such that every point x^0 from the region has the following properties:

1) the pair of strategies (ϕ_1^0, ϕ_2^0) of the form (2.3) exists, which forms the equilibrium situation and gives rise to the solution of the game $x(t) = x^0$, $t \in [0,1]$; 2) this strategies are such that every player i has the possibility to spend on the observation of the partner no more than the reserve, which he has, namely T_i. It means that the system of the intervals $[\tau_{j-1}^i, \tau_j^i] \in [0,1]$, $\tau_j^i - \tau_{j-1}^i = \Theta_j^i$, $j = 1, \ldots, m_i$, $d_i \le \Theta_j^i \le T_i$, $i = 1, 2$, exists which has the following property. For every k the number j will be found such, that $[t_k^i - d_i, t_k^i] \in [\tau_{j-1}^i, \tau_j^i]$ and under that $\sum_{j=1}^{m_i} \Theta_j^i \le T_i$. Here t_k^i are those moments which are defined in the strategies (ϕ_1^0, ϕ_2^0);

3) every other region D', which has the properties 1), 2), belongs to D_{T_1,T_2}.

In the paper the explicit expression of the border of the region D_{T_1,T_2} in terms of T_i and d_i, $i = 1, 2$ being found.

Let $D_{T_1} = \{x \in D / T_{10}(q_2(x)) \le T_1\}$, $D_{T_2} = \{x \in D / T_{20}(q_1(x)) \le T_2\}$.

The following lemma is true.

Lemma 3.1. $D_{T_1,T_2} = D_{T_1} \cap D_{T_2}$.

Proof. The set $D_{T_1} \cap D_{T_2}$ satisfies the conditions 1) and 2). D_{T_1,T_2} has to satisfy this conditions too and condition 3). That is $D_{T_1,T_2} \supset D_{T_1} \cap D_{T_2}$. Let prove the reverse inclusion. Let $x^0 \in D_{T_1,T_2}$, and let $[\tau_{j-1}^i, \tau_j^i]$, $j = 1, \ldots, m_i$, is the system of the intervals from the condition 2), $\tau_{j-1}^i - \tau_j^i = \Theta_j^i$, $\sum_{i=1}^{m_i} \Theta_j^i \le T_i$. It is known from the paper [7], that $T_{i0}(q^i(x^0)) = \min \sum_{j=1}^{m_i} \Theta_j^i \le T_i$. Therefore $T_{i0}(q^i(x^0)) \le T_i$, that is $x^0 \in D_{T_1} \cap D_{T_2}$. The proof of lemma 3.1 is finished.

If $T_1 < d_1$, then $D_{T_1} = \{x \in D / q_2(x) = 0\}$, since it is need not to control the second player when $q_2(x) = 0$.

If $T_1 = 1$, then $D_{T_1} = D$. (We assume that it is necessary to begin the observation at the moment of time $t = 0$ in order to become aware of the partner behaviour on the interval $[0, \tau]$, $\tau \in [0, d]$.)

When $T_1 = 1$ then it is possible to observe in continuous way.

If $T_1 = d_1$, then $D_{T_1} = \{x \in D / q_2(x) = 0\}$ too, since in case it is possible to receive information about the game going only one time at some moment of time t'.

If $q_2(x) \neq 0$, then it follows from the formula $t_k^i = 1 - (q^i)^k$ that it is necessary to receive information about the game going not at the moment t' only, but at some next moments of time too.

The analogous relationships take place for D_{T_2} too. It is necessary to perform the investigation of the function $T_{i0}(q^i)$ for determination D_{T_i} when $d_i < T_i < 1$.

So far as the presented below resonings are the same for $T_{10}(q_2)$ and $T_{20}(q_1)$ then the indexes i near the function $T_0(q)$ and the variable q will be omitted further till the Theorem 3.1.

So let us investigate the function $T_0(q) = n(q)d + (q)^{n(q)}$, where the function $n(q)$ accepts integer nonnegative meanings for every q, $0 < q < 1 - d$, and fits to the system

$$\begin{cases} (q)^{n(q)}(1-q) \leq d, \\ (q)^{n(q)-1}(1-q) > d. \end{cases} \tag{3.1}$$

1. For every $q \in (0, 1-d)$ the value $n(q)$ is uniquely determined because of the function $q^n(1-q)$ is a monotone decreasing function relatively n.

Really, let such q be found that $n_1(q)$, $n_2(q)$ meet the demands (3.1) and the inequality takes place $n_1(q) < n_2(q)$. Then the next inequality takes place:

$$(q)^{n_2(q)-1}(1-q) \leq (q)^{n_1(q)}(1-q) \leq d. \tag{3.2}$$

(The first inequality is true because of the decrease of the function $q^n(1-q)$ when n increase. The second inequality is true becauseof the acception that $n_1(q)$ fits to the system (3.1)).

So, $(q)^{n_2(q)-1}(1-q) \leq d$.

$n_2(q)$ fits to (3.1) under definition the function $n(q)$. Therefore the inequality $(q)^{n_2(q)-1}(1-q) > d$ has to be true. Because of this we have come to the contradiction. This fact proves that for every $q \in (0, 1-d)$ $n(q)$ is unique one.

2. Let define the concrete value $n(q)$ for every $q \in (0, 1-d)$.

The function $q^n(1-q)$ decreases relatively n and converges to zero evenly. Because of this the number N exists such that $q^N(1-q) < d$ for every $q \in [0,1]$ at once and $q_0^{N-1}(1-q_0) \geq d$ for some $q_0 \in [0,1]$. N depends on d only and does not depend on q.

2.1. For every d such that $1/4 < d < 1$, $N = 1$, because $\max\limits_{0 \leq q \leq 1} q(1-q) = 1/4$ when $q = 1/2$, $q^n(1-q)$ decreases when n increases and $1 - q > d$ for every $q \in [0, 1-d)$. Thereby $N = 1$ when $d \in (1/4, 1)$. Since $n(q) \geq 1$, so $n(q) = 1$ and $T_0(q) = d + q$ in this case.

2.2. Let $0 < d \leq 1/4$ then $N > 1$.

The equation $q^n(1-q) = d$ has no more than two values of roots when $n > 0$, n is integer. It has two values of roots when $0 < n < N$. May be, the values of roots coincides. It happens because of the function $q^n(1-q)$ strictly increases when $0 \leq q < n/n+1$, strictly decreases when $n/(n+1) \leq q \leq 1$ and $q^n(1-q) > q^{N-1}(1-q)$ when $n < N-1$. Besides $\hat{q}$, $\hat{q} \in [0,1]$ exists such that $\hat{q}^{N-1}(1-\hat{q}) \geq d$. This values of roots are distinguished by 0 and 1 because $d > 0$. If $0 < n < N$ then the equation $q^n(1-q) = d$ has one values of roots in the case when $n = N-1$ and $\max\limits_{q \in [0,1]} q^{N-1}(1-q) = (N-1/N)^{N-1}(1/N) = d$

only. In the rest of cases the equation $q^n(1-q) = d$, $q \in [0,1]$, $0 < n < N$, has two different values of roots.

Let the values of roots of the equation $q^n(1-q) = d$ be denoted as q_n, $\bar{q}_n$. Let $q_n \leq \bar{q}_n$.

2.3. The values of roots of the equation $q^n(1-q) = d$, $n = 1, \ldots, N-1$, are arranged in such order:

$$d < q_1 < q_2 < \cdots < q_{N-1} \leq \bar{q}_{N-1} < \cdots < \bar{q}_2 < \bar{q}_1 < 1 - d.$$

Let prove this statement.

$$q_1 = 1/2 - \sqrt{1/4 - d} > 1/2 - \sqrt{1/4 - d + d^2} =$$
$$= 1/2 - \sqrt{(1/2-d)^2} = 1/2 - (1/2 - d) = d,$$

$$\bar{q}_1 = 1/2 + \sqrt{1/4 - d} < 1/2 + (1/2 - d) = 1 - d.$$

So, $q_1 > d$, $\bar{q}_1 < 1 - d$.

Let prove that $q_{n-1} < q_n$, $1 < n < N-1$. Let us suppose that $q_{n-1} > q_n$ for some n. According to the property of the degree function $q_n^{n-1}(1-q_n) > q_n^n(1-q_n) = d$, the function $q^{n-1}(1-q)$ is continuous and takes the value 0 when $q = 0$, $d > 0$. Hence the value q_*, $q_* < q_n < q_{n-1}$, exists such that $q_*^{n-1}(1-q_*) = d$. But this contradicts to the facts, that the equation $q^{n-1}(1-q) = d$ has two values of roots and q_n is the smallest values of roots. Thereby $q_{n-1} \leq q_n$. The equality $q_{n-1} = q_n = \bar{\bar{q}}$ is possible when $\bar{\bar{q}}^n(1-\bar{\bar{q}}) = \bar{\bar{q}}^{n-1}(1-\bar{\bar{q}})$ only. That is $\bar{\bar{q}} = 0$, or $\bar{\bar{q}} = 1$. But as it was proved all values of roots belong to $(0, 1-d)$. Hence $q_{n-1} < q_n$. The proof of the inequality $\bar{\bar{q}}_{n-1} > \bar{\bar{q}}_n$ is analogous.

So, $d < q_1 < q_2 < \ldots < q_{N-1} \leq \bar{q}_{N-1} < \ldots < \bar{q}_2 < \bar{q}_1 < 1 - d$.

2.4. $q^n(1-q) \leq d$ when $q \leq q_n$, $q \geq \bar{q}_n$ and for every $q \in [0,1]$ when $n = N$.

$q^{n-1}(1-q) > d$ in such cases:

a) for $N - 1 > n - 1$ when $q_{n-1} < q < \bar{q}_{n-1}$,

b) for $n = N$ when $q_{N-1} \neq \bar{q}_{N-1}$,

c) for all $q < 1 - d$ when $n = 1$.

Hence $n(q) = N$ when $q_{N-1} < q < \bar{q}_{N-1}$;

$n(q) = N - 1$ when $q_{N-2} \leq q < q_{N-1}$, $\bar{q}_{N-1} \leq q < \bar{q}_{N-2}$;

....

$n(q) = 1$ when $0 < q \leq q_1$ and $\bar{q}_1 \leq q < 1 - d$.

Consequently the next theorem is proved:

Theorem 3.1. *$T_0(q)$ is expressed by the formula*

$$T_0(q) = \begin{cases} 0, & q = 0, \\ q + d, & 0 < q \le q_1, \\ q^2 + 2d, & q_1 < q \le q_2, \\ \cdots\cdots\cdots & \cdots\cdots\cdots \\ q^{N-1} + (N-1)d, & q_{N-2} < q \le q_{N-1}, \\ q^N + Nd, & q_{N-1} < q < \bar{q}_{N-1}, \\ q^{N-1} + (N-1)d, & \bar{q}_{N-1} \le q < \bar{q}_{N-2}, \\ \cdots\cdots\cdots & \cdots\cdots\cdots \\ q + d, & \bar{q}_1 \le q < 1 - d, \\ 1, & 1 - d \le q \le 1, \end{cases}$$

So, the concrete value $n(q)$ is clarified for every q.

For the following we have to prove the next lemma.

Lemma 3.2. *Let n and q are such values that $q^n(1-q) \le d$, $q^{n-1}(1-q) > d$, $0 < q < 1$, $d > 0$, n is integer positive number. Then $q^n + nd = \min\limits_m(q^m + md)$ for all integer positive m.*

Proof. So far as the function $q^n(1-q)$ decreases relatively n when $0 < q < 1$, then $\forall m \ge n$ $q^m(1-q) \le q^n(1-q) \le d$, that is $q^m(1-q) \le d$ or $q^{m+1} - q^m + d \ge 0$,

$$q^m + md - q^n - nd = q^m - q^n + (m-n)d = q^m - q^{m-1} + d + q^{m-1} -$$
$$- q^{m-2} + d + \ldots + q^{n+1} - q^n + d = \sum_{i=n}^{i=m-1} (q^{i+1} - q^i + d) \ge 0.$$

From the same reasons $\forall m < n$, $m > 0$ $q^m(1-q) \ge q^{n-1}(1-q) > d$, that is,

$$q^m - q^{m+1} - d > 0, \quad q^m + md - q^n - nd =$$
$$= q^m - q^n - (n-m)d = q^m - q^{m+1} - d + q^{m+1} - q^{m+2} - d + \ldots + q^{n-1} - q^n - d =$$
$$= \sum_{i=m}^{i=m-1} (q^i - q^{i+1} - d) > 0.$$

Lemma 3.2 is proved.

Lemma 3.3. *The function $T_0(q)$ is continuous and monotonically increasing relatively q when $q \in (0, 1-d]$.*

Proof. The continuity of function $T_0(q)$ when $q \in (0, 1-d]$ follows because of the continuity of function $q^n + nd$, $2 \le n \le N$, when $q \in (0, 1-d]$ and the next fact. For every n, $2 \le n \le N$, the equalities $q_{n-1}^{n-1} + (n-1)d = q_{n-1}^n + nd$, $\bar{q}_{n-1}^{n-1} + (n-1)d = \bar{q}_{n-1}^n + nd$ take place because $q_{n-1}^{n-1}(1 - q_{n-1}) = d$ and $\bar{q}_{n-1}^{n-1}(1 - \bar{q}_{n-1}) = d$.

Recall, that q_n, $\bar{q}_n$ denote the values of roots of equation $q^n(1-q) = d$, $q_n \le \bar{q}_n$. For the proof of monotone let notice that if q_*, q_{**} ($q_* < q_{**}$) belong to the same interval, for example $(q_{i-1}, q_i]$, then $T_0(q_*) < T_0(q_{**})$ because of increase the function $q^n + nd$ relatively q. Let q_*, q_{**} belong to the different intervals for example $q_* \in (q_{k-1}, q_k]$, $q_{**} \in [\bar{q}_m, \bar{q}_{m-1})$. Because $q_* \in (q_{k-1}, q_k]$, then $T_0(q_*) = q^k + kd$ and in this case k and q_* are connected by the relation

$$\begin{cases} q_*^k \quad (1 - q_*) \leq d \\ q_*^{k-1} \quad (1 - q_*) > d. \end{cases}$$

It permits to apply the lemma 3.2. The next relationships are $T_0(q_*) = q^k + kd = \min_n [q_*^n + nd] \leq q_*^m + md$. So far as $q_* < q_{**}$, then $q_*^m + md < q_{**}^m + md$ because of the increase of function $q^m + md$ relatively q. Besides $q_{**}^m + md = T_0(q_{**})$, since $q_{**} \in [\bar{q}_m, \bar{q}_{m-1})$ and the relations

$$\begin{cases} q_{**}^m \quad (1 - q_*) \leq d, \\ q_{**}^{m-1} \quad (1 - q_{**}) > d, \end{cases}$$

take place on this interval. Thus we have $T_0(q_*) < T_0(q_{**})$. The proof of strict increase of the function $T_0(q)$ is the same in the case when q_*, q_{**} belong to some other kinds of the intervals from the set $(0, q_1]$, $(q_{i-1}, q_i]$, $[\bar{q}_i, \bar{q}_{i-1}), \ldots, [\bar{q}_1, 1 - d)$, $i = 2, \ldots, N - 1$.

Thus, it is proved that $T_0(q)$ is continuous and strictly monotonically increase over $(0, 1 - d]$.

Theorem 3.2. *$D_{T_1} = \{x \in D/q_2(x) \leq q_{2T_1}\}$, where $q_{2T_1} = \max_{1 \leq n \leq [T_1/d_1]} \sqrt[n]{T_1 - nd_1}$ when $T_1 > d_1$ and $q_{2T_1} = 0$ when $T_1 \leq d_1$.*

Here [a] denotes the largest integer number which is less or equal a.

Proof. The function $T_{10}(q_2)$ is continuous, monotonically increases over $(0, 1 - d_1]$, $\lim_{q_2 \to 0} T_{10}(q_2) = d_1$, $T_{10}(1 - d_1) = 1$. Because of this unique q_{2T_1} exists such that $T_{10}(q_{2T_1}) = T_1$ for every $T_1 \in (d, 1]$. For every $q_2 \leq q_{2T_1}$ the inequality $T_{10}(q_2) \leq T_1$ takes place in the consequence of monotonically increase $T_{10}(q_2)$ and the equality $T_0(0) = 0$.

Because of this

$$D_{T_1} = \{x \in D/T_{10}(q_2(x)) \leq T_1\} = \{x \in D/q_2(x) \leq q_{2T_1}\}.$$

Let prove that $q_{2T_1} = \max_{1 \leq n \leq [T/d_1]} \sqrt[n]{T_1 - nd_1}$.

Let q_{2T_1} is such that $T_1 = (q_{2T_1})^j + jd_1$, $2 < j \leq N - 1$, and let q_2' exists such that $q_2' = \sqrt[m]{T_1 - md_1} > q_{2T_1}$, then $T_1 = md_1 + (q_2')^m$. $md_1 + (q_{2T_1})^m < md_1 + (q_2')^m = T_1$, that is $md_1 + (q_{2T_1})^m < jd_1 + (q_{2T_1})^j$ since $q_2' > q_{2T_1}$, the function $md_1 + (q_2')^m$ is increases relatively q_2. But this contradicts of the lemma 3.2, according to which $jd_1 + (q_{2T_1})^j = \min_{0<n}(nd_1 + (q_{2T_1})^n)$. (The conditions of lemma 3.2 for q_{2T_1} and j are fulfilled because of definition $T_{10}(q_2)$, which is equal to T_1, $T_1 = (q_{2T_1})^j + jd_1 = T_{10}(q_{2T_1}))$ in this case. It is possible to consider such n, that $n \leq [T_1/d_1]$. It follows from the requirement of nonnegativeness of the under root relation $T_1 - nd_1$ and the fact that q_2 is real and nonnegative.

$D_{T_1} = \{x \in D/q_2(x) = 0\}$ when $T_1 \leq d_1$. It may be write down as $D_{T_1} = \{x \in D/q_2(x) \leq q_{2T}(x)\}$, where $q_{2T_1}(x) = 0$.

The proof of the theorem is finished.

Consequence. $D_{T_1,T_2} = \{x \in D_2 / (q_2(x) \le q_{2T_1}) \wedge (q_1(x) \le q_{1T_2})\}$, where

$$q_{2T_1} = \begin{cases} \max\limits_{1 \le n \le [T_1/d_1]} \sqrt[n]{T_1 - nd_1}, & T_1 > d_1, \\ 0, & T_1 \le d_1, \end{cases}$$

$$q_{1T_2} = \begin{cases} \max\limits_{1 \le n \le [T_2/d_2]} \sqrt[n]{T_2 - nd_2}, & T_2 > d_2, \\ 0, & T_2 \le d_2. \end{cases}$$

Remark. *The condition "D_{T_1,T_2} is not empty" is the sufficient condition for the existence the equilibrium situation when limitations exist on the time of observation of the partner.*

Example.

$$M_1(x_1, x_2) = -x_1 + x_2; \quad M_2(x_1, x_2) = -x_1^2 - x_2; \quad -1 \le x_i \le 1, \quad i = 1, 2.$$

$$T_1 = 0,5, \quad T_2 = 1, \quad d_1 = d_2 = 0,1, \quad L_1 = L_2 = 0,$$

$$D = \{-1 \le x_i \le 1,\ i = 1, 2 / (x_2 > x_1) \wedge (x_2 < -x_1^2)\}.$$

$$q_{2T_1} = \max_{1 \le n \le [T_1/d_1]} \sqrt[n]{T_1 - nd_1} = a; \quad \mathit{0{,}58} < a < \mathit{0{,}59}, \quad n = 3.$$

$$q_2(x_1, x_2) = (-x_1^2 + 1 + x_1^2 + x_2)/(-x_1^2 + 1) = (x_2 + 1)/(-x_1^2 + 1) < a.$$

The boundary $D_{T_1} = D_{0,5}$ is situated among the curves

$$(x_2 + 1)/(-x_1^2 + 1) = \mathit{0{,}58}; \quad (x_2 + 1)/(-x_1^2 + 1) = \mathit{0{,}59}.$$

Further its part lies on the straight line $x_2 = x_1$.

References

[1] N.N. Krasovskii, A.I. Subbotin, *Positional Differential Games*, Nauka, Moscow, 1974, (Russian).

[2] A.B. Kurzhanskii, *Control and Observation in Circumstances of Uncertainty*, Nauka, Moscow, 1977, (Russian).

[3] L.A. Petrosjan, *The Differential Pursuit Games*, Leningrad State University, Leningrad, 1987, (Russian).

[4] A.F. Kleimenov, *The Nonantagonistic Positional Games*: Dissertation... Doctor of Physical and Mathematical Sciences, Institute of Mathematics and Mechanics of the Urals Branch of Academy of Sciences of the USSR, Sverdlovsk, 1991, (Russian).

[5] O. Pourtallier, G. Durand, B. Tomasin and G. Maynard de Lavalette, The Rabbit and the Hunter Game: an Application to a Close–In Weapon System, *5th Internat. Symp. on Dynamic Games and Applications: Preprint Volume* (1992), Geneva, 775–776.

[6] F.L. Chernous'ko, A.A. Melikjan, *The Game Problems of Control and Search*, Nauka, Moscow, 1978, (Russian).

[7] A.F. Kononenko, Statement of Problem. Model with Continuous Time, *Modern State of the Operation Researches Theory*, Nauka, Moscow, 1979, 173–179, (Russian).

[8] E.Z. Mokhonko, *Dynamics of Information Processes in Non-antagonistic Games*. 05.13.17 — Theoretical Foundations of Informatics. Dissertation... Doctor of Physical and Mathematical Sciences (DSc), Computing Center of Russian Academy of Sciences, Moscow, 1997, (Russian).

[9] E.Z. Mokhonko, The Influence of the Information Inexactitude on Retention of Situation of Equilibrium in Non-Antagonistic Differential Game, *Proceedings of Moscow Institute of Physics and Technology, Series of Aerophysics and Applied Mathematics*, Moscow Institute of Physics and Technology (MIPT), Moscow, 111–112, 1980, (Russian).

[10] E.Z. Mokhonko, *Information Processes in Non-antagonistic Differential Game*, Communications on Applied Mathematics, Computing Center of the USSR Academy of Sciences, 1989, (Russian).

In: Mathematics, Game Theory and Algebra Compendium ISBN 978-1-60456-659-8
Editor: Jacob H. Mathias, pp. 717-728

Chapter 60

VALUE FOR THE GAME WITH CHANGING COALITIONAL STRUCTURE

Leon A. Petrosjan and Svetlana I. Mamkina
Faculty of Mathematics and Mechanics
St. Petersburg State University
St. Petersburg, Russia

1. Introduction

Game-theoretic models considered in the modern game theory in principal can be divided into two classes: strategic and cooperative games. In strategic games the players choosing their strategies try to get maximal payoffs, in cooperative games it is assumed that players try to maximize the sum of their payoffs and the problem consists in the allocation of this maximal total payoff between them. There is a number of papers where the intermediate case is considered: the case when the cooperation is not full, and players form coalitions choosing strategies with the intention to maximize the payoff of the coalition to which they belong (see [1–4]). In this paper we consider games with perfect information and randomly changing coalitional partitions in some fixed vertices of the game tree. An algorithm for constructing the optimal subtree ("bunch") is proposed and the new value (the so called PMS-value) for such games introduced.

2. Dynamic Game with Changing Coalitional Partition

In what follows as basic model we shall consider the game in extensive form with perfect information, which includes the possibility of changing coalitional partitions in some fixed vertices of the game tree.

Definition 1. *A game tree is a finite oriented treelike graph K with the root x_0.*

We shall use the following notations. Let x be some vertex (position). We denote by $K(x)$ a subtree of the tree K with the root in x. We denote by $Z(x)$ the set of vertices directly following after x. The vertices y, directly following after x, are called alternatives in x ($y \in Z(x)$). The player who makes decision in x (who selects the next alternative position in x), will be denoted by $i(x)$. The choice of player $i(x)$ in position x will be denoted by $\overline{x} \in Z(x)$.

Let $N = \{1, ..., n\}$ — be the set of all players in the game. Under partition of the set N we shall understand the family of subsets $\Delta_k = \{S_j\}_{j=1}^{|\Delta_k|}$, such that $S_j \cap S_i = \emptyset$; $j \neq i$; $\bigcup S_j = N$. The set of all admissible partitions of N will be denoted by Δ.

Now we shall give the strict definition of the game with perfect information and changing coalitional partition.

Definition 2. *A game in extensive form with perfect information and changing coalitional partition $\Gamma(x_0)$ is a game tree K, with following additional properties :*

- *(I) The set of vertices (positions) is split up into $n+2$ subsets $P_1, P_2, ..., P_n, P_{n+1}, P_{n+2}$, which form a partition of the set of all vertices of the graph tree K. The vertices (positions) $x \in P_i$ are called players i personal positions, $i = 1, ..., n$; vertices (positions) $x \in P_{n+1}$ are called positions of the chance and vertices (positions) $x \in P_{n+2}$ are called terminal positions.*

- *(II) For each x the partition $\Delta(x) \in \Delta$ of the player set N is given, such that*

$$\begin{cases} \Delta(x) \equiv \Delta(y), & \forall y \in Z(x), \qquad \text{if } x \notin P_{n+1}, \\ \exists y \in Z(x): & \Delta(x) \not\equiv \Delta(y), \quad \text{if } x \in P_{n+1}. \end{cases}$$

- *(III) For each $y \in Z(x) = \{y_1, ..., y_r\}$, $x \in P_{n+1}$ the probability distribution $p(y_1), ..., p(y_r)$, $\sum\limits_{y \in Z(x)} p(y) = 1$, $r = |Z(x)|$ is defined.*

- *(IV) In each final vertex (position) the system of real numbers $h(w) = (h_1(w), ..., h_n(w))$, $w \in P_{n+2}$, $h_i(w) \geq 0$, $i = 1, ..., n$ is defined. Where $h_i(w)$ is the payoff of player i in the final vertex (position).*

Definition 3. *A strategy of player i is mapping $U_i(\cdot)$, which corresponds to each position $x \in P_i$ a unique alternative $y \in Z(x)$.*

The set of all strategies of player i is denoted by U_i.

In this setting we suppose that the player $i \in N$ in his position set $y \in P_i$ is playing in the interests of the coalition S_j, $(i \in S_j, S_j \in \Delta(x_j))$ to which he belongs , trying to maximize the payoff of this coalition. Suppose the players choose the strategy profile $U_1(\cdot), ..., U_n(\cdot)$. Then the game develops in the following way. The game starts in $x_0 \in P_{n+1}$, in which the coalitional partition $\Delta(x_0) \in \Delta$ is given. In this position x_0 the chance, according to the probability distribution defined in (III), is selecting the alternative $\overline{x}_1 \in Z(x_0)$, in which a coalitional partition $\Delta(\overline{x}_1) \in \Delta$ is defined. Suppose $\overline{x}_1 \in P_{i(\overline{x}_1)}$, then in position $\overline{x}_1$ the player $i(\overline{x}_1)$ makes a move choosing the alternative $\overline{x}_2 = U_{i(\overline{x}_1)}(\overline{x}_1) \in Z(\overline{x}_1)$, and acting in the interests of the coalition to which he belongs and which was realized after the chance move makes a choice in the initial position of the game $x_0 \in P_{n+1}$. Suppose that in $\overline{x}_2$ the coalitional partitions $\Delta(\overline{x}_2)$ and $\Delta(\overline{x}_1)$ coincide ($\Delta(\overline{x}_2) \equiv \Delta(\overline{x}_1)$). If $\overline{x}_2 \notin P_{n+1} \cup P_{n+2}$, then in position $\overline{x}_2$ player $i(\overline{x}_2)$ makes a move and chooses an alternative $\overline{x}_3 = U_{i(\overline{x}_2)}(\overline{x}_2) \in Z(\overline{x}_2)$, according to the interests of the coalition to which he belongs. If $\overline{x}_2 \in P_{n+1}$, then in position $\overline{x}_2$ the chance according to the probability distribution defined by (III) from definition 2 for position $\overline{x}_2$ is selecting the alternative $\overline{x}_3$. In the randomly selected position $\overline{x}_3 \in Z(\overline{x}_2)$ the coalitional partition $\Delta(\overline{x}_3) \in \Delta$ is defined which may or may not coincide with coalitional partition in $\overline{x}_2$. If on the stage k, $x_k \notin P_{n+1} \cup P_{n+2}$, then in x_k the player $i(\overline{x}_k)$ chooses an alternative $\overline{x}_{k+1} = U_{i(\overline{x}_k)}(\overline{x}_k) \in Z(\overline{x}_k)$, acting according to the interests of the coalition to which he belongs. If on the stage k the game finds itself in the position of the chance, i.e $\overline{x}_k \in P_{n+1}$, then in the position $\overline{x}_k$ an alternative $\overline{x}_{k+1}$ is chosen according to the probability distribution defined for this position $\overline{x}_k$ according to the condition (III) of the definition 2. In the chosen position $\overline{x}_{k+1} \in Z(\overline{x}_k)$ the coalitional partition $\Delta(\overline{x}_{k+1}) \in \Delta$ is defined, which may or may not coincide with coalitional partition defined for the position $\overline{x}_k$. The game ends after reaching the final position $\omega \in P_{n+2}$. This happens after a finite number of steps, since the game tree is finite. In the position ω, according to the condition (IV) of the definition 2, the payoffs of the players $(h_1(\omega), \ ..., \ h_n(\omega))$ are defined. Thus every strategy profile $U(\cdot) = (U_1(\cdot), \ ..., \ U_n(\cdot))$ defines the probability distribution over the set of final positions (because of the chance moves in $x \in P_{n+1}$), and thus the payoff $h_i(\omega)$ of the player i is a random variable. Consequently to each strategy profile $U(\cdot) = (U_1(\cdot), \ ..., \ U_n(\cdot))$ uniquely corresponds the vector of mathematical expectations of players payoffs $E_i(U_1(\cdot), \ ..., \ U_n(\cdot)) = E[h_i(\omega)]$, $i \in N$. Thus we can formulate our game as game in normal form $< N; \ U_1, \ ..., \ U_n; \ E_1, \ ..., \ E_n >$. But the normal form of the game is not very useful for the investigation of our game, since it takes not into account the belonging of players to different coalitions and the possibility of the transformation of the coalitions. This makes the use of classical solution concepts for the construction of optimal behavior practically impossible . In what follows we shall propose a new approach to the construction of the solution, which takes in account the existing coalitional structure of the game and the dynamics of its evolution.

3. The Algorithm for Aonstructing the Solution

As we mentioned before, we suppose that the player $i \in N$ in position $y \in P_i$ plays according to the interests of the coalition to which he belongs, i.e he strives to maximize the sum of the payoffs of players from coalition S_j, $(i \in S_j, \ S_j \in \Delta(x_j))$. In the same time to define the payoffs inside the coalition S_j the Shapley value is used, calculated for a specially defined characteristic function $V(R, x)$, $R \subset S_j$. Using this approach, we propose in this section an algorithm for constructing the solution in the game $\Gamma(x_0)$, which leads also to the calculation of the corresponding "bunch" of the paths in the game tree. The solution is constructed using the backward induction procedure from the terminal positions to the initial. The procedure is similar to the construction of a subgame perfect Nash equilibrium (see [6,7,9]) in the games with perfect information and also to the algorithm for the definition of the optimal trajectory in the games with partial cooperation (see [8]).

Before starting the investigation of the solution of $\Gamma(x_0)$, we introduce the notion of the length of the game $\Gamma(x_0)$. Under the length of the game $\Gamma(x_0)$ we shall understand the length of the maximal path in the game tree $K(x_0)$ (the number of vertices in the maximal path). Suppose that the length of $\Gamma(x_0)$ is equal to $T+1$. Consider the partition of the set of all possible positions in the game tree $K(x_0)$ on $T+1$ sets X_0, X_1, ..., $X_T : X_T = \{x_0\}$, where the set X_t consists of the positions which can be reached from the initial position after $T-t$ moves. The positions from X_t we shall denote by x_t, $t = 0, 1, ..., T$.

Consider now the set of positions of X_0. Since the length of the game is equal to $T+1$, then $X_0 \subset P_{n+2}$ and players payoffs are defined and are equal too $h_i(w)$, $w \in X_0$, $i = 1, ..., n$.

Step 1. Make a transition from the positions of X_0 to positions in the set $x_1 \in X_1$. If $x_1 \notin P_{n+1} \cup P_{n+2}$, then in x_1 the player $i(x_1)$ has to make a move. Suppose that $i(x_1)$ belongs to coalition $S_j(x_1)$; $S_j(x_1) \in \Delta(x_1)$. Here $\{S_k(x_1)\}$ is the set of coalitions constituting the coalitional partition $\Delta(x_1)$. The algorithm prescribes to each player $i(x_1)$ to choose with equal probabilities (equaprobable) each alternative $\overline{w} \in Z(x_1)$:

$$\max_{x \in Z(x_1)} \sum_{i \in S_j(x_1)} h_i(x) = \sum_{i \in S_j(x_1)} h_i(\overline{w}). \tag{1}$$

Denote by

$$\tilde{Z}_{i(x_1)}(x_1) = \arg \max_{x \in Z(x_1)} \sum_{i \in S_j(x_1)} h_i(x),$$

i.e.

$$\tilde{Z}_{i(x_1)}(x_1) = \left\{ y : \sum_{i \in S_j(x_1)} h_i(y) = \max_{x \in Z(x_1)} \sum_{i \in S_j(x_1)} h_i(x) \right\}.$$

The choice of the alternatives with equal probabilities means that we allow the use of mixed behaviour strategies. In [5] it is proved that the constructed strategies form a uniquely defined subgame perfect "indifferent" Nash equilibrium in the game where the elements of the corresponding coalitional partition are acting as players. The formula (1) defines the corresponding values of payoffs in a unique way.

If $x_1 \in P_{n+1}$ (x_1 is a chance move) and the game moves to the next positions $x \in Z(x_1)$ with probabilities $p(x)$, defined by the rules of the game, then the expected values of payoffs at x_1 are: $\sum_{x \in Z(x_1)} p(x) h_i(x)$, where $p(x)$ — is the probability of choosing the alternative (position) $x \in Z(x_1)$. If $x_1 \in P_{n+2}$, then the payoffs at x_1 are defined as $h_i(x_1)$, $i = 1, ..., n$. Since according to the algorithm in some cases the next alternative is chosen at random the proposed scheme does not define unique path (link), but defines a "bunch" or subtree.

Similarly one can construct the "bunch" with initial position (root) $x_1 \in X_1$ for every position x_1 of the set X_1. Thus on each subtree $K(x_1)$, $x_1 \in X_1$ (this subtree may consist from a unique position if $x_1 \in P_{n+2}$) the final position $\overline{x}_1$ of the constructed "bunch" is fixed in the game, or the probability distribution over $Z(x_1)$ (if the maximizing point in (1) is not unique, or $x_1 \in P_{n+1}$). Because of this behavior of players in subtrees $K(x_1)$, $x_1 \in X_1$, we are able to introduce the Bellmann functions $r_i^1 : X_1 \to R_+^1$, $i = 1, ..., n$, defined on the set X_1. Here r_i^1 – is the expected (predicted) payoff of player i in position

$x_1 \in X_1$, assuming that on the subtree $K(x_1)$ players act according to the constructed algorithm.

$$r_i^1(x_1) = \begin{cases} \frac{1}{|\tilde{Z}_{i(x_1)}(x_1)|} \sum\limits_{\overline{w} \in \tilde{Z}_{i(x_1)}(x_1)} h_i(\overline{w}), & \text{if} \quad x_1 \notin P_{n+2} \cup P_{n+1}, \\ h_i(x_1), & \text{if} \quad x_1 \in P_{n+2}, \\ \sum\limits_{w \in Z(x_1)} p(w) h_i(w), & \text{if} \quad x_1 \in P_{n+1} \end{cases} \quad (2)$$

(—A— is a number of elements in the set A) $i \in N$.

Step 2. Continue the motion towards the root of the game. Since at $x_2 \in X_2$ the algorithm may contain new elements we shall discuss this case in details. We define the behavior of players in position $x_2 \in X_2$. If $x_2 \notin P_{n+2} \cup P_{n+1}$, then in x_2 player $i(x_2) \in S_j(x_2)$ has to make a move. The algorithm orders player $i(x_2)$ to choose with equal probabilities each of the positions $\overline{x}_1$ satisfying the condition :

$$\max_{x \in Z(x_2)} \sum_{i \in S_j(x_2)} r_i^1(x) = \sum_{i \in S_j(x_2)} r_i^1(\overline{x}_1), \quad (3)$$

where r_i^1 is defined by (2).

Denote by

$$\tilde{Z}_{i(x_2)}(x_2) = \arg \max_{x \in Z(x_2)} \sum_{i \in S_j(x_2)} r_i^1(x_1).$$

The choice of the alternatives with equal probabilities means that we allow the use of mixed behaviour strategies. In [5] it is proved that the constructed strategies form a uniquely defined subgame perfect "indifferent" Nash equilibrium in the subgame where the elements of corresponding coalitional partition are acting as players.

If $x_2 \in P_{n+1}$ (at x_2 chance is making a choice), then since $\Delta(x_2)$ may not coincide with $\Delta(x_1)$, $x_1 \in Z(x_2)$, in the position x_2 we get an uncertainty in the players payoffs. To define the value of the function $r_i^2(x_2)$ in the position x_2 (in the root of the "bunch" in subtree $K(x_2)$), we have to single out the players payoffs from the payoff of the containing coalition., i.e. to define an allocation rule for the total coalitional payoff for each coalition $\{S_j(x_1)\}_{j=1}^{|\Delta(x_1)|}$. Such an allocation rule can be defined with the help of characteristic functions, formed for each element of the coalitional partition $\Delta(x)$, $x \in X_1$. The formation of such a characteristic function is explained in the next section. Now suppose that the characteristic function $v(x,\ R(S_k(x)))$, $R \subset S_k(x)$, $S_k(x) \in \Delta(x)$, $k = 1, \ldots, |\Delta(x)|$ is given. Choose an arbitrary coalition $S_k(x_1) \in \Delta(x_1)$. Consider a cooperative $|S_k(x_1)|$ player game $G(x_1,\ S_k(x_1))$ with c.f. $v(x,\ R(S_k(x_1))$, $R \subset S_k(x_1)$. The payoff of the maximal coalition $(S_k(x_1))$ in cooperative game $G(x_1,\ S_k(x_1))$ equals to:

$$v(x_1,\ S_k(x_1)) = \sum_{i \in S_k(x_1)} r_i^1(x_1). \quad (4)$$

Compute the Shapley value [10] in $G(x_1,\ S_k(x_1))$ with components $Sh_i(S_k(x_1))$, $i \in S_k(x_1)$, where

$$\sum_{i \in S_k(x_1)} Sh_i(S_k(x_1)) = v(x_1,\ S_k(x_1)).$$

And consider the Shaply value as an allocation rule of the total coalitional payoff $v_1(x_1,\ S_k(x_1))$. In a similar way we can find an optimal allocation for other coalitions from $\{\Delta(x_1) \backslash S_k(x_1)\}$.

Construct a vector

$$PMS(x_1) = (PMS_1(x_1),\ ...,\ PMS_n(x_1)), \tag{5}$$

where $PMS_i(x_1) = Sh_i(S_k(x_1))$, $i \in S_k(x_1)$, which defines the players payoffs for $1,\ ...,\ n$ in the case of $x_2 \in P_{n+1}$.

If $x_2 \in P_{n+2}$, the players payoffs in x_2 are simply equal to $h_i(x_2)$, $i = 1,\ ...,\ n$.

Thus we can define on X_2 functions $r_i^2 : X_2 \to R_1^+$, $i \in N$,

$$r_i^2(x_2) = \begin{cases} \frac{1}{|\tilde{Z}_{i(x_1)}(x_1)|} \sum\limits_{\overline{x}_1 \in \tilde{Z}_{i(x_1)}(x_1)} r_i^1(\overline{x}_1), & \text{if} \quad x_2 \notin P_{n+2} \cup P_{n+1} \\ h_i(x_2), & \text{if} \quad x_2 \in P_{n+2} \\ \sum\limits_{x_1 \in Z(x_2)} p(x_1) \cdot PMS_i(x_1), & \text{if} \quad x_2 \in P_{n+1} \end{cases} \tag{6}$$

where $p(x_1)$ are probabilities to choose the alternative x_1, $r_i^1(\overline{x}_1)$ is defined by (3) and $PMS_i(x_1)$ by (5) $(i \in N)$. Thus the "bunch" is constructed for every subtree $K(x_2)$.

The next steps are similar to steps 1 and 2. Consider now the step t. Let $x_t \in X_t$. Denote functions by $r_i^{t-1} : X_{t-1} \to R_+^1$, $i \in N$ that determine the players $i \in N$ payoffs after they behave according to the proposed algorithm in the positions $x_l \in X_l$, $l < t$.

Step t. Consider the position $x_t \in X_t$. If $x_t \notin P_{n+1} \cup P_{n+2}$, then in x_t, player $i(x_t) \in S_k(x_t) \in \Delta(x_t)$ is to make a choice. We prescribe him to select with equal probabilities any of the positions $\overline{x}_{t-1}$ satisfying the condition:

$$\max_{x \in Z(x_t)} \sum_{i \in S_k(x_t)} r_i^{t-1}(x) = \sum_{i \in S_k(x_t)} r_i^{t-1}(\overline{x}_{t-1}). \tag{7}$$

Denote by

$$\tilde{Z}_{i(x_t)}(x_t) = \arg \max_{x \in Z(x_t)} \sum_{i \in S_j(x_t)} r_i^{t-1}(\overline{x}_{t-1}) =$$

$$= \left\{ y : \sum_{i \in S_j(x_t)} r_i^{t-1}(y) = \max_{x \in Z(x_t)} \sum_{i \in S_j(x_t)} r_i^{t-1}(x) \right\}.$$

The choice of alternatives with equal probabilities means that we allow the use of mixed behaviour strategies. In [5] it is proved that the constructed strategies form a uniquely defined subgame perfect "indifferent" Nash equilibrium in the game where the elements of the corresponding coalitional partition are acting as players. This type of "indifferent" equilibrium we shall call PMN- equilibrium. It can be easily seen that in every game with perfect information there exist a unique PMN-equilibrium.

If $x_t \in P_{n+1}$, it is necessary to single out players payoffs from the coalitional payoff, i.e. to define an allocation of the coalitional payoff for every coalition $\{S_j(x_{t-1})\}_{j=1}^{|\Delta(x_{t-1})|}$. Select an arbitrary coalition $S_k(x_{t-1}) \in \Delta(x_{t-1})$. Consider a cooperative $|S_k(x_{t-1})|$ player game $G(x_{t-1},\ S_k(x_{t-1}))$ with c.f $v(x_{t-1},\ R(S_k(x_{t-1})),\ R \subset S_k(x_{t-1})$. The payoff of maximal coalition $(S_k(x_{t-1}))$ in the cooperative game $G(x_{t-1},\ S_k(x_{t-1}))$ is equal to:

$$v(x_{t-1},\ S_k(x_{t-1})) = \sum_{i \in S_k(x_{t-1})} r_i^{t-1}(x_{t-1}). \tag{8}$$

We shall consider the Shapely value with components $Sh_i(S_k(x_{t-1})),\ i \in S_k(x_{t-1})$, where

$$\sum_{i \in S_k(x_{t-1})} Sh_i(S_k(x_{t-1})) = v(x_{t-1},\ S_k(x_{t-1})),$$

as an optimal allocation of the coalitional payoff $v(x_{t-1},\ S_k(x_{t-1}))$. For the computation of the components of the Shapley value $Sh_i(S_k(x_{t-1})),\ i \in S_k(x_{t-1})$ we use the c.f $v(x_{t-1},\ R(S_k(x_{t-1})),\ R \subset S_k(x_{t-1}),\ S_k(x_{t-1}) \in \Delta(x_{t-1})$. In a similar way we can find the optimal allocation for other coalitions $\{\Delta(x_{t-1}) \backslash S_k(x_{t-1})\}$.

Define the vector

$$PMS(x_1) = (PMS_1(x_{t-1}),\ ...,\ PMS_n(x_{t-1})), \tag{9}$$

where $PMS_i(x_{t-1}) = Sh_i(S_k(x_{t-1})),\ i \in S_k(x_{t-1})$, which defines the payoffs of the players $1,\ ...,\ n$ in the case of $x_t \in P_{n+1}$. If $x_t \in P_{n+2}$, the payoffs of the players in x_t are defined as $h_i(x_t),\ i = 1,\ ...,\ n$.

Define the functions $r_i^t : X_t \to R_1^+,\ i \in N$ by formulas:

$$r_i^t(x_t) = \begin{cases} \frac{1}{|\tilde{Z}_{i(x_t)}(x_t)|} \sum\limits_{\overline{x}_{t-1} \in \tilde{Z}_{i(x_t)}(x_t)} r_i^{t-1}(\overline{x}_{t-1}), & \text{if} \quad x_t \notin P_{n+2} \cup P_{n+1} \\ h_i(x_t), & \text{if} \quad x_t \in P_{n+2} \\ \sum\limits_{x_{t-1} \in Z(x_t)} p(x_{t-1}) \cdot PMS_i(x_{t-1}), & \text{if} \quad x_t \in P_{n+1} \end{cases}$$

where $p(x_{t-1})$ is the probability to choose the alternative x_{t-1}, $r_i^{t-1}(\overline{x}_{t-1})$ is defined by (7) and $PMS_i(x_{t-1})$ by (9) $(i \in N)$. Thus the "bunch" and the payoffs on each subtree $K(x_t),\ x_t \in X_t$ are constructed.

The function $r_i^t(x_t)$ has the meaning of predicting player i's payoff in the position $x_t \in X_t$, assuming that in the subtree $K(x_t)$ the players will act according to the constructed algorithm.

Going downwards to the root of the game $\Gamma(x_0)$ and consequently determining the behavior of the players in the remaining sets X_τ, $(\tau = t+1, ..., T)$, we shall define the "bunch" consisting from the paths in a graph tree and the corresponding payoffs in $\Gamma(x_0)$. We call this "bunch" of paths "optimal bunch" in the game with changing coalitional partitions defined on the bases of PMS-vectors. Note that for each game we get a unique PMS-vector since the PMN-equilibrium constructed for corresponding subgames was uniquely defined.

4. Characteristic Functions for an Associated Cooperative Game

In this section we propose the way of defining the c.f. $v(x, R)$, $R \subset S_k(x)$ for $G(x,\ S_k(x))$. Denote by $\Gamma(x)$ the subgame of the game $\Gamma(x_0)$ starting in position x. When constructing the optimal bunch for $\Gamma(x_0)$ in section 2, the optimal behavior of player $i \in N$ was defined in each position P_i, $i \in N$. Denote by $U^*(\cdot) = (U_1^*(\cdot),\ ...,\ U_n^*(\cdot))$ the strategy profile defined in section 2 leading to the realization of optimal bunch in $\Gamma(x_0)$. The cooperative game $G(x, S_k(x))$ is constructed with the use of this strategies. Define the restriction of the strategy profile. Restriction of strategy profile in the subgame is this strategy profile considered on the positions of this subgame. Denote the restriction of U^* as $U_x^*(\cdot) = (U_{1x}^*(\cdot),\ ...,\ U_{nx}^*(\cdot))$. Let $\overline{\Gamma}(x)$ be the truncation of subgame $\Gamma(x)$, in which the choices of players $i \notin S_k(x)$ in personal positions are fixed according to U_{ix}^*. In the same way as we have done with optimal bunch we shall construct the c.f. $v(x, R)$, $R \subset S_k(x)$ using backward induction for $x \notin P_{n+1}$ (since for $x \in P_{n+1}$ our algorithm does not require the computation of the Shapley value). Consider the set of positions X_0. Since the length of the game is equal to $T+1$, then $X_0 \subset P_{n+2}$ and $v(x, R)$, $R \subset S_k(x)$ in this case is equal to $v(x,\ R) = \sum_{i \in R} h_i(x)$.

Step 1. Transfer from $x \in X_0$ to $x_1 \in X_1$ ($x_1 \notin P_{n+1}$), and consider the restriction $U_{x_1}^*(\cdot) = (U_{1x_1}^*(\cdot),\ ...,\ U_{nx_1}^*(\cdot))$ of strategy profile $U^*(\cdot)$ in subgame $\Gamma(x_1)$. Let $\overline{\Gamma}(x_1)$ be $\Gamma(x_1)$, where the choices of the players $i \notin S_k(x_1)$ are fixed according to strategy profile $U_{x_1}^*(\cdot)$. This means that the game $\overline{\Gamma}(x_1)$ is played among players from $S_k(x_1)$. For each subcoalition $R \subset S_k(x_1)$ consider an associated with $\overline{\Gamma}(x_1)$ zero-sum game $\overline{\overline{\Gamma}}(x_1,\ S_k(x_1))$ with two players: coalition R, as maximizing player (payoff of R is equal to the sum of payoffs of players in R), and the coalition $S_k(x_1) \backslash R$, as minimizing player (payoff of $S_k(x_1) \backslash R$ is equal to the payoff of coalition R with opposite sign). One can show that the payoff of each coalition R, cannot exceed the value of $v(x_1,\ S_k(x_1)) = \sum_{i \in S_k(x_1)} r_i^1(x_1)$

(see [2]), since by construction the coalition $S_k(x_1)$ gets the payoff $v(x_1, S_k(x_1))$ by using the best reply against the strategies $U^*_{ix_1}(\cdot)$ of players $i \notin S_k(x_1)$. Let $v(x_1,\ R)$ be the value of the game $\overline{\overline{\Gamma}}(x_1,\ S_k(x_1))$. With the help of $v(x_1,\ R)$, $R \subset S_k(x_1)$ the Shapley value in $\Gamma(x_1)$ is computed in the usual way.

Step t. Consider the position $x_t \in X_t$, $x_t \notin P_{n+1}$, and restriction $U^*_{x_t}(\cdot) = (U^*_{1x_t}(\cdot),\ ...,\ U^*_{nx_t}(\cdot))$ of the strategy profile $U^*(\cdot)$ in subgame $\Gamma(x_t)$, and the truncation $\overline{\Gamma}(x_t)$ of $\Gamma(x_0)$, in which the choices of players $i \notin S_k(x_t)$ in their personal positions are fixed according to $U^*_{x_t}(\cdot)$. Thus the game $\overline{\Gamma}(x_t)$ is the game among players in $S_k(x_t)$. For each subcoalition $R \subset S_k(x_t)$ consider an associated with $\overline{\Gamma}(x_t)$ zero-sum game $\overline{\overline{\Gamma}}(x_t, S_k(x_t))$ with two players; the coalition R, as maximizing player (the payoff of coalition R is equal to the sum of payoffs of players in R), and coalition $S_k(x_t)\backslash R$, as minimizing player (the payoff of $S_k(x_t)\backslash R$ is equal to the payoff of R with opposite sign). It can be shown that the payoff of coalition R cannot exceed the value of $v(x_t,\ S_k(x_t)) = \sum\limits_{i \in S_k(x_t)} r^t_i(x_t)$ (see [9]), since by construction the coalition $S_k(x_t)$ gets $v(x_t,\ S_k(x_t))$ by using the best reply strategies against strategies $U^*_{ix_t}(\cdot)$ chosen by $i \notin S_k(x_t)$. Let $v(x_t,\ R)$ be the value of $\overline{\overline{\Gamma}}(x_t,\ S_k(x_t))$. With the help of c.f $v(x_t,\ R)$, $R \subset S_k(x_t)$ the Shapley value in $\Gamma(x_t)$ can be computed in a usual way

Going further we can construct $v(x,\ R)$, $R \subset S_k(x)$ for all $x \notin P_{n+1}$ and the corresponding Shapley value. Note that the constructed c.f. $v(x,\ R)$, $R \subset S_k(x)$, $x \in X_t$ requires information about $U^*_{xt}(\cdot)$, but the last also requires information about $v(x,\ R)$, but for $x \in X_l$, $l < t$. Thus our construction is made correctly.

5. Example

Consider the game $\Gamma(x_0)$ with changing coalitional structure and game tree $K(x_0)$, represented on figure 1. Here $N = \{1,\ 2,\ 3\}$. The set off personal positions of player 1 is $P_1 = \{x_1,\ x_2,\ x_3,\ x_{23},\ x_{24},\ x_{25}\}$. The set of personal positions of player 2 is $P_2 = \{x_7,\ x_8,\ x_{11},\ x_{27},\ x_{29},\ x_{31}\}$. The set of personal positions of player 3 is $P_3 = \{x_5,\ x_{13},\ x_{33},\ x_{35},\ x_{37}\}$. The set of positions in which the chance makes the choice is $P_4 = \{x_0,\ x_{14},\ x_{39}\}$. The payoffs are written in terminal positions in columns, and in each column the first entry corresponds to the payoff of player 1, the second to the payoff of player 2 and so on. Suppose that the set of admissible coalitional partitions of player set $N = \{1,\ 2,\ 3\}$ has the form $\Delta = \{\Delta_1,\ \Delta_2,\ \Delta_3\}$, $\Delta_1 = \{\{1\},\ \{3,2\}\}$; $\Delta_2 = \{1,\ 2,\ 3\}$; $\Delta_3 = \{\{1,\ 3\},\ \{2\}\}$. Suppose in positions $x \in Z(x_0)$ following transition probabilities are defined $p(x_1)$, $p(x_2)$, $p(x_3)$, $p(x_1) + p(x_2) + p(x_3) = 1$, $p(x_1) = p(x_3) = \frac{1}{6}$; $p(x_2) = \frac{2}{3}$. Suppose also $\Delta(x_1) = \{\{1\},\ \{3,2\}\}$; $\Delta(x_2) = \{1,\ 2,\ 3\}$; $\Delta(x_3) = \{\{1,\ 3\},\ \{2\}\}$. Similarly in positions $x \in Z(x_{14})$ the transition probabilities $p(x_{20}) = p(x_{22}) = \frac{1}{6}$; $p(x_{21}) = \frac{2}{3}$, and coalitional partitions $\Delta(x_{20}) = \{\{1\},\ \{3,\ 2\}\}$; $\Delta(x_{21}) = \{1,\ 2,\ 3\}$; $\Delta(x_{22}) = \{\{1,\ 3\},\ \{2\}\}$ are defined.

In $x \in Z(x_{19})$ the transition probabilities $p(x_{23}) = p(x_{25}) = \frac{1}{6}$; $p(x_{24}) = \frac{2}{3}$ and coalitional partitions $\Delta(x_{23}) = \{\{1\}, \{3, 2\}\}$; $\Delta(x_{24}) = \{1, 2, 3\}$; $\Delta(x_{25}) = \{\{1, 3\}, \{2\}\}$ are given. Construct an optimal bunch in $\Gamma(x_0)$. The algorithm starts in terminal positions $x_{38}, x_{39}, x_{40}, x_{41}, x_{42}, x_{43}$. Consider subgames on subtrees $K(x_{33}), K(x_{35}), K(x_{37})$. $\Delta(x_{33}) = \Delta(x_{38}) = \Delta(x_{39})$. In x_{33} player 3 makes a choice. $3 \in \{3, 2\} \in \Delta(x_{33})$. $h_2(x_{38}) + h_3(x_{38}) = 5 > h_2(x_{39}) + h_3(x_{39}) = 4$. Thus in x_{33} player 3 chooses the alternative leading to x_{38}. Then $r^1(x_{33}) = (2, 3, 2)^\star$; $r^1(x_{31}) = (1, 1, 1)^\star$. $\Delta(x_{35}) = \Delta(x_{40}) = \Delta(x_{41})$.

Consider subgames on subtrees $K(x_{27})$, $K(x_{29})$, $K(x_{31})$. We have $\Delta(x_{27}) = \Delta(x_{31}) = \Delta(x_{33})$. In x_{24} moves player 1. $1 \in \{1\} \in \Delta(x_{23})$. Since $r_1^2(x_{26}) = 0 < r_1^2(x_{27}) = 2$, in x_{23} player 1 chooses x_{27}. Then $r^3(x_{23}) = (2, 3, 2)^\star$. Similarly we get: $r^3(x_{24}) = (3, 4, 1)^\star$; $r^3(x_{25}) = (1, 1, 1)^\star$.

Consider subgame on subtree $K(x_{19})$. In x_{19} chance makes a move. $\Delta(x_{23}) = \{\{3, 2\}, \{1\}\}$. Select the coalition $\{3, 2\} \in \Delta(x_{23})$ and consider a cooperative 2 person game $G(x_{23}, \{3, 2\})$ with c.f $v(x_{23}, R)$, $R \subset \{3, 2\}$. Computations give us the values of $v(x_{23}, R)$, $R \subset \{3, 2\}$. $v(x_{23}, \{3, 2\}) = 5$, $v(x_{23}, \{3\}) = 1$, $v(x_{23}, \{2\}) = 2$. Then in x_{23}: $Sh_2(\{3, 2\}, x_{23}) = 3$; $Sh_3(\{3, 2\}, x_{23}) = 2$; $\{\Delta(x_{23}) \setminus \{3, 2\}\} = \{1\}$.Then $Sh_1(\{1\}, x_{23}) = 2$. Construct $PMS(x_{23})$. $PMS_1(x_{23}) = Sh_1(\{1\}, x_{23}) = 2$, $PMS_2(x_{23}) = Sh_2(\{3, 2\}, x_{23}) = 3$, $PMS_3(x_{23}) = Sh_3(\{3, 2\}, x_{23}) = 2$. Thus $PMS(x_{23}) = (2, 3, 2)$. $\Delta(x_{24}) = \{1, 2, 3\}$. Consider cooperative 3-person game $G(x_{24}, \{1, 2, 3\})$ with c.f $v(x_{23}, R)$, $R \subset \{1, 2, 3\}$. Compute c.f $v(x_{23}, R)$, $R \subset \{1, 2, 3\}$. $v(x_{24}, \{1, 2, 3\}) = 8$, $v(x_{24}, \{1, 2\}) = 6$, $v(x_{24}, \{2, 3\}) = 5$, $v(x_{24}, \{1, 3\}) = 2$, $v(x_{24}, \{1\}) = 1$, $v(x_{24}, \{2\}) = 4$, $v(x_{24}, \{3\}) = 0$. Then in x_{24}: $Sh_1(\{1, 2, 3\}, x_{24}) = 2$; $Sh_2(\{1, 2, 3\}, x_{24}) = 5$; $Sh_3(\{1, 2, 3\}, x_{24}) = 1$; $PMS_1(x_{24}) = Sh_1(\{1, 2, 3\}, x_{24})$
$= 2$, $PMS_2(x_{24}) = Sh_2(\{1, 2, 3\}, x_{24}) = 5$, $PMS_3(x_{24}) = Sh_3(\{1, 2, 3\}, x_{24}) = 1$.

Thus $PMS(x_{24}) = (2, 5, 1)$. In similar way we get $PMS(x_{25}) = (\frac{1}{2}, 1, \frac{3}{2})$. Then $r_1^4(x_{19}) = p(x_{23}) \cdot PMS_1(x_{23}) + p(x_{24}) \cdot PMS_1(x_{24}) + p(x_{25}) \cdot PMS_1(x_{25}) = \frac{21}{12}$; $r_2^4(x_{19}) = p(x_{23}) \cdot PMS_2(x_{23}) + p(x_{24}) \cdot PMS_2(x_{24}) + p(x_{25}) \cdot PMS_2(x_{25}) = \frac{52}{12}$; $r_3^4(x_{19}) = p(x_{23}) \cdot PMS_3(x_{23}) + p(x_{24}) \cdot PMS_3(x_{24}) + p(x_{25}) \cdot PMS_3(x_{25}) = \frac{13}{12}$. Thus $r^4(x_{19}) = (\frac{3}{4}, 6, \frac{3}{4})^\star$ and $r^4(x_{16}) = (0, 1, 1)^\star$, $r^4(x_{17}) = (0, 0, 0)^\star$, $r^4(x_{18}) = (1, 4, 1)^\star$, $r^4(x_{20}) = (0, 12, 0)^\star$, $r^4(x_{21}) = (3, 0, 0)^\star$, $r^4(x_{22}) = (0, 0, 0)^\star$. Omit the detailed discussion of the following steps. Moving step by step downwards to x_0 we will reach x_0. Note, that according to the algorithm in positions x_5 player 3 and in position x_8 player 2 chooses the alternatives with equal probabilities. Since in x_0 the chance is making a move ($x_0 \in P_4$), $r^8 = (\frac{292}{108}, \frac{123}{108}, \frac{94}{108})^\star$, i.e. the payoffs along optimal bunch coincide with PMS vector (more precisely its mathematical expectation) for the whole game. The optimal bunch on the figure is denoted by bold line.

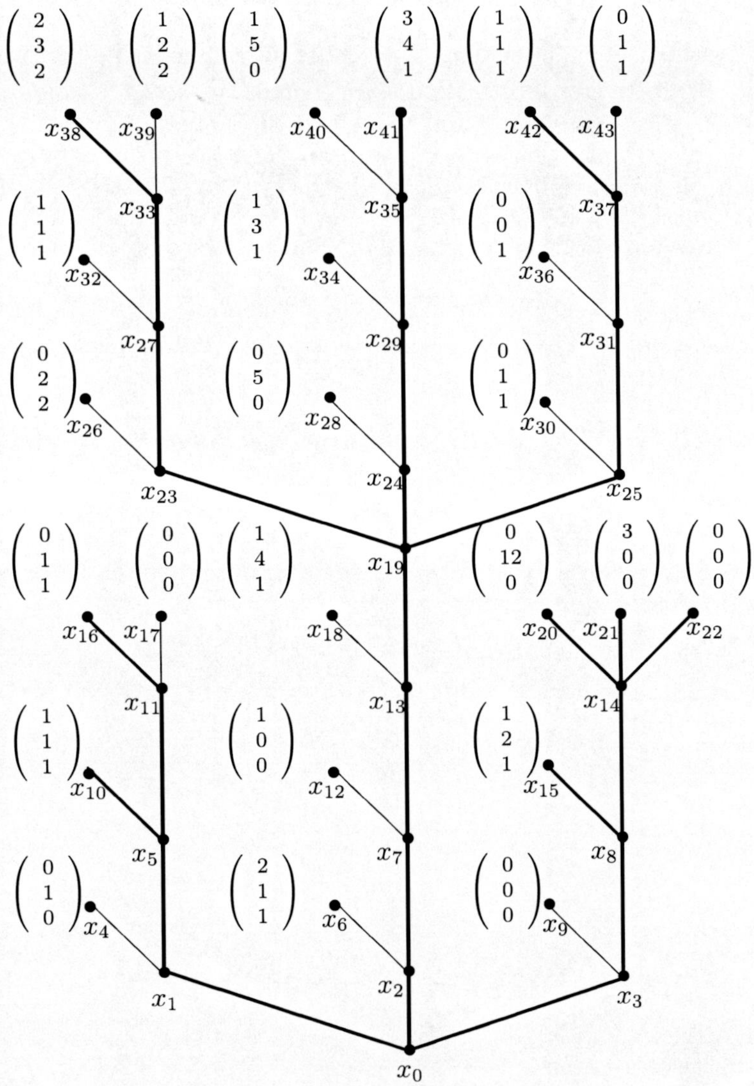

Figure 1.

References

[1] F. Bloch, Sequental Formation of Coalitions with Fixed Payoff Division, *Games and Economic Behaviour* **14** (1966), 90–123.

[2] R. Ono, S. Muto, Stability of Japan's Coalitions Cabinet, *Power and Fairness*, Ed. by M.J.Holer at al., Mohr Siebech, Germany, 2002, 17–25.

[3] G. Owen, A characterization of the Banzof-Coleman Index, *SIAM Journal of Applied Mathematics* **2** (1978), 315–327.

[4] G. Owen, Values of Games with a Priori Unions, *Mathematical Economy and Game Theory*, Ed. by R. Henn and O. Moeschlin., Berlin, 1977, 78–88.

[5] S.I. Mamkina, L.A. Petrosjan, The Set of all Subgame Perfect Nash Equilibria in Games With Perfect Information, *Proceedings of the 35th Conference on Stability and Control*, St. Petrsburg Univ. Press, St.Petersburg, 2004, 254–266 .

[6] J. Nash, Non-cooperative Games, *Ann. of Math.* **54** (1951), 286–295.

[7] L.A. Petrosjan, N.A. Zenkevich, *Game Theory*, World Sci., London, 1996.

[8] L.A. Petrosjan, D.A. Ayoshin, Values of Dynamic Games with Partial Cooperation, *Proceedings of the Institute of Mathematics and Mechanics* **6**, no. 1,2 (2000), Ekaterinburg, 160–172.

[9] L.A. Petrosjan, D.V. Kusyutin, *Games in Extencive Form: Optimality and Stability*, St.Petrsburg Univ. Press, St.Petersburg, 2000.

[10] L.S. Shapley, A Value for n-Person Games, *Contributions to the Theory of Games*, Ed. by H.W.Kuhn and A.W.Tucher., Princeton: Princeton University Press, 1953, 307–317.

In: Mathematics, Game Theory and Algebra Compendium ISBN 978-1-60456-659-8
Editor: Jacob H. Mathias, pp. 729-748

Chapter 61

THE REDISTRIBUTION PARADOX

Agnieszka Rusinowska
Department of Philosophy
Tilburg University, The Netherlands
Department of Mathematical Economics
Warsaw School of Economics
Warsaw, Poland
and Adrian van Deemen
University of Nijmegen
Nijmegen, The Netherlands

1. Introduction

When analyzing a voting situation, the important question arises how powerful actors are, and how to measure their voting power. In the course of about fifty years, a number of measures of voting strength, so called, power indices, have been proposed. The most important power indices to be found in the literature are the Shapley-Shubik index (Shapley and Shubik 1954, see also Shapley 1953), the normalized Banzhaf index (Dubey and Shapley 1979), the Penrose-Banzhaf index (Banzhaf 1965, Penrose 1946), the Deegan-Packel index (Deegan and Packel 1978, 1982), and the Holler index (Holler 1982, Holler and Packel 1983). These power indices have been reviewed, discussed and compared, for instance, in Felsenthal and Machover (1998). Apart from theoretical research, empirical research on power indices is of importance. In particular, one can find many empirical applications of power indices to the field of decision-making in the European Union, as well as to the analysis of national parliaments. In this paper, we use power indices to study parliamentary elections in Germany.

In general, a parliament can be modelled as a weighted voting game with political parties as players. The number of seats of a given party assigned to that party after an election is the weight of that player. Power indices applied to such a weighted voting game evaluate the distribution of power among the parties. In this paper, we apply this approach to the seat distribution arising from fifteen parliamentary elections in Germany. We analyze the German elections held in 1949, 1953, 1957, 1961, 1965, 1969, 1972, 1976, 1980, 1983, 1987, 1990, 1994, 1998, and 2002. For each of these fifteen seat distributions, we calculate the Shapley-Shubik index, the normalized Banzhaf and the Penrose-Banzhaf indices, the Deegan-Packel index, and the Holler index.

In the study and calculation of power indices, some unexpected and counterintuitive features emerged, called in the literature paradoxes of voting power. Well-known paradoxes of voting power are, in particular, the paradox of redistribution (Fischer and Schotter 1978, Schotter 1982), the paradox of new members (Brams 1975, Brams and Affuso 1976), the paradox of large size (Brams 1975), and the quarrelling paradox (Kilgour 1974). A theoretical analysis of voting power paradoxes and postulates for power indices is given, for instance, in Felsenthal and Machover (1995, 1998), and in Felsenthal, Machover and Zwicker (1998). In the literature, one can also find empirical research on paradoxes of power indices (see, for instance, Schotter 1982). In particular, the occurrence of voting power paradoxes in real politics has been investigated by Van Deemen and Rusinowska (2003), and Rusinowska (2001). Since there are paradoxes which are theoretically possible, but empirically do not occur frequently (for instance, Condorcet paradox), it is interesting to investigate how frequently a given paradox occurs in reality.

In this paper, we search for voting power paradoxes in German politics. Fifteen German elections will be analyzed and scrutinized on the occurrence of modified versions of the paradox of redistribution and the paradox of new members. Further, the frequencies of the appearance of these paradoxes will be calculated. We also check how frequently the power indices agree and disagree in the cases in which these paradoxes occur. Finally, we compare the frequencies calculated for the German data with the frequencies of the occurrence of the same paradoxes for the Dutch elections (see Van Deemen and Rusinowska 2003). For a clear description of the German electoral system, consider Farrell (2001).
One may define the aim of this paper as answering the following questions:

(1) Did the paradox of redistribution occur in the German elections in the period 1949-2002, and if so, how frequently?

(2) Did the paradox of new members occur in the German elections in the period 1949-2002, and if so, how frequently?

(3) How frequently do the power indices agree concerning the existence of the respective paradoxes?

(4) What conclusions can be drawn from the comparison of the occurrence of the analyzed paradoxes in the German data with the occurrence of these paradoxes in the Dutch elections?

By the paradox of redistribution and the paradox of new members mentioned in questions (1) and (2), respectively, we mean our modified versions of these paradoxes, defined in the next section.

The structure of this paper is as follows. Section 2 contains the theoretical background for analyzing the German elections. We briefly present some basic concepts of the theory of voting games, the definitions of power indices used in this paper, and the original formulations of the paradox of redistribution, and the paradox of new members. In our empirical research, we use slightly different versions of the paradoxes described for the reasons to be explained in Section 2. In Section 3, fifteen seat distributions in the German Parliament are studied. In Section 4, the paradox of redistribution is investigated empirically. We calculate the frequency of the occurrence of this paradox for the German Parliament, and we compare the results with the ones obtained for the Second Chamber in the Netherlands. Section 5 concerns the paradox of new members. We scrutinize the occurrence of this paradox in German politics, and we compare its frequency with the frequency of this paradox in Dutch politics. Section 6 discusses the agreement among the several indices concerning the occurrence of the paradoxes. Again, we compare the results obtained for the German politics with the results concerning the Dutch parliament. Finally, Section 7 contains some conclusions, and the Appendix contains the tables referred to.

2. Power Indices and Their Paradoxes

2.1. Voting Games and Power Indices

In this section, we briefly present the theoretical background. We start with the definition of a voting game, and next we give a short survey of most-known power indices. For a more extensive analysis of voting games see, for instance, Shapley (1962), Owen (1995), Van Deemen (1997), and Felsenthal and Machover (1998).

Let $N = \{1, 2, \ldots, n\}$ denote a nonempty (and finite) set of players. Any subset of N is called a *coalition*. A *voting game* G is an ordered pair $G = (N, W)$, where W is a set of coalitions, which satisfies the following assumptions:

(i) $W \neq \emptyset$

(ii) $\emptyset \notin W$

(iii) If $S \subset T$ and $S \in W$, then $T \in W$.

A member of W is called a *winning coalition*. The set of all coalitions not belonging to W is denoted by L, and a member of L is called a *losing coalition*. A *blocking* coalition is a losing coalition such that its complement is also losing.

A voting game is said to be *proper* if the complement of any winning coalition is losing, and it is *strong* if the complement of any losing coalition is winning. A voting game is called *decisive* if it is both proper and strong.

Player $k \in S$ is a *swinger* in a winning coalition S, if his or her removal from this coalition makes it losing. We refer to a winning coalition in which all players are swingers as a *minimal winning coalition*. In other words, a minimal winning coalition is a winning coalition such that no proper subset of it is winning.

A *dummy* can be characterized as a player who does not belong to any minimal winning coalition. A *veto-player* is a player who belongs to every minimal winning coalition. Finally, if one player forms the only minimal winning coalition, then this player is called a *dictator*. In other words, a dictator, if there is one, can be characterized as the sole player who is not a dummy.

One interesting and important class of voting games is the class of *weighted voting games*, in which a weight is assigned to each player indicating his voting strength in that game. We use the notation $[q; w_1, w_2, ..., w_n]$ to represent a weighted voting game. Here, q is the *quota* needed for a coalition to win, and w_k is the weight (number of votes/seats) of player k ($k = 1, 2, ..., n$). The quota q and the voting weights $w_1, w_2, ..., w_n$ are positive integers with $0 < q \leq \sum_{k=1}^{n} w_k$. A coalition S in a weighted voting game is winning if and only if the sum of the weights of its members is at least as large as q, that is,

$$S \in W \Leftrightarrow \sum_{k \in S} w_k \geq q. \tag{1}$$

A *power index* is a function ϕ that associates with each voting game (N, W) a vector $\phi = (\phi_1, ..., \phi_n)$, where the nonnegative real number ϕ_k is interpreted as a measure of the influence that player $k \in N$ can exert on the outcome. Many ideas about how to evaluate the distribution of power among the players have appeared. We give a short survey of the power indices which we use in this paper. We present only the basic formulae of these indices. For a more extensive analysis of the power indices see, for instance, Felsenthal and Machover (1998), Lucas (1982), Owen (1995), and Straffin (1982).

- The *Shapley-Shubik index* (Shapley and Shubik 1954) for the voting game $G = (N, W)$ is the vector $Sh(G) = (Sh_1(G), ..., Sh_n(G))$, given by

$$Sh_k(G) = \frac{p_k}{n!} \qquad (k = 1, 2, ..., n), \tag{2}$$

  where p_k is the number of orders in which player k is *pivotal*. A player is said to be pivotal for an order if it turns the coalition of players preceding him in that order into a winning coalition.

- The *normalized Banzhaf index* (Dubey and Shapley 1979) for the voting game $G = (N, W)$ is the vector $Bz(G) = (Bz_1(G), ..., Bz_n(G))$, defined by

$$Bz_k(G) = \frac{\eta_k}{\sum_{j \in N} \eta_j} \qquad (k = 1, 2, ..., n), \tag{3}$$

  where η_k is the number of winning coalitions in which player k is a swinger.

- The *Penrose-Banzhaf index* (Penrose 1946, Banzhaf 1965), also called the *non-normalized Banzhaf index* or the *absolute Banzhaf index* is the vector $PBz(G) = (PBz_1(G), ..., PBz_n(G))$, given by

$$PBz_k(G) = \frac{\eta_k}{\beta_k} = \frac{\eta_k}{2^{n-1}} \qquad (k = 1, 2, ..., n), \tag{4}$$

  where β_k denotes the total number of coalitions containing player k.

- The *Deegan-Packel index* (Deegan and Packel 1978, 1982) for the voting game $G = (N, W)$ is the vector $DP(G) = (DP_1(G), ..., DP_n(G))$, defined by

$$DP_k(G) = \frac{1}{m} \sum_{\{S \in M : k \in S\}} \frac{1}{s} \qquad (k = 1, 2, ..., n), \tag{5}$$

where M is the set of all minimal winning coalitions, m is the total number of minimal winning coalitions, and s is the number of players in S.

- The *Holler index* (Holler 1982, Holler and Packel 1983), also called the *Holler-Packel index* or the *public good power index*, is the vector
$H(G) = (H_1(G), ..., H_n(G))$, defined by

$$H_k(G) = \frac{m_k}{\sum_{j \in N} m_j} \qquad (k = 1, 2, ..., n), \tag{6}$$

where m_k denotes the number of minimal winning coalitions containing player k.

2.2. Voting Power Paradoxes

In this subsection, we will recapitulate some counterintuitive features of power indices as described and analyzed in the literature. These counterintuitive features are usually called *paradoxes of voting power.* We only describe the two voting power paradoxes which are considered in this paper, namely, the paradox of redistribution and the paradox of new members.

- *The paradox of redistribution* (Fischer and Schotter 1978, Schotter 1982) - This paradox appears when either a party's voting weight decreases and at the same time its voting power index increases, or when a party gains in terms of voting weight, but loses in voting power.
Let $G = [q; w_1, ..., w_n]$ and $G' = [q; w'_1, ..., w'_n]$ be two weighted voting games such that $\sum_{k=1}^{n} w_k = \sum_{k=1}^{n} w'_k$. A power index ϕ displays the *redistribution paradox* if

$$for\ some\ k,\ w'_k < w_k\ and\ \phi_k(G') > \phi_k(G). \tag{7}$$

- *The paradox of new members* (Brams 1975, Brams and Affuso 1976) -
This paradox appears when a new party joins the assembly and at least one 'old' party has greater voting power in this new situation than in the old one.
Let $G = [q; w_1, ..., w_n]$ be a voting game, and let $G' = [q'; w_1, ..., w_n, w_{n+1}]$ be a new game with a new party (n+1) joined. A power index ϕ displays the *paradox of new members* if

$$for\ some\ k \in N,\ \phi_k(G') > \phi_k(G). \tag{8}$$

The original definition of the paradox of redistribution above requires that both the total number of seats and the number of parties are fixed in the succeeding parliamentary periods. In this paper, we like to check the occurrence of this paradox for the German elections held between 1949 and 2002. For the whole period 1949-2002, none of the requirements mentioned is satisfied by the composition of the German Parliament. In the original definition of the paradox of new members, the weights of the old parties are held constant. Also this condition is not satisfied for the German Parliament. In order to be able to scrutinize the occurrence of the two paradoxes in reality, we define a modified version of each in the following way:

- **Modified paradox of redistribution** -
  Let $\widetilde{w_t}$ and ϕ_t denote the relative voting weight (that is, % of seats) in year t, and the power index of a party in year t, respectively. A power index ϕ displays the *modified redistribution paradox* if one of the following conditions (P1) or (P2) is satisfied:

  (P1): $\widetilde{w_t} > \widetilde{w_{t+k}}$ and $\phi_t \leq \phi_{t+k}$,

  (P2): $\widetilde{w_t} < \widetilde{w_{t+k}}$ and $\phi_t \geq \phi_{t+k}$.

- **Modified paradox of new members** -
  Let $\widetilde{w_t}$, ϕ_t, and P_t denote the relative voting weight in year t, the power index of a party in year t, and the set of all parties participating in an election in year t, respectively. A power index ϕ displays the *modified paradox of new members* if the following condition (P) is satisfied:

  (P): $\widetilde{w_t} \geq \widetilde{w_{t+k}}$ and $\phi_t < \phi_{t+k}$ and $P_t \subset P_{t+k}$.

Clearly, these definitions differ from the original formulations. First of all, since the total number of seats in the German Parliament was not constant for all parliamentary periods analyzed, we consider relative weights instead of real weights in the definitions of paradoxes (P1), (P2), and (P). Moreover, in the definitions of (P1) and (P2), we do not assume the number of parties to be constant. We could keep this assumption if we considered separately the German elections in 1961-1980, 1983-1987, and 1990-2002. However, since we wish to analyze the changes in the whole period 1949-2002, we cannot but drop the assumption about the fixed number of parties. Finally, also note that our formulation of the redistribution paradox differs from the original definition in adding the equality sign for the power indices ϕ_t and ϕ_{t+k} in (P1) and (P2).

In the definition of the modified paradox of new members, we do not assume that the weights of the old parties are held constant. The modified paradox of new members as defined by (P) occurs when there is a rise in the power of an incumbent party with less or equal relative weight after the entrance of new parties. Moreover, we add the condition that the set of all parties participating in an election in year t must be a subset of the set of all parties participating in an election in year $t + k$.

3. Power Indices for the German Parliament

The fifteen parliamentary elections in Germany we wish to scrutinize were held in 1949, 1953, 1957, 1961, 1965, 1969, 1972, 1976, 1980, 1983, 1987, 1990, 1994, 1998, and recently, in 2002. In this section, we study the fifteen seat distributions resulting from these elections. The total numbers of seats for the German Parliament in the periods mentioned differ from each other. This is why we have calculated the relative numbers of seats expressed as the percentages of the total number of seats for a given period. Table 1 in the Appendix presents the seat distribution matrix for the German Parliament in 1949-2002.

Both the relative weights and the numbers of seats are indicated in Table 1. For each party, and for each election in which the party participated, the first number is the relative weight, and the second number is the number of seats assigned to that party after the election.

The names of the German parties included in Table 1 are the following:
CDU/CSU - Christian Democratic Union/Christian Social Union
SPD - Social Democratic Party of Germany
FDP - Free Democratic Party
GRUNE - Alliance 90/The Greens
PDS - Party of Democratic Socialism
KPD/DKP - Communist Party
DRP/NPD - German Reich Party/National Democratic Party
DP - German Party
GB/BHE - Bloc of Expelles
BP - Bavaria Party
WAV - Reconstruction League
Zentrum - Centre Party
SSW - South Schleswig Voter's League

Using Table 1, one can check whether the weighted voting games representing the situation after the fifteen elections in Germany are proper, strong and, hence, decisive. It appears that all the fifteen games are proper, and all but one are strong (and hence decisive). The game representing the election in 1949 is the only one which is not strong, and therefore, is not decisive.
In order to see this, consider $S^* = \{CDU/CSU, KPD/DKP, DRP/NPD, DP, BP, Zentrum\}$. Since $\sum_{k \in S^*} w_k = 205$, and the quota $q = 206$, coalition S^* happens to be losing ($S^* \in L$), but also its complement $N \setminus S^*$ is losing, because $\sum_{k \in N \setminus S^*} w_k = 205$. It means that the 1949 game is not strong, and we say that coalition S^* is blocking. These results are in contrast to the Dutch situation. Van Deemen and Rusinowska (2003) found for that country that only three out of fifteen elections led to a proper and strong parliamentary game.

For each seat distribution, we have calculated the Shapley-Shubik index, the normalized Banzhaf index, the Penrose-Banzhaf index, the Deegan-Packel index, and the Holler index. The Shapley-Shubik and the normalized Banzhaf indices are the same for all but two seat distributions. They only differ from each other in 1949 and 1953. Table 2 in the Appendix shows the Shapley-Shubik and the normalized Banzhaf indices for the German Parliament.

Since the remaining power indices differ more frequently from each other, we have constructed separate tables for them. Tables 3, 4, and 5 in the Appendix present the absolute Banzhaf index, the Deegan-Packel index, and the Holler index, respectively.

The first conclusion we draw from Tables 2-5 concerns the appearance of dummies in German politics. For all five kinds of power indices used in this paper it is true that the power index of a player is equal to zero if and only if this player is a dummy (for the proof see, for instance, Van Deemen and Rusinowska 2003). Hence, we can easily conclude that dummy parties appeared in six out of fifteen seat distributions, that is, in 1953, 1957, 1990, 1994,

1998, and in 2002. Note that PDS is a dummy already since 1994. GRUNE was a dummy in 1990 and the Zentrum party in 1953. SPD, FDP, and DP were dummy parties in 1957. Analyzing the seat distributions since 1957, we find that CDU/CSU is the only party that never has been a dummy. Moreover, CDU/CSU was a dictator in 1957, and it is the only case of the occurrence of a dictator (and hence of a veto-player) in the German elections held between 1949 and 2002.

4. The Paradox of Redistribution in the German Parliament

We now scrutinize the occurrence of the redistribution paradox for the fifteen German parliamentary elections held between 1949 and 2002. The modified version of this paradox as defined by (P1) and (P2) (see Section 2.2) is used. We investigate the appearance of the paradox of redistribution for all German parties appearing more than once in Parliament in the period 1949-2002. There are seven of such parties: CDU/CSU, SPD, FDP, GRUNE, PDS, DP, and Zentrum. It appears that only for Zentrum, no redistribution paradox occurs. Tables 6-11 in the Appendix present the occurrence of the redistribution paradox for CDU/CSU, SPD, FDP, GRUNE, PDS, and DP, respectively. The symbols used in these tables mean the following:

P1 - paradox (P1) occurs for all power indices analysed (Sh, Bz, PBz, DP, H)
P2 - paradox (P2) occurs for all power indices analysed (Sh, Bz, PBz, DP, H)
PBP2 - paradox (P2) occurs only for the Penrose-Banzhaf index
HP1 - paradox (P1) occurs only for the Holler index
HP2 - paradox (P2) occurs only for the Holler index
DHP1 - paradox (P1) occurs only for the Deegan-Packel and the Holler indices
(DH)'P2 - paradox (P2) occurs only for the Shapley-Shubik index, the normalized Banzhaf index, and the Penrose-Banzhaf index

Note that the Shapley-Shubik index and the normalized Banzhaf index do *not* differ from each other with respect to the appearance of the redistribution paradox (both (P1) and (P2)) in German politics. Equally surprising, the Penrose-Banzhaf index does *not* differ from the Shapley-Shubik and the normalized Banzhaf indices with respect to the occurrence of paradox (P1).

In order to calculate the overall frequency of the paradox, we use the following method. First, we calculate the frequency of the paradox for each party mentioned. Let $nPiX$ denote the number of cases in which paradox (Pi) appears for power index X of a party, where $i = 1, 2$, and $X \in \{Sh, Bz, PBz, DP, H\}$. Let k be the number of elections in which a party participates. Since we like to analyze all possible cases in which the redistribution paradox might occur for a party, we have to check each pair of elections in which a party participates. Let l be the number of cases analyzed for a party. Hence, l is equal to the number of two-element combinations among k elements:

$$l = \binom{k}{2} = \frac{k!}{2!(k-2)!} = \frac{k(k-1)}{2}. \tag{9}$$

Furthermore, let $fPiX$ be the frequency of the paradox (Pi), $i = 1, 2$, for power index X of a party, and fPX be the frequency of the redistribution paradox for power index X of a party. Formally,

$$fP1X = \frac{nP1X}{l}, \qquad fP2X = \frac{nP2X}{l}, \qquad fPX = \frac{nP1X + nP2X}{l}. \tag{10}$$

The frequencies of the occurrences of paradoxes (P1) and (P2) for each kind of power index, and for all the seven German parties are presented in Table 12 (see the Appendix).

As was mentioned before, only for Zentrum no redistribution paradox occurs. For all remaining six parties, the frequencies of the occurrence of both paradoxes (P1) and (P2) are positive, for each power index analyzed. For GRUNE, PDS, and DP, if paradox (P1) or (P2) occurs in a given election, then it appears for all five kinds of power indices. Note that in Tables 9-11, the only notations used are P1 and P2. Hence, for each of these three parties (GRUNE, PDS, and DP), the frequencies of the occurrences of the paradox are the same for all power indices used. These frequencies are equal to 0,2 for GRUNE, 0,3333 for DP (for both (P1) and (P2)), and for PDS: 0,3333 and 0,5 for the occurrence of (P1) and (P2), respectively. Relatively low frequencies appear for the power indices of SPD (between 0,1429 and 0,1905). Paradoxes (P1) and (P2) occur relatively more frequently for FDP (the frequency is between 0,2476 and 0,4095), and for CDU/CSU (the frequency is between 0,0952 and 0,4476).

For the overall frequencies of the redistribution paradox, we will use the following notation. Let $TNPiX$ be the total number of cases in which paradox (Pi) appeared for power index X, $i = 1, 2$. Let L denote the total number of cases analyzed for all parties. Moreover, $FPiX$ is the frequency of paradox (Pi) for power index X, $i = 1, 2$, and finally, FPX denotes the frequency of the redistribution paradox for X. Formally, we can write

$$FP1X = \frac{TNP1X}{L}, \qquad FP2X = \frac{TNP2X}{L}, \qquad FPX = \frac{TNP1X + TNP2X}{L}. \tag{11}$$

The overall frequencies $FP1X$, $FP2X$, and FPX, where $X \in \{Sh, Bz, PBz, DP, H\}$, are presented in Table 13 (see the Appendix).

The main conclusion to be drawn from Table 13 is that the redistribution paradox occurs quite frequently in German politics. The frequency FPX is always greater than 0,45, and it is equal to 0,4588 for the Shapley-Shubik index and the normalized Banzhaf index, 0,4617 for the Deegan-Packel index, 0,4794 for the Penrose-Banzhaf index, and 0,5442 for the Holler index. Another conclusion is that the Holler index is the most sensitive index for the occurrence of both paradoxes (P1) and (P2) and, consequently, for the occurrence of the modified redistribution paradox. The Shapley-Shubik and the normalized Banzhaf indices are equally and the least sensitive.

Table 13 also presents a German-Dutch comparison, that is, it includes the frequencies $FP1$, $FP2$, and FP for the Second Chamber in the Netherlands in 1956-1998 (these results were presented in Van Deemen and Rusinowska 2003). For the Dutch elections, also the occurrence of a third version of the redistribution paradox, denoted by (P3) (Van Deemen and Rusinowska 2003), has been scrutinized. This version is described as follows:

(P3): $w_t = w_{t+k}$ and $\phi_t \neq \phi_{t+k}$,

where w_t and ϕ_t denote the voting weight in year t and the power index of a party in year t, respectively. Since paradox (P3) does not appear in the German data, we did not add it to our calculations. We conclude from Table 13, that the redistribution paradox appears much more frequently in German politics than in Dutch politics. For the Second Chamber in the Netherlands, the frequency FPX was between 0,1896 (for the normalized Banzhaf index) and 0,2395 (for the Holler index). Similarly to the German Parliament, the Holler index happens to be the most sensitive for the occurrence of the redistribution paradox in the Dutch data.

5. Paradox of New Members in German Politics

In this section we search for the occurrence of the modified paradox of new members (as defined by (P) in Section 2.2) for the German Parliament in 1949-2002. We use the following method. First, we identify all cases of entrances of new parties which satisfy the condition $P_t \subset P_{t+k}$, where P_t denotes the set of all parties which are elected in year t. There are 44 cases satisfying the condition $P_t \subset P_{t+k}$ (see Table 1). Hence, we compare the following 44 pairs of the elections:

1961-1983, 1961-1987, 1961-1990, 1961-1994, 1961-1998, 1961-2002,
1965-1983, 1965-1987, 1965-1990, 1965-1994, 1965-1998, 1965-2002,
1969-1983, 1969-1987, 1969-1990, 1969-1994, 1969-1998, 1969-2002,
1972-1983, 1972-1987, 1972-1990, 1972-1994, 1972-1998, 1972-2002,
1976-1983, 1976-1987, 1976-1990, 1976-1994, 1976-1998, 1976-2002,
1980-1983, 1980-1987, 1980-1990, 1980-1994, 1980-1998, 1980-2002,
1983-1990, 1983-1994, 1983-1998, 1983-2002,
1987-1990, 1987-1994, 1987-1998, 1987-2002.

For each of these 44 comparisons, we check whether the modified paradox of new members appears. We look for incumbent parties such that their relative weights did not increase, but at least one of their power indices increased after the entrance of new parties. We adopt the following symbols. Let nP be the number of cases in which the modified paradox of new members appeared for a power index of a party. Let m denote the number of cases analyzed for a party, that is, the number of cases from among 44 comparisons of the elections in which a party is an 'old' member participating in the elections compared. Moreover, let fP mean the frequency of the paradox of new members for a power index of a party, that is,

$$fP = \frac{nP}{m}. \tag{12}$$

For the German data, paradox (P) does not appear for the Holler index. For the Shapley-Shubik index, for both Banzhaf indices, and the Deegan-Packel index, paradox (P) occurs, and the results concerning its appearance are the same for all four kinds of power indices mentioned. Table 14 in the Appendix presents the frequencies of the paradox of new members for German parties.

Four incumbent parties have been found for the German elections: CDU/CSU, SPD, FDP, and GRUNE. For FDP and GRUNE, no paradox of new members occurred. The frequency of the appearance of this paradox for SPD is very low, and it is equal to 0,0455. The party which gained in terms of power most frequently after the entrance of new parties is CDU/CSU. The frequency of the paradox of new members for CDU/CSU is equal to 0,3409.

Comparing these results with the occurrence of the paradox of new members for Dutch parties (see Van Deemen and Rusinowska 2003), the conclusion is that the paradox occurs much more frequently in the Dutch Parliament than in the German Parliament. For the Second Chamber of the Dutch Parliament we have that in each cased analyzed, at least one incumbent party gained in terms of power after the entrance of new parties. For ten Dutch parties, the paradox of new members occurred for at least one power index, whereas, as was mentioned before, it appeared for only two German parties. Moreover, the greatest frequency of this paradox was equal to 0,8125 for the Dutch data, and only 0,3409 for the German data. Finally, as was also mentioned before, the results for the German elections happened to be exactly the same for the four kinds of power indices (except the Holler index), whereas the 'Dutch' results were different for all five power indices.

6. Agreements between Power Indices

In this section, we will investigate how frequently the power indices agree concerning the cases in which each paradox occurs. In order to investigate agreements between the different indices, we use the following method (Van Deemen and Rusinowska 2003). If for a party, paradox $(Pi) \in \{(P1), (P2), (P)\}$ appears for at least one of two different power indices ϕ, ϕ', then we define for a given party $nAPi(\phi - \phi')$ as the number of the cases in which the power indices (of the given party) ϕ and ϕ' agree concerning the occurrence of paradox (Pi). By analogy, let $nDAPi(\phi - \phi')$ be the number of the cases in which the power indices (of the given party) ϕ and ϕ' disagree about the occurrence of paradox (Pi). The *agreement index* for a party between ϕ and ϕ', denoted by $fAPi(\phi - \phi')$, is defined as the frequency of agreements between ϕ and ϕ' for a party concerning the cases in which (Pi) occurs. Formally,

$$fAPi(\phi - \phi') = \frac{nAPi(\phi - \phi')}{nAPi(\phi - \phi') + nDAPi(\phi - \phi')}, \tag{13}$$

where $nAPi(\phi - \phi') + nDAPi(\phi - \phi') \neq 0$.

We can also calculate the overall frequency of agreements for the German Parliament, i.e., considering the agreements of the power indices ϕ and ϕ' for all parties instead of just one party. We introduce the following notation:

$TNAPi(\phi - \phi')$ - the total number of cases in which the power indices ϕ and ϕ' agree concerning the occurrence of paradox (Pi)

$TNDAPi(\phi - \phi')$ - the total number of cases in which the power indices ϕ and ϕ' disagree concerning the occurrence of paradox (Pi)

$FAPi(\phi - \phi')$ - the agreement index between ϕ and ϕ' (frequency of agreements between ϕ and ϕ' concerning the cases in which (Pi) occurs)

$$FAPi(\phi - \phi') = \frac{TNAPi(\phi - \phi')}{TNAPi(\phi - \phi') + TNDAPi(\phi - \phi')}, \tag{14}$$

where $TNAPi(\phi - \phi') + TNDAPi(\phi - \phi') \neq 0$.

The agreement indices for each pair of power indices ϕ and ϕ', where $\phi, \phi' \in \{Sh, Bz, PBz, DP, H\}$ and $\phi \neq \phi'$, will be calculated for the occurrence of the paradoxes (P1), (P2), and (P). Hence, in total 30 agreement indices have to be calculated. The agreement indices are given in Table 15 (see the Appendix).

When analyzing Table 15, we can draw the following conclusions. Since the paradox of new members does not appear for the Holler index, each of the four agreement indices $FAP(\phi - H)$, where $\phi \in \{Sh, Bz, PBz, DP\}$ is equal to 0. Since the Shapley-Shubik index, the normalized Banzhaf index, the Penrose-Banzhaf index, and the Deegan-Packel index always agree about the occurrence of the paradox of new members, the remaining agreement indices FAP are equal to 1. Moreover, the Shapley-Shubik index and the normalized Banzhaf index always agree about the occurrence of both paradox (P1) and paradox (P2). Hence, $FAP1(Sh - Bz) = FAP2(Sh - Bz) = 1$. The agreement indices between the Shapley-Shubik index and the Penrose-Banzhaf index about (P1), as well as the agreement index between both Banzhaf indices about (P1) are equal to 1. The remaining agreement indices turn out to be very high - between 0,7024 and 0,9388. The index which agrees least frequently with the remaining power indices (except the Deegan-Packel index) is the Holler index. Nevertheless, the agreement indices between the Holler index and the other power indices about (P1) or (P2) are still pretty high, since they are always greater than 0,7.

Finally, we like to compare the results obtained for Germany with the results concerning the Dutch elections (Van Deemen and Rusinowska 2003). Besides the agreement indices for the German Parliament, Table 15 also shows the agreement indices about paradoxes (P1), (P2), and (P) for the Dutch data. The general conclusion from the German-Dutch comparison is that the agreement indices for the German Parliament are greater than for the Second Chamber in the Netherlands. Although each agreement index for the Dutch data is positive, the lowest one is equal to 0,3663, and there is no agreement index equal to 1. A similarity between the German and Dutch results concerns the Holler index. For both countries, the Holler index agrees least frequently with the other power indices (except the Deegan-Packel index), and the smallest agreement indices (concerning all paradoxes (P1), (P2), and (P)) are the agreement indices between the Holler index and the Penrose-Banzhaf index.

7. Conclusions

In the Introduction, we defined the aim of this paper as answering four questions. In this concluding section, we give the answers to the research questions as formulated in the Introduction.

(1) The modified paradox of redistribution occurred quite frequently in the German elections held between 1949 and 2002. The frequency of the redistribution paradox happens to be always greater than 0,45, and it is equal to 0,4588 for the Shapley-Shubik and the normalized Banzhaf indices, 0,4617 for the Deegan-Packel index, 0,4794 for the Penrose-Banzhaf index, and 0,5442 for the Holler index.

(2) The modified paradox of new members occurred in the German elections, but only rarely. Only two German parties (CDU/CSU and SPD) gained in terms of power after the entrance of new parties. The frequency of the paradox of new members is equal to 0,3409 for CDU/CSU, and 0,0455 for SPD. The results for all power indices used except the Holler index are the same. The paradox of new members does not appear for the Holler index.

(3) In general, power indices agree quite frequently concerning the existence of the respective paradoxes.

From an empirical point of view, there is hardly any difference between the Shapley-Shubik index and the normalized Banzhaf index for the German case. Since 1957, both indices always result in an identical power index matrix. Moreover, the Shapley-Shubik and the normalized Banzhaf indices always agree concerning the occurrence of both the redistribution paradoxes (P1) and (P2), and the paradox of new members (P). Each pair of power indices excluding the Holler index always agree concerning the appearance of the paradox of new members. Each agreement index concerning paradox (P1) is not smaller than 0,8142 (in particular, it is the agreement index between the absolute Banzhaf and the Holler indices). Each agreement index concerning (P2) is not smaller than 0,7024 (the agreement index between the absolute Banzhaf and the Holler indices).

(4) The redistribution paradox appears much more frequently in German politics than in Dutch politics. The overall frequency of the redistribution paradox for the German data is greater than twice the frequency of this paradox for the Dutch data for each power index analyzed. For the data of both countries, the Holler index happens to be the most sensitive one for the occurrence of the redistribution paradox. The paradox of new members occurs much more frequently in the Dutch Parliament than in the German one. This paradox appears for ten Dutch parties (with the greatest frequency equal to 0,8125) and only for two German parties (with the greatest frequency equal to 0,3409). Finally, the agreement indices about the occurrence of the paradoxes are, in general, greater for the German data than for the Dutch data.

Appendix

Table 1. Seat distribution (%) matrix for the German Parliament

year→ party↓	1949	1953	1957	1961	1965	1969	1972	1976	1980	1983	1987	1990	1994	1998	2002
CDU/ CSU	34.39 141	48.92 249	53.56 278	48.18 251	48.46 251	48.26 250	45.17 234	49.03 254	45.67 237	49.04 255	45.08 234	48.19 319	43.75 294	36.62 245	41.13 248
SPD	33.17 136	31.83 162	34.87 181	38.96 203	41.89 217	45.75 237	46.72 242	43.25 224	43.93 228	38.85 202	37.19 193	36.1 239	37.5 252	44.54 298	41.63 251
FDP	12.93 53	10.41 53	8.29 43	12.86 67	9.65 50	5.99 31	8.11 42	7.72 40	10.4 54	6.73 35	9.25 48	11.93 79	6.99 47	6.43 43	7.79 47
GRUNE										5.38 28	8.48 44	1.21 8	7.29 49	7.03 47	9.12 55
PDS												2.57 17	4.47 30	5.38 36	0.33 2
KPD/ DKP	3.66 15														
DRP/ NPD	1.22 5														
DP	4.15 17	2.95 15	3.28 17												
GB/ BHE		5.3 27													
BP	4.15 17														
WAV	2.92 12														
Zentr.	2.44 10	0.59 3													
SSW	0.24 1														
others	0.73 3														
Total (%)	100 410	100 509	100 519	100 521	100 518	100 518	100 518	100 518	100 519	100 520	100 519	100 662	100 672	100 669	100 603

Table 2. The Shapley-Shubik/normalized Banzhaf power index matrix for the German Parliament

year→ party↓	1949-Sh/ 1949-Bz	1953-Sh/ 1953-Bz	1957	1961	1965	1969	1972	1976	1980	1983	1987	1990	1994	1998	2002
CDU/ CSU	0.3155/ 0.3039	0.6/ 0.6364	1	0.3333	0.3333	0.3333	0.3333	0.3333	0.3333	0.5	0.5	0.5	0.5	0.1667	0.3333
SPD	0.2893/ 0.2833	0.1/ 0.0909	0	0.3333	0.3333	0.3333	0.3333	0.3333	0.3333	0.1667	0.1667	0.1667	0.1667	0.5	0.3333
FDP	0.2349/ 0.2615	0.1/ 0.0909	0	0.3333	0.3333	0.3333	0.3333	0.3333	0.3333	0.1667	0.1667	0.1667	0.1667	0.1667	0
GRUNE										0.1667	0.1667	0	0.1667	0.1667	0.3333
PDS												0.1667	0	0	0
KPD/ DKP	0.0298/ 0.0287														
DRP/ NPD	0.0095/ 0.0103														
DP	0.0365/ 0.0321	0.1/ 0.0909	0												
GB/ BHE		0.1/ 0.0909													
BP	0.0365/ 0.0321														
WAV	0.0254/ 0.0241														
Zentr.	0.0147/ 0.0161	0/ 0													
SSW	0.0024/ 0.0023														
others	0.0056/ 0.0057														
Total	1.0001/ 1.0001	1/ 1	1	0.9999	0.9999	0.9999	0.9999	0.9999	0.9999	1.0001	1.0001	1.0001	1.0001	1.0001	0.9999

Table 3. The absolute Banzhaf power index matrix for the German Parliament

year→ party↓	1949	1953	1957	1961	1965	1969	1972	1976	1980	1983	1987	1990	1994	1998	2002
CDU/CSU	0.5176	0.875	1	0.5	0.5	0.5	0.5	0.5	0.5	0.75	0.75	0.75	0.75	0.25	0.5
SPD	0.4824	0.125	0	0.5	0.5	0.5	0.5	0.5	0.5	0.25	0.25	0.25	0.25	0.75	0.5
FDP	0.4453	0.125	0	0.5	0.5	0.5	0.5	0.5	0.5	0.25	0.25	0.25	0.25	0.25	0
GRUNE										0.25	0.25	0	0.25	0.25	0.5
PDS												0.25	0	0	0
KPD/DKP	0.0488														
DRP/NPD	0.0176														
DP	0.0547	0.125	0												
GB/BHE		0.125													
BP	0.0547														
WAV	0.041														
Zentr.	0.0273	0													
SSW	0.0039														
others	0.0098														
Total	1.7031	1.375	1	1.5	1.5	1.5	1.5	1.5	1.5	1.5	1.5	1.5	1.5	1.5	1.5

Table 4. The Deegan-Packel power index matrix for the German Parliament

year→ party↓	1949	1953	1957	1961	1965	1969	1972	1976	1980	1983	1987	1990	1994	1998	2002
CDU/CSU	0.1342	0.4	1	0.3333	0.3333	0.3333	0.3333	0.3333	0.3333	0.375	0.375	0.375	0.375	0.2083	0.3333
SPD	0.1316	0.15	0	0.3333	0.3333	0.3333	0.3333	0.3333	0.3333	0.2083	0.2083	0.2083	0.2083	0.375	0.3333
FDP	0.175	0.15	0	0.3333	0.3333	0.3333	0.3333	0.3333	0.3333	0.2083	0.2083	0.2083	0.2083	0.2083	0
GRUNE										0.2083	0.2083	0	0.2083	0.2083	0.3333
PDS												0.2083	0	0	0
KPD/DKP	0.0995														
DRP/NPD	0.0688														
DP	0.0769	0.15	0												
GB/BHE		0.15													
BP	0.0769														
WAV	0.0823														
Zentr.	0.081	0													
SSW	0.0223														
others	0.0515														
Total	1	1	1	0.9999	0.9999	0.9999	0.9999	0.9999	0.9999	0.9999	0.9999	0.9999	0.9999	0.9999	0.9999

Table 5. The Holler power index matrix for the German Parliament

year→ party↓	1949	1953	1957	1961	1965	1969	1972	1976	1980	1983	1987	1990	1994	1998	2002
CDU/CSU	0.1161	0.3333	1	0.3333	0.3333	0.3333	0.3333	0.3333	0.3333	0.3333	0.3333	0.3333	0.3333	0.2222	0.3333
SPD	0.1071	0.1667	0	0.3333	0.3333	0.3333	0.3333	0.3333	0.3333	0.2222	0.2222	0.2222	0.2222	0.3333	0.3333
FDP	0.1339	0.1667	0	0.3333	0.3333	0.3333	0.3333	0.3333	0.3333	0.2222	0.2222	0.2222	0.2222	0.2222	0
GRUNE										0.2222	0.2222	0	0.2222	0.2222	0.3333
PDS												0.2222	0	0	0
KPD/DKP	0.1071														
DRP/NPD	0.0804														
DP	0.0893	0.1667	0												
GB/BHE		0.1667													
BP	0.0893														
WAV	0.0893														
Zentr.	0.0893	0													
SSW	0.0357														
others	0.0625														
Total	1	1.0001	1	0.9999	0.9999	0.9999	0.9999	0.9999	0.9999	0.9999	0.9999	0.9999	0.9999	0.9999	0.9999

Table 6. Redistribution paradox for CDU/CSU

	1953	1957	1961	1965	1969	1972	1976	1980	1983	1987	1990	1994	1998	2002
1949			PBP2	PBP2	PBP2	PBP2	PBP2	PBP2					(DH)'P2	PBP2
1953			HP1	HP1	HP1	HP1	P2	HP1	P2	HP1	HP1	HP1		HP1
1957														
1961				P2	P2	P1	P2	P1	HP2	P1	HP2	P1		P1
1965					P1	P1	P2	P1	HP2	P1	P1	P1		P1
1969						P1	P2	P1	HP2	P1	P1	P1		P1
1972							P2	P2	HP2	P1	HP2	P1		P1
1976								P1	HP2	P1	P1	P1		P1
1980									HP2	P1	HP2	P1		P1
1983										P1	P1	P1		HP1
1987											P2	P1		HP1
1990												P1		HP1
1994														HP1
1998														

Table 7. Redistribution paradox for SPD

	1953	1957	1961	1965	1969	1972	1976	1980	1983	1987	1990	1994	1998	2002
1949	DHP1	P2							(DH)'P2	(DH)'P2	(DH)'P2	(DH)'P2		
1953		P2												
1957														
1961				P2	P2	P2	P2	P2					HP2	P2
1965					P2	P2	P2	P2					HP2	P1
1969						P2	P1	P1					P1	P1
1972							P1	P1					P1	P1
1976								P2					HP2	P1
1980													HP2	P1
1983										P1	P1	P1		
1987											P1	P2		
1990												P2		
1994														
1998														HP1

Table 8. Redistribution paradox for FDP

	1953	1957	1961	1965	1969	1972	1976	1980	1983	1987	1990	1994	1998	2002
1949	HP1		P1	P1	P1	P1	P1	P1	DHP1	DHP1	DHP1	DHP1	DHP1	
1953				P1	P1	P1	P1	P1	P1	P1		P1	P1	
1957					P1	P1	P1		P1			P1	P1	P1
1961				P1	P1	P1	P1	P1						
1965					P1	P1	P1	P2			P2			
1969						P2	P2	P2	P2	P2	P2	P2	P2	P2
1972							P1	P2		P2	P2			
1976								P2		P2	P2			P2
1980											P2			
1983										P2	P2	P2	P1	P2
1987											P2	P1	P1	
1990												P1	P1	
1994													P1	P2
1998														P2

Table 9. Redistribution paradox for GRUNE

	1987	1990	1994	1998	2002
1983	P2		P2	P2	
1987			P1	P1	
1990					
1994				P1	
1998					

Table 10. Redistribution paradox for PDS

	1994	1998	2002
1990	P2	P2	
1994		P2	P1
1998			P1

Table 11. Redistribution paradox for DP

	1953	1957
1949	P1	
1953		P2

Table 12. Redistribution paradox for German parties

	CDU/CSU	SPD	FDP	GRUNE	PDS	DP	Zentrum
k	15	15	15	6	4	3	2
l	105	105	105	15	6	3	1
$nP1Sh$	34	15	37	3	2	1	0
$nP2Sh$	11	20	26	3	3	1	0
$nP1Bz$	34	15	37	3	2	1	0
$nP2Bz$	11	20	26	3	3	1	0
$nP1PBz$	34	15	37	3	2	1	0
$nP2PBz$	18	20	26	3	3	1	0
$nP1DP$	34	16	42	3	2	1	0
$nP2DP$	10	16	26	3	3	1	0
$nP1H$	47	17	43	3	2	1	0
$nP2H$	19	20	26	3	3	1	0
$fP1Sh$	0.3238	0.1429	0.3524	0.2	0.3333	0.3333	0
$fP2Sh$	0.1048	0.1905	0.2476	0.2	0.5	0.3333	0
$fP1Bz$	0.3238	0.1429	0.3524	0.2	0.3333	0.3333	0
$fP2Bz$	0.1048	0.1905	0.2476	0.2	0.5	0.3333	0
$fP1PBz$	0.3238	0.1429	0.3524	0.2	0.3333	0.3333	0
$fP2PBz$	0.1714	0.1905	0.2476	0.2	0.5	0.3333	0
$fP1DP$	0.3238	0.1524	0.4	0.2	0.3333	0.3333	0
$fP2DP$	0.0952	0.1524	0.2476	0.2	0.5	0.3333	0
$fP1H$	0.4476	0.1619	0.4095	0.2	0.3333	0.3333	0
$fP2H$	0.181	0.1905	0.2476	0.2	0.5	0.3333	0

Table 13. Redistribution paradox for German politics - Comparison with Dutch politics

country ↓	index X →	Sh	Bz	PBz	DP	H
Germany	L	340	340	340	340	340
Germany	$TNP1X$	92	92	92	98	113
Germany	$TNP2X$	64	64	71	59	72
Germany	$FP1X$	0.2706	0.2706	0.2706	0.2882	0.3324
The Netherlands	$FP1X$	0.0939	0.0881	0.0977	0.09	0.092
Germany	$FP2X$	0.1882	0.1882	0.2088	0.1735	0.2118
The Netherlands	$FP2X$	0.0977	0.1015	0.1169	0.1379	0.1475
Germany	FPX	0.4588	0.4588	0.4794	0.4617	0.5442
The Netherlands	FPX	0.1916	0.1896	0.2146	0.2279	0.2395

Table 14. Paradox of new members for the German Parliament (for Sh, Bz, PBz, and DP)

party ↓	nP	m	fP
CDU/CSU	15	44	0.3409
SPD	2	44	0.0455
FDP	0	44	0
GRUNE	0	8	0

Table 15. Agreement indices/German-Dutch comparison

Pi/country→	$P1$/Germany	$P1$/NL	$P2$/Germany	$P2$/NL	P/Germany	P/NL
$FAPi(Sh - Bz)$	1	0.7925	1	0.9259	1	0.8182
$FAPi(Sh - PBz)$	1	0.7544	0.9014	0.7077	1	0.8571
$FAPi(Sh - DP)$	0.9388	0.5238	0.9219	0.4302	1	0.5135
$FAPi(Sh - H)$	0.8142	0.5156	0.7662	0.4066	0	0.4595
$FAPi(Bz - PBz)$	1	0.7636	0.9014	0.7538	1	0.7941
$FAPi(Bz - DP)$	0.9388	0.55	0.9219	0.4706	1	0.5294
$FAPi(Bz - H)$	0.8142	0.541	0.7662	0.4396	0	0.5455
$FAPi(PBz - DP)$	0.9388	0.4203	0.831	0.3854	1	0.5
$FAPi(PBz - H)$	0.8142	0.4143	0.7024	0.3663	0	0.4595
$FAPi(DP - H)$	0.8673	0.9792	0.8194	0.8861	0	0.84

References

J. Banzhaf, Weighted voting doesn't work: a mathematical analysis, *Rutgers Law Review* **19** (1965), 317–343.

S.J. Brams, *Game theory and politics*, New York, Free Press, 1975.

S.J. Brams, P. Affuso, Power and size: a new paradox, *Theory and Decision* **7** (1976), 29–56.

S.J. Brams, W.F. Lucas and P.D. Straffin, (eds), *Political and related models*, Volume 2 in series *Models in applied mathematics*, New York: Springer, 1982.

J. Deegan, E.W. Packel, A new index of power for simple n-person games, *International Journal of Game Theory* **7** (1978), 113–123.

J. Deegan, E.W. Packel, *To the (minimal winning) victors go the (equally divided) spoils: a new index of power for simple n-person games*, In Brams, S.J., Lucas, W.F. and Straffin, P.D., o.c., 1982.

P. Dubey, L.S. Shapley, Mathematical properties of the Banzhaf power index, *Mathematics of Operations Research* **4** (1979), 99–131.

D.M. Farrell, *Electoral systems. A comparative introduction*, Hampshire: Palgrave Publishers Limited, 2001.

D.S. Felsenthal, M. Machover, Postulates and paradoxes of relative voting power. A critical re-appraisal., *Theory and Decision* **33** (1995), 195–229.

D.S. Felsenthal, M. Machover, *The measurement of voting power: theory and practice, problems and paradoxes*, London: Edward Elgar Publishers, 1998.

D.S. Felsenthal, M. Machover and W. Zwicker, The bicameral postulates and indices of a priori voting power, *Theory and Decision* **44** (1998), 83–116.

D. Fischer, A. Schotter, The inevitability of the paradox of redistribution in the allocation of voting weights, *Public Choice* **33** (1978), 49–67.

M.J. Holler, Forming coalitions and measuring voting power, *Political Studies* **30** (1982), 262–271.

M.J. Holler, E.W. Packel, Power, luck and the right Index, *Journal of Economics* **43** (1983), 21–29.

D.M. Kilgour, A Shapley value for cooperative games with quarreling, in: A. Rapoport (ed.) *Game theory as a theory of conflict resolution*, Boston: Reidel, 1974,193–206.

W.F. Lucas, *Measuring power in weighted voting games*, In Brams, S.J., Lucas, W.F. and Straffin, P.D., o.c., 1982.

G. Owen, *Game theory*, San Diego: Academic Press, 1995.

L.S. Penrose, The elementary statistics of majority voting, *Journal of the Royal Statistical Society* **109** (1946), 53–57.

A. Rusinowska, Paradox of redistribution in Polish politics, *Marie Curie Fellowships Annals (MCFA)*, **II** (2001), (in press).

A. Schotter, The paradox of redistribution: Some theoretical and empirical results, In Holler, M. J. (ed.) *Power, voting, and voting power*, Wurzburg-Wien: Physica-Verlag, 1982.

L.S. Shapley, A value for n-person games, *Annals of Mathematics Studies* **28** (1953), 307–317.

L.S. Shapley, Simple games: an outline of the descriptive theory, *Behavioral Science* **7** (1962), 59–66.

L.S. Shapley, M. Shubik, A method for evaluating the distribution of power in a committee system, *American Political Science Review* **48** (1954), 787–792.

P.D. Straffin, *Power indices in politics*, In Brams, S.J., Lucas, W.F. and Straffin P.D., o.c., 1982.

A.M.A. Van Deemen, *Coalition formation and social choice*, Boston: Kluwer Academic Publishers, 1997.

A.M.A. Van Deemen, A. Rusinowska, Paradoxes of voting power in Dutch politics, *Public Choice* **115** (2003), 109–137.

In: Mathematics, Game Theory and Algebra Compendium ISBN 978-1-60456-659-8
Editor: Jacob H. Mathias, pp. 749-763

Chapter 62

PLAYERS' INFORMATION IN TWO-PLAYER GAMES OF "SCORE SHOWDOWN"

Minoru Sakaguchi
Osaka University, Osaka, Japan

1. Two-Player Game of "Score Showdown"

Consider the two players I and II (sometimes they are denoted by 1 and 2, respectively). Let X_j (Y_j) be the random variable observed by I (II) at the j-th observation, $j = 1, 2$. We assume that X_1, X_2, Y_1, Y_2 are *i.i.d.* each with uniform distribution in $[0, 1]$.

The game is played as follows. I [II] observes $X_1 = x$ $[Y_1 = y]$ and chooses one of either A (*i.e.*, accepts the observed value) or R (*i.e.*, rejects his observed value and samples a new random variable).

The *score* for player I is defined by

$$S_1(X_1, X_2) = \begin{cases} X_1, \\ \varphi(X_1, X_2), \end{cases} \quad \text{if } X_1 \text{ is } \begin{cases} \text{accepted,} \\ \text{rejected and } X_2 \text{ is sampled,} \end{cases}$$

and the score $S_2(Y_1, Y_2)$ for II, is defined similarly, with X_is replaced by Y_is.

We call the game of "Keep-or-Exchange", "Risky Exchange" and "Showcase Showdown", when

$$\varphi(X_1, X_2) = X_2, X_2 I(X_2 > X_1), \text{ and } (X_1 + X_2) I(X_1 + X_2 \leq 1),$$

respectively. Here $I(e)$ is the indicator of the event e. For simplicity, we denote these games GKE, GRE and GSS, respectively. The name of GSS comes from Ref.[1].

After each player chooses his (or her) R or A, showdown is made, the scores are compared, and the player with the higher score than the opponent becomes the *winner*. Each player aims to maximize the probability of his (or her) winning.

We consider the three information types, under which the players decide their choices of either R or A.

I^{10-01} means that I observes $X_1 = x$, II observes $Y_1 = y$, and each player doesn't inform his observed value to his opponent.

I^{11-11} means that I observes $X_1 = x$, II observes $Y_1 = y$, and each player informs his observed value to his opponent.

I^{10-11} means that I observes $X_1 = x$, II observes $Y_1 = y$, and I informs his $X_1 = x$ to II, but II doesn't inform his $Y_1 = y$ to I.
The GKEs (GREs) under these three information types are solved in Sections 2, 3 and 4 (Sections 5,6 and 7). The most important difference between GKE and GRE is that "draw" occurs with positive probability in the latter, but it doesn't occur in the former.

In Section 9 we discuss the games GKE and GRE under information I^{10-11}, in which the "first-mover" I adopts some randomization in his strategy in order to restore his disadvantage. The results in Theorem 1~6, 3B and 6B are compared in Sections 8 and 9. Some open probrems in this area are mentioned in the final Section 10.

2. Game of "Keep-or-Exchange" under I^{10-01}

Suppose that the players' strategies have the form of

$$\begin{array}{ll} \text{I accepts (rejects) } X_1 = x, & \text{if } x > (<)\, a, \\ \text{II accepts (rejects) } Y_1 = y, & \text{if } y > (<)\, b. \end{array} \tag{2.1}$$

Let $M(a, b)$ donote I's winning probability when I and II choose a and b both in $[0, 1]$ respectively. Since "draw" (*i.e.*, players get an equal score) occurs with zero probability for $\forall(a, b) \in [0, 1]^2$, the game reduces to a continuous game with payoff function $M(a, b)$ on the unit square. We already know (Ref.[3 ; Theorem 2])

Theorem 1. *Solution to* GKE *under* I^{10-01} *is as follows : The game has a unique saddle point* (g, g) *and the saddle value* $M(g, g) = \frac{1}{2}$, *where* $g = \frac{1}{2}(\sqrt{5} - 1) \approx 0.61803$, *the golden bisection number.*

3. Game of "Keep-or-Exchange" under I^{11-11}

Define state (x, y) to mean that I observes $X_1 = x$, II observes $Y_1 = y$, and both players know both of x and y. The game reduces to the matrix game

$$\mathbf{M}(x, y) = \begin{array}{c} \\ R \\ A \end{array} \begin{array}{c} \begin{array}{cc} R & A \end{array} \\ \left[\begin{array}{cc} P(X_2 > Y_2) & P(X_2 > y) \\ P(x > Y_2) & I(x > y) \end{array} \right] \end{array} = \begin{pmatrix} 1/2 & \bar{y} \\ x & I(x > y) \end{pmatrix}, \tag{3.2}$$

where I, row player, is the maximizer and II, column player, is the minimizer.

Theorem 2. *Solution to* GKE *under* I^{11-11} *in state* (x, y) *is as follows ;*

Case	Optimal strategy-pairs	val.$\mathbf{M}(x,y)$
$y > x \vee \frac{1}{2}$	saddle point R-A	$\bar{y}$
$\frac{1}{2} > x \vee y$	R-R	$1/2$
$x > y \vee \frac{1}{2}$	A-R	x

(See Figure 1b in Section 8). *The winning probabilities are* $1/2, 1/2$.

Proof. By considering the two payoff matrices $\begin{pmatrix} 1/2 & \bar{y} \\ x & 0 \end{pmatrix}$ and $\begin{pmatrix} 1/2 & \bar{y} \\ x & 1 \end{pmatrix}$ coming from (3.1), we easily find the solution to the game in state (x, y) is given by the table in the theorem. Winning probability for each player is

$$P(W_1) = 1 - P(W_2) = \int_0^1 \int_0^1 \text{val.}\mathbf{M}(x,y)dxdy = \frac{1}{8} + \int_{1/2}^1 x^2 dx + \int_{1/2}^1 y\bar{y}dy = \frac{1}{2},$$

just as we can expect beforehand. □

4. Game of "Keep-or-Exchange" under $\mathbf{I}^{10-11}$

Define state $(x \mid Y_1)$ for I, to mean that I observes $X_1 = x$, and he doesn't know his rival's observed value of Y_1. Also define state $(y \mid x)$ for II to mean that II observes $Y_1 = y$ and is informed $X_1 = x$. Then the payoff matrix for II in state $(y \mid x)$ is

$$\mathbf{M}_2(y \mid x) = \begin{matrix} & R & A \\ R & P(X_2 < Y_2) & P(X_2 < y) \\ A & P(x < Y_2) & I(x < y) \end{matrix} = \begin{pmatrix} 1/2 & y \\ \bar{x} & I(x < y) \end{pmatrix}, \quad (4.1)$$

where I (row player) is the minimizer and II (column player) is the maximizer.

The payoff matrix for I in state $(x \mid Y_1)$ is

$$\mathbf{M}_1(x \mid Y_1) = \begin{matrix} & R & A \\ R & P(X_2 > Y_2) & P(X_2 > Y_1) \\ A & P(x > Y_2) & P(x > Y_1) \end{matrix} = \begin{pmatrix} 1/2 & \bar{Y_1} \\ x & I(x > Y_1) \end{pmatrix}. \quad (4.2)$$

We want to find the optimal strategy for each player. Assume that I's strategy is restricted to be under the

Assumption A. *Player I chooses* A (R), *if* $X = x > (\leq)\ a$, *for same* $a \in [0, 1]$ *which I must determine beforehand.*

In game theory we assume that players are intelligent, and each player should prepare for the optimal strategy employed by his opponent. Therefore if the Assumption A is open commonly to the players, player II as well as player I can compute the optimal dicision level a^* beforehand.

Theorem 3A. *Solution to* GKE *under* I^{10-11} *and Assumption A. The optimal choice for I is* $a^* = \sqrt{\frac{3}{8}} \approx 0.6124$. *The optimal strategy and the winning probability for II in state* $(y \mid x)$ *is ;*

Condition	II's optimal choice	II's winning prob.
$x < a^*$	A (R), if $y > (<)1/2$	$y \vee \frac{1}{2}$
$a^* < x < y$	A	*1*
$x > a^* \vee y$	R	$\bar{x}$

Players' winning probabilities are

$$P(W_1) = \frac{1}{3} + \frac{1}{4}\sqrt{3/8} \approx 0.4864,$$
$$P(W_2) = 1 - P(W_1) = \frac{2}{3} - \frac{1}{4}\sqrt{3/8} \approx 0.5136.$$

See Figure 1c in Section 8.

Proof. By (4.1), we easily find that II's optimal behavior in state $(y \mid x)$ is as shown by the table. So, we have

$$P(W_2) = \int_0^1 \int_0^1 \left\{ \left(\frac{1}{2} \vee y\right) I(x < a^*) + I(a^* < x < y) + \bar{x} I(x > a^* \vee y) \right\} dx dy \quad (4.3)$$
$$= \frac{5}{8} a^* + \frac{1}{2} \bar{a}^{*2} + \int_0^1 x \bar{x} dx = \frac{2}{3} - \frac{3}{8} a^* + \frac{1}{3} a^{*3}$$

which is a convex function with values $2/3$ at $a = 0$, $5/8$ at $a = 1$ and the minimal value $\frac{2}{3} - \frac{1}{4}\sqrt{\frac{3}{8}}$ at $a = \sqrt{\frac{3}{8}}$. □

Our result shows that the solution to GKE under information I^{10-11} and Assumption A is the same as in the sequential version of GKE (see Ref.[4 ; Theorem 1]). The reason is explained later in Section 7. Moreover we give another result Theorem 3B in Section 9.

5. Game of "Risky Exchange" under I^{10-01}

Suppose that players' strategies have the form of (2.1). Let $W_i, i = 1, 2$, be the event that player i wins. Let $M_i(a, b) = P(W_i \mid a, b), i = 1, 2$, denote the winning probability for player i, when players 1 and 2 choose a and b, respectively. Evidently we have for $\forall (a, b) \in [0, 1]^2$

$$P(\text{draw}) = P\{X_2 < X_1 < a, Y_2 < Y_1 < b\} = \frac{1}{4} a^2 b^2,$$

$$\sum_{i=1,2} M_i(a, b) + P(\text{draw} \mid a, b) = 1.$$

The game reduces to a continuous non-constant sum game with payoff functions $M_1(a, b)$ and $M_2(a, b)$, on the unit square. We already know (Ref.[5 ; Theorems 1 and 2]).

Theorem 4. *Solution to* GRE *under* I^{10-01} *is as follows. Let* $a^* \approx 0.54368$ *be a unique root in* $[0,1]$ *of the cubic equation*

$$a^3 + a^2 + a = 1.$$

The game has a unique equilibrium point (a^*, a^*) *and the equilibrium payoffs are*

$$M_1(a^*, a^*) = M_2(a^*, a^*) = \frac{1}{2}\left(1 - \frac{1}{4}a^{*^4}\right) \approx 0.48908$$

$$P(draw \mid a^*, a^*) = \frac{1}{4}a^{*^4} \approx 0.02184.$$

Note that in Theorem 1 (in Section 2), g is the unique root in $[0,1]$ of the quadratic equation $a^2 + a = 1$.

6. Game of "Risky Exchange" under I^{11-11}

Define state (x, y) as the same as in Section 3. Let $p_{AR}(x,y)$ $[q_{AR}(x,y)]$ denote the winning probability for I [II] when the players' choices are A by I and R by II in state (x, y). Other three probabilities $p_{RA}(x,y)$ $[q_{RA}(x,y)]$ *etc* are defined similarly. Also let $h_{RA}(x,y)$ *etc* denote the probability of draw, similarly. Hereafter, we shall sometimes omit the state description, for simplicity. We evidently have, for $\forall(x, y)$,

$$p_{AA} + q_{AA} + h_{AA} = 1 \tag{6.1}$$

and other three equations, and

$$h_{AA} = h_{AR} = h_{RA} = 0, \qquad h_{RR} = P(X_2 < x, Y_2 < y) = xy. \tag{6.2}$$

Furthermore we find that

$$\begin{aligned}
p_{AA} &= I(x,y), && (6.3)\\
p_{AR} &= P[\{y < Y_2 < x\} \cup \{Y_2 < y\}] = (x-y)I(x>y) + y, && (6.4)\\
p_{RA} &= P\{X_2 > x, X_2 > y\} = 1 - x \vee y, && (6.5)\\
p_{RR} &= P[(X_2 > x) \cap \{(X_2 > Y_2 > y) \cup (Y_2 < y)\}] && (6.6)\\
&= \frac{1}{2}(1+y^2) - xy - \frac{1}{2}(x-y)^2 I(x>y).
\end{aligned}$$

Therefore, from Eqs (6.1)∼(6.6), players in state (x, y) face the bimatrix game with the payoff bimatrix

$$\mathbf{M}(x,y) = \begin{array}{c|cc|cc|} & \text{R} & & \text{A} & \\ \hline \text{R} & p_{RR}, & 1 - xy - p_{RR} & 1 - x\vee y, & x \vee y \\ \text{A} & (x-y)I(x>y)+y, & \bar{y} - (x-y)I(x>y) & I(x>y), & I(x<y) \\ \hline \end{array} \tag{6.7}$$

$$= \begin{cases} \begin{array}{|cc|cc|} \hline \frac{1}{2}(1+y^2)-xy, & \frac{1}{2}(1-y^2) & \bar{y}, & y \\ \hline y, & \bar{y} & 0, & 1 \\ \hline \end{array} (\equiv \mathbf{M}(x,y \mid x \le y), \text{say}), & \text{if } x \le y; \\ \\ \begin{array}{|cc|cc|} \hline \frac{1}{2}(1-x^2), & \frac{1}{2}(1+x^2)-xy & \bar{x}, & x \\ \hline x, & \bar{x} & 1, & 0 \\ \hline \end{array} (\equiv \mathbf{M}(x,y \mid x > y), \text{say}), & \text{if } x > y. \end{cases}$$

Theorem 5. *Solution to* GRE *under* I^{11-11}, *in state* (x,y), *is as follows ;*

Case	Eq. strategy-pair	Eq, val. $\mathbf{M}(x,y)$
$y > x \vee (\sqrt{2}-1)$	saddle pt. R-A	$\bar{y},\ \ y$
$x < y < \sqrt{2}-1$	R-R	$\frac{1}{2}(1+y^2)-xy,\ \frac{1}{2}(1-y^2)$
$y < x < \sqrt{2}-1$	R-R	$\frac{1}{2}(1-x^2),\ \frac{1}{2}(1+x^2)-xy$
$x > y \vee (\sqrt{2}-1)$	A-R	$x,\ \ \bar{x}$

The winning probabilities for the players and the probability of draw are

$$P(\text{draw}) = \frac{1}{4}(17 - 12\sqrt{2}) \approx 0.00736,$$

$$P(W_1) = P(W_2) = \frac{1}{2}\{1 - P(\text{draw})\} = \frac{1}{8}(12\sqrt{2} - 13) \approx 0.49632.$$

(See Figure 1e.)

Proof. For the bimatrix $\mathbf{M}(x,y \mid x \le y)$ we note that

$$\begin{array}{ll} \frac{1}{2}(1-y^2) > (<)\, y, & \text{if } y < (>)\, \sqrt{2}-1 \\ \frac{1}{2}(1+y^2) - xy > (<)\, y, & \text{if } y < (>)\, 1 + x - \sqrt{2x+x^2} \equiv k(x), \text{say}. \end{array}$$

And $k(x)$ is convex, decreasing with $k(0) = 1$, $k(\sqrt{2}-1) = \sqrt{2}-1$, and $k(1) = 2-\sqrt{3} \approx 0.268$.

So, for the bimatrix $\mathbf{M}(x,y \mid x > y)$, we evidently have

$$\begin{array}{ll} \frac{1}{2}(1-x^2) > (<)\, x, & \text{if } x < (>)\, \sqrt{2}-1 \\ \frac{1}{2}(1+x^2) - xy > (<)\, x, & \text{if } x < (>)\, 1 + y - \sqrt{2y+y^2} = k(y). \end{array}$$

Therefore, by combining the above facts about $\mathbf{M}(x,y)$, we get the table in the theorem.

The probabilities we want to find are

$$\begin{aligned} P(\text{draw}) &= \int_0^{\sqrt{2}-1}\int_0^{\sqrt{2}-1} h_{RR}\,dxdy = \int_0^{\sqrt{2}-1}\int_0^{\sqrt{2}-1} xy\,dxdy = \frac{1}{4}(\sqrt{2}-1)^4 = \frac{1}{4}(17 \\ P(W_1) &= \iint_{0<x<y<\sqrt{2}-1} \left\{\frac{1}{2}(1+y^2) - xy\right\} dxdy + \iint_{0<y<x<\sqrt{2}-1} \frac{1}{2}(1-x^2)\,dxdy + \int_{\sqrt{2}-1}^{1} x^2 dx \\ &= \frac{1}{8}(4\sqrt{2}-5) + \sqrt{2} - 1 = \frac{1}{8}(12\sqrt{2}-13) \end{aligned}$$

and

$$P(W_2) = 1 - P(W_1) - P(\text{draw}) = 1 - \frac{1}{8}(12\sqrt{2}-13) - \frac{1}{4}(17-12\sqrt{2}) = \frac{1}{8}(12\sqrt{2}-13).$$

The result $P(W_1) = P(W_2)$ is consistent with our common sence. □

7. Game of "Risky Exchange" under I^{10-11}

Lemma 6.1. *Let* X *be a random variable uniformly distributed in* $[0, 1]$. *For* $\forall t \in [0, 1]$,

$$Z_{GSS}(t) = (t + X)I(t + X < 1) \qquad \textit{and} \qquad Z_{GRE}(t) = XI(X > t)$$

have the same distribution.

Proof. Let $F_{GSS}(z)$ and $F_{GRE}(z)$ be the c.d.f.s of $Z_{GSS}(t)$ and $Z_{GRE}(t)$, respectively. Then, clearly, $F_{GSS}(z) = F_{GRE}(z) = 0, \forall z \in [0, t)$. And, for $z \in (t, 1]$,

$$\begin{aligned} F_{GSS}(z) &= P\{(t + X)I(t + X < 1) \leq z\} \\ &= P[\{X < \bar{t} \text{ and } X \leq z - t\} \cup \{X \geq \bar{t}\}] = (z - t)^+ + t = z \end{aligned}$$

and

$$\begin{aligned} F_{GRE}(z) &= P\{XI(X > t) \leq z\} \\ &= P[\{X > t \text{ and } X \leq z\} \cup \{X \leq t\}] = (z - t)^+ + t = z. \qquad \square \end{aligned}$$

By considering Lemma 6.1, we find that GRE under information I^{10-11} and Assumption A (see Section 4), is identical to the sequential-move version of GSS. So, by referring to Ref.[3 ; Theorem 4] we obtain

Theorem 6A. *Solution to* GRE *under* I^{10-11} *and Assumption* A *is as follows. The optimal strategy for I is given by*
Choose A (R), *if* $x > (<)\ x_0 \approx 0.570$,
where x_0 *ia a unique root in* $(\sqrt{2} - 1, 1)$, *of the equation* $\bar{x}^4 = 2x(1 - 3x^2)$.
The optimal strategy for II in state $(y \mid x)$ *is given by*
Choose A (R), *if* $x < x_0$ *and* $y > (<)\ y_0(x)$,
Choose A (R), *if* $x > x_0$ *and* $y > (<)\ x$,
where $y_0(x) = (\sqrt{2} - 1)I(x \leq \sqrt{2} - 1) + (\bar{x}^2/(2x))\, I(\sqrt{2} - 1 < x < x_0)$.
Players' winning probabilities and probability of draw are

$$P(\text{draw}) \approx 0.0108, \quad P(W_1) \approx 0.4768, \quad P(W_2) \approx 0.5124.$$

(For the explicit expressions of $P(\text{draw})$ and $P(W_1)$, see Ref.[3 ; Theorem 4]. Also see Figure 1f in Section 8.)

8. Comparison between Theorem 1∼6.

The optimal choice-pairs in Theorem 1∼6 are shown by Figure 1a∼1f, in comparison. Here, GKE = game of Keep-or-Exchange, GRE = game of Risky Exchange, $P(D)$ =prob. of draw and $V_i = P(W_i), i = 1, 2$.

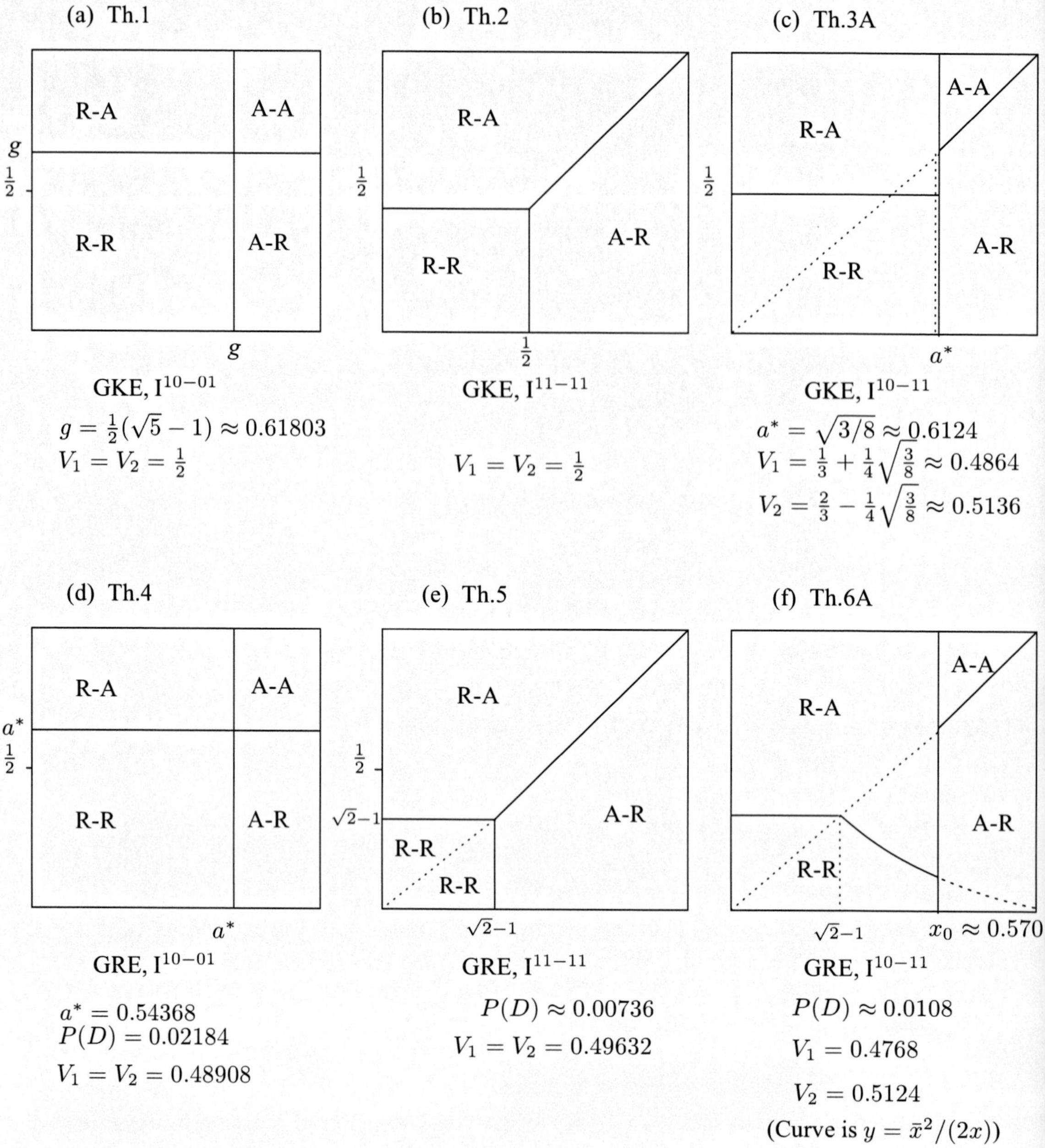

Figure 1. Optimal choice-pairs in GKE and GRE.

We abserve that (1) The decision threshould in GRE in each information type is smaller than that in GKE. (2) There doesn't exsit the optimal A-A pair under I^{11-11} ; and (3) In each of GKE and GRE under I^{10-11}, the border of the optimal A and R regions for II is more complicate than theose under I^{10-01} and I^{11-11}. And, we find that $P(\text{draw}) > 0$ and $P(W_1) < P(W_2)$.

9. More about Games under Information I^{10-11}.

Under information I^{10-11}, player I has an advantage over player I. It would seem natural that I would randomize his decision threshold in order to improve his disadvantage due to the leakage of his "hand" to his opponent. The situation is like in poker. See, for example, Ref.[2 ; Section 6]. Standing at this viewpoint, the next

Assumption B. *Player I, in state $X_1 = x$, chooses R if $x < \bar{a}$, chooses A if $x > a$ and employs the mixed strategy $\left(R, A\,;\ \dfrac{a-x}{a-\bar{a}} \cdot \dfrac{x-\bar{a}}{a-\bar{a}}\right)$, if $\bar{a} < x < a$, for some $a \in [\frac{1}{2}, 1]$ which he must determine beforehand,*

instead of **Assumption A** (stated in Section 4), is worth studying.

The best choice of a is not yet derived. Next two theorems show that the two extremes $a = \dfrac{1}{2}$ and $a = 1$ belong to the worst choices for I.

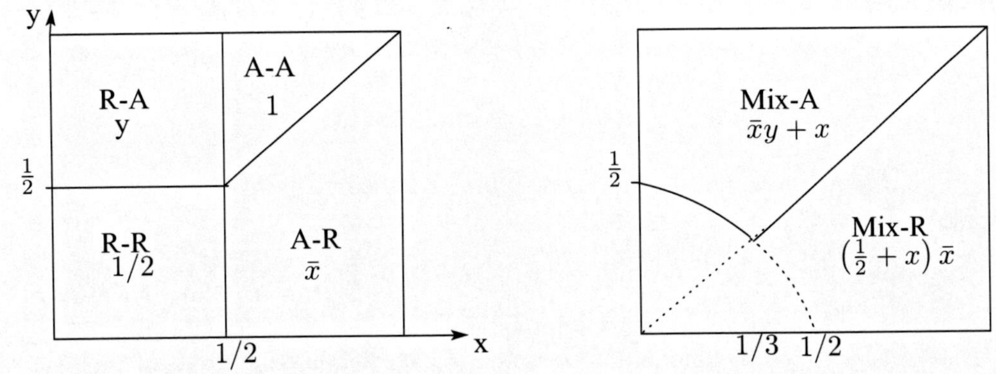

Case $a = 1/2$ Case $a = 1$

Figure 2. Optimal strategy-pair

(Mix. means I's mixed strategy $(R, A\,;\ \bar{x}, x)$. The curve is $\xi(x) = \frac{3}{2} + x - \bar{x}^{-1}$. II's winning prob. are mentioned therewith.)

Theorem 3B. *Solution to* GKE *under information* I^{10-11} *and Assumption* B.

(i). *Case $a = \dfrac{1}{2}$.*

The optimal strategy for II in state $(y \mid x)$ is ;

Choose A (R), *if $x < \dfrac{1}{2}$ and $y > (<)\ \dfrac{1}{2}$,*

Choose A (R), *if $x > \dfrac{1}{2}$ and $y > (<)\ x$,*

The winning probabilities are

$$P(W_1) = 1 - P(W_2) = \frac{23}{48} \approx 0.4792.$$

(ii). *Case $a = 1$.*

Players' optimal strategy-pairs and winning probabilities for II are as shown in Figure 2. We obtain

$$P(W_2) = 1 - P(W_1) = \frac{1}{6} + \frac{16}{81} + \left(-\frac{13}{1296} + \frac{1}{2}\log\frac{3}{2}\right) \approx 0.5569. \quad (9.1)$$

Proof. **(i)** Proof is easy by referring Theorem 3A and Eq.(4.3).

(ii) Since the payoff matrix for II in state $(y \mid x)$ is given by (4.1), II's winning prob. by choosing R (A) is the first (second) component of

$$(\bar{x}, x)\begin{pmatrix} 1/2 & y \\ \bar{x} & I(x<y) \end{pmatrix}. \quad (9.2)$$

For $x < y$, choosing A is better, if

$$\bar{x}y + x > \left(\frac{1}{2} + x\right)\bar{x},$$

or

$$y > \frac{1}{2} - x^2/\bar{x} = \frac{3}{2} + x\bar{x}^{-1} \equiv \xi(x), \text{ say.}$$

$\xi(x)$ is a concave decreasing function connecting the two points $(0, \frac{1}{2})$ and $(\frac{1}{3}, \frac{1}{3})$ in the triangle $0 < x < y < 1$.

Meanwhile, for $x > y$, choosing A is better, if $\bar{x}y > \left(\frac{1}{2} + x\right)\bar{x}$ *i.e.*, $y > \frac{1}{2} + x$ which is impossible. Therefore we obtain Figure 2.

$P(W_2)$ is obtained by calculating the integral over the unit square.

$$\int_0^1 \left(\frac{1}{2} + x\right) x\bar{x}dx \quad + \quad \int_{1/3}^1 dx \int_x^1 (\bar{x}y + x)dy \quad (9.3)$$
$$+ \int_0^{1/3} dx \left[\left(\frac{1}{2} + x\right)\bar{x}(\xi(x) - x) + \int_{\xi(x)}^1 (\bar{x}y + x)dy\right].$$

The first term is equal to $\frac{1}{6} \approx 0.16667$. The second term equals $\frac{16}{81} \approx 0.19753$. The third term is

$$\int_0^{1/3} dx \left[\frac{1}{2}x^3 - \frac{3}{2}x^2 - \frac{1}{8}x + \frac{3}{8} + \bar{x}^{-1}\left(\frac{1}{2}x^2 - \frac{1}{4}x + \frac{1}{4}\right)\right]$$
$$= -\frac{13}{1296} + \frac{1}{2}\log\frac{3}{2} \approx 0.19270.$$

Therefore we finally obtain

$$P(W_2) \approx 0.1667 + 0.19753 + 0.19270 \approx 0.55690. \qquad \square$$

Theorem 6B. *Solution to* GRE *under information* I^{10-11} *and Assumption* B.

(i). *Case* $a = \dfrac{1}{2}$.

The optimal strategy for II in state $(y \mid x)$ *is ;*

Choose A (R), *if* $x < \dfrac{1}{2}$ *and* $y > (<)\ \eta(x)$,

Choose A (R), *if* $x > \dfrac{1}{2}$ *and* $y > (<)\ x$,

where $\eta(x) = (\sqrt{2}-1)I(x < \sqrt{2}-1) + (\bar{x}^2/(2x))\, I\left(\sqrt{2}-1 < x < \frac{1}{2}\right)$.

Players' winning probabilities and probability of draw are

$$P(\text{draw}) \approx 0.0105, \quad P(W_2) \approx 0.5163 \quad \textit{and} \quad P(W_1) = 1 - P(W_2) - P(\text{draw}) \approx 0.4732.$$

(ii). *Case* $a = 1$.

Players' optimal strategy-pairs and the winning probabilities for II are as shown in Figure 3. *We have*

$$P(\text{draw}) \approx 0.0445, \quad P(W_1) < 0.4118 < 0.5417 < P(W_2).$$

and $\dfrac{1}{2}(P(W_1) + P(W_2)) \approx 0.4778$, *approximately. See Figure* 3.

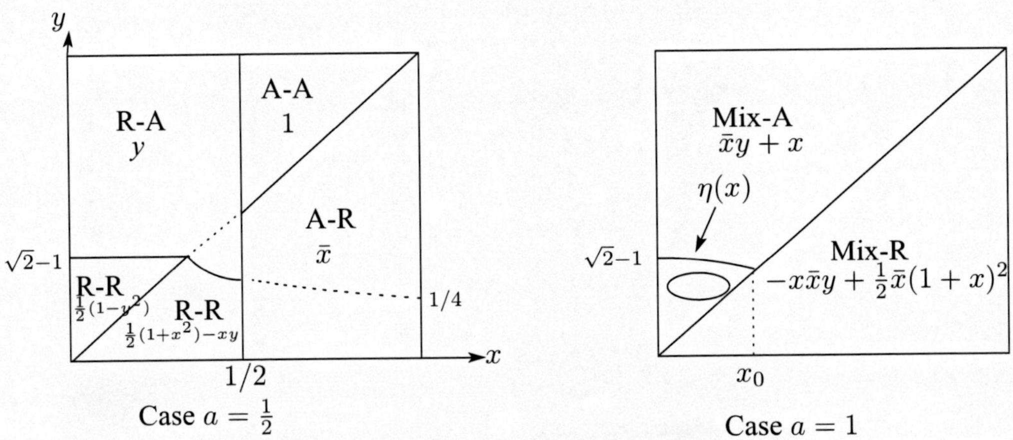

Figure 3. Optimal strategy-pair and II's winning probability.

(Here, $\eta(x) \equiv \left(\sqrt{1+\bar{x}^2} - 1\right)\bar{x}$, $x_0 \approx 0.3111$ is a unique root in $[0, 1]$ of the cubic equation $x^3 - x^2 - 3x + 1 = 0$. The part ⬭ is Mix.-R, $\frac{1}{2}(1+x) - xy - \frac{1}{2}\bar{x}y^2$.)

Proof of part **(i).** The payoff matrix for II in state $(y \mid x)$ is

$$\mathbf{M}_2(y \mid x) = \begin{array}{c} \\ R \\ A \end{array} \begin{array}{c} \begin{array}{cc} R & A \end{array} \\ \begin{pmatrix} q_{RR} & q_{RA} \\ q_{AR} & q_{AA} \end{pmatrix} \end{array} \tag{9.4}$$

where

$$q_{RR} = P[\{Y_2 < y\} \cap [\{Y_2 > X_2 > x\} \cup \{X_2 < x\}]] \tag{9.5}$$
$$= \frac{1}{2}(1-y^2)I(x<y) + \left\{\frac{1}{2}(1+x^2) - xy\right\} I(x>y),$$

$$q_{AR} = P(Y_2 > x \vee y) = 1 - x \vee y, \tag{9.6}$$

$$q_{RA} = P[\{y > X_2 > x\} \cup \{X_2 < x\}] = (y-x)I(y>x) + x = x \vee y, \tag{9.7}$$

$$a_{AA} = I(x<y). \tag{9.8}$$

From (9.4)∼(9.8), the following statements are evident ;

For $x < y \wedge \frac{1}{2}$, choosing A is better, if $q_{RA} = y > q_{RR} = \frac{1}{2}(1-y^2)$, *i.e.*, $y > \sqrt{2}-1$.

For $y < x < \frac{1}{2}$, choosing A is better, if $q_{RA} = x > q_{RR} = \frac{1}{2}(1+x^2) - xy$, *i.e.*, $y > \bar{x}^2/(2x)$.

For $y > x > \frac{1}{2}$, choosing A is better, if $q_{AA} = 1 > q_{AR} = \bar{y}$, *i.e.*, sure thing.

For $x > y \vee \frac{1}{2}$, choosing A is better, if $q_{AA} = 0 > q_{AR} = \bar{x}$, *i.e.*, impossible.

Combining the above all, we obtain Figure 3, Case $a = 1/2$, which shows players' optimal strategy-pairs and II's winning probabilities attached to each pair.

Winning prob. $P(W_2)$ is obtained by calculating the integral over the unit square.

$$\left[\int_0^{\sqrt{2}-1} \frac{1}{2}(1-y^2)y\,dy + \int_{\sqrt{2}-1}^{1/2} y^2 dy + \int_{1/2}^1 \left(\frac{3}{2}y - \frac{1}{2}\right) dy\right] \quad (9.9$$
$$+ \left[\int_0^{\sqrt{2}-1} dx \int_0^x \left\{\frac{1}{2}(1+x^2) - xy\right\} dy + \int_{\sqrt{2}-1}^{1/2} dx \left\{\int_0^{\bar{x}^2/(2x)} \left(\frac{1}{2}(1+x^2) - xy\right) dy\right.\right.$$
$$\left.\left. + \left(x - \frac{\bar{x}^2}{2x}\right)x\right\} + \int_{1/2}^1 x\bar{x}\,dx\right] \ (\equiv B_1 + B_2, \ \text{say, in this order}),$$

where B_1 (B_2) is the integral over the upper (lower) half of the unit square in Figure 3.

We find

$$B_1 = \frac{21}{16} - \frac{2}{3}\sqrt{2} \approx 0.36969.$$

The sum of the first and third terms in B_2 is

$$\int_0^{\sqrt{2}-1} \left\{\frac{1}{2}(1+x^2)x - \frac{1}{2}x^3\right\} dx + \left[\frac{x^2}{2} - \frac{x^3}{3}\right]_{1/2}^1 = \frac{5}{6} - \frac{1}{2}\sqrt{2} \approx 0.12623.$$

The second term in B_2 is

$$\int_{\sqrt{2}-1}^{1/2} \left(x^2 + \frac{\bar{x}^4}{8x}\right) dx = \frac{1}{8}\log\frac{1/2}{\sqrt{2}-1} + \left[-\frac{1}{2}x + \frac{3}{8}x^2 + \frac{1}{6}x^3 + \frac{1}{24}x^4\right]_{\sqrt{2}-1}^{1/2}$$
$$= -\frac{1}{8}\log(2\sqrt{2}-2) - \frac{499}{384} + \frac{11\sqrt{2}}{12} \approx 0.02041.$$

So finally, we get

$$B_1 + B_2 \approx 0.36969 + 0.12623 + 0.02041 \approx 0.51633. \tag{9.10}$$

Let D be the region in which R-R is the optimal pair in Figure 3, Case $a = 1/2$. Then

$$P(\text{draw}) = \int_0^1 \int_0^1 P\{(x,y) \in D\} P(X_2 < x, Y_2 < y) dx dy \tag{9.11}$$

$$= \int_0^1 \int_0^1 \left\{ I(x < \sqrt{2}-1, y < \sqrt{2}-1) + I\left(\sqrt{2}-1 < x < \frac{1}{2}\right) \frac{\bar{x}^2}{2x} \right\} xy dx dy$$
$$= \frac{1}{4}(\sqrt{2}-1)^4 + \frac{1}{4}\int_{\sqrt{2}-1}^{1/2} \bar{x}^2 dx = \frac{17}{4} - 3\sqrt{2} + \frac{1}{24}\left\{(2-\sqrt{2})^3 - \frac{1}{8}\right\} \approx 0.01053.$$

Eqs (9.9)~(9.11) give the proof of part (i). □

Proof of part (ii). II's winning prob. by choosing R (A) is the first (second) component of $(\bar{x}, x)\mathbf{M}_2(y \mid x)$. From Eqs (9.5)~(9.8) we can proceed as follows :
For $0 < x < y < 1$, choosing A is better, if

$$\bar{x}q_{RA} + xq_{AA} = \bar{x}y + x > \bar{x}q_{RR} + xq_{AR} = \bar{x} \cdot \frac{1}{2}(1-y^2) + x(1-y),$$

or equivalently,

$$y^2 + (2/\bar{x})\, y - 1 > 0,$$

that is,

$$y > \left(\sqrt{1+\bar{x}^2} - 1\right)/\bar{x} \ (\equiv \eta(x), \ \text{say}). \tag{9.12}$$

$\eta(x)$ is a concave , decreasing function connecting the two points $(0, \sqrt{2}-1)$ and (x_0, x_0) in the triangle $0 < x < y < 1$, where $x_0 \approx 0.311$ is a unique root in $[0,1]$ of the cubic equation $\sqrt{1+\bar{x}^2} - 1 = x\bar{x}$, *i.e.*,

$$x^3 - x^2 - 3x + 1 = 0. \tag{9.13}$$

Meanwhile, for $1 > x > y > 0$, choosing A is better, if

$$\bar{x}q_{RA} + xq_{AA} = x\bar{x} > \bar{x}q_{RR} + xq_{AR} = \bar{x} \cdot \left\{\frac{1}{2}(1+x^2) - xy\right\} + x\bar{x}$$

or equivalently,

$$y > \frac{1}{2}(x + x^{-1}),$$

which is impossible in the triangle $1 > x > y > 0$. Therefore we obtain Figure 3.

The winning prob. $P(W_2)$ is obtained by calculating the integral

$$\int_0^1 dx \int_0^x \left\{-x\bar{x}y + \frac{1}{2}\bar{x}(1+x)^2\right\} dy + \int_{x_0}^1 dx \int_x^1 (\bar{x}y + x) dy \tag{9.14}$$
$$+ \int_0^{x_0} dx \left[\int_x^{\eta(x)} \left(\frac{1+x}{2} - xy - \frac{1}{2}\bar{x}y^2\right) dy + \int_{\eta(x)}^1 (\bar{x}y + x) dy\right]$$

($\equiv B_1 + B_2 + B_3$, say, in this order).

Some computation gives

$$B_1 = \int_0^1 \left\{ -\frac{x^3\bar{x}}{2} + \frac{x\bar{x}}{2}(1+x)^2 \right\} dx = \frac{1}{2}\int_0^1 (x + x^2 - 2x^3)dx = \frac{1}{6} \approx 0.1667 \quad (9.15)$$

and

$$B_2 = \frac{1}{2}\int_{x_0}^1 \bar{x}(1+2x-x^2)dx = \frac{1}{2}\left\{\frac{3}{4} - \left(x_0 + \frac{1}{2}x_0^2 - x_0^3 + \frac{1}{4}x_0^4\right)\right\} \approx 0.2092. \quad (9.16)$$

Let R_2 be the rejection region for II in Figure 3. Then

$$\begin{aligned} P(\text{draw}) &= \int_0^1\int_0^1 \bar{x}P\{(x,y) \in R_2\} P(X_2 < x, Y_2 < y)dxdy \\ &= \int_0^1\int_0^1 \bar{x}\{I(x < x_0)\eta(x) + I(x > x_0)x\} xydxdy \\ &= \frac{1}{2}\left[\int_0^{x_0} x\bar{x}\eta(x)dx + \int_{x_0}^1 x^2\bar{x}dx\right]. \end{aligned}$$

The first and second integrals are

$$\begin{aligned} \int_0^{x_0} x\left(\sqrt{1+\bar{x}^2} - 1\right)dx &= \left[-\tfrac{1}{2}\left\{\bar{x}\sqrt{1+\bar{x}^2} + \log\left(\bar{x} + \sqrt{1+\bar{x}^2}\right)\right\} + \tfrac{1}{3}\left(1+\bar{x}^2\right)^{3/2}\right] \\ &\approx 0.0134 \end{aligned}$$

and

$$\int_{x_0}^1 (x^2 - x^3)dx = \frac{1}{12} - \frac{1}{3}x_0^3 + \frac{1}{4}x_0^4 \approx 0.0756,$$

respectively. Thus we have $P(\text{draw}) \approx \frac{1}{2}(0.0134 + 0.0756) = 0.0445$.

Finally we have to compute the thrid term of (9.11), *i.e.*, B_3. However we use an approximation in order to avoid the annoying computation for the integrals involving$\eta(x)$. Note that

$$(\bar{x}y + x) - \left(\frac{1+x}{2} - xy - \frac{1}{2}\bar{x}y^2\right) = \frac{1}{2}\bar{x}\left\{y^2 + (2/\bar{x})y - 1\right\} < 0, \quad \text{if } x < y < \eta(x) \text{ and}$$

and

$$B_3 > \left[\int_0^1 - \int_{x_0}^1\right] dx \int_x^1 (\bar{x}y + x)dy = \frac{3}{8} - 0.2092 \approx 0.1658 \quad \text{(by (9.16))}. \quad (9.17)$$

Therefore from (9.14)~(9.17), we obtain

$$P(W_2) = B_1 + B_2 + B_3 > \frac{1}{6} + 0.2092 + 0.1658 \approx 0.5417$$

and

$$P(W_1) = 1 - P(\text{draw}) - P(W_2) < 1 - 0.0445 - 0.5417 = 0.4138, \quad \text{approximately.}$$

This completes the proof of part (ii) of the theorem. □

10. Final Remark

Three-player games under various information are of interest. GKE under $I^{100-010-001}$ and $I^{100-110-111}$ are solved in Ref.[4 ; Theorem 3] and Ref.[4 ; Theorem 2], respectively. GRE under $I^{100-010-001}$ is solved in Ref.[6 ; Theorem 1] (The meaning of the information types in three-player games will be understood by referring to those in two-player games mentioned in Section 1). Several games, for example, GKE and GRE both under $I^{111-111-111}$ remain to be solved.

References

[1] P.R. Coe, W. Butterworth, Optimal stopping in the "Showcase showdown", *The American Statistician* **49** (1995), 271–275.

[2] A. Garnaev, *Search Games and Other Applications of Game Theory*, Springer, Berlin, 1999.

[3] M. Sakaguchi, Two-player games of "Score Showdown", *Scientiae Math. Japonicae* **60** (2004), 411–421.

[4] M. Sakaguchi, Three-player game of "Keep-or-Exchange", *Game Th. Appl.* **10** (2005), 148–160.

[5] M. Sakaguchi, Equilibrium in two-player games of "Showcase Showdown", *Scientiae Math. Japonicae* **61** (2005), 145–151.

[6] M. Sakaguchi, Equilibrium in the three-player game of "Risky Exchange", *Scientiae Math. Japonicae* **61** (2005), 335–344.

In: Mathematics, Game Theory and Algebra Compendium ISBN 978-1-60456-659-8
Editor: Jacob H. Mathias, pp. 765-784

Chapter 63

INFINITELY REPEATED SYMMETRIC 2×2-BIMATRIX GAMES

Alexei A. Semenishchev
Ural State University
Ekaterinburg
Russia

1. Introduction

Bimatrix games are sufficiently simple and, at the same time, powerful tool for analysis of the interaction between two players that could be animals, human beings or social institutions. The interests of the players are not antagonistic, i.e. each of them searches for the strategy that maximizes his personal profit. In the paper the simplest bimatrix games are considered: both players have two variants of action. Additionally, the property of symmetry of the game is required. The most widely known example of such games is the Prisoner's Dilemma [1]; other classes of games are considered as well.

In Section 3 a model of the infinitely repeated game with finite players' memory is constructed. The definitions of the strategies of the players are given. A graph of the game is suggested as the main analysis tool, and it is shown that the trajectory of the game has a cyclic structure. This property of the game allows defining payoff functions of the players in quite a simple way.

The problem of finding the Nash Equilibria [3] in the game is formulated in Section 4. The algorithm for the solution of this problem is suggested; results are shown in Appendix A.

Another important concept of non-antagonistic games, the Stackelberg solution [8], is analyzed in Section 5. The algorithm for finding an optimal leader's strategy in the repeated Prisoner's Dilemma, formulated in [6], has been adapted to solve any infinitely repeated symmetric bimatrix game. The optimal leader's strategy in the game with 1-step memory is described in Appendix B.

2. Symmetric 2×2 Bimatrix Games

Consider a bimatrix game with two strategies for each player. Denote these strategies by D and C, similarly to the Prisoner's Dilemma game [1], and the payoff matrices of the players – by A and B. A bimatrix game is called a *symmetric* one [9] if either payoff matrices satisfy the condition

$$B = A^T \tag{1}$$

or there exists such a substitution of the players' strategies that (1) is fulfilled. For example, the Prisoner's Dilemma game is symmetric because the payoff matrices of the players are [1]

$$A = \begin{pmatrix} R & S \\ T & P \end{pmatrix}, \quad B = \begin{pmatrix} R & T \\ S & P \end{pmatrix}.$$

We will also consider the *Battle of Sexes* [2] to be a symmetric game though its payoff matrices

$$A = \begin{pmatrix} 2 & -1 \\ -1 & 1 \end{pmatrix}, \quad B = \begin{pmatrix} 1 & -1 \\ -1 & 2 \end{pmatrix} \tag{2}$$

do not satisfy (1). Obviously, after the permutation of the first player's strategies new payoff matrices satisfy the condition of symmetry.

Symmetry of a bimatrix game means equal rights for both players, that is, the profit of the first player at the state (i, j) equals that of the second player at the state (j, i) for all $i, j \in \{D, C\}$. Since a symmetric bimatrix game can be prescribed by payoff matrix of the first player, it will be called a payoff matrix of such a game.

Here we deal with such a symmetric 2×2 bimatrix game whose payoff matrix's diagonal elements are different (i.e. $f_1(D, D) \neq f_1(C, C)$). A proper linear transformation reduces payoff matrix of such games to the form

$$A = \begin{pmatrix} \alpha & 1 \\ 0 & \beta \end{pmatrix}. \tag{3}$$

Therefore, the game is completely defined by two parameters and corresponds to some point on the plane (α, β). There exist 12 different types of such symmetric bimatrix games defined by order of parameters 0, 1, α, β (see for example [5]). Figure 1 demonstrates the partition of the plane (α, β) into domains corresponding to different game types.

For example, points of domain G_2 define the Hawk-Dove game. Points of the common bound of domains G_3 and G_4 correspond to the Battle of Sexes. Static Prisoner's Dilemma game is defined by the points of domain G_6; infinitely repeated Prisoner's Dilemma has additional restriction in this domain: $\beta > 1/2$.

3. Infinitely Repeated Bimatrix Games With Finite Players' Memory

Consider infinitely repeated symmetric bimatrix game; every step of this supergame is a bimatrix game defined by matrix (3). This approach has been found to be quite efficient while studying some classes of symmetric bimatrix games [1, 4, 7]. A strategy of a player

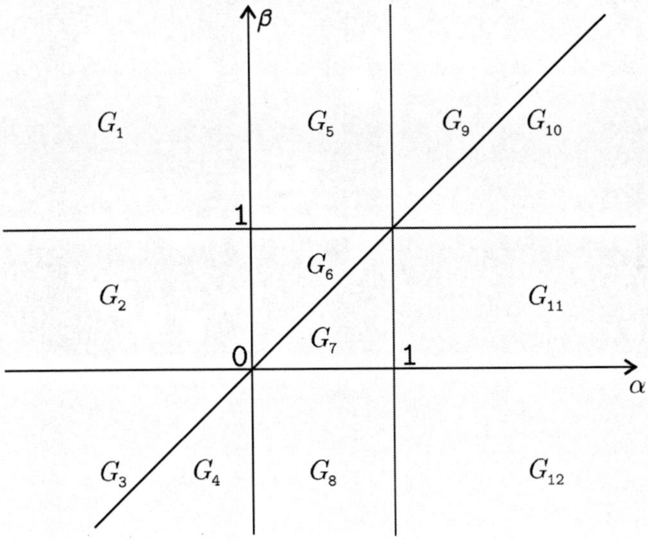

Figure 1. Different types of symmetric bimatrix games.

in such a game is some decision rule that determines player's choice at every step (see, for example, [1]). This decision rule can take into account all the information about game that is available at the current step.

We restrict the information used by the player's decision rule. Suppose the players to have m-step memory, that is, to choose the option D or C at the current step each player considers the results of m previous steps of the game. Integer $m \geq 0$ will be called a *memory length* (both players have the same memory length). The strategies of the players described above will be called *strategies with finite memory* or *strategies with m-step memory*. Examples of such strategies described in [1] are:

0. Strategies ALL C and ALL D can be regarded as strategies with 0-step memory.

1. Well-known strategy TIT FOR TAT is a strategy with 1-step memory.

2. Strategy TIT FOR TWO TATS, which is more remissive then TIT FOR TAT, has memory length 2.

For brevity we will use further the term 'game' instead of 'infinitely repeated symmetric bimatrix game with m-step memory of players'.

3.1. States of the Game

Under the suggested approach the player's choice is completely defined by the last m steps of the game. Therefore it is essential to use this previous history as the game state.

Definition 1. *We will call a game state at the current step an ordered set* $s = (s_{-m}, s_{-m+1}, \ldots, s_{-1})$, *where* s_{-i} *is the first player's profit* i *steps backward.*

A game state is defined by the profits of the first player only. The profits of the second player could be found if necessary using this information. Note that game states at steps $1, \ldots, m$ are undefined.

Denote a *set of all game states* by $\mathcal{S}$. Since $\forall\ i \in \overline{1,\ m}\ (s_{-i} \in \{0, 1, \alpha, \beta\})$ the set $\mathcal{S}$ consists of 4^m different states.

Define now a strategy of the player in the infinitely repeated game.

Definition 2. *A strategy of the player is some function that establishes correspondence between any possible game state and a choice of the player at this state:*

$$f : \mathcal{S} \mapsto \{C, D\} \tag{4}$$

Sets of all strategies of the first and the second player will be denoted by X and Y, respectively. We use here different notations for these sets because of some reasons that will be explained later. Definition of set $\mathcal{S}$ implies that both X and Y are finite and include 2^{4^m} different strategies.

3.2. Dynamics of the Game

Dynamics of the game is defined as follows. Let $s = (s_{-m}, s_{-m+1}, \ldots, s_{-1})$ is a current game state. Be given the strategies of both players $x \in X, y \in Y$, one can determine $p_1(x, y, s)$ which is the profit of the first player at state s. For example, if $x(s) = D$ and $y(s) = C$ then $p_1(x, y, s) = 1$ in accordance with payoff matrix (3). Then the game state that follows s is $t = (t_{-m}, t_{-m+1}, \ldots, t_{-1})$, where

$$\begin{aligned} t_{-m} &= s_{-m+1}, \\ t_{-m+1} &= s_{-m+2}, \\ &\vdots \\ t_{-2} &= s_{-1}, \\ t_{-1} &= p_1(x, y, s). \end{aligned} \tag{5}$$

Equations (5) are intended to be understood as follows. At every step of the infinitely repeated game all the profits a player keeps in mind shift one position left. The rightmost position of memory is occupied by the profit just obtained.

A trajectory of the game is a sequence of game states $\{s_i\}_{i=1}^{\infty} \subseteq \mathcal{S}$ that begins at some initial state s_1. Transitions between s_i and s_{i+1} for $i \geq 1$ are determined by the rule described above. Obviously, the trajectory depends both on initial state s_1 and on strategies of the players $x \in X, y \in Y$.

3.3. Graph of the Game

Since the game has a finite number of states and certain transitions between these states, it is convenient to illustrate the dynamics of the game at the oriented graph of following structure. The set of the game states $\mathcal{S}$ is regarded as a set V of the graph's nodes. Therefore every node $v \in V$ will be assigned to a set of coordinates, equal to coordinates of the corresponding game state s.

$$V = \{v = (v_{-m}, v_{-m+1}, \ldots, v_{-1}),\ v_{-i} \in \{0, 1, \alpha, \beta\}\} \tag{6}$$

Further, define the set of edges $E(\cdot, \cdot)$ of the graph in the situation when strategies of the players are undefined. Accordingly to the dynamics of the game, from every game state s there exist four available transitions to some other game states, depending on the choices of the players in s. We will treat these transitions as oriented edges coming out of the node v, which corresponds to the game state s. Analogously, it can be shown that for every node v there exist four edges coming into this node.

Definition 3. *Graph $G(\cdot, \cdot) = G(V, E(\cdot, \cdot))$ will be called graph of the infinitely repeated bimatrix game with m-step memory.*

Consider the following three subgraphs of graph $G(\cdot, \cdot)$. Subgraph $G(x, \cdot)$ is generated by strategy $x \in X$ of the first player, whereas the strategy of the second one is undefined. In this case for every node there exist two edges coming out of this node. Indeed, if the choice of the first player at the node v is determined, two possible choices of the second one give two variants of the result. Analogously one can consider subgraph $G(\cdot, y)$ which is generated by strategy $y \in Y$ of the second player under condition that the first player's strategy is undefined. Finally, be given the strategies of the players $x \in X, y \in Y$, one can find subgraph $G(x, y)$. For every node of this graph there exists the only outgoing edge.

Statement 1. *For arbitrary strategies of players $x \in X,\ y \in Y$ the graph $G(x, y)$ has at least one cycle.*

Proof. Indeed, the game is repeated an infinite number of times whereas number of nodes of the graph $G(x, y)$ equals 4^m, where m is the memory length. This implies that at some step the trajectory of the game will arrive at a node that has been passed earlier. From this moment onwards the trajectory of the game will pursue the achieved cycle. □

Graphs of the game with 1-step memory and 2-step memory are shown at Fig. 2. The strategies of the players are chosen at random. Choices of the first player are shown above the nodes, and those of the second player – below the nodes. The edges of the graph correspond to transitions between different game states. Heavily drawn edges mark cycles at the graph $G(x, y)$.

Accept the following notations. Denote the set of all cycles on the graph $G(x, y)$ by $Z(x, y)$. Symbols $Z(\cdot, \cdot)$, $Z(x, \cdot)$ and $Z(\cdot, y)$ will denote the sets of cycles on the graphs $G(\cdot, \cdot)$, $G(x, \cdot)$ and $G(\cdot, y)$, correspondingly.

Any of the sets $Z(\cdot, \cdot)$, $Z(x, \cdot)$ and $Z(\cdot, y)$ is not empty. Indeed, since $G(x, y)$ is a subgraph of $G(x, \cdot)$ (and $G(\cdot, y)$ and $G(\cdot, \cdot)$), the following inclusions are correct: $Z(x, y) \subseteq Z(x, \cdot)$, $Z(x, y) \subseteq Z(\cdot, y)$ and $Z(x, y) \subseteq Z(\cdot, \cdot)$. As it was proved above, the set $Z(x, y)$ is not empty, therefore the mentioned sets are not empty also.

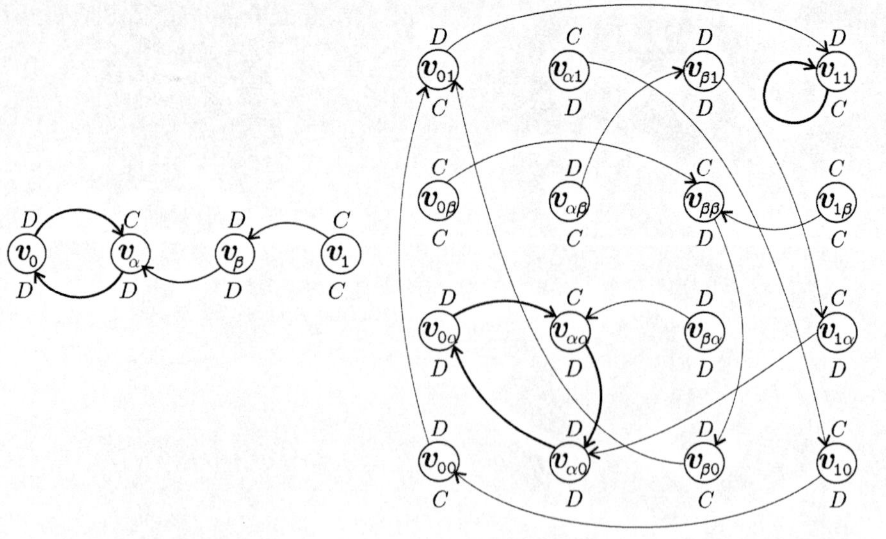

Figure 2. Graphs of the game with 1-step memory and 2-step memory.

3.4. Payoff Functions of the Players

There exist a number of different ways to determine the player's payoff in the infinitely repeated game. We will use here the Cesàro average:

$$f_i = \lim_{N\to\infty} \frac{1}{N} \sum_{n=1}^{N} f_i^n, \tag{7}$$

where $i = 1, 2$ is a number of a player and f_i^n is his profit at the n-th step of the game. As it was noted above, a trajectory of the game depends both on the strategies of the players $x \in X, y \in Y$ and on the initial game state s_1. Therefore $f_i = f_i(x, y; s_1)$.

Consider the simplest situation when graph $G(x, y)$ has the only cycle z. Then a trajectory of the game $\{s_1, s_2, \dots\}$ arrives at this cycle whatever the initial game state s_1 has been chosen. It can be easily proved that in this case players' payoff functions (7) are equal to the average profits at the cycle which are calculated as follows

$$M_1(z; \alpha, \beta) = \frac{k_\alpha \cdot \alpha + k_\beta \cdot \beta + k_1}{k_0 + k_\alpha + k_\beta + k_1}, \quad M_2(z; \alpha, \beta) = \frac{k_\alpha \cdot \alpha + k_\beta \cdot \beta + k_0}{k_0 + k_\alpha + k_\beta + k_1} \tag{8}$$

The following notations are used here: k_0 is the number of those transitions belonging to the cycle z, at which the first player obtains profit 0. These transitions correspond to those and only those edges of graph $G(x, y)$ that come into nodes $v = (v_{-m}, \dots, v_{-2}, 0)$ and belong to the cycle z. In other words, k_0 is a number of the nodes of the cycle z that have the last coordinate equal to 0. Similarly, figures $k_1,\ k_\alpha,\ k_\beta$ are defined as the number of nodes of kind $v = (v_{-m}, \dots, v_{-2}, 1)$, $v = (v_{-m}, \dots, v_{-2}, \alpha)$, $v = (v_{-m}, \dots, v_{-2}, \beta)$ at the cycle z. Numbers $k_0,\ k_1,\ k_\alpha,\ k_\beta$ will be called *multiplicities* of the profits $0,\ 1,\ \alpha,\ \beta$ at the cycle z.

Note the following property of functions (8). For any cycle z the difference $M_1(z; \alpha, \beta) - M_2(z; \alpha, \beta)$ is a constant that does not depend on the values of parameters α, β. Therefore, any cycle z at an arbitrary point (α, β) is either preferable for the first player or for the second one or both players have equal profits at this cycle.

Thus, every cycle z has a uniquely defined set of multiplicities $k = (k_0, k_\alpha, k_\beta, k_1)$. At the same time, there can exists a number of cycles that have the same set of multiplicities.

Definition 4. *A set of cycles having the same multiplicities* $k = (k_0, k_\alpha, k_\beta, k_1)$ *will be called a family of cycles defined by this set of multiplicities and will be denoted by* $Z_{k_0 k_\alpha k_\beta k_1}$.

It is easy to draw the conclusion that any of the multiplicities $k_0, k_\alpha, k_\beta, k_1$ is not greater than 4^{m-1} in a game with m-step memory. However, this condition is not sufficient for the existence of the cycle with a given set of multiplicities. Fix memory length m and consider the set of all cycles $Z(\cdot, \cdot)$ in the graph of the game with m-step memory.

Definition 5. *A set of nonnegative integer numbers* $k = (k_0, k_\alpha, k_\beta, k_1)$ *will be called an admissible set of multiplicities if there exists a cycle* $z \in Z(\cdot, \cdot)$ *that has such a set of multiplicities.*

In the game with 1-step memory any set $k = (k_0, k_\alpha, k_\beta, k_1)$, where each number is either 0 or 1, is an admissible set of multiplicities. In 2-step memory game there exists a criterion (see [6]) for a set $k = (k_0, k_\alpha, k_\beta, k_1)$ to be an admissible one. For the games with longer memory such criteria are unknown.

Now consider a trajectory of the game in the case when set $Z(x, y)$ includes more than one cycle. Since the initial game state can be chosen arbitrary, it is impossible to determine the cycle where the trajectory will come. That is why we consider the guaranteed profit of the player as his profit for given strategies x, y.

$$\begin{aligned} f_1(x, y; \alpha, \beta) &= \min_{z \in Z(x,y)} M_1(z; \alpha, \beta), \\ f_2(x, y; \alpha, \beta) &= \min_{z \in Z(x,y)} M_2(z; \alpha, \beta) \end{aligned} \tag{9}$$

As follows from (9), under fixed $x \in X$, $y \in Y$ the profits of the players $f_1(x, y; \alpha, \beta)$ and $f_2(x, y; \alpha, \beta)$ are piecewise linear function of parameters α, β.

4. Nash Equilibria in the Infinitely Repeated Bimatrix Game with m-step Memory

Consider a game with m-step memory under the condition that parameters α, β are fixed. Accordingly to definition [3], Nash equilibrium (NE) in this game is such pair of strategies $x^N \in X$, $y^N \in Y$ that for arbitrary $x \in X$, $y \in Y$ the following inequalities are fulfilled :

$$\begin{aligned} f_1(x, y^N; \alpha, \beta) \leq f_1(x^N, y^N; \alpha, \beta), \\ f_2(x^N, y; \alpha, \beta) \leq f_2(x^N, y^N; \alpha, \beta) \end{aligned} \tag{10}$$

Note that there can exist more then one NE in the game at a given point (α, β).

Now we formulate the problem of finding all NE at an arbitrary point (α, β).

Problem 1. *Given the memory length m, to find a partition of the plane of parameters α, β into domains $\{D_i\}_{i=1}^{n}$ such that in every domain D_i the set of Nash equilibria is the same for any point of this domain.*

4.1. Nash Algorithm of Finding the Nash Equilibria

As it was noted earlier, for a given memory length m each player has 2^{4^m} different strategies. Denote this number by r and enumerate the strategies of the players by numbers $0, \ldots, r-1$ by the following rule. Accept some conventional order of the game parameters, for example, $0, \alpha, \beta, 1$ (this order corresponds to the Prisoner's Dilemma game). Then enumerate nodes of the graph of game in lexicographic order: $v_0 = (0, \ldots, 0, 0)$, $v_1 = (0, \ldots, 0, \alpha)$, $v_2 = (0, \ldots, 0, \beta)$, ..., $v_{4^m-1} = (1, \ldots, 1)$. Thus, each strategy of the first player can be represented as a string of length r, consisting from characters D and C. A character at the i-th position of this string denotes the choice of the first player in the node v_i, $i = 0, \ldots, 4^m - 1$. For example, in the 1-step memory game the strategy ALL D has the code string $DDDD$ and the strategy TIT FOR TAT is defined as $DDCC$. Further, consider the characters D and C as digits of the binary number system (0 and 1, correspondingly). Such approach allows to present every strategy $x \in X$ by some number from the range $0, \ldots, r-1$.

Strategies of the second player are enumerated in a more sophisticated rule. Since the considered repeated game is constructed on the base of the symmetric game, it seems essential to require of it the property of *extended symmetry*:

$$\forall\, i, j \in \overline{0, r-1}\; \forall\, \alpha, \beta \in \mathbb{R}\; (f_1(i, j; \alpha, \beta) = f_2(j, i; \alpha, \beta)) \tag{11}$$

It can be proved that a proper way of enumerating the strategies of the second player exists for a game with arbitrary memory length. In a game with 1-step memory this method is reduced to the swap of the second player's choices in the nodes $v_0 = (0)$ and $v_3 = (1)$, following by the evaluation of obtained binary number. For example, the strategy TIT FOR TAT has number 3 ($DDCC$) if it is exploited by the first player. From the point of view of the second player, this strategy is coded by string $CDCD$, nevertheless it has the same number 3 in the set of the second player's strategies. If both players use this strategies, the set of cycles $Z(3, 3)$ consists of only three cycles with players' profits equal to $M_1(z_1; \alpha, \beta) = M_2(z_1; \alpha, \beta) = \alpha$, $M_1(z_2; \alpha, \beta) = M_2(z_2; \alpha, \beta) = \beta$ and $M_1(z_3; \alpha, \beta) = M_2(z_3; \alpha, \beta) = 1/2$, correspondingly.

To find NEs in the infinitely repeated bimatrix game with m-step memory, we consider it as a static bimatrix games with payoff matrices of players $P = P(\alpha, \beta)$ and $Q = Q(\alpha, \beta)$. Matrices P and Q are defined as follows:

$$P(\alpha, \beta) = (f_1(x, y; \alpha, \beta))_{x,y=0}^{r-1}, \quad Q(\alpha, \beta) = (f_2(x, y; \alpha, \beta))_{x,y=0}^{r-1} \tag{12}$$

The structure of these payoff matrices is rather complicated because each element of P and Q is a piecewise-linear function of variables α, β. Therefore, the values of the elements in different points (α, β) can be calculated under different formulas. The additional complexity arises because the sizes of payoff matrices increase fast when memory length m

increases. For example, the matrices P and Q have sizes 16×16 for the game with 1-step memory, and their sizes increase up to 65536×65536 for the game with 2-step memory. Nevertheless, under given parameters α, β matrices P and Q define bimatrix game, which is symmetric in virtue of (11). Standard methods can be invoked to find all NEs in this game. However, these methods don't appear to be applicable directly to the game with functional matrices (12). That is why we analyze instead a number of symmetric bimatrix games which are defined by matrices (11) for some points $\{(\alpha_i, \beta_i)\}_{i=1}^{n}$. The number and position of these points depend on memory length m.

Definition 6. *Points (α_1, β_1) and (α_2, β_2) will be called NE-equivalent points of infinitely repeated bimatrix game with m-step memory if the bimatrix game with payoff matrices $P(\alpha_1, \beta_1), Q(\alpha_1, \beta_1)$ has the same set of Nash equilibrium states as the bimatrix game with payoff matrices $P(\alpha_2, \beta_2), Q(\alpha_2, \beta_2)$.*

In terms of NE-equivalent points, Problem 1 can be formulated in the following way: to find a partition of the plane α, β into domains $\{D_i\}_{i=1}^{n}$, each of which consists of NE-equivalent points.

To solve Problem 1 we construct the following auxiliary set of lines L. Each cycle $z \in Z(\cdot,\cdot)$ defines the profit of the first player $M_1(z; \alpha, \beta)$ (see (8)), which is a linear function of parameters (α, β). The set L consists of all the following lines:

$$l = \{(\alpha, \beta) \in \mathbb{R} \mid M_1(z'; \alpha, \beta) = M_1(z''; \alpha, \beta), \quad z' \neq z''\} \tag{13}$$

The line l is not being defined for such cycles z', $z'' \in Z(\cdot,\cdot)$ that equation in (13) either has no solution or has solution coinciding with the entire plane of parameters α, β. Would the profits of the second player $M_2(c; \alpha, \beta)$ be used in (13), the set L becomes the same.

Lines of the set L have the following important property.

Statement 2. *If points (α_1, β_1) and (α_2, β_2) are not separated by any line of the set L, then these points are NE-equivalent.*

Proof. For the sake of brevity the following denotations will be used in this proof. The matrices $P(\alpha_1, \beta_1)$ and $Q(\alpha_1, \beta_1)$ will be denoted by P_1 and Q_1, respectively. The bimatrix game with matrices $P(\alpha_1, \beta_1)$ and $Q(\alpha_1, \beta_1)$ will be denoted by Γ_1. Similarly, denote the matrices $P(\alpha_2, \beta_2)$ and $Q(\alpha_2, \beta_2)$ by P_2 and Q_2, and the bimatrix game with these matrices – by Γ_2.

Suppose that some pair of strategies x^N, y^N constitutes the Nash equilibrium in the bimatrix game Γ_1. Prove that x^N, y^N are Nash equilibrium strategies in the bimatrix game Γ_2 as well.

Accordingly to the properties of NE strategies, the element $p_{x^N y^N}$ is the maximum of the y^N-th column of the matrix P_1, and the element $q_{x^N y^N}$ is the maximum of the x^N-th row of the matrix Q_1. Besides, the definition of the matrix $P(\alpha, \beta)$ implies

$$p_{x^N y^N} = \min_{z \in Z(x^N, y^N)} M_1(z; \alpha_1, \beta_1)$$

Let $p_{x^N y^N}$ achieves its value on some cycle $z' \in Z(x^N, y^N)$.

If we suppose that the pair x^N, y^N is not a NE in the point (α_2, β_2) then the following conclusion should be drawn: either the element $p_{x^N y^N}$ is not the maximum of the y^N-th column of the matrix P_2 or the element $q_{x^N y^N}$ is not the maximum of the x^N-th row of the matrix Q_2. The first assumption implies that there exists such a strategy $x^* \neq x^N$ that $p_{x^* y^N} > p_{x^N y^N}$ in the matrix P_2. Therefore, there exists such a cycle $z^* \in Z(x^*, y^N)$ that the following inequality is fulfilled:

$$M_1(z'; \alpha_1, \beta_1) < M_1(z^*; \alpha_2, \beta_2) \tag{14}$$

Since $M_1(z'; \alpha, \beta)$ and $M_1(z^*; \alpha, \beta)$ are continuous functions of variables α, β, there necessarily exists such a point (α^*, β^*) on the line between (α_1, β_1) and (α_2, β_2), that $M_1(z'; \alpha^*, \beta^*) = M_1(z^*; \alpha^*, \beta^*)$. This means that the point (α^*, β^*) belongs to some line of the set L, and this conclusion contradicts the condition that no line of the set L separates the points (α_1, β_1) and (α_2, β_2).

Analogously, one can prove that the element $q_{x^N y^N}$ remains the maximum of the x^N-th row in the matrix Q_2. Therefore, the pair of the strategies x^N, y^N is also the Nash equilibrium in the bimatrix game Γ_2.

Now prove the following fact: there are no NEs in the bimatrix game Γ_2, but those which exist in the game Γ_1.

Assume the contrary: the bimatrix game Γ_2 has some Nash equilibrium x^*, y^*, and this pair of strategies is not the NE in the game Γ_1. Consider the first part of this proof. One should exchange the notations (α_1, β_1) and (α_2, β_2), and the following conclusion will be drawn. If the pair x^*, y^* constitutes the Nash equilibrium state in the game Γ_2, it is the NE in the game Γ_1 also. □

The property of the lines of the set L allows to solve the Problem 1 using the following algorithm.

1. Constructing a partition of the plane (α, β) into the set of domains $\{F_i\}_{i=1}^{s}$, whose bounds are the lines of the set L.

2. Finding the Nash equilibria in the bimatrix game with matrices $P(\alpha_i, \beta_i)$, $Q(\alpha_i, \beta_i)$, where (α_i, β_i) is an arbitrary inner point of the domain F_i. Denote the set of these NE by N_{F_i}.

3. Joining those domains F_i whose sets N_{F_i} are identical.

The game with 1-step memory has been investigated using the described algorithm. The results of this investigation are shown in Appendix A.

5. Hierarchical Setting of the Game

In the previous section there were studied the Nash equilibria in the infinitely repeated bimatrix game with finite memory. The hierarchical setting of the game by Stackelberg [8] is another common way of solving the bimatrix games.

The setting is as follows. One of the players, called the *leader*, declares his strategy at the beginning of the game and undertakes to keep it till the end of the game. His partner, called the *follower*, analyzes the presented leader's strategy and determines his own strategy in order to maximize his profit. It is the known fact that the leader's profit in the hierarchical setting of the game is not less then his profit at any Nash equilibrium. That is why it is quite an interesting problem to solve the infinitely repeated bimatrix game in hierarchical setting.

Fix memory length m and assume the strategies of both players are enumerated as it is supposed in 4.1. Consider the first player to be a leader and the second one to be a follower. Given the leader's strategy $x \in X$ and values of the game parameters (α, β), the follower can define his best response not uniquely. Denote a set of best responses of the follower by $\Phi(x; \alpha, \beta)$:

$$\Phi(x; \alpha, \beta) = \{y \in Y, \mid y = \arg\max_{y \in Y} f_2(x, y; \alpha, \beta)\} \tag{15}$$

Remark 1. *The set $\Phi(x; \alpha, \beta)$ is defined and not empty for any strategy $x \in X$ and any point (α, β). The strategies of the set $\Phi(x; \alpha, \beta)$ could be found by the complete survey of the second player's strategies. It is possible because the set Y is finite.*

Define the guaranteed leader's profit as the minimal value of his profit for given strategy $x \in X$ and game parameters α, β:

$$\mu(x; \alpha, \beta) = \min_{y \in \Phi(x; \alpha, \beta)} f_1(x, y; \alpha, \beta) \tag{16}$$

Given the game parameters α, β, one can find the *optimal* leader's strategy that maximizes his guaranteed profit (16). In general, by the complete survey of the leader's strategies one can find a set of the optimal strategies in the point (α, β).

$$X^S(\alpha, \beta) = \{x \in X \mid x = \arg\max_{x \in X} \mu(x; \alpha, \beta)\} \tag{17}$$

If the set of optimal leader's strategies consists of more then one element, it can be represented by an arbitrary strategy $x^S(\alpha, \beta) \in X^S(\alpha, \beta)$. This reducing is correct because the leader's profits are the same for all the strategies $x \in X^S(\alpha, \beta)$, $y \in \Phi(x; \alpha, \beta)$.

The optimal leader's strategy depends on the values of the parameters α, β. Set the problem of finding this strategy for arbitrary pair of the game parameters.

Problem 2. *To construct the partition of the plane (α, β) into the set of domains $\{P_i\}_{i=1}^n$, each of which has been assigned to some optimal leader's strategy x_i^s.*

The partition $\{P_i\}_{i=1}^n$ that is the solution of Problem 2 will be called an *optimal partition*. Besides, we will require that both the leader's strategy x_i^s, and the cycle that defines the leader's profit, are the same in any point of domain P_i. If this condition is not fulfilled, we will replace domain P_i by the set of its subdomains, each of which satisfies the mentioned requirement.

5.1. Algorithm of Constructing the Optimal Partition

The algorithm described below solves Problem 2 in a finite number of steps. It consists of the choice of the initial partition followed by iterative improvement of the leader's strategy. The realization of the algorithm will be demonstrated for the game with 1-step memory.

5.1.1. Constructing the Initial Partition

At the initial stage of the algorithm the leader chooses his initial strategy arbitrarily. This strategy is the same in all the points of the plane (α, β). The choice of the initial strategy induces the initial partition.

Suppose the leader chooses the strategy ALL D=0 that means the choice of D at every node of the graph of the game.

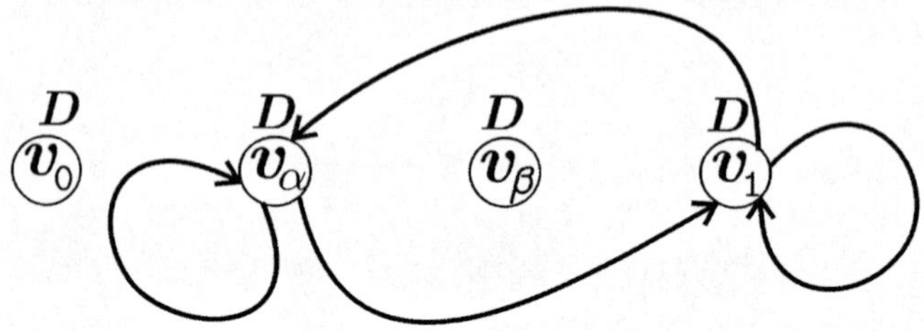

Figure 3. The strategy ALL D=0 and the set of cycles $Z(0, \cdot)$.

The set $Z(0, \cdot)$ consists of three cycles $v_\alpha, v_1, v_\alpha v_1$. The follower's profits on these cycles equal $\alpha, 0, \frac{\alpha}{2}$, respectively. His best response in the points of the half-plane $\alpha > 0$ is to choose the strategy ALL D that guarantees both his and the leader's profits equal α. In the points (α, β) for which $\alpha \leq 0$, the follower should choose the strategy ALL C in order to maximize his profit. In this case the follower's profit equals 0 whereas the leader's one equals 1.

Thus, the initial partition is constructed and the strategies of both the leader and the follower are defined in every domain.

The further process of solving Problem 2 consists in the complete survey of all the cycles on the graph of game. At every step of this process we consider the current partition of the plane (α, β) into set of the domains $\{P_i\}_{i=1}^{t}$. The leader's strategy in the domain P_i denote by x_i^S and his payoff function in this domain – by $f_i^S(\alpha, \beta)$. As it was noted above, this function is linear and it is the same for all the points of the domain P_i.

5.1.2. Choice of the Admissible Set of Multiplicities

If there is no any admissible set of multiplicities that has not been analyzed, the process is complete and the current partition is an optimal one. Otherwise, consider some admissible set of multiplicities $k^* = (k_0^*, k_\alpha^*, k_\beta^*, k_1^*)$ and the corresponding family of cycles $Z_{k_0^*, k_\alpha^*, k_\beta^*, k_1^*}$. The leader's profit on any of these cycles is the following:

$$f^*(\alpha, \beta) = \frac{k_\alpha^* \cdot \alpha + k_\beta^* \cdot \beta + k_1^*}{k_0^* + k_\alpha^* + k_\beta^* + k_1^*} \tag{18}$$

5.1.3. Improvement of the Leader's Strategy in the Domain P_i

Consider some domain P_i of the current partition of the plane (α, β). If the inequality

$$f^*(\alpha, \beta) > f_i^S(\alpha, \beta) \tag{19}$$

has a solution in the domain P_i, denote this solution by P_i^* and go to step 5.1.4., otherwise choose another domain P_i and repeat the current step.

If all the domains P_i have been analyzed, return to step 5.1.2.

5.1.4. Survey of the Cycles

The decidability of the inequality (19) means that the leader's profit on the cycles of the family $Z_{k_0^*, k_\alpha^*, k_\beta^*, k_1^*}$ is greater then his current profit in the subdomain $P_i^* \subseteq P_i$. Therefore one should analyze if these cycles can improve the leader's strategy.

Consider some cycle $z^* \in Z_{k_0^*, k_\alpha^*, k_\beta^*, k_1^*}$ and the subset of the leader's strategies $X^* = \{x \in X \mid z^* \in Z(x, \cdot)\}$. For each strategy $x^* \in X^*$ consider the following system of linear inequalities:

$$\begin{cases} M_2(z^*; \alpha, \beta) \geq M_2(z; \alpha, \beta), \\ \qquad z \in Z(x^*, \cdot) \backslash z^* \end{cases} \tag{20}$$

The inequalities of the system (20) should be read as follows: given the leader's strategy x^*, the follower would rather prefer the cycle z^* to any other cycle from $Z(x^*, \cdot)$. If this system has some solution $P_i^{**} \subseteq P_i^*$, the sufficient condition of improvement of the leader's strategy is the following:

$$\exists\, y^* \in Y : \; f_2(x^*, y^*; \alpha, \beta) = M_2(z^*; \alpha, \beta) \tag{21}$$

In other words, the follower must have an opportunity to achieve the profit proposed by the leader. The condition (21) together with (20) implies that either the set of cycles $Z(x^*, y^*)$ consists of the only cycle z^* or all the cycles of this set give the same profit to the follower (and, therefore, to the leader).

Survey all the strategies $x \in X^*$, and every case when both the system (20) has a solution and the condition (21) is fulfilled will be fixed by adding the pair (x^*, P_i^{**}) into a *list of the strategy improvement*. Enumerate all the pairs in this list: $L = \{(x^j, P_i^j)\}_{j=1}^{l}$.

If $\bigcup_{j=1}^{l} P_i^j = P_i^*$ then the set of multiplicities k^* is completely analyzed in the domain P_i. The leader's strategy is changed in the points of the subdomain P_i^*, and the process should return to step 5.1.2.

Otherwise, one should repeat the current step of the algorithm until all the cycles $z^* \in Z_{k_0^*, k_\alpha^*, k_\beta^*, k_1^*}$ are examined.

Problem 2 was solved for the game with 1-step memory. The optimal partition and the leader's strategies are shown in Appendix B.

Appendix A. Nash Equilibria in the Game With 1-step Memory

The solution of Problem 1 for the game with 1-step memory is shown below. This solution is being compared with Nash equilibria in the static symmetric 2×2 bimatrix games. The Pareto optimal strategies [2] in both cases are presented.

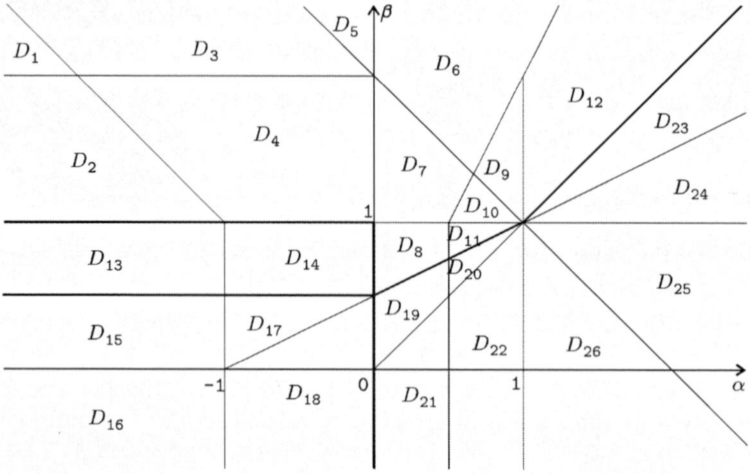

Figure 4. Nash equilibria in the game with 1-step memory.

The set of Nash equilibria is defined for every of domains $D_1, \ldots, D_{26}$. Different domains have different NE sets that are shown in Table 1. Since the game has the property of extended symmetry (11), some NE are omitted in order to reduce the amount of data. If the profit of the first player in the NE state $(i, j),\ i \leq j$ equals his profit in the state (j, i), we do not place the last one into Table 1. At the same time, there exist such NE that $f_1(i, j; \alpha, \beta) \neq f_1(j, i; \alpha, \beta)$ (for example, see those for domain D_1). In this case both pairs of the NE strategies are shown for clarity.

Table 1. Nash equilibria in the 1-step memory game

D_i	Nash equilibria	$f_1(\alpha, \beta)$	$f_2(\alpha, \beta)$
D_1	(0,12), (2,12)	1	0
	(3,14), (3,15), (6,6), (6,7), (7,11), (7,14), (7,15), (11,15), (15,15)	β	β
	(12,0),(12,2)	0	1
D_2	(0,12), (0,13), (2,12), (2,13), (4,9), (8,13)	1	0
	(3,14), (3,15), (6,6), (6,7), (7,11), (7,14), (7,15), (11,15)	β	β
	(9,4), (12,0), (12,2), (13,0), (13,2), (13,8)	0	1
D_3	(0,12), (2,12)	1	0
	(4,4)	$\frac{\alpha+\beta}{2}$	$\frac{\alpha+\beta}{2}$
	(3,13), (3,15), (6,6), (6,7), (7,11), (7,14), (7,15), (11,15), (15,15)	β	β
	(12,0), (12,2)	0	1

Table 1. Continued

D_i	Nash equilibria	$f_1(\alpha,\beta)$	$f_2(\alpha,\beta)$
D_4	(0,12), (0,13), (2,12), (2,13), (8,13)	1	0
	(4,4)	$\frac{\alpha+\beta}{2}$	$\frac{\alpha+\beta}{2}$
	(3,13), (3,15), (6,6), (6,7), (7,11), (7,14), (7,15), (11,15), (15,15)	β	β
	(12,0), (12,2), (13,0), (13,2), (13,8)	0	1
D_5	(4,4)	$\frac{\alpha+\beta}{2}$	$\frac{\alpha+\beta}{2}$
	(3,13), (3,15), (6,6), (6,7), (7,11), (7,14), (7,15), (11,15), (15,15)	β	β
D_6	(0,0), (0,2), (2,2)	α	α
	(4,4), (4,5), (5,12), (5,13), (13,13)	$\frac{\alpha+\beta}{2}$	$\frac{\alpha+\beta}{2}$
	(3,14), (3,15), (6,6), (6,7), (7,11), (7,14), (7,15), (11,15), (15,15)	β	β
D_7	(0,0), (0,2), (2,2)	α	α
	(4,4), (4,5)	$\frac{\alpha+\beta}{2}$	$\frac{\alpha+\beta}{2}$
	(3,14), (3,15), (6,6), (6,7), (7,11), (7,14), (7,15), (11,15), (15,15)	β	β
D_8	(0,0), (0,2), (2,2)	α	α
	(6,6), (6,7)	β	β
D_9	(0,0), (0,1), (0,2), (1,2), (1,8), (2,2)	α	α
	(4,4), (4,5), (5,12), (5,13), (13,13)	$\frac{\alpha+\beta}{2}$	$\frac{\alpha+\beta}{2}$
	(3,14), (3,15), (6,6), (6,7), (7,11), (7,14), (7,15), (11,15), (15,15)	β	β
D_{10}	(0,0), (0,1), (0,2), (1,2), (1,8), (2,2)	α	α
	(4,4), (4,5)	$\frac{\alpha+\beta}{2}$	$\frac{\alpha+\beta}{2}$
	(3,14), (3,15), (6,6), (6,7), (7,11), (7,14), (7,15), (11,15), (15,15)	β	β
D_{11}	(0,0), (0,1), (0,2), (1,2), (1,8), (2,2)	α	α
	(6,6), (6,7)	β	β
D_{12}	(0,0), (0,1), (0,2), (1,2), (1,8), (1,9), (2,2), (9,9)	α	α
	(4,4), (4,5), (5,12), (5,13), (13,13)	$\frac{\alpha+\beta}{2}$	$\frac{\alpha+\beta}{2}$
	(3,14), (3,15), (6,6), (6,7), (7,11), (7,14), (7,15), (11,15), (15,15)	β	β
D_{13}	(0,12), (0,13), (0,14), (0,15), (2,12), (2,13), (4,9), (4,11), (4,14), (4,15), (8,13), (8,15)	1	0
	(1,5)	$\frac{1}{2}$	$\frac{1}{2}$
	(6,6),(6,7)	β	β
	(9,4), (11,4), (12,0), (12,2), (13,0), (13,2), (13,8), (14,0), (14,4), (15,0), (15,4), (15,8)	0	1

Table 1. Continued

D_i	Nash equilibria	$f_1(\alpha,\beta)$	$f_2(\alpha,\beta)$
D_{14}	(0,12), (0,13), (0,14), (0,15), (2,12), (2,13), (8,13), (8,15)	1	0
	(1,5)	$\frac{1}{2}$	$\frac{1}{2}$
	(6,6),(6,7)	β	β
	(12,0), (12,2), (13,0), (13,2), (13,8), (14,0), (15,0), (15,8)	0	1
D_{15}	(0,12), (0,13), (0,14), (0,15), (2,12), (2,13), (4,9), (4,11), (4,14), (4,15), (8,13), (8,15)	1	0
	(1,5), (1,7), (3,5), (5,7)	$\frac{1}{2}$	$\frac{1}{2}$
	(6,6)	β	β
	(9,4), (11,4), (12,0), (12,2), (13,0), (13,2), (13,8), (14,0), (14,4), (15,0), (15,4), (15,8)	0	1
D_{16}	(0,12), (0,13), (0,14), (0,15), (2,12), (2,13), (4,9), (4,11), (4,14), (4,15), (6,9), (8,13), (8,15)	1	0
	(1,5), (1,7), (3,5), (5,7)	$\frac{1}{2}$	$\frac{1}{2}$
	(9,4), (9,6), (11,4), (12,0), (12,2), (13,0), (13,2), (13,8), (14,0), (14,4), (15,0), (15,4), (15,8)	0	1
D_{17}	(0,12), (0,13), (0,14), (0,15), (2,12), (2,13), (8,13), (8,15)	1	0
	(1,5), (1,7), (3,5), (5,7)	$\frac{1}{2}$	$\frac{1}{2}$
	(6,6)	β	β
	(12,0), (12,2), (13,0), (13,2), (13,8), (14,0), (15,0), (15,8)	0	1
D_{18}	(0,12), (0,13), (0,14), (0,15), (2,12), (2,13), (8,13), (8,15)	1	0
	(1,5), (1,7), (3,5), (5,7)	$\frac{1}{2}$	$\frac{1}{2}$
	(12,0), (12,2), (13,0), (13,2), (13,8), (14,0), (15,0), (15,8)	0	1
D_{19}	(0,0), (0,2), (2,2)	α	α
D_{20}	(0,0), (0,1), (0,2), (1,2), (1,8), (2,2)	α	α
D_{21}	(0,0), (0,2)	α	α
D_{22}	(0,0), (0,1), (0,2), (0,3), (1,2), (1,8), (3,8)	α	α
D_{23}	(0,0), (0,1), (0,2), (0,3), (1,2), (1,8), (1,9), (3,8), (9,9)	α	α
	(4,4), (4,5), (5,12), (5,13), (13,13)	$\frac{\alpha+\beta}{2}$	$\frac{\alpha+\beta}{2}$
	(6,6), (6,7), (7,11), (7,14), (7,15), (11,11), (11,15), (15,15)	β	β
D_{24}	(0,0), (0,1), (0,2), (0,3), (1,2), (1,8), (1,9), (3,8), (9,9)	α	α
	(4,4), (4,5), (5,12), (5,13), (13,13)	$\frac{\alpha+\beta}{2}$	$\frac{\alpha+\beta}{2}$
	(11,11), (11,15), (15,15)	β	β

Table 1. Continued

D_i	Nash equilibria	$f_1(\alpha,\beta)$	$f_2(\alpha,\beta)$
D_{25}	(0,0), (0,1), (0,2), (0,3), (1,2), (1,8), (1,9), (3,8), (9,9)	α	α
	(13,13)	$\frac{\alpha+\beta}{2}$	$\frac{\alpha+\beta}{2}$
D_{26}	(0,0), (0,1), (0,2), (0,3), (1,2), (1,8), (1,9), (3,8), (9,9)	α	α

Consider the Pareto optimal strategies of the players. In many of domains $D_1 \ldots D_{26}$ there exist a number of the Nash equilibria that are Pareto optimal. That is why we show only the profits of the players in this situations. Besides, since some domains have the same set of Pareto optimal strategies, we join them. The heavily-drawn lines on Fig. 4 divide the plane of the parameters α, β into four sets consisting of the following domains:

$D_1 - D_{12}$ The Pareto optimal profits of both the players equal β.

D_{13}, D_{14} In this domain there are three Pareto optimal situations with the players' profits $(1, 0)$, $(0, 1)$ and (β, β), correspondingly.

$D_{15} - D_{18}$ The Pareto optimal profits of the players in the points of this set are $(1, 0)$, $(0, 1)$ and $(\frac{1}{2}, \frac{1}{2})$.

$D_{19} - D_{26}$ In this set both the first and the second player get the profit α in the Pareto optimal situation.

The problem of finding the Nash equilibria in the static symmetric 2×2 bimatrix game with the payoff matrix (3) is solved as follows. The plane (α, β) is divided into four quadrants that have a common corner in the point $(0, 1)$, as it is shown on Fig. 5. The first quadrant is divided into domains D_I and D'_I that have the same set of NE but differ in the Pareto optimal strategies of the players.

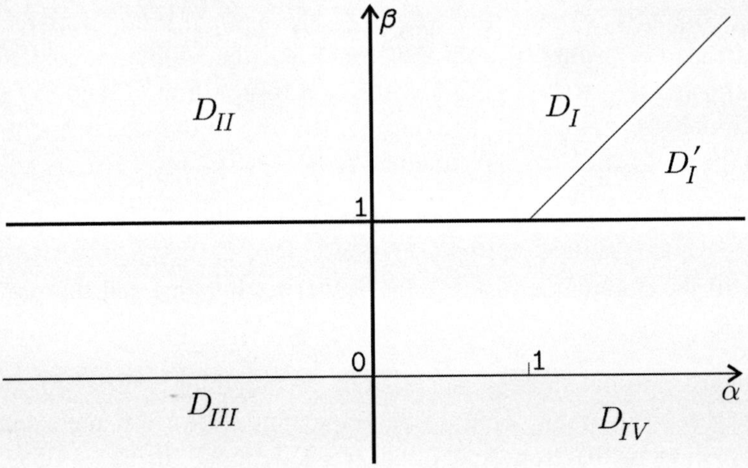

Figure 5. Nash equilibria in the static symmetric bimatrix game.

Table 2. Nash equilibria and their Pareto optimality in the static symmetric 2×2 bimatrix game

D_i	Nash equilibria	Pareto optimality
D_I	$(D, D): f_1 = f_2 = \alpha$	
	$(C, C): f_1 = f_2 = \beta$	Optimal
D'_I	$(D, D): f_1 = f_2 = \alpha$	Optimal
	$(C, C): f_1 = f_2 = \beta$	
D_{II}	$(C, C): f_1 = f_2 = \beta$	Optimal
D_{III}	$(D, C): f_1 = 1,\ f_2 = 0$	Optimal
	$(C, D): f_1 = 0,\ f_2 = 1$	Optimal
D_{IV}	$(D, D): f_1 = f_2 = \alpha$	Optimal

The Nash equilibrium strategies of the players and the Pareto optimal situation are shown in Table 2.

In comparison with static games, infinitely repeated symmetric bimatrix games with 1-step memory have some additional Pareto optimal strategies of the players.

In domains $D_{13},\ D_{14}$ (see Fig. 4) the players can get the Pareto optimal profits (β, β) unreachable in the static game. In domains $D_{15} - D_{18}$ there appears the opportunity to get the profits $(1/2, 1/2)$. In domains D_8, D_{11} the Pareto optimal profits of the players have increased: instead of the profits (α, α) in the static game, they can achieve the results (β, β) in the repeated bimatrix game with 1-step memory.

Appendix B. Stackelberg Solution of Repeated Symmetric Bimatrix Games With 1-step Memory

First, consider the Stackelberg solution of the static symmetric 2×2 bimatrix game with payoff matrix (3). For this, deal with the outcomes presented on Fig. 5 and in Table 2.

It is easy to find that in domains D_I and D_{II} the optimal leader's strategy is the choice of C. In this case the profits of both the leader and the follower equal β. The optimal strategy in domain D_{III} is D, the follower's best response is C, and his profit equals 0, whereas the leader's one equals 1. The optimal strategy of the leader in domains D_{IV} and D'_I is D, and the profits of both players equal α.

The structure of the optimal leader's strategy (i.e. the solution of Problem 2) in the 1-step memory game is shown on Fig. 6.

Examine all the domains $P_1 \ldots P_4$, the leader's strategies and the players' profits in them.

P_1 The leader's optimal strategy is TIT FOR TAT, and the profits of both players equal β. Note that this domain includes the trapezium $KLMN$ that consists of the points corresponding to the Repeated Prisoner's Dilemma. In terms of [1], the profit β means R, the reward for mutual cooperation. Thus, the problem of reaching the cooperation between players is solved in the considered setting of the game.

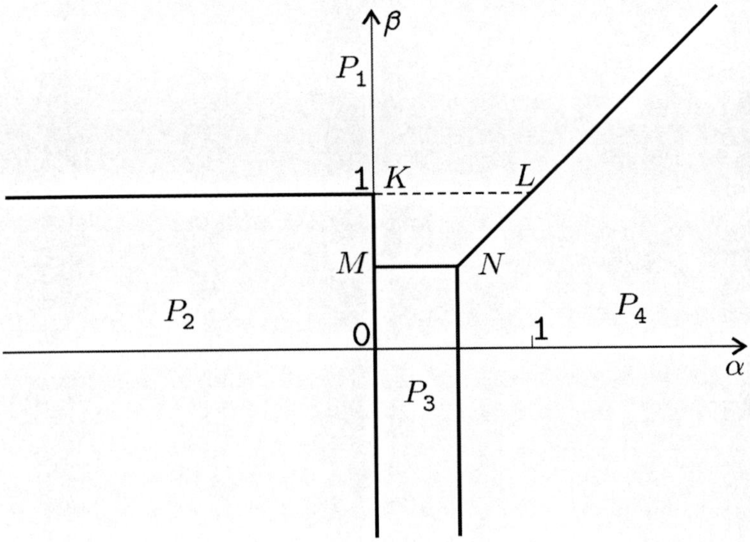

Figure 6. Solution of Problem 2 in the game with 1-step memory.

P_2 The leader exploits the strategy ALL D that guarantees him the profit equal to 1, whereas the follower's profit equals 0.

P_3 In this domain the optimal leader's strategy is TIT FOR TAT, like in domain P_1. The profits of both players equal $1/2$, which is better then β in the points of this domain.

P_4 The optimal leader's strategy here is ALL D. The profits of both the leader and the follower equal α.

It must be noted that in the repeated game with 1-step memory the improvement of the profits of both players takes place in domain P_3 and trapezium $KLMN$. Everywhere but in domain P_2 the leader's profit equals the follower's one. At the same time, the results of the investigation of the Repeated Prisoner's Dilemma with 2-step memory [6] demonstrate that the leader's profit can be greater than the follower's one. This tendency continues to the games with a longer memory.

The second remarkable fact is that the optimal leader's strategy in the game with 1-step memory is provided by only two kinds of behavior, TIT FOR TAT and ALL D. These two strategies are widely used in the majority of the investigations concerning the Repeated Prisoner's Dilemma and associated games [4, 5, 7].

Acknowledgements

The research was supported by the Russian Fund for Basic Researches (projects 01–01–06219, 03–01–00228).

References

[1] R. Axelrod, *The evolution of cooperation*, Basic Books, New York, 1984.

[2] R. D. Luce, H. Raiffa, *Games and Decisions*, Wiley, New York, 1957.

[3] J. Nash, Non-cooperative games, *Annals of Mathematics* **54** (1951), 286–295.

[4] M. Nowak, K. Sigmund, The alternating prisoner's dilemma, *Journal of Theoretical Biology Vol.* **168** (1994), 219–226.

[5] M. Posch, A. Pichler, and K. Sigmund, The efficiency of adapting aspiration levels, *IIASA Interim Report IR-98-103*, 1–27, 1998.

[6] A. Semenishchev, Solution of the repeated prisoner's dilemma with finite memory in hierarchical statement, *Game Theory and Applications Vol.* **6** (2001), 141–163.

[7] S. Smale, The prisoner's dilemma and dynamical systems associated to non-cooperative games, *Econometrica Vol.***48** (1980), 1617–1634.

[8] H. von Stackelberg, *The theory of the market economy*, Hodge, London, 1952.

[9] J. Weibull, *Evolutionary game theory*, MIT Press, Cambrige, MA, 1995.

In: Mathematics, Game Theory and Algebra Compendium ISBN 978-1-60456-659-8
Editor: Jacob H. Mathias, pp. 785-800

Chapter 64

NONSYMMETRIC CONSISTENT SURPLUS SHARING METHODS

Elena Yanovskaya
Faculty of Mathematics and Mechanics
St. Petersburg State University
St. Petersburg
Russia

1. Introduction

Fair allocation of costs or surplus among several agents participating in a joint venture is a central theme both of cooperative game theory and of public goods/public finance. There are two classes of allocation problems: cost allocation (or rationing) problems, in which the total sum to be distributed is less than the sum of agents' claims, and surplus sharing problems, where the total surplus exceeds the sum of individual claims. Given transferability of utility the question is: how should we divide the total surplus or cost among the agents? There is an excellent detailed survey of allocation problems [8].

The cost and surplus allocation problems are in some sense 'dual', and many allocation methods can be defined for problems from both classes. For example, the characterization of parametric methods for cost allocation problems due to Young [11] by continuity, symmetry, and consistency holds also for surplus sharing methods. In spite of the preference of some authors to begin with cost allocation problems, the surplus sharing problems seem to be simpler, because the condition of individual rationality (each agent should receive a value not less than his claim) implies nonnegativity of his share. The last condition, however, demands attention in rationing problems. So, in the paper only the surplus sharing problems will be investigated. The results can be transposed to the cost allocation problems as well.

The surplus sharing problems can be also treated as cooperative games with transferable utilities where intermediate coalitions are inessential. In such a game any symmetric and covariant solution divides equally the surplus (the difference between the total return, which is the value of the grand coalition, and the sum of individual claims, i.e. the individual values of the characteristic function). However in the context of allocation problems both axioms

– symmetry and covariance – become not so compelling as in the cooperative games. First, in view of nonnegativity of all claims and surplus, the covariance may be considered only in its weak version homogeneity. Second, the agents may have different exogenous rights – the different status of heirs, priorities between budget items etc. – and in such a case the symmetry axiom must be dropped. Just with the priority approach nonsymmetric allocation methods had been investigated [7]. Recently N.Naumova [9] has characterised nonsymmetric equal sacrifice methods for problems with claims unifying both the cost and the surplus sharing ones. For her model the key path independence property of allocation methods turns out to be too strong. For example, the uniform loss and uniform gain methods, which are path independent for cost and surplus sharing problems espectibely, are not continuously extended to the general problems with claims.

In paper parametric methods for surplus sharing problems are considered. Young [11] characterized symmetric parametric methods with help of three axioms: continuity, symmetry, and consistency. Each such a parametric method is determined by a function $f : \mathbb{R}_+ \times \mathbb{R}_+ \to \mathbb{R}_+$ such that the agent $i \in N$ in the problem with the set N of agent received the value $f(g_i, \lambda)$ where g_i is his claim, and the parameter λ is defined from the budget balance $\sum_{i \in N} f(g_i, \lambda) = T$, where T is the total revenue. It turns out (Theorem 1) that the strictly monotonic nonsymmetric parametric methods, determined by different strictly monotonic functions f_i, has the same characterization, of course not including symmetry with the only exception that the domain of the function f_i can be different: the parameter λ varies from 0 to some number $\Lambda(g_i) > 0$ which may be equal to infinity.

The equal sacrifice/gains methods for cost/surplus sharing problems are an important subset of the parametric ones. For symmetric case they are determined by a single continuous and strictly monotonic function $u(x)$ such that for the problem with claims $g_i, i \in N$ each agent receives the value x_i, where

$$u(x_i) - u(g_i) = u(x_j) - u(g_j) \ \forall i, j \in N \text{ and } \sum_{i \in N} x_i = T.$$

In another paper [12] Young obtained a characterization of equal sacrifice methods with help of axioms consistency, upper composition, strict resource monotonicity and strict ranking. For surplus sharing problems upper consistency means nothing but path independence. Axiom strict ranking implies symmetry, so in the proposed in this paper nonsymmetric extension of Young's theorem this axiom is replaced by strict demand monotonicity. Nonsymmetric equal gains methods are determined by continuous and strictly increasing functions u_i, which may be bounded or not, such that for a suprplus sharing problem with the set N of agents each agent receives x_i, where

$$u_i(x_i) - u_i(g_i) = u_j(x_j) - u_j(g_j) \ \forall i, j \in N, \text{ and } \sum_{i \in N} x_i = T.$$

The paper is organized as follows. In section 2 the definitions of surplus sharing methods and thir properties are given. Section 3 characterizes the nonsymmetric parametric surplus sharing methods (Theorem 1). Section 4 adds the path independence property to those considered in Section 3. Theorem 4 characterizes equal gains methods with help of axioms consistency, strict resource and demand monotonicity, and path independence. The key tool in the proof of this theorem is the known overlapping Gorman Theorem [1] about strict separability of social welfore functions.

2. Surplus Sharing Methods and Their Properties

Let $\mathcal{N}$ be a universal set of agents. For every finite subset $N \subset \mathcal{N}$ we denote by $\mathbb{R}^N$ the corresponding vector space, and by $\mathbb{R}^N_+, \mathbb{R}^N_{++}$ its nonnegative and positive orthants respectively.

A surplus sharing problem is a triple $\langle N, T; g\rangle$, where $N \subset \mathcal{N}$ – a finite subset of agents, $g \in \mathbb{R}^N_+$– their claims (investments), $T \geq 0$ – the total revenue to be distributed, and $\sum_{i \in N} g_i \leq T$.

Definition 1. *A surplus sharing method (rule) F is a mapping assigning to every problem $\langle N, T; g\rangle$ a solution vector $F(N,T;g) = (F_i(N,t;g)) \in \mathbb{R}^n_+$, satisfying the conditions $F(N,T;g) \geq g$, $\sum_{i \in N} F_i(N,T;g) = T$.*

We give here a list of the main properties of surplus sharing methods:

A surplus sharing method F is *continuous,* if for every N the function F is continuous in both variables T and g;

A surplus sharing method F is *anonymous*, if for every $N \subset \mathcal{N}$ and every permutation of agents $\pi : N \to N$ $\pi(F(N,T;g)) = F(N,T;\pi g)$;
F is *symmetric* or possesses *equal treatment property*, if the equality $g_i = g_j$, $i, j \in N$ implies $F_i(N,T;g) = F_j(N,T;g)$.

If the function F is not decreasing (increasing) in T, then the method F is *(strictly) resource monotonous*. If for all $i \in$ the functions F_i is not decreasing (increasing) in g_i, then the method F is *(strictly) claim monotonous*;

A method F is *consistent*, if for all $N \subset \mathcal{N}, i \in N$, and $j \in N \setminus i$ $F_j(N,T;g) = F_j(N \setminus i; T - F_i(N,T;g); g^i)$, where $g^i = (g_j)_{j \in N \setminus i}$. F is *pairwise consistent,* if for all $N \subset \mathcal{N}$, $i, j \in N$

$$F_{i(j)}\big(\{i,j\}, T - \sum_{k \neq i,j} F_k(N,T;g); (g_i, g_j)\big) = F_{i(j)}(N,T;g);$$

A method F satisfies *path independence*, if the equalities

$$F(N,T';g) = y, F(N,T;y) = x, \text{ where } T' < T, g < y < x,$$

imply $F(N,T;g) = x$;

A method F satisfies the property *(strict) ranking*, if

$$g_i \leq g_j,\ (g_i < g_j) \implies F_i(N,T;g) \leq F_j(N,T;g)\,(F_i(N,T;g) < F_j(N,T,g)).$$

A method F is *individually unbounded*, if for any problem $\langle N, T; g\rangle$

$$F_i(N,T;g) \to \infty \text{ as } T \to \infty \tag{1}$$

for all $i \in N$;
F is *almost individually unbounded*, if for any problem $\langle N, T; g\rangle$ property (1) holds for all $i \in N$ besides, probably, anyone.

Let $u_i : \mathbb{R}_+ \to \mathbb{R},\ i \in N$ be a collection of continuous and strictly monotonic functions.

Definition 2. *A method F is the equal gains method associated with $\{u_i\}, i \in N$, if $x = F(N,T;g)$ is the solution of the following system of equations:*

$$u_i(x_i) - u_i(g_i) = u_j(x_j) - u_j(g_j)\ \forall i,j \in N,\ \sum_{i \in N} x_i = T. \quad (2)$$

If the method F satisfies more anonymity, then in equations (2) all the functions u_i are equal: $u_i = u, i \in N$.

3. Parametric Methods

Young [11] proved that every symmetric, continuous, and pairwise consistent allocation method has a parametric representation and optimizes an additively separable objective function. He also showed tha the same result holds for surplus sharing methods as well.

Parametric methods for both cost and surplus sharing problems are generated by numerical *standards* [13], being functions asssociating with each agent, his claim, and gains (cost) a relative utility of this gain (cost). The parametric methods assign to agents the values equalizing the relative utilities.

We give the formal definition of parametric methods.

A surplus sharing method F is *parametric,* if there exist functions $f_i(g,\lambda)$, defined for all $i \in \mathcal{N},\ g,\lambda \geq 0$, non-decreasing in the second variable and satisfying the budget constraint $f_i(g,0) = g, \lim_{\lambda\to\infty} f_i(g,\lambda) = \infty$ such that for each problem $\langle N,g;T\rangle$ with $N \subset \mathcal{N}$

$$x = F(N,g;T) \implies x_i = f_i(g_i,\lambda), \quad (3)$$

where the parameter λ is defined from the equality $\sum_{i\in N} x_i = T$.

From the definition it follows that parametric methods are consistent.

In fact, consistency expresses a certain kind of independence of methods w.r.t. variations in populations. Given a method, consider some problem, apply the method to it, and imagine that some agents leaving it with their profirs. The requirement is that when the situation is re-evaluated from the viewpoint of the remaining agents, the method should award to each of them the same amount as it did initially. The parametric representations of parametric methods shows that these methods are indeed consistent.

In this section the cited result due to Young [11] is extended to nonsymmetric parametric methods under one additional strict resource monotonicity property. Note that (non strict) resource monotonicity of an allocation method follows from its continuity imposed as a condition in Young's theorem.

If a surplus sharing method F is not anonymous, but continuous and pairwise consistent (e.g. the priority methods [8]), then the functions $F_i(N,T;g), i \in N$ can be bounded or unbounded as $T \to \infty$. However, for each surplus sharing problem $\langle N,t;g\rangle$ at least for one agent $i \in N \quad F_i(N,T;g) \to_{T\to\infty} \infty$ because of the budget condition $\sum_{i\in N} F_i(N,T;g) = T$. Moreover, for the same agent i his profit $F_i(N,T;g)$ can increase to infinity together with T in one problem and be bounded in another one. Therefore the domains of functions f_i in the definition of parametric methods (3) can depend on g_i and

$i \in \mathcal{N}$. In order to avoid such cases N.Naumova [9] considered the individual unboundedness property (see section 2) demanding unboundedness the profits of both agents in any problems with $N = \{i, j\}$ when $T \to \infty$.

In this paper we shall not introduce such a condition on surplus sharing methods, but extend the definition of parametric methods admitting boundedness of the individual profits when the resource is unboundedly increasing.

A surplus sharing method F is *generalized parametric* if for all $i \in \mathcal{N}$ there exist the functions $f_i(g, \lambda)$, defined for $g \geq 0, \lambda \in [g, \Lambda_i(g))$, (where $\Lambda_i(g)$ may be equal to infinity), not decreasing in the second variable, satisfying the conditions $f_i(g, 0) = g, \lim_{\lambda \to \Lambda_i(g)} f(g, \lambda) = \infty$ such that

$$x = F(N, T; g) \iff x_i = f_i(g_i, \lambda) \text{ for all } i, \in N,$$

where $\lambda \in [0, \min_{i \in N} \Lambda_i(g_i))$ and is defined from the equality $\sum_{i \in N} x_i = T$.

Evidently, the generalized parametric methods are pairwise consistent and for any such a method F and two-person problem $\langle (i, j), T; (g_i, g_j) \rangle$ the inequality $\Lambda_i(g_i) < \Lambda_j(g_j)$ implies that $F_i((i, j), T; g\rangle \to_{T \to \infty} \infty$ and $F_j(\{i, j\}, T; g)$ is bounded as $T \to \infty$.

Theorem 1. *If a surplus sharing method F is continuous, pairwise consistent, and strictly resource monotonous, then it is generalized parametric.*

Proof. First, note that resource monotonicity of F implies its continuity in T. From the proof of Young's theorem it follows that symmetry, continuity, and pairwise consistency imply resource monotonicity.
In the absence of symmetry we put the property strict resource monotonicity.

Let a surplus sharing method F satisfy the conditions of the theorem. Then it may be defined by strictly increasing paths $L(g)$ in $\mathbb{R}^N$ getting out from the claim point such that

$$x \in L(g), \sum_{i \in N} x_i = T, \implies x = F(N, T; g). \tag{4}$$

Every such a path has a parametric representation, let

$$x_i = f_i^N(g, \lambda), \quad \lambda \in [0, \infty), \quad f_i(g, 0) = g_i, \tag{5}$$

where the functions f_i are continuously increasing in λ.

If a method F is pairwise consistent, then for each pair $i, j \in N$

$$(x_i, x_j) = F((i, j), T - \sum_{k \neq i, j} x_k; (g_i, g_j)).$$

From the definition of parametric methods (5) we get

$$x_i = f_i^{ij}(g_i, g_j; \lambda), \quad x_j = f_j^{ji}(g_i, g_j; \lambda), \quad x_i + x_j = T - \sum_{k \neq i, j} x_k, \tag{6}$$

where the functions f_i^{ij}, f_j^{ji} do not depend on N. Note that in representations (5),(6) the values of the parameter λ may be different.

For example, the values of one coordinate of the vector x may be put equal to λ, and the other coordinates are expressed through these values.

Let $\langle N, T; g\rangle$ be an arbitrary surplus sharing problem and let $x = F(N, T; g)$, $|N| \geq 3$. Then, similar to (6), we obtain the following representations for arbitrary $i, j, k \in N$:

$$\begin{aligned} x_i &= f_{ij}(g_j, g_i, x_j), \\ x_i &= f_{ik}(g_k, g_i, x_k), \\ x_k &= f_{kj}(g_j, g_k, x_j), \end{aligned} \qquad (7)$$

for some continuous functions f_{ij}, f_{ji}, f_{kj} increasing in the third variable. Put the expression for x_k in the second line. Then

$$x_i = f_{ik}(g_k, g_i, f_{kj}(g_j, g_k, x_j)). \qquad (8)$$

Compare (8) with the first line in (7). Both formulas express the dependence of x_i on x_j and in (7) x_i does not depend on g_k. Therefore the right hand of (8) does not depend on g_k as well, and the variables g_j, x_j are separated from g_i, i.e., we can write

$$x_i = \tilde{f}_{ik}(g_i, \psi_{kj}(g_j, x_j)) \qquad (9)$$

for some continuously increasing in the second variables functions $\tilde{f}_{1k}, \psi_{kj}$.

Since equalities (9) can be written for any $i, j, k \in N$, and since they do not depend on N, then for all $i, j \in \mathcal{N}$ the functions f_{ik}, ψ_{kj} do not depend on k, and we can write $f_{ik} = f_i$, $\psi_{kj} = \psi_j$. By strict monotonicity of f_i in the second variable there exists the converse:

$$\psi_j(g_j, x_j) = f_i^{-1}(g_i, x_i). \qquad (10)$$

Equalities (10) are valid for any pair of agents $i, j \in \mathcal{N}$, so $f_i^{-1} = \psi_i$ for all $i \in \mathcal{N}$ and we obtain

$$x = F(N, T; g) \implies f_i^{-1}(g_i, x_i) = \lambda \text{ for all } i \in N. \qquad (11)$$

However the domain of the parameter λ may depend on the set of agents N. The definition of functions f_i implies that if the profit $F_i(N, T; g)$ is bounded, then one of the following two cases may arise

1) $F_i((i,j), T; (g_i, g_j))$ is bounded in T for all problems $\langle (i,j), T; (g_i, g_j)\rangle$ and for all $j \in \mathcal{N}$ such that $g_j > 0$. Then for each $j \neq i$ $F_j((i,j), T; (g_i, g_j)) \to \infty$ as $T \to \infty$ and the method F, defined in (11), is parametric. Note that there may exists a unique $i \in \mathcal{N}$ possessing such a property. In fact, if there are at least two agents with bounded profits $F_i((i,j), T; g), F_j((i,j), T; g)$, then the methods F would not be well-defined for the problems with $N = \{i, j\}$ for sufficiently large T.

2) For some $k \in \mathcal{N}, g_i, g_j \geq 0$ $F_i((i,k), T, (g_i, g_k)) \to_{T\to\infty} \infty$. Comparing the problems with the agent sets (i,j) and (i,k), representation formula (11) implies that the value $f_i(g_i, \lambda)$ is bounded for all λ defining the method F for the problems with the agent set (i,j) and

$$F_i((i,k), T; (g_i, g_k)) = f_i(g_i, \lambda) \to \infty \text{ as } T \to \infty.$$

The last property of the function $f_i(g_i, \lambda)$ implies that in the problem $\langle (i,j), T; g\rangle$ the domain of λ defining the sharing vector $F((i,j), T; g)$ is contained in the similar domain for

the sharing vectors $F((i,k),T;g)$. Denote by
$\Lambda_{ij}(g_i,g_j) = \sup_T\{\lambda \mid f_i(g_i,\lambda) = F_i((i,j),T;g)$. Then

$$F_j((i,j),T;(g_i,g_j)) = f_j(g_j,\lambda) \to \infty \text{ as } \lambda \to \Lambda_{ij}(g_i,g_j)$$

and the function $f_j(g_j,\lambda)$ is defined for $\lambda \in [0,\Lambda(g_i,g_j))$. Since the function f_j does not depend on g_i, we have $\Lambda_{ij}(g_i,g_j) = \Lambda_j(g_j)$ for some function $\Lambda_j : \mathbb{R}_+ \to \mathbb{R}_+$.

Thus, under the condition of the case 2), for each $j \in \mathcal{N}$ and the function $f_j(g_j,\lambda)$ the domain of the parameter λ is some interval $[0,\Lambda_j(g_j))$, where $\Lambda_j(g_j)$ may be equal to infinity. This means that for every problem $\langle N,T;g\rangle$ $F_i(N,T;g) = f_i(g_i,\lambda)$, where $\lambda \in [0,\Lambda_N)$, $\Lambda_N = \min_{j\in N}\Lambda(g_j)$ and λ is defined from the equation $\sum_{i\in N} f_i(g_i,\lambda) = T$.

It is easy to note that the method F is well-defined, i.e., it assigns to each problem $\langle N,T;g\rangle$ a unique sharing vector $F(N,T;g)$ and is generalized parametric. □

On the picture three functions $f_i(g_i,\lambda), i = 1,2,3$ with different domains for the parameter λ are depicted.

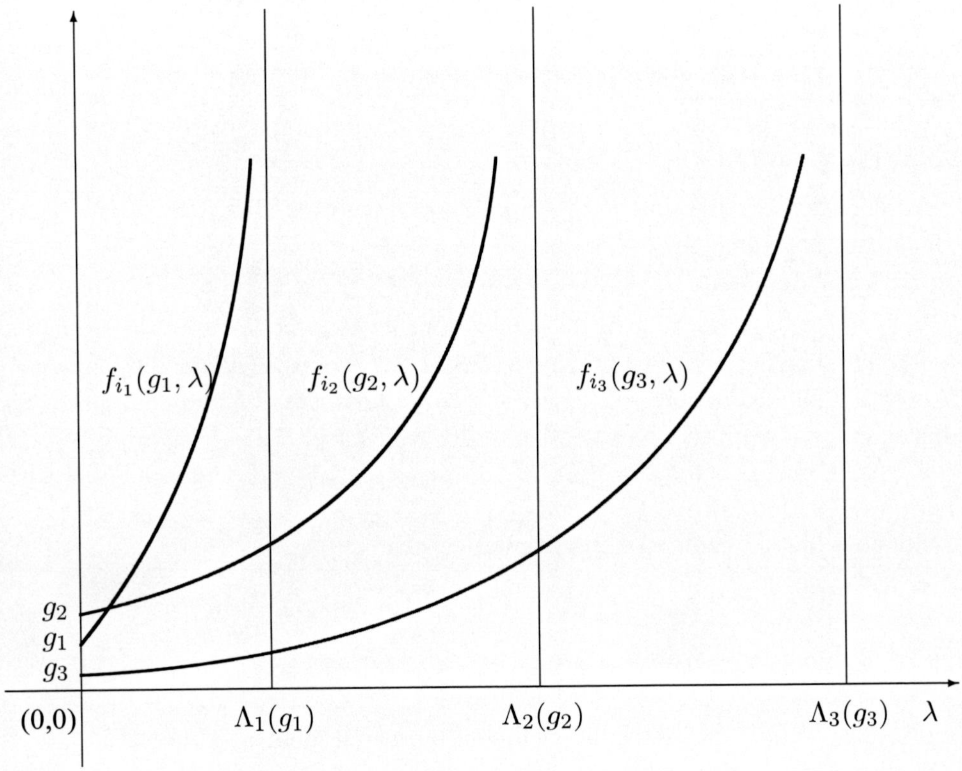

Figure 1. The functions defining the generalized method F for $N \subseteqq \{1,2,3\}$.

The well known *priority* methods are defined as follows: Let a strict ordering $\succ$ on the universal set of agents $\mathcal{N}$ be given such that the set $\mathcal{N}$ can be presented as

$$\mathcal{N} = i_1 \succ \ldots \succ i_k \succ \ldots.$$

For every surplus sharing problem $\langle N,T;g\rangle$ define the maximal agent $i(N)$ w.r.t. the ordering $\succ$ such that $i(N) \succ j$ for all $j \in N \setminus i(N)$. The priority method w.r.t. the ordering

$\succ$ is defined in the following way:

$$F_j(N,T;g) = \begin{cases} g_j, & \text{if } j \neq i(N), \\ T - \sum_{j\in N\setminus i(N)} g_j, & \text{if } j = i(N), \end{cases}$$

i.e., all the agents $j \neq i(N)$ receive only their claims and the agent $i(N)$ receives the whole surplus.

It is clear that the priority methods are pairwise consistent and continuous, though they are not strictly resource monotonic. Nevertheless, they are generalized parametric. Let us give their parametric representation. Let $\Lambda : \mathcal{N} \to \mathbb{R}_{++}$ be an arbitrary function representing the converse to $\succ$ relation:

$$i, j \in \mathcal{N}, i \succ j \iff \Lambda(i) < \Lambda(j).$$

For each $i \in \mathcal{N}$ define the function

$$f_i(g_i, \lambda) = \begin{cases} g_i & \text{for } \lambda \leq \Lambda(i-1), \\ \text{is arbitrary increasing from } g_i \text{ to } \infty & \text{for } \Lambda(i-1) < \lambda < \Lambda(i). \end{cases} \tag{12}$$

It is easy to check that the functions f_i in (12) give a parametric representation of the priority method w.r.t. $\succ$.

Strict resource monotonicity of F implies that the functions f_i are strictly increasing in λ, thus, they have the converse in λ :

$$\lambda = f_i^{-1}(x_i, g_i) = e_i(x_i, g_i), \quad x_i \geq g_i. \tag{13}$$

The functions $e_i : \mathbb{R}^2_+ \to \mathbb{R}$ may be called the *excess functions** by the analogy with the excess functions in cooperative game theory. An excess function assigns to each agent i, his claim g_i, and profit x_i his relative utility with the profit.

For the sharing vector $x = F(N,T;g)$ $e_i(x_i, g_i) = e_j(x_j, g_j) = \lambda$ for all $i, j \in N$, where $\lambda \in [0, \Lambda_i(g_{i0}))$, $\Lambda(g_{i0}) = \max_{i\in N} L(g_i)$.

Suppose that the method F is symmetric, then

$$g_i = g_j \implies x_i = F_i(N,T;g) = F_j(N,T;g) = x_j$$

which implies

$$e_i(g,x) = e_j(g,x) \implies e_i = e_j$$

for all $i, j \in N$, $N \subset \mathcal{N}$, thus, the method F turns out to be anonymous.

From Theorem 1 an optimization representation of parametric methods easily follows. The next theorem extends Young's Theorem 2 [11] to nonsymmetric nethods:

Theorem 2. *A method F satisfies the assumptions of Theorem 1 if and only if for each problem $\langle N,T;g\rangle$ the sharing vector $x^* = F(N,t;g)$ is a solution of the problem*

$$\min_{\substack{x\in\mathbb{R}^N_+ \\ x_i \geq g_i, \sum_{i\in N} x_i = T}} \sum_{i\in N} \int_{g_i}^{x_i} e_i(g_i, y)dy, \tag{14}$$

[0*] In [2] they are called *numerical standards*.

where $e_i = f_i^{-1}$ *and the functions* $f_i(g_i, \lambda)$, $g_i \geq 0, \lambda \in [0, \Lambda(g_i))$ *are the parametric representation of the method* F.

Proof. Sufficiency. Let x^* be a solution of problem 14. Since the excess values e_i are positive everywhere except for the diagonal where they are equal to zero, the objective function is strictly convex in all x_i. Thus, the minimum in (14) is unique and is attained in the point x^*, where $x_i^* > g_i$. The Lagrange conditions are necessary and sufficient for solving the problem (14):

$$e_i(g_i, x_i^*) = e_j(g_j, x_j^*) \quad \forall i, j \in N.$$

Now equalities (13) and Theorem 1 imply that $x^* = F(N, T : g)$.

The converse arguments together with Theorem 1 prove necessity. □

4. Path Independent Methods

In this section we consider strictly monotonous parametric methods satisfying more the path independence property. For such methods this property can be defined as follows: for all $g < y < x \in \mathbb{R}_{++}^N$

$$\begin{array}{l} e_i(g_i, x_i) = e_j(g_j, x_j), \\ e_i(g_i, y_i) = e_j(g_j, y_j) \end{array} \implies e_i(y_i, x_i) = e_j(y_j, x_j), \tag{15}$$

where e is the excess function corresponding to the method F.

Symmetric path independent methods were defined for the first time by Mill [3] in the context of taxation and then were characterized by Young [12] for cost allocation problems. It is easy to extend his result to surplus sharing problems. Then it is formulated as follows:

Theorem 3. [Young (1988)] *If a method* F *is consistent, path independent, satisfies strictly resource monotonicity and strict ranking, then it is an equal gains method (2).*

Note that strict ranking and strict resource monotonicity imply continuity, and continuity and strict ranking imply symmetry of the method F. H.Moulin in [8] formulated some open problems. One of them is a characterization of methods satisfying the weakened to 'non strict' monotonicity and ranking assumptions in the cited Young theorem. Such conditions would not imply symmetry. Therefore the problem is to describe all nonsymmetric, consistent, and path independent allocation methods.

Such methods were characterized by N.Naumova [9], however, she considered universal allocation problems unifying both the surplus and the cost ones. So, the allocation methods for such problems had to satisfy some additional properties.

The goal of this section is a characterization of consistent and path independent surplus sharing methods. The key tool for such a characterization is the Gorman theorem [1] on separable welfare functions. To apply the theorem it is necessary to add the properties of strict resource and claim monotonicity of the surplus sharing methods. Nevertheless, the main theorem of this section (Theorem 4), as it will be shown further, is a real extension of the Young theorem.

First, give the interrelations between some properties of the surplus sharing methods to be characterized.

Lemma 1. *If a method F is continuous, parametric, individually unbounded, and path independent, then the corresponding functions $f_i(g_i, \lambda)$ are non-decreasing in claims.*

Proof. Let us show that for two-agent problems with $N = \{i, j\}$ for any positive vector $x = (x_i, x_j) > 0$ there exist claims $g_i < x_i, g_j < x_j$ such that $x_i = f_i(g_i, \lambda), x_j = f_j(g_j, \lambda)$ for some $\lambda > 0$.

Since the functions $f_i(a, \lambda)$ are increasing in λ from a to infinity ($f_i(a, 0) = a$ for any $a \geq 0$,) then there exist $\lambda^1, \lambda^2 > 0$ such that $f_i(0, \lambda^1) = x_i, \;\; f_j(0, \lambda^2) = x_j$. If $\lambda_1 = \lambda_2$ then the claim has been proved. Let now $\lambda_1 < \lambda_2$. For any $g_i < x_i$ denote by $\lambda(g_i, x_i) > 0$ the solution of the equation $f_i(g_i, \lambda(g_i, x_i)) = x_i$. By continuity of F the functions f_i are continuous jointly in both variables in the domain $g_j \in [0, x_j], \lambda(g_j, x_j) \in [0, \lambda_2]$. Therefore, there is $g_j \in (0, x_j)$ such that $\lambda(g_j, x_j) = \lambda^1$ and we obtain $x_i = f_i(0, \lambda^1), x_j = f_j(g_j, \lambda^1)$. The claim has been proved.

If the method F is parametric and satisfies path independence, then for any claims $g^1, g^2 \in \mathbb{R}^N_+$ the paths defining the method $L_F(g_1), L_F(g_2)$ (4), having the parametric representations $x_i = f_i(g_i^1, \lambda), x_i = f_i(g_i^2, \lambda)$ respectively, either have no common points, or are sticking together, i.e., they coincide for sufficiently large λ.

Suppose that for some $j \in \mathcal{N}$ and λ^0 the function f_j is not monotonous in g_j. Then there exist $g_j^1 \neq g_j^2$, (let $g_j^1 < g_j^2$) such that $f_j(g_j^1, \lambda^0) = f_j(g_j^2, \lambda^0)$. This means that for any $i \in \mathcal{N}$ the paths $L_F(g_i, g_j^1)$, $L_F(g_i, g_j^2)$, defined parametrically by the functions f_i, f_j, are sticking together beginning with $\lambda = \lambda^0$. Since $f_j(g_j, 0) = g_j$ for all $g_j \geq 0$, then for any $g_j \in [g_j^1, g_j^2]$ the path $L_F(g_i, g)$ for sufficiently small λ is placed between the paths $l_F(g_i, g_j^1), L_F(g_i, g_j^2)$, and at $\lambda \geq \lambda_0$ they coincide. This means that for such λ's $f_j(g_j, \lambda) = f_j(g_j^1, \lambda) = f_j(g_j^2, \lambda)$. i.e.,the function f_j is monotonous in g_j.

Similarly, monotonicity of f_i can be proved. □

Relations (15) easily imply the following result:

Lemma 2. *If a method F is continuous, strictly resource monotonous, consistent, and path independent, then it is also claim monotonous (non strictly).*

In fact, (15) implies that the functions e_i are not increasing in g_i, i.e., the method F is claim monotonous.

An example of non strictly claim monotonous method satisfying the conditions of Lemma 2 is the following one:

$$F_i(N, T; g) = \max\{\lambda, \alpha\lambda + g\}, \; 0 < \alpha < 1. \tag{16}$$

If $\alpha = 0$, then the method is the uniform gain method [8] and if $\alpha = 1$, then it is an equal gains mehod corresponding to the utility functions $u_i(x) = x, \forall i \in \mathcal{N}$. It is clear that for $0 < \alpha < 1$ this method is strictly resource monotonous and non strictly claim monotonous.

Let us show interrelations between strict ranking and the strict claim monotonicity.

Lemma 3. *If a method F is consistent, satisfies strict resource monotonicity and strict ranking, then it is strictly claim monotonous.*

Proof. Consider two surplus sharing problems G, G' differing one from another only by claims of two agents:

$G = \langle N, T; \{g_1, \dots, g_i, \dots g_j = g_i + \varepsilon\}\rangle$; $G' = \langle N, T; \{g_1, \dots, g_i' = g_i + \varepsilon, \dots, g_j' = g_i\}\rangle$.

By strict ranking $F_i(G) = F_j(G') < F_j(G) = F_i(G')$. For other $k \neq i, j$ $F_k(G) = F_k(G')$. Let the agent j leave the problems G G' with the same gain $F_j(N, T; g), F_j(N, T; g')$ respectively. Then, by consistency, in the reduced problems G^j, G'^j the profits of all agents $k \neq i, j$ do not change, but the total sum in G_j are less than the total sum in G'^j. Therefore, the gain of agent i in G^j is less than his gain in G'^j:

$$F_i(G^j) < F_i(G'^j) = F_i(G^j) + \delta \tag{17}$$

for some $\delta > 0$.

Let now $G'' = \langle N \setminus j, T - F_j(G'); g^j\rangle$ be the problem differing from G'^j only by i-th claim, which is equal to g_i in G'' (in G'^j it was equal to $g_i + \varepsilon$.) From the problem G^j the problem G'' differs only by the total gain which is less in G'' than in G^j : $T - F_j(G') < T - F_j(G)$.

By strict resource monotonicity when we pass from G'' to G'^j the profits of all agents should strictly increase. Therefore, $F_i(G'') < F_i(G'^j)$ and (17) implies $F_i(G'') < F_i(G^j)$.

The last inequality proves strict claim monotonicity of F : □

Lemma 4. *If a method F is continuous, symmetric, consistent, path independent, and strictly claim monotonous, then it satisfies strict ranking.*

Proof. By Young's theorem [11] the method F is parametric, i.e., it is defined by a function $f(g_i, \lambda)$ such that $F_i(N, T; g) = f(g_i, \lambda)$, where $\lambda \in [0, \infty)$ (or $\lambda \in [0, \Lambda]$) is defined from the budget condition $\sum_{i \in N} f(g_i \lambda) = T$. For two-person cases ($N = \{i, j\}$) the paths $x_i = f(g_i, \lambda), x_j = f(g_j, \lambda)$ do not intersect for different claims (g_i, g_j), where g_i or $g_j = 0$, though they can stick together beginning from some λ's.

In the last case one of the three following possibilities may arise:

1) There exist $\lambda_0, \lambda_0' > 0$ such that $f(g_i, \lambda) = f(g_i', \lambda'), f(0, \lambda) = f(0, \lambda')$ for all $\lambda > \lambda_0, \lambda' > \lambda_0'$ and some $g_i \neq g_i' > 0$. This case contradicts strict claim monotonicity of F.

2) The case analogous to the case 1) where i and j are interchanged.

3) There are $\lambda, \lambda' > 0$ such that $f(0, \lambda) = f(g_i, \lambda'), f(g_j, \lambda) = f(0, \lambda')$. By symmetry of F this case may only happen on the diagonal, i.e., if $f(0, \lambda) = f(g_j \lambda) = f(0, \lambda') = f(g_j, \lambda')$, and we return to the cases 1) and 2).

Therefore, the paths defining the method F have no common points for different claims (g_i, g_j) when g_i or $g_j = 0$. This fact and strict claim monotonicity of F imply its strict ranking. □

Lemmas 3 and 4 imply that strict ranking of surplus sharing methods under assumptions of Young's Theorem 2 is stronger than strict claim monotonicity. Besides, symmetry implies the individual unboundedness property. Thus, the following lemma generalizes the Young Theorem:

Lemma 5. *If a surplus sharing method F is continuous, strictly claim and resource monotonous, almost individually unbounded, consistent, and path independent, then it is an equal gains method.*

Proof. Note that the conditions of the theorem do not imply symmetry of the method F, as those of Young's theorem do.

It is clear that the method F satisfies the conditions of Lemma 1 and by the Lemma there exist excess functions defining the method F by (13). Path independence implies relations (15), where the excess functions e_i are decreasing in g_i by strict claim monotonicity of F.

From almost individual unboundedness of F it follows that the domains of the excess functions e_i for all $i \in \mathcal{N}$ except for, perhaps, a unique $i = l$, are the following ones:

$$\{(g, x) \in \mathbb{R}^2_+ \mid g \leqq x\}. \tag{18}$$

Since the excess functions are defined up to monotonous transformations, we can suppose that their range is the whole nonnegative ray $\mathbb{R}_+$ for all $i \in \mathcal{N}$ (including l.) In the last part of the proof we consider the case when the domain of e_i are equal to (18) for all $i \in \mathcal{N}$. When boundedness of gains of the agent l is essential, it will be noted especially.

The first equation in (15) and strict monotonicity of the functions e_i in x_i imply

$$x_j = \varphi_{ij}(x_i, g_i, g_j) = e_j^{-1}(e_i(g_i, x_i), g_j), \tag{19}$$

for some functions φ_{ij}, increasing in the first and the third variables and decreasing in the second one. Then path independence conditions (15) can be rewritten as follows:

$$\varphi_{ij}(a, b, c) = \varphi_{ij}(a, t, \varphi_{ij}(t, b, c)) \tag{20}$$

for all $i, j, \; b \leq t \leq a$.

The functions $\varphi_{ij}(a, b, c)$ are defined on the domain $a \geq b$ of the positive orthant $\mathbb{R}^3_+$. In order to apply the Gorman theorem and to obtain separability of these functions it is necessary, first, to extend their domain up to the whole orthant and, second, to make them increasing in the second variable. For this purpose define the functions ψ_{ij} on $\mathbb{R}^3_+$ as follows:

$$\psi_{ij}(a, b, c) = \begin{cases} \varphi_{ij}(a, \frac{1}{b}, c), & \text{if } ab \geq 1, \\ \dfrac{1}{\varphi_{ij}(\frac{1}{a}, b, \frac{1}{c})}, & \text{if } ab \leq 1, \end{cases} \tag{21}$$

Since $\varphi_{ij}(a, a, b) \equiv b$ for all $i, j \in \mathcal{N}, \; a, b$ the functions ψ_{ij} are defined on the whole orthant $\mathbb{R}^3_+$, continuous, and increasing in each variable.

Besides, similar to the functions φ_{ij}, the second and the third variables of the functions ψ_{ij} are separated from the first one; and the first and second variables are separated from the third one. Therefore, by Gorman's theorem each subset of $\{a, b, c\}$ is separated from its complement, and the functions ψ_{ij} have the form

$$f^1_{ij}(\psi_{ij}(a, b, c)) = f^2_{ij}(a) + f^3_{ij}(b) + f^4_{ij}(c), \tag{22}$$

where $f^k_{ij}, \; k = 1, 2, 3, 4$ are continuous increasing functions.

Return to the functions φ_{ij}. Then equality (22) may be written as

$$\begin{aligned} f^1_{ij}(\varphi_{ij}(a, 1/b, c)) &= f^2_{ij}(a) + f^3_{ij}(b) + f^4_{ij}(c), \quad ab \geq 1, \\ f^1_{ij}\left(\frac{1}{\varphi_{ij}(1/a, b, 1/c)}\right) &= f^2_{ij}(a) + f^3_{ij}(b) + f^4_{ij}(c), \quad ab \leq 1. \end{aligned} \tag{23}$$

Put $h^3_{ij}(b) = f^3_{ij}(1/b)$. Then the first equality in (23) becomes φ_{ij}

$$f^1_{ij}(\varphi_{ij}(a,b,c)) = f^2_{ij}(a) + h^3_{ij}(b) + f^4_{ij}(c), \tag{24}$$

where the functions $f^1_{ij}, f^2_{ij}, f^4_{ij}$ are increasing and the function h^3_{ij} is decreasing.

Since the equality $x_i = g_i$ implies $x_j = g_j$, then for all $a, c > 0$

$$f^1_{ij}(c) = f^1_{ij}(\varphi_{ij}(a,a,c)) = f^2_{ij}(a) + h^3_{ij}(a) + f^4_{ij}(c),$$

that is,

$$f^1_{ij}(c) - f^4_{ij}(c) = f^2_{ij}(a) + h^3_{ij}(a). \tag{25}$$

Therefore, both hands of (25) are constant and $f^1_{ij} = f^4_{ij} = u_{ij},\ f^2_{ij} = -h^3_{ij} = v_{ij}$, where u_{ij}, v_{ij} are increasing functions.

Therefore, equalities (19) and (24) imply that

$$e_i(x_i, g_i) = \phi_{ij}(v_{ij}(x_i) - v_{ij}(g_i)) \tag{26}$$

for all $i, j \in N$ and some increasing functions ϕ_{ij}.

Put $f_{ij} = u_{ji} + v_{ij}$. Then for the solution vector $x = F(N, T, g)$

$$f_{ji}(x_j) - f_{ji}(g_j) = f_{ij}(x_i) - f_{ij}(g_i) \text{ for all } i \neq j. \tag{27}$$

Recall that the method F is parametric, hence,

$$e_i(x_i, g_i) = e_j(x_j, g_j) \text{ for all } i, j \in N. \tag{28}$$

Therefore, equalities (26), (27), and (28) imply that

$$e_i(x_i, g_i) = \phi_{ij}\big(f_{ij}(x_i) - f_{ij}(g_i)\big) \tag{29}$$

for all $i, j \in N, i \neq j$.

Rewrite equalities (29) using (27) twice: first for the same i and j, and then interchange them.

$$\begin{aligned} e_i(x_i, g_i) &= \phi_{ij}(f_{ij}(x_i) - f_{ij}(g_i)) = \\ &= \phi_{ij}(f_{ji}(x_j) - f_{ji}(g_j)); \\ e_j(x_j, g_j) &= \phi_{ji}(f_{ji}(x_j) - f_{ji}(g_j)) = \\ &= \phi_{ji}(f_{ij}(x_i) - f_{ij}(g_i)). \end{aligned} \tag{30}$$

Taking into account the equality $e_i(x_i, g_i) = e_j(x_j, g_j)$ which holds for the vectors x and g, from (30) it follows $\phi_{ij} = \phi_{ji}$ for all $i \neq j$.

The excess functions $e_i(x_i, g_i)$ defining a parametric method are determined up to a monotonous transformation – a parametric method equalizes the values of all e_i, – so for a unique pair i_0, j_0 we may put the function $\phi_{i_0 j_0}$ identical:

$$\begin{aligned} e_{i_0}(x_{i_0}, g_{i_0}) &= f_{i_0 j_0}(x_{i_0}) - f_{i_0 j_0}(g_{i_0}) \\ e_{j_0}(x_{j_0}, g_{j_0}) &= f_{j_0 i_0}(x_{j_0}) - f_{j_0 i_0}(g_{j_0}). \end{aligned} \tag{31}$$

Let $k \neq i_0, j_0$ be arbitrary. We have

$$\begin{aligned} e_k(x_k, g_k) &= \phi_{ki_0}(f_{ki_0}(x_k) - f_{ki_0}(g_k)) = \\ &= \phi_{i_0k}(f_{i_0k}(x_{i_0}) - f_{i_0k}(g_{i_0}) = \\ e_{i_0}(x_{i_0}, g_{i_0}) &= f_{i_0j_0}(x_{i_0}) - f_{i_0j_0}(g_{i_0}). \end{aligned} \tag{32}$$

The last equality in (32) implies that ϕ_{i_0k} is a linear function, $\phi_{i_0k}(t) = C_k t$, where C_k are positive constants. Thus, applying equalities (32) we obtain

$$e_k(x_k, g_k) = C_k(f_{ki_0}(x_k) - f_{ki_0}(g_k)), \tag{33}$$

where the functions $f_{ki_0}(x) \to_{x\to\infty} \infty$ for all $k \neq l$, $f_{li_0}(x) \to_{x\to A_l} \infty$, where $A_l = \sup_T F_l(N, g; T)$.

Equalities (32) and (33) mean that the method F is an equal gains method for the functions

$$u_i(x) = \begin{cases} f_{i_0j_0}(x) & \text{for } i = i_0, \\ f_{j_0i_0}(x) & \text{for } i = j_0, \\ C_i f_{ii_0}(x) & \text{for } i \neq i_0, j_0. \end{cases}$$

□

Lemmas 3 and 4 imply that Lemma 5 extends Young's Theorem 3, because the conditions of the Lemma together with symmetry imply all assumptions of the Theorem.

It remains to exclude individual unboundedness of the method F :

Theorem 4. *If a method F for surplus sharing problems is continuous, strictly claim and resource monotonous, consistent, and path independent, then it is an equal gains method.*

Proof. If a method F satisfies the assumptions of the Theorem, then it also satisfies the conditions of Theorem 2, so it is generalized parametric, i.e., there exists a collection of functions $f_i(g_i, \lambda)$ for all $i \in \mathcal{N}$, defined for $g_i \geq 0$, $\lambda \in [0, \Lambda(g_i))$, where $\Lambda((g_i)$ may be infinite.

Let us consider subsets $\mathcal{N}_\alpha \subset \mathcal{N}$ of agents such that for all surplus sharing problems with agent sets $N \subset \mathcal{N}_\alpha$ the method F is almost individually unbounded. Such subsets do exist, e.g., all two-agent sets possess this property. By Lemma 5 for each set $\mathcal{N}_\alpha$ there exist the functions $u_i^\alpha : [0, \infty) \to [0, \Lambda_i(0)), i \in \mathcal{N}_\alpha$, and also $u_l^\alpha : [0, A_l) \to [0, \infty)$ if $l \in \mathcal{N}_2$, where $A_l = \sup_\lambda f_l(g_l, \lambda)$, (by path independence, A_l does not depend on g_l) such that F is an equal gain method for the functions $\{u_i\}_{i \in \mathcal{N}_\alpha}$.

Since $\bigcup_\alpha \mathcal{N}_\alpha = \mathcal{N}$, the functions u_i^α exist for every $i \in \mathcal{N}$. However they may depend on the sets $\mathcal{N}_\alpha$. Fix an $i \in \mathcal{N}$, let $i = 1$ and denote $\mathcal{N}_j = \{1, j\}, j \in \mathcal{N} \setminus \{i\}$.. Then the corresponding functions $u_j^j, j \neq 1$ may be denoted as u_j. In each set $\mathcal{N}_j$ the functions u_1^j, u_j are defined up to identical affine transformations. Thus, we may put $u_1^j(0) = 0$ for all $j \in \mathcal{N} \setminus \{1\}$ and to normalize them such they satisfy the equalities

$$f_1(g_1, u_1^j(x_1) - u_1^j(g_1)) = x_1. \tag{34}$$

for all $j \in \mathcal{N} \setminus \{1\}$, $g_1, x_1, x_1 > g_1$.

The right hand of (34) does not depend on g_1, so is the left hand and we can put $g_1 = 0$. Then $f_1(0, u_i^j(x_1)) = x_1$, and

$$u_1^j(x_1) = f^{-1}(0, x_1) = u_1(x_1) \text{ for all } j \in \mathcal{N} \setminus \{1\},$$

where the inverse function is taken in the second variable.

Thus, we obtained the functions u_i for all $i \in \mathcal{N}$. It remains to show that the method F is an equal gains method with respect to the collection of functions $\{u_i\}_{i \in \mathcal{N}}$.

Let $\langle N, T; g \rangle$, $N \subset \mathcal{N}, 1 \in N$ be an arbitrary problem and $F(N, T, g) = x$. As it has been already proved, the method F is an equal gains method for each problem $\{1, j\}, T; g\rangle, j \in \mathcal{N} \setminus \{1\}$. Therefore, by its consistency

$$u_1(x_1) - u_1(g_1) = u_j(x_j) - u_j(g_j) \text{ for all } j \in N \setminus \{1\}. \tag{35}$$

Equalities (35) imply that F is an equal gains method w.r.t. the functions $\{u_i\}_{i \in \mathcal{N}}$ for the problems $\langle N, T; g \rangle$ for any $N \subset \mathcal{N}, 1 \in N$. By consistency of F it is also the same method for the problems $\langle N, T; g \rangle$ for $\not\ni i$ as well. □

To conclude with, note that the condition strict claim monotonicity is essential and can not be weakened as the following example shows.

For each $\alpha \in [0, 1]$ consider the method F^α with the parametric representaton

$$f_\alpha(g, \lambda) = \max\{g + (1 - \alpha)\lambda, \lambda\}.$$

If $\alpha = 0$, then it is the *constrained egalitarian method* $F_i^0(N, T; g) = g_i + \dfrac{T - \sum_{i \in N} g_i}{n}$. It is an equal gains method w.r.t. $u_i(x) = x \, \forall i$.

If $\alpha = 1$, then F^1 is the *uniform gain method*, with the parametric representation $f_i^1(g, \lambda) = \max\{g, \lambda\}$. It is neither strictly resource nor strictly claim monotonous.

For any $\alpha \in (0, 1)$ the method F^α is consistent, path independent, strictly resource monotonous, and not strictly claim monotonous. It is not an equal gains method.

$$f_\alpha(g, \lambda) = \max\{g + (1 - \alpha)\lambda, \lambda\}.$$

If $\alpha = 0$, then it is the *constrained egalitarian method* $F_i^0(N, T; g) = g_i + \dfrac{T - \sum_{i \in N} g_i}{n}$. It is an equal gains method w.r.t. $u_i(x) = x \, \forall i$.

If $\alpha = 1$, then F^1 is the *uniform gain method*, with the parametric representation $f_i^1(g, \lambda) = \max\{g, \lambda\}$. It is neither strictly resource nor strictly claim monotonous.

For any $\alpha \in (0, 1)$ the method F^α is consistent, path independent, strictly resource monotonous, and not strictly claim monotonous. It is not an equal gains method.

References

[1] W.M. Gorman, The structure of utility functions, *Rev.Econ.Stud.* **35** (1968), 369–390.

[2] M. Kaminski,'Hydraulic' rationing, *Math. Soc. Sci.* **40** (2000), 131–155.

[3] J.S. Mill, On liberty, In: M.Warnock, ed., *Utilitarianism*, Glasgow, Fontana Press, 1859.

[4] H. Moulin, The separability axiom and equal sharing methods, *J. Econ. Theory* **36** (1985), 120–148.

[5] H. Moulin, Equal or proportional division of a surplus, and other methods, *Int. J. Game Theory* **16** (1987), 161–186.

[6] H. Moulin, Axioms of cooperative decision making, In: *Econometric Society Monographs* **15** (1988), Cambridge University Press, Cambridge.

[7] H. Moulin, Priority rules and other asymmetric rationing methods, *Econometrica* **68** (2000), 643–684.

[8] H. Moulin, Axiomatic cost and surplus sharing methods, In: *Handbook on Social Choice and Welfare*, K.J.Arrow, A.K.Sen, and K.Suzumura (Eds.), Elsevier Science B.V., 2002, 289–357.

[9] N.I. Naumova, Nonsymmetric equal sacrifice solutions for claim problems, *Math. Soc. Sci.* **43** (2002), 1–18.

[10] L.S. Shapley, A value for n-person games, Contributions to the Theory of Games II, *Annals of Mathematical Studies* **28** (1953), Princeton University Press, 307–317.

[11] H.P. Young, On dividing an amount according to individual claims or liabilities, *Math. Oper. Res.* **12** (1987), 397–414.

[12] H.P. Young, Distributive justice in taxation, *J. Econ. Theory* **44** (1988), 321–335.

[13] H.P. Young, *Equity in Theory and Practice*, Princeton University Press, Princeton, 1994.

INDEX

C

F

G

H

I

J

K

L

M

N

O

P

Q

R

S

T

U

V

W

Y

Z